Optimization in Solving Elliptic Problems

Optimization in Solving Elliptic Problems

by
Eugene G. D'yakonov

Steve McCormick
Editor of the English Translation

CRC Press
Boca Raton   New York   London   Tokyo

Library of Congress Cataloging-in-Publication Data

D'iakonov, E. G. (Evgenii Georgievich)
    [Minimizatsiia vychislitel 'noi raboty.   English]
    Optimization in solving elliptic problems / Eugene G. D'yakonov ;
    editor of the English translation, Steve McCormick.
        p.   cm.
    Includes bibliographical references and index.
    ISBN 0-8493-2872-1 (alk. paper)
    1. Differential equations, Elliptic – – Asymptotic theory.
    I. McCormick, S. F. (Stephen Fahrney), 1944-   . II. Title.
    QA377.D57513 1995   95-351
    515'.353 – – dc20   CIP

Preface

Minimization of computational work for elliptic and related problems, including boundary value problems for strongly elliptic systems and corresponding eigenvalue problems, is one of the most important and difficult problems of computational mathematics. Since the problem is closely connected with asymptotically optimal approximations involving a minimal number $N \equiv N(\varepsilon)$ of degrees of freedom to get a desired accuracy $\varepsilon > 0$ of the approximation, it has many common features with classical optimization problems of approximation theory and numerical integration.

A precise setting for these two problems and a significant advance in their solutions are connected first of all with the names of Kolmogorov, Nikolskii, Sobolev, and Bakhvalov (see [32, 39, 388, 460, 477]). But it was quite clear from the very beginning that discretizations of elliptic problems involve one additional and extremely difficult problem of solving linear and nonlinear grid systems with a very large number of unknowns N. Therefore, not only questions of grid approximations, but also questions of constructing effective direct and iterative methods (algorithms) for arising systems, are very important, in particular, connected with the required computational work (with the computational complexity).

The first results in this direction leading to iterative methods for some model difference systems with almost asymptotically optimal work estimate $W = O(N \ln N |\ln \varepsilon|)$ (with the truly optimal estimate for $\ln W$ if $\varepsilon \asymp N^{-k}$, $k > 0$) were discovered by Douglas, Peaceman, and Rachford (see [405, 138]) in 1955, 1956 and became known as alternating direction iteration methods (ADI methods). They showed that attempts to minimize the computational work for solution of difference systems can really lead to nearly optimal asymptotic behavior of W and highly efficient practical algorithms. These and other results led in the beginning of the 1960s to posing the problem as one of asymptotic minimization of computational work for classes of elliptic problems, solutions of which should be elements of a compact set M in a considered Sobolev space G (the typical example for second-order problems is $G = W_2^1(\Omega)$, $M = \{u : \|u\|_{W_2^{1+\gamma}(\Omega)} \leq K\}$, where the parameter $\gamma > 0$ characterizes the solution's additional smoothness). Approximation properties of these compact sets can be described in several ways, but we prefer to use $N_0(\varepsilon)$-widths in the sense of Kolmogorov (see citeogan-r,pinkus,ti), as suggested by Babushka and Sobolev in 1965 (see citebab-s), which is especially suitable for projective-grid methods (finite element methods).

The asymptotic behavior of $N_0(\varepsilon)$ defines the optimal one of the minimal number of degrees of freedom $N(\varepsilon) \asymp N_0(\varepsilon)$ and the optimal asymptotic behavior of $W(\varepsilon) \asymp N_0(\varepsilon)$ of the required computational work to get such accuracy. Construction of computational algorithms with $W(\varepsilon) \asymp N_0(\varepsilon)$ is the main objective of this book. It is based on the author's work [150–206] and is achieved under fairly general conditions in general dimensions. To obtain such results, special modifications of finite element and difference methods are used for which effective iterative methods with preconditioning, symmetrization, and coarse grid continuation can be applied. It may be said that the book provides a justification (to a some degree) of the Kolmogorov-Bakhvalov hypothesis that the optimal asymptotic characteristics of numerical methods for a correct elliptic boundary value problem, from the point of view of accuracy and required computational work, should be determined by estimates of ε-entropy or similar information characteristics of the corresponding compact sets.

This book is intended for people interested in contemporary theory of numerical methods for elliptic problems and its applications.[1] To widen the circle of possible readers, some necessary facts from functional analysis are given as well. In this respect, the Introduction plays a very significant role and contains also an overview of the future results. Among the topics considered in detail, we mention: a priori estimates, existence of solutions of nonlinear systems, correctness of problems, estimates of accuracy of projective and difference methods, a posteriori estimates of accuracy, algebraic-geometrical methods for constructing special quasiuniform and composite grids, fast elliptic solvers, and convergence of iterative methods with model operators for discrete analogs of elliptic, saddle-point, and spectral (eigenvalue) problems. The iterative methods include preconditioned iterative methods with spectrally equivalent operators, multigrid methods, two-stage methods with inner iterations, decomposition methods, and fictitious grid domain methods. Incorporation of iteration parameters and use of these methods as inner iterations in the Newton-Kantorovich scheme are also considered.

This book consists of an introduction and nine chapters. The Intro-

---

[1] Basic results described in this book were used by the author in his lectures at Moscow State University (Department of Computational Mathematics and Cybernetics).

duction is actually a special chapter and gives a simplified representation of the obtained results. It also contains some general and necessary facts from Hilbert and Sobolev space theory, from N-widths theory, from correctness theory for elliptic boundary value problems, and from the theory of Rayleigh-Ritz and general projective methods. Similar facts related to Stokes type systems and spectral problems are given in the first paragraphs of Chapters 7 and 9.

Chapter 1 is devoted to general questions of numerical methods for correct operator equations in a Hilbert space, with special emphasis on projective methods. Convergence of iterative methods with model symmetric operators is investigated, including the case with perturbations and possible coarse grid continuation. Schur compliments and their use in construction of preconditioners are discussed as well.

Chapter 2 deals mainly with projective-grid methods associated with the triangulation of d-dimensional regions. Estimates of convergence, multilevel splittings of finite element subspaces, estimates of angles between subspaces, and spectral equivalence of projective-grid and difference operators are discussed as well.

In Chapter 3, analysis of the effective iterative methods mentioned above is given for linear and nonlinear grid systems.

Chapter 4 provides some algorithms for construction of special classes of grids and triangulations.

Chapter 5 is devoted to ultimate analysis of the required computational work in the case of coarse grid continuation and includes the general case of a correct elliptic equation, some nonlinear equations and systems, and some elasticity problems.

Chapter 6 describes similar results for difference methods. Special attention is paid to analysis based on the use of the inner product in Sobolev space $W_2^2(Q)$. It enables one to obtain the desired estimates of the computational work for rather strong nonlinear equations and systems. The case of fourth-order equations and systems, including some plate and shell problems, is also described.

Chapter 7 is devoted to asymptotically optimal algorithms for elliptic problems with linear constraints, like Stokes and Navier-Stokes systems in two and three dimensions. Elasticity problems with a small parameter are also considered and special attention is paid to algorithms with estimates of the required computational work independent of this parameter.

In Chapter 8, fourth-order elliptic problems are considered. The central focus is on the methods associated with reduction of the original problem to boundary value problems for second-order systems of Stokes type.

Chapter 9 is devoted to estimates of convergence of projective-grid methods and to the analysis of modified gradient methods in the case of spectral (eigenvalue) problems involving elliptic and strongly elliptic operators. Estimates of computational work are obtained that are of the same type as for elliptic boundary value problems. The case of problems with linear constraints is also included.

Many algorithms considered in this book have found practical applications; by way of an illustration of their capabilities, calculations for elasticity problems and for problems in the theory of nonlinear shells are briefly described. Fortunately, many of the obtained algorithms, considered best from the above theoretical point of view, are often very instructive for designing effective algorithms for many types of modern computers.

All basic mathematical statements are given with sufficiently complete proofs. The almost self-contained exposition is typical of most of the chapters, with the probable exception of Chapter 5, where some statements with proofs similar to those used elsewhere are given only with brief indications. Since the many mathematical results discussed in this book need concentrated attention of the reader, it is important to allow the possibility to consider each section as separately as possible. Therefore, enumeration of theorems, lemmas, figures, and formulas refers to the given section unless preceded by one or two additional numbers, which then indicate the appropriate section and chapter. For example, Theorem 0.1.2 references Theorem 2 of Section 1 of Chapter 0. The same is true with respect to formulas, but they are always given with the corresponding number of the paragraph.

At this point, I would like to express my gratitude to all my colleagues who influenced my understanding of questions discussed in the book. With special gratitude, I recollect the role of S. L. Sobolev, who directed my interest to optimization of numerical methods when I was his postgraduate student. I am also very grateful to N. S. Bakhvalov, Yu. A. Dubinskii, A. V. Knyazev, G. M. Kobelkov, V. S. Ryaben'kii, G. L. Siganevich, V. D. Valedinskii, E. E. Tyrtyshnikov, and A. A. Zlotnik for their comments on the text of the book in Russian.

With regard to the present extended and updated version of the book in English, I would like to express my deep gratitude to Steve McCormick, an editor of the text in English, whose comments were always very instructive and to the point. Moreover, he was one of the initiators who motivated me to undertake the task. A great help to me in the preparation of Latex files was my daughter Barbara.

This work was partially supported by the Independent Research and Development Program at Ecodynamics Research Associates, Inc., under the project management of Dr. Stanly Steinberg.

The final text was completed while visiting the University of Colorado at Denver by invitation of A. Knyazev, J. Mandel, and Steve McCormick.

Editor's Preface

While Gene D'yakonov seldom ventured outside of the Soviet Union before its dissolution, news of his groundbreaking work regularly made its rounds in the global mathematical community. Yet, until now, several of his results have been virtually inaccessible to the West. This text is therefore all the more important in that, in addition to many important developments of other leading mathematicians, it contains most of the major results of Gene's own highly productive mathematical career. It is much more than a simple translation of his earlier Russian book, however: this book includes lots of new results, especially those that fill previous gaps in the theory of asymptotically optimal algorithms for elliptic boundary value problems. Major results cover accuracy estimates for finite difference and *projective-grid* (finite element) methods, development of topologically equivalent grids and triangulations, founding theory for convergence of a broad spectrum of iterative methods based on preconditioning and symmetrization, and a host of applications that include fluid flow and elasticity. An especially compelling aspect of this book is its treatment of algorithm optimization in the context of *N-widths of compact sets in Sobolev spaces and the Kolmogorov-Bakhvalov hypothesis*. It is destined to become a classic in the literature on optimal numerical methods for elliptic partial differential equations.

To quote Professors N. S. Bakhvalov and G. I. Marchuk in their comments on the Russian version:

This unique book was written by one of the most recognized authorities on optimization in numerical methods and contains a series of new results of extremely high mathematical level, which have not only fundamental but also utmost practical importance. Its appearance is a significant event in contemporary computational mathematics. The book must be recommended to the attention of specialists in numerical methods and a wider circle of readers.

According to the Scientific Board of Moscow State University:

The book contains a series of results obtained by D'yakonov of extremely high scientific level and which have become a basis for a new scientific di-

rection in the theory of numerical methods. This direction is connected with construction of asymptotically optimal algorithms for elliptic problems and has not only fundamental importance but also the greatest applied one. Many of the methods suggested by D'yakonov are widely used in computational practice.

In the Bulletin of Highest Qualification Commission of Russia, Bakhvalov writes:

The book itself is a remarkable example of how many deep ideas and results of classical mathematics find the most important applications in the theory of optimization of computational algorithms and in practical solution of many difficult applied problems. The book must be recommended to the attention not only of specialists in computational mathematics, but also to those in functional analysis, differential equations, approximation theory, and computational geometry. Several chapters may be recommended to all interested in questions of optimization of numerical methods.

An ability essential to any pioneer is a deep understanding of the significance and potential implications of new developments. I gained a personal appreciation of Gene's remarkable abilities in this regard when I first met him in the summer of 1992 at Moscow State University. It had been common knowledge in the field that the first modern multigrid methods were developed by Fedorenko and Bakhvalov in the 1960s. But we had also understood multigrid to be only a theoretical curiosity then, with no one appreciating its practical importance until Brandt's revolutionary work of the 1970s. I was very surprised to find this to be a misconception. During my visit, I had asked Gene to translate segments of the summary of his earlier book [162] written in 1966. A sentence of that translation jumped out at me off the page:

We have considered here iterative methods based on the idea of using spectrally equivalent operators. Some recent results published by Bakhvalov and Fedorenko should be mentioned (they analyzed methods associated with a sequence of nested grids). It must be expected that the combination of both ideas will yield iterative methods that are not only asymptotically optimal but also extremely effective for solving hard practical problems.

Seeing this and recognizing its historical significance, I gained an instantaneous admiration of his foresight, which I maintained throughout my reading of the material contained in the following pages. As I read, I began to realize that I would learn not only many important technical matters, but also insights into the future directions of computational mathematics, where his work is no doubt taking us.

The Author

Eugene G. D'yakonov, born in Nevel, U.S.S.R., is a well-known specialist on numerical methods for partial differential equations. He is a Professor at the Moscow State University in the Department of Computer Mathematics and Cybernetics, and is a member of the American Mathematical Society. He received a M.Sc. in 1957 from the Moscow State University and a Ph.D. from the Institute of Applied Mathematics, the Russian Academy of Sciences, in 1963.

The most important results of his investigations are related to the theory of accuracy of grid (finite element and difference) methods for elliptic linear and nonlinear systems and spectral problems with elliptic operators, the theory of effective iterative methods for algebraic problems of a large order, and similar topics for systems with linear constraints such as Stokes and Navier-Stokes systems. Several important principles on construction and investigation of grid methods and corresponding iterative algorithms belong to him. One of them is connected with the fundamental idea of using spectrally equivalent grid operators in preconditioned iterative methods; the beginning of its wide recognition dates back to the 1960s after publication of his pioneer paper in 1961 (see [150]). There is also an international recognition of his investigations of ADI and splitting methods for multidimensional nonstationary problems of mathematical physics. He has been invited to speak at many international conferences.

Some difficult nonlinear problems of mathematical physics have been successfully solved on the basis of effective computational algorithm proposed by Professor D'yakonov. He has authored over 110 papers and three books (in Russian).

The Editor

Steve McCormick is a Professor of Applied Mathematics at the University of Colorado at Boulder. He received his Ph.D. in mathematics at the University of Southern California in 1971, with a thesis about abstract theory for iterative methods in Hilbert spaces. He has since been on the faculty of the Claremont Colleges, Colorado State University, and the University of Colorado at Denver. His research interests include multigrid methods, multilevel adaptive refinement, least squares methods for partial differential equations, eigenvalue problems, inverse problems, structural analysis, particle transport, computational fluid dynamics, and high performance computing.

Basic Notation

1. General.

$\forall, \exists, \Rightarrow, \Leftrightarrow$ are standard logical symbols;

$x \in M, x \notin M, A \subset B, A \cup B, A \cap B, A \setminus B, A \times B$ are standard set theory notations;

$\equiv$ is equality by definition;

$\emptyset \equiv$ emptyset;

$\{x\} \equiv$ a set of elements x;

$\{x : \ldots\} \equiv$ a set of x with the given property;

$F : A \mapsto B \equiv$ a mapping of the set A into the set B;

$F^{-1} \equiv$ the inverse mapping when F is one-to-one;

$F\{A\} \equiv \{y : \exists x \in A \text{ such that } y = F(x)\} \equiv$ the image of A under the mapping F;

$F^{-1}\{M\} \equiv \{x : F(x) \in M\}$;

$\mathbf{R} \equiv$ the set of real numbers;

$[k] \equiv$ the integer part of the number k;

$\mathbf{R}^d \equiv$ the Euclidean space of vectors $x \equiv [x_1, \ldots, x_d]$ with standard linear operations, with inner product $(x, y) \equiv x_1 y_1 + \ldots + x_d y_d$, and norm $|x| \equiv (x, x)^{1/2}$;

$e_r \equiv [\underbrace{0, \ldots, 0}_{r-1}, 1, \underbrace{0, \ldots, 0}_{d-r}] \equiv$ the rth coordinate unit vector in $\mathbf{R}^d$;

$A \in \mathbf{R}^{m \times n} \equiv$ matrix $A \equiv [a_{i,j}]$ with elements $a_{i,j} \in \mathbf{R}$ having m rows and n columns;

$A^T \equiv$ the transpose of the matrix A; $A^T = [a_{j,i}] \in \mathbf{R}^{n \times m}$;

$D(B; n) \equiv$ the block diagonal matrix whose all diagonal blocks are a square matrix B and the number of them is n;

$\Omega \equiv$ a bounded domain (region) in the Euclidean space $\mathbf{R}^d$; in the case of a model region, the symbol Q is often used instead of Ω;

$\partial \Omega \equiv \Gamma \equiv$ the boundary of Ω;

$\bar{\Omega} \equiv \Omega \cup \partial \Omega \equiv$ the closure of Ω;

$|\Omega| \equiv$ the measure of Ω in $\mathbf{R}^d$;

$|\Gamma|_{(d-1)} \equiv$ the $(d-1)$-dimensional measure of Γ consistent with the measure in $\mathbf{R}^d$;

$$(u, v)_{0, \Omega} \equiv \int_\Omega uv d\Omega;$$

$K \neq K(h) \Leftrightarrow$ the constant K does not depend on h;

$f(h) \asymp g(h) \Leftrightarrow$ there exist positive constants κ_0 and κ_1 such that $\kappa_0 \leq f(h)/g(h) \leq \kappa_1$;

$W = O(N) \Leftrightarrow$ there exists a positive constant κ such that $|W| \leq \kappa |N|$;

$f(N) = o(N) \Leftrightarrow \lim_{N \to \infty} f(N)/N = 0$;

$\delta(i; j) \equiv$ the Kronecker symbol: $\delta(i; j) = 1$ if $i = j$ and $\delta(i; j) = 0$ otherwise;

$0 \equiv$ the null number, vector, function, matrix, operator;

$\text{supp} f \equiv$ the closure of the set of points at which the function f does not vanish;

$\arg \min \Phi \equiv \{u : \Phi(u) \leq \Phi(v), \forall v\} \equiv$ the set of elements minimizing the functional Φ.

2. Normed linear spaces.

lin $\{\psi_1, \ldots, \psi_N\}$ ≡ the lineal (the linear span) of elements $\psi_1, \ldots, \psi_N$;

dim S ≡ the dimension of the subspace S;

$S_1 \oplus S_2$ ≡ the direct sum of the subspaces S_1 and S_2;

$\|u\|_U$ ≡ the norm of the element u from a normed linear space U;

$\text{dist}_U\{u; M\} \equiv \inf\limits_{v \in M} \|u - v\|_U$ ≡ the distance from u to the set M;

$S_U(v; r) \equiv \{v : \|u - v\|_U \leq r\}$ ≡ the closed ball in U with center v and radius $r > 0$;

$S_U(r) \equiv S_U(0, r)$;

$\mathcal{L}(U; F)$ ≡ the normed linear space of linear bounded operators L mapping U into F;

$$\|L\|_{U \mapsto F} \equiv \sup_{v \neq 0} \frac{\|Lv\|_F}{\|v\|_U};$$

$\mathcal{L}(U) \equiv \mathcal{L}(U; U)$;

$\|L\|_U \equiv \|L\|_{U \mapsto U}$;

I ≡ the identity operator;

Ker $L \equiv \{v : Lv = 0\}$ ≡ the kernel (null-space) of the operator L;

Im $L \equiv L\{U\}$ ≡ the image (range) of the operator L;

L'_v ≡ the Gateaux derivative of a differentiable operator L evaluated at the point v:

$$L'_v z = \lim_{t \to +0} \frac{L(v + tz) - L(v)}{t}, \quad \forall z; \quad L'_v \in \mathcal{L}(U; F).$$

3. Euclidean and Hilbert spaces.

G or H ≡ a Euclidean or Hilbert space with the inner product of elements u and v denoted by (u, v); the symbol H is preferred when the space is $\mathbf{R}^N$;

$G_1 \times \cdots \times G_k$ ≡ the Descartes product of the Euclidean or Hilbert spaces $G_1, \ldots, G_k$, which is a Euclidean or Hilbert space with elements $u \equiv [u_1, \ldots, u_k]$ and inner product $(u, v) \equiv (u_1, v_1) + \cdots + (u_k, v_k)$;

$\hat{G}_h$ ≡ a finite-dimensional subspace of G associated with the projective method under consideration;

$S^\perp \equiv \{v : (v, u) = 0, \forall u \in S\}$;

$b(u; v) \equiv (L(u), v)$ ≡ a bilinear or quasibilinear form mapping $G^2 \equiv G \times G$ into $\mathbf{R}$ and associated with the linear or nonlinear operator $L : G \mapsto G$;

G^* ≡ the linear space of bounded linear functionals l mapping G into $\mathbf{R}$;

A^* ≡ the adjoint operator to $A \in \mathcal{L}(G_1; G_2)$: for all $u \in G_1$ and $v \in G_2$, $(Au, v)_{G_2} = (u, A^*v)_{G_1}$;

A_s ≡ the symmetric (self-adjoint) part of $A \in \mathcal{L}(G) : A_s \equiv 2^{-1}(A + A^*)$;

A_a ≡ the antisymmetric (skewsymmetric) part of $A \in \mathcal{L}(G) : A_a \equiv 2^{-1}(A - A^*)$;

$\mathcal{L}^+(G)$ ≡ the set of linear, symmetric, and positive definite operators in $\mathcal{L}(G)$; for an operator $L \in \mathcal{L}^+(G)$, there exists a positive constant ν such that $L \geq \nu I$;

$H(B)$ ≡ the Euclidean space differing from the Euclidean space H only by inner product defined by $B \in \mathcal{L}^+(H)$, namely

$$(u, v)_{H(B)} \equiv (u, v)_B \equiv (Bu, v)_H \equiv (Bu, v);$$

$S_B(u; r) \equiv S_{H(B)}(u; r)$;

$\|L\|_B \equiv \|L\|_{H(B) \mapsto H(B)}$;

$\lambda(A)$ ≡ an eigenvalue of A;

sp $A \equiv \{\lambda(A)\}$ ≡ the spectrum of A;

$r(A) \equiv \max |\lambda(A)|$ ≡ the spectral radius of A;

$\theta_G(S; S') \equiv \|P - P'\|$ ≡ the gap between the subspaces S and S' of G, where P and P' are orthogonal projectors on S and S', respectively;

$\theta_B(S; S') \equiv \theta_{H(B)}(S; S')$.

4. Spaces of functions mapping a region $\Omega \subset R^d$ into R.

$C(\Omega) \equiv$ the Banach space of real-valued continuous functions on Ω with norm $\|u\|_{C(\Omega)} \equiv \sup_x |u(x)|$;

$D_r u \equiv u_{x_r} \equiv \frac{\partial u}{\partial x_r}$;

$u_{x_r, x_l} \equiv D_r D_l u$;

$D^\alpha u \equiv D_1^{\alpha_1} \ldots D_d^{\alpha_d} u$ $(\alpha \equiv [\alpha_1, \ldots, \alpha_d]$, $|\alpha| \equiv \alpha_1 + \cdots + \alpha_d$ is the order of the derivative;

$\nabla u \equiv [D_1 u, \ldots, D_d u] \equiv \text{grad } u$;

$$|\nabla u| \equiv [(D_1 u)^2 + \cdots + (D_d u)^2]^{1/2};$$

$C_0^\infty(\Omega) \equiv$ the space of functions such that, together with all of their derivatives, are continuous for all $x \in \mathbf{R}^d$ and have supports belonging to Ω;

$L_p(\Omega)$ (for $1 \le p < \infty$)$\equiv$the Banach space with norm

$$\|u\|_{L_p(\Omega)} \equiv (|u|^p, 1)_{0,\Omega}^{1/p} \equiv |u|_{0,p};$$

$L_2(\Omega) \equiv$ the Hilbert space with the inner product $(u, v) \equiv (u, v)_{0,\Omega}$;

$W_p^m(\Omega)$ (for $1 \le p < \infty$ and $m = [m]$) $\equiv$ the Sobolev space of functions having generalized derivatives up to order m such that

$$\|u\|_{W_p^m(\Omega)} \equiv [\sum_{|\alpha| \le m} \frac{|\alpha|!}{\alpha!} (|D^\alpha u|^p, 1)_{0,\Omega}]^{1/p} < \infty,$$

where $\alpha! \equiv \alpha_1! \times \cdots \times \alpha_d!$;

$\overset{0}{W}_p^m(\Omega) \equiv$ the subspace formed by completing $C_0^\infty(\Omega)$ in $W_p^m(\Omega)$;

$W_2^m(\Omega) \equiv G^{(m)} \equiv$ the Hilbert space with the inner product

$$(u, v)_{G(m)} \equiv \sum_{l=0}^m (u, v)_{l,\Omega};$$

$$(u, v)_{l,\Omega} \equiv \sum_{|\alpha|=l} \frac{l!}{\alpha!} (D^\alpha u, D^\alpha v)_{0,\Omega}$$

is (for $l \ge 1$) a semi-inner product associated with the seminorm

$$|u|_{l,\Omega}^2 \equiv \sum_{|\alpha|=l} \frac{l!}{\alpha!} |D^\alpha u|_{0,\Omega}^2;$$

$$(u, v)_{1,\Omega} \equiv (D_1 u, D_1 v)_{0,\Omega} + \cdots + (D_d u, D_d v)_{0,\Omega};$$

$$|u|_{1,\Omega} \equiv (|\nabla u|^2, 1)_{0,\Omega}^{1/2} = \left(\int_\Omega [(D_1 u)^2 + \cdots + (D_d u)^2] d\Omega \right)^{1/2};$$

$|u|_{2,\Omega} \equiv ((D_1^2 u)^2 + 2(D_1 D_2 u)^2 + (D_2^2 u)^2, 1)_{0,\Omega}^{1/2}$ (for $d = 2$);

$W_2^1(\Omega; \Gamma_0)$ (when Γ_0 is a closed subset of Γ consisting of a finite number of pieces belonging to smooth $(d-1)$-dimensional surfaces and $|\Gamma_0|_{(d-1)} > 0$) $\equiv$ the Hilbert space formed by completing the preHilbert space with the inner product $(u, v)_{m,\Omega}$ and whose elements are smooth functions which vanish in some vicinities of Γ_0;

$W_2^{m+\gamma}(\Omega) \equiv G^{(m+\gamma)}$ (for $m = [m] \geq 0, 0 < \gamma < 1$) $\equiv$ the Hilbert space formed by completing the preHilbert space with the inner product

$$(u,v)_{m,\Omega} \equiv (u,v)_{m,\Omega} + \int_\Omega \int_\Omega \sum_{|\alpha|=m} \frac{m!}{\alpha!} \frac{(D^\alpha u(x) - D^\alpha u(y))(D^\alpha v(x) - D^\alpha v(y))}{|x-y|^{(d+2\gamma)}} dx dy$$

and whose elements are smooth functions on Ω;
$G^{1/2}(\Gamma) \equiv$ the Hilbert space of traces on Γ of elements of $G^{(1)}$ with the norm

$$\|u\|_{1/2,\Gamma}^2 \equiv |u|_{0,\Gamma}^2 + |u|_{1/2,\Gamma}^2;$$

$$|u|_{1/2,\Gamma}^2 \equiv \int_\Gamma \int_\Gamma \frac{|u(x) - u(y)|^2}{|x-y|^{d+1}} dx dy;$$

$\mathrm{rot}\, w \equiv [D_2 w, -D_1 w]$ (for $d = 2$)

5. Projective-grid methods (PGMs, FEMs).

$\hat{\Omega}_h \equiv$ an approximation to $\bar{\Omega}$ generated by the grid with the parameter h and consisting of a finite set of cells;
$\hat{\Gamma} \equiv$ an approximation to $\partial\Omega \equiv \Gamma$;
$T_h(\hat{\Omega}) \equiv$ a triangulation (simplicial partition) of $\hat{\Omega}$;
$T_{c,h}(\hat{\Omega}) \equiv$ a composite triangulation of $\hat{\Omega}$ (union of standard triangulations of parts of $\bar{\Omega}$ in its partition under consideration);
$T_{c,h}^{(l)}(\hat{\Omega}) \equiv$ a composite triangulation of the lth level of refinement;
$\Omega_h \equiv$ the set of grid nodes (elementary nodes when domains with slits are considered)
P_i associated with basis functions $\hat{\psi}_i(x)$ of the finite element subspace $\hat{G}_h$;
$T \in T_h \Leftrightarrow T$ is a simplex from the triangulation $T_h(\hat{\Omega})$;
$T \in T_{c,h}(\hat{\Omega}) \Leftrightarrow T$ is a simplex from the composite triangulation $T_{c,h}(\hat{\Omega})$;
$T \in T_{c,h}^{(l)}(\hat{\Omega}) \Leftrightarrow T$ is a simplex from the composite triangulation $T_{c,h}^{(l)}(\hat{\Omega})$ of the lth level of refinement;
$L_h \equiv [b(\hat{\psi}_j; \hat{\psi}_i)] \equiv$ the matrix associated with the given bilinear form and the basis of $\hat{G}_h$ (a projective analog of the linear operator $L \in \mathcal{L}(G)$ defined by the bilinear form $(Lu,v)_G = b(u;v)$ for all u and v);
$B_h, J_h, \Lambda_h \equiv$ some model operators.

6. Difference methods associated with parallelepiped grids.

$h \equiv [h_1,\ldots,h_d] \equiv$ the vector of step sizes of the parallelepiped grid with grid nodes $x_i \equiv [i_1 h_1,\ldots,i_d h_d]$ and the multiindex $i \equiv [i_1,\ldots,i_d]$;
$\|h\| \equiv (h_1^2 + \cdots + h_d^2)^{1/2}$;
$\Pi_i \equiv [i_1 h_1, (i_1+1)h_1] \times \cdots \times [i_d h_d, (i_d+1)h_d] \equiv$ a cell of the grid;
$\|h\| \equiv h_1 \times \cdots \times h_d \equiv$ the volume of a cell of the grid;
$I_r x_i \equiv x_i + h_r e_r, r = 1,\ldots,d$;
$I_{-r} x_i \equiv x_i - h_r e_r, r = 1,\ldots,d$;
$u_i \equiv u(x_i) \equiv$ the value of the grid function at the node x_i;
$I_{-r} u_i \equiv u(x_i - h_r e_r)$;
$I_r u_i \equiv u(x_i + h_r e_r)$;
$\partial_r u_i \equiv h_r^{-1}[I_r u_i - u_i]$;
$\bar{\partial}_r u_i \equiv h_r^{-1}[u_i - I_{-r} u_i]$,
$\hat{\partial}_r u_i \equiv (2h_r)^{-1}[I_i u_i - I_{-r} u_i]$,
$\tilde{\partial}_r u_i \equiv$ any of the differences $\partial_r u_i, \bar{\partial}_r u_i$, or $\hat{\partial}_r u_i$;
$\tilde{\partial}^\alpha u_i \equiv \tilde{\partial}_1^{\alpha_1} \times \cdots \times \tilde{\partial}_d^{\alpha_d} u_i$;
$\tilde{\partial}^0 u_i \equiv u_i$;

$\Lambda_r u_i \equiv -\bar{\partial}_r \partial_r u_i = -h_r^{-2}[I_{-r}u_i - 2u_i + I_r u_i]$;
$\Omega_h \equiv$ the set of nodes associated with unknown values of the grid function in the difference system.

In the case of finite (vanishing outside $\bar{\Omega}$) grid functions, the following inner products and norms are used:
$(\bar{\partial}^\alpha u, \bar{\partial}^\alpha v)_0 \equiv ||h|| \sum_{x_i} \bar{\partial}^\alpha u_i \bar{\partial}^\alpha v_i$;
$|u|_{0,p} \equiv (||h|| \sum_{x_i} |u_i|^p)^{1/p}, 1 \le p < \infty$;
$|u|_{m,p} \equiv (\sum_{|\alpha| \le m} |\partial^\alpha u|_{0,p}^p)^{1/p}$.

7. Spaces of vector-functions mapping Ω into $\mathbf{R}^k$.
$u \equiv [u_1(x), \ldots, u_k(x)], \ x \in \bar{\Omega}$;
$\vec{u} \equiv [u_1(x), \ldots, u_d(x)]$ (for $x \in \mathbf{R}^d$);
div $\vec{u} \equiv D_1 u_1 + \cdots + D_d u_d$;
$\partial^\alpha u \equiv [\partial^\alpha u_1, \ldots, \partial^\alpha u_k]$;
$W_2^m(\Omega) \equiv W_2^{m_1}(\Omega) \times \cdots \times W_2^{m_k}(\Omega) \ (m = [m_1, \ldots, m_k])$;
$\hat{u} \equiv [\hat{u}_1(x), \ldots, \hat{u}_k(x)], \ x \in \hat{\Omega} \equiv$ a progective (finite element) approximation to u;
$u_h \equiv u \equiv [u_1, \ldots, u_k] \equiv$ a grid function mapping Ω_h into $\mathbf{R}^k$;
$u_i \equiv [u_1(x_i), \ldots, u_k(x_i)]^T \equiv [u_1, \ldots, u_k]_i^T$;
$\bar{\partial}^\alpha u_i \equiv [\bar{\partial}^\alpha u_1, \ldots, \bar{\partial}^\alpha u_k]_i$;
$|u|_{m,\Omega} \equiv [|u_1|_{m,\Omega}^2 + \cdots + |u_k|_{m,\Omega}^2]^{1/2}$;
$||u||_{m+\gamma,\Omega} \equiv [||u_1||_{m+\gamma,\Omega}^2 + \cdots + ||u_k||_{m+\gamma,\Omega}^2]^{1/2}$.

8. Iterative methods for grid systems.
$N \equiv N_h \equiv$ the number of the unknowns in the system $L_h(u_h) = f_h, u_h \in H, f_h \in H$;
$u_h \equiv \mathbf{u} \equiv u \equiv [u_1, \ldots, u_N]^T$;
$u^n \equiv$ the nth iterate (approximation at the nth iteration);
$z^n \equiv u^n - u \equiv$ the error at the nth iteration;
$r^n \equiv L_h(u^n) - f \equiv$ the residual at the nth iteration ($r^n = L_h z^n$ when L is linear);
$\tau \equiv$ an iteration parameter;
$J_h \asymp B_h \Leftrightarrow$ the grid operators J_h and B_h are spectrally equivalent.

9. Computational algorithms.
$a \equiv$ computational algorithm;
$a(v) \equiv$ the output of the algorithm a with the initial vector v;
$\epsilon > 0 \equiv$ the desired accuracy;
$W(\epsilon) \equiv W_a(\epsilon) \equiv$ the required computational work in algorithm a for obtaining ϵ accuracy.

10. Iterative methods for algebraic eigenvalue (spectral) problems.
$Lu = \lambda M u$ refers to the original grid spectral problem with $L = L^* > 0, M = M^* > 0$;
$\lambda_1, \ldots, \lambda_N$ denote eigenvalues enumerated in increasing order;
$\nu_1, \ldots, \nu_l$ is the set of all different eigenvalues enumerated in increasing order;
$U(\nu_j) \equiv$ the eigensubspace associated with the eigenvalue $\nu_j, j = 1, \ldots, l$;
$\lambda(u)$ (if $u \ne 0$) $\equiv$ the Rayleigh quotient defined by

$$\lambda(u) \equiv \frac{(Lu, u)}{(Mu, u)};$$

$u^n \equiv$ the nth iterate (an approximation to the desired eigenfunction obtained at the nth iteration);
$r^n \equiv Lu^n - \lambda(u^n)Mu^n \equiv$ the residual at the nth iteration;
$Mu = sLu$ refers to the original grid spectral problem with $M = M^*, L = L^* > 0$;

$s_1, \ldots, s_N$ denote eigenvalues enumerated in decreasing order $(s_1 \geq s_2 \geq \cdots \geq s_N)$;
$t_1, \ldots, t_l$ is the set of all different eigenvalues enumerated in decreasing order;
$s(u)$ (if $u \neq 0$) $\equiv$ the Rayleigh quotient defined by

$$s(u) \equiv \frac{(Mu, u)}{(Lu, u)}$$

$(s(u) = (\lambda(u))^{-1}$ if $M > 0)$;
$r_0^n \equiv Mu^n - s(u^n)Lu^n \equiv$ the residual at the nth iteration.

Contents

Introduction

§ 1. Modern formulations of elliptic boundary value problems

1.1. Variational principles of mathematical physics. Mathematical physics is a fundamental language of modern science, and one of its most powerful tools. This importance was clear to the genius Galileo nearly four centuries ago, and is now commonly understood. All branches of classical mathematical physics like acoustics, elasticity, heat conduction, hydrodynamics, magnetics, metereology, oceanography, atomic physics, and engineering provide many important and often difficult mathematical problems. These problems are usually posed in terms of partial differential equations, frequently as *elliptic boundary value problems* in particular, which are very interesting in themselves and, to some degree, lead to an understanding of other types of problems. Similar equations are encountered very often in the relatively new mathematical branches of chemistry, biology, medicine, sociology, and ecology, e.g.. Therefore, it is reasonable to develop elliptic boundary value problems from a unified mathematical point of view, ignoring their origin and nonmathematical formulations. Moreover, it will be very fruitful to pose these problems and to discuss possible ways to approximate them in the most suitable and clear mathematical language developed in this century, which is based on the fundamental notions and principles of functional analysis (see, e.g., [18, 27, 117, 124, 326, 337, 418, 459, 489]).

For clarity, we start by considering one of the simplest models of elliptic boundary value problems—the *Dirichlet problem* for *Poisson's equation* in a bounded region $\Omega \in \mathbf{R}^2$ with a sufficiently smooth boundary Γ. Its classical formulation is: find a function $u(x) \equiv u(x_1, x_2)$, defined and continuous on the closure $\bar{\Omega}$, such that u has continuous second-order derivatives $u_{x_r x_r}$ in Ω ($r = 1, 2$), it satisfies the Poisson equation

$$-\Delta u \equiv -u_{x_1 x_1} - u_{x_2 x_2} = g(x), \quad x \in \Omega, \tag{1.1}$$

with a given continuous function $g(x)$, and it vanishes on the boundary of the region:

$$u \mid_\Gamma = 0. \tag{1.2}$$

Such formulations of elliptic boundary value problems have certain conceptual advantages, especially when these problems are considered for the first time. But even slight complications yield enough unsettled questions to leave one feeling dissatisfied. For example, for a discontinuous function $g(x)$, which may appear in many practical problems, there is a necessity to change the formulation on lines of discontinuity of $g(x)$.

This feeling of dissatisfaction deepens when we recollect that the problem under consideration is connected to the *variational problem* of minimizing the *energy functional*

$$\Phi(v) \equiv \int_\Omega [v_{x_1}^2 + v_{x_2}^2 - 2vg]d\Omega \equiv (|\nabla v|^2 - 2vg, 1)_{0,\Omega} \tag{1.3}$$

(here and elsewhere, we prefer to write $(u, v)_{0,\Omega}$ instead of $\int_\Omega uvd\Omega$). More precisely, the variational problem is to find

$$u = \arg\min\{\Phi(v); \ \forall v \in V\}, \tag{1.4}$$

that is, to find $u \in V$ such that $\Phi(u) \leq \Phi(v)$, $\forall v \in V$, where the *space of admissible functions* V consists of continuously differentiable functions on $\bar{\Omega}$ that vanish on Γ. V is a *real linear space* with the usual operations of summation and scalar multiplication (only such real linear spaces will be considered in this book). Following the classical Euler's approach, assume that u satisfies (1.4), that is, $\Phi(u) \leq \Phi(w), \forall w \in V$. Given a function $v \in V$, we introduce the function $\varphi(t) \equiv \Phi(u+tv)$, which is quadratic in the real variable t. Now since $\varphi(t) \geq \varphi(0)$, we have $\varphi'(t) \mid_{t=0} = 0$. Therefore, $(u_{x_1}, v_{x_1})_{0,\Omega} + (u_{x_2}, v_{x_2})_{0,\Omega} = (g, v)_{0,\Omega}$, $\forall v \in V$. If it is assumed that u has continuous second-order derivatives, then integration by parts on the left-hand side of this relation yields $(-\Delta u - g, v)_{0,\Omega} = 0$ and, hence, (1.1). So (1.1) is just *Euler's equation* for the variational problem (1.4), which was formulated under weaker assumptions on the admissible functions.

Similar variational problems for the theory of elasticity will be considered in Chapter 5; many examples from other branches of mathematical physics can readily be given. We will be able to study them in a unified form if we consider V as a linear space with an *(energy) inner product* (u, v), defined for the above-mentioned V by

$$(u, v)_V \equiv (u, v)_{1,\Omega} \equiv (u_{x_1}v_{x_1} + u_{x_2}v_{x_2}, 1)_{0,\Omega} \equiv (u, v)_L. \tag{1.5}$$

We recall axioms of an inner product: $(u, v) = (v, u)$, $(t_1u_1 + t_2u_2, v) = t_1(u_1, v) + t_2(u_2, v)$, $\|u\|^2 \equiv (u, u) \geq 0, \|u\| = 0 \Leftrightarrow u = 0$. (Here and elsewhere, 0 may denote a null scalar, vector, function, operator, or matrix.) From these axioms, it is easy to obtain the following important properties of every inner product: [1]

$$(u, v) \leq \|u\|\|v\|, \quad \|u + v\| \leq \|u\| + \|v\|, \tag{1.6}$$

$$(u, v) = 0 \Leftrightarrow \|u + v\|^2 = \|u\|^2 + \|v\|^2, \tag{1.7}$$

$$\|u + v\|^2 + \|u - v\|^2 = 2[\|u\|^2 + \|v\|^2]. \tag{1.8}$$

From this short excursion into basic functional analysis, we return to our original representative variational problem. Realization of the importance of such variational problems as mathematical models of physical states and processes is an outstanding scientific achievement, obtained from the efforts of many great scientists, and over many centuries. [2]

The first few years of the twentieth century may be regarded as the beginning of a modern period of understanding how variational problems should be properly formulated, ushered in by Hilbert's famous lecture at the congress of mathematicians (Paris, 1900), in which he underlined the

---

[1] The first inequality in (1.6) is the *Cauchy's inequality* (sometimes it is referred to as the *Cauchy's-Schwartz inequality*) and it is equivalent to the second one in (1.6), which is known as the *triangle inequality*); (1.7) is the generalized Pythagorean theorem and its converse; and (1.8) is the *parallelogram property*. It will be useful later to observe that (1.6)–(1.8) remain true for the more general *semi-inner product*, which is an inner product in every way except that it may not have the property that $(u, u) = 0$ implies $u = 0$ (in which case $\| \ \|$ is called a *seminorm*).

Recall that any linear space V with $\dim V < \infty$ is a *Euclidean space*; if V is an infinite-dimensional linear space, then we refer to it as a *preHilbert space*—it may be called a *Hilbert space* under some additional assumptions described in the following subsection. We also recall that every linear space with an inner product is a *normed linear space* : $\|v\| \geq 0$, $\|v\| = 0 \Leftrightarrow v = 0$, $\|tv\| = |t|\|v\|$, and $\|u + v\| \leq \|u\| + \|v\|$. Moreover, each normed linear space is a particular case of a *metric space* with distance between u and v defined by $\rho(u; v) \equiv \|u - v\|$. Finally, for V with an inner product, we use V^* to denote the linear space of bounded linear functionals l mapping V into $\mathbf{R}$ with norm $\|l\| \equiv \sup |l(v)|/\|v\|$, $v \neq 0$.

[2] The first insights that nature in all its manifestions chooses the easiest paths can be attributed to the time of Aristotle. Much later, classical works of Fermat, Bernoulli, Leibnitz, Euler, Lagrange, and other remarkable scientists of the seventeenth and eighteenth centuries led to the formulation of some *variational principles of mathematical physics*, especially the *Lagrange variational principle*, which may be considered to be directly responsible for the first appearance of problems like (1.4) (see [124, 326]). The impression made by such perfect physical laws on contemporaries of Lagrange was so strong, in fact, that many considered them as a mathematical proof of the existence of God. In the first half of the nineteenth century, it was believed by many mathematicians, most notably Dirichlet and Riemann, that solutions of problems like (1.4) must exist, and that existence proofs were only waiting to be discovered. Later came a period of disappointment caused by Weierstrass' example of a variational problem of similar type that proved to have no solution.

prime importance of the justification of variational principles in connection with an appropriate extension of the admissible space V. [3]

The solution of this famous problem of Hilbert turned out to be a rather complicated matter, and had taken a number of decades more and the efforts of many outstanding mathematicians to be clearly understood (see [124, 372, 326, 459, 418]). To explain the main difficulty, we need additional facts about problem (1.4) and the space V from (1.5). For this V, consider the norms $\|u\|_V \equiv |u|_{1,\Omega}$, $\|u\|^2_{L_2(\Omega)} \equiv |u|^2_{0,\Omega} \equiv (u^2, 1)_{0,\Omega}$. The well-known (see [18, 117]) *Poincaré-Steklov inequality* holds:

$$|u|_{0,\Omega} \leq \gamma |u|_{1,\Omega} \tag{1.9}$$

for some constant γ. This means that, for a fixed $g \in L_2(\Omega)$ and

$$l(v) \equiv (g, v)_{0,\Omega}, \tag{1.10}$$

we have $\|l\| \leq \gamma |g|_{0,\Omega}$ and

$$\Phi(v) = |v|^2_{1,\Omega} - 2l(v) \geq \inf \Phi(v) \equiv d > -\infty. \tag{1.11}$$

Hence, we can find a sequence $\{u^n\} \subset V$ with the property that $\Phi(u^n) \to d$ as $n \to \infty$. Such a sequence, following the famous publication of Courant, Friedrichs, and Lewy [123], is called a *minimizing sequence* (see [117, 124, 326]). Given $\varepsilon > 0$, we thus have $\Phi(u^n) < d + \varepsilon^2/4$ for all sufficiently large n, which we write as $n > N(\varepsilon)$. Then, for $n > N(\varepsilon)$ and $p \geq 1$, using (1.8) we have $\|u^{n+p} - u^n\|^2_V = 2[\Phi(u^{n+p} + \Phi(u^n)] - 4\Phi([u^{n+p} + u^n]/2) < \varepsilon^2$. Such a sequence is called a *Cauchy's sequence* or a *fundamental sequence*. Thus, each minimizing sequence is a fundamental one, which raises the main question: Is it a convergent sequence or not? Recall that, in the case $V = \mathbf{R}$, notions of convergent and fundamental sequences are equivalent (this statement is known as the *Cauchy's criterion* for convergence). The same is true in the case $V = \mathbf{R}^N$, but the situation is completely different for preHilbert spaces, and this was the main difficulty in the study of variational problems at the beginning of this century.

1.2. Variational problems in a Hilbert space. Suppose now a preHilbert space V at hand has two additional features: every fundamental sequence in it is a convergent one and there exists a sequence of elements $a_1, a_2, \ldots$, such that, for every element $v \in V$, it is possible to choose a subsequence a_{n_k} convergent to v. Then V is called a Hilbert space. The first property is the most difficult to satisfy and is called completeness of

---

[3] This was the essence of his famous twentieth problem. At about this time, the major importance of these principles on computation was demonstrated by Rayleigh, and later by Ritz.

the metric space; the second property characterizes separability of the space and usually does not present any real difficulty, since the desired sequence can be chosen consisting of continuous and smooth functions.

Theorem 1. *Let G be a Hilbert space with norm $\|v\| \equiv \|v\|_G$. Suppose $l \in G^*$ and define the energy functional $\Phi : G \mapsto \mathbf{R}$ by $\Phi(v) \equiv \|v\|^2 - 2l(v)$. Then problem (1.4) with $V = G$ has a unique solution.*

Proof. The proof follows directly from the classical *Riesz representation theorem* (see [18, 117, 292]): for any $l \in G^*$ there exists a unique $f \in G$ such that, $l(v) = (f, v)$, $\forall v \in G$, and $\|f\| = \|l\|$. Then the bilinearity of the inner product allows us to write $\Phi(v) = \|v - f\|^2 - \|f\|^2$, and the theorem follows by observing that the unique minimizer of Φ must be $u = f$. $\square$

In the same way, the following theorem is also proved.

Theorem 2. *Let conditions of Theorem 1 be satisfied. Let u' be the solution of the same problem but with l' replacing l. Then $\|u - u'\| = \|l - l'\|$.*

Consider a Hilbert space G and $G^2 \equiv G \times G$, the *Descartes product*. Then a mapping $b : G^2 \mapsto \mathbf{R}$ is a *bilinear form* defined on G^2 (see [18, 117]) if $b(u; v)$ is linear in the variable $u \in G$ for each fixed $v \in G$ and it is linear in $v \in G$ for each fixed $u \in G$.

Theorem 3. *Let G be a Hilbert space and $b(u; v)$ be a bilinear form defined on G^2 such that, for all $u, v \in G$,*

$$b(u; v) = b(v; u), \tag{1.12}$$

$$|b(u; v)| \leq \delta_1 \|u\| \|v\|, \tag{1.13}$$

$$I(u) \equiv b(u; u) \geq \delta_0 \|u\|^2, \ \delta_0 > 0. \tag{1.14}$$

Then problem (1.4) with $V = G, l \in G^*$, and

$$\Phi(v) \equiv b(v; v) - 2l(v) \tag{1.15}$$

is equivalent to the problem of finding a unique $u \in G$ such that

$$b(u; v) = l(v), \ \forall v \in G. \tag{1.16}$$

Proof. Note that the bilinear form under consideration is in fact an inner product, so it induces a new geometry on the linear space G. Let $G(L)$ denote this new space under the new inner product

$$(u, v)_L \equiv b(u; v) \tag{1.17}$$

and corresponding norm $\|u\|_L \equiv (u, u)_L^{1/2}$. Then by (1.13) and (1.14) we have $\delta_0 \|u\|^2 \leq b(u; u) \leq \delta_1 \|u\|^2$, $\forall u \in G$. Hence, each fundamental sequence in $G(L)$ is a fundamental one in G, and they converge to the same

element. This implies that $G(L)$ is a Hilbert space and $l \in (G(L))^*$. Therefore, again using the Riesz theorem, we infer the existence of an element
$f_L \in G$ such that $l(v) = (f_L, v)_L$ and $\Phi(v) = \|v\|_L^2 - 2(f_L, v)_L$. Hence,
Theorem 1 applies and the unique solution of (1.16) is $u = f_L$. $\square$ [4]

**1.3. Completion of a preHilbert space and basic properties
of Sobolev spaces.** Theorems 1–3 have emerged as a result of the efforts of mathematicians of this century, enabled by the creation of a new
mathematical tool and language—functional analysis. This development
was pioneered primarily by Hilbert, Friedrichs, Sobolev, and Mikhlin (see
[326, 373, 459, 418]). The main hurdle that this theory had to overcome to
justify the variational principle was, as it is now understood, to pass from
the given preHilbert space to an appropriate Hilbert space, that is, to obtain the desired completeness. This hurdle is similar to that of passing from
the rational numbers to the reals. In this respect, the most promising construction was suggested by Cauchy, and it was used by Hausdorff for completion of general metric spaces (recall that every linear space with an inner
product is a metric space with a distance $\rho(u; v) \equiv \|u - v\|$). In the *Haus-
dorff completion*, we consider all fundamental sequences $\{u^n\} \equiv \mathbf{u}$ and call
two fundamental sequences $\mathbf{u}$ and $\mathbf{v}$ *equivalent* if $\lim_{n \to \infty} \|u^n - v^n\| = 0$.
This does give a true equivalence relation: $\mathbf{u} \sim \mathbf{u}$; $\mathbf{v} \sim \mathbf{u}$ implies $\mathbf{u} \sim \mathbf{v}$;
and $\mathbf{u} \sim \mathbf{v}$ and $\mathbf{v} \sim \mathbf{w}$ imply $\mathbf{u} \sim \mathbf{w}$. Therefore, in the set of all
such fundamental sequences, $\mathbf{u}$, we can consider the *equivalence classes*
$Cl(\mathbf{u}) \equiv \mathbf{U} \equiv \{\mathbf{v} : \mathbf{v} \sim \mathbf{u}\}$. The set of these classes is clearly a preHilbert
space under the definitions

$$\alpha \mathbf{U} + \beta \mathbf{V} \equiv Cl(\alpha \mathbf{u} + \beta \mathbf{v}) \text{ and } (\mathbf{U}, \mathbf{V}) \equiv \lim_{n \to \infty} (u^n, v^n), \qquad (1.18)$$

where $\mathbf{u}, \mathbf{v}$ are representatives of $\mathbf{U}, \mathbf{V}$, respectively, and $\alpha, \beta \in \mathbf{R}$. [5]

---

[4] Properties (1.12)–(1.14) are often referred to by the respective terminology symmetry, boundedness, and coercitivity. The Hilbert space $G(L)$, often referred to as the
energy space, was probably first suggested by Friedrichs. The idea of choosing appropriate geometries in this way is of fundamental importance, not only in the analysis of
elliptic boundary value problems but also in the construction of effective iterative methods for solving their discretizations. It is straightforward to generalize Theorem 3 to
the case where the minimum in (1.4) over V is replaced by the minimum over a *linear
manifold* $\varphi + G' \equiv \{v : v = \varphi + w, w \in G'\}$, where G' is a *subspace* of G (that is, G'
contains each linear combination of its elements and may itself be regarded as a Hilbert
space). This case is typical of inhomogeneous Dirichlet conditions for (1.1), where v on
the boundary Γ must coincide with a given function φ. If the function ϕ is given as an
element of G, then the simple change of variables $v = \varphi + v'$ leads to the case considered
in Theorem 3. Similar elegant theorems, dealing with minimization on a given bounded,
closed, and nonempty subset of G and leading in the terminology of Lions to *variational
inequalities* (see [109, 235, 280, 246]), are now widely known.

[5] More difficult is the proof that this space is complete (see [292, 511]). Note that the

Of prime importance to us is the *Sobolev completion*, which is the special case of a Hausdorff completion based on the *Sobolev norm*. We begin with a more general form designed for functional normed spaces and leading to particular *Banach spaces*, which are complete normed linear spaces. We confine ourselves to a sketch of the basic notions and theory, and refer the reader elsewhere for details and an extensive treatment of the subject (see [3, 67, 292, 382, 459]). We begin with the fundamental Banach space $L_p(\Omega)$, $p \geq 1$, where Ω is a bounded region in $\mathbf{R}^d$ with the *Lipschitz boundary* Γ (see [3, 67, 256]). Consider a set of continuous functions $u(x)$ on $\bar{\Omega}$ such that

$$\|u\|_{L_p(\Omega)} \equiv (|u|^p, 1)_{0,\Omega}^{1/p} \equiv |u|_{0,p} < \infty, \qquad (1.19)$$

where the integral over Ω is understood in the Riemann sense. By the well known *Hölder inequality*

$$|(u, v)|_{0,\Omega} \leq |u|_{0,p}|v|_{0,q}, \qquad (1.20)$$

with $1/p + 1/q = 1$, it is possible to prove that this set is a normed linear space, the completion of which (see [3, 423]) is the Banach space $L_p(\Omega)$. [6]

Let $C_0^\infty(\Omega)$ denote the space of real-valued functions of $x \in \mathbf{R}^d$ that are continuous together with their derivatives of any order and that have support belonging to the region Ω (denote supp f as the closure of the set of points where the function f does not vanish). For given $u, w \in L_1(\Omega)$ and for a given index $1 \leq r \leq d$, if $(u, D_r\phi)_{0,\Omega} = -(w, \phi)_{0,\Omega}$, $\forall \phi \in C_0^\infty(\Omega)$, where $D_r\phi \equiv \frac{\partial \phi}{\partial x_r}$, then w is called a *generalized derivative in the Sobolev sense* of the function u with respect to x_r; similarly, if

$$(u, D^\alpha\phi)_{0,\Omega} = (-1)^{|\alpha|}(w, \phi)_{0,\Omega}, \ \forall \phi \in C_0^\infty(\Omega), \qquad (1.21)$$

with a multi-index $\alpha \equiv [\alpha_1, \ldots, \alpha_d]$, $D^\alpha\phi \equiv D_1^{\alpha_1}\ldots D_d^{\alpha_d}\phi$, and $|\alpha| \equiv \alpha_1 + \ldots + \alpha_d$, then w is a generalized derivative of the order $|\alpha|$. It is

---

subspace consisting of those classes containing a *stationary sequence* (that is, a sequence of the form $\{u, u, \ldots\}$ for some $u \in V$) is isometric to V.

[6] This completion leads to rather dramatic changes in the nature of elements of the spaces, which may be difficult for a reader without the proper mathematical background: sometimes it will be necessary to abandon the notions of Riemann integrals and continuous functions in favor of the more general Lebesgue integrals and measurable functions. It is important to keep in mind that, when speaking about a function, we will mean in fact a set of *equivalent functions*: two functions are equivalent if they differ only on a set of measure zero. Other changes will occur when we consider Sobolev spaces: usual derivatives will be replaced by generalized derivatives in a Sobolev sense which we will define. Nonetheless, all of the above mentioned notions carry over to this more general setting provided we have reasonably good types of regions and functions, in particular, approximating functions. So we will outline only a few central concepts and results of this remarkable branch of the modern theory of functions.

easy to show that $c_1 D^\alpha u_1 + c_2 D^\alpha u_2 = D^\alpha [c_1 u_1 + c_2 u_2]$ for any constants c_1 and c_2 if $D^\alpha u_1$ and $D^\alpha u_2$ exist.

The *Sobolev space* $W_p^m(\Omega)$, with an integer $m = [m] \geq 0$, consists of all functions $u \in L_p(\Omega)$ such that $D^\alpha u \in L_p(\Omega)$ for all α such that $|\alpha| \leq m$; the norm in $W_p^m(\Omega)$ is given by

$$\|u\|_{W_p^m(\Omega)} \equiv [\sum_{|\alpha| \leq m} \frac{|\alpha|!}{\alpha!} (|D^\alpha u|^p, 1)_{0,\Omega}]^{1/p}, \tag{1.22}$$

where $\alpha! \equiv (\alpha_1!) \dots (\alpha_d!)$ and $D^\alpha u = u$ if $|\alpha| = 0$.[7] Here and below we assume that Γ satisfies certain additional conditions, such as being piecewise smooth. Then, $W_p^m(\Omega)$ is also a completion of the space of m-times continuously differentiable functions on $\bar{\Omega}$ in the metric connected with the norm (1.22). For $p = 2$, we have the especially important case of the Hilbert space $W_2^m(\Omega) \equiv G^{(m)}$. (This space is often denoted by H^m, but we instead reserve the notation H for Euclidean spaces.) Then the Sobolev norm and corresponding seminorm are given, respectively, by

$$\|u\|_{m,\Omega}^2 = \sum_{k=0}^{m} |u|_{k,\Omega}^2 \quad \text{and} \quad |u|_{k,\Omega}^2 \equiv \sum_{|\alpha|=k} \frac{k!}{\alpha!} |D^\alpha u|_{0,\Omega}^2, \tag{1.23}$$

independent of the *Descartes coordinate system* (see [460]).

Equivalent Norm Theorem. [8] *Let $G = W_2^m(\Omega)$ and suppose we are given functionals $l_1, \dots, l_k$ in G^* with the following property: any polynomial Q of degree at most $m - 1$ that satisfies the conditions $l_i(Q) = 0, i \in [1, k]$, must be $Q = 0$. Then the norm (1.22) on G is equivalent to the norm*

$$|u|_{m,p} \equiv (\sum_{|\alpha|=m} |D^\alpha u|^p, 1)_{0,\Omega}^{1/p} + \sum_{i=1}^{k} |l_i(u)|. \tag{1.24}$$

$C(\Omega)$ *Embedding Theorem. For the space $W_p^m(\Omega)$ with $pm > d$, there exists a constant K such that every $u \in W_p^m(\Omega)$ is an element of $C(\Omega)$ and $\|u\|_{C(\Omega)} \equiv \sup |u(x)| \leq K \|u\|_{W_p^m(\Omega)}$.* [9]

---

[7] There is also the widely known approach of constructing a similar space on the basis of the Schwartz theory of distributions (see [338]).

[8] This theorem (see [67, 292, 456, 459]) will be extremely useful for obtaining estimates of the accuracy of various interpolations in Sobolev spaces and, eventually, for estimates of the accuracy of various discretization methods.

[9] It should be emphasized that u here relates to an appropriate representative of the class of functions equivalent to u, as is usually understood in all similar statements about elements of Sobolev spaces (see [3, 67, 456, 459]).

To formulate the next theorem (see [3, 67, 456, 459]), which deals with two important cases simultaneously, denote by Γ' either Ω if $s = d$ or an intersection of Ω and an s–dimensional and (piecewise) smooth surface S; in particular, for $s = d - 1$, Γ' may be the boundary $\partial\Omega \equiv \Gamma$. Now if u is a continuous function on $\bar{\Omega}$, then the *trace* of u on Γ' is defined to be simply the restriction of u to Γ', that is, the function equal to u at all points of Γ'. In the general case, we approximate u by a sequence $\{u^n\}$ of continuous functions from $W_p^m(\Omega)$ so that $\lim_{n\to\infty} \|u - u^n\|_{W_p^m(\Omega)} = 0$, and consider traces of u^n on Γ'. Then the following theorem asserts that the traces of u^n constitute a fundamental sequence in $L_q(\Gamma')$ and its limit serves as the trace of u. To avoid further notation, we denote the trace of u simply by u. For an equivalent definition of the trace of a function, see [3, 67, 292].

$L_q(\Gamma')$ **Embedding Theorem.** *Suppose that* $1 \leq s \leq d$, $1 < p \leq q < \infty$, *and* $0 \leq m - d/p + s/q$. *Then, for each* $u \in W_p^m(\Omega)$, *there exists a* trace *on* Γ' *such that*

$$\|u\|_{L_q(\Gamma')} \leq K\|u\|_{W_p^m(\Omega)} \tag{1.25}$$

with constant K independent of u.

The given theorems enable us to use simplified norms for some subspaces of $W_p^m(\Omega)$. As an example, take $p = 2$ and define Γ_0 as a closed subset of $\Gamma \equiv \partial\Omega$ consisting of a finite number of pieces belonging to smooth $(d-1)$–dimensional surfaces. By $|\Gamma|_{(d-1)}$ we denote the $(d-1)$-dimensional measure defined on Γ that is consistent with the measure on $\mathbf{R}^d$ (for $d = 2, |\Gamma|_{(1)}$ is simply the length of Γ). When $|\Gamma_0|_{(d-1)} > 0$, we define a Hilbert space as the completion of the preHilbert space that is comprised of smooth functions that vanish in the vicinity of Γ_0 and that is equipped with the inner product $(u, v)_{m,\Omega}$ (see (1.23)). In particular, for $m = 1$ and sufficiently well behaved Ω and Γ_0, it is possible to prove that this Hilbert space coincides with the subspace of $W_2^1(\Omega)$ consisting of functions with zero traces on Γ_0 (in (1.24) we can take $k = 1$ and $l_1(u) \equiv (u, 1)_{0;\Gamma_0}$). So, in this case, we may replace $\|u\|_{W_2^1(\Omega)}$ by the equivalent but simpler norm $|u|_{1,\Omega}$. In the case $\Gamma_0 = \Gamma$, this space, which we denote by $G_0^{(1)}$, is a completion of the space of functions vanishing in the vicinity of the boundary. The formulated embedding theorems imply the boundedness of the embedding operators that map the space under consideration into $C(\Omega)$ and $L_q(\Gamma')$. The strengthened variants of these theorems, with $m - d/p + s/q > 0$, maintain that the operators are not only bounded, but also *compact* (that is, they map each bounded sequence into a sequence containing a convergent subsequence or, said differently, they map bounded sets into precompact sets; see [3, 67, 292, 459]).

In the theory of elliptic boundary value problems, especially in the case of regions with nonsmooth boundary, of fundamental importance are generalized Sobolev spaces $W_p^m(\Omega)$, where $m = [m] + \gamma$, $[m] \geq 0$ is an integer, and $0 < \gamma < 1$.[10] We confine ourselves to the case $p = 2$, which leads to the Hilbert space $G^{(m)}$ defined as the completion of smooth functions under the inner product

$$(u, v)_{m,\Omega} \equiv (u, v)_{[m],\Omega} +$$

$$\int_\Omega \int_\Omega \sum_{|\alpha|=[m]} \frac{[D^\alpha u(x) - D^\alpha u(y)] \times [D^\alpha v(x) - D^\alpha v(y)]}{|x - y|^{(d+2\gamma)}} dx\, dy. \qquad (1.26)$$

An analogous space $W_2^m(\Gamma)$ can be introduced for the boundary. Its most important case in the sequel will be for $m = 1/2$, which defines the space $G^{1/2}(\Gamma)$ with the norm $\|u\|_{1/2,\Gamma}^2 \equiv |u|_{0,\Gamma}^2 + |u|_{1/2,\Gamma}^2$, where

$$|u|_{1/2,\Gamma}^2 \equiv \int_\Gamma \int_\Gamma |x - y|^{-(d+1)} |u(x) - u(y)|^2 dx\, dy.$$

Direct and Inverse Trace Theorems for $W_2^1(\Omega)$. There exist positive constants K and K' such that, for each $u \in G^{(1)}$, its trace on Γ belongs to $G^{1/2}(\Gamma)$ and $\|u\|_{G^{1/2}(\Gamma)} \leq K\|u\|_{G^{(1)}}$. Conversely, for each given function $u \in G^{1/2}(\Gamma)$, there exists a function $v \in G^{(1)}$ such that its trace on Γ coincides with u and $\|v\|_{G^{(1)}} \leq K'\|u\|_{G^{1/2}(\Gamma)}$. [11]

Equivalent norm and embedding theorems remain true for general m and a wider class of regions that do not necessarily have Lipschitz boundaries. For instance, the unit circle with a cut defined by one radial line is permitted. However, our next theorem depends critically on having a Lipschitz boundary.

Extension Theorem. For any region Ω with Lipschitz piecewise smooth boundary, there exists a bounded linear operator $E \in \mathcal{L}(W_p^m(\Omega); W_p^m(\mathbf{R}^d))$ such that Eu is an extension of any $u \in W_p^m(\Omega)$, that is, $Eu(x) = u(x)$ if $x \in \Omega$ (see [67]).

---

[10] They were introduced and investigated by Besov, Gagliardo, Nicol'skii, Slobodetckii, and Yakovlev (see [3, 68, 237, 338, 506]). These investigations led to an understanding of the Dirichlet principle via the discovery of a remarkable criterion for a function defined on the boundary to be the trace of some function in $G^{(1)} \equiv W_2^1(\Omega)$. To appreciate the significance of this criterion, it is worth noting that not even continuity of a function on Γ is sufficient for it to be the trace of a function in $G^{(1)}$, as was demonstrated by Hadamard ([459]).

[11] We can replace $|u|_{1/2,\Gamma}^2$ by $|u|_{1/2}^2 \equiv \sum_{r=1}^d \int_\Gamma \int_t (\frac{u(x+te_r)-u(x)}{t})^2 d\Gamma\, dt$ (it was shown by Burenkov, Gagliardo, Yakovlev (see [68, 237]) that these norms are equivalent). Works of Andreev and Dryja (see [10, 141]), which are devoted to the use of discrete analogs of $G^{1/2}(\Gamma)$, underlined the real significance of such spaces from the computational point of view. During the past decade, a substantial effort has been spent on studying the use of these grid spaces as a basis for *domain decomposition methods* (see, e.g., [84, 130, 142, 384, 457] and references therein).

Finally, we note that similar results hold for $p = \infty$ (see [3]).

1.4. Generalized solutions of elliptic boundary value problems.
We began this book by considering a representative elliptic boundary value problem in its classical form (1.1)–(1.2), then developing its variational form (1.4)–(1.15) in an appropriate Hilbert space. The solution of the variational problem is usually referred to as the *generalized* or *weak solution* of the original problem (see [18, 326, 459]). Most suitable for us, as we have already mentioned, are the *projective formulations* like (1.16), both as a means of specifying the problem and as a substantial step towards construction of effective numerical methods for its approximate solution. Similar formulations can be found in the classical works of Euler, Bubnov, and Galerkin, but, of course, without Hilbert spaces. The formulations we develop are connected with such names as Vishik, Lax, and Milgram (see [18, 117, 326, 371, 493]. Theorem 5 below is often referred to as the *Lax-Milgram theorem*. But first we emphasize the possibility to use *operator equation formulations*.

Theorem 4. Let G be either a Hilbert or Euclidean space. Suppose that $b(u; v)$ is a bilinear form on G^2 that satisfies (1.13) and that $l(v) \equiv (f, v)$ for some $f \in G$. Then (1.16) is equivalent to the linear operator equation

$$Lu = f, \qquad (1.28)$$

where the operator L is defined by the relation $(Lu, v) \equiv b(u; v)$, $\forall u, \forall v \in G$, and satisfies $\|L\| \leq \delta_1$. If (1.12) and (1.14) are also satisfied, then L is symmetric and $\delta_0 I \leq L \leq \delta_1 I$.

Proof. The proof is readily obtained if we again use the Riesz theorem and, for fixed u and each v, write $b(u; v) = (w, v)$ and define $w \equiv Lu$. That L is a linear operator follows from $b(tu + t'u'; v) = tb(u; v) + t'b(u'; v)$. Writing $v = Lu$, we have $\|Lu\|^2 = b(u; Lu) \leq \delta_1 \|u\| \|Lu\|$, which gives the estimate for $\|L\|$. The remaining assertions are immediate. $\square$

Theorem 5. Let G be a Hilbert or Euclidean space. Suppose that $b(u; v)$ is a bilinear form satisfying conditions (1.13) and (1.14) and that $l(v) = (f, v)$ for a $f \in G$. Then problem (1.16) is equivalent to the linear operator equation (1.28) with such an invertible operator L that $\|L^{-1}\| \leq \delta_0^{-1}$.

Proof. If we could prove that L^{-1} exists, then it would be a linear operator and the conditions of the theorem would imply $\delta_0 \|u\|^2 \leq (Lu, u) = (f, u) \leq \|f\| \|u\|$. Replacing u by $L^{-1}f$, we could then deduce $\delta_0 \|L^{-1}f\| \leq \|f\| \|L^{-1}f\|$, from which would follow $\|L^{-1}f\| \leq \delta_0^{-1} \|f\|$. So it suffices to prove the existence of L^{-1}, which can be done in the simplest way if we replace equation (1.28) by the equivalent equation $u = Q(u)$, where $Q(u) \equiv u - \tau(Lu - f)$ and $\tau > 0$. It easy to see that, for small enough

$\tau > 0$, we have $(1 - 2\tau\delta_0 + \tau^2\delta_1^2)^{1/2} \equiv q < 1$. Then, for all u and v, we have

$$\|Q(v) - Q(u)\|^2 = \|v - u\|^2 - 2\tau(Lv - u, v - u) + \tau^2\|Lv - u\|^2 \le q^2\|v - u\|^2.$$

Hence, Q is a *contraction operator*, and the *contraction mapping princi-ple* (see [292, 317]) guarantees that the equation $u = Q(u)$ has a unique solution. Thus, L^{-1} must exist. $\square$ [12]

Consider an example of applying Theorem 5. With the space $W_2^1(\Omega; \Gamma_0)$ defined above, suppose $|\Gamma_0|_{(d-1)} > 0$ and let $\Gamma_1 \equiv \Gamma \setminus \Gamma_0$ and $\|u\| \equiv |u|_{1,\Omega}$. Define the bilinear form

$$b(u; v) \equiv b^{(0)}(\Omega; u; v) + (\sigma, uv)_{\Gamma_1}, \tag{1.30}$$

where

$$b^{(0)}(\Omega; u; v) \equiv \sum_{r,l=1}^{d} (a_{rl}, D_l u D_r v)_{0,\Omega}$$

$$+ \sum_{r=1}^{d} (b_r v D_r u - b_r' u D_r v, 1)_{0,\Omega} + (c, uv)_{0,\Omega}, \tag{1.31}$$

$(u, v)_{\Gamma_1} \equiv (u, v)_{0,\Gamma_1}$, and the coefficients a, b, b', c, σ are assumed to be piece-wise continuous and bounded, which guarantees the existence of all integrals in (1.30) and (1.31).

Lemma 1. *Under the above conditions, the form (1.30) satisfies (1.13).*

Proof. It suffices to prove that each term on the right-hand sides of (1.30) and (1.31) are bounded in absolute value from above by $K\|u\|\|v\|$. For example, let $X_r \equiv |(b_r v, D_r u)_{0,\Omega}|$, $r \in [1, d]$. Then

$$X_r \le \sup |b_r(x)|(|v|, |D_r u|)_{0,\Omega} \le \sup |b_r(x)|\|v\|_{0,\Omega}|u|_{1,\Omega},$$

and, by virtue of (1.9), we deduce that $X_r \le \|u\|\|v\|$. For estimation of the term involving σ, it is possible to apply (1.25) and write

$$|u|_{0,\Gamma_1} \le \gamma_1 |u|_{1,\Omega}. \tag{1.32}$$

Other terms are treated analogously. $\square$

For verifying (1.14), we introduce $\vec{b}(x) \equiv [b_1(x) - b_1'(x), \dots, b_d(x) - b_d'(x)]$, $|\vec{b}| \equiv \sup |\vec{b}(x)|$, $|c_-| \equiv \sup |c(x)|$, $|\sigma_-| \equiv \sup |\sigma(x)|$, where the supre-mum for $|\vec{b}(x)|$ is taken over all $x \in \Omega$ and the other two are taken over x such that $c(x) < 0$ and $\sigma(x) < 0$, respectively. I_d will denote the identity matrix in $\mathbf{R}^{d \times d}$ and ν_0 will denote a positive constant.

---

[12] The contraction property of Q was used in a more general case in [155, 315, 408]. It is easy to see that $\tau = \delta_0/\delta_1^2$ minimizes the contraction constant q.

Lemma 2. Let the matrix $A \equiv 2^{-1}[a_{r,l}(x) + a_{l,r}(x)]$ be such that

$$A \geq \nu_0 I_d, \quad \forall x \in \Omega, \tag{1.33}$$

and $\nu_0 - |c_-|\gamma^2 - |\sigma_-|\gamma_1^2 - \gamma|\vec{b}| \equiv \nu_0 - \nu' \equiv \delta_0 > 0$. Then (1.14) holds.

Proof. It is easy to see that $b(u; u) \geq \nu \sum_{r=1}^{d} |D_r u|_{0,\Omega}^2 -$ $|\sum_{r=1}^{d}(b_r - b_r')u, D_r u)_{0,\Omega}| - |c_-|\|u\|_{0,\Omega}^2 - |\sigma_-|\|u\|_{0,\Gamma_1}^2$. We can estimate the terms with $|c_-|$ and $|\sigma_-|$ using (1.19) and (1.32). To estimate the term $X \equiv \sum_{r=1}^{d}((b_r - b_r')u, D_r u)$, we introduce an inner product on the space $(L_2(\Omega))^d$ and use the Cauchy inequality (1.6), which implies that $|X| \leq [(\sum_{r=1}^{d}(|b_r - b_r'|^2, u^2)_{0,\Omega}^{1/2}|u|_{1,\Omega}$. Thus, $|X| \leq |\vec{b}||u|_{0,\Omega}\|u\| \leq |\vec{b}|\gamma\|u\|^2$ and (1.14) holds. $\square$

Theorem 5 together with the subsequent lemmas leads to the conclusion that problem (1.16), (1.30)–(1.33) has a unique solution and is equivalent to equation (1.28) with L having bounded inverse L^{-1}. Such linear problems are referred to as *correct* ones. [13]

§ 2. Projective-grid methods (finite element methods)

2.1. Rayleigh-Ritz method. Consider the variational problem from Theorem 1.3. At the center of our attention from now on will be questions on the theory of numerical methods for such and more general problems.

---

[13] This notion was introduced by Hadamard and plays a fundamental role in the theory of operator equations in general and of elliptic boundary value problems in particular. It implies that (1.28), for every given $f \in G$, has a solution that is unique and depends continuously on f. We shall discuss this subject in Chapter 1 in connection with nonlinear problems and their approximations. It should be emphasized that only boundary conditions of Dirichlet type on Γ_0 were incorporated in the structure of the space G (they are often referred to as *essential boundary conditions*). The role of the remainder of the boundary was quite different and was connected with the bilinear form and the linear functional, which in the general case might have taken in the form $l(v) \equiv (g, v)_{0,\Omega} + \sum_{r=1}^{d}(g_r, v)_{0,\Omega} + (g', v)_{0,\Gamma_1}$ with given functions $g, g_r \in L_2(\Omega)$ and $g' \in L_2(\Gamma_1)$. In the classical setting, boundary conditions on Γ_1 should be specified and are referred to as *natural boundary conditions*; they include conditions of *Neumann* type. It is also very important to note that the operator L in (1.28), defined by the relation $(Lu, v) \equiv b(u; v)$, $\forall u, v \in G$, (see Theorem 4), is not a standard differential operator. For example, if $b(u; v) = (u, v)$ then L is the identity operator in G although this bilinear form was associated with problem (1.1), (1.2). Along the same lines we see that form (1.30) with $\sigma = 0$ corresponds to the mixed boundary value problem, with boundary conditions of Dirichlet and Neumann type on Γ_0 and Γ_1, respectively, for the elliptic equation $L_d u \equiv -\sum_{i,l=1}^{d} D_r(a_{rl}D_l u) + \sum_{r=1}^{d}(b_r D_r u + D_r(b'u)) + cu = g$ (in this case $l(v) = (f, v) = (g, v)_{0,\Omega}$). Finally, conditions that we imposed here on the coefficients could have been significantly more general (see [326] and § 5.1). Moreover, there are many special cases that simplify the development. For instance, if $\Gamma_0 = \Gamma$ and $D_r(b_r - b_r') \leq 0$ for all r, then $X \geq 0$.

One of the numerical methods of fundamental theoretical importance, and at the same time serving as a key structure of many modern and widely used computational algorithms, is the *Rayleigh-Ritz method*. The basic idea is very natural and simple: the original variational problem posed in the Hilbert space G is restricted to finite-dimensional subspaces:

$$\hat{u} = \arg\min \Phi(\hat{v}), \quad \hat{v} \in \hat{G}, \tag{2.1}$$

where $\hat{G} \equiv \hat{G}_N$, with $\dim \hat{G}_N = N$, are subspaces of G and are expected to produce proper approximations $\hat{u} \equiv \hat{u}_N$ to the solution u in G which we are eager to find.[14] To deal practically with the finite-dimensional linear spaces $\hat{G}$, we need to choose *bases* for them (see [468, 509]) and we emphasize right away that this step deserves special attention.

To see this from a historical perspective, note that if $\hat{\psi}_1, \ldots, \hat{\psi}_N$ is the chosen basis for $\hat{G}$[15], then for each $\hat{v} \in \hat{G}$ there is a unique expansion

$$\hat{v} = v_1 \hat{\psi}_1 + \ldots + v_N \hat{\psi}_N. \tag{2.2}$$

Returning to the problem of establishing accuracy of the method, the following two theorems (see [27, 117, 255, 372, 379, 418]) state a strong connection between the problem under discussion and the classical problem of the *best approximation in a Hilbert space* (see [41, 117, 255, 372]).

Theorem 1. Let G be a Hilbert space, $\Phi(v) \equiv \|v\|^2 - 2l(v)$, $l \in G^*$, and $l(v) = (f, v)$. Let problem (1.4) with $V = G$ be approximated by problems (2.1) of the Rayleigh-Ritz method. Then each problem (2.1) has a unique solution $\hat{u}$, which is the best approximation to u from the subspace $\hat{G}$.

Proof. Since $\Phi(v) = \|v - f\|^2 - \|f\|^2$, it is obvious that each solution of (2.1) is the best approximation to $f = u$ (see Theorem 1.1) from $\hat{G}$, which is unique. $\square$

The solution of (2.1) can be found from the orthogonal decomposition

$$u = Pu + (u - Pu), \quad Pu \in \hat{G}, \quad (u - Pu) \perp \hat{G},$$

where P is the *orthoprojector* (orthogonal projection operator) from G onto $\hat{G}$ and $\hat{u} = Pu$ is the *orthogonal projection* of u onto $\hat{G}$.

---

[14] This idea was known to Euler, at least in certain special cases.

[15] In many of the first applications of the method at the beginning of this century, the full range of values $N = 1, 2, \ldots$ were being used, and the chosen basis for $\hat{G}_N$ was simply the first N elements of a given sequence of functions. This meant that $\hat{G}_N \subset \hat{G}_{N+1}$ and that the functions $\hat{\psi}_i$ did not depend on N. In modern variants of the method, only certain values of N are used: subspaces $\hat{G}$ and, consequently, their dimensions N are defined by some grid parameter h (in the simplest case of a cubical grid, h is the mesh size), and the functions $\hat{\psi}_i$ depend on h. One of the major advances of this new structure, as we shall see, is that it allows for basis functions with local support (i.e., they are nonzero only in small neighborhoods), so the resulting discrete systems are often very sparse and less inherently complicated.

It is easy to understand that this theorem also gives a way to estimate the variation of $\hat{u}$ which is induced by a variation of f: if we replace f by f', then the new solution $\hat{u}' \in \hat{G}$ of (2.1) satisfies $\|\hat{u} - \hat{u}'\| \leq \|f - f'\|$, because the fact that P is an orthoprojector implies

$$\|Pu - Pu'\| \leq \|u - u'\|. \tag{2.3}$$

Letting $\hat{u}$ be the best approximation from $\hat{G}$ to a given u, then the distance from u to $\hat{G}$ is defined by dist $\{u; \hat{G}\} \equiv \|u - \hat{u}\|$.

Theorem 2. *Let G be a Hilbert or an Euclidean space and suppose a bilinear form $b(u; v)$ is defined on G^2 such that conditions* (1.12)–(1.14) *are satisfied. Let $\Phi(v) = b(v; v) - 2l(v)$ (see* (1.15)) *with $l(v) = (f, v)$. Let $K \equiv (\delta_1/\delta_0)^{1/2}$ and variational problem* (1.4) *with $V = G$ be approximated by problems* (2.1) *of the Rayleigh-Ritz method. Then each problem* (2.1) *has a unique solution $\hat{u}$ and*

$$\|\hat{u} - u\| \leq K \text{dist}\{u; \hat{G}\}. \tag{2.4}$$

Proof. As in the proof of Theorem 1.3, it is extremely effective to use the Hilbert space $G(L)$ with the inner product defined by $b(u; v)$. Then the situation is just what we considered above, and $\hat{u} = P_L u$, $\|\hat{u} - u\|_L = \text{dist}_{G(L)}\{u; \hat{G}\}$, where P_L is the orthoprojector (in the sense of $G(L)$) onto $\hat{G}$. Since $\delta_0\|v\|^2 \leq \|v\|_L^2 \leq \delta_1\|v\|^2$, estimate (2.4) is valid. $\square$

Similar to (2.3), we have $\|\hat{u} - \hat{u}'\| \leq \|f - f'\|$.

It is very important that problem formulations (1.4), (2.1) be replaceable by projective ones. For (1.4), we have already used such a formulation (see (1.16)). For (2.1), we note that it is equivalent to: find $\hat{u} \in \hat{G}$ such that

$$b(\hat{u}; \hat{v}) = l(\hat{v}), \quad \forall \hat{v} \in \hat{G}. \tag{2.5}$$

An operator formulation for (2.1), analogous to (1.28) for (1.4), is: find $\hat{u} \in \hat{G}$ such that

$$\hat{L}\hat{u} = \hat{f}, \tag{2.6}$$

where $\hat{L} \equiv PLP$, $\hat{f} \equiv Pf$, and P is the orthoprojector from G onto $\hat{G}$.

From a computational point of view, it is of vital importance to formulate (2.1), (2.5), and (2.6) algebraically, that is, to write a system of linear equations for the vector $\mathbf{u}$ of components u_i defined from (2.2) with $\hat{v}$ replaced by $\hat{u}$. More precisely, expansion (2.2) constitutes an *isomorphism* between $\hat{G}$ and the space $\mathbf{R}^N \equiv H$ comprised of vectors

$$\mathbf{v} \equiv [v_1, \ldots, v_N]^T. \tag{2.7}$$

The most direct way to such a system leads from (2.5) rewritten in the form: find $\hat{u} = \sum_{j=1}^{N} u_j \hat{\psi}_j$ such that $b(\hat{u}; \hat{\psi}_i) = l(\hat{\psi}_i)$, $i = 1, \ldots, N$. The vector $\mathbf{u} \equiv u_h$ is then the solution of the linear system

$$\mathbf{Lu} = \mathbf{f}, \tag{2.8}$$

where [16]

$$\mathbf{L} \equiv L_h \equiv [b(\hat{\psi}_j; \hat{\psi}_i)] \tag{2.9}$$

and

$$\mathbf{f} \equiv [l(\hat{\psi}_1), \ldots, l(\hat{\psi}_N)]^T. \tag{2.10}$$

2.2 Bubnov-Galerkin method and projective methods. Projective formulations (2.5), (1.16) enable us to apply similar approximations for more general problems (see Theorem 1.5) without symmetry condition (1.12) or corresponding variational form (1.4). These are modern variants of *Bubnov-Galerkin methods*, very often referred to as *projective methods* (see [117, 317]), and formulated in a very simple manner: in problem (1.16) we replace u, v by elements of the chosen subspace $\hat{G}$ (see (2.5)).

Theorem 3. *Let the conditions of Theorem 1.5 be satisfied and consider problem (1.16) approximated by corresponding problems (2.5). Then each of these approximate problems has a unique solution $\hat{u}$ and estimate (2.4) of the accuracy of the projective method holds.* [17]

Proof. Existence and uniqueness of the solutions follow from Theorem 1.5. Comparing (2.5) and (1.16) with $v = \hat{v}$, we see that $b(\hat{u} - u; \hat{w}) = 0$, for all $\hat{w} \in \hat{G}$. Hence,

$$b(\hat{u} - u; \hat{u} - u) = b(\hat{u} - u; \hat{v} - u).$$

Estimating the left-hand side of the latter equality from below and the right-hand side from above, we obtain

$$\delta_0 \|\hat{u} - u\|^2 \leq \delta_1 \|\hat{u} - u\| \|\hat{v} - u\|, \tag{2.11}$$

which leads directly to (2.4). □

Therefore, if the set of subspaces $\hat{G}_N$ satisfies

$$\lim_{N \to \infty} \text{dist}\{u; \hat{G}_N\} = 0, \quad \forall u \in G, \tag{2.12}$$

---

[16] The matrix L_h here is the *Gram matrix* of the chosen basis for $\hat{G}$ in the Hilbert space $G(L)$. Solution of such systems with very large N and some special matrices L_h will be one of the main focal points of this book.

[17] Theorem 3 is often referred to as Cea's theorem (see [117, 244]).

then we can be sure that *the projective method converges* (in the case of (2.11), it is said that the *set of subspaces* $\hat{G}_N$ *approximates* G). Moreover, it is clear that in order to estimate dist $\{u; \hat{G}_N\}$, it suffices to estimate $\|u - \hat{w}\|$ for a suitably chosen $\hat{w}$ (e.g., $\hat{w}$ can be chosen as the function in $\hat{G}$ that interpolates u on every grid cell, which allows for local estimation of the error (see § 5)).

We have already emphasized that the algebraic form of the method is crucial from a computational point of view and it depends critically on the choice of a basis for $\hat{G}_N$. But, in this respect, everything is done in the same way as it was done in the previous subsection, including the resultant system (2.8), (2.9), and (2.10), except that we cannot now consider L_h as a Gram matrix ($L_h^T \neq L_h$).

2.3. Projective-grid methods (finite element methods). We are now in a position to regard and analyze the modern variants of projective methods widely known as *finite element methods* (FEMs), also known as: *Galerkin, variational-difference, variational-grid, and projective-grid methods* (PGMs) (see [95, 244, 324, 351, 379, 390, 470, 489]). [18]

Some particular cases of PGMs, especially those associated with rectangular grids, have a lot in common with *difference methods* (finite difference methods) and are referred to as *projective-difference* or *variational-difference* methods. They will be considered in detail in Subsection 2.4. This important relationship between these methods was stressed by Courant in 1943 (see [122, 320]), but not given much attention until two decades later. [19] The first formulations in the mid 1960's of FEMs (see [320, 390,

---

[18] They may be regarded as particular cases of projective methods based on a given grid (mesh), with parameter h for the closure $\bar{\Omega}$ of the original region and for the elliptic boundary value problem under consideration; each function $\hat{v}$ (see (2.2)) is specified at every grid cell by a finite number of parameters like its values at cell *vertices (nodes, grid points)*; and instead of N and $\hat{G}_N$, it is reasonable to write N_h and $\hat{G}_h$, which stresses their dependence on the grid. Usually, the basis function $\hat{\psi}_i$ also depends on h, and its *support* consists of just a few cells of the grid, although some basis functions may have nonlocal support, especially for problems with singularities in their solutions. Thus, the term *projective-grid methods* (PGMs), suggested by Mikhlin, seems most appropriate for the mathematical nature of these methods.

[19] Moreover, some similar subspaces $\hat{G}_h$ had been considered by many mathematicians even earlier (e.g., *piecewise bilinear functions*, associated with rectangular grids, were used by Sobolev [458]; papers of Courant, Friedrichs, and Lewy should be mentioned as well; the one-dimensional case dealing with piecewise linear functions may be traced even to Euler and Leibnitz; the two-dimensional case can be found in [444]). These particular spaces of continuous piecewise polynomial functions are now often referred to as examples of *spline spaces* (see [5, 117, 446]). So we see that, from the mathematical point of view, the methods under consideration may not be regarded as markedly new. It seems reasonable that the lack of interest in these methods before the 1960's stemmed mostly from the necessity to solve large and complicated (compared to the difference case) systems of equations.

470, 518]) as methods of discretization, which were suitable for cases of regions with complicated geometry, gained wide recognition, due largely to the emergence of electronic computers, the first advances in numerical solution of elliptic boundary value problems, and the realization that complicated geometry is a major obstacle for difference methods. It was soon understood (see [88, 234, 394, 470]) that these methods have the same nature as the variational-difference method of Courant, and this stimulated a new wave of attention. Today, it is clear that no great gulf exists between difference and FEMs: many results, first obtained for one, have found adequate analogs for the other.[20]

We return now to consider perhaps the simplest but nonetheless extremely important case of PGMs based on a *triangulation* of a closed planar region.

    Figure 1. Regular triangulations of triangle and rectangle.

We assume, for the time being, that the boundary $\Gamma \equiv \partial\Omega$ consists of a finite number of closed line segments, as is the case for a polygonal domain. Approximations of general regions will be investigated later in Chapters 2, 4, and 5.

By a triangulation of the region Ω, more accurately of its closure $\bar{\Omega}$, [21] we mean a partition $T_h(\bar{\Omega})$ of $\bar{\Omega}$ into a finite set of triangles $T_k \in T_h(\bar{\Omega})$ such that no two different triangles have common inner points and no side of one triangle can be a part of a side of another, i.e., different triangles may only have one common vertex or one common side. In other words, we may say that a *triangular grid* is being used, or, in terms of FEMs, that

---

[20] In this respect, we recall that optimal estimates of accuracy (which we discuss in § 5) were first obtained for FEMs (PGMs), but as early as the late 1960s they were proved for difference approximations as well (see [164, 173, 424]). An enlightening example is the coincidence of the usual 13-point difference approximation of the biharmonic operator with the discrete operator for a specially designed PGM, shown by Strelkov [472]. But probably the most important changes took place in our attitude towards certain sophisticated iterative methods, which we shall be discussing briefly in § 3 and, in more detail, in Chapters 1 and 3. These methods were first designed for difference systems and only in the early 1970's were they extended to finite element systems (see [13, 137, 165, 166]). Currently, especially in case of multidimensional problems, such methods have become indispensable tools for solution of many of the large linear and nonlinear grid systems arising in practice, and they are the main hope for dealing with the more difficult systems that will appear in the future.

[21] In the following, for simplicity, we sometimes refer to triangulations of regions instead of their closures. It should be clear from context what is meant.

we are working with *triangular elements*.

Very often the required triangular grid is obtained through partitioning of each cell of a given *quadrilateral grid* into two triangles; e.g., for a given cell of a *rectangular grid*, drawing one of its diagonals will give a partition of the cell; if all diagonals of rectangular cells are parallel, the obtained triangulation will be called a *regular triangulation* (see Figure 1).

The examples of such grids and triangulations given below (see Figures 2–8) contain only a very small number of cells; in practical problems, we often encounter cases of more complicated regions with on the order of 10^3 to 10^6 cells.

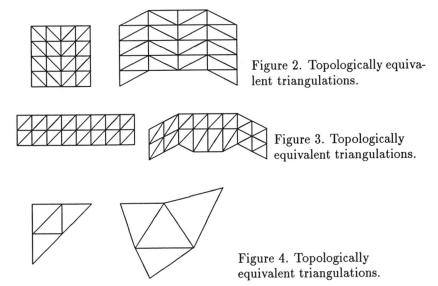

Figure 2. Topologically equivalent triangulations.

Figure 3. Topologically
equivalent triangulations.

Figure 4. Topologically
equivalent triangulations.

All of these illustrations are nonetheless instructive and suggest simple algorithms for constructing similar triangulations with more cells; with the exception of Figure 8, these triangulations may be referred to as *quasiuniform triangulations* with parameter h (called the linear size), that is, there exist positive constants $\kappa_0, \kappa_1, \kappa_2, \kappa_3$ such that

$$\kappa_0 h \leq l \leq \kappa_1 h, \quad \kappa_1 h^2 \leq s \leq \kappa_2 h^2,$$

where l stands for the length of any side of any triangle from $T_h(\bar{\Omega})$ and s stands for the area of any such triangle. Figure 8 attempts to illustrate the case of a triangulation with local refinement, and the parameter h here must be a vector providing the information about all triangles.

In Chapter 4, we discuss algorithms for automatic construction of some types of triangulations. For the time being, we confine ourselves to the

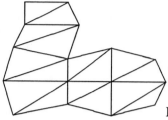

Figure 5. Triangulation of polygonal type.

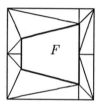 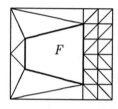

Figure 6. Triangulations of
doubly-connected regions.

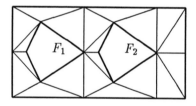

Figure 7. Triangulations of
a triply-connected region.

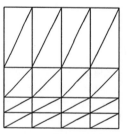 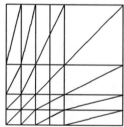 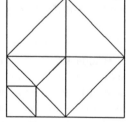

Figure 8.
Triangu-
lations
with local
refinement.

extremely useful notion of *topologically equivalent triangulations*. More precisely, we say that triangulations $T_h(\bar{\Omega})$ and $T_h(\bar{Q})$ are topologically equivalent if there exists a one-to-one mapping, continuous with respect to both directions, of $\bar{\Omega}$ onto $\bar{Q}$ such that each elementary triangle from $T_h(\bar{\Omega})$ is mapped onto a corresponding triangle from $T_h(\bar{Q})$. Notice that, in this case, we may speak about continuous *piecewise affine mappings*, which are a subject of a branch of topology called *simplicial topology* (see [430]). Figures 2–4 provide simple examples of topologically equivalent triangulations, which generally exist in great variety.

Of special significance are cases of triangulations of a *model region Q* such that $\bar{Q}$ consists of a finite number of triangles, each of which is half of a unit square with sides parallel to coordinate axes, and $T_h(\bar{Q})$ consists of triangles, all vertices of which are nodes of a square grid with mesh size $h \equiv 1/p$. These cases are notable because each node P may be characterized by a vector parameter $i \equiv [i_1, i_2]$, where i_1, i_2 are integers such that $P \equiv P_i \equiv [i_1 h, i_2 h]$. [22]

Suppose now that somehow we have obtained the desired triangulation $T_h(\bar{\Omega})$, with the Dirichlet boundary Γ_0 covered by a collection of some sides of elementary triangles, and what we wish next is to construct an associated subspace $\hat{G}_h$ of the Hilbert space $G \equiv W_2^1(\Omega; \Gamma_0)$. Then all vertices of triangles (nodes of the grid) belonging to $\bar{\Omega}$ and not belonging to Γ_0 constitute a set Ω_h of, say, N_h nodes. With each node P from Ω_h, we associate a basis function $\hat{\psi}_P(x)$ defined by the following conditions: $\hat{\psi}_P(x) = 1$ if $x = P$, $\hat{\psi}_P(x) = 0$ if x coincides with any point from Ω_h different from P, and the restriction of $\hat{\psi}_P(x)$ to an arbitrary triangle from $T_h(\bar{\Omega})$ is a linear function.

It is clear that the nodal basis function $\hat{\psi}_p(x)$ defined in this way is continuous on $\bar{\Omega}$, vanishing on Γ_0, and nonzero only on the triangles for which node P is a vertex. This means that the support, S_P, of each $\hat{\psi}_P$ consists of triangles with the common vertex P. Inside each elementary triangle T, the first partial derivatives of $\hat{\psi}_P$ exist and are constant (nonzero for $T \subset S_P$ and vanishing outside of S_P). So it is not difficult to verify (using integration by parts) that

$$(\hat{\psi}_P, D_r \varphi)_{0,\Omega} = -(D_r \hat{\psi}_P, \varphi)_{0,\Omega}, \quad \forall \varphi \in C_0^\infty(\Omega). \qquad (2.13)$$

To this end, it suffices to represent the integral over Ω as a sum of integrals over triangles T and carry out integration by parts for them; terms on each side of T either cancel each other (if the side belongs to two triangles)

---

[22] We remark also that there are some very interesting algorithms for constructing a triangulation with a given set of nodes (see [215, 287, 377, 445]) based on the notions of *Delaunay triangulation* and of *Dirichlet tesselation*.

or are equal to zero since $\varphi|_\Gamma = 0$. This means that the usual derivative $D_r\hat{\psi}_P$, existing only inside each T, may be regarded as a generalized derivative in the Sobolev sense, defined for almost all $x \in \Omega$. Therefore, $\hat{\psi}_P \in W_2^1(\Omega; \Gamma_0) = G$. [23]

None of the $\hat{\psi}_P$ defined above can be a linear combination of the others. Thus, they constitute a basis for the space $\hat{G}_h \equiv \hat{G} = \text{lin}\{\hat{\psi}_P\}$, where lin (lineal) stands for the linear span of the given system of functions. $\hat{G}$ is characterized as the space of functions that are continuous on $\bar{\Omega}$, vanishing on Γ_0, and linear on each triangle in $T_h(\Omega)$.

If we index the nodes by $i = 1, 2, \ldots, N$ in some order and write $\hat{\psi}_i$ instead of $\hat{\psi}_P$ and S_i instead of S_P, then system (2.8) takes the form

$$u_1 b(\hat{\psi}_1; \hat{\psi}_1) + \ldots + u_N b(\hat{\psi}_N; \hat{\psi}_1) = l(\hat{\psi}_i), \quad i = 1, \ldots, N, \qquad (2.14)$$

with integrals actually being taken over $S_i \cap S_j$ for computation of $b(\hat{\psi}_j; \hat{\psi}_i)$. So, in the case of (1.30), we have to find integrals over T of functions of the form $a_{r,l}, b_r\hat{\psi}_i, b'_r\hat{\psi}_j, c\hat{\psi}_j\hat{\psi}_i$ and integrals over some sides of T of functions like $\sigma\hat{\psi}_j\hat{\psi}_i$ (similar integrals are needed for the calculation of $l(\hat{\psi}_i)$).[24]

2.4. The simplest projective-grid operators. For triangulations like those shown in Figures 1 and 2, it is convenient to obtain the desired grid operators L_h using vector indexes $i \equiv [i_1, i_2]$, with $u_i \equiv u_P$. The closed side defined by nodes P_i and P_j will be denoted by $[P_iP_j]$. We will also be using the following difference notation: $e_1 \equiv [1, 0], e_2 \equiv [0, 1]$, $I_{-r}u_i \equiv u(P_i - h_re_r), I_ru_i \equiv u(P_i + h_re_r), \partial_ru_i \equiv h_r^{-1}[I_ru_i - u_i], \bar{\partial}_ru_i \equiv h_r^{-1}[u_i - I_{-r}u_i], \Lambda_ru_i \equiv -\bar{\partial}_r\partial_ru_i = -h_r^{-2}[I_{-r}u_i - 2u_i + I_ru_i], r = 1, 2$; sometimes we shall prefer to write index i after parenthesis, e.g.: $[I_{-r}u - 2u + I_ru]_i \equiv I_{-r}u_i - 2u_i + I_ru_i$.

Lemma 1. *Let $\bar{\Omega} \equiv \bar{Q}$ be the rectangle $[0, l_1] \times [0, l_2]$ and $G \equiv \overset{o}{W}{}_2^1(Q)$, and suppose the rectangular grid with nodes $P_i \equiv [i_1h_1, i_2h_2]$ and mesh sizes $h_r \equiv l_r(N_r+1)^{-1}, r = 1, 2$ is used. Let each cell of the grid be triangulated by one of its diagonals and let the associated piecewise linear basis functions $\hat{\psi}_i$ be defined with $i_r \in [1, N_r]$. Let $b(u; v) \equiv (u, v)_{1,Q}, l(v) \equiv (g, v)_{0,Q}$. Then discrete system (2.14) can be rewritten in the form*

$$h_1h_2(\Lambda_1u + \Lambda_2u)_i = f_i, \quad P_i \in Q_h, \qquad (2.15)$$

---

[23] In the case of one-dimensional problems, analogous basis functions are the so-called *hat functions*, defined as follows: $\hat{\psi}_i(t) = |t - ih|/h$ if $t \in [ih - h, ih + h]$ and zero otherwise.

[24] There exist useful formulas (see [117]) for integration of polynomials over triangles. Very often (see [518]), a special procedure dealing with a *local stiffness matrix* is used to obtain all required integrals with respect to a fixed given triangle, and these local matrices are then assembled to obtain L_h.

$$u_i = 0, \quad P_i \in \partial Q. \tag{2.16}$$

Proof. Let $X \equiv (D_1\hat{u}, D_1\hat{\psi}_i)_{0,S_i} = (D_1\hat{u}, D_1\hat{\psi}_i)_{\Pi_{i,1,0}} + (D_1\hat{u}, D_1\hat{\psi}_i)_{\Pi_{i,1,1}}$, where $\hat{u} = \sum_{P_i \in \Omega_h} u_i\hat{\psi}_i$ and $\Pi_{i,1,0}$ and $\Pi_{i,1,1}$ are the unions of two triangles with the common sides $[P_{i_1-1,i_2}P_i]$ and $[P_iP_{i_1+1,i_2}]$, respectively. Since $\hat{u}$ and $\hat{\psi}_i$ on each of these triangles are linear, we see that $D_1\hat{u} = \bar{\partial}_1 u_i, D_1\hat{\psi}_i = h_1^{-1}$ on $\Pi_{i,1,0}$, and $D_1\hat{u} = \partial_1 u_i, D_1\hat{\psi}_i = -h_1^{-1}$ on $\Pi_{i,1,1}$. Thus, $X = h_1h_2\Lambda_1 u_i$, which gives the first term on the left-hand side of (2.15). In the same manner, $Y \equiv (D_2\hat{u}, D_2\hat{\psi}_i)_{0,S_i}$ is found, which completes the proof. $\square$

Note that this projective-grid operator Λ_h coincides with the widely used difference operator multiplied by h_1h_2, which was pointed out by Courant in 1943 for the case when all diagonals of the cells are parallel (in the case of a *regular triangulation*). The proof used above was suggested in [177] and is rather fruitful for some types of problems, as we shall see in Section 4. It may be easily generalized to more complicated problems. For example, instead of the considered bilinear form, if we take the *diffusion form*

$$b(u; v) \equiv (a_1, D_1uD_1v)_{0,\Omega} + (a_2, D_2uD_2v)_{0,\Omega} \tag{2.17}$$

with coefficients $a_r(x)$ being constants with respect to the triangulation $T_h(\bar{\Omega})$ (here $\Omega = Q$), then (2.15) should be replaced by

$$L_hu_i \equiv -h_1h_2[\bar{\partial}_1(Y_1a_1\partial_1 u) + \bar{\partial}_2(Y_2a_2\partial_2 u)]_i, \quad P_i \in \Omega_h, \tag{2.18}$$

where Y_ra_r denotes the arithmetic mean of the (constant) values of a_r in the two triangles having the common side along which the difference $\partial_r u$ is taken. In a number of cases, it is useful to extend coefficients like $a_r(x)$ to all $x \in \mathbf{R}^2$ by assigning $a_r(x)$ the value zero outside $\bar{\Omega}$ (this is usually referred to as a *finite extension*). On the basis of this, it is not difficult to prove the next theorem.

Theorem 4. Let each cell of a rectangular grid be split into two triangles as in Lemma 1, and let $\bar{\Omega}$ be covered by a finite number of these triangles such that the Dirichlet part Γ_0 of its boundary is covered by a finite number of their sides. Let $G \equiv W_2^1(\Omega; \Gamma_0)$ and consider the bilinear form defined on G^2 by (2.17). Let Ω_h be the set of nodes of the grid belonging to $\bar{\Omega}$ but not to Γ_0. Then representation (2.18) is valid provided we extend a_r by setting $a_r(x) = 0$ for all x outside $\bar{\Omega}, r = 1, 2$.

By virtue of (2.17), it is easy to find a representation of L_hu_i in terms of the values u_i, with $P_i \in \Omega_h$. For example, in the case of $G = G^{(1)}$ and Figure 1, with $A_1 \equiv [0,0]$ and $A_2 \equiv [l_1, 0]$, it is easy to verify that

$$L_h u_i = -h_1 h_2 [\bar{\partial}_1 (Y_1 a_1 \partial_1 u)]_i - h_1 [Y_2 a_2 \partial_2 u]_i \text{ if } P_i \in (A_1 A_2), \qquad (2.19)$$

$$L_h u_i = -h_2 [Y_1 a_1 \partial_1 u]_i - h_1 [Y_2 a_2 \partial_2 u]_i \text{ if } P_i = A_1, \qquad (2.20)$$

where $(A_1 A_2)$ denotes the open side from A_1 to A_2. In the particular case that a_r is constant on Ω, formulas (2.19), (2.20) take the form $L_h u_i = 2^{-1} a_1 h_1 h_2 \Lambda_1 u_i - h_1 a_2 \partial_2 u_i, P_i \in (A_1 A_2)$, and $L_h u_i = -2^{-1} [a_1 \partial_1 u + a_2 \partial_2 u]_i, P_i = A_1$. The addition in the bilinear diffusion form (2.17) of a reaction term $(\sigma, uv)_\Gamma$ leads to a change in the form of $L_h u_i$ only for $P_i \in \Gamma$, namely, the added term $(\sigma, \hat{u} \hat{\psi}_i)_{\Gamma \cap S_i}$, where $\Gamma \cap S_i$ consists in general of two sides of the triangles. For constant σ, the required integrals can be found by *Simpson's rule*: the integral over $[0, 1]$ of a product of two linear functions $\hat{u}$ and $\hat{\psi}$, with $\hat{\psi}(0) = 1$ and $\hat{\psi}(1) = 0$, is just $(2u_0 + u_1)/6$. Therefore, e.g.,

$$L_h u_{0,0} = -[a_1 \partial_1 u_{0,0} + a_2 \partial_2 u_{0,0}]/2 + \sigma/6[h_1(u_{1,0} + 2u_{0,0}) + h_2(u_{0,1} + 2u_{0,0})],$$

$$L_h u_i = a_1/2 h_1 h_2 \Lambda_1 u_i - h_1 a_2 \partial_2 u_i + \sigma h_1 Y_{1,\Gamma} u_i, \quad P_i \in (A_1 A_2),$$

with $Y_{1,\Gamma} u_i \equiv [I_{-1} u_i + 4u_i + I_1 u_i]/6, P_i \in (A_1 A_2)$.

Thus, for the cases considered, the structure of grid operator L_h is very similar to that of a difference operator (with the simplified reaction term σu_i). This underscores the close relationship between PGMs (FEMs) and difference methods, and will be of great help in constructing some effective iterative methods for solution of systems like (2.8). We shall return to the subject in Section 3.1, being content here just to recognize that system (2.15), (2.16), after elimination of the values $u_i = 0$ on the boundary and renumbering of remaining values, can be rewritten in the matrix form

$$\Lambda \mathbf{u} \equiv \Lambda_1 \mathbf{u} + \Lambda_2 \mathbf{u} = \mathbf{f} \qquad (2.21)$$

with $\mathbf{u} \in \mathbf{R}^N, N \equiv N_1 N_2$,

$$\Lambda_r = \Lambda_r^T > 0, \ r = 1, 2, \quad \Lambda_1 \Lambda_2 = \Lambda_2 \Lambda_1. \qquad (2.22)$$

Commutability of the operators in (2.22) is remarkable and rare, and, due to a well known theorem of the linear algebra (see [328]), it ensures existence of an orthonormal basis of $\mathbf{R}^N$ comprised of common eigenvectors of these symmetric matrices (operators) (for the particular case under consideration, it was shown in 1959 by Birknoff and Varga ([70]) that this property holds only in the case of a rectangular region). For operators in (2.21), this basis can even be given in explicit form:

$$e_k(x_1, x_2) \equiv (2h_1/l_1)^{1/2} \sin[\pi k_1 x_1 l_1^{-1}] (2h_2/l_2)^{1/2} \sin[\pi k_2 x_2 l_2^{-1}], \quad (2.23)$$

$k \equiv [k_1, k_2]$, $k_r \in [1, N_r]$, $r = 1, 2$ (see [232, 440]), which is similar to the Poisson expression for eigenfunctions of Laplace's operator; the corresponding eigenvalues are
$\lambda_k \equiv 4h_1 h_2 [h_1^{-2} \sin^2[\pi k_1 h_1 (2l_1)^{-1}] + h_2^{-2} \sin^2[\pi k_2 h_2 (2l_2)^{-1}]]$, and they belong to $[\alpha_0, \alpha_1]$, with $\alpha \equiv \alpha_1/\alpha_0 \asymp h^{-2}$ if $h_1 \asymp h_2 \asymp h$, i.e., $\exists\, k_0 > 0$ and $k_1 > 0$ such that $k_0 \leq \alpha h^2 \leq k_1$.

2.5. Composite grids and triangulations; local grid refinement.
So far we have considered FEMs (PGMs) in which all grid nodes $P_i \in Q_h$ (that is, the nodes belonging to the closed region $\bar{Q}$ and not to $\Gamma_{Q,0}$) have had equal treatment and have been represented in the basis for $\hat{G}_h$ by their own basis functions $\hat{\psi}_P$. Refinement of a grid or triangulation in subregions where the solution of the problem may change very rapidly has always been regarded as a potentially reasonable way to improve the accuracy and efficiency of the basic method. Figure 8 illustrates one possible refinement approach that preserves the standard triangulation (see also [49, 512]). One of the main disadvantages of such refinement is evident: it leads to rather complicated grids. As an alternative, we consider a very popular *local grid refinement* procedure (see [29, 80, 142, 205, 224, 347, 362, 363, 365]) that is based on *composite grids* and *composite triangulations* (for difference schemes, see e.g., [330, 494]).

It seems reasonable to describe any such grid or triangulation not as a stationary structure, but as a result of a process of refinement consisting of, say, p levels. To start with, we need an initial triangulation $T^{(0)}(\bar{Q}) \equiv T^0$, a closed subset $\bar{Q}^{(1)} \subset \bar{Q}$ consisting of a number of triangles from $T^{(0)}$, and an integer $t_1 > 1$, which will be called the refinement ratio. Then each triangle from $T^{(0)}$ is uniformly subdivided into t_1^2 smaller triangles (which can be readily done by subdividing each side of the given triangle into t_1 equal parts and drawing straight lines through the points of division parallel to the other sides). [25]

The collection of cells of the old triangulation not belonging to $Q^{(1)}$ and of the new smaller cells belonging to $\bar{Q}^{(1)}$ define the composite triangulation $T^{(1)}$ of the first level. If $p > 1$, then an analogous procedure is carried out with respect to $\bar{Q}^{(2)} \subset \bar{Q}^{(1)}$ and the refinement ratio $t_2 > 1$, and so on. Subsets $\bar{Q}^{(l)} \subset \bar{Q}^{(l-1)}$ with $l = 2, \ldots, p$ can often be taken such that $\bar{Q}^{(l)}$ does not have common points with the closure of $\bar{Q}^{(l-2)} \setminus \bar{Q}^{(l-1)}$; under this assumption, lines separating $\bar{Q}^{(l)}$ and $\bar{Q}^{(l-1)} \setminus \bar{Q}^{(l)}$ consist of sides of

---

[25] When dealing with an approximation of the space $W_2^1(Q; \Gamma_{Q,0})$, it is natural to assume that $\Gamma_{Q,0}$ is covered by a number of sides of the triangles of $T^{(0)}$.

triangles from $T^{(l-1)}$ subdivided into t_l equal parts, and we shall refer to
these lines as *cutting lines*, for they define the decomposition

$$\bar{Q} = F_1 \cup F_2 \cup \ldots \cup F_{p+1}, \qquad (2.24)$$

with F_l being the closure of $\bar{Q}^{(l-1)} \setminus \bar{Q}^{(l)}$, $l = 1, \ldots, p$, $F_{p+1} \equiv Q^{(p)}$, and
$\bar{Q}^{(0)} \equiv \bar{Q}$ (these closed sets may only have common points on their bound-
aries). Thus, after the pth local refinement of the initial triangulation, we
come to the composite triangulation $T^{(p)} \equiv T_{c,h}(\bar{Q})$ consisting of standard
triangulations of subsets $F_1, \ldots, F_{p+1}$ and representing a particular case of
composite grids with local refinement.

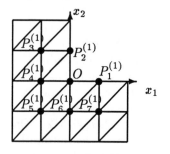
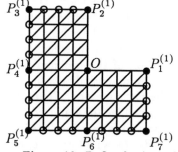

Figure 9. Initial triangulation.   Figure 10: Refined triangulation.

Figures 9–11 illustrate such local regular refinement of a uniform grid
for L-shaped region $\bar{Q}$: if $P_1 = [1, 0]$ and $P_2 = [0, 1]$, then $\bar{Q}^{(1)} \equiv \bar{Q} \cap \{x :
|x_r| \leq q, r = 1, 2\}$ and $t_1 = 4$; the refinement for the subregion is shown in
Figure 10. The resulting composite triangulation is shown in Figure 11.

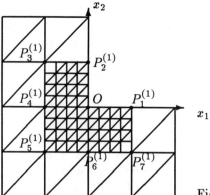

Figure 11. Composite triangulation.

Given a composite triangulation $T_{c,h(\bar{Q})}$, we can define an associated
finite element space $\hat{G}_h \equiv \hat{G}_{c,h}$ as the subset of $W_2^1(Q; \Gamma_{Q,0})$ consisting

of functions $\hat{v}$ that are continuous on $\bar{Q}$ and linear on each triangle $T \in T_{c,h}(\bar{Q})$. This immediately implies that some vertices P_i' of the triangles in $T_{c,h}(\bar{Q})$ but not in $\Gamma_{Q,0}$ do not correspond to degrees of freedom v_i of $\hat{v}$ and basis functions $\hat{\psi}_{Q,i}$. Indeed, such vertices, which we call *seminodes* (also called *slave nodes*), are found only on cutting lines and can be defined as new vertices, that is, as appearing on these lines as the result of a corresponding local refinement; values of $\hat{v}$ then must be defined through linear interpolation from the values of $\hat{v}$ at the two neighboring vertices (nodes). More precisely, if $[P_0^{(l-1)} P_1^{(l-1)}]$ is the side of the triangle in $T^{(l-1)}$ containing a seminode P_i' and this triangle is subdivided into t_l^2 smaller triangles, then this side contains exactly $t_l - 1$ new vertices P_i'. We renumber them to get $\vec{P}_i' = \vec{P}_0^{(l-1)} + i/t_l(\vec{P}_1^{(l-1)} - \vec{P}_0^{(l-1)})$, where $i \in [1, t_l - 1$ and $\vec{P}$ denotes the radius-vector of the point P. In this case, we must have

$$\hat{v}(P_i') = \hat{v}(P_0^{(l-1)}) + i t_l^{-1}[\hat{v}(P_1^{(l-1)}) - \hat{v}(P_0^{(l-1)})], \quad i \in [1, t_l - 1]. \quad (2.25)$$

For example, Figure 10 contains three seminodes P_i' on the side $[P_3^{(1)} P_2^{(1)}] \equiv [P_0^{(0)} P_1^{(0)}]$, which together with other possible seminodes are marked by circles. As usual, we denote by Q_h the set of nodes P_i corresponding to the basis functions $\hat{\psi}_{Q,i}$. These functions can be defined in the same way as in Subsection 3 and can differ from the old ones only for P_i belonging to the cutting lines. Figures 12 and 13 show supports of such functions associated with the nodes $P_3^{(1)}$ and $P_4^{(1)}$, respectively: it is easy to see that they can be described as the closures of the old supports of the similar functions defined before the refinement, except that the smaller triangles without vertices on cutting lines are deleted.

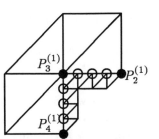

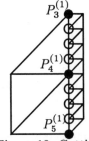

Figure 12. Cutting line corner basis function support.

Figure 13. Cutting line side basis function support.

We complete this section by determining the changes in the structure of equations (2.15) due to extension to the case of Q in Figure 11 and its triangulation $T_{c,h}(\bar{Q})$. It is clear that (2.15) holds for any node belonging to just one level (i.e., not belonging to any cutting line) provided the appropriate mesh size is used in the expression for $\Lambda_r u_i$ and the values of u at any seminodes in the formulas are replaced according to the linear interpolation formula (2.25). But such elimination of values of the grid function at the seminodes makes sense only for the ultimate form of the matrix. This is in contrast to the desire to describe the structure of the grid operator in general terms, especially if we want to compute the vector $L_h \mathbf{u}$ for a given vector $\mathbf{u}$, which is a matter of importance for implementing iterative methods. For such situations, it is reasonable to calculate values of $\hat{v}$ (see (2.2),(2.7)) at all seminodes via interpolation of the type (2.25) and use them for calculation of $L_h \mathbf{u}$. Observe that such a procedure is attractive from the point of view of obtaining independent but alike calculations on subregions F_l (see (2.24)). Thus, we may describe ways for obtaining $L_h \mathbf{u}$ that use not only the vector $\mathbf{u}$, but also the vector u consisting of values of the grid function $\hat{u}$ at nodes and seminodes of $\bar{Q}$. The most significant changes to (2.15) occur for nodes on the cutting lines. Consider as typical illustrations cases of nodes $P_3^{(1)}$ and $P_4^{(1)}$ from Figures 12 and 13 and let h_1, h_2 denote the mesh sizes before refinement, say, with ratio t_l. Then, for $P_i \equiv P_3^{(1)}$, we have

$\Lambda_1 u_i \equiv h_1^{-1} h_2^{-1} (D_1 \hat{u}, D_1 \hat{\psi}_i)_{0,Q} = -h_1^{-1} \partial_1 u_i (1/2 + 1/2 t_l^{-1}) + h_1^{-1} \bar{\partial}_1 u_i$, because $D_1 \hat{u} = \partial_1 u_i$ and $D_1 \hat{\psi}_i = -h_1^{-1}$ on the t_l^{-1} small triangles having a side belonging to $[P_3^{(1)} P_2^{(1)}]$ and area $h_1 h_2 2^{-1} t_l^{-2}$. Similarly, $\Lambda_2 u_i = -h_2^{-1} \partial_2 u_i + h_2^{-1} \bar{\partial}_2 u_i (1/2 + 1/2 t_l^{-1})$. Thus, $L_h \mathbf{u}_i = h_1 h_2 [\Lambda_1 u_i + \Lambda_2 u_i]$ with these new operators Λ_1, Λ_2 given above. In the same way for $P_i \equiv P_4^{(1)}$, we have

$$A_2 \mathbf{u}_i = h_1 h_2 [-h_2^{-1} \partial_2 u_i (1/2 + (2 t_l)^{-1}) + h_2^{-1} \bar{\partial}_2 u_i (1/2 + (2 t_l)^{-1})].$$

However, to determine $A_1 \mathbf{u}_i \equiv (D_1 \hat{u}, D_1 \hat{\psi}_i)_{0,Q}$, note that $D_1 \hat{u}, D_1 \hat{\psi}_i$ may have different values at the different small triangles (see Figure 13). These values are just the difference approximations calculated with the new grid size $h_1' \equiv t_l^{-1} h_1$ over correspondingly small horizontal intervals. We denote these differences for $\hat{u}$ by $\partial_1' u_i$ (for the node $P_i \equiv P_4^{(1)}$) and by $\partial_1' u_{i,j_2}'$ with $|j_2| \in [1, t_l - 1]$ (for seminodes numbered in the natural way on the sides $[P_i P_3^{(1)}]$ and $[P_i P_5^{(1)}]$, respectively). Then, for $\hat{\psi}_i$, the corresponding differences are simply the numbers $t_l, t_l - 1, \ldots, 1$ multiplied by $-h_1^{-1}$, and we may write $A_1 \mathbf{u}_i = -h_1^{-1} (Y_2 \partial_1' u)_i + h_1^{-1} \bar{\partial}_1' u_i$, with $(Y_2 \partial_1' u)_i \equiv (t_l)^{-2} [\partial_1' u_i + \sum_{|j_2|=1}^{t_l-1} (t_l - |j_2|) \partial_1' u_{i,j_2}']$. Since $t_l + 2(t_l - 1) + \ldots + 2 = t_l^2$, $(Y_2 \partial_1' u)_i$ gives an average value of $D_1 \hat{u}$ with respect to all small triangles, as shown in Figure 13 (for Figures 12 and 13, $t_l = 4$). Thus, we see that, even for systems associated with composite grids with local refinement, it is possible to obtain fairly simple formulas for grid operators, at least for model regions.

§ 3. Methods of solution of discretized problems; asymptotically optimal and nearly optimal preconditioners

3.1. Specificity of grid systems; direct methods. Consider a linear elliptic boundary value problem discretized by a *grid method*, by which we mean any method connected with a chosen infinite sequence of grids (e.g., FEMs, difference methods, their variations known as *box, control volume finite element*, or *finite volume element methods* (see [53, 106, 268, 275, 366, 335]), and some *spectral* and *collocation methods* (see [107, 108, 236, 276, 414])). Then the ultimate product of such a procedure is a hopefully reasonable sequence of linear *grid systems* of form (2.8) with unknown vectors $\mathbf{u} \equiv u_h \equiv [u_1, \ldots, u_N]^T$, which in this section will be denoted simply by u. Because N, the number of unknowns, is considered to be an element of an increasing sequence of positive integers, we are especially interested in properties of the matrices $L_h \equiv L$ for large N. The N-dimensional vectors u, i.e., the *grid functions* defined at the grid nodes, may be regarded as elements of the N-dimensional Euclidean space $H \equiv H_h \equiv \mathbf{R}^N$ equipped with the simplest inner product $(u, v) \equiv (u, v)_{\mathbf{R}^N} \equiv u_1 v_1 + \ldots + u_N v_N$. [26] Here and henceforth, for a given Euclidean space H, let $\mathcal{L}(H)$ denote the space of linear operators mapping H into itself and let I denote the identity operator in $\mathcal{L}(H)$. We shall also make use of the notation $\mathcal{L}^+(H) \equiv \{B : B \in \mathcal{L}(H), B = B^* > 0\}$, which is the subset of $\mathcal{L}(H)$ of symmetric (self-adjoint) positive operators. [27]

For $B \in \mathcal{L}^+(H)$, the space $H(B)$ will denote the new Euclidean space that differs from H only in that is uses the following inner product:

$$(u, v)_B \equiv (Bu, v)_H \equiv (Bu, v), \quad \|u\|_B \equiv (u, u)_B^{1/2}. \qquad (3.1)$$

We will have occasion to regard L_h as a mapping $H(B_1) \mapsto H(B_2)$ $(B_r \in \mathcal{L}^+(H), r = 1, 2)$. We shall thus write

$$\|L\|_{H(B_1) \mapsto H(B_2)} \equiv \sup_{v \neq 0} \frac{\|Lv\|_{B_2}}{\|v\|_{B_1}},$$

---

[26] In this book the notion of Euclidean space is connected with a linear finite-dimensional space equipped with any inner product; possible forms of the inner product will be discussed below.

[27] The finite-dimensionality of Euclidean spaces means that there is no need to distinguish *positive operators* B defined by the property $(Bv, v) > 0, \forall v \neq 0$, and *positive definite operators* defined by the stronger condition $B \geq \delta I$, with fixed positive constant δ. For an infinite-dimensional Hilbert space G, these notions are not equivalent, and we will take $\mathcal{L}^+(G)$ to mean the set of positive definite operators.

or, simply, $\|L\|_B$ and $\|L\|$ for the cases $B_1 = B_2 = B$ and $B = I$, respectively. We shall also use $\lambda(L)$ to denote any eigenvalue of L and $\mathrm{sp}\,L \equiv \{\lambda(L)\}$ to denote the *spectrum of L*.

In this section, we confine ourselves primarily to the symmetric positive case that $L_h \equiv L = L^* > 0$ and that computation of the vector Lv, for any given v, can be done in $W = O(N)$ arithmetic operations. [28]

There are various algorithms for solving linear systems. From a general point of view, they might be subdivided in two main classes: *direct methods* and *iterative methods*. Given an invertible matrix L and a vector f, direct methods use a finite sequence of operations to obtain the exact solution of (2.8), assuming no rounding errors.

The number of required arithmetic operations for, say, algorithm a will be denoted by $W_a(N)$, and will serve as a measure of the required *computational work*. [29]

---

[28] The latter property is typical for PGMs, difference methods, and some of their variations because they lead to *sparse matrices*, that is, to matrices L_h with $O(N)$ nonzero elements (see [240, 248, 469]). This is generally not the case for spectral methods.

Sparsity of matrices is a very important and pleasant feature of grid methods because of its potential for reducing computational work. On the other hand, another very important but very unpleasant feature of these methods is the extremely wide range of eigenvalues of L_h: as we saw in Subsection 2.4, for $L_h \equiv \Lambda$ from (2.21), $\mathrm{sp}\,L_h \subset [\alpha_0, \alpha_1]$, where $\alpha_0 > 0$ and $\alpha \equiv \alpha_1/\alpha_0 \asymp h^{-2}$ even for the optimal bounds. We shall see below that it is exactly this quantity α that determines the rate of convergence of classical iterative methods, and this rate therefore degrades severely as $h \to 0$. The *condition number* of an invertible matrix A, defined by $\mathrm{cond}\,A \equiv \|A\|\|A^{-1}\|$ (introduced by Turing (see [226, 505, 509])), is roughly equal to α for the case under consideration, and so the stability of the obtained approximate solution of the system with respect to rounding errors should also degrade rapidly as the number of unknowns increases. This rather general feature of discretizations of elliptic operators must be properly considered in the development and analysis of solution techniques.

[29] It is well known that the classical *Gauss elimination method* yields $W_a(N) = O(N^3)$ in general, although at present some algorithms achieve $W_a(N) = O(N^k)$ with $k \approx 5/2$ (see [401]). It remains an open question in the class of direct methods for general problems as to what is the optimal asymptotic value of $W_a(N)$, in other words, what is the *computational complexity of the problem* (see [479, 480]). We must also keep in mind that the influence of rounding errors might be rather dramatic for these methods (see [505]). More encouraging is that *band matrices* with bounded bandwidth can be treated by Gauss elimination at an obviously optimal asymptotic cost of $W_a(N) \asymp N$. This holds, e.g., for one-dimensional elliptic grid problems and has been exploited extensively by many similar algorithms (see [252, 440]) for difference systems like $\Lambda_r u = f$ (see (2.21)). For two-dimensional problems, therefore, it is reasonable to try to find a numbering scheme for the nodes and their unknowns that leads to the minimal bandwidth. Such attempts have led to the development for special classes of matrices of certain modifications of the Gauss method like the *Cholesky triangular factorization* (see [240]). Unfortunately, one can prove (see [164]) that for systems like (2.15), (2.16) with $h_r \asymp h, r = 1, 2$, it is impossible to obtain a smaller bandwidth than $l \asymp h^{-1}$. This explains why, for the Cholesky triangular factorization of such matrices for more general regions and grids, we have only estimates of the type $W_a(N) = O(N^{5/2})$ (see [146, 240]).

Finally, our discussion of direct methods would not be complete without mentioning that, for certain model problems like (2.15), (2.16) defined on regular grids for rectangles or cubes (parallelepipeds), there exist some remarkable *fast direct methods* that were developed from the efforts of many mathematicians (see [51, 46, 119, 159, 164, 281, 421, 440, 447, 474, 502]). These methods will be discussed in more detail in § 3.1. What is important to us now is that they are characterized by estimates of the type

$$W = O(N(\ln N)^r), \; r = 0, 1, \tag{3.2}$$

$$W = O(N \ln \ln N), \tag{3.3}$$

and that in one way or another they are connected with *separation of variables* and use of the so-called *fast discrete Fourier transform* (see § 3.1 and [41, 120, 440]). The first appearance of such methods that attain (3.2) with $r = 1$ (in the multidimensional case) were in the late 1960's (see [38, 159, 281]). [30]

In certain instances, below we will deal with similar asymptotic estimates for the solution of the system $Lu = f$ with a given f. In such cases, the operators L will usually be denoted by either Λ or B.

3.2. Classical iterative methods. Consider linear system (2.8) with

$$L \equiv L_h \in \mathcal{L}^+(H) \tag{3.4}$$

and

$$\mathrm{sp}\, L \subset [\alpha_0, \alpha_1], \alpha_0 > 0, \tag{3.5}$$

and the *method of simple iteration* of the form

$$u^{n+1} = u^n - \tau(Lu^n - f) \equiv u^n - \tau r^n. \tag{3.6}$$

Here, u^n is the current approximation to the solution u (*nth iterate*), u^{n+1} is the new approximation (($n + 1$)th iterate), u^0 is a given initial approximation, $\tau > 0$ is an iteration parameter that must be specified, and

Nevertheless, these variants of standard direct methods have been instrumental in solving many practical grid systems with moderately large N, especially for two-dimensional problems.

[30] In 1975, Bank and Rose suggested more sophisticated direct methods, calling them *marching algorithms*, that employed some form of block Gaussian elimination. They attained the optimal case of $r = 0$ in (3.2), but unfortunately used elimination procedures that were exponentially unstable in the sense of growth in roundoff error. Later modifications of these methods achieved the optimal or near optimal estimate (3.3) (see [51, 46, 69, 440, 447]).

$r^n \equiv Lu^n - f$ is the *residual* at the nth iteration.[31] In contrast to direct methods, iterative methods do not generally produce the desired solution, but rather a sequence of hopefully increasingly better approximations to it, and the iterative method is said to be *convergent* if $\lim_{n \to \infty} \|u^n - u\| = 0$. To analyze such convergence, we let $z^n \equiv u^n - u$, the *error at the nth iteration*, and note that (3.6) is equivalent to $z^{n+1} = R(\tau)z^n$, where $R(\tau) \equiv I - \tau L$ is the so-called *error reduction operator*. Thus, $\|z^{n+1}\| \leq \|R(\tau)\| \|z^n\|$, so it seems reasonable to attempt to minimize $\|R(\tau)\|$ with respect to τ, i.e., take the operator $R(\tau)$ as close to 0 as possible. Now, for any symmetric operator A in a Euclidean space, we have $\|A\| = max|\lambda(A)| \equiv r(A)$, which can be readily verified via expansions with respect to an *orthonormal basis* of H comprised of eigenvectors of A (see [226, 378, 468, 497]). Then,

$$\|R(\tau)\| \leq q(\tau) \equiv \max_{t \in [\alpha_0, \alpha_1]} |1 - \tau t|,$$

where $q(\tau) = \max\{|1 - \tau\alpha_0|, |1 - \tau\alpha_1|\}$ and $q(\tau) < 1$ if $0 < \tau < 2/\alpha_1$. Hence, it is reasonable to minimize $q(\tau)$, which is achieved by $\tau^* \equiv 2(\alpha_0 + \alpha_1)^{-1}$, yielding $\|R(\tau^*)\| \leq q \equiv (\alpha - 1)/(\alpha + 1) < 1$, where again $\alpha \equiv \alpha_1/\alpha_0$. Hence, if the initial guess u^0 has error z^0 that satisfies $\|z^0\| \leq K_0$ and a desired *accuracy* or *tolerance* $\varepsilon > 0$ is given, then $\|z^m\| \leq \varepsilon$ holds with $m = [|\ln(\varepsilon K_0) \ln q| + 1]$ ([k] denotes the integer part of k). For our model problem (2.15), (2.16), (2.21), the eigenvectors of Λ are given in (2.23). Hence, $\Lambda_r e_k = \lambda_{r,k_r} e_k$ with $\lambda_{r,k_r} \equiv 4h_1 h_2 h_r^{-2} \sin^2(\pi k_r h_r (2l_r)^{-1})$. For $h_1 = h_2 = h$, we see that $c_0 h^2 \leq \lambda(L) \leq 8$ with $c_0 \approx \pi^2(l_1^2 + l_2^2)$ and that the constants used cannot be improved as $h \to 0$. Hence, $q \geq 1 - c_1 h^2$ for some $c_1 > 0$, and the required number of iterations of (3.6) is $m \asymp h^{-2}|\ln \varepsilon|$ (there exist constants $c_3 \geq c_2 > 0$ such that $c_2 \leq mh^2|\ln \varepsilon|^{-1} \leq c_3$). Thus, to obtain the desired ε-accuracy, we must perform $W \asymp h^{-4}|\ln \varepsilon|$ arithmetic operations, which far exceeds $N \asymp h^{-2}$.

We can improve this estimate by using more sophisticated classical methods like *Richardson's iteration*, named for its inventor (see [422]). The basic idea is to allow for new parameters τ_n for each iteration in order to improve overall error reduction. The scheme is given by

$$u^{n+1} = u^n - \tau_n r^n, n = 0, \ldots, k - 1. \tag{3.7}$$

Its error satisfies $z^{n+1} = (I - \tau_n L)z^n$, which implies that $z^k = Z_k z_0$, where $Z_k \equiv (I - \tau_0 L) \cdots (I - \tau_{k-1} L)$ is the error reduction operator. The optimal strategy for choosing the iteration parameters $\tau_0, \ldots, \tau_{k-1}$ consists

---

[31] This method is also sometimes referred to as the *method of Jacobi*, who applied it to particular L with specific τ.

in minimizing $\max_{\lambda \in [\alpha_0, \alpha_1]} |Q_k(\lambda)|$, where $Q_k(\lambda) \equiv (1 - \tau_0 \lambda) \ldots (1 - \tau_{k-1} \lambda)$. [32]

The first iterative methods for difference systems that achieved the nearly optimal estimate

$$W = O(N \ln N |\ln \varepsilon|) \tag{3.8}$$

were discovered by Douglas, Peaceman, and Rachford (see [405, 138]), and became known as *alternating direction iteration (ADI) methods.* [33]

3.3. Iterative methods with spectrally equivalent operators; optimal preconditioning. Success of the ADI methods, as well as the fast direct methods developed later, stemmed from taking advantage of properties (2.22), (2.23), which seldom hold in practice. The development of algorithms with the optimal asymptotic values of $\ln W$ for more general problems was impelled by two basic concepts: *model operators* and *inner iterations* (see [150] and later publications [153, 154, 155, 159, 160, 162, 163, 166, 177, 135, 258, 259, 260, 496, 497]).

Definition. Consider linear grid operators $L_h \equiv L$ and $B_h \equiv B$ in $\mathcal{L}^+(H)$, where the corresponding Euclidean space $H \equiv H_h$ is defined for all admissible grid parameters $h \in \{h\}$. Then operators L and B are called

---

[32] Recognized by several researchers (see [41, 226, 232, 355, 497]) in the 1950's, this optimal choice of parameters is determined by roots of the *Chebyshev polynomial* of degree k: $T_k(t) \equiv \cos(k \arccos t)$. This will be further discussed in Chapter 1. Such an optimal choice leads to the estimate $k \asymp \alpha^{1/2} |\ln \varepsilon|$, which yields some improvement. However, for our model problem (2.15),(2.16), we are still left with the rather excessive estimate $W = O(h^{-3} |\ln \varepsilon|)$. It was recognized almost immediately (see [232]) that, for large α and k, the numerical stability of Richardson's iteration depends critically on the order of the roots of $T_k(t)$. The optimal ordering was discovered by Lebedev and Finogenov in 1971 (see [355, 440]). We also note that attempts to use more sophisticated *gradient methods* like *conjugate gradients*, as well as *Gauss-Seidel* and *successive overrelaxation (SOR) methods* developed in the 1950's (see [271, 277]), led to the same estimates for W. Thus, all classical iterative methods tended to exhibit rather poor asymptotic estimates of convergence and computational complexity. The modern variant of relaxation methods developed by Frankel and Sheldon (see [271]), known as *symmetric successive overrelaxation methods*, attains the asymptotic estimate $W = O(h^{-5/2} |\ln \varepsilon|)$. It is notable that a similar idea of symmetrization was suggested in 1950 by Aitken for the general Gauss-Seidel method (see [6]).

[33] For $\varepsilon \asymp N^{-k}, k > 0$, we have $\ln W = \ln N + o(\ln N)$, which implies that the asymptotically optimal estimate for $\ln W$ is obtained; recall that $f(N) = o(\ln N)$ implies that $\lim_{N \to \infty} \frac{f(N)}{\ln N} = 0$. Such work laid the foundation for further inventions and improvements in the field (we shall return to the discussion of the subject in Chapter 3) and has been used with great success on many important model problems. Probably even more significant was the clear and definite confirmation that attempts to minimize computational work for solving difference systems do lead to nearly optimal asymptotic estimates of W and very efficient practical algorithms. Interestingly enough, these methods are attracting renewed attention because of their suitability for modern vector and parallel supercomputers (see [131, 288, 398]).

spectrally equivalent if there exist positive constants δ_0 and δ_1, independent of h, such that

$$\delta_0(Bv, v) \le (Lv, v) \le \delta_1(Bv, v), \quad \forall v \in H. \tag{3.9}$$

We write this relationship as $L \asymp B$, noting that it clearly satisfies all of the conditions to be an equivalence, as mentioned, e.g., in Subsection 1.3. Reference to a spectrum is made here for two reasons. First, if $\dim H = N$ and $\lambda_i(L)$ denotes the ith eigenvalue of L in the increasing order (and similarly for $\lambda_i(B)$), then (3.9) implies that

$$\delta_0 \lambda_i(B) \le \lambda_i(L) \le \delta_1 \lambda_i(B), \quad i = 1, \dots, N, \tag{3.10}$$

which follows from Fisher's theorem (see Theorem 9.1.1). The second especially important reason is given in the following lemma.

Lemma 1. *Let H be a Euclidean space, B, L belong to $\mathcal{L}^+(H)$, and*

$$\delta_0 B \le L \le \delta_1 B, \quad \delta_0 > 0. \tag{3.11}$$

Then $A \equiv B^{-1}L$ is a symmetric operator when it is regarded as a mapping of the Euclidean space $H(D)$ (with either $D = B$ or $D = L$) into itself, and

$$\mathrm{sp}\,(B^{-1}L) \subset [\delta_0, \delta_1]. \tag{3.12}$$

Proof. We have $X \equiv (Au, v)_B = (B(B^{-1}Lu), v) = (Lu, v), \ \forall u, \forall v.$ Hence, $X = (u, Lv) = (u, Av)_B$. Similarly, $(Au, v)_L = (L(B^{-1}L)u, v) = (B^{-1}Lu, Lv) = (u, Av)_L$. Consider any eigenvector w of A: $Aw = \lambda w$. Then $Lw = \lambda Bw$, $\lambda = (Lw, w)(Bw, w)^{-1}$, and, since (3.11) and (3.9) are equivalent, it follows that $\lambda \in [\delta_0, \delta_1]$. $\square$

Lemma 1 implies that when $L \asymp B$ with constants δ_0, δ_1, the spectrum of $A \equiv B^{-1}L$ belongs to the finite interval $[\delta_0, \delta_1]$ and $A \in \mathcal{L}^+(H(D))$ with $D = B$ or L (or other more complicated choice whose practical value is unclear). Moreover, the iterative method

$$Bu^{n+1} = Bu^n - \tau_n(Lu^n - f) \tag{3.13}$$

applied to (2.8) is equivalent to the method $u^{n+1} = u^n - \tau_n(Au^n - B^{-1}f)$ (of type (3.7)) applied to the *preconditioned system*

$$Au \equiv B^{-1}Lu = B^{-1}f \tag{3.14}$$

with the symmetric and positive operator A ($A \in \mathcal{L}^+(H(D))$, where $D = B$ or L), the spectrum of which satisfies (3.12). In particular, method (3.13) with $\tau_n = \tau \equiv 2(\delta_1 + \delta_0)^{-1}$, which may be regarded as a *modified method of the simple iteration,* yields the estimates $\|I - \tau B^{-1}L\|_D \leq [\delta - 1]/[\delta + 1]$, $\|z^k\|_D \leq \varepsilon\|z^0\|_D$ if $k \asymp |\ln\varepsilon|$, where $\delta \equiv \delta_1/\delta_0$, $D = B$ or L. [34]

Modifications of other classical methods like Richardson's iteration and gradient methods can be used with success (they will be discussed in detail in § 1.3). The above operator B is often referred to as a *model operator* or a *preconditioner,* and its choice is of fundamental importance. In particular, if B is such that $L \asymp B$ and any system $Bv = g$ can be solved with estimates of W given by (3.2), then the modified method of the simple iteration (see (3.13)) gives a way to obtain an ε-approximation to the solution of (2.18) at a cost of

$$W = O(N(\ln N)^r|\ln\varepsilon|). \tag{3.15}$$

Therefore, for constructing asymptotically optimal algorithms for solving linear grid systems (2.8) with $L \in \mathcal{L}^+(H)$, i.e., with $L = L^* > 0$, it suffices to determine a model operator B with the two above mentioned properties. More precisely, we shall call B an *optimal preconditioner* if it is spectrally equivalent to the operator L and algorithms with estimates $W = O(N)$ are known for solving systems $Bv = g$ with a given g; we shall call it a *nearly optimal preconditioner* if solutions of the above systems require estimates (3.2) with $r \geq 0$ and $\delta \equiv \delta_1/\delta_0 = O((\ln N)^s)$ with $s \geq 0$. In accordance with this, we shall be speaking about *optimal preconditioning* or *nearly optimal preconditioning.* [35]

---

[34] It is clear that δ is, in fact, a bound on the condition number of the operator $A = B^{-1}L$: $\delta \geq \text{cond}\{B^{-1}L\}$, and that this estimate is exact for optimal values of δ_0 and δ_1 equal to the minimal and maximal eigenvalues of the generalized eigenvalue problem $Lu = \lambda Bu$, respectively.

[35] We will return to constructing optimal and nearly optimal preconditioners later (this problem will be discussed in detail in Chapters 2 and 3); right now we make some comments about Lemma 1, used for $D = B$ in the author's paper [150] (see also [154, 155, 162]) in connection with constructing model operators B spectrally equivalent to a grid operator L_h, which was obtained through the difference approximation of the first boundary value problem in a d-dimensional cube in the case of an elliptic equation of $2m$th order ($m \geq 1$) with variable coefficients; the consideration of such equations was especially instructive since, for them, $\alpha \asymp h^{-2m}$ and $\delta \asymp 1$. From a mathematical point of view, Lemma 1 is very close to a similar result by Kantorovich, dealing with the case $D = B$ and differential elliptic operators (see [292]); he applied it to the modified method of *steepest descent* (to be discussed in § 1.3) with special attention to elliptic equations with polynomial coefficients and polynomial iterates. There are also common points with notions of *alike operators* (see [373]).

We should also emphasize again the role of Friedrich's idea of using an appropriate inner product in the given Hilbert space, which was discussed in § 0.2. Moreover, Lemma 1 relates to the more general concept of improving the given system $Lu = f$ by passing to system (3.14), with B^{-1} constructed to approximate L^{-1}. The goal is to obtain $\lambda(B^{-1}L)$ as close to 1 as possible (see [110, 226]); the importance of using a symmetric B and $\|B^{-1}L\|$ was stressed in [102, 150, 264, 277].

At the present time, a variety of fruitful approaches now exists to constructing optimal and nearly optimal preconditioners. While the entire Chapter 3 and some sections of

Thus, for $L = L^* > 0$, many effective iterative methods exist and are united by the use of Lemma 1. But what can we do with more general problems? Answers differ and depend on what kind of generalization we have in mind. For $L = L^* \geq 0$ with Ker $L \equiv \{v : Lv = 0\}$, the answer is almost obvious (see [162, 351, 440]) and is given by the following lemma.

Lemma 2. *Let* $L = L^* \geq 0, \dim Ker\, L \geq 1$, *and* $B = B^* > 0$, *and suppose inequalities (3.9) hold for all* v *orthogonal in the sense of* $H(B)$ *to* Ker L. *Then the operator* $A \equiv B^{-1}L$ *is symmetric as an element of* $\mathcal{L}(H(B))$ *and*

$$\mathrm{sp}\ (B^{-1}L) \subset \{0 \cup [\delta_0, \delta_1]\}. \tag{3.16}$$

Proof. The proof of symmetry of A is the same as for Lemma 1. Therefore, the space $H(B)$ is an *orthonormal sum* of Ker A and Im $A \equiv \{Av : v \in H\}$ (see [64, 226, 328]). [36] Note that Ker $A =$ Ker L, and that all eigenvectors corresponding to nonzero eigenvalues belong to Im A. Thus, for such eigenvectors, we have $Av = \lambda v$, $(Lv, v) = \lambda(Bv, v)$, which, by virtue of (3.9), implies (3.16). $\square$ [37]

Chapters 5, 6, and 8 will be devoted to this problem, we just mention briefly such basic and widely applied concepts as:

- Select a simple enough part $\Lambda \equiv \Lambda_\Omega$ of the grid operator $L_h \equiv L_\Omega$ such that, for solving a system $\Lambda v = g$ with an arbitrary given g, some fast direct methods are available (see § 3.1), and yet Λ represents a principal part of L_h in some sense.

- Choose Λ so that effective inner iterations for systems $\Lambda v = g$ can be found leading to the so-called two-stage preconditioners B of the form $B = \Lambda(I - Z_k)^{-1}$, where k is the number of the performed inner iterations and Z_k is the corresponding error reducing operator (see § 3.4).

- Instead of Λ_Ω, choose a similar operator Λ_Q defined on a grid topologically equivalent to the original one but for a simpler model region Q (see the next § 4).

- A partition of the given $\bar{\Omega}$ or of a model closed region $\bar{Q}$ can be used to construct B (domain decomposition preconditioners and methods of Schwarz type (see § 3.5)).

- On the basis of a partition of the given grid and splitting of the original finite element space (for $\bar{\Omega}$ or $\bar{Q}$), use multigrid to construct model operators B (see § 3.7); other fruitful algebraic multigrid constructions of model operators are also available (see [89, 291, 350, 362]).

- If a sufficiently simple region $\bar{\Pi}$ can be obtained from $\bar{Q}$ by adding $\bar{F}$ (F is often referred to as a fictitious domain, $\bar{\Pi} = \bar{Q} \cup \bar{F}$), then B_Q can be constructed via B_Π (see § 3.6).

[36] Ker A and Im A are sometimes denoted by $N(A)$ and $R(A)$.
[37] For $L = L^* \geq 0$ there exists a possibility to use a model operator $B = B^* \geq 0$ satisfying (3.9) (if positive constants δ_0 and δ_1 do not depend on the grid, we may again call these operators spectrally equivalent). This case can be reduced to the original case by restricting the operators to the common invariant subspace Im $L=$Im B.

3.4. Symmetrizations of systems. For general systems (2.8), including nonlinear problems, a modified method of the simple iteration can be used, as we will see in § 1.3; some applications of the Richardson method are also known (see [386]).

A very promising alternative is to reduce general linear systems to the *normal equations* $L^* L u = L^* f$ with symmetric positive operators $L^* L$ by the well-known *Gauss transformation*, also known as as the *Gauss symmetrization of the system*, which is equivalent to applying the *least squares method* based on minimization of $\|Lv - f\|^2$. This idea, together with the construction of nearly optimal preconditioners for the symmetrized difference systems, was investigated in [156, 162] (see also § 6.2), but it was immediately understood that it could work well only for elliptic problems that exhibited regularity estimates of the form $\|u\|_{W_2^2(\Omega)} \leq K|f|_{0,\Omega}$. Other combinations of preconditioning and symmetrization were examined. For example, an alternative to (3.14) is the system $LB^{-1}v = f$, which in symmetrized form is $B^{-1}L^* LB^{-1}v = B^{-1}L^* f$. But such approaches did not generally lead to any essentially new results. The reason, as it became clear only in the 1980's (see [198, 199]), lies in the use of inappropriate geometry: Gauss symmetrization deals with operators as mappings of a Euclidean space into itself, but the geometry of this space is not consistent with the geometry of the original Hilbert space and the chosen PGM subspace. A more relevant approach consists in regarding L as an element of $\mathcal{L}(H(B_2); H(B_1^{-1}))$, with $B_r = B_r^* > 0, r = 1, 2$, for which we have the following lemma.

Lemma 3. *Let L be considered as an element of $\mathcal{L}(H(B_2); H(B_1^{-1}))$. Then its conjugate (or adjoint) operator L' is given by $L' = B_2^{-1}L^* B_1^{-1}$.*

Proof. We have $X \equiv (Lu, v)_{B_1^{-1}} = (u, L^* B_1^{-1}v) = (B_2 u, L'v), \forall u, \forall v$. Hence, $X = (u, L'v)_{B_2}$, from which the lemma follows. $\square$

Thus, the symmetrization defined by the chosen pair of spaces leads to the system

$$Au \equiv B_2^{-1}L^* B_1^{-1}Lu = B_2^{-1}L^* B_1^{-1}f, \tag{3.17}$$

where $A \in \mathcal{L}^+(H(B_2))$ provided L is invertible. Now suppose $B \in \mathcal{L}^+(H)$ and consider the three choices $B_1 = I, B_2 = B^2$; $B_1 = B^2, B_2 = I$; $B_1 = B_2 = B$. This yields the respective symmetrized operators $A = B^{-2}L^* L$; $A = L^* B^{-2}L$; $A = B^{-1}L^* B^{-1}L$. In § 1.3, we will obtain estimates of their localized spectra of the form

$$\mathrm{sp}\, A \subset [\delta_0, \delta_1], \ \delta_0 > 0, \tag{3.18}$$

with $\delta_0 \neq \delta_0(h), \delta_1 \neq \delta_1(h)$, meaning that $L^* L \asymp B^2$; $L^* B^{-1}L \asymp I$; $L^* B^{-1}L \asymp B$. We emphasize that the last relationship is equivalent to the natural conditions

$$\|L\|_{H(B) \mapsto H(B^{-1})} \leq \delta_1^{1/2} \text{ and } \|L^{-1}\|_{H(B^{-1}) \mapsto H(B)} \leq \delta_0^{-1/2}, \qquad (3.19)$$

which were discussed in [154, 155] and written in the form

$$\delta_0 \|v\|_B^2 \leq \|Lv\|_{B^{-1}}^2 \leq \delta_1 \|v\|_B^2, \qquad \forall v \in H. \qquad (3.20)$$

As we shall see in § 4, these conditions may be regarded as a consequence of the correctness of the original elliptic boundary value problem and, in fact, were established in [164] for difference methods and in [166] for PGMs and FEMs. A number of relevant investigations (see [16, 82, 218, 221, 244, 253, 277, 353, 355, 432, 440]) should also be mentioned. Some of these studies considered the *generalized least squares method* (or *generalized method of minimal residuals*) for converting the system $Lu = f$ into a variational problem of the type $u = \arg\min \|Lu - f\|_D^2$, with an operator $D = D^* > 0$. Such schemes have been known for a long time (see, e.g., [64, 277]) and were used even earlier by Courant for constructing strongly convergent minimizing sequences (see [124]). In our case, this approach leads to the symmetrized system

$$L^* D L u = L^* D f, \qquad (3.21)$$

which, with improper choice of D, can significantly increase the condition number of the matrix. Thus, for (3.19), it seems reasonable to use $D \asymp B^{-1}$, as suggested in [164]. [38]

3.5. Coarse grid continuation (multigrid acceleration of the basic iterative algorithm). As we have seen, optimal preconditioning and appropriate symmetrization of grid systems may lead to algorithms for finding ε-approximations of their solutions that yield computational work estimates of type (3.15) with $r = 0$. Now it seems prudent to choose this ε, which governs the accuracy in the approximation of the grid system solution, to be on the same order of accuracy of approximation of the given original problem in a Hilbert space, and which is often of the form $\varepsilon \asymp N^{-k}$ for some fixed $k > 0$. (The form of discretization error will be discussed in more detail in § 5.) For such cases, (3.15) with $r = 0$ becomes

$$W = O(N \ln N). \qquad (3.22)$$

To eliminate the multiplier $\ln N$ in these estimates is not a simple matter, especially for the case of curvilinear boundaries. We will deal with this

---

[38] Note that taking such a D and using it to further precondition (3.21) leads to (3.17) with $B_1 = B_2 = B$. We emphasize that the same result is obtained without preconditioning by an appropriate symmetrization in the sense of Lemma 3, again with $B_1 = B_2 = B$.

problem in Chapter 5 in detail and with a simpler variant in § 1.4. [39]

3.6. Some nonelliptic applications. The concept of a model operator or preconditioner has gained growing recognition as a very useful tool for solution of a variety of large systems of equations that have nothing to do with elliptic boundary value problems. [40]

§ 4. Invariance of operator inequalities under projective approximations

4.1. Rayleigh-Ritz method and Gram matrices. Returning to system (2.8) for the Rayleigh-Ritz method, we shall treat it as an operator equation in the Euclidean space $H \equiv \mathbf{R}^N$. Recall that the matrix $L \equiv L_h$, defined by (2.9), is the Gram matrix for the chosen basis of the subspace $\hat{G} \subset G(L)$. The Euclidean space of elements $\hat{v} \in \hat{G}$ under the inner product inherited from $G(L)$ will be denoted by $\hat{V}$, so that $(\hat{u}, \hat{v})_{\hat{V}} = b(\hat{u}; \hat{v})$.

Lemma 1. *If conditions* (1.12)–(1.14) *are satisfied, then* $L_h \in \mathcal{L}^+(H)$ *and the Euclidean spaces* $\hat{V}$ *and* $H(L_h)$ *are isometric.*

Proof. Expansions (2.2) give an isomorphism between and $H \equiv \mathbf{R}^N$

---

[39] For now, we just emphasize the usefulness in this context of the classical *method of continuation with respect to a parameter* (see [64]) for solution of systems that involve the given parameter. For systems associated with grids obtained via p successive global refinements of an initial grid, as suggested by many authors especially in applications, a natural choice for the parameter is the index $l \in [0, p]$ of the level of refinement. That is, the basic idea of the *multigrid acceleration of the iterative algorithm* at hand is to provide an initial guess on the given grid by applying first the iterative method to coarser levels of refinement. Such analysis for methods of type (3.13) with $B \asymp L$ was carried out in [181, 187], showing that for successive applications of the basic iteration on grids with the levels of refinement $l = 0, 1, \ldots, p$, only a finite number of iterations on each level are needed to obtain the desired accuracy. Similar estimates for certain multigrid methods (referred to as *full multigrid methods*) were obtained for the first time by Bakhvalov (see [37]). Note that coarse grid continuation is used even in the classical Newton interpolation formula (see [64]).

[40] While this topic is beyond our present scope, we only mention a few typical results. First, several papers (see [113, 112, 482, 483]) deal with circulant preconditioners for systems with *Toeplitz matrices*. Second, a class of important grid problems arises in the theory of *queueing problems* (see, e.g., [111, 297]). On the basis of *Markovian queueing networks* with overflow capacity, it is possible to describe steady-state probability distributions as solutions of the *Kolmogorov balance equations*. The resulting $N \times N$ homogeneous linear system $Lu = 0$ has solutions corresponding to different states in the network and are identified as values of the grid function u at nodes of a cubical grid in Euclidean space of dimension equal to the number of queues. The nonsymmetric matrix L is such that dim Ker $L = 1$, and its nonzero values correspond to those of the simplest difference Laplacian with boundary conditions analogous to the Neumann type. Thus, a natural choice for solving our system is to apply an iterative method to its appropriate symmetrization, using a model operator that is spectrally equivalent to the corresponding grid Laplacian. This will be analyzed in § 1.3 for symmetric nonnegative model operators.

(see (2.2), (2.7)). Hence,

$$b(\hat{u}; \hat{v}) = (L_h \mathbf{u}, \mathbf{v}), \forall \hat{u} \in \hat{G}, \forall \hat{v} \in \hat{G}, \tag{4.1}$$

which implies that $L_h = L_h^*$. Taking $\hat{v} = \hat{u}$ in (4.1) yields

$$\|\hat{u}\|_{\hat{V}} \equiv \|\hat{u}\|_L = \|\mathbf{u}\|_{L_h}, \tag{4.2}$$

which shows that the isomorphism is an isometry. $\square$.

Similarly, it is easy to verify that expansions (2.2) provide an isometry between the Euclidean spaces $\hat{G}$ and $H(\Lambda) \equiv H(J)$, where

$$\Lambda \equiv J \equiv [(\hat{\psi}_j, \hat{\psi}_i)_G] \tag{4.3}$$

is the Gram matrix of the same basis, but now in the original inner product of the Hilbert space G. We are using here two letters for this matrix to stress that it corresponds, on the one hand, to a particular (model) case of the operator L_h and, on the other hand, to a projective approximation of the identity operator $I \in \mathcal{L}(G)$ (see, e.g., (2.21), which shows that J is not the identity matrix). The next elementary lemma, which is very well known (see [166, 373]), nevertheless plays a fundamental role in the translation of certain operator inequalities in a Hilbert space into an algebraic language.

Lemma 2. *Let conditions (1.12)–(1.14) be satisfied and let L_h and J be defined by (2.9), (4.3). Then $\delta_0 J \le L_h \le \delta_1 J$, $L_h \asymp J$.*

Proof. From (1.12)–(1.14) and Lemma 1, for an arbitrary $\hat{u}$, we have $\delta_0\|\hat{u}\|_G^2 \le \|\hat{u}\|_{\hat{V}}^2 \le \delta_1\|\hat{u}\|_G^2$. Hence, $\delta_0\|\mathbf{u}\|_J^2 \le \|\mathbf{u}\|_{L_h}^2 \le \delta_1\|\mathbf{u}\|_J^2$. $\square$

We also recall a few important properties of the matrices in $\mathcal{L}^+(H) \equiv \{B : B = B^* > 0\}$ that will be of a later use. First, it can be shown (see [226, 239]) that each $B \in \mathcal{L}^+(H)$ admits a unique operator $B^{1/2} \in \mathcal{L}^+(H)$ such that $(B^{1/2})^2 = B$ and, hence, $\|v\|_B = \|B^{1/2}v\|, \forall v \in H$. The change of variables $B^{1/2}v \equiv u$ then allows us to rewrite (3.9) in the form $\delta_0 I \le B^{-1/2}LB^{-1/2} \le \delta_1 I$.

Lemma 3. *If A and B belong to $\mathcal{L}^+(H)$ and $A \ge B$, then $B^{-1} \ge A^{-1}$.*

Proof. Consider the operator $C = B^{-1}A$. Then $C \in \mathcal{L}^+(H(B))$ and $C \ge I$. Thus, $C^{-1} = A^{-1}B \le I$ and, for all v, $(A^{-1}Bv, v)_B \le (v, v)_B$. The change of variables $Bv = u$ leads to $(A^{-1}u, u) \le (B^{-1}u, u)$. $\square$

Lemma 4. *Let $A = A^* > 0$ and $B = B^* \ge 0$. Then*

$$B \le A \Leftrightarrow \|B^{1/2}A^{-1/2}\| \le 1.$$

Proof. We have $(Bu, u) \le (Au, u) \Leftrightarrow \|B^{1/2}A^{-1/2}v\|^2 \le \|v\|^2$, where $v \equiv A^{1/2}u$. Since u can be any vector and the operator $A^{1/2}$ is invertible, the set of the corresponding v is just the given Euclidean space. $\square$

Lemma 5. If $B \in \mathcal{L}^+(H)$, then

$$\|v\|_{B^{-1}} = \max_{u \neq 0}\{|(u, v)|/\|u\|_B\}, \quad \forall v \in H.$$

Proof. For all u, v, we have $|(|u, v)| = |(B^{1/2}u, B^{-1/2}v)| \leq \|u\|_B\|v\|_{B^{-1}}$, with equality if $u = Bv$. $\square$ [41]

The next lemma, which is well known (see [272, 505]), will often be of use to us.

Lemma 6. If $A \in \mathcal{L}(H_1; H_2)$ and $B \in \mathcal{L}(H_2; H_1)$, then

$$\text{sp } (AB) \setminus 0 = \text{sp } (BA) \setminus 0.$$

Proof. We sketch a simple proof in the most important case of $H_1 = H_2 = H = \mathbf{R}^N$. The proof will follow if we can show that, for any $\lambda \neq 0$, the operators $AB - \lambda I$ and $BA - \lambda I$ are either both invertible or both singular. Without loss of generality, we consider only the case $\lambda = 1$. Now, to this end, assume that $AB - I$ is invertible and write $(AB - I)^{-1} = C$. Then $I = (AB - I)C$ and $ABC = I + C$, so that $R \equiv (BA - I)(BCA - I) = B(I + C)A - BCA - BA + I = I$. Thus, $BA - I$ is invertible and $(BA - I)^{-1} = BCA - I$. $\square$ [42]

Lemma 7. Suppose $A = A^* \geq B = B^* \geq 0$. Then $A^{1/2} \geq B^{1/2}$.

Proof. Consider the case $A > 0$. Suppose that the assertion is not true. Then there exists u such that $(A^{1/2}u, u) < (B^{1/2}u, u)$, and the largest eigenvalue λ_N of the problem $B^{1/2}v = \lambda A^{1/2}v$ must be such that $\lambda_N > 1$. Then, for the corresponding eigenvector $v_N \equiv w$ with $B^{1/2}w = \lambda_N A^{1/2}w$, we have $(Bw, w) = (B^{1/2}w, B^{1/2}w) = \lambda_N^2(Aw, w)$. Hence, $(Bw, w) > (Aw, w)$, a contradiction with the condition. Hence, the assertion must be true for $A > 0$. If $A \geq 0$, we introduce $A_n \equiv A + 1/nI > 0$, where $n = 1, 2, \ldots$. Then $A_n^{1/2} \geq B^{1/2}$, and in the limit we obtain the desired inequality. $\square$ [43]

4.2. Projective approximations of operators. Consider now the general case of invertible operators $\hat{L} \equiv PLP$ and $L_h \equiv [b(\hat{\psi}_j; \hat{\psi}_i)]$, which

---

[41]Lemma 5 explains why estimates of the form (3.19) will often be used in this book: for the model equation $Bv = f$, use of $\|f\|_{B^{-1}}$ leads to the equality $\|v\|_B = \|f\|_{B^{-1}}$, so the estimate for L^{-1} in (3.10) may be regarded as optimal (we estimate the chosen norm of the solution via the weakest norm of the right-hand side). Similar estimates became standard in the theory of elliptic equations after the appearance of Weyl's estimates (see [326, 371]) that apply when the right-hand sides are divergences of elements of a given vector field.

[42]If $H_1 = H_2$, then sp $(AB) = $ sp (BA). Moreover, $\|A\|^2 = r(A^*A)$ and $\|A^*\|^2 = r(AA^*)$, where $r(B) \equiv \max|\lambda(B)|$ is the *radius of the spectrum* of B. Hence, Lemma 6 leads to the well-known relation $\|A\| = \|A^*\|$ for arbitrary $A \in \mathcal{L}(H)$.

[43]The more general conclusion $A^\alpha \geq B^\alpha$ with $0 \leq \alpha \leq 1$ is widely known as the *Heinz inequality* (see [318, 296]).

were earlier defined by the relations

$$b(\hat{u}; \hat{v}) = (\hat{L}\hat{u}, \hat{v})_G = (L_h \mathbf{u}, \mathbf{v})_H, \quad \forall \hat{u} \in \hat{G}, \forall \hat{v} \in \hat{G},$$

(see (2.2), (2.7), and (2.9)).

Theorem 1. *With J defined by (4.3), we have*

$$\|\hat{L}\| = \|L_h\|_{H(J) \mapsto H(J^{-1})} = \|J^{-1/2} L_h J^{-1/2}\| \tag{4.4}$$

and

$$\|(\hat{L})^{-1}\| = \|L_h^{-1}\|_{H(J^{-1}) \mapsto H(J)} = \|J^{1/2} L_h^{-1} J^{1/2}\|. \tag{4.5}$$

Proof. Define $q(\hat{u}) \equiv \|\hat{L}\hat{u}\|(\|\hat{u}\|)^{-1}$ for $\hat{u} \neq 0$. Then

$$\|\hat{L}\| = \max_{\hat{u} \neq 0} q(\hat{u}) \text{ and } \|\hat{L}^{-1}\| = \max_{\hat{u} \neq 0} [q(\hat{u})]^{-1}.$$

We first show that for each function $\hat{u} \in \hat{G}$ and the corresponding vector $\mathbf{u}$ (see (2.2) and (2.7)), we have

$$\|\hat{L}\hat{u}\| = \|L_h \mathbf{u}\|_{J^{-1}}. \tag{4.6}$$

To this end, notice that $X \equiv \|\hat{L}\hat{u}\| = \max_{\hat{v} \neq 0} |(\hat{L}\hat{u}, \hat{v})|/\|\hat{v}\|$. Thus,

$$X = \max_{\hat{v} \neq 0} \frac{|b(\hat{u}; \hat{v})|}{\|\hat{v}\|} = \max_{\mathbf{v} \neq 0} \frac{|(L_h(\mathbf{u}, \mathbf{v})|}{\|\mathbf{v}\|_J} = \|L_h \mathbf{u}\|_{J^{-1}},$$

and (4.6) holds. This implies that $q(\hat{u}) = \|L_h \mathbf{u}\|_{J^{-1}}/\|\mathbf{u}\|_J$ and $\max_{\hat{u} \neq 0} q(\hat{u}) = \|L_h\|_{H(J) \mapsto H(J^{-1})}$, from which (4.4) follows. Using again (4.6) and the changes of variables $\hat{L}\hat{u} = \hat{v}$ and $L_h \mathbf{v} = \mathbf{f}$, we can verify in much the same way that

$$\|\hat{L}^{-1}\| = \max_{\mathbf{u} \neq 0} \frac{\|\mathbf{v}\|_J}{\|L_h \mathbf{v}\|_{J^{-1}}} = \max_{\mathbf{f} \neq 0} \frac{\|L_h^{-1} \mathbf{f}\|_J}{\|\mathbf{f}\|_{J^{-1}}},$$

from which (4.5) follows. $\square$

Here again we emphasize the importance of inequalities

$$\|L_h^{-1}\|_{H(J^{-1}) \mapsto H(J)} \leq K_0 \tag{4.7}$$

and

$$\|L_h\|_{H(J) \mapsto H(J^{-1})} \leq K_1, \tag{4.8}$$

with constants K_0 and K_1 independent of h. They will be very useful later, especially for obtaining inequalities (3.19) with model operators $B \asymp J$.

Theorem 2. Let conditions (1.13), (1.14) *be satisfied. Then inequalities* (4.7), (4.8) *with* $K_0 = \delta_0^{-1}, K_1 = \delta_1$ *are valid.*

Proof. Inequalities (1.13), (1.14) in combination with Theorem 1.5 for projective problem (3.5) imply that $\|\hat{L}\| \leq \delta_1$, $\|(\hat{L})^{-1}\| \leq \delta_0^{-1}$. These inequalities together with (4.4), (4.5) lead to (4.7), (4.8). □

Estimate (4.7) is a consequence of (1.14). It is very important to find more general conditions for this fundamental estimate. In order to do so, we recall the next well-known *theorem about perturbations of an invertible operator* (see [292, 295, 317]).

Theorem 3. Let G be a Hilbert or a Euclidean space, $A \in \mathcal{L}(G)$ an invertible operator with $\|A^{-1}\| = a < \infty$, and $A' \in \mathcal{L}(G)$ a perturbed operator with $a\|A - A'\| \equiv q \leq 1$. Then A' is invertible and $\|(A')^{-1}\| \leq a(1-q)^{-1}$.

Proof. The simplest way to prove this theorem is via the contraction mapping principle, which we applied already in Theorem 1.5. Given arbitrary f, consider an operator equation $A'u = f$ replaced by its equivalent $u = Cu \equiv -A^{-1}(A' - A)u + g$, where $g \equiv A^{-1}f$. Since $\|Cu - Cv\| \leq q\|u - v\|$, then the operator C defines a contraction mapping in the complete metric space. Thus, the given equation has a unique solution u that satisfies $\|u\| \leq q\|u\| + \|g\|$, implying that $\|(A')^{-1}f\| \leq (1-q)^{-1}a\|f\|$. This proves the desired inequality. □

Theorem 4. Let G denote a Hilbert space and consider an invertible operator $L \in \mathcal{L}(G)$, with $\|L^{-1}\| \leq K$, split into the sum $L = L_1 + L_2$, where L_2 is a compact linear operator and L_1 corresponds to a bilinear form satisfying conditions (1.13), (1.14) *with constants $\delta_{1,1}, \delta_{0,1} > 0$ instead of δ_1, δ_0. Let $\{\hat{G}_N\}$ be a sequence of subspaces approximating the Hilbert space G (see* (2.12)) *and consider the projective operators $\hat{L}, L_h$ defined by* (2.5),(2.8). *Then there exists a constant $K_0 > 0$ such that, for sufficiently large N, the operators $\hat{L}, L_h$ are invertible and*

$$\|(\hat{L})^{-1}\| \leq K_0, \quad \|L_h^{-1}\|_{H(J^{-1}) \mapsto H(J)} \leq K_0.$$

Proof. The assumptions on L_1 imply that $\|L_1^{-1}\| \leq \delta_{0,1}^{-1}$, $\|\hat{L}^{-1}\| \leq \delta_{0,1}^{-1}$. Writing $\hat{L} = \hat{L}_1 A_0$, where $A_0 \equiv I + \hat{L}_1^{-1}\hat{L}_2$, then $\hat{L}$ is invertible if and only if the operator $A_0 \in \mathcal{L}(G_N)$ is, which is what we now show. Following [317, 372], we prove that the extended operator $A' \equiv I + \hat{L}_1^{-1}\hat{L}_2 P \in \mathcal{L}(G)$ has a bounded inverse for large enough N (P is the orthoprojector onto $\hat{G}_N$ so that $A'v = A_0 v$ for all $v \in \hat{G}_N$). We regard A' as a perturbation of $A \equiv I + L_1^{-1}L_2 = L_1^{-1}L$ with $A^{-1} = L^{-1}L_1$, $\|A^{-1}\| \leq K\delta_{1,1}$. Note that $Q_N \equiv A' - A = (R_N + R'_N)L_2$, where $R_N \equiv \hat{L}_1^{-1} - L_1^{-1}$ and $R'_N \equiv \hat{L}_1^{-1}(P - I)$. For arbitrary f, from (2.12) and Theorem 2.3 we may assert that

$$\lim_{N \to \infty} \|(R_N + R'_N)f\| = 0,$$

which means that $\{Q_N\}$ is a sequence of operators that converge pointwise to 0 (see [341]). Since L_2 is a compact operator, the set of all $L_2 v$ with $\|v\| \leq 1$ is compact. Hence, from a well-known theorem of functional analysis (see [341]), the sequence Q_N is convergent to 0 in the usual norm sense, that is, $\lim_{N \to \infty} \|Q_N\| = 0$. If N is large enough and $\|Q_N\|\|A^{-1}\| \leq q < 1$, then Theorem 3 allows us to write $\|(A')^{-1}\| \leq (1-q)^{-1}\|A^{-1}\|$ and $\|\hat{L}^{-1}\| \leq (1-q)^{-1} K \delta_{1,1} \delta_{0,1}^{-1} \equiv K_0$. This and (4.5) lead to (4.7). □

4.3. Spectral equivalence of grid operators defined on topologically equivalent triangulations. We start by considering two topologically equivalent triangulations: $T_h(\bar{\Omega})$ of the original region Ω with the Lipschitz boundary and $T_h(\bar{Q})$ of a chosen *model region Q*. [44] We assume that $T_h(\bar{Q})$ is a collection of triangles obtained from a given square grid with mesh size h by standard subdivision of each grid cell into two equal triangles (choice of the two possible diagonals can be made cell by cell). Let $\mathrm{II} : \bar{Q} \mapsto \bar{\Omega}$ be a continuous piecewise-affine mapping (so that the image of each triangle $T' \in T_h(\bar{Q})$ is a corresponding triangle $T \in T_h(\bar{\Omega})$, that is, $\mathrm{II}\{T'\} = T$ and $\mathrm{II}^{-1}\{T\} = T'$). This one-to-one correspondence between the closed regions written in the form $x = \mathrm{II}z$, $x \in \bar{\Omega}$, $z \in \bar{Q}$, defines an isomorphism between the finite element spaces $\hat{G}_{\Omega,h}$ and $\hat{G}_{Q,h}$ of functions

$$\hat{u}_\Omega(x) = \sum_{i=1}^N u_i \hat{\psi}_{\Omega,i}(x) \text{ and } \hat{u}_Q(z) = \sum_{i=1}^N u_i \hat{\psi}_{Q,i}(z), \quad (4.9)$$

where u_i correspond to the same values of the functions $\hat{u}_\Omega$ and $\hat{u}_Q$ at the equivalent nodes with the index i and $\hat{\psi}_{\Omega,i}, \hat{\psi}_{Q,i}$ are usual continuous piecewise linear basis functions (see (2.14), (2.15)). Let the positive piecewise constant functions $a(x), a_Q(z)$ have the same constant values at inner points of corresponding triangles $T, T'(T = \mathrm{II}\{T'\})$, and define the following grid operators (matrices):

$$\Lambda_\Omega \equiv \Lambda_{\Omega,h} \equiv [\sum_{r=1}^2 (a(x), \frac{\partial \hat{\psi}_{\Omega,j}}{\partial x_r} \frac{\partial \hat{\psi}_{\Omega,i}}{\partial x_r})_{0,\Omega}] \quad (4.10)$$

and

$$\Lambda_Q \equiv \Lambda_{Q,h} \equiv [\sum_{r=1}^2 (a_Q(z), \frac{\partial \hat{\psi}_{Q,j}}{\partial z_r} \frac{\partial \hat{\psi}_{Q,i}}{\partial z_r})_{0,Q}]. \quad (4.11)$$

---

[44] Recall that we sometimes refer to triangulations of regions instead of their closures.

We have not been specific about the boundary conditions or the sets $\Omega_h = \Pi\{Q_h\}$ so that we could maintain generality of the analysis: the only question of importance is whether N is the number of all grid points belonging to $\bar{Q}$ or only those belonging to the complement of $\Gamma_{Q,0} \equiv \Pi^{-1}\{\Gamma_0\}$. For the former case, we have $\Lambda_Q = \Lambda_Q^* \geq 0$, and, for the latter, we have the stronger assertion $\Lambda_Q \in \mathcal{L}^+(H)$.

For an arbitrary triangle $T \equiv P_1 P_0 P_2$, we choose P_0 to denote the vertex corresponding to the right angle of the corresponding triangle T' and define

$$\mu(T) \equiv [S_1 + S_2](2S_{1,2})^{-1}, \tag{4.12}$$

where $S_1 \equiv |P_0P_1|^2$, $S_2 \equiv |P_0P_2|^2$, $S_{1,2} \equiv |[\vec{P_0P_1}, \vec{P_0P_2}]|$.[45] Evidently, $2\mu(T) \geq 1$, with equality when T is half of a square. Thus, the quantity

$$\mu(T_h(\bar{\Omega})) \equiv \max \mu(T), \quad T \in T_h(\bar{\Omega}), \tag{4.13}$$

may be of use to define a measure of deformation that the ideal triangulation $T_h(\bar{Q})$ must undergo to become $T_h(\bar{\Omega})$ (see § 2.2).

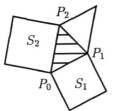

Figure 1. Relevant
element areas.

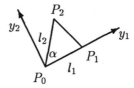

Figure 2. Relevant
element angle and side lengths.

Theorem 5. *For each $h \in \{h\}$, assume we are given two topologically equivalent triangulations $T_h(\bar{\Omega})$ and $T_h(\bar{Q})$ with grid operators $\Lambda_\Omega, \Lambda_Q$ defined by formulas (4.10), (4.11). Suppose there exists a $\mu < \infty$ such that*

$$\sup \mu(T_h(\bar{\Omega})) \leq \mu. \tag{4.14}$$

Then $\Lambda_\Omega, \Lambda_Q$ are spectrally equivalent operators and

$$\delta_{0,\Omega}\Lambda_Q \leq \Lambda_\Omega \leq \delta_{1,\Omega}\Lambda_Q, \quad \delta_{1,\Omega} \equiv \mu + (\mu^2 - 1)^{1/2} \equiv \delta_{0,\Omega}^{-1}. \tag{4.15}$$

Proof. Expansions (2.2) with (2.7) and (4.2) lead to

$$(\Lambda_\Omega \mathbf{u}, \mathbf{u}) = (a(x), |\nabla u|^2)_{0,\Omega}, \quad (\Lambda_Q \mathbf{u}, \mathbf{u}) = (a_Q(z), |\nabla v|^2)_{0,Q},$$

---

[45] S_r is the area of the square with side $P_0P_r, r = 1, 2$, and $S_{1,2}$ is the area of the parallelogram with sides P_0P_1, P_0P_2 (see Figures 1 and 2).

$$\text{where } |\nabla u|^2 \equiv \sum_{r=1}^{2} [\frac{\partial \hat{u}_\Omega}{\partial x_r}]^2, \text{ and } |\nabla v|^2 \equiv \sum_{r=1}^{2} [\frac{\partial \hat{u}_Q}{\partial z_r}]^2.$$

Integrals over Ω, Q may be regarded here as sums of integrals over all possible triangles T, T', which is a simple but popular and useful notion based on *additivity with respect to subregions* of the corresponding bilinear forms. Thus, to obtain the desired inequalities, it suffices to do so simply for $|\hat{u}_\Omega|_{1,T}^2$ and $|\hat{u}_Q|_{1,T'}^2$. A second useful notion to apply here is to select suitable parameters. Functions $\hat{u}_\Omega, \hat{u}_Q$ are completely defined by their equal values u_k at nodes (vertices of triangles) $P_k, P_k', k = 0, 1, 2$, but choosing them as parameters is not so satisfactory. A more reasonable choice is to use differences $u_1 - u_0 \equiv t_1, u_2 - u_0 \equiv t_2$, because this leads leads to the simplest form

$$|\hat{u}_Q|_{1,T'}^2 = 2^{-1}(t_1^2 + t_2^2).$$

Hence, $X \equiv |\hat{u}_\Omega|_{1,T}^2 = 2^{-1}S_{1,2}|\text{grad } \hat{u}_\Omega|^2$. It is well known that $|\text{grad } u|$ does not depend on the choice of Descartes coordinates. Therefore, we may work with the system of coordinates y_1, y_2 such that the origin of the system coincides with P_0 and the abscissa (y_1-axis) goes along side P_0P_1. Writing $|P_0P_1| = l_1, |P_0P_2| = l_2, \angle P_1P_0P_2 = \alpha$, and $\frac{\partial}{\partial y_r} \equiv D_r$ (see Figure 2), then $D_1 u = l_1^{-1}t_1, l_2^{-1}t_2 = \cos \alpha D_1 u + \sin \alpha D_2 u$ and

$$2X = S_{1,2}[t_1^2 l_1^{-2} + (\sin \alpha)^{-2}(t_2 l_2^{-1} - \cos \alpha t_1 l_1^{-1})^2].$$

Thus,

$$X = 2^{-1}[S_{1,2}^{-1}(S_2 t_1^2 + S_1 t_2^2) - 2t_1 t_2 \cot \alpha].$$

Therefore, in order to obtain bounds for $X[2^{-1}(t_1^2 + t_2^2)]^{-1}$, it suffices to find eigenvalues of the matrix

$$\mathbf{A} \equiv \begin{bmatrix} S_2 S_{1,2}^{-1} & -\cot \alpha \\ -\cot \alpha & S_1 S_{1,2}^{-1} \end{bmatrix}.$$

These eigenvalues are roots of the equation $\lambda^2 - \lambda(S_1 + S_2)S_{1,2}^{-1} + 1 = 0$, that is, $\lambda_1 = (f(\mu(T)))^{-1}, \lambda_2 = f(\mu(T))$, where $f(t) \equiv t + (t^2 - 1)^{-1/2}$, which is monotonically increasing for $t \geq 1$. Hence,

$$\delta_{0,\Omega}|\hat{u}_Q|_{1,T'}^2 \leq |\hat{u}_\Omega|_{1,T}^2 \leq \delta_{1,\Omega}|\hat{u}_Q|_{1,T'}^2, \quad \forall T.$$

Summing up of these inequalities leads to (4.15).□ [46]

---

[46]Condition (4.14) is equivalent to the requirement that there exists a $\beta_0 \in (0, \pi/2)$ such that $0 < \beta_0 \leq \beta \leq \pi - \beta_0$ for every angle β of every triangle in $T_h(\bar{\Omega})$.

Consider now some useful generalizations of Theorem 5, starting with

$$b_\Omega(u; v) \equiv \sum_{T \in T_h(\bar\Omega)} [a(T)(u, v)_{1,T} + c(T)(u, v)_{0,T}] + \sum_l \sigma(l)(u, v)_{0,l}, \quad (4.17)$$

where l denotes a triangle side, summation with respect to l is carried out for all such sides with $\sigma(l) > 0$, and $a(T), c(T), \sigma(l)$ denote nonnegative constants. Let also a bilinear form be defined by a similar expression with T', l' instead of T, l but with the same constants at corresponding pairs T, T' and l, l' under the mapping Π^{-1} (that is, under the transformation of variables $z = \Pi^{-1}(x)$).

Theorem 6. *For the family of topologically equivalent triangulations* $T_h(\bar\Omega), T_h(\bar Q)$ *and basis functions* $\hat\psi_{\Omega,i}, \hat\psi_{Q,i}$, *define the operators*

$$\Lambda'_\Omega \equiv [b_\Omega(\hat\psi_{\Omega,j}; \hat\psi_{\Omega,i})], \quad \Lambda'_Q \equiv [b_Q(\hat\psi_{Q,j}; \hat\psi_{Q,i})]. \quad (4.18)$$

Assume that there exist constants $k_1 \geq k_0 > 0, k'_1 \geq k'_0 > 0$ *such that, for all* T *with* $c(T) > 0$ *and all* l *with* $\sigma(l) > 0$, *the following inequalities hold:*

$$k_0 h^2 \leq 2|T| \leq k_1 h^2, \quad k'_0 h \leq |l|_{(1)} \leq k'_1 h. \quad (4.19)$$

Let $\delta_0 \equiv \min\{\delta_{0,\Omega}, k_0, k'_0\}, \delta_1 \equiv \max\{\delta_{1,\Omega}, k_1, k'_1\}$. *Then*

$$\delta_0 \Lambda'_Q \leq \Lambda_\Omega \leq \delta_1 \Lambda'_Q. \quad (4.20)$$

Proof. As in the proof of Theorem 5, we see that $(\Lambda'_\Omega u, u) = b_\Omega(\hat u_\Omega; \hat u_\Omega)$, $(\Lambda'_Q u, u) = b_Q(\hat u_Q; \hat u_Q)$, and we need to examine only additional terms of the type $c(T)|\hat u_\Omega|^2_{0,T}$ and $\sigma(l)|\hat u_\Omega|^2_{0,l}$ with $c(T) > 0, \sigma(l) > 0$. There is no need to introduce new parameters for them, and the mapping $x = \Pi(z)$ that is affine on T transforms them into corresponding terms for $b_Q(\hat u_Q; \hat u_Q)$ multiplied by $|T|$ or $|l|_{(1)}$. This together with (4.19) and (4.15) leads to (4.20).□

Consider now the rather unusual bilinear form

$$b'_\Omega(u; v) \equiv b_\Omega(u; v) + \sum_l d(l)(u, v)_{1,l}, \quad (4.21)$$

where $d(l) \geq 0$ denotes a constant on l and $(u, v)_{1,l} \equiv (u'(s), v(s))_{0,l}$, with $u'(s)$ denoting the first derivative of u with respect to the arclength parameter s. Strictly speaking, such bilinear forms have almost nothing to do with classical elliptic boundary value problems of second order, for the space $W_2^1(Q)$ may contain functions without the above mentioned derivatives. Nevertheless, a very important class of problems, e.g., from the theory of plates and shells with stiffened edges, may be transformed to systems of Stokes type with analogous bilinear forms (similar problems will be considered later in § 8.4). In any event, in addition to the conditions of Theorem 6, we must make the assumption that (4.19) also holds for all l for which $d(l) > 0$. Then, a term of the type $X \equiv |\hat{u}_\Omega|^2_{1,l}$, after the local change of the parameter $s = s_0 + s'h|l|^{-1}_{(1)}$, leads to $X = h|l|^{-1}_{(1)}|\hat{u}_Q|^2_{1,l'}$ with $l' = \Pi^{-1}l, h|l|^{-1}_{(1)} \in [(k'_1)^{-1}, (k'_0)^{-1}]$. Thus, the assertion of Theorem 6 holds with δ'_r replacing $\delta_r, r = 1, 2$, where $\delta'_0 \equiv \min\{\delta_0, \min(k'_1)^{-1}\}$, and $\delta'_1 \equiv \max\{\delta_1, \max(k'_0)^{-1}\}$. [47]

4.4. Spectral equivalence of grid operators defined on composite triangulations with local refinements. Consider now the very important and widely used variants of FEMs involving composite grids with local refinements, as introduced at the end of § 2 (see Figures 9–13 from § 2). For such variants the condition numbers of the related matrices rapidly increase with further refinement (see [54]), so the problem of constructing effective preconditioners becomes especially important (see [54, 85, 362]).

Recall that a composite triangulation consists of distinct triangles with no common inner points; however, a side of one triangle may be a part of a side of another, and vertices of all such triangles not belonging to the Dirichlet boundary can either be grid nodes or seminodes, at which values of grid functions must be specified via interpolation procedures like (2.25). Figure 11 illustrates such regions and triangulations.

Theorem 7. Let $T_{c,h}(\bar{\Omega})$ and $T_{c,h}\bar{Q}$ be composite topologically equivalent triangulations and, for all $P_i \in \Omega_h$, consider the basis functions $\hat{\psi}_{\Omega,i}, \hat{\psi}_{Q,i}$ defined as in Subsection 2.5 (see Figures 12 and 13) and the operators $\Lambda'_\Omega, \Lambda'_Q$ defined via formulas (4.18) from the Theorem 6, but with these basis functions. Assume that condition (4.19) is satisfied. Then inequalities (4.20) remain true.

Proof. It suffices to compare $X \equiv b_\Omega(\hat{u}_\Omega; \hat{u}_\Omega)$ and $X' \equiv b_Q(\hat{u}_Q; \hat{u}_Q)$ for the corresponding finite elements functions defined on the composite topologically equivalent triangulations under consideration. To do so, we represent X, X' as sums of terms we considered in the proofs of Theorems 5 and 6. The separate terms only involve corresponding pairs T, T' and l, l', and do not depend on other triangles and sides. Therefore, all reasoning from the proofs of Theorems 5 and 6 may be applied to establish (4.20). □

---

[47]Similar estimates can be obtained for problems associated with two-dimensional manifolds, including, e.g., the surface of a cube or tetrahedron, or the composition of several polygons having a common side but belonging to different planes. What really matters is that the given bilinear forms are sums of the corresponding bilinear forms defined for the triangles on the manifold at hand. Elliptic boundary value problems on such two-dimensional manifolds are often found in engineering applications.

§ 5. *N*-widths of compact sets and optimal numerical methods for classes of problems

For an elliptic boundary value problem on a bounded domain $\Omega \in \mathbf{R}^d$ that can be regarded as a correct operator equation (1.28) in a Hilbert space $G \subset W_2^1(\Omega) \equiv G^{(1)}$, our main goal is to find suitable approximations to its solution u. More precisely, given a prescribed tolerance $\varepsilon > 0$, we want to find a function v such that $\|v - u\| \leq \varepsilon$, in which case v is called an *ε-approximation* to u. Since ε may be rather small, this can inherently take a lot of computation, and we must select our strategy for obtaining such v with the case $\varepsilon \to 0$ in mind. Thanks to fundamental results obtained by many investigators concerning properties of solutions of elliptic boundary value problems (see [256, 312]), we can often be sure that u is in the subset $M \equiv M_K$ of $G^{(m+\gamma)} \equiv W_2^{m+\gamma}(\Omega)$ of functions that satisfy

$$\|u\|_{m+\gamma,\Omega} \leq K, \quad m = [m] \geq 1, 0 < \gamma \leq 1. \tag{5.1}$$

Sometimes—very often only after many computational experiments are done—it becomes clear that additional properties of the solution should be taken into account. In other words, we may not want to worry about being efficient for finding every possible solution in M, but only those that exhibit properties that are specific to the application at hand. Nevertheless, even if we are ready to return to the problem and reexamine our selection of the appropriate numerical method, it seems reasonable to start by assuming that we are dealing with the *class of problems*, denoted by Cl, such that all solutions are in M and each $u \in M$ is a solution of at least one of the problems from this class (e.g., we may regard problems (1.28) with fixed L but with f ranging over $L\{M\}$).

5.1. Approximations of compact sets and criteria for optimality of computational algorithms. The set M of functions satisfying (5.1) is a *compact set* in the space G, that is, each sequence of elements of M contains a convergent subsequence to an element of M. More importantly, this notion of a compact set in a Hilbert space or in a complete metric space (in the general case) yields a remarkable approximation property: a set M in a Hilbert space is compact if only if, for each positive number ε, there exists a finite set of elements $a_1(\varepsilon), \ldots, a_{N_\varepsilon'}(\varepsilon)$ in M such that, for every $u \in M$, there exists an element a from this finite set that satisfies $\|u - a\| \leq \varepsilon$. [48] Thus, $N_\varepsilon \equiv N_\varepsilon(M)$ defines the minimal number

---

[48] This statement is the classical *Hausdorff criterion of compactness* (see [292]); any finite set that satisfies this approximation property is called an *ε-net* for M; and every such set with the minimal number N_ε is called an *optimal ε-net of the compact set.*

of elements in a table approximating the compact set with ε-accuracy, and its asymptotic behavior as $\varepsilon \to 0$ is important to understand.

For certain metric spaces and compact sets, well known asymptotic estimates are given in terms of $\log_2 N_\varepsilon$, which is called the ε-*entropy of the compact set* (see [41, 477]) because it characterizes the minimum amount of information necessary to specify an element of an ε-net.

Publications of Kolmogorov in the 1950's (see [409, 477]), which were devoted to asymptotic analysis of the ε-entropy of some sets of functions, attached special significance to this notion from the point of view of optimal approximation of functions from a given compact set. They motivated the original attempts to construct asymptotically optimal numerical methods for solving of elliptic boundary value problems, first suggested by Bakhvalov in the late fifties for Dirichlet problems in a planar region with smooth boundary for the harmonic (Laplace) equation (see [38, 39, 40]). These results, together with the first asymptotically optimal algorithms for solving difference elliptic systems (considered briefly in § 3 and, of course, with remarkable results of Nikolskii, Sobolev, and Bakhvalov on the theory of *optimal quadrature formulas*; see [388, 460]), led to the formulation of the *Kolmogorov-Bakhvalov hypothesis*, which states that the optimal asymptotic characteristics (of accuracy and computational complexity) of numerical methods for a correct elliptic boundary value problem should be determined by the ε-entropy or similar information characteristics of the corresponding compact sets. More precisely, it states that the ε-entropy must coincide with $\log_2 N(\varepsilon)$ and with $\log_2 W(\varepsilon)$, where $N(\varepsilon)$ is the minimal number of degrees of freedom for obtaining ε-approximation of any $u \in M$ in the asymptotically optimal numerical method and $W(\varepsilon)$ is the computational work required by the asymptotically optimal algorithm to obtain ε-approximations to solutions of arising systems.

Instead of ε-entropy, a currently more popular measure of these characteristics, especially in connection with PGMs, is N-*width in the sense of Kolmogorov* (see [409, 477]), as suggested by Babushka and Sobolev in 1965 (see [32]). The N-width of M is denoted by $\pi_N \equiv \pi_N[M; G]$ and defined by

$$\pi_N \equiv \inf_{V_N} \sup_{u \in M} \|u - Pu\|,$$

where V_N denotes an arbitrary subspace of G with $\dim V_N \leq N$ and P denotes the orthoprojector of the Hilbert space G onto V_N.

In what follows, $\varepsilon > 0$ will denote the desired accuracy of approximation and $N_0(\varepsilon)$ will correspond to the minimal (optimal) value of N that achieves the estimate $\pi_N \leq \varepsilon$. Suppose that for a given numerical method associated with approximations from the subspace V_N, we wish to obtain an ε-approximation to the solution u of the given problem from the class

Cl defined above. To achieve this accuracy for all possible problems, that is, for all $u \in M$, we must conclude that $N \geq N_0(\varepsilon)$. Therefore, comparison of asymptotic values of N and $N_0(\varepsilon)$ is of fundamental importance for judgments about the possible accuracy of numerical methods and their computational characteristics (see [32, 170, 183, 187, 394]), especially in connection with the required *computational work* or *computational complexity* (see [479, 480]), which had been already used in § 3. [49]

Definition 1. A method of discretization of problems from the given class Cl, *dealing with the parameter* h *and the subspaces* $\hat{G}_h \subset G$, $\dim G_h \equiv N_h$, *is called asymptotically optimal with respect to accuracy if there exists* $\kappa > 0$ *such that, for each problem from the class* Cl *with the solution* u, *its discrete analog has a unique solution* $\hat{u}_h$ *that satisfies the estimate* $\|\hat{u}_h - u\| \leq \kappa \pi_N$. (In this case, the given estimate of accuracy is often referred to as an *asymptotically optimal estimate of accuracy*.)

It readily follows from this definition that the asymptotically optimal methods noted above may give ε-approximations for the solutions of the problems from the class Cl only for sufficiently large $N \geq \kappa_2 \varepsilon^{-d/\gamma}$ with $\kappa_2 > 0$. This value of N determines the number of the unknowns in system (2.8) and an obvious lower bound for $W(\varepsilon)$ for any possible algorithm leading to the ε-approximations v of their solutions; here and in what follows, we speak about approximations in the sense of the Euclidean space $H(J)$ (see (4.3)), with the norm consistent with the norm of the subspace $\hat{G}_h$ (see Lemma 4.1). Therefore, if we denote by $\hat{v}_h$ the corresponding element of $\hat{G}_h$ (see (2.2)), then we must have: $\|\hat{v}_h - u\|_G = O(\varepsilon)$.

Definition 2. A computational algorithm leading to ε-*approximations in the sense of the Hilbert space* G *of the solutions* u *of problems from the class* Cl *is called asymptotically optimal if it is characterized by an estimate of the type* $W_a(\varepsilon) \leq \kappa_3 N_0(\varepsilon) |\ln \varepsilon|$.

Definition 3. A computational algorithm leading to ε-*approximations in the sense of the Hilbert space* G *of the solutions* u *of problems from the*

---

[49] Recall that our measure of computational work is denoted by W_a and corresponds to the number of arithmetic operations required by the given algorithm a. These arithmetic operations are considered in our analysis to be performed by an imaginary computer with infinite precision when our attention is on the study of the computational complexity. But, of course, such analysis must be supplemented by a more realistic study that takes into account rounding errors (as we do in § 1.4). Moreover, we must reexamine our judgment about the algorithm if we begin to consider possible implementations on modern vector and parallel computers. Fortunately, the best algorithms obtained from the above theoretical point of view are often very instructive for designing effective algorithms for many types of concrete computers (see, e.g., [28, 85, 130, 131, 250, 257, 288, 362, 398, 510]). Therefore, we may proceed in our analysis of optimal algorithms with the sound hope that it is not only interesting by itself, but also may be of great practical value for solving hard problems arising today and in the foreseeable future.

class Cl is called nearly asymptotically optimal *or* logarithmically optimal *if it is characterized by the estimate* $W_a(\varepsilon) \le \kappa_3 N_0(\varepsilon)|\ln \varepsilon|^r$ *with* $r > 0$.

For the compact set $M \subset G \equiv G^{(1)}$ defined via (5.1), it will be shown that $\pi_N \asymp N^{-\nu/d}$ with $\nu \equiv m + \gamma - 1$, i.e., $\exists \, \kappa_1 \ge \kappa_0 > 0$ such that

$$\kappa_0 N^{-\gamma/d} \le \pi_N[M; G] \le \kappa_1 N^{-\gamma/d}; \tag{5.2}$$

estimates (5.2) are also valid in the case of $G = W_2^1(\Omega; \Gamma_0)$ and of vector-functions with components satisfying (5.1). When (5.2) applies, the nearly optimal algorithms are thus characterized by the estimates

$$W(\varepsilon) \le K \varepsilon^{-d/\nu} |\ln \varepsilon|^r. \tag{5.3}$$

The construction of algorithms with estimates (5.3) $(r = 0, r = 1)$ is precisely the problem of *minimization of computational work,* which is the main subject of this book and which has attracted much attention of mathematicians over the last half of this century.

We will consider a sufficiently wide variety of elliptic boundary value problems, linear and nonlinear, in Chapter 5; in Chapter 7 and 8, we will pay special attention to the case of elliptic problems with linear constraints like div $u = 0$, which is typical for the Stokes and Navier-Stokes problems; in Chapter 9, we will be dealing with spectral (eigenvalue) problems associated with elliptic operators. But for all such problems, the desired asymptotically optimal algorithms will be constructed on the basis of PGMs satisfying two conditions:

1) they must be asymptotically optimal methods with respect to the accuracy;

2) they must generate grid systems for which effective iterative algorithms can be constructed with estimates of type (3.8) and (5.3). [50]

5.2. Estimates of N-widths in spaces like $W_2^1(\Omega)$. In what follows, the compact set M in the space $G \equiv G^{(1)}$ is defined either by condition (5.1) with $\nu \equiv m + \gamma - 1$ or by the more general conditions

$$\|u\|_{1+\nu, \Omega_s} \le K_s, \quad s \in [1, p], \tag{5.4}$$

where $\bar{\Omega}_1, \ldots, \bar{\Omega}_p$ constitute a decomposition of $\bar{\Omega}$ and each of the subregions satisfies the same assumptions about the boundary that were imposed on the boundary of Ω.

---

[50] We stress again that the first condition has a lot in common with similar conditions in the theory of approximation of functions and in the theory of numerical integration, and is therefore easier to satisfy. Moreover, it can be accomplished in many ways. But the second condition, dealing with the necessity to solve large grid systems, is the main feature of this objective that determines its difficulty and that will be, without doubt, the subject of much further investigation.

Lemma 1. *Let a set* $M \subset G \equiv W_2^1(\Omega)$ *consist of functions satisfying conditions* (5.4). *Then* M *is compact in* G.

Proof. Consider an arbitrary sequence $\{u^n\} \subset M$ and corresponding sequences $\{u_s^n\}$ consisting of restrictions u_s^n of functions u^n to Ω_s, $s \in [1, p]$. Then, on the basis of embedding theorems for the spaces $W_2^{1+\nu}(\Omega_s)$, we may assert that there exists a subsequence (for simplicity, we identify it with the given sequence) such that: u_s^n converge to u_s in the sense of the Hilbert space $W_2^1(\Omega_s)$ and, at the same time, u^n converges weakly to u in the Hilbert space $W_2^1(\Omega) = G$. Since

$$\|u\|_{1,\Omega}^2 = \|u_1\|_{1,\Omega_1}^2 + \ldots + \|u_p\|_{1,\Omega_p}^2,$$

then we may conclude that $\{u^n\}$ is a fundamental sequence in the Hilbert space G and, therefore, converges to $u \in G$. $\square$

We will make use of a cube $Q \equiv [0, a]^d, a > 0$, and of the following orthonormal basis of the space $L_2(Q)$:

$$e_k(x) \equiv (2/a)^{d/2} \sin[\pi k_1 x_1 / a] \ldots \sin[\pi k_d x_d / a]$$

with $k \equiv [k_1, \ldots, k_d]$ being a vector with integer components (this system of functions also constitutes an orthogonal basis of the Hilbert space $\overset{\circ}{W}_2^1(Q)$).

For each $u \in L_2(Q)$, we have $u = \sum_k a_k e_k$, $a_k = (u, e_k)_{0,Q}$. We regard these functions as defined for all $x \in \mathbf{R}^d$ and will use the *Weyl space* $W^{(r)}$ with $r \geq 0$ and the norm

$$|u|_{r,a}^2 \equiv \sum_k a_k^2 [k_1^{2r} + \ldots + k_d^{2r}]$$

(see [104, 387]). Note that there exists a constant $\delta_{1,r}$ such that

$$\|u\|_{r,Q}^2 \leq \delta_{1,r}' \|u\|_{r,a}^2, \quad \forall u \in W^{(r)}.$$

Denote by $W_2^r(Q; \rho)$ the subspace of $W_2^r(Q)$ consisting of functions vanishing in a given ρ-vicinity of ∂Q, that is, $u(x) = 0$ for almost all x with $\text{dist}\{x; \partial Q\} \leq \rho$ ($\rho > 0$). Then, for $\rho > 0$ sufficiently small, there exist constants $\delta_{1,r}' \geq \delta_{0,r}' > 0$ such that

$$\delta_{0,r} |u|_{r,a}^2 \leq \|u\|_{r,Q}^2 \leq \delta_{1,r} |u|_{r,a}^2, \quad \forall u \in W_2^r(Q; \rho)$$

(see [104]); note that the need to appeal to $W_2^r(Q; \rho)$ instead of $W_2^r(Q)$ is essential only for $r = [r] + 1/2$ and is connected with a loss of additivity of the space $W_2^{1/2}(Q)$ with respect to subregions (see § 2). We shall also use the subspace S_N with $\dim S_N \equiv N \equiv n^d$ defined as a linear span of functions e_k with $k_s \in [1, n], s \in [1, d]$.

Lemma 2. *For the subspace S_N, there exists $K(\mu)$ such that*
$\|u\|_{1+\nu,Q} \leq K(\nu)n^\nu\|u\|_{1,Q}$, $\forall u \in S_N$.

Proof. We have $\|u\|_{1+\nu,Q} \leq \delta_{1,1+\nu}|u|_{1+\nu,a}$. Since

$$|u|_{1+\nu,a}^2 = \sum_k a_k^2[k_1^{2(1+\nu)} + \ldots + k_d^{2(1+\nu)}] \leq K' \sum_k a^2[k_1^2 + \ldots + k_d^2]$$

with

$$K' \equiv \max_{1 \leq k_s \leq n, s=1,\ldots,d} \frac{\sum_k a_k^2[k_1^{2(1+\nu)} + \ldots + k_d^{2(1+\nu)}]}{\sum_k a_k^2[k_1^2 + \ldots + k_d^2]}$$

and $\sum_k a_k^2[k_1^2 + \ldots + k_d^2] \leq K'K''\|u\|_{1,Q}^2$, we may write

$$|u|_{1+\nu,a}^2 \leq K'K''\|u\|_{1,Q}^2.$$

To estimate K', it suffices to use the elementary inequality
$(a_1 + a_2)/(b_1 + b_2) \leq max\{1/b_1; a_2/b_2\}$, $\forall a_i > 0$, $\forall b_i > 0, i = 1,2$, which
gives $K' = n^{2\nu}$. $\square$

Theorem 1. *Let the compact set M be defined via (5.1) in the Hilbert
space $G \equiv G^{(1)}$. Then there exists $\kappa_0 > 0$ such that*

$$\pi_N[M; G] \geq \kappa_0 N^{-\nu/d}. \tag{5.5}$$

Proof. It suffices to confine ourselves to the case $N = n^d$. Let V_N be an
arbitrary subspace of G with dim $V_N \leq N$. Define $V_{N;Q}$ as the subspace of
restrictions of functions from V_N to a cube Q belonging to Ω such that dist
$\{\partial Q; \partial\Omega\} \geq 2\rho > 0$. Then, in any subspace S_P of G with $P \equiv (n+1)^d$, there
must exist a function v with $\|v\|_{1,Q} = 1$ orthogonal in the Hilbert space
$W_2^1(Q)$ to $V_{N,Q}$ such that $\|v\|_{1+\nu,Q} \leq K(\nu)(n + 1)^\nu$. Let $p(u) \in G^{(1+\nu)}$
denote the extension of $u \in W_2^{1+\nu}(Q)$ with $p(ku) = kp(u)$,

$$\|p(u)\|_{1+\nu,\Omega} \leq \kappa_4\|u\|_{1+\nu,Q}.$$

Then, for this $v \in S_P$, we may take $k = K\kappa_4^{-1}\|v\|_{1+\nu,\Omega}^{-1} \geq \kappa_5 n^{-\nu}$ with
$\kappa_5 > 0$. This gives $p(kv) \equiv u \in M$. Then,

$$\|u - z\|_{1,\Omega} \geq \|Kv - z\|_{1,Q} \geq \|kv\|_{1,Q} \geq \kappa_5 n^{-\nu}, \forall z \in V_N,$$

and (5.5) holds. $\square$

Theorem 2. *Let $W_2^1(\Omega; \partial\Omega) \subset G \subset G^{(1)}$ and the set M be defined by
(5.4). Then (5.5) holds.*

Proof. Consider a subregion Ω_s and a ball $S \equiv S(x_0; r_0) \subset \Omega$ with
a sufficiently small radius $r_0 > 0$ such that dist$[S; \partial\Omega_s] \geq 2\rho > 0$. As in

the above proof, for each $u \in W_2^{1+\nu}(S)$, we define an extension $p(u) \in W_2^{1+\nu}(\Omega)$. Let $\varphi \in C_0^\infty(S(x_0, r_0 + \rho))$ (see (1.21)) be a standard function of the type used in a *partition of unity for a region* (see [67, 387]), that is, $\varphi(x) = 1$ if $x \in S$ and $\varphi(x) \geq 0$. Then, multiplying $p(u)$ by φ gives a function $v \equiv p_0(u) \in \overset{o}{W_2^1}(\Omega)$ with $\|v\|_{1+\nu,\Omega} \leq \kappa_6 \|u\|_{1+\nu,S}$, where κ_6 is a constant not depending on $u \in W_2^{1+\nu}(S)$. If functions $u \in W_2^{1+\nu}(S)$ are elements of the compact set $M_s(\kappa_7)$ defined by the inequality $\|u\|_{1+\nu,\Omega_s} \leq \kappa_7$ and we take a sufficiently small κ_7 leading to the inequalities $\kappa_6\kappa_7 \leq K_r, r = 1, \ldots, p$ (see(5.4)), then the extension $p_0(u)$ used above gives an element of the compact set M under consideration. We denote the set of these extensions as $p_0\{M_s(\kappa_7)\}$. Then, for an arbitrary subspace V_N of G, we may consider a corresponding subspace $V_{N,s}$ of restrictions of its elements to Ω_s and write

$$\|p_0(u) - Pu\| \geq \|u - [Pu]_{\Omega_s}\|_{1,\Omega_s} \geq \|u - P_s u\|_{1,\Omega_s},$$

where $[v]_{\Omega_s}$ denotes the restriction of $v \in G$ to Ω_s and P_s stands for the orthoprojector of the Hilbert space $W_2^1(\Omega_s)$ onto the subspace $V_{N,s}$. This implies that $\pi_N[M; G] \geq \pi_N[M_s; W_2^1(\Omega_s)]$. The right-hand side of this inequality may now be estimated by virtue of (5.5). $\square$

Theorem 3. Let a Hilbert space G be a subspace of vector-functions $\bar{u} \equiv [u_1, \ldots, u_k]$ such that

$$[\overset{o}{W_2^1}(\Omega)]^k \subset G \subset [W_2^1(\Omega)]^k. \tag{5.6}$$

Let a compact set M in G be defined by conditions

$$\|u_l\|_{1+\nu,\Omega_s} \leq K_s, \quad s \in [1, p], l \in [1, k]. \tag{5.7}$$

Then estimate (5.5) holds.

Proof. In the compact set M, consider the subset of vectors $\bar{u} \equiv [u_1, 0, \ldots, 0]$, which constitutes a compact set to which Theorem 2 applies. Using (5.6), (5.7), we may then assert that $\pi_N[M; G]$ is not smaller than the N-width from Theorem 2. $\square$ [51]

5.3. Optimality of projective-grid methods. For simplicity, we confine ourselves to the very instructive example of PGM using the most basic triangulation $T_h(\bar{Q}) \equiv T_h(\bar{\Omega}))$ that consists of triangles T such that each T is a half square with side $h(T)$ (in the case of a quadratic grid,

---

[51] Theorems 1–3 can be generalized in much the same way to the case of compact sets in the Hilbert space $G^{(m)}$ with $m = 0, 1, 2, \ldots$, and also to the corresponding spaces $G \subset G^{(m_1)} \times \ldots \times G^{(m_k)}$ of vector-functions. These estimates were used in the author's publications [183, 187] and have some points in common with results of [394, 409].

$h(T) = h$). Estimates (2.4) might be easily specified to find an estimate
of $\|\hat{v} - u\|$ for a suitable $\hat{v} \in \hat{G}_h$. In the case of $u \in G^{(1+\nu)}$ with $\nu > 0$,
the embedding theorem into $C(Q)$ applies because $2(1 + \nu) > 2$. Thus, we
construct

$$\text{int } u \equiv \hat{v} \equiv \sum_{P_i \in Q_h} u(P_i)\hat{\psi}_{P_i}(x) \tag{5.8}$$

(see (2.2)). Note that, on each triangle T, int u is just the linear interpolant
of u. Then, for $z(x) \equiv \hat{v}(x) - u(x)$, we may write

$$\|z\|_{1,Q}^2 = \sum_T \|z\|_{1,T}^2 \tag{5.9}$$

and make use of the fact that, on each $T \equiv \triangle P_1 P_0 P_2$, we have
$z(P_k) = 0, k = 0, 1, 2$ and $z \in W_2^{1+\nu}(T)$.

In what follows, we assume that $\nu = 1$ and consider the standard
(reference) triangle R with vertices $A_1(1; 0), A_0(0; 0), A_2(0; 1)$. Let V be
a Hilbert space consisting of functions $u \in W_2^2(R)$ satisfying conditions
$u(A_i) = 0, i = 0, 1, 2$ with the inner product

$$(u, v)_V \equiv [(D_1^2 u, D_1^2 v)_{0,R} + 2(D_1 D_2 u, D_1 D_2 v)_{0,R} + (D_2^2 u, D_2^2 v)_{0,R}]. \tag{5.10}$$

It is easy to verify that the norm of u in the Hilbert space V, denoted
by $|u|_{2,R}$, is equivalent to the standard norm $\|u\|_{2,R}$ in the Hilbert space
$W_2^2(R)$ (see Equivalent Norm Theorem in § 2). Thus, there exist constants
K_0, K_1 such that

$$|u|_{0,R} \leq K_0 \|u\|_V, \quad |u|_{1,R} \leq K_1 \|u\|_V, \quad \forall u \in V. \tag{5.11}$$

Furthermore, on the basis of the well known properties of symmetric compact operators, which will be discussed partly in Section 9, it is possible
to show that the optimal (minimal) values of K_0, K_1 in (5.11), denoted by
K_0^*, K_1^*, exist and can be calculated as solutions of the variational problems

$$(K_0^*)^{-1/2} = \min_{v \in V}\{|v|_{0,R}^2 |v|_V^{-2}\}, \quad (K_1^*)^{-1/2} = \min_{v \in V}\{|v|_{1,R}^2 |v|_V^{-2}\}. \tag{5.12}$$

These are just eigenvalue problems for an elliptic fourth-order operator (see
[117, 164, 394] and § 2.3). For example, the constant K_1^* from (5.12) was
calculated in [12, 453] (by different numerical methods) to be 0.4888; hence,
we may use $K_1 = 1/2$. Some interesting estimates of similar constants from
below have been obtained in [473], which relate to the theory of packing

and covering for the compact sets under consideration (see [427]). In any event, having obtained estimates (5.11) for the reference triangle, we may apply a one-to-one affine mapping of this triangle T (actually given by $x'_r = h(T)x_r, r = 1, 2$), which leads to estimates

$$|u|_{0,T} \leq K_0 h_T^2 |u|_{2,T}, \quad |u|_{1,T} \leq K_1 h_T |u|_{2,T}. \tag{5.13}$$

Therefore,

$$\|z\|_{1,Q}^2 \leq \sum_T \{[h_T^2 K_1^2 + h_T^4 K_0^2] |u|_{2,T}^2\}, \tag{5.14}$$

which implies that $\|z\|_G \leq K'h|u|_{2,Q}$ when all $h(T) = h$, as is the case, e.g., for system (2.15), (2.16). Therefore, to get ε-accuracy, we may take $h \asymp \varepsilon$ and the number of the unknowns in the system is $N \asymp h^{-2}$. This leads to $\varepsilon \asymp N^{-1/2}$, and comparison of this estimate with (5.2) shows that the PGM under consideration is an asymptotically optimal method with respect to accuracy. It should be noted that estimates of accuracy

$$\|u - \hat{u}\|_G \leq Kh^\nu \|u\|_{1+\nu,Q} \tag{5.15}$$

with $0 < \nu \leq 1$ can be obtained for similar but more general PGMs associated with so-called quasiuniform triangulations and will be considered for d-dimensional elliptic boundary value problems in Chapter 2. [52] If

$$G \equiv G^{(1)} \equiv W_2^1(Q), \tag{5.16}$$

then we have

$$\|u\|_{0,l_h}^2 \leq \gamma'^2 h|u|_{1,Q}^2 + h^{-1}|u|_{0,Q}^2, \quad \forall u \in G. \tag{5.17}$$

We now return to questions of the optimality of PGMs from the point of view of the required computational work for finding proper ε-approximations

---

[52]The procedure used for obtaining (5.15) is typical for many estimates connected with small regions: they are derived, first of all, for some standard regions, then a proper h-transformation of variables is used to obtain estimates for the small regions at hand. Of course, such estimates with small parameters have been long and widely used in many branches of mathematics. In fact, the same procedure serves with the frequently used *Bramble-Hilbert lemma* (see [117]) to obtain estimates like (5.15) of the accuracy of FEMs. But we shall prefer to use the standard procedure described above because, first of all, it emphasizes the importance of appropriate subspaces and the classical theorem about equivalent norms in Sobolev spaces, and, secondly, it might be applied for proving estimates with no connection to an approximation property. For example, in case of $G \equiv W_2^1(Q; \Gamma_0)$ with $|\Gamma_0|_{(1)} > 0$, we have the standard inequality $\|u\|_{0,l} \leq \gamma_1 |u|_{1,G}$ with $l \subset Q$ denoting a smooth arc. The contraction of this standard region by a factor of h^{-1} ($h > 0$ is a small parameter) leads to Q_h, l_h and the estimate $\|u\|_{0,l_h} \leq \gamma_1 h^{1/2} |u|_{1,Q_h}$.

to solutions of systems (2.14). If these systems with N unknowns take the model form (2.15), (2.16) with properties (2.21), (2.22), then it is possible to apply fast direct methods mentioned in § 3 (see also § 3.1) with asymptotically optimal estimates (3.2), (3.3) of the computational work W (optimal estimates correspond to $r = 0$ in (3.2)). Thus, for these model problems with the simplest geometry of the original region, asymptotically optimal algorithms can be constructed on the basis of the simplest PGMs (FEMs) and rather sophisticated direct methods that exploit the properties like (2.22). In case of more general grid systems, iterative methods of type (3.13) and their generalizations for the symmetrized grid system (3.17) can be of great help in obtaining algorithms with estimates

$$W = O(N \ln^r N |\ln \varepsilon|), r \geq 0. \qquad (5.18)$$

Such methods will be considered in detail in Chapters 1 (§ 3–§ 5), 2 (§ 4), 3, and 5, and will be one of our main concerns. We emphasize that, at the present time, we have a fairly rich variety of nearly optimal iterative methods, yielding estimates of (5.18) with $r > 0$, and optimal ones, yielding estimates (5.18) with $r = 0$. Nevertheless, it is not an exaggeration to say that construction of such methods is, and will continue for many years to be, one of the most important central problems of computational mathematics. Note that we already outlined a way in § 3, in the case of $\varepsilon \asymp N^{-s}, s > 0$ (for the examples above considered, we may take $\varepsilon \asymp N^{-1/2}$), to eliminate the multiplier $\ln \varepsilon$ in (5.18) through coarse grid continuation; this will be investigated further in § 1.4 and 5.1. [53]

Often, problems occur with piecewise smooth functions defining the functional l from (2.5) (see (1.33)). Sometimes, it is reasonable to take the information about l as an element of G^*. If, e.g., $l(\psi) = -(D_1 g_1 + D_2 g_2, \psi)_{0,\Omega}$ with $\psi \in \overset{o}{W}{}_2^1 (\Omega)$, then integration by parts yields $l(\psi) = (g_1, D_1\psi)_{0,\Omega} + (g_2, D_2\psi)_{0,\Omega}$. This form now makes sense even for $g_r \in L_2(\Omega)$ and, if $D_r g_r \in L_2(\Omega)$, it gives a sufficient reserve of smoothness for approximating values of function g_r using its averaged over T values. (We have no such reserve for approximating $D_r g_r \in L_2(\Omega)$.)

---

[53] There is also the simpler but nevertheless important problem of estimating the errors appearing in the result of approximations of integrals over triangles $T \in T_h(\bar{\Omega})$ necessary for obtaining matrices L_h and vectors f from (2.8)–(2.10). Recall that, if we integrate polynomials, we may make use of a number of known (see [117]) formulas for such integrals. In the general case, we may approximate the given coefficients (functions) by polynomials on T and estimate necessary perturbations of L_h and f in a way we will consider later in § 1.2. It is also possible to take into account these approximation errors (which we will do later) in the same manner as with rounding-off errors: to this end, we have to carry out an analysis of iterative methods with perturbations, which will be done in § 1.4.

Chapter 1

General theory of numerical methods for operator equations

§ 1. General questions

1.1. General notions. In this section, we start by considering the most general and important notions of the theory of numerical methods, which are equally applicable to all reasonable approximations of a given operator equation, including all types of grid methods.[1]

Consider Banach spaces U and F (see Subsection 0.1.2), a linear operator L mapping U into F, and a given element $f \in F$. Then the problem of finding $u \in U$ such that $Lu = f$ is called *correct (well posed)* if the bounded inverse operator L^{-1} exists: $\|L^{-1}\|_{U \mapsto F} \leq K < \infty$ (recall that the inverse of a linear operator is always linear). In other words, correctness of the given linear problem implies that it has a unique solution u for each given right-hand side f that is uniformly continuous in f (replacing f by $\tilde{f}$ yields a new solution $\tilde{u}$ with $\|u - \tilde{u}\| \leq K\|f - \tilde{f}\|$). To tackle nonlinear problems in a general setting, we have to localize u and f by requiring that they belong to closed subsets S and S_F of the spaces U and F, respectively. In the general case, we write the operator equation in the form

$$L(u) = f, \quad u \in S, \tag{1.1}$$

---

[1] There are many investigations devoted to similar topics (see, e.g., [11, 18, 64, 292, 419, 229, 475, 485]), although the notions and statements used here have been taken mainly from the author's publications [164, 165, 166].

with $f \in S_F$.[2]

We will call (1.1) *a correct problem on S_F* if, for each $f \in S_F$, it has a unique solution u that is uniformly continuous in f.

The essence of a numerical method for solving (1.1) is to replace it by a sequence of algebraic problems

$$L_h(u_h) = f_h, \quad u_h \in S_h, \quad h \in \{h\}. \tag{1.2}$$

Here we use h as a parameter taken from a normed space, with norm $|h|$, and assume that $\{h\}$ does not contain 0 but does contain a sequence converging to 0 (for the simplest grid methods on cubical grids, the parameter h is simply the mesh size and $|h| = h$). Vectors $u_h \equiv \mathbf{u} \in \mathbf{R}^{N_h}$ and $f_h \equiv \mathbf{f} \in \mathbf{R}^{N_h}$ in (1.2) are considered as elements of normed linear spaces U_h and F_h with $\dim U_h = \dim F_h = N_h \to \infty$ as $|h| \to 0$, $S_h \in U_h$, and the operator L_h maps S_h into F_h. We will later return to the analysis of problems (1.2) for given U_h and F_h; but, for the time being, we assume that each has a unique solution in U_h and try to specify what we mean by a numerical method for the given problem. We will speak about a *numerical method for the operator equation* if we have not only the set of algebraic problems (1.2), but also a way to compare solutions u and u_h.

The most general way to compare elements of U and U_h is to compare their images in a special normed space Q_h under certain linear operators $q_h \in \mathcal{L}(U; Q_h)$ and $p_h \in \mathcal{L}(U_h; Q_h)$, where q_h satisfies

$$\lim_{|h| \to 0} \|q_h u\|_{Q_h} = \|u\|_{U_0}, \quad \forall u \in U. \tag{1.3}$$

Here, the space U is embedded in a weaker normed space U_0 (in the simplest case, $U_0 = U$). Then the set of $z_h \equiv p_h u_h - q_h u$ can be considered as the *error* or *accuracy of the method*, and we call the method *convergent* if

$$\lim_{|h| \to 0} \|z_h\|_{Q_h} = 0. \tag{1.4}$$

(Note that (1.3) implies uniqueness of the limit.) The numerical method is convergent with *order* $k > 0$ if $\|z_h\|_{Q_h} \leq K|h|^k$.

Let all S_{F_h} be such that, for each $f_h \in S_{F_h}$, we have a unique solution $u_h \equiv L_h^{-1}(f_h)$. We say that the numerical method is *correct on sets S_{F_h}* if, for L_h^{-1} mapping S_{F_h} into S_h, there exists a constant K such that $\|L_h^{-1}(f_{1,h}) - L_h^{-1}(f_{2,h})\|_{U_h} \leq K\|f_{1,h} - f_{2,h}\|_{F_h}$, $K \neq K(h)$, for each $f_{k,h} \in S_{F_h}, k = 1, 2$. Of course, correctness of the method on the above

---

[2] Usually, S and S_F will be closed balls $S_U(v; r_1)$ and $S_F(f; r_2)$ in U and F, respectively, and we will assume that $0 \in S, L(0) = 0$. The case $r = \infty$ is allowed, which obviates localization and is often appropriate for mildly nonlinear problems.

mentioned sets implies that inverse operators L_h^{-1} (defined on these sets) satisfy *the Lipschitz condition* with the same constant K and, for the linear case, $\|L_h^{-1}\|_{F_h \mapsto U_h} \leq K$ or, in other words, for every solution u_h of the linear problem $L_h u_h = f_h$ with a given f_h, the *a priori estimate*

$$\|u_h\|_{U_h} \leq K\|f_h\|_{F_h}, \quad K \neq K(h), \tag{1.5}$$

holds. [3]

1.2. A general convergence theorem. The analysis of convergence deals with the three spaces U, U_h, and Q_h, and the above mentioned operators q_h and p_h mapping the first two respective spaces into Q_h. [4]

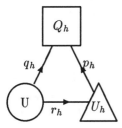

Figure 1. Operators connecting basis spaces.

We define *the approximation error* of the method with respect to a solution u of (1.1) as the set of $\zeta_h \equiv L_h(r_h u) - f_h$; it is said that *the approximation (the approximation with order $k > 0$) is valid* if

$$\lim_{|h| \to 0} \|\zeta_h\|_{U_h} = 0, \quad (\|\zeta_h\|_{U_h} \leq K|h|^k).$$

Everywhere below, $\xi_h(u) \equiv p_h r_h u - q_h u$ and $\|p_h\| \equiv \|p_h\|_{U_h \mapsto Q_h}$.

Theorem 1 (see [164, 165]). Let problems (1.2) for each $f_h \in S_{F_h}$ have unique solutions $u_h \in U_h$. Suppose that $L_h(r_h u) \in S_{F_h}$ and inequality (1.5)

---

[3] The choice of the normed spaces U_h and F_h deserves special discussion, but here we just emphasize the significance and usefulness of Theorem 0.4.1 and Lemmas 0.4.1 and 0.4.2 in case the spaces U and F are the same Hilbert space G: we will describe some fruitful approaches in Section 2 to the analysis of correctness in the case of Euclidean spaces U_h and F_h. It should be pointed out that sometimes, instead of correctness of the method, one prefers to speak about *stability of the method*, and numerical methods with weak stability (with K depending on h) can be useful.

[4] Our goal here is not a discussion of possible reasonable choices of these spaces and operators but a general scheme of investigation that would enable us to deal with rather difficult problems. Hence, it makes sense to introduce one more family of operators $r_h \in \mathcal{L}(U; U_h)$ analogous to the operators widely used in difference methods: for a continuous function $u \in U$, $r_h u$ is usually a vector of values of u at all considered grid points; if we have to deal with more complicated functions like those from the space $L_2(\Omega)$, then we can take a vector of averaged values, computed with respect to small cubes centered around the grid points (e.g., Steklov's middle values, which will be discussed in Section 2.3; other types of averaging are also possible).

is valid. Then the accuracy of the method satisfies

$$\|z_h\|_{Q_h} \leq K\|p_h\|\|\zeta_h\|_{F_h} + \|\xi_h(u)\|_{Q_h};$$

in the case that Q_h is a Euclidean space and $(\xi_h(u), p_h(r_h u - u_h))_{Q_h} = 0$, we obtain the sharper estimate

$$\|z_h\|^2_{Q_h} \leq (K\|p_h\|\|\zeta_h\|)^2 + \|\xi_h(u)\|^2_{Q_h}. \tag{1.6}$$

Proof. Since $z_h = p_h u - q_h u = (p_h u_h - p_h r_h u) + (p_h r_h u - q_h u)$, we have $\|z_h\|_{Q_h} \leq \|p_h(u_h - r_h u)\|_{Q_h} + \|\zeta_h\|_{Q_h}$ or $\|z_h\|_{Q_h} \leq \|p_h\|\|u_h - r_h u\|_{U_h} + \|\zeta_h\|_{Q_h}$. Since $L_h(r_h u) \in S_{F_h}$, we can apply (1.5), which gives $\|u_h - r_h u\|_{U_h} \leq K\|\zeta\|_{F_h}$. Substitution into the previous inequality leads to the first desired inequality.[5] (1.6) now follows from (0.1.7). □

Two important cases are simplified by use of just two of three spaces for each h. These two cases are characterized by the choices $Q_h = U$ and $Q_h = U_h$. The first case with a Hilbert space $G = U = F$ and the operator q_h defined by $q_h u = u$ is typical of projective methods (see [18, 292, 317]) that approximate the given operator equation by a sequence of finite-dimensional operator equations in subspaces $\hat{G}_h \subset G$:

$$\hat{L}(\hat{u}) = \hat{f}, \ \hat{u} \in \hat{G}_h. \tag{1.7}$$

They can be reduced to algebraic problems (1.2) after selection of bases for these subspaces and the use of corresponding expansions of $\hat{u}$ (see 0.2.2); the vectors u_h are just the coefficients of these expansions. Sometimes, as we have seen in Section 0.2, it is possible to use operators r_h so that $\zeta_h = 0$, and all we need is to estimate the energy norms of $u - r_h u$. But, as we shall see below, for projective methods it is very effective to take $Q_h = \hat{G}_h$ and to use Theorem 1. Sometimes it is useful to regard Q_h as a subspace of U_0, $G \in U_0$. The second important case with $Q_h = U_h, p_h u_h \equiv u_h, q_h = r_h$ is typical for many difference methods [64, 229, 378, 435]; to get the desired estimate of accuracy, we have to carry out an analysis of correctness (stability) of the difference systems and of the approximation error. [6]

---

[5] A similar result was obtained later by R. Temam [475].

[6] Both analyses depend crucially on the choices of norms defined on the spaces U_h and F_h, as is now widely understood (see, for example, [166, 229, 435]). But, for difference methods and especially for more general variants of grid methods, very often a better analysis can be carried out by treating Q_h as a subspace either in U or in U_0 ([166, 424, 439, 475]). Currently, special attention is paid to *domain decomposition methods* using different types of approximation in different parts of the given region. For such methods, as was pointed out in [166], the spaces Q_h can be different from U and U_h and, in the general case, can coincide with subspaces of a space U_0 containing U.

1.3. Estimates of accuracy of projective methods. We have already introduced estimates of the desired type in Section 0.2. Cea's Theorem 0.2.3 is remarkably simple and can be readily generalized to nonlinear problems like (1.1) and (1.7). To do so, we again use their projective formulations based on the *quasibilinear form* $b(u; v)$ defined for all $u \in S, v \in G$ by $b(u; v) \equiv (L(u), v)$. If the functional $l \in G^*$ is such that $l(v) = (f, v)$, we can formulate the problems mentioned above (the original one and its approximation) in the form: find respective u and $\hat{u}$ such that

$$b(u; v) = l(v), \quad \forall v \in G, u \in S, \tag{1.8}$$

and

$$b(\hat{u}; \hat{v}) = l(\hat{v}), \quad \forall v \in \hat{G}, \quad \hat{u} \in \hat{S}. \tag{1.9}$$

The approximate finite-dimensional problem (1.9) is equivalent to the algebraic one (1.2) with $u_h \equiv \mathbf{u} \in \mathbf{R}^{N_h}$. We emphasize again the importance of the proper choice of the Euclidean spaces (see Lemma 0.4.1)

$$U_h \equiv H(\Lambda), \quad F_h \equiv H(\Lambda^{-1}), \quad \Lambda \equiv J, \tag{1.10}$$

if we are concerned with investigation of the nonlinear systems (1.2) and effective iterative methods for finding their solutions. Localizations for u and $\hat{u}$ of the form

$$S \equiv S_G(r), \hat{S} \equiv S \cap \hat{G} \tag{1.11}$$

can easily be obtained, with $r = \delta_0^{-1}\|l\|$, if

$$b(v; v) \geq \delta_0\|v\|^2, \quad \forall v \in G, \quad \delta_0 > 0 \tag{1.12}$$

(usually referred to as definiteness of the form). Indeed, it is easy to see that $\delta_0\|u\|^2 \leq b(u; u) = l(u) \leq \|l\|\|u\|$.

    Theorem 2. Let problems (1.8) *and* (1.9) *with balls* S *and* $\hat{S}$ *from* (1.11) *have solutions* u *and* $\hat{u}$, *and suppose for all* v *in* S *and for all* $w \in G$ *that*

$$b(v; v - u) - b(u; v - u) \geq \sigma_0\|v - u\|^2, \quad \sigma_0 > 0, \tag{1.13}$$

and

$$|b(v; w) - b(u; w)| \leq \sigma_1\|v - u\|\|w\|. \tag{1.14}$$

Suppose also that $K = \sigma_1/\sigma_0$. *Then* $\|\hat{u} - u\| \leq K\text{dist}\{u; \hat{G}\}$.

    Proof. It follows from (1.8) and (1.9) that $b(\hat{u}; \hat{v}) - b(u; \hat{v}) = 0$, $\forall \hat{v} \in \hat{G}$. Therefore,

$$b(\hat{u}; \hat{u} - u) - b(u; \hat{u} - u) = b(\hat{u}; \hat{v} - u) - b(u; \hat{v} - u). \tag{1.15}$$

We may regard $\hat{v}$ as an element of $\hat{S}$ and estimate the left-hand side from below using condition (1.13). The right-hand side of (1.15) is estimated from above with the help of (1.14). Hence, from (1.15), it follows that

$$\sigma_0\|\hat{u} - u\|^2 \le \sigma_1\|\hat{u} - u\|\|\hat{v} - u\|. \tag{1.16}$$

Let Pu be the best approximation in $\hat{G}_h$ to the solution u. Since $\|u\|^2 = \|Pu\|^2 + \|u - Pu\|^2$, then $\|Pu\| \le \|u\| \le r$. Therefore, replacement of $\hat{v}$ in (1.16) by Pu yields the desired estimate. $\square$ [7]

Now we turn to an analysis of more difficult problems, assuming only their correctness.

Theorem 3. *For a linear operator equation $Lu = f$ with invertible L, suppose that projective approximations $L_h u = f_h$ are such that L_h^{-1} exist and*

$$\|L_h^{-1}\|_{H(J^{-1}) \mapsto H(J)} \le K_0, \quad \|L_h\|_{H(J) \mapsto H(J^{-1})} \le K_1$$

(see (1.10)). Let $P_h \equiv P$ be the orthoprojector of G onto $\hat{G}_h$. Then

$$\|\hat{u} - Pu\|_G \le K_0 K_1 \|Pu - u\|_G \tag{1.17}$$

and

$$\|\hat{u} - u\|_G \le (1 + K_0^2 K_1^2)^{1/2}\|Pu - u\|_G. \tag{1.18}$$

Proof. Using the standard formulations (1.2), (1.9), we may write

$$b(\hat{w}; \hat{v}) = l(\hat{v}) + b(\hat{w}; \hat{v}) - b(u; \hat{v}), \quad \forall \hat{v} \in \hat{G}.$$

Taking into account (1.9), we see that

$$b(\hat{u}; \hat{v}) - b(\hat{w}; \hat{v}) = b(u; \hat{v}) - b(\hat{w}; \hat{v}). \tag{1.19}$$

---

[7] The particular case of this theorem with conditions (1.13) and (1.14) holding for all u, v, and w is the best known (see [96, 484, 489]) and corresponds to the case of a nonlinear *strongly monotone operator* L satisfying the Lipschitz condition with constant σ_1. Then (1.13) and (1.14) can be rewritten in the form $(L(v), v - u) - (L(u), v - u) \ge \sigma_0\|v - u\|^2$ and $\|L(v) - L(u)\| \le \sigma_1\|v - u\|$ (we have taken $w = L(v) - L(u)$). Under such conditions, it is even possible to prove that problem (1.1) with $S = U$ always has a unique solution. In fact, equations $L(u) - f = 0$ and $u = u - \tau(L(u) - f)$ with $\tau > 0$ are equivalent and, for a small τ, the operator A defined by $A(u) \equiv u - \tau(L(u) - f)$ is a *contraction*, i.e., there exists q such that $0 < q < 1$ and $\|A(v) - A(u)\| \le q\|v - u\|$, $\forall u, \forall v$. Then the *contraction mapping principle*, one of the milestones of functional analysis, states that the equation $u = A(u)$, linear or nonlinear, has a unique solution; moreover, it is a limit of iterations $u^{n+1} = A(u^n)$, $n = 0, 1, \dots$ as n tends to infinity [292, 341]. For many important problems of mathematical physics conditions (1.13) and (1.14) are satisfied for all u and v from a ball $S_G(r)$; such localizations were effectively used, e.g., in classical investigations by Ladyzenskaya [323] devoted to Navier-Stokes problems. As it was shown in [403, 426], when strong a priori estimates are valid, it is often possible to pass to a new problem (with the same solution u) such that these conditions are satisfied for all u and v.

We first rewrite (1.19) in terms of the Euclidean space H with vectors $\mathbf{u}, \mathbf{v}, \mathbf{w}$ instead of $\hat{u}, \hat{v}, \hat{w}$, respectively. Then the left-hand side of (1.19) is $(L_h(\mathbf{u}), \mathbf{v}) - (L_h(\mathbf{w}), \mathbf{v})$ with inner products in the sense of H. Hence, (1.19) is equivalent to $L_h(\mathbf{u}) - L_h(\mathbf{w}) = \zeta$, where

$$(\zeta, \mathbf{v}) = b(u; \hat{v}) - b(\hat{w}; \hat{v}). \tag{1.20}$$

Therefore, $\mathbf{u} - \mathbf{w} = L_h^{-1}\zeta$ and

$$\|\mathbf{u} - \mathbf{w}\|_J \leq K_0 \|\zeta\|_{J^{-1}}. \tag{1.21}$$

In (1.20) take $\mathbf{v} = J^{-1}\zeta$. Then

$$\|\zeta\|_{J^{-1}}^2 \leq K_1 \|\hat{w} - u\|_G \|\hat{v}\|_G = K_1 \|\hat{w} - u\|_G \|\mathbf{v}\|_J.$$

Hence, $\|\zeta\|_{J^{-1}}^2 \leq K_1 \|Pu - u\| \|\zeta\|_{J^{-1}}$, which when substituted into (1.21) gives the estimate equivalent to (1.17). To prove (1.18), it suffices to note that $\hat{u} - u = (\hat{u} - Pu) + (Pu - u)$ and to use orthogonality of $Pu - u$ and $\hat{G}$. $\square$ [8]

Theorem 4. Let u be a solution of nonlinear problem (1.1). Let problems (1.9) with $\hat{S} = S_G(u; r) \cap \hat{G}_h$ be written in the form (1.2), and assume that the operators L_h are such that method (1.2) is correct on $S_{F_h} \equiv L_h\{S_h\}$. Suppose that K_0 is the constant in the Lipschitz condition for L_h^{-1} on S_{F_h}, and that

$$\|b(v; z) - b(v'; z)\| \leq K_1 \|v - v'\| \|z\|, \quad \forall v \in \hat{S}, \forall v' \in \hat{S},$$

and $Pu \equiv w \in \hat{S}$. Then estimates (1.17), (1.18) remain true.

Proof. Using the quasi-bilinear form b, we may start from the same equalities (1.19), (1.20). It is easy to see that the estimate for ζ remains true. Further, elements $L_h(\mathbf{u})$ and $L_h(\mathbf{w})$ belong to S_{F_h}. Therefore, (1.21) holds and leads to the desired estimates. $\square$

1.4. Estimates using supplementary Hilbert spaces. [9]

---

[8] Estimate (1.17) corresponds to the case of Theorem 1 with the space $Q_h = \hat{U}_h$ and $r_h(u) = \mathbf{w}, p_h(u_h) = \hat{u}, q_h(u) = Pu$. Similarly, for (1.18), the difference is only in the choices $Q_h = U, q_h(u) = u$. With regard to the conditions used above, it is useful to recall Theorems 1–3 from Section 0.4, which clearly indicate that these conditions are fairly general for linear problems. As for nonlinear problems, we are now in a position to formulate and prove a similar theorem using the notation $L\{S\} \equiv \{Lv : v \in S\}$ for the image of the set S.

[9] So far we have considered convergence theorems for projective methods, that, like Theorem 0.2.3, do not contain the principal result of Rayleigh-Ritz methods theory given by Theorem 0.2.2 for $L \in \mathcal{L}^+(G)$ (the set of linear, bounded, symmetric, and positive definite operators mapping the given Hilbert space G into itself). This is a very serious drawback, especially when constants like $\delta_1(\delta_0)^{-1}$ or $K_0 K_1$ are comparatively large, as is

Returning to our wish to improve the given theorems, we assume that

$$L = A + L_0, \tag{1.22}$$

where A is a principal part of L in the sense that

$$k_0 I \leq A \leq k_1 I, \quad k_0 > 0, \tag{1.23}$$

and L_0 is a minor part of L in the sense that there exists a supplementary Hilbert space G_0 with the norm $\|u\|_0$ such that G is embedded into G_0 and

$$|(L_0 v, w)| \leq k_2 \|v\| \|w\|_0, \quad \forall v \in G, \forall w \in G. \tag{1.24}$$

For example, in the diffusion-convection case with G a subspace of $W_2^1(\Omega)$, it is reasonable to take A to be the diffusion part of L, L_0 the convection part, and $G_0 = L_2(\Omega)$.

Theorem 5. *Let conditions of Theorem 3 with respect to* L, L_h *and conditions* (1.22)–(1.24) *be satisfied. Let* P_A *be the orthoprojector of the Hilbert space* $G(A)$ *with the inner product* $(u, v)_A \equiv (Au, v)$ *onto* $\hat{G}_h$ *and suppose*

$$\|P_A u - u\|_0 \leq k_3 |h|^s \|P_A u - u\|_A, \quad s > 0. \tag{1.25}$$

Let $k_4 \equiv k_0^{1/2} k_2 k_3$. *Then*

$$\|\hat{u} - u\|_A \leq \left(1 + (K_0 k_4 |h|^s)^2\right) \|P_A u - u\|_A. \tag{1.26}$$

Proof. We rewrite (1.19) in the form $b(\hat{u} - \hat{w}; \hat{v}) = b_A(u - \hat{w}; \hat{v}) + (L_0(u - \hat{w}), \hat{v})$, where b_A is the bilinear form corresponding to the operator A and $\hat{w} \equiv P_A u$. Using properties of the chosen orthoprojector, we may replace (1.19) by $b(\hat{u} - \hat{w}; \hat{v}) = (L_0(u - \hat{w}), \hat{v})$ and pass to (1.20) with $(\zeta, \mathbf{v}) = (L_0(\hat{w} - u), \hat{v})$. Therefore, if we use the projective approximation $A_h \in \mathcal{L}^+(H)$ defined for the operator A, then $\|\mathbf{u} - \mathbf{w}\|_{A_h} = \|L_h^{-1}\zeta\|_{A_h}$. Since $\|\mathbf{v}\|_{A_h} \leq (k_1)^{1/2} \|\mathbf{v}\|_J$, we have

$$\|\mathbf{u} - \mathbf{w}\|_{A_h} \leq k_1^{1/2} \|L_h \zeta\|_J \leq K_0 k_1^{1/2} \|\zeta\|_{J^{-1}}.$$

Analogous to what we have done before, we have

$$\|\zeta\|_{J^{-1}} \leq K_2 \|P_A u - u\|_0 \leq k_2 k_3 |h|^s \|P_A u - u\|_0.$$

---

the case, e.g., for diffusion-convection problems with dominating convection. Note that, from this point of view, even the classical Theorem 0.1.2 does not promise too much in the case of large K, and one is justified in attempts to construct new grid methods with better estimates. Some relatively new methods of this type for elasticity problems will be considered in Chapter 7, and many investigators have focused much attention on the diffusion-convection problems mentioned above (see, e.g., [253, 451, 490]), although much remains to be clarified in this direction.

Thus,

$$\|P_A u - \hat{u}\|_A = \|\mathbf{u}\|_{A_h} \leq k_4 |h|^s \|P_A u - u\|_A,$$

from which (1.26) follows immediately. □ [10]

Theorem 6. *Let the conditions of Theorem 3 with respect to L, L_h and conditions (1.22)–(1.24) be satisfied. Then*

$$\|\hat{u} - u\|_A \leq (1 + K_0)\|P_a u - u\|_A + K_0 k_1^{1/2} k_2 \|P_a u - u\|_0. \qquad (1.27)$$

Proof. For $\hat{w} \equiv P_a u$, we have (1.20) with

$$(\zeta, \mathbf{v}) = b_A(u - \hat{w}; \hat{v}) + (L_0(u - \hat{w}); \hat{v})$$

and $\|\zeta\|_{J^{-1}} \leq K_0\|P_a u - u\|_A + k_2\|P_a u - u\|_0$. Thus,

$$\|\hat{u} - P_a u\|_A \leq K_0\|P_a u - u\|_A + K_0 k_1^{1/2} k_2 \|P_a u - u\|_0.$$

This and the triangle inequality lead to (1.27). □

Theorem 7. *Let the conditions of Theorem 3 with respect to L, L_h and conditions (1.22)–(1.24) be satisfied. Then*

$$\|\hat{u} - u\|_A \leq (1 + K_0)\|P_0 u - u\|_A + K_0 k_1^{1/2} k_2 \|P_0 u - u\|_0. \qquad (1.28)$$

Proof. It suffices to repeat the proof of Theorem 6 with $\hat{w} \equiv P_0 u$. □[11]

§ 2. Conditions for correctness of discrete problems

2.1. A priori estimates of solutions. Given a fixed h, consider the corresponding discrete (grid) system (1.2), which we rewrite as the operator equation

$$L(u) = f, \ u \in S, \qquad (2.1)$$

in the Euclidean space $H \equiv \mathbf{R}^N$. We want to be sure that this equation has a solution. If it can be transformed to the form $u = A(u)$, where A

---

[10] Constant k_3 in (1.25) may depend on u; in many elliptic problems with $G_0 = L_2(\Omega), G \subset W_2^1(\Omega)$, it is possible to take $s = 1$ ([117, 394]).

Sometimes it seems reasonable to take projection operators other than P and P_A. We consider two such choices. The first deals with the operator P_a (approximating operator) defined by assigning the values of $Pu \in \hat{G}_h$ at the grid points to be the values of u there under the assumption that u is a continuous function (it will be especially useful for PGMs like those considered in the Introduction (see also § 2.3)). The second defines P_0 as the orthoprojector of the Hilbert space G_0 onto $\hat{G}_h$.

[11] We obtained Theorems 5–7 as improvements of Theorem 0.2.3. In a similar manner, it is possible to generalize Theorems 2 and 4 to nonlinear problems, but this is left to the reader as an exercise.

is a contraction operator mapping the closed set S into itself, then this is a very satisfying achievement since it yields even more than we desired— uniqueness of the solution and a constructive way to obtain a good approximation to it. Very often this will be the case in our investigations, but sometimes it will pose a real mathematical problem, the solution of which will be significantly facilitated if we can find a way to prove the existence of u and localize it. In this respect, it is difficult to overestimate the importance and beauty of *topological principles* like that given by the following well-known theorem (see [235, 323, 341]).

Brower Fixed Point Theorem. Let S be a nonempty, bounded, closed, and convex set in a Euclidean space. Assume that a continuous (on S) operator A maps S into itself. Then there exists at least one $u \in S$ such that $A(u) = u$, called the fixed point of A. [12]

We are concerned with applications of this theorem to the Euclidean spaces like $H(B)$ with $B \in \mathcal{L}^+(H)$, having in mind, first of all, operators B that are spectrally equivalent to operators J of projective methods considered in Section 0.4 and that defined norms $\|u\|_J$ equal to $\|\hat{u}\|$ in the given subspace $\hat{G}$ of the original Hilbert space G. [13]

Theorem 1. Let $B \in \mathcal{L}^+(H)$, $S \equiv S_B(r)$, and $r > 0$, and suppose we are given an operator L that is continuous on S and such that, for all $v \in S$,

$$(L(v), v) \geq \delta \|v\|_B^2, \quad \delta > 0. \tag{2.2}$$

Suppose $f \in S_{B^{-1}}(\delta r)$. Then operator equation (2.1) has a solution, and each solution $u \in S$ satisfies $\|u\|_B \leq \delta^{-1} \|f\|_{B^{-1}}$.

Proof. We introduce the function $Q(v) \equiv B^{-1}(L(v) - f)$, which obviously is continuous on S. Given v with $\|v\|_B = r$, we obtain (see Lemma 0.4.2) $(Q(v), v)_B = (L(v), v) - (B^{-1/2}f, B^{1/2}v)$. Since $\|B^{-1/2}f\| = \|f\|_{B^{-1}}$, it is easy to verify that $(Q(v), v)_B \geq \delta r^2 - r\|f\|_{B^{-1}} \geq 0$. If we assume that S contains no solutions of (2.1), this would imply that $\|Q(v)\|_B > 0$ for all $v \in S$ and that the function $P(v) \equiv -rQ(v)\|Q(v)\|_B^{-1}$ is continuous on S. Moreover, for $\|v\| \leq r$, we see that $\|P(v)\|_B = r$, which means that P maps S into itself and that the Brower theorem for P applies. Therefore, assuming no solutions of (2.1) exist leads to the existence of $v \in S$ such that $P(v) = v$. But, in this case,

$$(P(v), v)_B = \|v\|_B^2 = r^2 > 0,$$

---

[12] Brower's theorem was generalized by Schauder for the case of a Hilbert space under the assumption that L is not only continuous, but also compact, that is, it maps each bounded sequence into a precompact set (see subsection 0.1.3). Brilliant results for differential equations are connected with the *Leray-Schauder topological principle* (see [19, 323]).

[13] Exposition in this section primarily follows the author's publications [164, 165, 166].

which leads to a contradiction with the estimate

$$(P(v), v)_B = -r(Q(v), v)_B \|Q(v)\|_B^{-1} \le 0.$$

Therefore, (1.2) must have a solution. Then, for such u, we may apply standard reasoning: since $(L(u), u) = (f, u)$, we estimate the left-hand side from below by $\delta \|u\|_B^2$ and the right-hand side from above by $\|B^{-1/2}f\| \|B^{1/2}u\|$, and we come to the conclusion that

$$\delta \|u\|_B^2 \le \|f\|_{B^{-1}} \|u\|_B, \tag{2.3}$$

which gives the necessary a priori estimate. $\square$

A special case of this theorem deals with

$$L = R + P, \tag{2.4}$$

with linear R satisfying $R = R^* \ge \delta B$ and P continuous on the whole space H satisfying the *antisymmetry* property ([145, 323])

$$(Pv, v) = 0, \quad \forall v \in H \tag{2.5}$$

(for linear P, $P^* = -P$). Next we analyze more general cases under the assumption that the principal linear part of L, denoted by R, is an invertible operator. To this end, we give two lemmas (see [162]), which use the notation that $\lambda(A)$ denotes any eigenvalue of the operator A.

Lemma 1. Assume that $R = A - c_0 I$ with $A = A^*$, $\lambda(A) \ge \lambda_0 > 0$, $c_0 \ge 0$, and $|\lambda(R)| \ge d > 0$. Let $\delta_0 \equiv g(c_0 + d)$, $\delta_1 = 1$ for $\lambda_0 > c_0 - d$, and $\delta_1 \equiv \max\{1; g(\lambda_0)\}$ for $\lambda_0 \le c_0 - d$ with $g(t) \equiv (1 - c_0/t)^2$. Then

$$\delta_0 \|Av\|^2 \le \|Rv\|^2 \le \delta_1 \|Av\|^2, \ \forall v \in H. \tag{2.6}$$

Proof. Since we deal here with a symmetric operator A, we may use expansions of v with respect to an orthonormal basis of H consisting of its eigenvectors $z_1, z_2, \ldots, z_N$, that is, $Az_i = \lambda_i z_i$, where $(z_i, z_j) = 0$ if $i \ne j$, and $\|z_i\| = 1$ for all i, j. Let $v = \sum_{i=1}^N v_i z$. Then

$$\|Av\|^2 = \sum_{i=1}^N (\lambda_i v_i)^2, \ \|Rv\|^2 = \sum_{i=1}^N ((\lambda_i - c_0)v_i)^2.$$

Hence, $\min g(t) \|Av\|^2 \le \|Rv\|^2 \le \max g(t) \|Av\|^2$, where minimum and maximum are taken with respect to all t such that $t \ge \lambda_0$, $|t - c_0| \ge d$. It is easy to verify that $\min g(t) = g(c_0 + d) = \delta_0$. As for the estimate $\max g(t) \le \delta_1$, it suffices to note that, if $c_0 > d$, then $g(c_0 + d) \le g(c_0 - d)$, which completes the proof. $\square$

Lemma 2. *Let* $A = A^* > 0$, $M = M^* > 0$, $\lambda_0 > 0$, $R \equiv A - c_0 M$, *where* $c_0 \geq 0$ *and* $|\lambda(M^{-1}R)| \geq d > 0$. *Then*

$$\delta_0\|v\|_A^2 \leq \|Rv\|_{A^{-1}}^2 \leq \delta_1\|v\|_A^2, \tag{2.7}$$

with the same constants δ_0, δ_1 *as in Lemma 1.*

Proof. It is easy to see that the operators $M^{-1}A$, $M^{-1}R$ are symmetric as elements of $\mathcal{L}(H(M))$ (see Lemma 0.3.1). We take an orthonormal basis for the space $H(M)$ consisting of eigenvectors $z_1, z_2, \ldots, z_N$ of $M^{-1}A$, that is, $Az_i = \lambda_i M z_i$, $\forall i$, and proceed as in the proof of Lemma 1 using expansions of v with respect to this basis. This yields inequalities

$$\delta_0\|M^{-1}Av\|_M^2 \leq \|M^{-1}Rv\|_M^2 \leq \delta_1\|M^{-1}Av\|_M^2,$$

which are equivalent to operator inequalities in the space $\mathcal{L}(H)$:

$$\delta_0 AM^{-1}A \leq RM^{-1}R \leq \delta_1 AM^{-1}A.$$

Note that $AM^{-1}A$ is positive. Therefore, Lemma 0.3.1 again implies that $\omega \equiv \mathrm{sp} \ \{(AM^{-1}A)^{-1}(RM^{-1}R)\} = \mathrm{sp} \ \{A^{-1}M(A^{-1}M)(M^{-1}R)M^{-1}R\}$ belongs to the interval $[\delta_0, \delta_1]$. It is easy to verify that multipliers $A^{-1}M$ and $M^{-1}R$ commute, so ω is the spectrum of $A^{-1}RA^{-1}R$, which is a positive operator considered as an element of $\mathcal{L}(H(A))$. Consequently, in the space $\mathcal{L}(H)$, we may write $\delta_0 A \leq RA^{-1}R \leq \delta_1 A$, which leads to (2.7). $\square$ [14]

Now we return to the nonlinear problem (2.1) with $L = R + \mathcal{P} = A - c_0 I$ and consider two very similar theorems dealing with spaces $H(A)$ and $H(B)$, respectively.

Theorem 2. *Let the conditions of Lemma 2 be satisfied,* $B \in \mathcal{L}^+(H)$, $A \geq \delta B$, $\delta > 0$, $S = S_B(r)$. *Suppose* $\mathcal{P}$ *is a continuous operator on* S *such that, for all* $v \in S$,

$$\|\mathcal{P}(v)\|_{A^{-1}} \leq \mu\|v\|_B, \ \mu < \rho \equiv \delta^{1/2}d(d + c_0)^{-1}. \tag{2.8}$$

Suppose that $\|f\|_{A^{-1}} \leq r(\rho - \mu) \equiv r'$. *Then (2.1) with* L *from (2.4) has a solution and, for each solution* u, *we have* $\|u\|_B \leq (\rho - \mu)^{-1}\|f\|_{A^{-1}}$.

Proof. We begin with the case $\|f\|_{A^{-1}} < r'$ and introduce a family of vector fields continuous on S given by

$$Q_\nu(v) \equiv v + \nu R^{-1}(\mathcal{P}(v) - f). \tag{2.9}$$

Q_ν depends continuously on the parameter $\nu \in [0, 1]$. Then

$$\|Q_\nu(v)\|_B \geq \|v\|_B - \nu\|R^{-1}\mathcal{P}(v)\|_B - \nu\|R^{-1}f\|_B \equiv X(v).$$

---

[14] In a similar manner, we can consider $c_0 < 0$, but this is not of much interest because R is always positive for this case.

Since $\|v\|_A \geq \delta^{1/2}\|v\|_B$ and $\|R^{-1}v\|_A \leq (1 + c_0d^{-1})\|v\|_{A^{-1}}$ (see (2.7)), then, for $\|v\|_B = r$, we have

$$X(v) \geq [1 - \delta^{-1/2}(1 + c_0d^{-1})\mu]r - \delta^{-1/2}(1 + c_0d^{-1})\|f\|_{A^{-1}}$$

and, therefore, $\|Q_\nu(v)\|_B \geq \rho^{-1}[(\rho - \mu)r - \|f\|_{A^{-1}} > 0$. This means that none of the fields can have zero values $Q_\nu(v)$ on the boundary of S, and we may say (see [316]) that *homotopy on the boundary of the ball* of the fields $Q_0(v)$ and $Q_1(v)$ holds and that *rotations of the fields* are nonzero since $Q_0(v) = v$. Therefore, theorems of topology ([316]) connected with rotations of vector fields assure that each field $Q_\nu(v)$ must have a zero point in S. Thus, there exists u such that $Q_1(u) = 0$, $L(u) = f$, and

$$\|u\|_B \leq \|R^{-1}\mathcal{P}(u)\|_B + \|R^{-1}f\|_B \leq \rho^{-1}(\mu\|u\|_B + \|f\|_{A^{-1}}),$$

which leads to the desired a priori estimate. The remaining case $\|f\|_{A^{-1}} = r'$ can be reduced to the one considered if we notice that (2.1), with $f_n \equiv (1 - 1/n)f$, $n = 0, 1, \ldots$ instead of f, yield solutions u_n and they form a set that contains a sequence convergent to a solution of (2.1) as $n \to \infty$. $\square$

We could have formulated this theorem for the simple case $A = B$ and $\delta = 1$, but in its present general form, the following theorem is a direct consequence.

Theorem 3. *Let $B \in \mathcal{L}^+(H)$, $S = S_B(r)$, L be given by (2.4) with a linear operator R such that $\|R^{-1}\|_{H(B^{-1}) \mapsto H(B)} \leq K_0$, and $\mathcal{P}$ be a continuous operator on S such that $\|\mathcal{P}(v)\|_{B^{-1}} \leq \mu\|v\|_B$, for all $v \in S$. Suppose that $K_0\mu < 1$ and $\|f\|_{B^{-1}} \leq rK_0^{-1}(1 - \mu K_0) \equiv r_0$. Then (2.1) has a solution and any solution u satisfies $\|u\|_B \leq K_0(\mu r + \|f\|_{B^{-1}})$.* [15]

Sometimes it will be of use to work with the space $H(A^2)$, especially when it serves as an approximation of the Sobolev space $W_2^2(\Omega)$.

Theorem 4. *Let conditions of Lemma 1 be satisfied, the operator L in (2.4) be such that the operator $\mathcal{P}$ is continuous on $S \equiv S_{A^2}(r)$, and*

$$\mathcal{P}(v), R(v) \geq -\delta\|Av\|^2, \quad \delta_0 > \delta \geq 0, \ \forall v \in S. \tag{2.10}$$

Suppose that $\|f\| \leq r\delta_1^{-1/2}(\delta_0 - \delta) \equiv r'$. Then (2.1) has a solution and each solution u satisfies $\|Au\| \leq \delta_1^{1/2}(\delta_0 - \delta)^{-1}\|f\|$.

Proof. As in the proof of Theorem 2, we restrict our analysis to the case $\|f\| < r'$. Then, for $\|Av\| = r$, we have

$$(RQ_\nu(v), Rv) = \|Rv\|^2 + \nu(\mathcal{P}(v), Rv) - \nu(f, Rv).$$

---

[15] The above notion of the rotation of a vector field is equivalent to the notion of the *degree of a mapping with respect to* 0 used in the Leray-Schauder principle (see [19, 316]).

The right-hand side is estimated from below by $(\delta_0 - \delta)\|Av\|^2 - \|f\|\|Rv\| > 0$. Therefore, the rotations of the fields $Q_0(v)$ and $Q_1(v)$ are equal, and the same reasoning leads to the existence of $u \in S$ such that $Q_1(u) = 0$. For each solution, we have $(\delta_0 - \delta)\|Au\|^2 \leq (L(u), Ru) \leq \|f\|\|Ru\|$, which gives the desired a priori estimate. $\square$

2.2. Theorems of correctness. Theorems 1–4 have enabled us to maintain the existence of solutions of (1.2) for each f from the corresponding ball S_F. We now study the sensitivity of u with respect to perturbations of f, which we do by way of the two problems

$$L(u_k) = f_k, \quad u_k \in S, \quad f_k \in S_F, \quad k = 1, 2. \tag{2.11}$$

Theorem 5. Let conditions of Theorem 1 be satisfied and suppose, for all v, v' in S, that

$$(L(v) - L(v'), v - v') \geq \delta_0\|v - v'\|_B^2, \quad \delta_0 > 0. \tag{2.12}$$

Let $S_F \equiv S_{B^{-1}}(\delta r), K \equiv \delta_0^{-1}$. Then solutions of (2.11) always exist and satisfy

$$\|u_2 - u_1\|_B \leq K\|f_2 - f_1\|_{B^{-1}}. \tag{2.13}$$

Proof. Theorem 1 guarantees existence of solutions u_k of (2.11). Since $(L(u_2) - L(u_1), u_2 - u_1) = (f_2 - f_1, u_2 - u_1)$, we have $\delta_0\|u_2 - u_1\|_B^2 \leq \|f_2 - f_1\|_{B^{-1}}\|u_2 - u_1\|_B$. The latter inequality leads to (2.13). $\square$ [16]

The following three theorems deal with nonlinear problems without a strong monotonicity assumption; they can be regarded as special modifications of the well-known *perturbation principle* of Banach-Kantorovich of an invertible linear operator (see [292]).

Theorem 6. Let conditions of Theorem 2 be satisfied and suppose, for all v, v' in S, that

$$\|\mathcal{P}(v) - \mathcal{P}(v')\|_{A^{-1}} \leq \mu'\|v - v'\|_B, \quad \nu' < \rho.$$

Then problems (2.11), with $S_F \equiv S_{A^{-1}}(r(\rho - \mu))$, have unique solutions and estimate (2.13) with $K \equiv \delta_0^{-1}$ holds.

Proof. Since $u_2 - u_1 = R^{-1}(f_2 - f_1) - R^{-1}(\mathcal{P}(u_2) - \mathcal{P}(u_1))$, then $\|u_2 - u_1\|_B \leq \delta^{-1/2}(\|R^{-1}(f_2 - f_1)\|_A + \|R^{-1}[\mathcal{P}(u_2 - \mathcal{P}(u_1)])\|_A) \equiv X$. Thus, $X \leq \delta^{-1/2}(c_0 + d)d^{-1}\|f_2 - f_1\|_{A^{-1}} + \delta^{-1/2}(c_0 + d)d^{-1}\mu'\|u_2 - u_1\|_B$, and $(1 - \mu'\rho^{-1})\|u_2 - u_1\|_B \leq \rho^{-1}\|f_2 - f_1\|_{A^{-1}}$, which leads to (2.13). $\square$

Next is a similar theorem whose proof is even simpler, and therefore omitted.

---

[16] Operators that satisfy (2.12) are said to be *strongly monotone operators;* see [101, 238, 337, 484, 489, 493]; the proof we used typifies the theory used in their analysis.

Theorem 7. Let conditions of Theorem 3 be satisfied and suppose, for all v, v' in S, that $\|\mathcal{P}(v) - \mathcal{P}(v')\|_{B^{-1}} \leq \mu\|v - v'\|_B$, where $K_0\mu < 1$. Then problems (2.11), with $S_F \equiv S_{B^{-1}}(r_0)$, have solutions and for them estimate (2.13) with $K \equiv K_0(1 - K_0\mu')^{-1}$ holds.

The next theorem is a direct consequence of Theorem 4.

Theorem 8. Let conditions of Theorem 4 be satisfied and suppose, for all v, v' in S, that $(\mathcal{P}(v) - \mathcal{P}(v'), R(v - v')) \geq -\delta'\|A(v - v')\|^2$, with $\delta_0 - \delta' > 0$. Then problems (2.11), with S_F being a ball in H defined by the condition $\|f\| \leq r'$, have solutions and the estimate

$$\|A(u_2 - u_1)\| \leq K\|f_2 - f_1\| \tag{2.14}$$

with $K \equiv \delta_1^{1/2}(\delta_0 - \delta')^{-1}$ holds.

These theorems have dealt with problem (1.2) for a fixed h and H. If we want to transform them into correctness theorems, all we need to do is require that the appropriate constants do not depend on h. These theorems also enable us to investigate readily the influence of perturbations of operators on obtained solutions.

Theorem 9. Let $S_F = S_F(r)$ and suppose problems (2.11) have unique solutions that satisfy the correctness estimate $\|u_2 - u_1\|_U \leq K\|f_2 - f_1\|_F$. Suppose the perturbed problem $\tilde{L}(\tilde{u}) = \tilde{f}$, with the same S, S_F, has a unique solution $\tilde{u}$ satisfying $\|f - \tilde{f}\|_F + \|L(\tilde{u}) - \tilde{L}(\tilde{u})\|_F \leq r$. Then

$$\|u - \tilde{u}\|_U \leq K\|f - \tilde{f}\|_F + K\|\tilde{L}(\tilde{u}) - L(\tilde{u})\|_F.$$

Proof. It suffices to note that $\tilde{u}$ satisfies the equation $L(\tilde{u}) = f + \zeta$, where $\zeta \equiv L(\tilde{u}) - \tilde{L}(\tilde{u}) + \tilde{f} - f$ and $\|\zeta\|_F \leq r$. □ [17]

Linear problems with $S = U, S_F = F$ lead to a significant simplification of the given theorems. Our next theorem is a reformulation of Theorem 8 and serves as an illustration of such a simplification.

Theorem 10. Let $L = R + \mathcal{P}$ with linear R and $\mathcal{P}$. Suppose the conditions of Lemma 1 are satisfied for R and that

$$(\mathcal{P}v, Rv) \geq -\delta\|Av\|^2, \delta_0 > \delta, \ \forall v \in H.$$

Then L is invertible and $\|L^{-1}\|_{H \mapsto H(A^2)} \leq \delta_1^{1/2}[\delta_0 - \delta]^{-1}$.

---

[17] We have analyzed cases of problems with localizations like $\|u\|_U \leq r$, to which the more general localization $u \in S(v_0; r)$ can be reduced by the transformation $v = v' + v_0$. From a theoretical point of view, we can even take an unknown solution u as the center v_0 of the ball. Sometimes, approximations to u, obtained, e.g., as solutions of *linearized problems* or approximations to them, may lead to reasonable v_0 (we will discuss linearizations in the next subsection and in Section 3).

2.3. Derivatives of nonlinear operators. We will need basic facts from theory of *differentiable operators* mapping $S_U(v; r)$ into a Banach space F. Element v will constitute an operator $L'_v \in \mathcal{L}(U; F)$.

An operator L'_v is called the *Gateaux derivative of L in v* if, for each z (see, e.g., [19, 235, 244, 341, 484]), $\lim_{t \to 0}[L(v + tz) - L(v)]/t = L'_v z$. Similarly, L'_v is the *Fréchet derivative* if, for each z, $\lim_{\|z\|_U \to 0} \|L(v + z) - L(v) - L'_v z\|_F \|z\|_U^{-1} = 0$. It is known that if the Gateaux derivative exists in the vicinity of v and depends on v continuously, then the Fréchet derivative exists at v and they coincide (see, e.g., [244, 341]). Moreover, for a linear operator L, we have $L'_v = L$. Finally, if L'_v and $A'_{L(v)}$ exist then the composition $A(L)$ has a derivative and $(AL)'_v = A'_{L(v)} L'_v$. Of special importance to us will be the case $U = F = \mathcal{R}^N$ with the operator L defined by $L(u) = [f_1(u), \ldots, f_N(u)]^T$, where $u \equiv [u_1, \ldots, u_N]$ and the functions $f_i(u)$ have the usual derivatives $\frac{\partial f_i}{\partial u_j}$ that are continuous in the vicinity of v. Then L'_v is the *Jacobian matrix*, that is, $L'_v \equiv [\partial f_i / \partial u_j]$; in this case, there is no necessity to specify which derivative we have in mind since both coincide.

Many basic notions and theorems of classical mathematical analysis have become extremely useful and to some degree routine tools in dealing with nonlinear problems in modern functional spaces. For example, under proper understanding of the integral, the classical *Newton-Leibnitz formula* for $L(v + h) - L(v)$ holds; from it, we can conclude (see [19, 292, 341]) that if L'_v depends continuously on $v \equiv u + tz, 0 \leq t \leq 1$, then the *Lipschitz inequality*

$$\|L(u + z) - L(u)\|_F \leq \|L'_v\|_{U \mapsto F} \|z\|_U \tag{2.15}$$

holds. We will use this remarkable inequality often; it is sometimes referred to as the *formula of finite increments* because it may be regarded as a generalization of the classical *Lagrange formula of finite increments*. Now, if additionally we assume that $\|L'_v - L'_u\|_{U \mapsto F} \leq l\|v - u\|_U$, then it is possible to prove that

$$\|L(u + v) - L(u) - L'_u z\|_U \leq \frac{l}{2}\|z\|_U^2. \tag{2.16}$$

This inequality may be regarded as a generalization of a particular case of the classical *Taylor formula* (see [481]). This formula means that L'_u may be regarded, in the accuracy sense, as the best linearization of the nonlinear operator L at u.

Lemma 3. *Let $H = U = F$ be a Euclidean or Hilbert space and suppose that an operator L has a Gateaux derivative L'_{u+tz} at each point $u + tz$, with u fixed and $t \in (0, 1)$. Then, for every w, there exists $\theta \in (0, 1)$ such that*

$$X \equiv (L(u + z) - L(u), w) = (L'_{u+\theta z}, w). \tag{2.17}$$

Proof. Let $g(t) \equiv (L(u + tz) - L(u), w)$ with $t \in (0, 1)$. Then $g'(t) = (L'_{u+tz}z, w)$, $g(0) = 0$, $g(1) = X$. Therefore, the classical Lagrange formula gives $X = g(1) - g(0) = g'(\theta)$ which leads to (2.17). $\square$

We conclude this exposition of the essentials of nonlinear differentiable operator theory by stressing the fact that very often we do not need explicit derivatives—with a quasibilinear form $b(u; v)$, it may suffice to find a bilinear form $b'(u; v) \equiv \lim_{t \to 0}[b(u + tz; w) - b(u; w)]/t$ that corresponds to L'_u.

2.4. Theorems of correctness for differentiable operators. In case of differentiable operators, many conditions in the above theorems, including those in Section 1, can be specified on the basis of Lemma 3. For example, conditions (2.12) will be satisfied if, for all $v \in S$, we have $L'_{v,s} \geq \delta_0 B$, $\delta_0 > 0$, where $L'_{v,s}$ is the symmetric part of the linear operator L'_v. Sometimes, it is easy to specify even the statements of the theorems. As an example, we give a modification of Theorem 7.

Theorem 11. *Let $B = B^* > 0$ and $S \equiv S_B(u^0; r')$, and suppose that L is a continuously differentiable operator on S such that, for all $v, u, u + z$ in S,*

$$\|(L'_v)^{-1}\|_{H(B^{-1}) \mapsto H(B)} \leq K_0, \tag{2.18}$$

and

$$\|L(u + z) - L(u) - L'_u z\|_{B^{-1}} \leq k\|z\|_B^{1+\alpha}, \alpha > 0. \tag{2.19}$$

Let $\mu' \equiv k(2r')^\alpha$, $K_0\mu' < 1$, and $S_F \equiv S_{B^{-1}}(L(u^0); r'K_0^{-1}(1 - \mu'K_0))$. Then the conclusion of Theorem 7 remains true.

Proof. For $R \equiv L'_{u^0}$, we define $\mathcal{P}(v) \equiv L(v) - L(u^0) - R(v - u^0)$ and

$$Q_\nu(v) \equiv v - u^0 + \nu R^{-1}[\mathcal{P}(v) + L(u^0) - f],$$

with $f \in S_F$ and $\nu \in [0, 1]$ as in the proof of Theorem 2. Then, for $\|v - u^0\|_B = r'$, we have $\|Q_\nu(v)\|_B \geq r' - K_0(\|\mathcal{P}(v)\|_{B^{-1}} + \|L(u^0) - f\|_{B^{-1}})$. Therefore, existence of solutions of (2.11) follows from the same reasoning used in the proof of Theorem 2. For solutions u_1, u_2, we see that $L'_{u_2}(u_2 - u_1) + L(u_2) - L(u_1) - L'_{u_2}(u_2 - u_1) = f_2 - f_1$. Thus,

$$\|u_2 - u_1\|_B \leq K_0(\|f_2 - f_1\|_{B^{-1}} + k\|u_2 - u_1\|_B^{1+\alpha}),$$

which leads immediately to (2.13). $\square$

The following well-known theorem about *implicit functions* will often be of use to us.

The Implicit Function Theorem. *Suppose L is continuously differentiable on $S \equiv S(u_0; r)$ and there exists $(L'_{u_0})^{-1} \in \mathcal{L}(F; U)$. Then, in a vicinity of $f_0 \equiv L(u_0)$, there exists a continuous inverse operator L^{-1} and $L^{-1}(f_0) = u_0$ (see [341]).*

§ 3. Iterative methods with model symmetric operators

We return now to the problem that arose in Sections 0.3–0.5 while we were discussing possible approaches to constructing asymptotically optimal algorithms for solving linear grid systems. [18]

3.1. Estimates of rates of convergence in the Euclidean space $H(B)$ of the modified method of simple iteration. We consider here in detail the most essential questions in the theory of iterative methods with model operators, starting with an analysis of the modified method of the simple iteration for nonlinear systems (2.1) regarded as operator equations in the Euclidean space H. For a given model operator $B \in \mathcal{L}^+(H)$, [19] the modified method of the simple iteration takes the form

$$Bu^{n+1} = Bu^n - \tau(L(u^n) - f). \tag{3.1}$$

In what follows, we suppose that there exists a solution u of (2.1), $z^n \equiv u^n - u$ will denote the error at the nth iteration, and balls $S_B(u; r)$ will be denoted by $S_r, r > 0$.

Theorem 1. Let u be a solution of (2.1) and $S \subset S_r$, and suppose, for all $u + z \in S_r$, that

$$(L(u + z) - L(u), z) \geq \delta_0(\|z\|_B)\|z\|_B^2 \tag{3.2}$$

and

$$\|L(u + z) - L(u)\|_{B^{-1}}^2 \leq \delta_1(\|z\|_B)\|z\|_B^2. \tag{3.3}$$

Let method (3.1) be applied with $0 < \tau < 2\delta_0/\delta_1$ and an arbitrary initial iterate $u^0 \in S_r$. Then it converges and $\|z^n\|_B \leq [\rho(\tau; r)]^n\|z_0\|_B$, where $\rho(\tau; r) \equiv \rho(\tau) \equiv [1 - 2\tau\delta_0 + \tau^2\delta_1]^{1/2} < 1$, $\rho(\tau) \geq \rho(\tau^*) = [1 - \delta_0^2/\delta_1]^{1/2}$, and $\tau^* \equiv \delta_0/\delta_1$.

---

[18] The idea was to develop iterative methods based on model symmetric (self-adjoint) operators, denoted here by B. Key questions on their design deserve special attention, which we consider in Chapter 3. However, what we are interested in now is estimates of the rate of convergence of such methods, that is, estimates of error reduction as a function of the number of the iterations (for one iteration, we prefer to speak about *convergence factor*.) Exposition here is based on author's publications [150, 153, 154, 155, 156, 159, 160, 162, 163, 164, 165, 166, 170, 182, 187, 191, 198, 199, 200]. For similar work, see [82, 101, 135, 244, 252, 258, 260, 315, 407, 434, 440, 442, 497, 507].

[19] Conditions on the operators L and B will be imposed on some closed balls $S_B(v; r)$ with a radius $r > 0$, and in order not to repeat similar formulations, we shall require that everywhere below $\delta_k(t), \sigma_k(t)$ denote nonnegative functions on $[0, r]$ that are nonincreasing if $k = 0$ and nondecreasing if $k \geq 1$; their minimal (for $r = 0$) or maximal (for $r = 1$) values at the end point $t = r$ will be denoted by $\delta_k, \sigma_k, k = 0, 1$, and we shall assume that $\delta_0 > 0$, $\sigma_0 > 0$.

Proof. In order to make use of (3.2) and (3.3), we have to show that all iterates u^n remain in S_r. To this end, we take an arbitrary z with $\|z\|_B \leq r$ and define $A(z) \equiv z - \tau B^{-1}[L(u+z) - L(u)]$. Then,

$$\|A(z)\|_B^2 = \|z\|_B^2 - 2\tau(L(u+z) - L(u), z) + \tau^2\|B^{-1}[L(u+z) - L(u)]\|_B^2.$$

Conditions (3.2), (3.3) together with the restrictions imposed on τ give $\|A(z)\|_B^2 \leq \rho(\tau; \|z\|_B)\|z\|_B^2 \leq (\rho(\tau))^2 r^2 < r^2$. Hence, all u^n belong to S_r. Further, for the errors at the nth and $(n+1)$th iterations, the standard recurrence relation

$$z^{n+1} = z^n - \tau B^{-1}[L(u+z^n) - L(u)] = A(z^n)$$

holds. Hence, $\|z^{n+1}\|_B \leq \rho(\tau)\|z^n\|_B$, which leads to the desired estimate. Optimality of the choice $\tau = \tau^*$ is straightforward. $\square$ [20]

Theorem 2. Let u be a solution of (2.1) with $S \subset S_r$ and suppose the operator L is continuously differentiable at every point $u + z \in S_r$ such that, for all z' and w, $(L'_{u+z}z', z') \geq \delta_0(\|z\|_B)\|z'\|_B^2$ and $|(L'_{u+z}z', w)| \leq [\delta_1(\|z\|_B)]^{1/2}\|z'\|_B\|w\|_B$. Let τ and u^0 be the same as in Theorem 1. Then the assertion of Theorem 1 remains true.

Proof. By virtue of (2.17), for some $\theta \in (0, 1)$, we may write $(L(u+z) - L(u), z) = (L'_{u+\theta z}z, z) \geq \delta_0(\theta\|z\|_B)\|z\|_B^2$. Hence, (3.2) holds. Similarly, $|(L(u+z) - L(u), v)| = |(L'_{u+\theta z}z, v)| \leq [\delta_1(\|z\|_B)]^{1/2}\|z\|_B\|v\|_B$, $\forall v$ (see (1.17)). This and Lemma 4.5 yield (3.3). Hence, the conditions of Theorem 1 are satisfied. $\square$ [21]

Theorem 3. Let u be a solution of (2.1) and $S \subset S_r$, and suppose L is a continuously differentiable operator at each point $u + z \in S_r$ such that,

$$\sigma_0(\|z\|_B)\|z'\|_B^2 \leq ((L'_{u+z})_s z', z') \leq \sigma_1(\|z\|_B)\|z'\|_B^2, \quad \forall z', \tag{3.4}$$

and

$$\|(L'_{u+z})_a z'\|_{B^{-1}}^2 \leq \sigma_2(\|z\|_B)\|z'\|_B^2, \quad \forall z'. \tag{3.5}$$

---

[20] The uniqueness of the solution of (2.1) is a direct consequence of (3.2); convergence of the method can be also expressed by saying that the solution is an attractor of all elements of the ball S_r. The theorem can be found in [159, 163] and is based on earlier work (see [155, 315, 408, 493]) in which conditions of the type (2.12), (1.14) that are more restrictive than (3.2), (3.3) were used. Such conditions can be further simplified in the case of differentiable operators.

[21] It is possible to get more accurate, but at the same time more complicated, estimates on the basis of information about the symmetric and antisymmetric parts of L'_v, the derivative of the operator L. These parts have the respective forms $(L'_v)_s \equiv 2^{-1}[L'_v + (L'_v)^*]$, $(L'_v)_a \equiv 2^{-1}[L'_v - (L'_v)^*]$. Such estimates were for the first time suggested by Gunn [260] for linear problems; we consider somewhat better estimates from [155, 162] under significantly more general conditions.

Let the method (3.1) be applied with $u^0 \in S_r$ and $0 < \tau < 2(\sigma_1 + \sigma_2/\sigma_0)^{-1}$. Then it converges and

$$\|z^{n+1}\|_B \le \rho_1(\tau; \|z^n\|_B)\|z^n\|_B \le \rho_1(\tau; r)\|z^n\|_B, \qquad (3.6)$$

where $\rho_1(\tau; t) \equiv \{1 - 2^{-1}[\sigma_0(t) + \sigma_1(t)]^2 + \tau^2 \sigma_2(t)\}^{1/2} + \tau 2^{-1}(\sigma_1(t) - \sigma_0(t))$, $\rho_1(\tau; r) \equiv \rho_1(\tau) \ge \rho_1(\tau^*)$, with the optimal iteration parameter $\tau^* \equiv 2[\sigma_0 + \sigma_1 - (\sigma_1 - \sigma_0)\sigma_2^{1/2}(\sigma_2 + \sigma_0\sigma_1)^{-1/2}][4\sigma_2 + (\sigma_0 + \sigma_1)^2]^{-1}$.

Proof. By virtue of (2.15), we get the basic recurrence relation $\|z^{n+1}\|_B \le \|R\|_B \|z^n\|_B$, where $R \equiv I - \tau B^{-1} L'_v$, $v \equiv u + \theta z^n$, and, as we may assume, $\|z^n\|_B \le r$. Hence, for proving (3.6), it suffices to show that

$$\|Rw\|_B \le \rho_1(\tau; \|z^n\|_B)\|w\|_B. \qquad (3.7)$$

Letting $R = R_1 + R_2$ with $R_1 \equiv (1 - \alpha)I - \tau B^{-1}(L'_v)_s$, $R_2 \equiv \alpha I - \tau B^{-1}(L'_v)_a$ and using (0.1.6), we may write $\|Rw\|_B \le \|R_1 w\|_B + \|R_2 w\|_B$. Here, $\alpha \in [0, 1]$ is an artificial parameter that will be selected to minimize an estimate from above for the right-hand side of this inequality. Letting $\sigma'_k \equiv \sigma_k(\|z\|_B)$, then (3.4) and Lemma 0.3.1 give

$$\sigma'_0 B \le (L'_v)_s \le \sigma'_1 B, \quad \|R_1\|_B \le \max_{t \in [\sigma'_0, \sigma'_1]} |1 - \alpha - \tau t|.$$

Since $B^{-1}(L'_v)_a \in \mathcal{L}(H(B))$ is antisymmetric, then $\|R_2 w\|_B^2 = \alpha^2 \|w\|_B^2 + \tau^2 \|(L'_v)_a\|_{B^{-1}}^2 \le [\alpha^2 + \tau^2 \sigma_2']\|w\|_B^2$. Hence, we may conclude that, for every α,

$$\|R\|_B \le \rho(\tau; \alpha) \equiv \max\{|1 - \alpha - \tau \sigma'_0|; |1 - \alpha - \tau \sigma'_1|\} + [\alpha^2 + \tau^2 \sigma'_2]^{1/2}.$$

Now we are in a position to take advantage of the presence of the free parameter and to choose $\alpha \in [0, 1]$ to minimize the function $\rho(\tau; \alpha)$ with respect to α. For this purpose, we analyze properties of the function $\rho(\tau; \alpha)$ with $\tau > 0, 0 \le \alpha \le 1$, and introduce two subsets Q_1 and Q_2 of the set of possible pairs $[\tau, \alpha]$: Q_1 will be the subset with $1 - \alpha - \tau \sigma'_0 \ge \max\{0; -(1 - \alpha - \tau \sigma'_1)\}$ and Q_2 with $1 - \alpha - \tau \sigma'_0 < 0$. Since $1 - \alpha - \tau \sigma'_0 \ge 1 - \alpha - \tau \sigma'_1$, we may replace $\rho(\tau; \alpha)$ by

$$\rho(\tau; \alpha) = \rho_{1,0}(\tau; \alpha) \equiv 1 - \alpha - \tau \sigma'_0 + [\alpha^2 + \sigma'_2 \tau^2]^{1/2} \text{ if } [\tau; \alpha] \in Q_1,$$

$$\rho(\tau; \alpha) = \rho_{2,0}(\tau; \alpha) \equiv -(1 - \alpha - \tau \sigma'_1) + [\alpha^2 + \sigma'_2 \tau^2]^{1/2} \text{ if } [\tau; \alpha] \in Q_2.$$

Further, $\rho_{1,0}(\tau; \alpha)$ on Q_1 is a decreasing function of α and $\rho_{2,0}(\tau; \alpha)$ on Q_2 is an increasing function of α. Then, for $\tau \le 2(\sigma'_0 + \sigma'_1)^{-1}$, we take $\alpha = 1 - 2^{-1}(\sigma'_1 + \sigma'_0)$ and obtain $\|R\|_B \le \rho_1(\tau; t)$ with $t \equiv \|z^n\|_B$. For $\tau \ge 2(\sigma'_0 + \sigma'_1)^{-1}$, the best choice is $\alpha = 0$, which leads to

$\|R\|_B \le \tau[\sigma_1' + (\sigma')_2^{1/2}] - 1 = \rho_{2,0}(\tau;t)$ with $t = \|z^n\|_B$. It is easy to see that $\rho_{2,0}(\tau;t) \ge \rho_{2,0}(2(\sigma_0' + \sigma_1')^{-1};t)$, which enables us to deal only with the case $\tau \le 2(\sigma_0' + \sigma_1')^{-1}$. Therefore, for

$$\tau < 2[\sigma_1 + \sigma_2/\sigma_0]^{-1} \le 2[\sigma_1' + \sigma_2'/\sigma_0']^{-1},$$

estimate (3.7) holds with $\rho_1(\tau;t) < 1$ and leads to (3.6), since $\rho_1(\tau;t) \le \rho_1(\tau;r) = \rho_1(\tau)$. Minimization of the function $\rho_1(\tau)$ with respect to $\tau \in (0, 2(\sigma_0 + \sigma_1)^{-1})$ is again straightforward and leads to $\tau = \tau^*$. $\square$ [22]

    Theorem 4. *Let $L \in \mathcal{L}(H)$ and $B \in \mathcal{L}^+(H)$, and suppose there exist positive constants δ_0, δ_1 such that $(Lz, z) \ge \delta_0\|z\|_B^2$; $|(Lv, z)| \le \delta_1^{1/2}\|v\|_B\|z\|_B$, $\forall z, \forall v$. Then L is invertible. If, additionally in method (3.1), the iteration parameter $\tau \in (0, 2\delta_0/\delta_1)$ is used, then $\|R\|_B \le \rho(\tau) < 1$ for $R \equiv I - \tau B^{-1}L$ and $\rho(\tau) \equiv [1 - 2\tau\delta_0 + \tau^2\delta_1]^{1/2}$.*

    Theorem 5. *Let $L \in \mathcal{L}(H)$, $B \in \mathcal{L}^+(H)$, and $\sigma_0 B \le L_s \le \sigma_1 B$, and, for all v, z, suppose that $|(L_a v, z)| \le \sigma_2^{1/2}\|v\|_B\|z\|_B$. Let τ and $\rho_1(\tau)$ be defined as in Theorem 3. Then $\|R\|_B \le \rho_1(\tau) < 1$.*

    3.2. Estimates of the rate of convergence in the Euclidean space $H(B^2)$. [23] In the sequel, $2\delta_0/\delta_1 \equiv \tau'$.

    Theorem 6. *Let u be a solution of (2.1), $B \in \mathcal{L}^+(H)$, and $S \subset S_r' \equiv S_{B^2}(u;r)$, and suppose, for all $u + z \in S_r'$, that $(L(u + z) - L(u), Bz) \ge \delta_0(\|Bz\|)\|Bz\|^2$ and $\|L(u + z) - L(u)\|^2 \le \delta_1(\|Bz\|)\|Bz\|^2$. Then method (3.1) with $\tau \in (0, \tau')$ and $u^0 \in S_r'$ converges and $\|Bz^n\| \le (\rho(\tau))^n\|Bz^0\|$.*

    Theorem 7. *Let u be a solution of (2.1), $B \in \mathcal{L}^+(H)$, $S \subset S_r' \equiv S_{B^2}(u;r)$. Suppose $(L(u + z) - L(u), Bz) \ge \delta_0(\|Bz\|)\|Bz\|^2$ and*

---

[22] The case $\sigma_2 = 0$ gives $\tau^* = 2/(\sigma_0 + \sigma_1)$, $\rho_1(\tau^*) = (\sigma - 1)/(\sigma + 1)$ with $\sigma \equiv \sigma_1/\sigma_0$ (see [155, 442, 440]), that is, after replacement of σ by δ, we have the same optimal estimate that was obtained in § 0.3. This is, of course, a very pleasant result at least from a theoretical point of view. If the functions $\sigma_k(t), k = 0, 1, 2$, can be found explicitly (such situations will be described in Chapter 5), then we can use an increasing sequence $\{\tau_n\}$ of iteration parameters with τ_n determined by constants $\sigma_k(r_n), k = 0, 1, 2$, where $r_n \ge \|z^n\|_B$. The same is applicable to method (3.1) with τ replaced by τ_n under the conditions of Theorem 1. Simpler cases of these theorems are obtained when the δ_k, σ_k in conditions (3.2)–(3.5) are assumed constant for all u, z. It is then possible to replace the equation $L(u) = f$ by the equivalent one $u = R(u) \equiv u - \tau B^{-1}[L(u) - f]$ and to estimate $\|R(u) - R(v)\|_B^2 = \|u - v - \tau B^{-1}[L(u) - L(v)]\|_B^2$ in the same manner as in the proofs of these theorems. This implies that R is a contraction in the Euclidean space $H(B)$ and that method (3.1) converges for each u^0. Furthermore, when $L \in \mathcal{L}(H)$, that is, L is linear, then the given theorems lead to their simplest variants as follows.

[23] We have already used the Euclidean space $H(B^2)$ with $B \in \mathcal{L}^+(H)$ for analysis of correctness of grid systems (see Theorem 2.8). Usefulness of such an approach will be clarified in Chapter 6. For the time being, following [156, 164], we analyze convergence of method (3.1) rewritten in the equivalent form $B^2(u^{n+1} - u^n) = -\tau[BL(u^n) - Bf]$. In other words, we apply Theorems 1 and 2, regarding B^2 as a model operator for system $BL(u) = Bf$, which obviates the proof of our next three theorems.

$\|L(u + z) - L(u)\|^2 \leq \delta_1(\|Bz\|)\|Bz\|^2$. *Let L be continuously differentiable on S'_r such that, for all $u + z \in S'_r$ and all v, $(L'_{u+z}v, Bv) \geq \delta_0(\|Bz\|)\|Bz\|^2$, $\|L'_{u+z}v\|^2 \leq \delta_1(\|Bz\|)\|Bv\|^2$. Then method (3.1) with $\tau \in (0, \tau')$ and $u^0 \in S'_r$ converges as in Theorem 6.*

Theorem 8. Let $L \in \mathcal{L}(H)$ and $B \in \mathcal{L}^+(H)$, and suppose there exist positive constants δ_k, $k = 0, 1$, such that, for all v, $\delta_0\|Bv\|^2 \leq (Lv, Bv)$ and $\|Lv\|^2 \leq \delta_1\|Bv\|^2$. Then L is an invertible operator and if $\tau \in (0, \tau')$ and $R \equiv I - \tau B^{-1}L$, then $\|R\|_{B^2} \leq \rho(\tau) < 1$. [24]

3.3. Condition numbers of symmetrized linear systems; generalizations for nonlinear problems. The modified method of simple iteration for linear problems with $L \in \mathcal{L}^+(H)$ may be replaced by significantly more effective iterative methods with model operators like the modified Richardson or gradient methods (in particular, the modified conjugate gradient methods), as we noted already in § 0.3. We delay discussion of these methods in detail until later, because there is a necessity to continue studying variants of the modified method of simple iteration for nonlinear problems, which reduces in the linear case to the classical method of simple iteration applied to symmetrized linear problems (0.3.17). But, to do this, we should prove that localizations of the spectrums of the symmetrized linear operators given in (0.3.18)–(0.3.20) hold (see [155, 156, 162, 164, 198, 199]).

Lemma 1. Let $L \in \mathcal{L}(H), B \in \mathcal{L}^+(H), A_1 \equiv B^{-2}L^*L$, and $A_2 \equiv L^*B^{-2}L$. Then localization (0.3.18) with $A \equiv A_1$ holds if and only if

$$\delta_0\|Bv\|^2 \leq \|Lv\|^2 \leq \delta_1\|Bv\|^2, \quad \forall v, \tag{3.8}$$

localization (0.3.18) with $A \equiv A_2$ holds if and only if

$$\delta_0\|Bv\|^2 \leq \|L^*v\|^2 \leq \delta_1\|Bv\|^2, \quad \forall v. \tag{3.9}$$

Proof. Conditions (3.8) may be rewritten in the operator form $\delta_0 B^2 \leq B^2 A_1 L^*L \leq \delta_1 B^2$. By virtue of Lemma 0.3.1 and the subsequent remark, (3.8) is equivalent to (0.3.18) with $A = A_1$. Similarly, (3.9) is equivalent to the requirement sp $(B^{-2}LL^*) \subset [\delta_0, \delta_1]$. By virtue of Lemma 0.4.6, sp $(B^{-2}LL^*) = $ sp $(L^*B^{-2}L) = $ sp A_2. Hence, (3.9) and the condition sp $A_2 \subset [\delta_0, \delta_1]$ are equivalent. $\square$

Note that (3.8), (3.9) with $\delta_0 > 0$ are equivalent to the corresponding conditions

$$\|L\|_{H(B^2) \mapsto H} \leq \delta_1^{1/2}, \quad \|L^{-1}\|_{H \mapsto H(B^2)} \leq \delta_0^{-1/2}, \tag{3.8'}$$

---

[24] Theorem 8 is based on Theorem 4, so it is possible to modify Theorem 5 to apply to the space $H(B^2)$, but the general complexity of the operator $(BL)^* = L^*B$ makes such a result less meaningful.

$$\|L^*\|_{H(B^2)\mapsto H} \le \delta_1^{1/2}, \quad \|(L^*)^{-1}\|_{H\mapsto H(B^2)} \le \delta_0^{-1/2}. \qquad (3.9')$$

In the case of $A = A_3 \equiv B^{-1}L^*B^{-1}L \in \mathcal{L}^+(H(B))$, the analysis is exactly the same as for the general case $A \equiv B_2^{-1}L^*B_1^{-1}L \in \mathcal{L}^+(H(B_2))$ (see (0.3.17)) with $B_r \in \mathcal{L}^+(H), r = 1, 2$.

Lemma 2. *Let $B_r \in \mathcal{L}^+(H), r = 1, 2$, and $A \equiv B_2^{-1}L^*B_1^{-1}L$, where L is an invertible operator. Then condition (0.3.18) is equivalent to*

$$\|L\|_{H(B_2)\mapsto H(B_1^{-1})} \le \delta_1^{1/2}; \quad \|L^{-1}\|_{H(B_1^{-1})\mapsto H(B_2)} \le \delta_0^{-1/2}.$$

Proof. The latter conditions may be rewritten in the form $\delta_0\|v\|_{B_2}^2 \le \|Lv\|_{B_1^{-1}}^2 \le \delta_1\|v\|_{B_2}^2$, $\delta_0 > 0, \forall v \in H$. Hence, they are equivalent to the inequalities $\delta_0 B_2 \le L^*B_1^{-1}L \le \delta_1 B_2$, $\delta_0 > 0$. Again using Lemma 0.3.1 and the subsequent remark, we conclude that these conditions are equivalent to (0.3.18). □ [25] Thus, inequalities (0.3.19), (3.8'), (3.9') play the key role, not only in the analysis of correctness of the original problem, but also in the construction of effective iterative methods with model operators. They will be analyzed for some grid analogs of elliptic boundary value problems in Chapters 5–8. For now, we consider a simple but important illustration.

Lemma 3. *Let A and M be symmetric positive operators, and the operator $L \equiv A - c_0 M$ be such that $|\lambda(M^{-1}L)| \ge d > 0$ and $\lambda(M^{-1}A) \ge \lambda_0 > 0$. Suppose c_0, λ_0, and d define the constants $\delta_0 \equiv \delta_0', \delta_1 \equiv \delta_1'$ in the same way as in Lemma 2.1. Let $B \in \mathcal{L}^+(H)$ be a model operator such that $\delta_{0,1}B \le A \le \delta_{1,1}B$ with $\delta_{0,1} > 0$. Then inequalities (0.3.20) with $\delta_0 \equiv \delta_0'\delta_{0,1}/\delta_{1,1}$, $\delta_1 \equiv \delta_1'\delta_{1,1}/\delta_{0,1}$ hold.*

Proof. By virtue of Lemma 2.2 with $R \equiv L$, for each v, we have $\delta_0'\|v\|_A^2 \le \|Lv\|_{A^{-1}}^2 \le \delta_1'\|v\|_A^2$. Taking into account the conditions imposed on B, we write

$$\delta_0'\delta_{0,1}\|v\|_B^2 \le \|Lv\|_{A^{-1}}^2 \le \delta_1'\delta_{1,1}\|v\|_B^2.$$

Lemma 0.4.3 implies that $\delta_{1,1}^{-1}B^{-1} \le A^{-1} \le \delta_{0,1}^{-1}B^{-1}$, which leads to the desired inequalities. □

Now we are in a position to analyze the convergence of the following three variants (see [162, 164, 199, 200, 202]) of the modified method of simple iteration:

$$B^2[u^{n+1} - u^n] = -\tau(L'_{u^n})^*[L(u^n) - f], \qquad (3.10)$$

---

[25] Of course, this lemma with $A = A_3$ and its conditions are equivalent to (0.3.19) and (0.3.20). Notice also that Lemma 1 may be regarded as a consequence of Lemma 2, since the choice $B_1 = I, B_2 = B^2$ leads to $A = A_1$ and the choice $B_2 = I, B_1 = B^2$ leads to $A = A_2$. Thus, as we just have seen, the *condition number* of the operator $A \in \mathcal{L}^+(H(B_2))$, is determined by the condition number of the original operator $L \in \mathcal{L}(H(B_2), H(B_1))$. (Relevant papers were cited in § 0.3.)

$$u^{n+1} - u^n = -\tau(L'_{u^n})^* B^{-2}[L(u^n) - f], \tag{3.11}$$

$$B[u^{n+1} - u^n] = -\tau(L'_{u^n})^* B^{-1}[L(u^n) - f)]. \tag{3.12}$$

For linear L, these variants reduce to the usual simple iteration applied to the symmetrized system (0.3.17) with $A = A_1, A_2, A_3$, respectively. Since all of these cases may be analyzed in much the same way, we confine ourselves to a detailed proof of Theorem 11 for (3.12), and content ourselves only with stating Theorems 9 for (3.10) and 10 for (3.11); another reason is that we shall give a proof of Theorem 11 dealing with an iterative method more general than (3.12) with two model operators $B_r \in \mathcal{L}^+(H), r = 1, 2$ and which may include all cases (3.10)–(3.12). In what follows, for simplicity of presentation, instead of the functions $\delta_k(t), k = 0, 1, 2$ like those used in Theorems 1–3, 6, and 7, we consider constants $\delta_k, k = 0, 1, 2$, and an iterative parameter τ such that

$$0 \le \delta_2 < \delta_0^2, \quad 0 < \delta_0 \le \delta_1, \quad 0 < \tau < 2[\delta_1 + \delta_2^{1/2}]^{-1}, \tag{3.13}$$

$$\rho_2(\tau) \equiv \max\{|1 - \tau\delta_0|; |1 - \tau\delta_1|\} + \tau\delta_2^{1/2} < 1, \tag{3.14}$$

$\rho_2(\tau) \ge \rho(\tau^*) = 1 - 2[\delta_0 - \delta_2^{1/2}](\delta_1 + \delta_0)^{-1}$, where $\tau^* \equiv 2(\delta_0 + \delta_1)^{-1}$.

   Theorem 9. *Let u be a solution of (2.1), $B \in \mathcal{L}^+(H)$, and $S \subset S_r \equiv S_{B^2}(u; r)$, and suppose the operator L is continuously differentiable on S_r such that, for all v, $u + z \in S_r$, and $\theta \in (0, 1)$, $\delta_0\|Bv\|^2 \le \|L'_{u+z}v\|^2 \le \delta_1\|Bv\|^2$, $\|(L'_{u+\theta z} - L'_{u+z})v\|^2 \le \delta_2/\delta_1\|Bv\|^2$. Let (3.13) hold, $u^0 \in S_r$, and $D \equiv B^2$. Then method (3.13) converges and*

$$\|z^{n+1}\|_D \le \rho_2(\tau)\|z^n\|_B, \tag{3.15}$$

with $\rho_2(\tau)$ from (3.14).

   Theorem 10. *Let u be a solution of (2.1), $B \in \mathcal{L}^+(H)$, and $S \subset S_r \equiv S(u; r)$, and suppose the operator L is continuously differentiable on S_r such that, for all v, $u + z \in S_r$, and $\theta \in (0, 1)$, $\delta_0\|Bv\|^2 \le \|(L'_{u+z}v)^*\|^2 \le \delta_1\|Bv\|^2$ and $\|(L'_{u+\theta z} - L'_{u+z})^*v\|^2 \le \delta_2/\delta_1\|Bv\|^2$. Let (3.13) be satisfied, $u^0 \in S_r$, and $D \equiv I$. Then method (3.11) is convergent and estimate (3.15) is valid.*

   Theorem 11. *Let u be a solution of (2.1), $B_r \in \mathcal{L}^+(H), r = 1, 2$, and $S \subset S_r \equiv S_{B_2}(u; r)$, and suppose the operator L is continuously differentiable on S_r such that, for all v, $u + z \in S_r$, and $\theta \in (0, 1)$,*

$$\delta_0\|v\|_{B_2}^2 \le \|(L'_{u+z}v)\|_{B_1^{-1}}^2 \le \delta_1\|v\|_{B_2}^2, \tag{3.16}$$

$$\|(L'_{u+\theta z} - L'_{u+z})v\|_{B_1^{-1}}^2 \le \delta_2/\delta_1\|v\|_{B_2}^2. \tag{3.17}$$

Let (3.13) *be satisfied,* $u^0 \in S_r$, *and* $D \equiv B_2$. *Then the iterative method*

$$B_2[u^{n+1} - u^n] = -\tau(L'_{u^n})^* B_1^{-1}[L(u^n) - f] \qquad (3.12')$$

is convergent and estimate (3.15) *is valid.*

Proof. For an arbitrary $u^n \in S_r$, by virtue of (2.15), there exists $\theta \in (0,1)$ such that $\|z^{n+1}\|_{B_2} \le \|R\|_{B_2}\|z^n\|_{B_2}$, where $R \equiv I - \tau B_2^{-1}(L'_{u^n})^* B_1^{-1} L'_{u^n+\theta z^n} \equiv R_1 + R_2$, $R_1 \equiv I - \tau B_2^{-1}(L'_{u^n})^* B_1^{-1} L'_{u^n}$, and $R_2 \equiv \tau B_2^{-1}(L'_{u^n})^* B_1^{-1}[L'_{u^n} - L'_{u^n+\theta z^n}]$. The operator R_1 is symmetric if it is regarded as an element of $\mathcal{L}(H(B_2))$. Hence, conditions (3.16) lead to $\|R_1\|_{B_2} \le \max\{|1 - \tau\delta_0|; |1 - \tau\delta_1|\}$. As for R_2, we may write

$$\|R_2 v\|_{B_2} \le \tau \|B_2^{-1/2}(L'_{u^n})^* B_1^{-1/2}\| \|B_1^{-1/2}[L'_{u^n} - L'_{u^n-(1-\theta)z^n}]v\|.$$

This together with (3.17) and the inequality $\|B_2^{-1/2}(L'_{u^n})^* B_1^{-1/2}\| \le \delta_1^{1/2}$, which is a consequence of (3.16) (see § 0.4), lead to $\|R_2\|_{B_2} \le \rho_2(\tau)$ (see (3.14)). Therefore, $u^{n+1} \in S_r$, and the basic recurrence estimate (3.15) holds. □

On the basis of these theorems, it is possible to carry out an analysis of convergence of the important class of methods obtained with L'_{u^n} in (3.10)–(3.12) replaced by a fixed L'_{u^n} for several successive iterations. Another interesting case is of the operator L taken in the form $L = A + \mathcal{P}$, where A is a linear invertible operator and $\mathcal{P}$ is in some sense a subdominant nonlinear operator. Then the iterative method

$$B_2[u^{n+1} - u^n] = -\tau A^* B_1^{-1}[Au^n + \mathcal{P}(u^n) - f], \qquad (3.18)$$

with two model operators $B_r \in \mathcal{L}^+(H), r = 1, 2$ might apply. [26]

3.4. A posteriori estimates. Suppose that some numerical process we use produces a vector w with a fairly *residual* $r(w) \equiv L(w) - f$ in some appropriate norm. It would then be useful to draw some conclusions about the unknown solution u and the error $w - u$, that is, to obtain some *a posteriori estimates*. Such estimates are often based either on the correctness theorems or on the contraction mapping principle, and although any of our earlier theorems can be used for this purpose, we confine ourselves to Theorem 1 (see [159]).

Theorem 12. *Let* $B \in \mathcal{L}^+(\mathcal{H})$ *and* $S \equiv S_B(w; r/2)$, *and suppose the operator* L *is such that, for all* $u \in S$ *and* $u + z \in S$, *conditions* (3.2), (3.3) *are satisfied. Let* $\tau^* \equiv \delta_0/\delta_1$ *and* $\rho^* \equiv [1 - \delta_0^2/\delta_1]^{1/2}$, *and suppose*

---

[26] For example, suppose u is a solution of (2.1) with $S \subset S_r \equiv S_{B_2}(u; r)$ and, for all v and all $u + z \in S_r$, that we have $\delta_0\|v\|_{B_2}^2 \le \|Av\|_{B_1^{-1}}^2 \le \delta_1\|v\|_{B_2}^2$ and $\|\mathcal{P}(u + z) - \mathcal{P}(u)\|_{B_1^{-1}}^2 \le \delta_2/\delta_1\|z\|_{B_2}^2$. Then, in the case of (3.13) and $u^0 \in S_r$, (3.15) again holds.

w satisfies $\|L(w) - f\|_{B^{-1}} \leq r(1 - \rho^*)[2\tau^*]^{-1}$. *Then problem* (2.1) *has a unique solution.*

Proof. Let $R(v) \equiv v - \tau^* B^{-1}[L(v) - f]$, $\forall v \in S$. Then $\|R(v) - R(w)\|_B \leq r_1 + r_2$, where $r_1 \equiv \|v - w - \tau^* B^{-1}[L(v) - L(w)]\|_B \leq \rho^* r/2$ (see the proof of Theorem 1) and $r_2 \equiv \tau^* \|L(w) - f\|_{B^{-1}} \leq r/2(1 - \rho^*)$. Hence, $r_1 + r_2 \leq r/2$ and R maps S into itself. But, for all v and v' from S, $\|R(v) - R(v')\|_B \leq \rho^* \|v - v'\|_B$, so R is a contraction mapping of S into itself and it must have a unique fixed point u, which is the unique solution of (2.1). □

We return now to method (3.1), yielding $\{u^n\}$, and draw some conclusions about the above assumptions and the iteration parameters. Starting with the linear case, we write $z^n \equiv u^n - u$, $r^n \equiv Lu^n - f = Lz^n$, $R \equiv I - \tau B^{-1}L$, $\mathcal{R} \equiv I - \tau L B^{-1}$. Since $z^{n+1} = Rz^n$ and $r^{n+1} = \mathcal{R}r^n$, then the respective operators R and $\mathcal{R}$ will be called the *error reduction* and *residual reduction operators* for method (3.1).

Theorem 13. *The error reduction and residual reduction operators for* (3.1) *satisfy* $\|R\|_B = \|\mathcal{R}\|_{B^{-1}}$, $\|R\|_{B^2} = \|\mathcal{R}\|$. [27]

Proof. The equality $\|R\|_B = \|B^{1/2}RB^{-1/2}\|$ (see § 0.4) implies that

$$\|R\|_B = \|I - \tau B^{-1/2}LB^{-1/2}\| = \max_{v \neq 0}\{\|B^{-1/2}\mathcal{R}B^{-1/2}v\|/\|v\|\}.$$

The change of variables $v = B^{-1/2}u$ leads to the first desired equality. Similarly, for the Euclidean space $H(B^2)$, we have $\|R\|_{B^2} = \|B^1RB^{-1}\| = \|I - \tau LB^{-1}\| = \|\mathcal{R}\|$, which gives the second desired equality. □ [28]

---

[27] This fact was used in [164] for an analysis of the modified gradient method, based on minimization of $\|L(v) - f\|^2_{B^{-1}}$.

[28] This theorem together with Theorems 4 and 5 imply the recurrence estimates $\|r^{n+1}\|_{B^{-1}} \leq \rho \|r^n\|_B$, where ρ denotes either $\rho(\tau)$ (see Theorem 4) or $\rho_1(\tau)$ (see Theorem 5). The vectors r^n and $B^{-1}r^n$ must be computed in the iteration process (3.1), so it is a simple matter to find $(B^{-1}r^n, r^n) = \|r^n\|^2_{B^{-1}}$. In other words, we may regard the sequence $\{b^n \equiv \|r^n\|_{B^{-1}}\}$ as readily available and use it to control our choice of τ. For example, ignoring rounding errors, if we encounter inequality $b^{n+1}/b^n > \rho$, then our suppositions on τ, even with the optimal and unknown constants δ'_k and σ'_k, were not justified, so we must use smaller values of τ. In particular, for Theorem 4 with the requirement $0 < \tau < 2\delta'_0/\delta'_1$ and the parameter $\tau \equiv \tau'$ that is used, we may conclude that $\tau' \geq 2\delta'_0/\delta'_1$; a similar estimate with $\|r^n\| \equiv b^n$ may be of help in the case of Theorem 8. Cases of iterative methods for symmetrized linear systems will be analyzed later (see Theorem 18), but for now we turn to nonlinear problems. Now there is no simple connection between the behaviors of z^n and r^n, so it might be instructive to consider the sequence $\{b^n \equiv \|u^{n+1} - u^n\|_D\}$ with $D = B, B^2, I$, in accordance with the Euclidean space in which contraction of the corresponding operator was proved. More precisely, if we consider as a typical method (3.1), we may write $u^{n+1} - u^n = R(u^n) - R(u^{n-1})$ with $R(v) \equiv v - \tau B^{-1}L(v)$; if we know that the operator R with proper τ must be a contraction mapping of S_r into itself (in the sense of the Euclidean space $H(B)$), then we may take $D = B$ and write $b^{n+1} \leq \rho b^n$; if this inequality for the parameter τ that is

3.5. Modifications of Richardson's iteration. We have already emphasized the efficiency of the modified Richardson method for linear systems with $L = L^* > 0$ under reasonable preconditioning (see (0.3.13)). We have also emphasized the attraction of the appropriate symmetrization of a linear system with a general invertible operator (see (0.3.17), (0.3.18), Lemmas 1 and 2), which will enable us again to use the Richardson method. We now examine the convergence estimates in more detail. For simplicity, consider the most general case of the symmetrized system (0.3.17), rewritten here as $Au = g \equiv B_2^{-1} L^* B_1^{-1} f$, and the modified Richardson method of the form

$$u^{n+1} - u^n = -\tau_n B_2^{-1} L^* B_1^{-1} [L(u^n) - f], \quad n = 0, 1, \ldots, k-1, \quad (3.19)$$

with two model operators $B_r \in \mathcal{L}^+(H), r = 1, 2$. In what follows, we make use of the operators

$$A \equiv B_2^{-1} L^* B_1^{-1} L, \quad \mathcal{A} \equiv L B_2^{-1} L^* B_1^{-1}, \quad D \equiv L^* B_1^{-1} L, \quad (3.20)$$

and $D' \equiv L B_2^{-1} L^*$. If $\exists\, L^{-1}$ then $D \in \mathcal{L}^+(H)$ and $D' \in \mathcal{L}^+(H)$.

Lemma 4. *For an arbitrary $L \in \mathcal{L}(H)$, the operators defined in (3.20) are such that A is symmetric as a mapping of the Euclidean space $H(B_2)$ into itself and $\mathcal{A}$ is symmetric as a mapping of the Euclidean space $H(B_1^{-1})$ into itself; the spectrums of A and $\mathcal{A}$ coincide; and if the operator L is invertible, then $A \in \mathcal{L}^+(H(B_2))$, $A \in \mathcal{L}^+(H(D))$, $A \in \mathcal{L}^+(H(B_1^{-1}))$, and $A \in \mathcal{L}^+(H(D')^{-1})$.*

Proof. The relation $(L^* B_1^{-1} L)^* = L^* B_1^{-1} L$ and Lemma 0.3.1 imply the asserted properties of A. For $\mathcal{A}$, it suffices to note that $(\mathcal{A}u, v)_{B_1^{-1}} = (u, \mathcal{A}v)_{B_1^{-1}}$ and apply Lemma 0.3.1. The spectra of $A = (B_2^{-1} L^* B_1^{-1})L$ and of $\mathcal{A} = L(B_2^{-1} L^* B_1^{-1})$ coincide by virtue of Lemma 0.4.6. □

Recall that these spectra belong to the same interval $[\delta_0, \delta_1]$, with $\delta_0 > 0$ specified in Lemma 2. To carry out the analysis of convergence of method (3.19), we will make use of: the standard error $z^n \equiv u^n - u$; the residual $r^n \equiv Lu^n - f = Lz^n$; the relations $z^{n+1} = [I - \tau_n A]z^n$, $r^{n+1} = [I - \tau_n \mathcal{A}]r^n$, $z^k = Z_k z_0$, $r^k = Z_{r,k} r^0$ with $Z_k \equiv Q_k(A)$, $Z_{r,k} \equiv Q_k(\mathcal{A})$, where $Q_k(\lambda) \equiv (1 - \tau_0 \lambda) \cdots (1 - \tau_{k-1} \lambda)$; and some relatively unusual relations dealing with the residual $\xi^n \equiv Au^n - f$ of the symmetrized system $Au = g$ and having the form

$$\xi^n = B_2^{-1} L^* B_1^{-1} r^n, \quad \xi^{n+1} = [I - \tau_n A]\xi^n, \quad \xi^k = Z_k \xi^0.$$

---

used is not fulfilled, then $\tau \geq 2\delta_0'/\delta_1'$, and we should take a smaller τ and reexamine our information about the corresponding δ_k and σ_k.

The points $\cos \pi(2i+1)/(2k)$ with $i \in [0, k-1]$ are roots of the classical Chebyshev polynomial $T_k(t) \equiv \cos(k \arccos t)$ of degree k, written here for $|t| \leq 1$. The basic set of iteration parameters are determined by

$$\{\tau_n^{-1}\} = \{t_i\}, \quad t_i \equiv \varphi(\cos(\pi(2i+1)/(2k))), \quad i \in [0, k-1], \qquad (3.21)$$

where the function $\varphi(t) \equiv 2^{-1}[\delta_1 + \delta_0 + (\delta_1 - \delta_0)t]$ maps the line segment $[-1, 1]$ onto the given one $[\delta_0, \delta_1]$. Note also (see [41, 232, 271, 252]) that

$$T_k(t) = \frac{[t + (t^2 - 1)^{1/2}]^k + [t - (t^2 - 1)^{1/2}]^k}{2}, \quad |t| \geq 1, \qquad (3.22)$$

$q_k \equiv [T_k((\delta_1 + \delta_0)/(\delta_1 - \delta_0))]^{-1} < 1$, $|T_k(-t)| = |T_k(t)|$, and $T_k(t) \geq 0$ if $t \geq 1$. [29]

Theorem 14. *Let $B_r \in \mathcal{L}^+(H), r = 1, 2$, and $A \equiv B_2^{-1}L^*B_1^{-1}L$. Suppose that sp $A \subset [\delta_0, \delta_1]$ with $\delta_0 > 0$. For the system $Lu = f$, consider a cycle of k iterations of the method (3.19) applied with iteration parameters defined by (3.21). Then the reduction operators Z_k and $Z_{r,k}$ are symmetric as elements of $\mathcal{L}(H(E))$ and $\mathcal{L}(H(B_1^{-1}))$, respectively, and*

$$\|Z_k\|_E = \|Z_{r,k}\|_{B_1^{-1}} \leq q_k; \qquad (3.23)$$

if the set $\{t_i\}$ from (3.21) is replaced by the set $\{t_i^+\}$, where $t_i^+ \equiv \varphi^+(\alpha_i)$ with $\alpha_i \equiv \cos \pi(2i+1)/k$ and $i \in [0, k-1]$ (for even $k = 2m$), $|i-m| \in [0, m]$ (for odd $k = 2m+1$), and

$$\varphi^+(t) \equiv 2^{-1}[\delta_1 + \delta_0 - (\delta_1 - \delta_0)t], \qquad (3.21')$$

then the reduction operators $Z_k \equiv Z_k^+$ and $Z_{r,k} \equiv Z_{r,k}^+$ are symmetric nonnegative mappings of the respective Euclidean spaces $H(E)$ and $H(B_1^{-1})$ into themselves, and $0 \leq Z_k^+ \leq q_k^+ I$, $0 \leq Z_{k,r}^+ \leq q_k^+ I$, where $q_k^+ \equiv 2q_k(1 + q_k)^{-1} < 1$.

Proof. The reduction operators are polynomials of the respective operators A and $\mathcal{A}$, which are symmetric in the sense specified in Lemma 4. Thus, the desired symmetries of the reduction operators hold, and we need

---

[29] In further constructions of model operators, we will use inner iterations based on the modified Richardson method. As was fairly recently noticed (see [22, 191, 202]), it is reasonable in such situations to use, instead of the classical polynomials $T_k(t)$, some polynomials $T_k^+(t)$, leading to the error reduction operators $Z_k \equiv Z_k^+ \geq 0$ as mappings of the Euclidean space $H(E)$ into itself. (The case with $Z_k \leq 0$ has not found reasonable applications.) Thus, we try to unite similar statements about the convergence of (3.19) into a single theorem.

to prove only the given inequalities. The polynomial $Q_k(\lambda) \equiv Q$ in the case of the parameters from (3.21) takes the well-known form

$$Q = T_k \left(\frac{2\lambda - \delta_1 - \delta_0}{\delta_1 - \delta_0} \right) [T_k \left(-\frac{\delta_1 + \delta_0}{\delta_1 - \delta_0} \right)]^{-1} = q_k T_k \left(\frac{2\lambda - \delta_1 - \delta_0}{\delta_1 - \delta_0} \right),$$

with the standard Chebyshev polynomial $T_k(t)$ as in the classical theory of the Richardson method. Therefore, norms of the symmetric operators Z_k and $Z_{r,k}$ are estimated by $\max\limits_{\lambda \in [\delta_0, \delta_1]} |Q_k(\lambda)| \le q_k$, and the first part of the theorem is proved. Consider now the set $\omega \equiv \{\pi(2i+1)/k\} \subset [0, 2\pi]$. For the case $k = 2m$, ω does not contain π and all its points are displaced on $(0, 2\pi)$ symmetrically with respect to π. Therefore, the set $\{\alpha_i\}$ from (3.21') is a twice repeated set of the points $\alpha_0, \dots, \alpha_m$, which are the roots of the polynomial $T_m(t) = \cos(m \arccos t)$, $|t| < 1$. Moreover, each of these α_i is a two-multiple root of the polynomial $T_k(t) + 1 \equiv T_k^+(t)$. Therefore, for the case (3.21') with $k = 2m$, we deal with a nonnegative polynomial $Q_k(\lambda) \equiv Q_k^+(\lambda)$ on $[\delta_0, \delta_1]$, which can be written in the form

$$Q_k(\lambda) = \left(T_m \left(\frac{\delta_1 + \delta_0 - 2\lambda}{\delta_1 - \delta_0} \right) \right)^2 \left(T_m \left(\frac{\delta + 1}{\delta - 1} \right) \right)^{-2}$$

or $Q_k(\lambda) = q_k^+ [T_k(\frac{\delta_1 + \delta_0 - 2\lambda}{\delta_1 - \delta_0}) + 1]$. Hence,

$$\max\limits_{\lambda \in [\delta_0, \delta_1]} |Q_k(\lambda)| = \frac{2}{T_{2m}(\frac{\delta+1}{\delta-1}) + 1} = \frac{2q_k}{1 + q_k} = q_k^+.$$

Similarly, for the case $k = 2m + 1$, the set of α_i with $i \in [0, 2]m$, consists of the twice repeated points $\cos(\pi(2i+1)/(2m+1))$ with $i \in [0, m-1]$ and of the single point $\alpha_m = -1$. In other words, these points are roots of the polynomial $T_k(t) + 1 = T_k^+(t)$. Since $\varphi^+(-1) = \pi$, it is easy to see that $Q_k(\lambda) = (1 - \lambda/\delta_1)[r_m(\lambda)]^2$, which is nonnegative on $[\delta_0, \delta_1]$. Moreover,

$$Q_k(\lambda) = T_k(\frac{\delta_1 + \delta_0 - 2\lambda}{\delta_1 - \delta_0]}) + 1 [T_k(\frac{\delta + 1}{\delta - 1}) + 1]^{-1} = q_k^+ [T_k(\frac{\delta_1 + \delta_0 - 2\lambda}{\delta_1 - \delta_0}) + 1]$$

for all k, and the same reasoning as above is applicable for deriving the necessary estimates. $\square$ [30] There is also a possibility, in the case (3.21), to

---

[30] It is notable that, for the case $k = 2m$, we might have used the standard mapping of $[-1, 1]$ onto $[\delta_0, \delta_1]$ defined by the function $\varphi(t)$; it would have led to twice repeated cycles composed of m iterations with the standard m iteration parameters from (3.21). Also, for the case $k = 2m + 1$, we might have used $2m$ iteration parameters defined by the m points $[\varphi(\cos \pi(2i+1)/(2m+1))]^{-1}$ with $i \in [0, m]$ and the remaining parameter $\tau_n = 1/\delta_1$. Recall that from the computational point of view, especially when either

come to the final iterate u^k in another way using recurrence relations of the type (see [41, 440])

$$u^{n+1} - u^n = \omega_n \omega_{n-1}[u^n - u^{n-1}] - \frac{2(1 + \omega_n \omega_{n-1})}{\delta_1 + \delta_0}[Au^n - g],$$

where $n = 1, \ldots, k-1$, $u^1 - u^0 = -\tau_0[Au^0 - g]$, $\tau_0 \equiv 2/[\delta_1 + \delta_0]$, $\omega_n = [2\frac{\delta_1 + \delta_0}{\delta_1 - \delta_0} - \omega_{n-1}]^{-1}$ if $n \geq 1$ and $\omega_0 = [\delta_1 - \delta_0][\delta_1 + \delta_0]^{-1}$. This relation is based on the important property of Chebyshev polynomials, namely, that they satisfy the recurrence relation $T_n(t) = 2tT_{n-1}(t) - T_{n-2}(t)$ ($n \geq 2$, $T_0(t) = 1, T_1(t) = t$), which yields u^n with the remarkable property $z^n = Z_n z^0$, $\|Z_n\|_E \leq q_n$, and $\lim_{n \to \infty} \omega_n = (\delta^{1/2} - 1)/(\delta^{1/2} + 1)$. Obviously, this algorithm might also be used in the case of iterative parameters from (3.21') if $k = 2m$ (we simply twice repeat the above described procedure for (3.21) with $k = m$). Also, it might be used to accelerate the modified method of simple iteration (see [271]). Note that the important constant q_k may be rewritten in the form $q_k = 2\rho^k(1 + \rho^{2k})^{-1}$ with $\rho \equiv (\delta^{1/2} + 1)/(\delta^{1/2} - 1)$, $\delta \equiv \delta_1/\delta_0$ (it is not difficult to verify that this ρ may be replaced by the inverse value $\rho' \equiv 1/\rho$). The given expressions for q_k and q_k^+ enable us obtain the desired ε-accuracy ($0 < \varepsilon < 1$) using sufficiently large k. It is not difficult to verify that such k are characterized by the relations

$$q_k \leq \varepsilon \Leftrightarrow k \geq \frac{\ln 1/\varepsilon + [(\ln 1/\varepsilon)^2 - 1]^{1/2}}{\ln \rho} \equiv X(\varepsilon), \quad q_k^+ \leq \varepsilon \Leftrightarrow k \geq X(\varepsilon'),$$

with $\varepsilon' \equiv \varepsilon(2 - \varepsilon)$, or by their simplified versions

$$k \geq \frac{\ln 2/\varepsilon}{\ln \rho} \quad \Rightarrow \quad q_k \leq \varepsilon, \quad k \geq \frac{\ln 2/\varepsilon'}{\ln \rho} \quad \Rightarrow \quad q_k^+ \leq \varepsilon'.$$

Note also that $\ln \rho = \ln[1 + \frac{2/\delta^{-1/2}}{1 - \delta^{-1/2}}]$ and $\ln(\rho)^{-1} \sim \delta^{1/2}$ as $\delta \to \infty$.

Returning now to particular cases of (3.19), we emphasize again that this covers three variants for symmetrized system (0.3.17) with $A = A_1 \equiv$

---

k or δ are comparatively large and the influence of the rounding errors is significant, we should pay special attention to the order in which the elements of the sets $\{t_i\}$ and $\{t_i^+\}$ are being used as inverses of the iteration parameters. A fairly simple rule for appropriate enumeration that leads to numerical stability for the case $k = 2^m$ can be described recursively (see [41, 332, 355, 440]) as follows: if, for $k = 2^{m-1}$, the permutation $[i_0^{m-1}, i_1^{m-1}, \ldots, i_{2^{m-1}-1}^{m-1}]$ defines the stable enumeration of the set $\{t_i\}$ from (3.21), then the similar permutation in the case $k = 2^m$ must be

$$[i_0^{m-1}, 2^m - 1 - i_0^{m-1}, i_1^{m-1}, 2^m - 1 - i_1^{m-1}, \ldots, i_{2^{m-1}-1}^{m-1}, 2^m - 1 - i_{2^{m-1}-1}^{m-1}].$$

$B^{-2}L^*L, A = A_2 \equiv L^*B^{-2}L$, and $A = A_3 \equiv B^{-1}L^*B^{-1}L$, and it is even possible to include formally the method (0.3.17) with $L \in \mathcal{L}^+(H)$ under the choice $B_1 = B, B_2 = L$ or $B_2 = B, B_1 = L$. Thus, in the case (0.3.13) with the iteration parameters from (3.21), Theorem 14 and Lemma 0.3.1 imply the estimates

$$\|z^k\|_E \leq q_k\|z_0\|_E, \quad \|r_k\|_{B_1^{-1}} \leq q_k\|r^0\|_{B_1^{-1}} \tag{3.24}$$

with $E = B$ or $E = L$, where q_k should be replaced by q_k^+ if the iteration parameters from (3.21′) are used (observe that $A = B^{-1}L \in \mathcal{L}^+(H(B))$ is a consequence of Lemma 0.3.1). The cases with A_1 and A_2 are connected with Gauss symmetrization of the operator L as an element of $\mathcal{L}(H(B^2); H)$ and $\mathcal{L}(H; H(B^{-2}))$, respectively. For them, (3.19) with $B_1 = I, B_2 = B^2$ and $B_1 = B^2, B_2 = I$ takes the form

$$B^2(u^{n+1} - u^n) = -\tau_n L^*(Lu^n - f), \text{ or } u^{n+1} - u^n = -\tau_n L^*B^{-2}(Lu^n - f).$$

The most important and general case of (3.19) with $B_1 = B_2 = B$ (see Theorems 0.4.1 and 0.4.4 and (0.3.19) with $B \asymp J$) may be rewritten as

$$B(u^{n+1} - u^n) = -\tau_n L^*B^{-1}(Lu^n - f). \tag{3.25}$$

3.6. Use of orthogonalization. Having in mind possible generalizations of the modified Richardson methods for cases of general symmetric operators $L = L^*$, we start with the simplest case of a nonnegative operator $L = L^* \geq 0$ with Ker $L \equiv \{v : Lv = 0\}$. Then, as is known from linear algebra: the Euclidean space H is an orthogonal sum of the subspaces Ker L and Im $L \equiv \{f : \exists u : Lu = f\}$; the system $Lu = f$ has a solution if and only if $f \in$ Im L; and all solutions may differ only so that $u - u' \in$ Ker L. Note that, for $B = B^* > 0$, we have: Ker $(B^{-1}L) =$ Ker L, Im $(B^{-1}L) = \{v : (Bv, z) = 0, \forall z \in$ Ker $L\}$, and, to make our problem with $f \in$ ImL correct, we may use formulations like (2.1) with S being either Im L or Im A.

   Theorem 15. *Let* $L = L^* \geq 0, B = B^* > 0, S \equiv Im\ (B^{-1}L)$, *and* $f \in Im\ L$. *Suppose, for all* $v \in S$, *that condition* (0.3.9) *with* $\delta_0 > 0$ *is satisfied. Then problem* (2.1) *has a unique solution and, for method* (0.3.13) *with* $u^0 \in S$ *and iteration parameters from* (3.21), *we have the estimates* $\|z^k\|_B \leq q_k\|z^0\|_B, \|r^k\|_{B^{-1}} \leq q_k\|z^0\|_{B^{-1}}$.

   Proof. Conditions on S and f guarantee existence of the unique solution u that coincides with the solution of $Au \equiv B^{-1}Lu = g \equiv B^{-1}f, u \in S$. Since $u^0 \in S$, we have $z^0 \in S$. Note that S is the linear span of eigenvectors of the symmetric operator A (regarded as an element of $\mathcal{L}(H(B))$ that do not belong to Ker A. Therefore, z^0 may be represented as a linear combination of eigenvectors of A corresponding to eigenvalues from the interval $[\delta_0, \delta_1]$, and (3.24) applies .□ [31]

Theorem 16. *Let L be an invertible symmetric operator and suppose several of its eigenvectors $y_1, \ldots, y_m$ corresponding to eigenvalues $\lambda_1, \ldots, \lambda_m$ are known. Let these vectors be an orthonormal basis of the subspace $U_m \equiv lin\ \{y_1, \ldots, y_m\}$ in the sense of the Euclidean space H. Consider the orthoprojector $P^\perp$ defined by $P^\perp f \equiv f - (f, y_1)y_1 - \ldots - (f, y_m)y_m$. For all v orthogonal in the Euclidean space $H(B)$ to U_m, suppose that*

$$\delta_0(Bv, v) \le (LP^\perp v, P^\perp v) \le \delta_1(Bv, v). \qquad (3.26)$$

For system $Lu = f$, consider the iterative method

$$B(v^{n+1} - v_n) = -\tau_n[P^\perp LP^\perp v^n - P^\perp f], \quad n = 0, \ldots, k-1, \qquad (3.27)$$

applied with the iteration parameters from (3.21). Then $\|\xi^k\|_B \le q_k\|\xi^0\|_B$ and $\|\zeta^k\|_{B^{-1}} \le q_k\|\zeta^0\|_{B^{-1}}$, where $\xi^n \equiv v^n - P^\perp u$ and $\zeta^n \equiv P^\perp(LP^\perp v^n - f)$.

Proof. It is easy to see that $u = u_f + v$, where

$$u_f \equiv \lambda_1^{-1}(f, y_1) + \ldots + \lambda_m^{-1}(f, y_m)$$

and $v = P^\perp u \perp U_m, Lv = P^\perp f$. Thus, $Cv \equiv P^\perp LP^\perp v = P^\perp f$. Note that $C^* = C$ and Ker $C = U_m$, and that the method (0.3.13) for the system with the operator C is actually the iterative method under consideration. Note again that the operator $A \equiv B^{-1}C$ is symmetric as a mapping of the Euclidean space $H(B)$ into itself, that Ker $A = U_m$, and that sp $A \subset \{0 \cup [\delta_0, \delta_1]\}$ (see (3.26)). Therefore, the desired estimates follow directly from Theorem 15. □

It is not difficult to generalize Theorem 16 to the case dim Ker $L > 0$, but we shall concentrate instead on a new case associated with a model operator $B = B^* \ge 0$, as suggested in [182]. Namely, we investigate iterations (0.3.13) assuming that

$$L = L^* \ge 0,\ B = B^* \ge 0,\ \text{Ker } L = \text{Ker } B \oplus Q_0,\ \text{Ker } B \perp Q_0, \qquad (3.28)$$

---

[31] Similarly, if we take the operator A from (3.20) and $S \equiv \text{Im } A = \{v : (B_1 v, z) = 0, \forall z \in \text{Ker } L = \text{Ker } A\}$, in much the same way we can estimate the rate of convergence of method (3.19). Consider now the most interesting case of an invertible symmetric operator L. There are two possible approaches to the problem. One is based on the information that the spectrum of L belongs to a union of two intervals not containing the point 0: sp $L \subset ([-\delta_2, -\delta_3] \cup [\delta_0, \delta_1])$ with positive $\delta_k, k \in [0, 3]$; a special choice of iteration parameters is necessary with sufficiently large k (see [355]). The second approach needs more detailed information about eigenvectors of L corresponding to negative eigenvalues (we consider some effective methods for this in Chapter 9) and is a very useful when dealing with a number of systems with the same operator L. Since this approach eliminates some of the eigenvectors, it may work well even in the case of positive operators (see [187]).

where $\perp$ is understood in the sense of the Euclidean space H. Obviously, $\text{Im } L \subset \text{Im } B \equiv H'$ and the subspace H' is invariant with respect to both B and L; restrictions of these operators to H' will be denoted by B' and L', respectively. Of course, $B' \in \mathcal{L}^+(H')$. We also will be making use of the orthoprojector P of the Euclidean space H onto $\text{Ker } B$ and of the orthoprojector $P^\perp \equiv I - P$.

Lemma 5. *Let conditions* (3.28) *be satisfied and suppose, for all* $v \in S_B \equiv \{v : v \perp \text{Ker } B \text{ and } (Bv, z) = 0, \forall z \in Q_0\}$, *that inequalities* (0.3.9) *are valid with* $\delta_0 > 0$. *Then the Euclidean space* $H'(B')$ *has an orthonormal basis composed of eigenvectors of the operator* $A' \equiv (B')^{-1}L'$, *all eigenvalues of* A' *belong to* $\{0 \cup [\delta_0, \delta_1]\}$, *and the number of zero eigenvalues coincides with* $\dim Q_0$.

Proof. We have $L' = (L')^* \geq 0$, $\text{Ker } L' = Q_0$, and $\text{Im } (B')^{-1}L' = \{v : v \in H' \text{ and } (Bv, z) = 0, \forall z \in Q_0\}$. Thus, (0.3.9) holds for all $v \in \text{Im } [(B')^{-1}L']$, and the conclusions follow from Lemma 0.3.2. $\square$

We turn now to analysis of iterative method (0.3.13) for problem (2.1) with

$$f \in \text{Im } L, \quad S \equiv \{v : (B'v, z) = 0, \forall z \in \text{Ker } L\}. \tag{3.29}$$

Recall that the solution with the minimal norm is called the *normal solution* of the system (see [7, 226, 328]).

Theorem 17. *Let conditions of Lemma 5 be satisfied and suppose method* (0.3.13) *is applied to problem* (2.1), (3.29) *with the set of iteration parameters defined by* (3.21). *Then, for arbitrary* u^0, *the systems involving* u^{n+1} *are solvable and*

$$(B\xi^k, \xi^k) \leq q_k^2 (B\xi^0, \xi^0) \text{ and } \|r^k\|_{(B')^{-1}} \leq q_k \|r^0\|_{(B')^{-1}}$$

with $\xi^n \equiv P^\perp u^n - u$.

Proof. Since $f \in \text{Im } L \subset H'$ and $\text{Ker } B \subset \text{Ker } L$, then $g(v) \equiv Bv - \tau_n(Lv - f) \in H'$ for all v. Hence, $g(u^n) = g(P^\perp u^n)$, and $Bu^{n+1} = g(u^n)$ is solvable. The set of solutions of the fixed system is a plane in the Euclidean space H consisting of the points $w \equiv u^{n+1} + z$, $\forall z \in \text{Ker } L$, where u^{n+1} is any solution of the system. Therefore, $P^\perp u^{n+1} \in H'$ is the normal solution of the system, and it easy to see that the sequence $P^\perp u^n$ coincides with the sequence v^n if $v^0 = P^\perp u^0$ and $B'[v^{n+1} - v^n] = -\tau_n[L'v^n - f]$. This shows that Theorem 17 is a consequence of Theorem 16. $\square$ [32]

---

[32] Given an operator B, we will often make use of its *pseudoinverse operator* (see [7, 328, 415]), [33] denoted by $B^\dagger$, which maps each $g \in H$ on $B^\dagger g$, where $B^\dagger g$ is the normal solution of the system $B^*Bv = B^*g$ (for $B = B^* \geq 0$, we have $B^\dagger g = (B')^{-1}P^\perp g$). It is then possible to obtain a generalization of Theorem 17 dealing with method (3.25) where B^{-1} is replaced by $B^\dagger$ (it is again important that (3.28) must be satisfied, so we may make use of the fact $\text{Ker } L^*B^\dagger L = \text{Ker } L$ when $\text{Im } L \cap \text{Ker } L = 0$, which is the

Lemma 6. *Let $L = L^* \geq 0, B = B^* \geq 0$, and $M = M^* > 0$, and suppose $\mathrm{Ker}\, L$ is an orthogonal sum in the Euclidean space $H_1 \equiv H(M)$ of $\mathrm{Ker}\, B$ and a subspace Q_0. Let $S_M \equiv \{v : (Mv, z) = 0, \forall z \in \mathrm{Ker}\, B\}$ and $S_1 \equiv \{v : v \in S_M, (Bv, z) = 0, \forall z \in Q_0\}$, and suppose $(0.3.9)$ with $\delta_0 > 0$ is satisfied for all $v \in S_1$. Then: S_M is an invariant subspace of the operators L_1 and B_1; regarded as a Euclidean space G with the inner product $(u, v)_G \equiv (Bu, v)$, it has an orthonormal basis $\{\psi_i\}$ composed of eigenvectors such that $L_1 \psi_i = \lambda_i \psi_i$; and for these λ_i the statement of Lemma 5 remains true.*

Proof. Note that $S_M = \mathrm{Im}\, B_1$, $(Bu, v) = (B_1 u, v)_M$, and $L_1 \psi_i = \lambda_i B_1 \psi_i \Leftrightarrow L \psi_i = \lambda_i B \psi_i$. Hence, S_M is invariant with respect to L_1 and B_1. If we denote by L_1' and B_1' the corresponding restrictions of the operators L_1 and B_1, then Lemma 5 applies. $\square$

On the basis of this lemma, one may obtain a generalization of Theorem 17 with $P^\perp$ being the orthoprojector of the Euclidean space $H(M)$; the corresponding norms of residuals $r_1^n \equiv M^{-1} r^n$ will then coincide with $(B^\dagger r^n, r^n)^{1/2}$. The important case dealing with a generalization of Theorem 16 will be discussed separately in § 5.2.

3.7 Adaptation of iteration parameters. We discuss here a very practical aspect of applying above iterations dealing with the need to use approximate and sometimes even wrong constants δ_0 and δ_1 (see [164, 198, 271]). The optimal values of these constants will be denoted by δ_0^* and δ_1^*; e.g., in the case of $(0.3.9)$, they correspond to the minimal and maximal eigenvalues of the operator $B^{-1} L$. The positive constants δ_0, δ_1 corresponding to the iteration parameters from (3.21), $(3.21')$ we denote now by $\delta_0^{(m)}, \delta_1^{(m)}$. Our wish is to reexamine and possibly improve these constants after performing a cycle consisting of k iterations under consideration. The main tool for this will be estimates of norms of residuals r^k. More precisely, suppose the iteration parameters $\tau_0, \ldots, \tau_{k-1}$ defined by (3.21) with given $\delta_k \equiv \delta_k^{(m)}, k = 0, 1$ are used in the iterations (3.19) and define $t^{(m)} \equiv [\delta_1^{(m)} + \delta_0^{(m)}][\delta_1^{(m)} - \delta_0^{(m)}]^{-1}$. Suppose this cycle of iterations yielded the residual r^k. Define the actual observed reduction factor $\rho_k^{(m)} \equiv \|r^k\|_D / \|r^0\|_D$ with $D \equiv B_1^{-1}$. Compute $\alpha_k^{(m)} \equiv \rho_k^{(m)} / q_k = \rho_k^{(m)} T_k(t^{(m)})$ (a ratio of the observed reduction factor and the theoretical factor that assumes the used estimates are valid). If $\alpha_k^{(m)} \leq 1$, then there is no reason to reject the given

---

case if $L = L^*$). An important case is when, instead of eigenvectors y_i of the operator L, eigenvectors ψ_i of the operator $L_1 \equiv M^{-1} L$ with $M = M^* > 0$ are known; such a case is typical for finite element approximations of elliptic eigenvalue (spectral) problems which are described in Chapter 9. This case is reduced to the one considered by replacing the original system $Lu = f$ by the system $L_1 u = g \equiv M^{-1} f$ with a symmetric operator L_1 regarded as an element of $\mathcal{L}(H_1)$ and with the Euclidean space $H_1 \equiv H(M)$ and $\mathrm{Ker}\, L_1 = \mathrm{Ker}\, L$. The same applies to the operator $B_1 \equiv M^{-1} B$ with $\mathrm{Ker}\, B_1 = \mathrm{Ker}\, B$. We present here only an analog of Lemma 5.

constants, because the obtained result is in accordance with the theoretical estimate. In this case, we may continue our work with these constants and perform a new cycle of either k or even $2k$ iterations. If the contrary case $\alpha_k^{(m)} > 1$ holds, and if we are sure that this is not a result of rounding errors, then we should reexamine these constants, which we do here using the inverse function $T_k^{(-1)}(t)$ defined for $t \geq 1$ (for such t, the polynomial $T_k(t)$ is monotonically increasing): if $\lambda_m' > 1$ is such that $T_k(\lambda_m') = t$, then this number may be found in the explicit form $\lambda_m' = [a^2 + 1]/(2a)$, where $a^k \equiv t + [t^2 - 1]^{1/2} \equiv x$ is a solution of the equation $x + x^{-1} = 2t$.

Theorem 18. If $\rho_k^{(m)} \geq 1$, then

$$\delta_1^* \geq 2^{-1}[\delta_1^{(m)} + \delta_0^{(m)} + \lambda_m'(\delta_1^{(m)} - \delta_0^{(m)})]. \tag{3.30}$$

If $\rho_k^{(m)} < 1$, $\alpha_k^{(m)} > 1$, and $\delta_1^* \leq \delta_1^{(m)}$, then

$$\delta_0^* < 2^{-1}[\delta_1^{(m)} + \delta_0^{(m)} - \lambda_m'(\delta_1^{(m)} + \delta_0^{(m)})]. \tag{3.31}$$

Proof. Recall that $r^k = Q_k(\mathcal{A})r^0$ and $\mathcal{A}$ (see (3.20)) is symmetric and sp $\mathcal{A} \subset [\delta_0^*, \delta_1^*]$ (see Lemma 0.4.1). If we use the standard expansions of r^k, r^0 with respect to the orthonormal basis for the Euclidean space $H(B_1^{-1})$ consisting of eigenvectors of $\mathcal{A}$, then we may write $a_i^{(k)} = Q_k(\lambda_i)a_i^{(0)}$, where λ_i stands for the ith eigenvalue of $\mathcal{A}$ and a_i^n denotes a corresponding coefficient in the expansion of r^n, and

$$\|r^n\|_{B_1^{-1}}^2 = \sum_i [a_i^{(n)}]^2. \tag{3.32}$$

Since Q_k was defined via the Chebyshev polynomial $T_k(t)$, it is monotonically decreasing on $[0, \delta_0^{(m)}]$ from 1 to $[T_k(t^{(m)}]^{-1}$, and the function $|Q_k(\lambda)|$ is monotonically increasing if $\lambda \geq \delta_1^{(m)} \equiv \delta_1$. Thus, in the case $\rho_k^{(m)} \geq 1$, we must have $\delta_1 \leq \delta_1^*$ (meaning that δ_1 was not a proper upper bound for eigenvalues of the operator A) and $|Q_k(\delta_1)| \leq |Q_k(\delta_1^*)|$. From (3.32), it follows that the strongest divergence may occur only for components a_i with $\lambda_i = \delta_i^*$. Thus, $\|r_k\|_{B_1^{-1}} \geq |Q_k(\delta_1^*)|\|r_0\|_{B_1^{-1}}$,

$$q_k T_k\left(\frac{2\delta^* - \delta_1 - \delta_0}{\delta_1 - \delta_0}\right) \geq \rho_k^{(m)} = q_k \alpha_k^{(m)}, \quad T_k\left(\frac{2\delta_1^* - \delta_1 - \delta_0}{\delta_1 - \delta_0}\right) \geq \alpha_k^{(m)},$$

and we deduce that $(2\delta_1^* - \delta_1 - \delta_0)(\delta_1 - \delta_0)^{-1} \geq \lambda_m'$ and that (3.30) holds. Similarly, if $\delta_1 \geq \delta_1^*$, then $|a_i^{(k)}|$ with all i must decrease in comparison with $|a_i^{(0)}|$, with the weakest decrease corresponding to $\lambda_i = \delta_0^*$. Therefore, the second variant mentioned in the theorem leads to $\|r_k\|_{B_1^{-1}} \leq$

$|Q_k(\delta_0^*)|\|r_0\|_{B_1^{-1}}$. Thus, $T_k([\delta_1 + \delta_0 - 2\delta_0^*][\delta_1 - \delta_0]^{-1}) > \alpha_k^{(m)}$ and (3.31) holds. $\square$

A similar theorem holds in the case of the modified Richardson method leading to $Z_k \geq 0$.

Theorem 19. Let constants $\delta_1 \geq \delta_0 > 0$ define both the set $\{t_i^+\}$ as in Theorem 14 and the iteration parameters $\tau_0, \ldots, \tau_{k-1}$ in the modified Richardson method by the condition $\{\tau_n^{-1}\} = \{t_i^+\}$. Suppose the cycle of k iterations lead to the residual r^k with $\rho_k^{(m,+)} \equiv \|r^k\|_{B_1^{-1}}/\|r^0\|_{B_1^{-1}}$ and $\alpha_k^{(m,+)} \equiv \rho_k^{(m,+)}/q_k^+$. If $\rho_k^{(m,+)} \geq 1$, then

$$\delta_1^* \geq 2^{-1}[\delta_1^{(m)} + \delta_0^{(m)} + \lambda_m^{(1)}(\delta_1^{(m)} - \delta_0^{(m)})], \qquad (3.30')$$

where $\lambda_m^{(1)} \equiv T_k^{(-1)}(\alpha_k^{(m,+)} + (-1)^{k-1})$. If $\rho_k^{(m,+)} < 1$, $\alpha_k^{(m,+)} > 1$, and $\delta_1^* \leq \delta_1^{(m)}$, then

$$\delta_0^* < 2^{-1}[\delta_1^{(m)} + \delta_0^{(m)} - \lambda_m^{(0)}(\delta_1^{(m)} + \delta_0^{(m)})], \qquad (3.31')$$

where $\lambda_m^{(0)} \equiv T_k^{(-1)}(\alpha_k^{(m,+)} - 1)$.

Proof. Reasoning is much the same as in the above proof. We only have to note that, in the first case, we have

$$|Q_k^+(\delta_1^*)| = q_k^+|T_k(\frac{\delta_1 + \delta_0 - 2\delta_1^*}{\delta_1 - \delta_0}) + 1| \geq \rho_k^{(m)}.$$

Thus, $|T_k(\frac{\delta_1+\delta_0-2\delta_1^*}{\delta_1-\delta_0}) + 1| \geq \alpha_k^{(m)}$ and $T_k(\frac{2\delta_1^*-\delta_1-\delta_0}{\delta_1-\delta_0}) \geq \alpha_k^{(m)} + (-1)^{k-1}$, which leads to (3.31'). The second case leads to $T_k(\frac{\delta_1+\delta_0-2\delta_0^*}{\delta_1-\delta_0}) + 1 \geq \alpha_k^{(m)}$, and (3.31') holds. $\square$ [34]

3.8. Modified gradient methods. We confine ourselves here to the case of an invertible operator L and the system $Lu = f$ transformed to $Au = g$ (see(3.20)); recall that this includes the case of the preconditioned system (0.3.14) with $A \equiv B^{-1}L$. Then, by virtue of Lemma 4 and Theorem 0.2.1, the solution u of this problem coincides with the solution of the

---

[34] The obtained approximate constants can be increased and decreased somewhat and serve as new more accurate ones. Usually it is important to find first of all dependable approximations to δ_1^*. For this, as we have seen, divergence of the method is very instructive; in such cases, it is reasonable to keep the initial iterate. At subsequent stages, when we try to improve convergence, there is no such need. Various strategies of increasing the number of iterations in the cycle or of using a three-level algorithm are possible (see [271]). Similar algorithms with adaptation of iterative parameters in (0.3.13) with $B \times L$ were used with success in the code of Orehov (Moscow State University, Chair of Theory of Elasticity, 1979) dealing with numerical solution of some complicated grid systems in the theory of shells.

variational problem $u = \arg\min\Phi(v)$ with the functional written in either of two forms:

$$\Phi(v) \equiv (Av, v)_{B_2} - 2(g, v)_{B_2}, \tag{3.33}$$

or

$$\Phi(v) = \|v - u\|_D^2 - \|u\|_D^2, \tag{3.34}$$

where $D \equiv L^* B_1^{-1} L = B_2 A$. The equivalent form

$$\Phi(v) = \|Lv - f\|_{B_1^{-1}}^2 - \|f\|_{B_1^{-1}}^2 \tag{3.35}$$

may also be used. [35]

The classical *method of the steepest descent* selects τ to minimize the function $\varphi(\tau) \equiv \Phi(u^n - \tau\xi_n)$. Since, for all τ, we have $z^{n+1} = z^n - \tau\xi^n$ and $r^{n+1} = r^n - \tau A r^n$ (see (3.20)), then the desired choice of τ leads to minimization of $\|z^n - \tau\xi^n\|_D^2$ and $\|r^n - \tau A r^n\|_{B_1^{-1}}^2$, and, therefore, must correspond to orthogonality of $z^n - \tau\xi^n$ and ξ^n in the Euclidean space $H(D)$ and of $r^n - \tau A r^n$ and $A r^n$ in the Euclidean space $H(B_1^{-1})$. We therefore conclude that the optimal τ is

$$\tau = \frac{\|B_1^{-1} r^n\|_{D'}^2}{\|A r^n\|_{B_1^{-1}}^2} = \frac{\|L^* B_1^{-1} r^n\|_{B_2^{-1}}^2}{\|A r^n\|_{B_1^{-1}}^2},$$

and the method of the steepest descent may be regarded as the *method of minimal errors* in the Euclidean space $H(D)$ and simultaneously as the *method of minimal residuals* in the Euclidean space $H(B_1^{-1})$, as noticed by Petryshyn (see [407]) for particular cases.

We emphasize the fundamental importance of another point of view, namely, that the considered methods in fact use Rayleigh-Ritz approximations with one-dimensional planes (straight lines) in H. This gives rise to the well-known generalizations dealing with k-dimensional planes

$$X_k \equiv u^n + U_k(u_n) \equiv \{v : v = u^n + z, z \in U_k(u_n)\},$$

where the subspace

$$U_k(u_n) \equiv \mathrm{lin}\{\xi^n, A\xi^n, \ldots, A^{k-1}\xi^n\}$$

---

[35] The *gradient of the functional* at the point u of the Euclidean space H is defined as vector $w(u)$ such that $\lim_{t\to\infty}[\Phi(u + tz) - \Phi(u)]/t = (w(u), z)$, and it corresponds to the direction of fastest increase of the functional. If for some reasons we wish to work with the Euclidean space $H(B)$, then the gradient in the sense of this Euclidean space is simply $w_B(u) = B^{-1}w(u)$. Consequently, the gradient of Φ at the point u^n in the sense of the Euclidean space $H(B_2)$ is $\xi^n = Au^n - g$, and the classical *gradient method* of minimization of functional (3.33) takes the form $u^{n+1} = u^n - \tau\xi^n$, $\tau > 0$.

is usually referred to as the *Krylov subspace*. Specifically, in the k-step
method of steepest descent, the new iterate u^{n+1} is chosen as the best (in
the sense of the Euclidean space $H(D)$) approximation to the solution u
among elements of the plane X_k, that is, among vectors

$$v \equiv u^n - \alpha_0 \xi^n - \alpha_1 A \xi^n - \ldots - \alpha_{k-1} A^{k-1} \xi^n$$

with k free parameters α_i. They correspond to the residuals

$$r^n - \alpha_0 L \xi^n - \alpha_1 L A \xi^n - \ldots - \alpha_{k-1} L A^{k-1} \xi^n = r',$$

where $r' \equiv r^n - \alpha_0 \mathcal{A} r^n - \alpha_1 \mathcal{A}^2 r^n - \ldots - \alpha_{k-1} \mathcal{A}^{k-1} r^n$. Therefore, the best
choice must also define the best approximation to r^n among vectors r'.

  Theorem 20. Let conditions of Theorem 14 with respect to the operators
$L, B_r, r = 1, 2$ *be satisfied and suppose the k-step method of steepest descent
for solution of the system* $Au = g$ *is used. Then*

$$\|z^{n+1}\|_D \le q_k \|z^n\|_D, \quad \|r^{n+1}\|_{B_1^{-1}} \le q_k \|r^n\|_{B_1^{-1}}. \tag{3.36}$$

  Proof. The main tool to obtain estimates of the rate of convergence of
the iterative variational methods is their comparison with the corresponding
Richardson iteration with the same initial iterate. Thus, we consider a given
u^n as initial guess v^0 and perform an imaginary cycle of k iterations (3.19)
with the iterative parameters from (3.21) defined by the optimal constants
δ_0^* and δ_1^*; the last iterate in this modified Richardson method is denoted
by v^{n+1}. Then Theorem 14 implies that $\|v^{n+1} - u\|_D \le q_k \|z^n\|_D$. Note
that $v^{n+1} \in u^n + U_k(u^n)$ and that u^{n+1} must be the best approximation to
u in the sense of the Euclidean space $H(D)$ among elements of the plane
X_k. Hence, $\|u^{n+1} - u\|_D \le \|v^{n+1} - u\|_D \le q_k \|z^n\|_D$. Likewise, comparison
of $\|r^{n+1}\|_{B_1^{-1}}$ and $\|L v^{n+1} - f\|_{B_1^{-1}}$ gives the desired estimate for r^{n+1}. $\square$

  These estimates refer to the worst possible case, and very often better
results are observed especially in the few first iterations. Besides, cases of
clustering of the eigenvalues, that is, when large groups of them almost
coincide, are very favorable to these methods (as is sometimes the case for
certain optimal preconditioners).

  Very effective algorithms currently exist for obtaining the desired best
approximations u^{n+1} among elements of the Krylov subspaces like $U_k(u^n)$
when k is relatively small. They are often referred to as the *modified conjugate gradient* or *Lanczos methods*, and are based on recurrence orthogonalization procedures for constructing an orthonormal basis in the sense of
the Euclidean space $H(D)$ of the subspace at hand (see [41, 64, 244, 252,
277, 353, 432, 433]). They are especially remarkable in that they generate intermediate approximations from subspaces with dimensions equal to

$1, 2, \ldots, k - 1$. If we regard the operator A as a mapping of the Euclidean space $H(B_2) \equiv G$ or $G(A) = H(D)$ into itself (see (3.20)), then we may apply the standard formulas of the method with symmetric and positive operators (see [41, 244, 252, 412, 440]) and write $\xi^0 \equiv Au^0 - g$, $p^0 \equiv \xi^0$,

$$\xi^{m+1} \equiv \xi^m - \alpha_m Ap^m, \; u^{m+1} \equiv u^m - \alpha_m p^m, \; p^{m+1} \equiv \xi^m - \beta_m p^m, \quad (3.38)$$

$$\alpha_m \equiv (\xi^m, p^m)_{B_2} \|p^m\|_D^{-2}, \quad \beta_m \equiv -\|\xi^{m+1}\|_{B_2}^2 \|\xi^m\|_{B_2}^{-2}. \quad (3.39)$$

Also, we have

$$(\xi^m, p^m)_{B_2} = (L^* B_1^{-1} r^m, p^m) = \|\xi^m\|_{B_2}^2 = \|L^* B_1^{-1} r^m\|_{B_2^{-1}}^2, \; m = 0, \ldots,$$

vectors $p^0, \ldots, p^{m-1}$ constitute an orthonormal (in the sense of the Euclidean space $H(D)$) basis of the corresponding Krylov subspace

$$U_k(u_n) \equiv \lim\{\xi^0, A\xi^0, \ldots, A^{k-1}\xi^0\};$$

and $(\xi^i, \xi^j)_{B_2} = 0$ if $i \neq j$ and $= 1$ if $i = j$. Smallness of $\|r^m\|_{B^{-1}}$ may serve as a signal to stop the calculations. Recall that the considered modified conjugate gradient method contains three basic variants connected with symmetrizations of three types (see Lemmas 1–3). It also contains the most known variant corresponding to the case $L = L^* > 0$ and the preconditioned system $B^{-1}Lu = B^{-1}f$ (see [41, 278, 244, 252, 440, 497]).

Consider briefly the case $L = L^* \geq 0$ and $B = B^* \geq 0$ from Lemma 5 and, [36] in particular,

$$\delta_0 B \leq L \leq \delta_1 B, \quad \delta_0 > 0. \quad (3.40)$$

If we use the same $S_B = \text{Im } B = H'$, L', B' from Lemma 5 and Theorem 17 and note that sp $(B')^{-1}L' \subset [\delta_0, \delta_1]$, then the modified conjugate gradient method with $u_0 \in H'$ can be rewritten in the form

$$\xi^{m+1} \equiv \xi^m - \alpha_m (B')^{-1} Lp^m, \; u^{m+1} \equiv u^m - \alpha_m p^m, \; p^{m+1} \equiv \xi^m - \beta_m p^m, \quad (3.41)$$

where $\xi^0 = (B')^{-1} r^0$, $p^0 \equiv \xi^0$,

$$\alpha_m \equiv (\xi^m, p^m)_B \|p^m\|_L^{-2}, \quad \beta_m \equiv -\|\xi^{m+1}\|_B^2 \|\xi^m\|_B^{-2}. \quad (3.42)$$

---

[36]Similar optimal model operators B will be constructed in Chapter 3, e.g., for the Neumann problem for a second-order elliptic equation and for the second boundary value problems in the theory of elasticity.

Observe that $(\xi^m, p^m)_B = (r^m, p^m)$ and $\|\xi^m\|_B^2 = (r^m, \xi^m)$. Then, applying the proofs of Theorems 17 and 20, we obtain our next theorem.

Theorem 21. Suppose the conditions of Lemma 5 are satisfied, $u^0 \in H'$, and the above modified conjugate gradient method is applied. Then $\|z^{n+1}\|_L \leq q_k \|z^n\|_L$.

Some modified gradient methods may be constructed for nonlinear problems as well. For illustration, we confine ourselves to the case of the variational problem in the Euclidean space $H(B)$ of the form

$$u = \arg \min_v \Phi_3(v) \equiv \|Lv - f\|_{B^{-1}}^2 \qquad (3.43)$$

with a continuously differentiable operator L. Then the modified conjugate gradient method takes the form $\eta^0 \equiv (L'_{u^0})^* B^{-1} r^0$, $p^0 \equiv B^{-1} r^0$, $u^{m+1} \equiv u^m - \tau_m p^m$, where $m \geq 0$, $\tau_m = \arg \min \Phi(u^m - \tau p^m)$;

$$\eta^{m+1} \equiv (L'_{u^n})^* B^{-1} r^m, \quad p^{m+1} \equiv B^{-1} \eta^{m+1} - \beta^m p^m,$$

with $\beta^m \equiv -[(\eta^{m+1}, \eta^{m+1} - \eta^m)_{B^{-1}}]\|\eta^m\|_{B^{-1}}^{-2}$ (see [244, 412]). Of course, this optimal τ can be found only approximately and there are many different ways to do it. Other methods deal with similar variational problems but under additional simplifications. For example, in the case of problem (3.43) with a linear operator L, we may try to find $u^{n+1} = u^n - \tau B^{-1} r^n$ such that $\tau = \arg \min_\tau \|r^n - \tau LB^{-1} r^n\|_{B^{-1}}^2$. This method is sometimes referred to as the *modified method of minimal residuals,* and it gives

$$\tau = (LB^{-1} r^n, B^{-1} r^n)\|LB^{-1} r^n\|_{B^{-1}}^{-2}.$$

Obviously, estimates of Theorem 13 for $\|r^{n+1}\|_{B^{-1}}$ remain valid.

To summarize our results about the modified gradient methods, we conclude that they have both assets and liabilities compared with the modified Richardson methods leading to linear reduction error operators. Richardson methods are obviously simpler and very competitive when we use them (with sufficently accurate bounds for the corresponding spectrum) as inner iterations in some sophisticated constructions of model operators; there is also the possibility of applying Chebyshev polynomials with weights (see [355]) to take into account some additional information about the spectrum. On the other hand, modified gradient methods do not need to have a priori information about the spectrum and very often in the first few iterations demonstrate radical improvements in the accuracy of initial approximations. For some complicated problems, it might be useful to apply a combined strategy and to pass to the modified Richardson method when covergence becomes slow.

3.9. Nonsymmetric model operators. Here, we consider now some important and in many respects unresolved aspects of using nonsymmetric model operators. As we have already emphasized in § 1, problems with dominant antisymmetric part L_a of the linear operator L can be found in many branches of mathematical physics. If we use a nonsymmetric model operator B and replace the original system $Lu = f$ by $B^{-1}Lu = B^{-1}f$, then it seems natural to symmetrize it in the Euclidean space H and to work with the system

$$Au \equiv L^*(B^{-1})^*B^{-1}Lu = L^*(B^{-1})^*B^{-1}f. \qquad (3.44)$$

If we note that $(B^{-1})^*B^{-1} = (BB^*)^{-1}$ and introduce $D \equiv (BB^*)^{1/2}$, then we may rewrite it as $Au \equiv L^*D^{-2}Lu = L^*D^{-2}f$ with a symmetric model operator D. Thus, of paramount importance are estimates (3.9), (3.9') with B^2 replaced by D^2, since they lead to inequalities $\delta_0 I \leq L^*D^{-2}L \leq \delta_1 I$. Moreover, the rate of convergence of the modified Richardson method

$$u^{n+1} - u^n = -\tau_n L^*(B^*)^{-1}[B^{-1}(Lu^n - f)] \qquad (3.45)$$

is completely defined by these estimates. A similar generalization holds for Theorem 10. But the major obstacle for such problems is the need to construct model operators B leading not only to the desired estimates, but also to easily solved systems $Bv = g$; in the case of (3.44), (3.45), we even must have easily solved systems with the operator B^*; and the same applies to attempts to construct operators B such that inequalities (3.8), (3.8') are valid and to use iterations

$$B^*B(u^{n+1} - u^n) = -\tau_n L^*(Lu^n - f),$$

or their generalizations of type (3.10) with B^2 replaced by B^*B. Although much attention (see [225, 250, 440, 490]) has been paid to this topic, the main developments are no doubt yet to occur. For difficult problems, use of an artificial parameter $\lambda \in [0, 1]$ before L_a or a corresponding nonlinear part of the operator might be useful. In this case, we deal the parameterized system $L_s u + \lambda L_a u = f$, and the method of continuation with respect to this parameter can perhaps be combined with coarse grid continuation (see § 4).

§ 4. Solution of grid systems and asymptotic estimates of the required computational work

4.1. Estimates of the computational work in the modified method of the simple iteration. Theorems of convergence for iterative methods (3.1), (3.10)–(3.12) indicate that

$$\|z^{n+1}\|_D \leq \rho\|z^n\|_D, \quad \rho < 1, \qquad (4.1)$$

where $z^n \equiv u^n - u$, $u^n \in S_D(u; r)$ and $D = B, B^2, I$. Rewriting these methods in the generic form $u^{n+1} = F(u^n)$, then to carry out a more general analysis of convergence, we consider *perturbed iterative methods* of the form $v^{n+1} = F(v^n) + \zeta^n$, and the basic recurrence relation

$$\|v^k - u\|_D \le \rho \|v^n - u\|_D + \|\zeta^n\|_D.$$

Here, ζ^n may appear as a result of rounding errors or some other approximations made in the implementation of the iterative method (e.g., when inner iterations have been applied for approximate solution of certain systems; in § 3.4 we shall discuss other ways to deal with such problems on the basis of an appropriate change in the form of the model operator).

If all $v^n \in S_D(u; r)$, then this recurrence estimate leads to

$$\|v^k - u\|_D \le \rho^k \|v^0 - u\|_D + (1 - \rho)^{-1} \max_n \|\zeta^n\|_D.$$

The goal here is to analyze in detail what such a procedure requires of computational work to ensure that the desired accuracy is achieved. This means that we have to pay special attention to the dependence on h (i.e., on the grid) both of the number of unknowns $N = \dim H$ and of the convergence factor bound ρ in (4.1); as usual, we denote by $W(a) \equiv W$ the number of arithmetic operations performed by a given algorithm a; output of the algorithm starting with an initial vector v is denoted by $a(v)$.

Theorem 1. For method (3.1), suppose that the conditions of either Theorem 3.1 or 3.2 are satisfied, confirming estimate (4.1) with $D = B$ when $u^n \in S_r \equiv S_B(u; r)$. Consider algorithm a_1 for finding ε_L- approximations in the Euclidean space $H(B^{-1})$ to the vector $L(v)$, with an arbitrary given vector $v \in S_r$, and suppose $W(a_1) \le W_1(N, \varepsilon_L)$. Consider algorithm a_2 for finding $\varepsilon_{B^{-1}}$-approximations in the Euclidean space $H(B)$ to solutions of

$$Bv = g \tag{4.2}$$

with an arbitrary g with $\|g\|_{B^{-1}} \le 2r$, and suppose $W(a_2) \le W_2(N, \varepsilon_{B^{-1}})$. Assume $\varepsilon_\xi \equiv \varepsilon_{B^{-1}} + \tau \varepsilon_L \le r(1 - \rho)$ and

$$\varepsilon' \equiv \varepsilon - \varepsilon_\xi (1 - \rho) - 1 > 0. \tag{4.3}$$

Suppose algorithm a is constructed by using a_1 and a_2 with the modified method of the simple iteration (3.1) as follows. Given a vector $v^n \in S_r$, define a new iterate $v^{n+1} \equiv a_3(v^n)$ by the following rules:

1. *Find $a_1(v^n) \equiv L(v^n) + \xi_L^n$ with $\|\xi_L^n\|_{B^{-1}} \le \varepsilon_L$.*

2. *Find $g \equiv \tau[L(v^n) + \xi_L^n - f]$.*

3. For system (4.2), find $a_2(g) \equiv w \equiv v + \xi_{B-1}^n$ with $\|\xi_{B-1}^n \leq \varepsilon_{B-1}$.

4. Find $v^{n+1} = v^n - w$.

Then $K(\varepsilon, \rho)$ iterations of algorithm a produce a vector v satisfying

$$\|v - u\|_D < \varepsilon \tag{4.4}$$

with computational work $W(a) \leq K(\varepsilon, \rho)[3N + W_1(N, \varepsilon_L) + W_2(N, \varepsilon_{B-1})]$, where

$$K(\varepsilon, \rho) \equiv [\frac{|\ln \frac{\varepsilon'}{r}|}{|\ln \rho|} + 1]. \tag{4.5}$$

Proof. It is a simple matter to verify that

$$B(v^{n+1} - v^n) = -\tau(L(v^n) - f) + \xi^n, \tag{4.6}$$

where $\xi^n \equiv B\xi_{B-1}^n - \tau\xi_L^n$. Hence, $\|v^{n+1} - u\|_B \leq \|v^n - u - \tau B^{-1}(L(v^n) - L(u))\|_B + \|\xi^n\|_{B-1} \leq \rho\|v^n - u\|_B + \varepsilon_\xi \leq \rho r + \varepsilon_\xi \leq r$ and all v^{n+1} in (4.6) belong to S_r if $v^0 \in S_r$. Therefore,

$$\|v^{n+1} - u\|_B \leq \rho^{n+1}r + \frac{\varepsilon_\xi}{1 - \rho}. \tag{4.7}$$

Since $B(u^{n+1} - v^n) = g_1 \equiv -\tau(L(v^n) - f)$ and $\|g_1\|_{B-1} \leq \|u^{n+1} - u\|_B + \|v^n - u\|_B \leq (1 + \rho)r$, we have $\|g\|_{B-1} \leq (1 + \rho)r + \tau\varepsilon_L < 2r$. Therefore, algorithms a_1 and a_2 require the computational work specified in the conditions of the theorem, and the computational work in one iteration of algorithm a is estimated as

$$W(a_3) \leq 3N + W_1(N, \varepsilon_L) + W_2(N, \varepsilon_{B-1}).$$

Inequalities (4.7) and (4.3) indicate that we can achieve (4.4) with $v \equiv v^k$ after performing $k = K(\varepsilon; \rho)$ iterations (recall that $[x]$ refers to an integer part of x). □

The analysis of the perturbed iterative methods (3.10)–(3.12) may be carried out in much the same way. For this reason, we confine ourselves to the case of iterative method (3.12).

Theorem 2. *For the method (3.12), suppose the conditions of Theorem 3.11 with $B_1 = B_2 = B$ are satisfied, confirming estimate (4.1) with $D = B$ where $u^n \in S_r \equiv S_B(u; r)$. Consider the respective algorithms a_1 and a_2 for finding ε_L-approximations in the Euclidean space $H(B^{-1})$ to the vector $L(v)$ with a given vector $v \in S_r$ and for finding ε_{B-1}-approximations in the Euclidean space $H(B)$ to solutions of systems (4.2) with*

$$\|g\|_{B-1} \leq \max\{\delta_1^{1/2}r + \varepsilon_L; \delta_1^{1/2}(\delta_1^{1/2}r + \varepsilon_L + \varepsilon_{B-1}) + \varepsilon'\} \equiv r' \tag{4.8}$$

that satisfy the estimates $W(a_1) \le W_1(N, \varepsilon_L)$ and $W(a_2) \le W_2(N, \varepsilon_{B^{-1}})$. Consider algorithm a_3 for finding ε^*-approximations in the Euclidean space $H(B^{-1})$ to $(L'_v)^* w$ with an arbitrary given $v \in S_r$ and w with $\|w\|_B \le \delta_1^{1/2} r + \varepsilon_L + \varepsilon_{B^{-1}} \equiv r^*$, and suppose $W(a_3) \le W_3(N, \varepsilon^*)$. Assume that

$$\varepsilon_\xi \equiv \tau\{\varepsilon_{B^{-1}} + \varepsilon^* + \delta_1^{1/2}(\varepsilon_{B^{-1}} + \varepsilon_L)\} \le \frac{r}{1-\rho} \qquad (4.9)$$

and that (4.3) is satisfied. Suppose algorithm a is constructed by using a_1, a_2, and a_3 in the method (3.12) as follows. Given a vector $v^n \in S_r$ define a new iterate $v^{n+1} \equiv a_4(v^n)$ by the following rules:

1. *Find $a_1(v^n) \equiv L(v^n) + \xi_L^n$ with $\|\xi_L^n\|_{B^{-1}} \le \varepsilon_L$.*

2. *Find $g \equiv a_1(v^n) - f$.*

3. *For (4.2), find $a_2(g) \equiv w^n \equiv B^{-1}g + \xi_1^n$ with $\|\xi_1^n\|_{B^{-1}} \le \varepsilon_{B^{-1}}$.*

4. *Find $w^* \equiv a_3(w^n) \equiv (L'v)^* w^n + \eta^n$ with $\|\eta^n\|_{B^{-1}} \le \varepsilon^*$.*

5. *For (4.2) with $g = w^*$, find $\zeta^n \equiv a_2(w^*) \equiv B-1w^n + \xi_2^n$ with $\|\xi_2^n\|_{B^{-1}} \le \varepsilon_{B^{-1}}$.*

6. *Find $v^{n+1} = u^n - \tau\zeta^n$.*

Then $K(\varepsilon, \rho)$ iterations of algorithm a produce a vector v satisfying (4.4) with $D = B$ and required computational work bounded according to $W(a) \le K(\varepsilon, \rho)\{3N + W_1(N, \varepsilon_L) + W_3(N, \varepsilon^*) + 2W_2(N, \varepsilon_{B^{-1}})\}$, where $K(\varepsilon, \rho)$ is given by (4.5).

Proof. The rules of algorithm a_4 produce $v^{n+1} = a_4(v^n)$ at the nth iteration and it is a simple matter to verify that

$$v^{n+1} - v^n = -\tau\{B^{-1}(L'_{v^n})^* B^{-1}(L(v^n) - f)\} + \xi^n, \qquad (4.10)$$

with $\xi^n = -\tau[\xi_2^n + B^{-1}\eta^n + B^{-1}(L'_{v^n})^*(\xi_1^n + B^{-1}\xi_L^n)]$ and

$$X \le \|\xi_2^n\|_B + \|\eta^n\|_{B^{-1}} + \|B^{-1/2}(L'_{v^n})^* B^{-1/2}\|(\|\xi_1^n\|_B + \|\xi_L^n\|_{B^{-1}}), \quad (4.11)$$

where $X \equiv \|\xi^n\|_B / \tau$. The operator $B^{-1/2}(L'_{v^n})^* B^{-1/2}$ is the adjoint of $B^{-1/2}L'_{v^n}B^{-1/2}$, so their norms coincide with $T \equiv \|L'_{v^n}\|_{H(B) \to H(B^{-1})} \le \delta_1^{1/2}$. Hence, $\|\xi^n\|_B \le \varepsilon_\xi$ (see (4.9), (4.11)) and $\|v^{n+1} - u\|_B \le \rho\|v^n - u\| + \varepsilon_\xi \le \rho r + \varepsilon_\xi \le r$ (see(4.1), (4.10)), whence (4.7) follows with $D = B$. What remains to be done is to justify applicability of algorithms a_1, a_2, and a_3 with estimates of the computational work specified in the conditions of the theorem. Since all $v \in S_r$, then applicability of algorithm a_1 is obvious;

regarding algorithms a_2 and a_3, we note that
$\|g\|_{B^{-1}} \le \|B^{-1/2} L'_{u+\theta(v^n-u)} B^{-1/2}\| \|v^n - u\| + \varepsilon_L$ and, hence, $\|g\|_{B^{-1}} \le \delta_1^{1/2} r + \varepsilon_L \le r'$ with r' from (4.8). Similarly,

$$\|w^n\|_B \le \|g\|_{B^{-1}} + \varepsilon_{B^{-1}} \le r^*, \quad \|w^*\|_{B^{-1}} \le \delta_1^{1/2} \|w^n\|_B + \|\eta^n\|_B \le r'.$$

Therefore, all necessary localizations for the algorithms are proved. □

Theorem 3. Let $\varepsilon \asymp N^{-\alpha}$ and $\alpha > 0$, and suppose algorithms a_1, a_2 from Theorem 1 and a_1, a_2, a_3 from Theorem 2 lead to ε-approximations with estimates of computational work $W \le Kg(N)$. Suppose ρ in (4.1) is independent of N. Then the algorithms a constructed in these theorems obtain ε-approximations in the sense of the Euclidean space $H(D)$ to the solution of (2.1) with estimates $W(a) \le K_1 g(N) \ln N$.

Proof. The proofs given for Theorems 1 and 2 indicate that permissible bounds for the errors at various stages of the respective algorithm a, that is, $\varepsilon_L, \varepsilon_{B_{-1}}, \varepsilon^*$, may be of the same asymptotic order as the required accuracy ε. Thus, each of the iterations requires computational work with estimates $W \le K_2 g(N)$. Since ρ in (4.1) does not depend on h or N, the number of performed iterations is estimated as $k = O(|\ln \varepsilon|) = O(N)$. □ [37]

4.2. Modified classical iterative methods with spectrally equivalent model operators. We start with linear operators.

Definition 1. Operators $L \in \mathcal{L}^+(H)$ and $B \in \mathcal{L}^+(H)$ are connected by *relationship C^0* if they are spectrally equivalent.

Definition 2. Operators $L \in \mathcal{L}(H)$ and $B \in \mathcal{L}^+(H)$ are connected by relationship $C^{0,0}$ if the conditions of Theorem 3.4 are satisfied with constants δ_0 and δ_1 independent of h. [38]

Definition 3. Operators $L \in \mathcal{L}(H)$ and $B \in \mathcal{L}^+(H)$ are called connected by *relationship $C^{0,1}$* if the conditions of Theorem 3.5 are satisfied with constants σ_r independent of h, $r = 0, 1, 2$.

In the case of nonlinear operators L, we formulate more general relationships $C^k(u, r)$ and $C^{0,m}(u, r)$ with $r > 0$ being a constant independent of h. They enable us to construct the modified method of the simple iteration with the solution $u \equiv u_h$ of (2.1) being an attractor of all $u \in S_r$.

---

[37] We stress that Theorems 1–3 with estimates of W from (0.3.2), (0.3.3) lead to nearly asymptotically optimal algorithms for solution of systems (2.1); similar algorithms can be obtained on the basis of the iterative methods investigated in § 3.

In the following subsection, we formulate conditions for independence of ρ in (4.1) with respect to h for some of these methods in terms of special relations between the given operator $L \equiv L_h$ and a model grid operator $B \equiv B_h$, considered as mappings of the Euclidean space $H \equiv H_h$ into itself. Recall that we deal with $h \in \{h\}$ corresponding to H_h with $\dim H_h \to \infty$ as $|h| \to 0$.

[38] If the conditions of Theorem 3.8 are satisfied, then the operators BL and B^2 are connected by the same relationship $C^{0,0}$.

Definition 4. Operators L and $B \in \mathcal{L}^+(H)$ are called connected by *relationship* $C^0(u, r)$ if the conditions of Theorems 3.1 or 3.2 are satisfied with functions $\delta_0(t)$ and $\delta_1(t)$ independent of h.

It is clear that similar conditions of Theorems 3.7 or 3.6 imply that operators BL and B^2 are connected by the relationship $C^0(u, r)$.

Definition 5. Operators L and $B \in \mathcal{L}^+(H)$ are connected by *relationship* $C^{0,1}(u, r)$ if the conditions of Theorem 3.3 are satisfied with functions $\sigma_r(t), r = 0, 1, 2$, independent of h. [39]

Definition 6. Operators L and $B \in \mathcal{L}^+(H)$ are called connected by *relationship* $C^1(u, r)$ if the conditions of Theorems 3.9 are satisfied, confirming the validity of (4.1) with ρ independent of h and $D = B^2$. More precisely, the functions $\delta_r(t), r = 0, 1, 2$, must be independent of h and the function $\delta_2(t)$ must be sufficiently small.

Similar specifications are assumed to hold for the next two definitions.

Definition 7. Operators L and $B \in \mathcal{L}^+(H)$ are called connected by *relationship* $C^2(u, r)$ if the conditions of Theorems 3.10 are satisfied, confirming the validity of (4.1) with ρ independent of h and $D = I$.

Definition 8. Operators L and $B \in \mathcal{L}^+(H)$ are called connected by *relationship* $C^3(u, r)$ if the conditions of Theorems 3.11 are satisfied, confirming the validity of (4.1) with ρ independent of h and $D = B$. [40]

The spectral equivalence of any of the four pairs of operators L and B, L^*L and B^*B, $L^*B^{-2}L$ and I, $L^*B^{-1}L$ and B enables us to use modifications of the classical iterative methods discussed in § 3. Their rates of convergence are determined by the quotient (condition number) $\delta \equiv \delta_1/\delta_0$ and do not depend on h. In the case of the modified Richardson method (3.19) with any of the three main variants $A = A_1, A = A_2$, or $A = A_3$, ε-accuracy in (4.4) is achieved if

$$k \approx [\frac{\delta^{1/2}}{2|\ln \frac{2}{\varepsilon r}|}] \tag{4.12}$$

under the condition that $\|u - u^0\|_D \leq r$. If we use a fixed number k of the iterations in a cycle and repeat these cycles m times, we can again obtain

---

[39] It is easy to see that if L and B are connected by the relationship $C^{0,1}(u, r)$, then they are also connected by the relationship $C^0(u, r)$; in the case of a continuously differentiable operator L, the converse is also valid. Both of these relationships give sufficient conditions for convergence of the method (3.1); what is really important from the practical point of view is a selection of the iterative parameter τ.

[40] We stress again that relationships $C^k(u, r)$ enable us to construct methods (3.1), (3.10)–(3.12) that lead to a contraction mapping on certain balls $S_B(u; r)$ with respect to their centers. Under additional conditions, it is even possible to construct contraction mappings of the above balls into themselves. In particular, for linear operators L, it is possible to obtain contraction mappings of the whole spaces H and relationships $C^k(u; r)$ become simpler relationships C^k.

(4.4) with

$$m = [|\log_{q_k}(\varepsilon r)| + 1]. \tag{4.13}$$

The same applies to the modified conjugate gradient methods with cycles consisting of k iterations. Of course, estimates (4.12), (4.13) are valid only when we neglect rounding errors; otherwise a special investigation of these perturbed methods must be carried out; and we have already cited some of the important existing results in § 3 (see also [374]).

4.3. Continuation methods (multigrid acceleration of a basic iterative algorithm). A very important class of stationary problems in mathematical physics in general and in structural mechanics in particular is characterized by the presence of a parameter, say, $\lambda \in [0, 1]$. As a result of discretization, the corresponding finite dimensional problems take the form

$$L(u(\lambda), \lambda) = f(\lambda), \quad u(\lambda) \in S, \quad 0 \leq \lambda \leq 1, \tag{4.14}$$

and we are especially interested in tracing their *solution branches*, that is, in sufficiently accurate approximations of the solution curve $u(\lambda), \lambda \in [0, 1]$. This problem also takes a discretized form if we confine ourselves to a finite number of problems

$$L_l(u_l) \equiv L(u_l; \lambda_l) = f_l, \quad u_l \in S_l, \quad l = 0, 1, \ldots, p, \tag{4.15}$$

defined on a one-dimensional grid

$$0 = \lambda_0 < \lambda_1 < \ldots < \lambda_p = 1 \tag{4.16}$$

on the original interval $[0, 1]$. [41]

Continuation methods for solving (2.1) are very often used with respect to an artificial parameter λ introduced in such manner that in (4.15), (4.16) $L(v, 1) = L(v), f_p = f$, and the initial problem (4.15) with $l = 0$ is easily solved. Such methods are extremely useful in situations where sufficiently

---

[41] There are some complicated cases when small changes in a parameter may result in rather dramatic changes in solutions; they usually are referred to as *bifurcation problems*, and have attracted a lot of attention (see [8, 64, 300, 380]). Simpler situations lead to the expectation that solutions must be continuously dependent on the parameter and, consequently, an obtained good approximation to the solution $u(\lambda_l)$ may serve as a reasonable initial guess (iterate) in an iterative method used for solving the system with the next parameter value λ_{l+1}. Such procedures are usually called *continuation methods*, or *predictor-corrector methods* because they use predicted initial iterates taken from the problem using the previous parameter value and correct it by performing a number of iterations, say k_{l+1}, of the iterative method at hand; sometimes, when there is more substantial smoothness of solutions with respect to the parameter, some *extrapolation procedures* involving a few computed approximations might be useful (see [8, 64, 214, 397]).

accurate approximations to the solutions are necessary to ensure the convergence of the applied iterative method, and, in particular, to trace a desired solution branch.

Classical continuation methods work with the same Euclidean space H and change only the form of the operator equations (4.15). While solving hard grid systems, it seems reasonable to connect the parameter with the grid itself. This very natural idea, which has been used in many applications for a long time, is often referred to as coarse grid continuation (see § 0.3). The associated parameter l in (4.15) characterizes the parameter h_l from the chosen set $\{h\}$ in the grid systems (2.1) and (1.2). More precisely, we suppose that all grids with the parameter $l = 0, 1, \ldots, p$ are quasiuniform and parameterized by the numbers

$$h_l = 2^{-l} h_0 = 2^{p-l} h, \quad h \equiv h_p, \quad l = 0, \ldots, p. \tag{4.17}$$

Problem (4.15) with the grid parameter h_l is usually called a problem of the lth *level of refinement*. We regard this problem as an operator equation in a Euclidean space H_l with $S_l \subset H_l$ and assume that

$$N_l = \dim H_l \asymp h_l^{-d} = h^{-d} 2^{(l-1)d}, \tag{4.18}$$

where d stands for the dimension of the Euclidean space $\mathbf{R}^d$ containing the original region associated with the given elliptic boundary value problem. For each such problem, we may apply a certain iterative method (let us say a basic iterative algorithm) leading after k iterations to the estimate

$$\|v_l^k - u_l\|_{D_l} \leq q^k \|v_l^0 - u_l\|_{D_l} + \frac{\nu_l}{1-q}, \tag{4.19}$$

with the number $q < 1$ responsible for the rate of convergence of the method, with the number $\nu_l \geq 0$ connected to possible perturbations (see Theorems 1 and 2), and with the operator $D_l \in \mathcal{L}^+(H_l)$. Let $W(N_l, \nu_l)$ serve as an upper bound on the number of arithmetic operations required to perform one iteration of the method and I_l^{l+1} be a linear operator mapping the Euclidean space H_l onto the Euclidean space H_{l+1} associated with the grid of the $(l+1)$th level of refinement (usually, $I_l^{l+1} v_l$ is computed by some interpolation procedure). Then coarse grid continuation (the multigrid acceleration of the basic iterative algorithm) is determined by an initial iterate $v_0 \in H_0$, a number k of iterations performed on each level, and relations

$$v_{l+1}^0 = I_l^{l+1} v_l \in S_{l+1}, \quad l = 0, \ldots, p-1, \quad v_l \equiv v_l^k. \tag{4.20}$$

In what follows, we regard the original problem (1.1) in the Hilbert space and its discretizations (4.16) as correct ones with solutions $u \in G, u_l \in H_l$;

systems (4.16) correspond to systems (1.2) and, for each H_l, we define an operator $p_l \in \mathcal{L}(H_l; G)$ such that $\hat{v}_l \equiv p_l v_l \in G$; and we write $\|u\|_G \equiv \|u\|$. Suppose that

$$\|u - \hat{u}_1\| \leq Kh_l^\gamma, \quad \gamma > 0, \tag{4.21}$$

$$\|\hat{v}_1 - p_{l+1}(I_l^{l+1} v_l)\| \leq K_1 h_l^\gamma, \quad \forall v_l \in S_l, \quad \nu_l \leq K_2 h_l^\nu, \tag{4.22}$$

and that the operators D_l are such that

$$0 < \alpha_0 \leq \frac{\|v_l\|_{D_l}}{\|\hat{v}_l\|} \leq \alpha_1, \quad \forall v_l \neq 0. \tag{4.23}$$

We also require that

$$q^k \frac{\alpha_1}{\alpha_0} \leq \rho^k \equiv \sigma < 2^{-\gamma}, \quad \sigma^p \frac{K_0}{\alpha_0} \leq \frac{\varepsilon}{4}, \quad \frac{K_3 h^\gamma}{1 - \sigma 2^\gamma} \leq \frac{\varepsilon}{4}, \tag{4.24}$$

with $K_3 \equiv \sigma[K + 2^\gamma(K_1 + K)] + (K_2(1-q))/\alpha_0$. It is not difficult to verify that, given numbers $q, \alpha_1, \alpha_0, \gamma, K, K_0, K_1, K_2$, we can select the numbers k and p such that inequalities (4.24) are satisfied with $k \asymp 1, p \asymp |\ln \varepsilon|$, that is, there are $p \asymp |\ln \varepsilon|$ levels of refinement and only a finite number k of iterations on each level.

Theorem 4. *Let $\varepsilon > 0$ be the desired accuracy for finding the solution u of the original problem (1.1) in the Hilbert space G. Suppose, for the discretized problems (4.15), that we use a basic iterative method on each level that satisfies (4.19) with computational work $W(N_l) \leq K' N_l |\ln h_l|^r$, $r \geq 0$ for each iteration. Let conditions (4.21–(4.24) be satisfied with $k \asymp 1, p \asymp |\ln \varepsilon|$. Then multigrid acceleration of this iterative method yields the estimate $\|\hat{v}_p - u\| \leq \varepsilon$ with computational work*

$$W = O(\varepsilon^{-d/\gamma} |\ln \varepsilon|^r) = O(h^{-d} |\ln h|^r).$$

Proof. From (4.19)–(4.24), we have $x_{l+1} \equiv \hat{v}_{l+1} - \hat{u}_{l+1} \leq \sigma \eta_{l+1} + \nu_{l+1} [\alpha_0(1-q)]^{-1}$, where $\eta_{l+1} \equiv \|p_{l+1}(I_l^{l+1} v_l) - \hat{u}_{l+1}\|$. We estimate η_{l+1} by

$$\eta_{l+1} \leq \|p_{l+1}(I_l^{l+1} v_l) - \hat{v}_l\| + x_l + \|\hat{u}_l - u\| + \|u - \hat{u}_{l+1}\|.$$

Thus, $\eta_{l+1} \leq x_l + (K_1 + K)h_l^\gamma + Kh_l^\gamma$, which yields the key recurrence relation

$$x_{l+1} \leq \sigma x_l + K_3 h_{l+1}^\gamma. \tag{4.25}$$

Since $h_p^\gamma + \sigma h_{p-1}^\gamma + \ldots + \sigma^{p-1} h_1^\gamma \leq h^\gamma (1 - \sigma 2^\gamma)^{-1}$, then multiplying (4.25) by σ^{p-l} and taking sums of the obtained inequalities implies that

$$x_p \leq \sigma^p K_0 / \alpha_0 + K_3 h^\gamma (1 - \sigma 2^\gamma)^{-1} \leq \varepsilon/2.$$

This leads to the estimate $\|\hat{v}_p - u\| \leq \|\hat{u}_p - u\| + \varepsilon/2 \leq \varepsilon$. Estimation of the required computational work gives (see [181, 183]) $W \leq K_4[h^{-d}|\ln h|^r + (2h)^{-d}|\ln 2h|^r + \ldots + h_0^{-d}|\ln h_0|^r]$. Thus, $W \leq K_5 h^{-d}[|\ln h|^r(1 + 2^{-d} + 4^{-d} + \ldots) + (\ln 2)^r[2^{-d} + 2^r 4^{-d} + \ldots]] \leq K_6 h^{-d}|\ln h|^r$, and the desired estimates hold. □ [42]

§ 5. Block elimination of unknowns; Schur matrices; cooperative operators

5.1.   Block elimination of unknowns.   Block elimination of unknowns, one of the major tools of linear algebra, is in one form or another being widely used in many contemporary direct and iterative methods. Thus, it makes sense to study separately the most general results for this method, with particular emphasis on application to constructing effective preconditioners in the case of grid systems generated by PGMs.

Consider a linear system $Au = f$ with $u \in H \equiv \mathbf{R}^n, f \in H$, and $A \in \mathbf{R}^{n \times n}$, [43] and suppose that it may be written in the block form

$$Au \equiv \begin{bmatrix} A_{1,1} & A_{1,2} \\ A_{2,1} & A_{2,2} \end{bmatrix} \begin{bmatrix} u_1 \\ u_2 \end{bmatrix} = \begin{bmatrix} f_1 \\ f_2 \end{bmatrix}, \tag{5.1}$$

where $u \equiv [u_1, u_2]^T \in H \equiv H_1 \times H_2, u_r \in H_r, \dim H_r \equiv n_r, r = 1, 2, n_1 + n_2 = n$, and the block $A_{1,1} \in \mathbf{R}^{n_1 \times n_1}$ is invertible and such that $A_{1,1}v = g$ is easily solved. Then

$$A_{1,1}u_1 = f_1 - A_{1,2}u_2, \tag{5.2}$$

and elimination of u_1 leads to the system

$$S_2(A)u_2 \equiv S_2 u_2 \equiv (A_{2,2} - A_{2,1}A_{1,1}^{-1}A_{1,2})u_2 = g_2, \tag{5.3}$$

where

$$g_2 \equiv f_2 - A_{2,1}A_{1,1}^{-1}f_1. \tag{5.4}$$

The original system is equivalent to system (5.3), (5.2) with g_2 from (5.4), which also may be written in the block-triangular form

$$A_t u \equiv \begin{bmatrix} A_{1,1} & A_{1,2} \\ 0 & S_2 \end{bmatrix} \begin{bmatrix} u_1 \\ u_2 \end{bmatrix} = \begin{bmatrix} f_1 \\ g_2 \end{bmatrix}.$$

Thus, the elimination procedure (sometimes referred to as the *tearing* or *bordering method*) may be reduced to the following stages:

---

[42] Various generalizations of this theorem are possible with different numbers k_l instead of k, different refinement ratios, and other estimates of the computational work like $W(N) = O(h^{-d} \ln |\ln h|)$.

[43] Here we prefer to write n, n_r instead of N, N_r.

1. Solve the system $A_{1,1}v_1 = f_1$ and evaluate the vector $g_2 = f_2 - A_{2,1}v_1$.

2. Solve the matrix equation $A_{1,1}X = A_{1,2}$ (involving the solution of n_2 systems with the same matrix $A_{1,1}$ and right-hand sides coinciding with the $(n_1 + k)$th column of A, $k = 1, \ldots, n_2$, which is perfectly suited for implementation on vector and parallel computers).

3. Find the matrix $S_2 = A_{2,2} - A_{2,1}X$.

4. Solve the system $S_2 u_2 = g_2$.

5. Evaluate the vector $g_1 = f_1 - A_{1,2}u_2$.

6. Solve the system $A_{1,1}u_1 = g_1$.

Required computational work may be characterized as $W \approx (n_2 + 2)W_1 + W_2$, where W_1 and W_2 are upper bounds for the computational work required to solve the systems with matrices $A_{1,1}$ and S_2, respectively. [44]

5.2. Basic properties of Schur matrices. It will be important to analyze properties of the Schur matrix S_2 on the basis of given properties of the original matrix A. To this end, for each $u_2 \in H_2$, we introduce a *prolongation operator* defined by

$$p(u_2) \equiv [p_1(u_2), u_2]^T \in H,$$

where $p_1(u_2) \equiv -A_{1,1}^{-1}A_{1,2}u_2 \in H_1$.

Lemma 1. *For each $u_2 \in H_2$ and each $v \equiv [v_1, v_2]^T \in H$, we have representations: $(S_2 u_2, v) = (Ap(u_2), v)$; $(S_2 u_2, u_2) = (Ap(u_2), p(u_2))$; if, additionally, $A^* = A, A_{1,1} > 0$, then*

$$(S_2 u_2, u_2) = \min_{u_1}(Au, u). \tag{5.5}$$

---

[44] The described procedure may be connected with a *block-triangular factorization of the matrix*

$$A = \begin{bmatrix} A_{1,1} & 0 \\ A_{2,1} & S_{2,2} \end{bmatrix} \begin{bmatrix} I_1 & A_{1,1}^{-1}A_{1,2} \\ 0 & I_2 \end{bmatrix}.$$

The matrix $S_2(A) \equiv S_2 \equiv A/A_{1,1}$ is often referred to as the *Schur complement* to the block $A_{1,1}$, or simply the *Schur matrix*.

We emphasize that this algorithm as posed is well suited only for blocks with a relatively small number n_2, and in many contemporary variants, dealing with large n_2, stages 2–4 are replaced by a separate procedure leading directly to $u_2 = S_2^{-1}g_2$ without computing the matrix S_2. This is achieved by the application of some inner iterations (we shall discuss this topic in detail in § 3.4, now we only note that finding u_2 will amount to performing of a chosen number of inner iterations). The same applies to the systems with $A_{1,1}$; especially attractive are blocks such that the solutions of systems involving them can be reduced to independent solution of smaller subsystems.

Proof. Since $(Au, v) = (A_{1,1}u_1 + A_{1,2}u_2, v_1) + (A_{2,1}u_1 + A_{2,2}u_2, v_2)$, in the case of $u = p(u_2)$, we may write

$$(Ap(u_2), v) = (A_{2,2}u_2, v_2) + (A_{2,1}u_1, v_2) = (S_2u_2, v_2).$$

If $A_{1,1} > 0, A_{1,2}^T = A_{2,1}$, and we wish to minimize

$$(Au, u) = (A_{1,1}u_1, u_1) + 2(A_{1,2}u_2, u_1) + (A_{2,2}u_2, u_2)$$

with respect to u_1 for a given u_2, then Theorem 0.1.3 is applicable, and we may write $p_1(u_2) = \arg\min_{u_1}(Au, u)$. Hence, (5.5) holds. $\square$

In this section, we shall be dealing only with symmetric matrices A.

Lemma 2. *Let $A = A^* > 0$ and suppose, for all u_2, that*

$$(Ap(u_2), p(u_2)) \geq \delta_0(A_{2,2}u_2, u_2), \quad \delta_0 > 0. \tag{5.6}$$

Then $S_2 \in \mathcal{L}^+(H_2)$ and

$$\delta_0 A_{2,2} \leq S_2 \leq \delta_1 A_{2,2}, \quad \delta_1 = 1. \tag{5.7}$$

Proof. Since $A_{r,l}^* = A_{r,l}$ for all r and l, we have $S_2^* = S_2$. Moreover, $(S_2u_2, u_2) = (Ap(u_2), p(u_2))$, whence (5.7) follows directly. $\square$

Lemma 3. *If $A^* = A$ and $A_{1,1} > 0$, then $A \geq 0 \Leftrightarrow S_2 \geq 0$; $A > 0 \Leftrightarrow S_2 > 0$.*

Proof. It suffices to show that $A \geq 0, A > 0$ if $S_2 \geq 0, S_2 > 0$, respectively. But both statements follow directly from (5.5) if we note that $u_2 \neq 0 \Leftrightarrow p(u_2) \neq 0$ (see [7]). $\square$

In the case $A_{1,1} > 0$, making use of the prolongation operator p, we define $\text{Ker}_2 A \equiv \{u_2 : p(u_2) \in \text{Ker } A\}$.

Lemma 4. *Let $A = A^* \geq 0$ and $A_{r,r} > 0$ $(r = 1, 2)$, and suppose inequality (5.6) is satisfied for all u_2 such that $A_{2,2}u_2 \perp \text{Ker } S_2$ in the Euclidean space H_2. Then*

$$\delta_0(A_{2,2}u_2, u_2) \leq (S_2u_2, u_2) \leq (A_{2,2}u_2, u_2), \quad \forall u_2 \in \text{Im }(A_{2,2}^{-1}S_2).$$

Proof. The operator $A_{2,2}^{-1}S_2$ is symmetric as an element of $\mathcal{L}(H_2(A_{2,2}))$. Note that $\text{Ker } S_2 = \text{Ker }(A_{2,2}^{-1}S_2)$, so $\text{Im }(A_{2,2}^{-1}S_2) = \{u_2 : A_{2,2}u_2 \perp \text{Ker } S_2\}$. All $u_2 \in \text{Im }(A_{2,2}^{-1}S_2)$ are such that (5.6) holds, and Lemma 2 applies. $\square$

Usually we reduce solving system (5.1) to solving (5.3) with the Schur matrix S_2. But in some situations (which we discuss in § 3.6 when we deal with the iterative fictitious grid region method), it becomes reasonable to solve system (5.3) by solving special cases of systems (5.1).

Lemma 5. *Suppose operators A and $A_{1,1}$ are invertible. Then system (5.3) with arbitrary $g_2 \in H_2$ has a unique solution u_2 coinciding with the* *second component of the solution* $u \equiv [u_1, u_2]^T \in H$ *of system (5.1) with the right-hand side such that*

$$f_1 = 0, \quad f_2 = g_2. \tag{5.8}$$

Proof. Take f from (5.8). Then system (5.1) has a unique solution that must satisfy (5.2)–(5.4) with $g_2 = f_2$. Therefore, u_2 is a solution of the system $S_2 u_2 = g_2$ under consideration. Finally, it has only one solution (otherwise, the operator A would not have been invertible). □

Lemma 6. Suppose the operator $A_{1,1}$ is invertible and system (5.3) with a given $g_2 \in H_2$ has a solution u_2. Then system (5.1) with the right-hand side from (5.8) has a solution $u = [p_1(u_2), u_2]$.

Proof. Given u_2, we define u_1 from (5.2) with $f_1 = 0$. Then $u_1 = p_1(u_2)$, and the obtained vector $u = [u_1, u_2]^T$ is a solution of (5.1). □

Lemma 7. Suppose the operators A and B in $\mathcal{L}(H)$ have the block structure defined by (5.1) and are such that

$$A = A^* \geq 0, \ B = B^* \geq 0, \ \delta_0 B \leq A \leq \delta_1 B, \ \delta_0 > 0.$$

Let $A_{1,1} > 0, B_{1,1} > 0$. Then, for the Schur matrices $S_2(A) \equiv A/A_{1,1}$ and $S_2(B) \equiv B/B_{1,1}$, we have

$$\delta_0 S_2(B) \leq S_2(A) \leq \delta_1 S_2(B). \tag{5.9}$$

Proof. The desired estimates (5.9) are equivalent (see Lemma 1) to

$$\delta_0(Bu_B, u_B) \leq (Au, u) \leq \delta_1(Bu_B, u_B), \quad \forall u_2, \tag{5.10}$$

where $u \equiv p(u_2)$ and $u_B \equiv [w_1, u_2]^T$ with $w_1 \equiv -B_{1,1}^{-1}B_{1,2}u_2$. Now recall that $(Bu_B, u_B) = \min\limits_{u_1}(Bu, u)$ and $(Ap(u_2), p(u_2)) = \min\limits_{u_1}(Au, u)$ (see Lemma 1). Therefore, $X \equiv \delta_0(Bu_B, u_B) \leq \delta_0(Bu, u)$, and the conditions of Lemma 7 imply that $X \leq (Au, u) \leq (Au_B, u_B) \leq \delta_1(Bu_B, u_B)$, that is, (5.10) and, therefore, (5.9) holds. □ [45]

5.3. Schur complements in the case of Gram matrices. Suppose that G is either a Hilbert space or a Euclidean space. Consider a subspace $\hat{G}$ such that $\hat{G} \equiv \hat{G}_1 \oplus \hat{G}_2$ with basis $\hat{\psi}_1, \ldots, \hat{\psi}_{n_1+n_2}$,

$$\hat{G}_1 \equiv \text{lin } \{\hat{\psi}_1, \ldots, \hat{\psi}_{n_1}\}, \text{ and } \hat{G}_2 \equiv \text{lin } \{\hat{\psi}_{n_1+1}, \ldots, \hat{\psi}_{n_1+n_2}\}.$$

Denote by $A, A_{1,1}, A_{2,2}$ the Gram matrices defined by these three respective bases. Then, for the matrix A and its block-diagonal part D, we may write

$$A \equiv \begin{bmatrix} A_{1,1} & A_{1,2} \\ A_{2,1} & A_{2,2} \end{bmatrix}; \quad D \equiv \begin{bmatrix} A_{1,1} & 0 \\ 0 & A_{2,2} \end{bmatrix}. \tag{5.11}$$

---

[45] The same result may be deduced from the similar statement: if $A^* = A \geq B^* = B \geq 0, A_{1,1} > 0, B_{1,1} > 0$ and $A \geq B$, then $S_2(A) \geq S_2(B)$.

For our analysis, we make use of the *angle between subspaces* $\hat{G}_1$ and $\hat{G}_2$ defined as the biggest angle $\alpha \in [0, \pi/2]$ that, for all elements $\hat{u}_r \in \hat{G}_r$ with $r = 1, 2$, satisfies

$$|(\hat{u}_1, \hat{u}_2)| \leq \cos \alpha \|\hat{u}_1\|\|\hat{u}_2\| \qquad (5.12)$$

(of course, the biggest angle corresponds to the minimal value of possible $\cos \alpha$ satisfying (5.11); also note that if the angle $\alpha(\hat{u}_1; \hat{u}_2)$ between nonzero elements $\hat{u}_1$ and $\hat{u}_2$ is defined in a standard way as $\cos \alpha(\hat{u}_1; \hat{u}_2) \equiv |(\hat{u}_1, \hat{u}_2)|(\|\hat{u}_1\|\|\hat{u}_2\|)^{-1}$, then the angle between the subspaces is the minimal angle between their elements). [46]

   Theorem 1. *Let α be the angle based on an inner or a semi-inner product between the subspaces $\hat{G}_1$ and $\hat{G}_2$, bases of which define matrices A and D from (5.11). Let $s^2 \equiv 1 - \cos^2 \alpha$. Then*

$$(1 - \cos \alpha)D \leq A \leq (1 + \cos \alpha)D, \qquad (5.13)$$

and

$$s^2(A_{r,r}u_r, u_r) \leq (Au, u), \quad r = 1, 2, \quad \forall u. \qquad (5.14)$$

   Proof. Take any $\hat{u}_1 \equiv u_1\hat{\psi}_1 + \ldots + u_{n_1}\hat{\psi}_{n_1} \in \hat{G}_1$ and

$$\hat{u}_2 \equiv u_{n_1+1}\hat{\psi}_{n_1+1} + \ldots + u_{n_1+n_2}\hat{\psi}_{n_1+n_2} \in \hat{G}_2.$$

These expansions define the vectors

$$\mathbf{u_1} \equiv [u_1, \ldots, u_{n_1}]^T, \ \mathbf{u_2} \equiv [u_{n_1+1}, \ldots, u_{n_1+n_2}]^T, \ \mathbf{u} \equiv [u_1, \ldots, u_{n_1+n_2}]^T.$$

Then it is easy to see that

$$A_{1,2}\mathbf{u_2} = [(\hat{\psi}_1, \hat{u}_2), \ldots, (\hat{\psi}_{n_1}, \hat{u}_2)]^T, \quad (A_{1,2}\mathbf{u_2}, \mathbf{u_1}) = (\hat{u}_1, \hat{u}_2).$$

Hence, $(A\mathbf{u}, \mathbf{u}) = (\hat{u}_1, \hat{u}_1) + (\hat{u}_2, \hat{u}_2) + 2(\hat{u}_1, \hat{u}_2)$. Note now that

$$2|(\hat{u}_1, \hat{u}_2)| \leq \cos \alpha \{\|\hat{u}_1\|^2 + \|\hat{u}_2\|^2\}.$$

Therefore, inequalities (5.13) are valid (see [21, 183, 347]). Inequality (5.14) is a fairly obvious consequence of the above representation of $(A\mathbf{u}, \mathbf{u})$ and the estimate $2|\cos \alpha\|\hat{u}_1\|\|\hat{u}_2\|\| \leq (\cos \alpha\|\hat{u}_r\|)^2 + \|\hat{u}_l\|^2$ with $r \neq l$. $\square$

---

[46] These notions may be directly generalized for spaces with semiinner products, which were described in § 0.1 (recall that the inner product axiom $\|u\| = 0 \Rightarrow u = 0$ does not necessarily hold). As we shall see later, these generalizations of Gram matrices and angles will be very useful even when we wish to estimate angles in standard Euclidean spaces for some finite element subspaces through local analysis connected with separate cells of the grid. A Gram matrix is nonnegative in the case of a semi-inner product; if it is actually positive, then the corresponding subspace may be regarded as a Euclidean space.

In what follows, we deal only with semiinner products and subspaces for which the block $A_{1,1} > 0$, so it will be possible to use the standard notion of an orthoprojector P_1 on the subspace $\hat{G}_1$: for each $u \in G$, $P_1 u$ is the unique element of $\hat{G}_1$ such that $(u - P_1 u, \hat{v}_1) = 0$, $\forall \hat{v}_1 \in \hat{G}_1$; of course, P_1 is a linear operator.

Theorem 2. *Let P_1 be an orthoprojector on the subspace $\hat{G}_1$ defined above and let $P_1^\perp \equiv I - P_1$, $P_1^\perp \hat{\psi}_i \equiv \hat{\varphi}_i$. Then the Schur complement $S_2(A)$ in the case of the Gram matrix A from (5.11) is a new Gram matrix generated by the functions $\hat{\varphi}_{n_1+1}, \ldots, \hat{\varphi}_{n_1+n_2}$ and*

$$(S_2(A)u_2, u_2) = \|\hat{u}_2 - P_1 u_2\|^2. \tag{5.15}$$

Proof. Denote by Y_k the kth column of the matrix $A_{1,2}$, that is,

$$Y_k = [(\hat{\psi}_1, \hat{\psi}_{n_1+k}), \ldots, (\hat{\psi}_{n_1}, \hat{\psi}_{n_1+k})]^T.$$

Define the vector $X_k \equiv A_{1,1}^{-1} Y_k \equiv [x_1, \ldots, x_{n_1}]^T$. Then the function $x_1 \hat{\psi}_1 + \ldots + x_{n_1} \hat{\psi}_{n_1}$ is the best approximation in $\hat{G}_1$ to $\hat{\psi}_{n_1+k} \in \hat{G}_2$ and coincides with $P_1 \hat{\psi}_{n_1+k}$. Note that lth row R_l of the matrix $A_{2,1}$ has the form

$$R_l = [(\hat{\psi}_{n_1+l}, \hat{\psi}_1), \ldots, (\hat{\psi}_{n_1+l}, \hat{\psi}_{n_1})],$$

implying $R_l X_k \equiv c_{l,k} = (\hat{\psi}_{n_1+l}, P_1 \hat{\psi}_{n_1} + k)$. Thus, for the element $s_{l,k}$ of the matrix $S_2(A)$, we have $s_{l,k} = (\hat{\psi}_{n_1+l}, \hat{\psi}_{n_1+k}) - c_{l,k} = (\hat{\psi}_{n_1+l}, \hat{\varphi}_{n_1+k})$. Note that $(\hat{v}_1, \hat{\varphi}_{n_1+k}) = 0$ for every $\hat{v}_1 \in \hat{G}_1$. Then

$$s_{l,k} = ((I - P_1)\hat{\psi}_{n_1+l}, \hat{\varphi}_{n_1+k}) = (\varphi_{n_1+l}, \varphi_{n_1+k}),$$

and we obtain the desired representation of $S_2(A)$. Now, making use of properties of Gram matrices, we observe that

$$Z \equiv (S_2(A)u_2, u_2) = \|u_{n_1+1}\hat{\varphi}_{n_1+1} + \cdots + u_{n_1+n_2}\hat{\varphi}_{n_1+n_2}\|^2.$$

Thus, $Z = \|\sum_{i=1}^{n_2} u_{n_1+i}(I - P_1)\hat{\psi}_{n_1+i}\|^2 = \|(I - P_1)\hat{u}_2\|^2$ and (5.15). $\square$

Lemma 8. *In addition to the Gram matrix A, consider a new Gram matrix $\bar{A}$ of the same block structure defined by a new basis $\bar{\psi}_1, \ldots, \bar{\psi}_{n_1+n_2}$, where $\bar{\psi}_i = \hat{\psi}_i$ if $i \leq n_1$ and $\bar{\psi}_l = \hat{\psi}_i - \hat{v}_i$ with $\hat{v}_i \in \hat{G}_1$ if $i > n_1$. Then*

$$S_2(A) = S_2(\bar{A}). \tag{5.16}$$

Proof. Since $(I - P_1)\bar{\psi}_i = (I - P_1)\hat{\psi}_i$, this lemma is a direct corollary of Theorem 2. $\square$

    Lemma 9. Suppose the conditions of Theorem 1 are satisfied and let
$s^2 \equiv 1 - \cos^2 \alpha = \sin^2 \alpha$. *Then*

$$s^2 A_{2,2} \leq S_2(A) \leq A_{2,2}. \qquad (5.17)$$

Proof. On the basis of (5.15), we conclude that

$$(S_2 u_2, u_2) \geq \text{dist}^2[\hat{u}_2; \hat{G}_1] \geq s^2 \|\hat{u}_2\|^2 = s^2 (A_{2,2} u_2, u_2).$$

This and the obvious inequality $\|(I - P_1)\hat{u}_2\| \leq \|\hat{u}_2\|$ give (5.17). □
    5.4. Cooperative model operators.[47]
    *Theorem 3. Assume that the subspaces $\hat{G}_1$ and $\hat{G}_2$, associated with the
grid parameter h, are such that*

$$\|P_1^{\perp} \hat{u}_2\| \geq \kappa_0 h^{\lambda}, \lambda \geq 0, \kappa_0 > 0, \quad \forall \hat{u}_2 \in \hat{G}_2, \|\hat{u}_2\| = 1. \qquad (5.18)$$

Then there exist numbers $\kappa_r > 0, r = 1, 2, 3$, such that

$$\kappa_1 h^{2\lambda} D \leq A \leq \kappa_2 D, \qquad (5.19)$$

and

$$\kappa_3 h^{\lambda} A_{2,2} \leq A_{2,2} - A_{2,1} A_{1,1}^{-1} A_{1,2} \leq A_{2,2}. \qquad (5.20)$$

    Proof. Let $\alpha(u_2)$ denote the angle between an arbitrary given function
$\hat{u}_2 \in \hat{G}_2$ and the subspace $\hat{G}_1$, and α denote the angle between the subspaces
$\hat{G}_1$ and $\hat{G}_2$ (of course, $\alpha \leq \alpha(u_2)$). Then $\sin \alpha(\hat{u}_2) \geq \kappa_0 h^{\lambda}$. Hence,
$\sin \alpha \geq \kappa_0 h^{\lambda}$ and $\cos \alpha \leq [1 - \kappa_0^2 h^{2\lambda}]^{1/2}$. Now (5.11) and (5.16) lead to
(5.19), (5.20) with $\kappa_1 = \kappa_0^2$, $\kappa_2 = 2$, $\kappa_3 = \kappa_0^2$. □
    Conditions (5.18) follow from the simpler conditions

$$\|\varphi_{n_1+k}\| \geq \kappa_0 h^{2\lambda}, k = 1, \ldots, n_2, \qquad (5.21)$$

when the system of the functions $\hat{\varphi}_{n_1+1}, \ldots, \hat{\varphi}_{n_1+n_2}$ is orthogonal. This
makes sense if the subspace $\hat{G}_1$ is a standard finite element space and $\hat{G}_2$

---

    [47] Operators (matrices) A of block structure (5.1) that lead to relatively easily solved
systems with matrices $A_{1,1}$ and $S_2(A)$ in the role of preconditioners have long been
used by many authors (see, e.g., [21, 119, 152, 155, 174, 183, 186, 187, 199, 309, 452]).
Some preconditioners (generalized splitting operators) of such a structure were suggested
in [152] (see § 3.2); optimal preconditioners can be found in [155, 174, 183]. Such
operators are widely applied in many contemporary variants of domain decomposition
and multigrid methods (see [24, 80, 71, 128, 203, 263, 307, 322, 289, 347, 498]), and we
consider some of them in Chapter 3. They may be referred to as *model block-triangular
factorized operators* or *cooperative operators*. The latter terminology emphasizes the
connection with game theory (joint efforts of both players are required to obtain the
desired solution, each making decisions associated with the Euclidean space H_1 or H_2).
We first consider a theorem from [183] helpful for the analysis of problems with dim $H_2 =
n_2 \times 1$.

is a linear span of several singular functions such that their supports do not have common points (we shall analyze such cases in Chapter 5, where similar optimal preconditioning will be described).

Now, given the operator $A = A^* \geq 0$ from (5.11), possibly with complicated $A_{1,1}$ or $S_2(A)$, we try to find a model cooperative operator $B = B^* \geq 0$ of the same block structure,

$$B \equiv \begin{bmatrix} B_{1,1} & B_{1,2} \\ B_{2,1} & B_{2,2} \end{bmatrix}, \tag{5.22}$$

and comparatively simple $B_{1,1}$ and $S_2(B) \equiv B_{2,1}B_{1,1}^{-1}B_{1,2}$. Of course, the case of a block-diagonal B is the simplest one and requires only the inequalities connecting the corresponding blocks of B and D (see (5.11)). We assume that

$$B_{1,2} = A_{1,2} = B_{2,1}^T, \ A_{1,1} \leq B_{1,1} \leq \frac{1}{1 - q_1^+}A_{1,1}, \ B_{2,2} = D_{2,2} + A_{2,1}B_{1,1}^{-1}A_{1,2},$$
$$\tag{5.23}$$

with $0 < q_1^+ < 1$, and that systems with $B_{1,1}$ and $S_2(B) = D_{2,2}$ are easy to solve. Such B can also be written in the form

$$B = \begin{bmatrix} B_{1,1} & 0 \\ 0 & D_{2,2} \end{bmatrix} \begin{bmatrix} I_1 & 0 \\ D_{2,2}^{-1}A_{2,1} & I_2 \end{bmatrix} \begin{bmatrix} I_1 & B_{1,1}^{-1}A_{1,2} \\ 0 & I_2 \end{bmatrix}, \tag{5.24}$$

which was used in [174] for construction of some optimal preconditioners. We shall also make use of an operator C with blocks

$$C_{1,1} = B_{1,1}, \quad C_{1,2} = A_{1,2}, \quad C_{2,1} = A_{2,1}, \quad C_{2,2} = A_{2,2}. \tag{5.25}$$

Lemma 10. *Let conditions* (5.22), (5.23), (5.25) *and those of Theorem 1 with* $s^2 = 1 - \cos^2 \alpha > 0$ *be satisfied. Then*

$$A \leq C \leq \xi_1 A, \quad s^2 A_{2,2} \leq S_2(C) \leq A_{2,2} \tag{5.26}$$

with

$$\xi_1 \equiv 1 + \frac{q_1^+}{s^2(1 - q_1^+)}.$$

Proof. Since $B_{1,1} \geq A_{1,1}$, then $C \geq A$. For $X \equiv ((C - A)u, u)$, we have

$$X = ((B_{1,1} - A_{1,1})u_1, u_1) \leq q_1^+(1 - q_1^+)^{-1}(A_{1,1}u_1, u_1).$$

Then, by virtue of (5.14), we obtain $C \leq \xi_1 A$. We have

$$(S_2(C)u_2, u_2) = (A_{2,2}u_2, u_2) - (B_{1,1}^{-1}A_{1,2}u_2, A_{1,2}u_2).$$

Since $B_{1,1}^{-1} \leq A_{1,1}^{-1}$, we see that $S_2(C) \geq S_2(A)$. Now, applying (5.17), we conclude that all inequalities in (5.26) are valid (similar statements can be found in [189, 191, 22]). □

Theorem 4. *Let conditions of the above lemma be satisfied and suppose there exist constants*

$$0 < \kappa_0 \leq 1 \leq \kappa_1 \tag{5.27}$$

such that

$$\kappa_0 S_2(B) \leq S_2(C) \leq \kappa_1 S_2(B). \tag{5.28}$$

Then

$$\kappa_0 B \leq C \leq \kappa_1 B \tag{5.29}$$

and

$$\frac{\kappa_0}{\xi_1} B \leq A \leq \kappa_1 B. \tag{5.30}$$

Proof. Operators B and C have equal blocks except for $B_{2,2}$ and $C_{2,2}$. Hence, $S_2(C) - S_2(B) = A_{2,2} - B_{2,2}$. If $S_2(C) \geq S_2(B)$, then $A_{2,2} \geq B_{2,2}$ and $C \geq B$. Thus, we may consider only the case $0 < \kappa_0 < 1$, for which

$$\frac{1}{\kappa_0} C_0 - B \geq 0 \Leftrightarrow (\frac{1}{\kappa_0} - 1) \begin{bmatrix} B_{1,1} & A_{1,2} \\ A_{2,1} & \frac{\kappa_0}{1-\kappa_0}(\frac{1}{\kappa_0} A_{2,2} - B_{2,2}) \end{bmatrix} \geq 0,$$

with the last inequality equivalent (see Lemma 3) to

$$\frac{\kappa_0}{1 - \kappa_0}(\frac{1}{\kappa_0} A_{2,2} - B_{2,2}) - A_{2,1} B_{1,1}^{-1} A_{1,2} \geq 0,$$

which is in turn equivalent to $\kappa_0 S_2(B) \leq S_2(C)$ (see (5.28)). Therefore, the left inequality in (5.29) with κ_0 is proved. The inequality with κ_1 follows from this and the observation $S_2(B) \geq \kappa_1^{-1} S_2(C)$. Finally, (5.30) follows directly from (5.29) and (5.26). □ [48]

---

[48] This theorem can be found in [203, 204] and has some points in common with similar statements from [22]. It will be very useful for some multigrid constructions of optimal preconditioners (see Chapter 3) based on appropriate recurrent splittings of finite-element subspaces (see [24, 50, 347, 517, 512]). We remark finally that these operators might be very useful for certain difficult practical problems when we are interested, not in asymptotic characteristics, but instead want to work with systems in the Euclidean spaces H_1 and H_2 separately (see, e.g., [1, 28, 86, 231, 257]).

Chapter 2

Projective-grid methods for second-order elliptic equations and systems

§ 1. Projective-grid methods associated with triangulation of the region and piecewise polynomial functions

1.1. Topologically equivalent grids and triangulations. In the introduction we exposed some basic notions and results related to the theory of PGMs (FEMs) in the two-dimensional case. Here we discuss a similar topic but for the more difficult d-dimensional case, $d \geq 2$. For this we need to specify notions of grids and triangulations constructed for a given region. Recall that, for $d = 2$ and $d = 3$, geometrical pictures and Descartes coordinate systems involve use of sets of points beside standard notions of vectors and operations with them. Thus, we shall be able to work in the general linear space V in the same geometric manner as in the usual physical space $\mathbf{R}^3$, if we use axioms of Weyl and think of V not only as a collection of elements (vectors) v as above, but also as a collection of points P_v (see [463]) such that: every ordered pair of points P_u and P_v defines a unique element (vector) $[P_u, P_v]$, (P_u is its beginning, P_v is its end); each point P_u and each $z \in V$ define a unique point P_v that allows for the identification $[P_u, P_v] = z$; and if $[P_u, P_v] = z_1$ and $[P_v, P_w] = z_2$, then $[P_u, P_w] = z_1 + z_2$. In this case, fixing a point O as the origin, we can identify elements $[O, P_v]$ with v as in classical analytical geometry in $\mathbf{R}^3$ with direction-magnitude vectors. According to this, in what follows, we treat $\mathbf{R}^d$ as a point-vector space (affine-vector space) with the above axioms.

Using the point O as the origin of a coordinate Descartes system with the standard orthonormal basis, consisting of vectors $e_1, \ldots, e_d$, we have a one-to-one correspondence between points M with coordinates $[x_1, \ldots, x_d] \equiv x$ and vectors $\vec{OM} \equiv x_1 e_1 + \ldots + x_d e_d$, being elements of the Euclidean space. Hence, we may use the standard notions of inner points of a set, of domains (regions), of their boundaries, and of closures of domains in the same way as in the case of $\mathbf{R}^3$. We understand a *grid* to be a certain partition of the closure of the given region into a finite set of cells, such that each *cell* is the closure of a certain subregion and these different subregions have no common inner points. Of course, above all, we are interested in grids with the cells of the simplest form.

For a given set of j points $P_0, \ldots, P_j$, we define their convex hull $T \equiv P_0 \ldots P_j \equiv \{M : \vec{OM} = \sum_{i=0}^{j} \alpha_i \vec{OP_i}, \alpha_1 \geq 0, \sum_{i=0}^{j} \alpha_i = 1\}$. Such set of points is called a *polyhedron*. Its particular and very important case, corresponding to points $P_0, \ldots, P_j$ with $j \leq d$ and linearly independent vectors $\vec{P_0 P_1}, \ldots, \vec{P_0 P_j}$, is called a j-dimensional simplex $T \equiv [P_0 P_1 \ldots P_j]$; each point P_i is a vertex of T; for $1 \leq k \leq j-1$, k-dimensional faces of T are k-dimensional simplexes with $k+1$ different vertices $P_{i_1}, \ldots, P_{i_k}$ from the given set of P_i; 1-dimensional faces are called edges. Of course, for $d = 1, 2, 3$ the simplex T is an interval, a triangle, and a tetrahedron, respectively.

In what follows, we shall be dealing with special d-dimensional simplexes, which may be referred to as *regular simplicial parts of cubes* in $\mathbf{R}^d$. More precisely, consider a cube $Q_a \equiv \{x : 0 \leq x_i \leq a, i \in [1, d], a > 0\}$ with 2^d vertices $[\alpha_1, \ldots, \alpha_d]$, where each number α_i is either 0 or a; given the direction $[1, 1, \ldots, 1]$ of one of its diagonals, we take $d+1$ points:

$$P_0 \equiv 0, P_1 \equiv P_0 + a e_{j_1}, P_2 \equiv P_1 + a e_{j_2}, \ldots, P_d \equiv P_{d-1} + a e_d, \qquad (1.1)$$

where all $j_1, \ldots, j_d$ are different integers and each $j_r \in [1, d]$. Then these points $P_0, \ldots, P_d$ define a d-dimensional simplex $T \equiv [P_0 P_1 \ldots P_d]$; the cube Q_a contains $d!$ such different congruent simplexes; and one of them can match another by an isometric mapping (see [125, 168, 169, 471]). Collection of these $d!$ simplexes constitutes a *regular triangulation of the cube* defined by the chosen diagonal.[1] Figure 1 illustrates such partition of a triangular prism into three simplexes, where $P_0 P_3$ corresponds to the above diagonal. Any such simplexes and similar ones corresponding to other $2^{d-1} - 1$ directions of diagonals of the cube Q_a may be called a *regular simplicial part of a cube*.

---

[1] Similar triangulations can be readily constructed in the case of a parallelepiped .

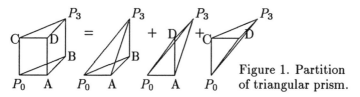

Figure 1. Partition of triangular prism.

As we did for $d = 2$, here we assume that the triangulation $T_h(\bar{\Omega})$ of a bounded region Ω in the Euclidean space $\mathbf{R}^d$ constitutes a set of simplexes such that common parts of any two different simplexes may be: the empty set, a vertex, a common edge, a common 2-dimensional face, and so on (in other words, we deal with a *simplicial grid*; each cell of the grid is a simplex). Of course, in the case of a general $\bar{\Omega}$, we have to deal either with triangulations $T_h(\hat{\Omega}_h)$ of their approximations $\hat{\Omega}_h$ or with some simplexes replaced by more general figures, which will be referred to as quasisimplexes (e.g., curvilinear triangles if $d = 2$) to be specified below.

Many triangulations, in particular, composite ones which will be considered below, can be described as a result of some refinement procedures applied to separate cells. For $d = 2$ and a given integer t (refinement ratio), it is a simple matter to partition a triangle T into t^2 equal subtriangles (by subdividing each side into t equal parts and drawing through the points of division straight lines parallel to other sides of the triangle; Figure 2 corresponds to $t = 3$. Figure 3 represents a 3-dimensional simplex (tetrahedron) $T \equiv [P_0ABP_3]$, which must be partitioned into 3^3 smaller congruent simplexes with edges reduced by a third and with each face subdivided into 3^2 subtriangles. [2]

For regions Ω of a complicated form, it is very important to simplify information about them and about the corresponding grids on the basis of some model regions and grids. For example, suppose that $T_h(\bar{Q})$ is a triangulation of the closure of a model given domain $Q \subset \mathbf{R}^d$ with Lipschitz boundary. Suppose also that there exists an one-to-one mapping

---

[2] In the general case, we can obtain the desired partition of a 3-dimensional simplex by subdividing each edge into t equal parts and drawing through points of division planes parallel to coordinate (affine) ones containing the pairs of vectors: $a_1, a_2; a_1, a_3$, and a_2, a_3 with $a_1 \equiv P_0\vec{P}_1, a_2 \equiv P_1\vec{P}_2, a_3 \equiv P_2\vec{P}_3$.

Similar general partition of a d-dimensional simplex into t^d congruent subsimplexes can be proved to exist, but we simply take it as an assumption. The global refinement of a given triangulation with the refinement ratio t can be carried out for all cells separately if $d \leq 3$. It is not the case if $d \geq 4$, since there exist different partitions of a 3-dimensional simplex and they must coincide on common 3-dimensional faces of neighboring cells. Also, if for $d \geq 3$ we wish to construct the desired triangulation starting from a cubical grid, we cannot use arbitrary diagonals and triangulations defined by them of different cubical cells because we may come to different partitions of the same square (a common face of two cells).

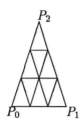

Figure 2. Partition
of triangle.

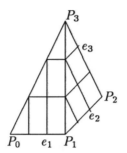

Figure 3: Partition of tetrahedron.

$$z = \Pi^{-1}(x) \equiv \hat{\Pi}(x) \equiv Z(x), \quad x = \Pi(z), \tag{1.2}$$

continuous with respect to both directions, of $\bar{\Omega}$ onto $\bar{Q}$ (in other words, $\bar{\Omega}$ and $\bar{Q}$ are topologically equivalent). Note that here, of course, x and $z \equiv \hat{x}$ refer to points $[x_1, \ldots, x_d]$ and $[z_1, \ldots, z_d]$ from $\bar{\Omega}$ and $\bar{Q}$, respectively,

$$\hat{\Pi}(x) \equiv [\hat{\Pi}_1(z), \ldots, \hat{\Pi}_d(x)]^T \equiv [Z_1(x), \ldots, Z_d(x)]^T,$$

$$\hat{\Pi}^{-1}(x) \equiv [\hat{\Pi}_1^{-1}(z), \ldots, \hat{\Pi}_d^{-1}(z)]^T \equiv \Pi(z) \equiv [\Pi_1(z), \ldots, \Pi_d(z)]^T.$$

Then we can define on $\bar{\Omega}$ a grid with cells T_Ω being the images of the simplexes $T_Q \in T_h(\bar{Q})$, that is, $T_\Omega \equiv \Pi\{T_Q\}$. In the general case, these cells and grids can be referred to as *quasisimplexes*, and quasitriangulations. Similarly, the corresponding images of cubical cells and cubical grids can be called *quasicubes* and *quasicubical grids*. The most remarkable case corresponds to all quasisimplexes being standard simplexes. Then we have *topologically equivalent triangulations* $T_h(\bar{Q})$ and $T_h(\bar{\Omega})$: each elementary simplex T_Q from $T_h(\bar{Q})$ is mapped to a corresponding simplex T_Ω from $T_h(\bar{\Omega})$ (in this case, Π from (1.2) is a one-to-one piecewise affine mapping, that is, $\Pi(\hat{x}) = \Pi_{T_Q}\hat{x} + \hat{x}_{0,T_Q}$, where Π_{T_Q} is an invertible matrix). Observe that, in this case, a quasicube consists of a finite number of simplexes; it will be called an *affine quasicube* (for $d = 3$, the image of each face of a cube consists of two triangles with a common side but possibly belonging to different planes).

There is a great variety of model regions Q. In the case of $d = 2$ and a simply connected $\bar{\Omega}$ with Lipschitz boundary, we can choose a triangle, a rectangle, a union of a several rectangles and triangles, and so on. Sometimes it is even suitable to use not the planar $\bar{Q}$, but a region $\bar{Q}$ belonging to some two-dimensional manifold like a *Riemann surface* (which are extremely important for representation of multivalued functions of the complex variable or of two real variables). For example, quasisquare grids

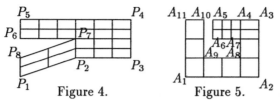

Figure 4. Figure 5.

Grids topologically equivalent to square grids on Riemann's surfaces.

for $\bar{\Omega}$ depicted in Figures 4 and 5 are topologically equivalent to square grids for two regions $\bar{Q} \equiv \bar{Q}_1 \cup \bar{Q}_2$, with $\bar{Q}_1$ and $\bar{Q}_2$ belonging to two intersecting planes with the line of intersection containing the image of $[P_2 P_7]$ and $[A_7 A_4]$, respectively, and with $\bar{Q}_2$ being rectangles (they correspond to the images of $[P_1 P_2 P_7 P_8]$ and of $[A_6 A_7 A_4 A_5]$). Such surfaces will be very useful for regions with boundaries containing slits (see Subsection 6). Also, if we take equal numbers of grid points on $[P_8 P_7]$ and $[P_6 P_7]$ (see Figure 4), then the information about a grid function defined on such a grid differs from the information about the function defined on a rectangular grid only for nodes on $[P_8 P_7]$. This may be useful for a simplification of the data about the grid functions under consideration. Under the presence of periodic conditions, grids on cylindrical and toroidal surfaces might be of help; grids on closed two-dimensional manifolds might be represented through grids either on a sphere or on the surface of a cube, and so on. The natural and simplest choices for the model $\bar{Q}$, in the case of $d = 3$ and a simply connected $\bar{\Omega}$ with Lipschitz boundary, are a cube, a prism, a tetrahedron, a union of several cubes and tetrahedrons (regular simplicial parts of cubes), and so on. The latter choice (with a multi-connected $\bar{Q}$) might be used for a multi-connected Ω as well. Some three-dimensional manifolds in the four-dimensional Euclidean space (generalizations of Riemann surfaces) might be useful. [3]

1.2. Composite triangulations. In § 0.2 for $d = 2$, we discussed advantages of composite triangulations $T_{c,h}(\bar{\Omega})$ composed of cells of trian-

---

[3] Various types of grids and triangulations that are topologically equivalent to model ones now are widely used in practice. A problem of constructing such grids and triangulations, including composite triangulations with local refinement, will be considered in more detail in Chapter 4. For the moment, we just emphasize that our interest in them is connected not only with the obvious simplification of all information about the grid (see [478, 518]), but above all with the possibility to construct effective model operators and asymptotically optimal algorithms for a wide class of elliptic boundary value problems (see [175, 176, 177, 178, 179, 187]). It seems very interesting that, in the role of a basic space unit from a computational point of view, we may now use the same as in d-dimensional geometry—the simplex (the corresponding asymptotically optimal algorithms will be described in Chapters 3 and 5).

gulations for closed subregions $F_i, i \in [1, p_1]$, where

$$\bar{\Omega} = F_1 \cup F_2 \cup \ldots \cup F_{p_1+1} \qquad (1.3)$$

and every two of these closed sets may have common points only on their boundaries. In the same way, we can deal with $d \geq 3$. [4]

1.3. Piecewise linear and piecewise polynomial functions. Suppose that the closure $\bar{Q}$ of a given model domain Q in the Euclidean space $\mathbf{R}^d$ consists of a finite number of d-dimensional simplexes $T_{0,k}$; their collection defines a standard triangulation $T^{(0)}(\bar{Q})$. In approximating the space $W_2^1(Q; \Gamma_0)$ on the basis of this triangulation we usually identify the set of grid nodes $Q_h^{(0)}$ with the set of vertices of the simplexes $T_{0,k}$ not belonging to Γ_0; we consider this space as a Hilbert space G with the inner product $(u, v)_{1,Q}$. Define the basic spline (finite element) subspace $\hat{G}_{gl}$ of G, associated only with the initial global triangulation, as

$$\hat{G}^{(0)} \equiv \hat{G}_{gl} \equiv \{\hat{u} : \hat{u} = \sum u_i \hat{\psi}_i^{(0)}(x), \ P_i^{(0)} \in Q_h^{(0)}\}. \qquad (1.4)$$

For $l \geq 0$, we refer to the given composite triangulation $T^{(l)}$ as an old one and define the new composite triangulation via the $(l+1)$th level of local refinement. More precisely, we use $Q_h^{(l+1)}$ as a set of vertices $P_i^{(l+1)}$ of simplexes $T_{l+1,k}$ from $T^{(l+1)}(\bar{Q})$ such that $P_i^{(l+1)}$ is in correspondence with the basis continuous piecewise linear function $\hat{\psi}_i^{(l+1)}(x)$ such that $\hat{\psi}_i^{(l+1)}(P_i^{(l+1)}) = 1$, $\hat{\psi}_i^{(l+1)}(P_j^{(l+1)}) = 0$ for $i \neq j$ and $\hat{\psi}_i^{(l+1)}(x)$ is linear on each simplex $T_{l+1,k}$ from the triangulation $T^{(l+1)}(\bar{Q})$. Observe that the set $Q_h^{(l+1)}$ contains $Q_h^{(l)}$ and some new vertices of simplexes from $T^{(l+1)}$.

---

[4] Any such grid or triangulation can be obtained as a result of a refinement process consisting of p levels of global refinement and p_1 levels of local ones. We describe only the local refinements and start by considering an initial triangulation $T^{(0)}(\bar{\Omega}) \equiv T^0$, a closed subset $\bar{\Omega}^{(1)} \subset \bar{\Omega}$ consisting of a number of cells from $T^{(0)}$, and an integer $t_1 > 1$ (refinement ratio). Next, each simplex from $T^{(0)}$ contained in $\bar{\Omega}^{(1)}$ is partitioned into a set of t_1^d equal smaller simplexes. Cells of the old triangulation not belonging to $\Omega^{(1)}$ and of all new smaller cells belonging to $\bar{\Omega}^{(1)}$ define the composite triangulation $T^{(1)}$ of the first level. If $p_1 > 1$, then a similar procedure is carried out with respect to $\bar{\Omega}^{(2)} \subset \bar{\Omega}^{(1)}$ and the refinement ratio $t_2 > 1$, and so on. Subsets $\bar{\Omega}^{(l)} \subset \bar{\Omega}^{(l-1)}$ with $l \in [2, p_1]$ usually can be taken such that $\bar{\Omega}^{(l)}$ have no common points with the closure of $\bar{\Omega}^{(l-2)} \setminus \bar{\Omega}^{(l-1)}$; under this assumption, the $(d-1)$-dimensional surfaces separating $\bar{\Omega}^{(l)}$ and $\bar{\Omega}^{(l-1)} \setminus \bar{\Omega}^{(l)}$ consist of faces of simplexes from $T^{(l-1)}$ subdivided into $(t_l)^{d-1}$ equal parts. Thus, after all of these consequent p_1 local refinements of the initial triangulation, we come to a composite triangulations with local refinement $T^{(p_1)} \equiv T_{c,h}(\bar{\Omega})$ consisting of standard triangulations of subsets (see (1.3)) $F_1, \ldots, F_{p_1+1}$, being a particular case of more general composite grids with local refinement. Here, $F_{p_1+1} = \bar{\Omega}^{(p_1)}$, F_{p_1} is a closure of $\bar{\Omega}^{(p_1-1)} \setminus \Omega^{(p_1)}$ and so on. Refinements in different separate parts of F_i with different numbers t can also be useful.

Other new vertices are referred to as seminodes, and values of the function $\hat{u}$ at them are obtained through linear interpolation on old simplexes from $T^{(l)}$ containing the corresponding ones from $T^{(l+1)}$. Thus,

$$\hat{G}^{(l)} \equiv \{\hat{u} : \hat{u} = \sum u_i \hat{\psi}_i^{(l)}(x), \ P_i^{(l)} \in Q_h^{(l)}\} \qquad (1.5)$$

with $l \in [0, p_1]$. Along with the basis $\hat{\psi}_i^{(l+1)}(x)$ for $\hat{G}^{(l+1)}$ with $l \in [0, p_1 - 1]$, consider a new basis $\bar{\psi}_i^{(l+1)}(x)$ with

$$\bar{\psi}_i^{(l+1)} \equiv \hat{\psi}_i^{(l+1)} \text{ for } P_i^{(l+1)} \in Q_h^{(l+1)} \setminus Q_h^{(l)} \qquad (1.6)$$

and

$$\bar{\psi}_i^{(l+1)} \equiv \hat{\psi}_i^{(l)} \text{ for } P_i^{(l+1)} \in Q_h^{(l)}, \qquad (1.7)$$

assuming that the numbers of nodes from $Q_h^{(l+1)} \setminus Q_h^{(l)}$ are less than those of nodes from $Q_h^{(l)}$. The indicated choice of the new basis (often referred to as the hierarchical basis, see [34, 50, 512, 517]) leads to splittings

$$G^{(l+1)} = G_1^{(l+1)} \oplus G_2^{(l+1)}, \quad l \in [0, p-1], \qquad (1.8)$$

where

$$\hat{G}_2^{(l+1)} \equiv \hat{G}^{(l)}, \qquad (1.9)$$

$$\hat{G}_1^{(l+1)} \equiv \{\hat{u} \in \hat{G}^{l+1} : \hat{u}(P_i^{(l)}) = 0\}, \qquad (1.10)$$

and $l \in [0, p-1]$. This yields the following splitting of the space:

$$\hat{G}_h = \hat{G}_{gl} \oplus \hat{G}_{loc}, \qquad (1.11)$$

with

$$\hat{G}_{loc} \equiv \hat{G}^{(1)} \oplus \cdots \oplus \hat{G}^{(p_1)} \qquad (1.12)$$

(see (1.4), (1.5), (1.8)–(1.10)).

Recall that a very suitable formula for linear interpolation and for defining a linear function $\hat{u}$ on a d-dimensional simplex $T \equiv [P_0 P_1 \ldots P_d]$ is based on the use of so-called *barycentric coordinates* $\lambda_0, \lambda_1, \ldots, \lambda_d$, which are defined for a point M by the conditions

$$\sum_{i=1}^{d} \lambda_i \vec{P_0 P_i} = \vec{P_0 M}, \quad \lambda_0 = 1 - \sum_{i=1}^{d} \lambda_i. \qquad (1.13)$$

The formula for a linear function $\hat{u}$ takes the form (see [117])

$$\hat{u}(M) = \sum_{i=0}^{d} \hat{u}(P_i)\lambda_i. \qquad (1.14)$$

It is clear that more general approximating subspaces $\mathcal{J}^{(m)} \equiv \hat{G}_h$ can be used on the basis of *Lagrangian interpolating polynomials* of degree $m \geq 2$ on simplexes (see [28, 117, 181, 314, 345]). To construct such a polynomial $\tilde{u}^{(m)}(M)$ on a simplex T, we subdivide it into m^d subsimplexes and use values of the function at all of their vertices. If $d = 2, m = 2$, and triangle $T \equiv [P_1 P_0 P_2]$, we can write

$$\hat{u}^{(2)}(M) = \sum_{i=0}^{d} u(P_i)\lambda_i(2\lambda_i - 1) + 4 \sum_{i<j} u(P_{i,j})\lambda_i\lambda_j, \qquad (1.15)$$

where $P_{i,j}$ refers to the middle point of the edge $[P_i P_j]$.

For $d = 2$, we also describe the polynomial of mth degree using affine coordinates $\xi - 1, \xi_2$ of the point M such that $\vec{P_0 M} = \xi_1\vec{e_1} + \xi_2\vec{e_2}$, where $\vec{e_1} \equiv m^{-1}\vec{P_0 P_1}, \vec{e_2} \equiv m^{-1}\vec{P_0 P_2}$. If values of the function at the points with affine coordinates $[i_1|\vec{e_1}|, i_2|\vec{e_2}|], 0 \geq i_r \geq m, i_1 + i_2 \leq m$ are denoted by $u_{i_1,i_2}, \partial_1 u_{i_1,i_2} \equiv |e_1|^{-1}(u_{i_1+1,i_2} - u_{i_1,i_2}), \partial_2 u_{i_1,i_2} \equiv |e_2|^{-1}(u_{i_1,i_2+1} - u_{i_1,i_2})$, and $\partial_1^0\partial_2^0 u \equiv u$, then (see [64])

$$\hat{u}^{(m)}(\xi_1, \xi_2) = \sum_{k=0}^{m} \sum_{i_1+i_2=k} \partial_1^{i_1}\partial_2^{i_2} u_{0,0}(i_1!i_2!)^{-1} \times \xi_1(\xi - |e_1|)$$

$$\times \cdots \times [\xi_1 - (i_1 - 1)|e_1|]\xi_2(\xi_2 - |e_2|)\ldots[\xi_2 - (i_2 - 1)|e_2|]. \qquad (1.16)$$

1.4. Polylinear functions and their generalizations. For the simplest case of a cubical or parallelepiped grid with cells

$$\Pi_i \equiv [i_1 h_1, (i_1 + 1)h_1] \times \cdots \times [i_d h_d, (i_d + 1)h_d]$$

the space $\tilde{G}_h^{(1)}$ of continuous and piecewise polylinear functions (see [117, 324, 394, 458, 470]) is being used. Such functions can be written in the form

$$\hat{u} = \left\{ \prod_{r=1}^{d} [I + (x_r - x_{r,i_r})\partial_r] \right\} u. \qquad (1.17)$$

More general spaces $\tilde{G}_h^{(m)}$ are associated with functions that are continuous on Ω (on $\bar{\Omega}$ for the domain with Lipschitz boundary) and piecewise polynomial on Π_i. More precisely, each such function on a cell Π_i is a polynomial of degree m with respect to each variable x_r if other variables are fixed. Using the differences on the parallelepiped grid refined with the ratio m and coordinates

$$\xi_1 \equiv \frac{x_1 - x_{1,i_1}}{mh_1}, \ldots, \xi_d \equiv \frac{x_d - x_{d,i_d}}{mh_d},$$

we can write these functions in the form

$$u = \left\{ \prod_{r=1}^{d} (I + \xi_r \partial_r + \ldots + (m!)^{-1} \xi_r (\xi_r - 1) \ldots [\xi_r - (m-1)] \partial_r^m) \right\} u. \quad (1.18)$$

1.5. Prism grids; cylindrical coordinates. Consider now

$$\bar{\Omega} \equiv [\bar{\Omega}_2] \times [0 \le z \le l] \subset \mathbf{R}^3, \quad (1.19)$$

with Ω_2 a bounded region lying in the coordinate plane x, y with a given triangulation $T_h(\bar{\Omega}_2)$. For $T_k \in T_h(\bar{\Omega}_2)$, we define a prism

$$Z_{k,j} \equiv T_k \times [z_j \le z \le z_{j+1}], \quad (1.20)$$

where

$$0 = z_0 < z_1 < \ldots < z_{N_3+1} = l$$

correspond to nodes of a one-dimensional grid on $[0, e]$. For given values of the function u at the vertices of each $Z_{k,j}$, we can represent the desired function on the prism as

$$u(x, y, z) = \sum_{i=0}^{2} \lambda_i [u(P_i, z_j) + (z - z_j) \partial_z u(P_i, z_j)], \quad (1.21)$$

where P_0, P_1, P_2 refer to the vertices of T_k and $\lambda_0, \lambda_1, \lambda_2$ are barycentric coordinates of the point $[x, y] \in T_e$. Such functions are continuous on $\bar{\Omega}$ and belong to $G^{(1)}$. [5]

1.6. Boundary value problems on regions with non-Lipschitz boundary. Here we reexamine the basic notions of PGM theory for the case of a region Ω with non-Lipschitz boundary $\partial \Omega \equiv \Gamma$. We assume, however, that the closed region $\bar{\Omega}$ has Lipschitz boundary $\partial \bar{\Omega}$ and that

$$\partial \Omega = \partial \bar{\Omega} \cup \Gamma_s, \quad (1.22)$$

where Γ_s is a set of a finite number of slits $\Gamma_{s,k}$ with each a connected $(d-1)$-dimensional closed domain (with sufficiently smooth boundary) belonging to a smooth $(d-1)$-dimensional surface in $\mathbf{R}^d$. Also, we assume that, for

---

[5] Such spaces and their generalizations are especially useful for $\bar{\Omega}$ being solids of rotation, when x, y, z are cylindrical coordinates. Possible generalizations are based on m levels of refinement of the triangulation and each interval $[z_j, z_{j+1}]$, leading to a partition of the prism $Z_{k,j}$ into m^3 congruent subprisms. Then, given values of the function at their vertices, we can define functions of x and y for all $z = z_j + kh, k \in [0, m]$ via Lagrangian polynomials of degree m with respect to x and y. Then the desired function from $G^{(1)}$ can be obtained on each prism $Z_{k,j}$ through interpolation with respect to z.

each point $P \in \Gamma_s \backslash \partial \Gamma_s$, there exists a closed ball, with center at P and with small enough radius, such that it is partitioned by Γ into a finite number, say k_P, of closed connected subregions $\bar{S}_j(P), j \in [1, k_P]$, such that each of them has Lipschitz boundary. For the domain Ω depicted in Figure 6, Γ consists of the boundaries of a trapezoid, a triangle, and two vertical and one horizontal straight line segments, Γ_s consists of these segments, and $\partial \bar{\Omega}$ is the union of the boundaries of a trapezoid and the triangle.

Figure 6. Region with slits.

As an example of a three-dimensional region, we can take an open cube and the set of interior slits being a union of several triangles such that any two different triangles either have no common points or have a common side. [6]

Figure 7. Simple slit.   Figure 8. Double slit.

To apply Sobolev space theory to regions with slits, we must pay special attention to the basis notions and theorems (see [67, 68]). For example, we may speak of the spaces $W_2^m(\Omega)$ with $m = [m] \geq 1$ as completions of spaces

---

[6] The principal difference between regions with Lipschitz boundary and with non-Lipschitz boundary can be easily understood if we consider the space of functions defined on the domain Ω such that each of them has continuous bounded derivatives. Suppose that $P \in \Gamma$ and we have a sequence $\{P_n\}$ convergent to P. Then, for each function u from the above class, provided Ω has Lipschitz boundary, we obtain the sequence $\{u(P_n)\}$ that converges to a limit that does not depend on the choice of the sequence. Therefore, we may extend $u(x)$ to a continuous function on $\bar{\Omega}$. In contrast, for Ω with the non-Lipschitz boundary, the sequence $\{u(P_n)\}$ must be convergent only when $P \notin \Gamma_s$ or $P \in \partial \Gamma_s$; otherwise, it might be nonconvergent; but if $P \in \Gamma_s \backslash \partial \Gamma_s$ and all P_n belong to the same subregion $S_j(P)$, it will again be convergent, with the limit independent of the choice of the sequence in this subregion. Therefore, in the latter case, we may obtain k_P different limit values, which we can take as values $u_j(P)$ of functions $u_j(x)$ continuous on $\bar{S}_j(P), j = 1, \ldots, k_P$, and coinciding with u on $\Omega \cap S_j(P)$. For example, for Figure 6, $k_P = 2$ if $P \in \Gamma_s \backslash \partial \Gamma_s$ and P is not the point P_0 of intersection of horizontal and vertical slits, for which $k_{P_0} = 4$. For the simple slit depicted in Figure 7, $k_P = 2$ for its inner points, and it is convenient to depict it as a double slit (see Figure 8). Note that this is also in accordance with the treatment of the slit from an engineering standpoint as a mathematical model of a crack (a fracture) with a small but positive width, in a shell, plate, and so on.

of smooth functions defined on Ω in the corresponding normed space and apply embedding theorems from § 0.2, provided that slits preserve the so-called *cone condition* or its generalizations (see [67]) and we agree to use different traces of functions on the same slit (e.g., for the space $W_2^1(\Omega)$ with Ω containing the simple slit from Fig. 7, going along the x_2-axis, the two traces $u_+(x)$ and $u_-(x)$ are defined by restricting $u(x)$ to those parts of Ω with $x_1 > 0$ and $x_1 < 0$, respectively). But the Extension Theorem does not hold (see [b-i-n]); instead, we can use theorems dealing with extensions to larger regions that are either regions with continuations of some old slits or, in general case, belong to some Riemann surfaces (or their generalizations). For example, if we consider a unit circle with slit $\Gamma_s \equiv \{[t, 0], 0 \le t \le 1\}$, then it is easy to see that an extension of a function $u \in W_2^m(\Omega)$ to the function in $W_2^m(\mathbf{R}^2 \setminus [0, \infty))$ is possible, where $[0, \infty)$ is the positive x_1-semi-axis. Therefore, in working with the Hilbert space $G = W_2^1(\Omega; \Gamma_0)$ for such regions, we can include parts of double slits into Γ_0 (e.g., for Figure 8, the left part of the double slit may belong to Γ_0). The case

$$\Gamma_s \subset \Gamma_0 \tag{1.23}$$

is the simplest, and the theorems from § 0.2 apply since the space G can be regarded as a subspace of $G' \equiv W_2^1(\Omega'; \Gamma_0 \setminus \Gamma_s)$, where Ω' is a domain with Lipschitz boundary such that $\bar{\Omega}' = \bar{\Omega}$.

Returning now to PGMs, the case (1.23) needs no alterations of the basic notions and constructions of PGMs. But if (1.23) does not hold, when we speak about a grid (triangulation) of $\bar{\Omega}$ (noncomposite) and of the set of nodes $P \in \Gamma_s \setminus \partial\Gamma_s$, then each must be regarded as a *multinode* in accordance with the number k_P of regions $S_j(P)$ defined above and the same number of possible different limit values $\hat{u}(P^{(j)}), j \in [1, k_P]$ (the same applies to seminodes in case of composite triangulations). Each of these elementary nodes $P^{(j)} \in \partial\Gamma_s, P \notin \Gamma_0$, must be in one-to-one correspondence with the basis function. Moreover, these basic functions are continuous only on Ω and the corresponding domains $S_j(P)$ and are actually k_P-valued functions if considered on $\bar{\Omega}$; in other words, supports of these functions must belong only to one $\bar{S}_j(P)$ (crossing slits is not allowed).

Splitting of the nodes on a simple slit is depicted at Figure 8, and it is notable that even different grids (maybe composite ones) might be used at different subregions $S_j(P)$ leading to a composite triangulation of the region. By the above reasoning, when we speak about topological equivalence of grids, it makes sense instead of grids (triangulations) for $\bar{\Omega}$ to consider those for regions on some Riemann surfaces (and their generalizations), as was mentioned with respect to the grids depicted in Figures 4 and 5. After making the above adjustments of notions of nodes and basis functions, and denoting the corresponding set of nodes (elementary nodes) associated with basis functions by Ω_h, we can use again expansions of type (0.2.2), rewritten in the form

$$\hat{u} = \sum_{P_i \in \Omega_h} u_i \hat{\psi}_{P_i}(x) \qquad (1.24)$$

and apply results concerning error estimates of projective methods and resulting algebraic systems. Here we meet a very interesting problem dealing with generalizations of spaces with a fractional index to the class of domains under consideration since. To define these spaces through completions of the spaces of smooth functions on Ω becomes very restrictive and does not allow us to work with functions having different limit values on different edges of slits.[7] An effective approach to the construction of these spaces is based on partitions of $\bar{\Omega}$ into subregions with Lipschitz boundaries, the use of conditions of (0.5.4) and some additional matching conditions for subdomains having common parts of their boundaries which do not belong to Γ (see [67, 68, 506]); another way to introduce them is based on partitions of Ω with overlapping (see [68, 103, 459]). We mostly avoid such spaces (§ 3.5 will be an exception, where such a definition of the space $W_2^{1/2}(\Gamma)$ is used), since for our error analysis conditions of type (0.5.4) are quite satisfactory; note that estimates (0.5.5) of N-widths from below remain true. [8]

1.7. Use of symmetry of the solution. If the original elliptic boundary value problem in a Hilbert space G is associated with a region having a symmetry property and some additional conditions on given coefficients and functions are satisfied, then the corresponding symmetry property of the solution can be expected. Taken into account for construction of the numerical method, it might lead to a significant reduction of the number of unknowns and of computational work. For example, if the region Ω is symmetric with respect to the $(d-1)$-dimensional plane $x_1 = 0$, then the solution of the original boundary value problem for the elliptic equation under consideration may be an even or odd function with respect to x_1, that is, either

$$u(-x_1, x_2, \ldots, x_d) = u(x_1, x_2, \ldots, x_d) \qquad (1.25)$$

---

[7] Such an approach probably works well for regions on Riemann surfaces like those for Figures 4 and 5.

[8] We conclude with a few other computational advantages of Riemann surfaces. For example, a simply connected planar region with a simple slit cannot be topologically equivalent to an open rectangle. But if we replace the simple slits by double ones and the region by a corresponding one on a Riemann surface, then such an equivalence is possible. Even the closed domain on such a surface depicted in Figure 6 (with double slits) can be topologically transformed into a rectangle. This remark might be useful in constructing grids of the desired simplicity. Also, we can use completely independent triangulations in different $S_j(P)$, and even different approximations of slits might be of help if we deal, e.g., with a curvilinear slit.

or

$$u(-x_1, x_2, \ldots, x_d) = -u(x_1, x_2, \ldots, x_d). \tag{1.26}$$

In general, following [500], we associate the symmetry property of the region with an isometric mapping m of $\mathbf{R}^d$ onto itself such that $m^r = I$ for some integer $r \geq 1$ and

$$m(x) \in \bar{\Omega} \text{ if } x \in \bar{\Omega}; \tag{1.27}$$

on the basis of (1.27), we define an invertible *symmetry operator* $S \in \mathcal{L}(G)$ and corresponding *symmetric subspace*

$$G_{(s)} \equiv \{v : v \in G, Sv = v\}, \tag{1.28}$$

which may be considered as a new Hilbert space (with the old inner product). Then, if we can prove that

$$u \in G_{(s)}, \tag{1.29}$$

where u is the solution of the problem under consideration, then we consider a reformulation that deals only with corresponding subspaces $\hat{G}_{(s),h} \subset G_{(s)}$. Note that for the above considered example of symmetry with respect to the plane $x_1 = 0$, we have $m(x) = [-x_1, x_2, \ldots, x_d]$, and $Su(x) = u(m(x))$ or $Su(x) = -u(m(x))$ for (1.25) and $Su(x) = -u(m(x))$ for (1.26). For a system of equations with $u(x) \equiv [u_1(x), \ldots, u_k(x)]$, the operator S may take a more complicated form since various combinations of odd and even components are possible (we shall consider such cases in Chapter 5–7). Consider the original problem given in the form (1.18) rewritten here as

$$b(u; v) = l(v), \ \forall v \in G, \forall u \in U. \tag{1.30}$$

Theorem 1. Let problem (1.30) have a unique solution and suppose that

$$Sw \in U, \forall w \in U, \tag{1.31}$$

$$b(Sw; Sv) = b(w; v), \ \forall w \in U, \forall v \in G, \tag{1.32}$$

$$l(Sv) = l(v), \quad \forall v \in G. \tag{1.33}$$

Then (1.29) holds.

   Proof. Conditions (1.31)–(1.33) imply that $b(Su; Sv) = l(Sv), \ \forall v \in G$. Since $S \in \mathcal{L}(G)$ and is invertible, then the set of all Sv is just the Hilbert space G. Thus, $b(Su; v) = l(v), \forall v \in G$, and Su is a solution of the problem. By uniqueness, (1.29) is valid. $\square$

It is obvious that Theorem 1 holds for projective approximations (see (1.1.9) with $\hat{S}$ replaced by $\hat{U}$) provided that

$$S\hat{w} \in \hat{U}, \ \forall \hat{w} \in \hat{U} \equiv \hat{U} \equiv \hat{G} \cap U. \tag{1.34}$$

Hence, $S\hat{w}$ must be an element of $\hat{G}$. [9]

§ 2. Linear homotopy and change of space variables in constructing projective-grid methods

2.1. Change of variables in projective-grid methods. Mappings of given regions Ω into simpler model regions Q were defined by the transformation (1.2) of original variables $x \equiv [x_1, \ldots, x_d]$ into variables z, associated with points of Ω and Q, respectively. This transformation yields an isomorphism between linear spaces of functions defined on Ω and Q, with the following correspondence between their elements:

$$u_Q \equiv Zu, \quad u \equiv \Pi u_Q, \tag{2.1}$$

where $u_Q(z) = u(x)$ if z and x satisfy (1.2) (we prefer here to speak about functions defined on domains to emphasize a possibility of slits and different limit values of functions for points on them). Now we are interested in conditions on these transformations leading to the isomorphism of the spaces $W_2^m(\Omega) \equiv G^{(m)}$ and $W_2^m(Q)$ $(m > 0)$ with the estimates

$$K_0 \|u_Q\|_{m,Q} \leq \|u\|_{m,\Omega} \leq K_1 \|u_Q\|_{m,Q}, \quad K_0 > 0 \tag{2.2}$$

---

[9] We restrict ourselves to the symmetry of the triangulation with respect to the plane $x_1 = 0$. For two-dimensional problems, construction of such a triangulation presents no problem. For $d \geq 3$, if we construct the desired triangulation (or a composite triangulation in the general case) for that part of Ω whose points satisfy the condition $x_1 \geq 0$, and if we use its symmetric image for that part with $x_1 \leq 0$, then $(d-1)$-dimensional simplicial partitions of the corresponding part of the plane will coincide for both triangulations (composite triangulations). This implies that the union of all simplexes of both triangulations (composite triangulations) yields the desired symmetric triangulation (composite triangulation) of $\bar{\Omega}$ (note that in this case the plane $x_1 = 0$ contains no seminodes). Therefore, for such PGMs we can obtain symmetric subspaces $\hat{G}_{(s),h}$ of functions in $\hat{G}_h$ satisfying condition (1.25) or (1.26). What remains to be verified is only that $S\hat{w} \in U$ if $\hat{w} \in U$. This is usually the case for standard localizations. Note finally that (1.32) may hold in cases that (1.33) does not. This is the situation if, e.g., $l(v) \equiv (f(x), v(x))_{0,\Omega}$, where the function f is neither odd nor even with respect to x_1. Then we can make use of the representation $f(x) = f_1(x) + f_2(x)$, where $2f_1(x) \equiv f(x_1, x_2, \ldots, x_d) - f(-x_1, x_2, \ldots, x_d)$, $2f_2(x) \equiv f(x_1, x_2, \ldots, x_d) + f(-x_1, x_2, \ldots, x_d)$. For linear problems, then, the solution is the sum of solutions of two problems defined by right-hand sides $f_1(x)$ and $f_2(x)$. The same holds for the corresponding grid systems, and this can be effectively exploited in the solution process (see, e.g., [134, 168]).

for all $u \in G^{(m)}$ and $u_Q = Zu$ (the equality $u_Q(z) = u(x)$ is taken to mean for almost all x). Such an isomormphism is denoted as

$$W_2^m(\bar{\Omega}) \leftrightarrow W_2^m(Q), \qquad (2.3)$$

and we write the Jacobian matrices in the form $Z_x' \equiv [\frac{\partial Z_i}{\partial x_j}]$, $\Pi_z' \equiv [\frac{\partial \Pi_i}{\partial z_j}]$ with their determinants $\det Z_x'$ and $\det \Pi_z$; we also write $m' \equiv \max\{1, m\}$ if $m = [m]$ and $m' \equiv [m] + 1$ if $m > [m]$. We formulate an important general lemma, the proof of which can be found in [67, 387].

Lemma 1. *Suppose that the functions $Z_j, j = 1, \ldots, d$, from (1.2) have continuous uniformly bounded derivatives up to the order m' and that $|\det \Pi_z'| \geq \delta > 0$ for all $z \in Q$. Suppose $u \in G^{(m)}$. Then $u_Q \in W_2^m(Q)$ and there exists a constant $K_0 > 0$ such that the inequality from (2.2) involving K_0 holds for all $u \in G^{(m)}$; under similar conditions on the functions Π_j and $\det Z_x'$, there exists a constant $K_1 > 0$ such that the corresponding inequality from (2.2) holds for all $u \in G^{(m)}$; the generalized derivatives of u and u_Q of the order $k \leq [m]$ are connected by the same equalities as the derivatives in the classical sense under the change of variables $x = \Pi(z)$, but interpreted now as holding for almost all x and z.*

This lemma enables us to write

$$\mathrm{grad}_x u \equiv \nabla_x u = (Z_x')^* \nabla_z u_Q, \quad \nabla_z u_Q = (\Pi_z')^* \nabla_x u, \qquad (2.4)$$

where $\nabla_x u \equiv [\frac{\partial u}{\partial x_1}, \cdots, \frac{\partial u}{\partial x_d}]^T$, $\nabla_z u_Q \equiv [\frac{\partial u_Q}{\partial z_1}, \cdots, \frac{\partial u_Q}{\partial z_d}]^T$, $A^* \equiv A^T$, and

$$|u|_{1,\Omega}^2 = (|(Z_x')^* \nabla_z u_Q|^2, |\det \Pi_z'|)_{0,Q},$$

$$|u_Q|_{1,Q}^2 = (|(\Pi_z')^* \nabla_x u|^2, |\det Z_x'|)_{0,\Omega}. \qquad (2.5)$$

Conditions on the functions Π_j and Z_j might be essentially weakened, which can be important in constructing PGMs, that is, in constructing corresponding grids and subspaces for regions with a complicated geometry. Trying to move to a simpler model region on the basis of transformation (1.2), it seems reasonable to partition (cut) $\bar{\Omega}$ into a union of a finite number of blocks (supercells, superelements) $\bar{S}_{\Omega,k}, k \in [1, k_0]$ (different domains S_i and S_j do not have common points), and to construct the desired mappings of these blocks onto corresponding model blocks $S_{Q,k}$ of $\bar{Q}$ separately, provided of course that these mappings coincide on common parts of the boundaries of two neighboring blocks (*matching of blocks*) (see [117, 177, 178, 180, 379, 390, 478, 518]).

Lemma 2. *Let $\bar{S}_{Q,k}$ and $\bar{S}_{\Omega,k}, k \in [1, k_0]$, be corresponding blocks from partitions of $\bar{Q}$ and $\bar{\Omega}$ under the continuous transformation $x = \Pi z$ on Q (see (1.2)) and such that all functions $\Pi_i(z), i \in [1, d]$, have continuous* and bounded first partial derivatives on each domain $S_{Q,k}$. Suppose that $|\det \Pi'_z| \geq \delta > 0$ and that a function $\psi_{i,Q}$ is also continuous on Q and has continuous and uniformly bounded first partial derivatives on each domain $S_{Q,k}$. Then

$$\psi_i \equiv \psi_{i,\Omega} \equiv \Pi \psi_{i,Q} \in G^{(1)}. \tag{2.6}$$

Proof. The functions ψ_i inside each $S_k \equiv \Pi\{S_{Q,k}\}$ have continuous first derivatives that are uniformly bounded. Also, $\det Z'_x \det \Pi'_z = 1$. Hence, we may apply (2.5) and conclude that $\psi_i \in W^1_2(S_k)$. These functions are continuous on Ω; their limit values are defined on Γ but may be different for any point on a slit. Thereby, we obtain (2.6). $\square$ [10]

2.2. Standard quasitriangles and quasisquares. There are many ways to transform the geometry of the given region. In choosing the form of a transformation, one must keep in mind its ultimate use—namely, the grid approximations and resulting grid systems that will be obtained. It is therefore reasonable to use the simplest transformations of classical topology, such as *linear homotopy* (see [66, 115, 333, 379]). Of course, these transformations can be effectively applied only for certain standard blocks, and it is very important to define them properly and especially to be able to partition the given region into a set of such blocks. For two-dimensional manifolds, the most natural choice of a block is a triangle with two straight sides and one curvilinear side like region $A_1 A_3 A_4 \equiv T'$ with the curvilinear side $\Gamma_3 \equiv A_4 A_3$ shown in Figure 1.

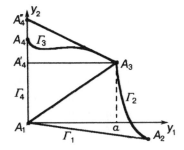

Figure 1. Quasisquare.

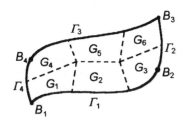

Figure 2. Partition of region.

We refer to it as a *quasitriangle* since it can be readily transformed into the ordinary triangle $T \equiv \triangle A_1 A_3 A_4$ by the linear homotopy, which in a suitable Descartes coordinate system with y_2-axis going along $[A_1 A_4]$, for a fixed y_1, is just a transformation mapping one straight line segment into

---

[10] If the set of $\psi_{i,Q}$ is a basis for a space $\hat{G}_{Q,h}$, then the set of ψ_i is a basis for the corresponding space $\hat{G}_{\Omega,h}$. But of course the latter space may be rather complicated and is determined by the form of the transformation Π.

another (note that straight sides of $\mathcal{T}$ coincide with those of T and that the exact form of the transformation will be given below). The similar transformation of the quasitriangle $T'' \equiv A_1 A_2 A_3$ with the curved side Γ_2 (the arc $A_2 A_3$) into the triangle $\triangle A_1 A_2 A_3$ enables us to construct a desired transformation of a *quasisquare* $A_1 A_2 A_3 A_4$ (see Figure 1) into a quadrilateral consisting of $\triangle A_1 A_2 A_3$ and $\triangle A_1 A_3 A_4$ (the case of $A_2 A_3 \parallel A_3 A_4$ is permitted). Below we use the the standard notation $f \in C^m[0, a]$, referring to a function defined on the segment $[0, a]$, with first continuous uniformly bounded m derivatives inside the segment (sometimes we write $f \in C^m$ even for functions of several variables). Also, we use the following definition (see [176, 177, 178]): a quasitriangle $\mathcal{T} \equiv A_1 A_2 A_3$ is called a *standard quasitriangle of the order m* if there exists a Descartes coordinate system $[y_1, y_2]$ with y_2-axis going along $[A_1 A_4]$ and such that $A_1 = [0, 0]$, $A_4 = [0, f(a)]$, $A_3 = [a, f(a)]$, with $a > 0, f(a) > 0$, and the curve $A_1 A_3$ defined by the equation

$$y_2 = f(y_1), \quad 0 \le y_1 \le a, \quad f \in C^{m+1},$$

and there exist triangles $A_1 A_3 A_4'$ and $A_1 A_3 A_4''$ with the vertices A_4' and A_4'' on the ray $A_1 A_4$ such that $\triangle A_1 A_2 A_4' \subset T' \subset \triangle A_1 A_2 A_4''$.

A simply connected bounded closed region T' with boundary Γ partitioned by the points $A_1, A_2, A_3,$ and A_4 into a set of four arcs $\Gamma_r, r \in [1, 4]$ (see Figure 1), is called a *standard quasisquare of order m* if two neighboring arcs (say, Γ_1 and Γ_4) are closed straight line segments and the segment $[A_1 A_3]$ subdivides $\mathcal{T}$ into two standard quasitriangles, say T' and T'', of order m. Figure 2 illustrates a possible partition of the given $\bar{\Omega}$ into a union of standard quasisquares and quasitriangles.

For T', let the line segments $[A_1 A_3], [A_4 A_3], [A_4' A_3],$ and $[A_4'' A_3]$ belong to the straight lines defined by $y_2 = \lambda y_1$, $y_2 - f(a) = \lambda_1(y_1 - a)$, $y_2 - f(a) = \lambda_1'(y_1 - a)$, and $y_2 - f(a) = \lambda_1''(y_1 - a)$, respectively. Then the linear homotopy $t = Z(y)$ mapping the quasitriangle T' into the $\triangle \, T'$ (sides $[A_1 A_4]$ and $[A_1 A_3]$ are fixed and the arc $\overparen{A_4 A_3}$ is replaced by the chord $[A_4 A_3]$) is given by the transformation $t_2 = \lambda t_1 + (y_2 - \lambda y_1) p_1(y_1) \equiv \eta(y_1, y_2)$, $t_1 = y_1$, where $p_1(y_1) \equiv (y_1 - a)(f(y_1) - \lambda y_1)^{-1}(\lambda_1 - \lambda)$. The inverse transformation $y = \Pi(t)$ is defined by $y_1 = t_1$, $y_2 = \lambda t_1 + (t_2 - \lambda t_1) p_2(t_1) \equiv \zeta(t_1, t_2)$, where $p_2(t_1) \equiv (f(t_1) - \lambda t_1)[(\lambda_1 - \lambda)(t_1 - a)]^{-1}$. In the sequel, $a_1 \equiv (\lambda - \lambda_1)(\lambda - \lambda_1'')^{-1}$, $a_2 \equiv (\lambda - \lambda_1)(\lambda - \lambda_1')^{-1}$, $\mathcal{T} \equiv T'$.

Lemma 3. Suppose $\mathcal{T}$ is a standard quasitriangle of order $m \ge 0$. Then, for all its inner points $[y_1, y_2]$, we have

$$|\frac{\partial \eta}{\partial y_1}| \le \kappa_{1,0}, \quad |\frac{\partial \eta}{\partial y_2}| \le \kappa_{0,1}, \quad \det Z_y' \in [a_1, a_2], \tag{2.8}$$

$|\frac{\partial^{k_1+k_2}\zeta}{\partial t_1^{k_1}\partial t_2^{k_2}}| \le \bar{\kappa}_{k_1,k_2}$, $1 \le k_1 + k_2 \le m+1$, *with the constants specified in the proof.*

Proof. For all possible λ (not necessarily a positive number), we have $\lambda > \lambda_1' \ge \lambda_1 \ge \lambda_1''$ and it is easy to see that $\kappa_{0,1} = \frac{\lambda-\lambda_1}{\lambda-\lambda_1'}$ and

$$\frac{\partial \eta}{\partial y_2} = \frac{\partial(t_1,t_2)}{\partial(y_1,y_2)} = p_1(y_1) \in [\frac{\lambda-\lambda_1}{\lambda-\lambda_1'}, \frac{\lambda-\lambda_1}{\lambda-\lambda_1'}].$$

Also, $|1 - p_1(y_1)| \le \max\{\frac{\lambda_1-\lambda_1''}{\lambda-\lambda_1'}; \frac{\lambda_1'-\lambda_1}{\lambda-\lambda_1'}\} \equiv K_0$. Since $\frac{\partial \eta}{\partial y_1} = \lambda(1 - p_1(y_1)) + \frac{y_2-\lambda y_1}{f(y_1)-\lambda y_1}[(\lambda_1 - \lambda)(1 - p_1(y_1)) - (f'(y_1) - \lambda_1)p_1(y_1)]$, then $\kappa_{1,0} = K_0 \max\{|\lambda|; |\lambda_1|\} + \kappa_{0,1}\|f'(y_1) - \lambda_1\|_C$. Similarly,

$$\frac{\partial \zeta}{\partial t_2} = p_2(t_1) \in [\frac{\lambda-\lambda_1'}{\lambda-\lambda_1}, \frac{\lambda-\lambda_1''}{\lambda-\lambda_1}], \text{ and } \bar{\kappa}_{0,1} = \frac{\lambda-\lambda_1''}{\lambda-\lambda_1}.$$

For $k_2 \ge 2$, $\bar{\kappa}_{k_1,k_2} = 0$, and we need to analyze only the most difficult case of $\frac{\partial^{k_1+1}\zeta}{\partial t_1^{k_1}\partial t_2} = p_2^{(k_1)}(t_1)$. To estimate $|p_2^{(k_1)}(t_1)|$, we use the Taylor expansion

$$f(t_1) = \sum_{r=0}^{k_1} \frac{f^{(r)}(a)}{r!}\xi^r + \varphi_{k_1}(\xi),$$

where $\xi \equiv t_1 - a$ and

$$\varphi_{k_1}(\xi) \equiv \frac{1}{k_1!}\int_0^\xi f^{(k_1+1)}(a+\nu)(\xi-\nu)^{k_1}d\nu.$$

For $p_2^{(k_1)}(t_1) = \frac{1}{\lambda_1-\lambda}\frac{\partial^{k_1}\varphi_{k_1}(\xi)}{\partial\xi^{k_1}} \equiv X$, using $C_k^r \equiv \frac{k!}{r!(k-r)!}$, we have

$$X = \frac{1}{\lambda_1 - \lambda}\sum_{r=0}^{k_1} C_{k_1}^r \frac{(-1)^r}{\xi^{r+1}}\int_0^\xi f^{(k_1+1)}(a+\nu)(\xi-\nu)^r d\nu.$$

Hence, $\bar{\kappa}_{k_1,1} = \frac{\|f^{k_1+1}(t_1)\|_C}{\lambda-\lambda_1}\sum_{r=0}^{k_1} C_{k_1}^r \frac{1}{r+1}$. Similarly, we obtain

$$\bar{\kappa}_{1,0} = \|f'(t_1) - \lambda\|_C[\frac{\lambda-\lambda_1''}{\lambda-\lambda_1} + \max\{|\lambda - \frac{(\lambda-\lambda_1')^2}{\lambda-\lambda_1}|; |\lambda - \frac{(\lambda-\lambda_1'')^2}{\lambda-\lambda_1}|\}],$$

$$\bar{\kappa}_{k_1,0} = \|f^{k_1}(t_1)\|_C \left(\frac{\lambda-\lambda_1''}{\lambda-\lambda_1} + \frac{\max\{|\lambda_1'|; |\lambda_1''|\}}{\lambda-\lambda_1}\sum_{r=1}^{k_1}\frac{1}{r}C_{k_1}^r\right)$$

(for $k_1 \ge 2$), which completes the proof. $\square$

Lemma 4. *Let T be a standard quasitriangle of the order $m \geq 0$. Then*

$$\delta_0 |u|^2_{1,T} \leq |u_Q|^2_{1,T} \leq \delta_1 |u|^2_{1,T}, \quad \forall u \in W^1_2(T), \tag{2.9}$$

$$\delta_0 \equiv \min_{y \in T} \left\{ \kappa_0(y) \frac{(\lambda_1 - \lambda)(y_1 - a)}{f(y_1) - \lambda y_1} \right\}, \tag{2.10}$$

$$\delta_1 \equiv \max_{y \in T} \left\{ \kappa_1(y) \frac{(\lambda_1 - \lambda)(y_1 - a)}{f(y_1) - \lambda y_1} \right\}, \tag{2.11}$$

$\kappa_0(y)$ *and* $\kappa_1(y)$ *are the minimal and maximal eigenvalues of the matrix*

$$A(y) \equiv \begin{bmatrix} 1 & a_{1,2} \\ a_{1,2} & a_{2,2} \end{bmatrix}, \quad a_{1,2} \equiv \frac{\partial \eta}{\partial y_1}, \ a_{2,2} \equiv |\nabla_y \eta|^2. \tag{2.12}$$

Proof. From (2.5), with $Q = T$, we see that

$$|u_T|^2_{1,T} = (|(Z'_y)^* \nabla_y u|^2, |\det Z'_y|)_{0,T}. \tag{2.13}$$

Since $A(y) = (Z'_y)^* Z'_y$, then (2.9)–(2.11) follow from (2.13) and (2.8). $\square$ [11]
In the sequel, we use $C^k_r \equiv \frac{r!}{k!(r-k)!}$,

$$|u^{(r)}_T|^2 \equiv \sum_{k=0}^{r} C^k_r \left(\frac{\partial^r u}{\partial y_1^k \partial y_2^{r-k}} \right)^2, \quad |u^{(r)}_T|^2 \equiv \sum_{k=0}^{r} C^k_r \left(\frac{\partial^r u_T}{\partial t_1^k \partial t_2^{r-k}} \right)^2. \tag{2.14}$$

Were we not interested in the nature of the constants, then the following lemma could be formulated as a trivial consequence of (2.3) and Lemma 1. However, we prove it in its current form (suggested in [183]), because for affine mappings it yields optimal bounds for $|u^{(r)}_T|$ and reduces to

$$|u^{(r)}_T| = |u^{(r)}_T| \tag{2.15}$$

if only transformations of the Descartes coordinate system are used (see also [460]).

Lemma 5. *Let T be a standard quasitriangle of the order $m \geq 1$ and $r \leq m + 1$. Then for all its inner points $[y_1, y_2]$, we have*

$$|u^{(r)}_T| \leq K_{r,1} |u^{(1)}_T|^2 + \ldots + K_{r,r} |u^{(r)}_T|^2; \tag{2.16}$$

if T is a triangle and the transformation $y = \Pi(t)$ is affine defined by the matrix Π, then

$$|u^{(r)}_T| \leq \|\Pi\|^r |u^{(r)}_T|, \ |u^{(r)}_T| \leq \|\Pi^{-1}\|^r |u^{(r)}_T|. \tag{2.17}$$

[11] In the same manner, we can estimate $|u|^2_{1,T}$.

Proof. We drop the indices T and $\mathcal{T}$ and write $\mathrm{grad}_y u \equiv \mathrm{grad}_y^{(1)} u$. Suppose that, for $k \geq 2$,

$$\mathrm{grad}_y^{(k)} u \equiv \left[\begin{array}{c} \mathrm{grad}_y^{(k-1)} \frac{\partial u}{\partial y_1} \\ \mathrm{grad}_y^{(k-1)} \frac{\partial u}{\partial y_2} \end{array} \right] \in \mathbf{R}^{2^k}.$$

Note that the relation $C_{s+1}^{k+1} = C_s^k + C_s^{k+1}$ and the induction imply that $|\mathrm{grad}_y^{(k)} u|^2 = |u_{\mathcal{T}}^{(k)}|^2$. We also use the vector $\mathrm{grad}_y^{(k),l} u$ containing l blocks equal to $\mathrm{grad}_y^{(k)} u$, and similarly we define $\mathrm{grad}_t^{(k)} u$ and $\mathrm{grad}_t^{(k),l} u$. It then for (2.16) suffices to prove, for $r \leq m+1$, that

$$|\mathrm{grad}_t^{(r)} u| \leq \sum_{l=1}^{r} K_{r,l} |\mathrm{grad}_y^{(r),2^{r-l}} u|. \tag{2.18}$$

Concerning the connection between $\mathrm{grad}_t^{(r)} u \equiv U^r$ and $\mathrm{grad}_y^{(r)} u$, we have

$$U^1 = \mathrm{grad}_t^{(1)} u = A_{1,1} \mathrm{grad}_y^{(1)} u, \quad A_{1,1} \equiv B \equiv \left[\begin{array}{cc} 1 & \frac{\partial \zeta}{\partial t_1} \\ 0 & \frac{\partial \zeta}{\partial t_2} \end{array} \right],$$

where B corresponds to $(\Pi_z')^*$ from (2.4). But, for $r \leq m$,

$$U^r = A_{r,r} \mathrm{grad}_y^{(r)} u + A_{r,r-1} \mathrm{grad}_y^{(r-1),2} u + \cdots + A_{r,1} \mathrm{grad}_y^{(1),2^{r-1}} u, \tag{2.19}$$

where $D(B; n)$ denotes the block diagonal matrix whose all diagonal blocks are the square matrix B and the number of them is n. Then we obtain the similar expression for U^{r+1}, with
$A_{r+1,r+1} \equiv D(A_{r,r}; 2) P_{r+1,r+1} D(B; 2^r) P_{r+1,r+1}^{(1)}$,

$$A_{r+1,k} \equiv \left[\begin{array}{cc} (A_{r,k})_{t_1} & 0 \\ 0 & (A_{r,k})_{t_2} \end{array} \right] + D(A_{r,k-1}; 2^2) P_{r+1,k} D(B; 2^r) P_{r+1,k}^{(1)}$$

if $2 \leq k \leq r$, and $A_{r+1,1} \equiv \left[\begin{array}{cc} (A_{r,1})_{t_1} & 0 \\ 0 & (A_{r,1})_{t_2} \end{array} \right]$,

where P refers to permutation matrices (of norm 1), $(A)_{t_s}$ refers to the derivative of the matrix A with respect to t_s, $s = 1, 2$, and elements of the matrices $A_{r,l}$, $r > 1$, contain only derivatives of order less than r of elements of B and, in accordance with Lemma 3, are uniformly bounded. Hence, (2.19) holds for all $r \leq m+1$ and leads to

$$|\text{grad}_t^{(r)}u|^2 \le \sum_{l=1}^{r} K_{r,l}|\text{grad}_y^{(l),2^{r-l}}u|^2 \qquad (2.20)$$

yielding inequality (2.18) and (2.16). (If elements of B are constants, then (2.20) takes the form $|\text{grad}_t^{(r)}u| \le \|B\|^r|\text{grad}_y^{(r)}u|$ (see (2.17)); if B is an orthogonal matrix, we then have $|U^r| \le |\text{grad}_y^{(r)}u|$. Since the reverse inequality also holds, we obtain (2.15).) □ [12]

2.3. Linear and central homotopy in space. Linear homotopy is easily applicable for the multi-dimensional case $d \ge 3$. We restrict ourselves to consideration only of the illustrative case $d = 3$.

For a given simplex $T = A_0 B_0 B_1 B_2$, introduce a Descartes coordinate system $[y_1, y_2, y_3]$ with origin at point A_0 and y_3-axis along vector $\vec{A_0 B_0} \equiv \vec{l}$ (see Figure 3).

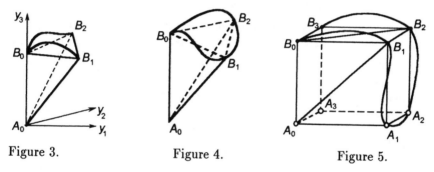

Figure 3.   Figure 4.   Figure 5.

Let the orthogonal projections of the vertices B_0, B_1, and B_2 onto the coordinate plane $y_3 = 0$ be denoted by B_0', B_1', and B_2'. Let $\triangle B_0 B_1 B_2 \equiv \triangle'$. Suppose we replace the face $S_0 \equiv \triangle B_0 B_1 B_2$ by a surface S defined by the function

$$y_3 = f(y_1, y_2), \quad [y_1, y_2] \in \triangle, \quad f \in C^{m+1}, m \ge 0$$

and $B_k \in S, k \in [0, 2], [B_1 B_2] \subset S, S \cap \triangle A_0 B_1 B_2 = [B_1 B_2]$. Then the resulting closed region (solid) T is referred to as a *standard quasisimplex of type* $[\vec{l}, m]$. If linear equations $y_3 = g(y_1, y_2)$ and $y_3 = f_0(y_1, y_2)$ correspond to planes containing $\triangle A_0 B_1 B_2$ and $\triangle B_0 B_1 B_2$, respectively, then the linear homotopy transforming T into simplex T is given by the equations

---

[12]For $u \in G^{(m)}(\Omega)$, we can approximate it by a sequence of smooth functions u_n and integrate corresponding inequalities from Lemmas 3–5 for them over the regions under consideration. Taking into account the estimates for det Z_y', we can then pass to the limit and obtain the corresponding estimates for norms in the given Sobolev spaces (we shall return to consideration of similar estimates for piecewise affine mappings later).

$$t_1 = y_1, \quad t_2 = y_2, \quad t_3 = g + \frac{f_0 - g}{f - g}(y_3 - g) \equiv \xi(y)$$

(here, of course, g, f_0, and f depend on $y \equiv [y_1, y_2, y_3]$). [13] The restriction $[B_1 B_2] \subset S$ is rather severe: even a region as a ball cannot be partitioned into a union of standard quasisimplexes of the above type. A more general transformation is a *central homotopy* or *perspective transformation*. To describe it, consider barycentric coordinates $\lambda \equiv [\lambda_0, \lambda_1, \lambda_2]$ of points $M_0 \in S_0$. Then each point on the ray along $\vec{A_0 M_0}$ is characterized by the four-dimensional vector $M' \equiv [\lambda, r] \equiv [\lambda_0, \lambda_1, \lambda_2, r]$, where $r = |A_0 M|$. Suppose S_0 and S are defined by equations $r = \varphi_0(\lambda) > 0$ and $r = \varphi(\lambda) > 0$, where $\varphi \in C^{m+1}$, $m \geq 0$, $\varphi(e_i) = \varphi_0(e_i), i \in [1,3]$. We define a *standard quasisimplex T of order m* as the locus of points M such that

$$\vec{A_0 M} = r \vec{A_0 M'}, \quad M' \in S_0, \quad 0 \leq r \leq \varphi(\lambda).$$

Then the perspective transformation of T' into the simplex T is given by the relations $\lambda_T = \lambda$, $r_T = r \frac{\varphi_0(\lambda)}{\varphi(\lambda)} \equiv \xi(\lambda)$. This means that we again deal with the linear homotopy, but now for every fixed λ. Note that it is widely used for the perspective representation of functions of two variables to depict the function so that its properties are most easily visualized. It is not very difficult to express this transformation in terms of Descartes coordinates. For example, if $A_0 = [0,0,0]$ and S_0 belongs to the plane $x_3 = |A_0 B_0|$, then it can be written in the form

$$t_i = y_i [y_3^2 + f_0^2]^{1/2} [y_3^2 + f^2]^{-1/2}, \quad i = 1, 2, 3, \tag{2.21}$$

where $y_3 = f_0(y_1, y_2)$ defines the plane containing the triangle S_0. The next lemma is an easy consequence of (2.21) and Lemma 1.

Lemma 6. *Let T be a standard quasisimplex of order $[m]$, $m > 0$, and consider the equation*

$$y_3 = f(y_1, y_2), \ f \in C^{m+1}, \tag{2.22}$$

---

[13] It is possible to obtain generalizations of Lemmas 3 and 4 dealing with planes P_{-1} and P_1, containing the straight line segment $[B_1 B_2]$ and defined by the equations $y_3 = f_{-1}(y_1 y_2)$ and $y_3 = f_1(y_1, y_2)$, with $g \leq f_{-1} \leq f \leq f_1$. Also, for generalizations of Lemmas 4 and 5, it is reasonable to use matrices

$$B \equiv A_{1,1} \equiv \begin{bmatrix} 1 & 0 & \frac{\partial \zeta}{\partial t_1} \\ 0 & 1 & \frac{\partial \zeta}{\partial t_2} \\ 0 & 0 & \frac{\partial \zeta}{\partial t_3} \end{bmatrix}, \ \Pi' \equiv \begin{bmatrix} 1 & 0 & \frac{\partial \eta}{\partial y_1} \\ 0 & 1 & \frac{\partial \eta}{\partial y_2} \\ 0 & 0 & \frac{\partial \eta}{\partial y_3} \end{bmatrix}, \ A(y) \equiv \Pi'(\Pi')^*.$$

for S. Then (2.21) *leads to* (2.3), *with* $\Omega = T$ *and* $Q = T$. [14]

2.4. Piecewise affine mappings. Let $R \equiv T$ denote a d-dimensional reference simplex with vertices $P_0' \equiv 0$, $P_1' \equiv P_0' + e_1, \ldots, P_d' \equiv P_{d-1}' + e_d \equiv [1, 1, \ldots, 1]$ and let the simplex T denote its image under the mapping

$$x = \Pi z + a_0, \tag{2.23}$$

where $x \equiv [x_1, \ldots, x_d]^T \in T$, $z \equiv [z_1, \ldots, z_d]^T \in R$. If $a_j \equiv [a_{1,j}, \ldots, a_{d,j}]^T$ is the jth column of the matrix Π, $j \in [1, d]$, that is, if $\Pi = [a_1 \ldots a_d]$, then the vectors $a_1, \ldots, a_d$ are linearly independent and the vertices of T are

$$P_0 \equiv a_0, \; P_1 \equiv P_0 + a_1, \ldots, P_d \equiv P_{d-1} + a_d. \tag{2.24}$$

It is also clear that a given simplex T with vertices from (2.24) can be obtained as the result of the transformation (2.23) applied to R. For $u(x) \equiv u_T$ defined on T, let $u_R \equiv u(\Pi z + a_0)$, which is defined on R. For smooth enough functions, we introduce (see (2.14)) $|u_T^{(r)}|^2 \equiv \sum_{|\alpha|=r} \frac{r!}{\alpha!}|D_z^\alpha u|^2$ and $|u_T^{(r)}|^2 \equiv \sum_{|\alpha|=r} \frac{r!}{\alpha!}|D_x^\alpha u_T|^2$.

Lemma 7. Suppose R and T are the simplexes defined above. Then (2.17) *holds; if additionally Π is an orthogonal matrix Π, then $|u_T^{(r)}| = |u_T^{(r)}|$.*

Proof. The proof of Lemma 5 applies with following changes. We define $\text{grad}_x u \equiv \text{grad}_x^{(1)} u_T$ and, for $k \geq 2$, define the d^k-dimensional vector

$$\text{grad}_y^{(k)} u \equiv \begin{bmatrix} \text{grad}_y^{(k-1)} \frac{\partial u_T}{\partial x_1} \\ \vdots \\ \text{grad}_x^{(k-1)} \frac{\partial u_T}{\partial x_d} \end{bmatrix}.$$

We can prove that $|\text{grad}_x^{(k)} u_T|^2 = |u_T^{(k)}|^2$ by mathematical induction. Consider, e.g., the term $(D_1 D^\alpha u_T)^2$ with $|\alpha| = k$ in the expression for $|u_T^{(k+1)}|^2$ with coefficient $a_1 \equiv \frac{(k+1)!}{(\alpha_1+1)!\alpha_2!\ldots\alpha_d!}$. The same term in $|\text{grad}_x^{(k+1)} u_T|^2$ corresponds to the d terms $(D_1 D^\alpha)^2$, $(D_2 D^{\alpha+e_1-e_2})^2, \ldots, (D_d D^{\alpha+e_1-e_d})^2$, with respective coefficients: $b_1 \equiv \frac{k!}{\alpha_1!(\alpha_2)!\ldots(\alpha_d)!}$, $b_2 \equiv \frac{k!}{(\alpha_1+1)!(\alpha_2-1)!\ldots\alpha_d!}, \ldots,$ $b_d \equiv \frac{k!}{(\alpha_1+1)!\alpha_2!\ldots(\alpha_d-1)!}$. Since $a_1 = b_1 + \cdots + b_d$, we obtain the desired equality. Now there is no need for the vectors $\text{grad}_x^{(k),l} u_T$ ((2.18) and (2.19) take

---

[14] Six standard quasisimplexes with the common vertex A_0 generate some standard quasicubic cells, e.g., with only three possible curvilinear faces $B_0 B_1 B_2 B_3$, $A_3 A_2 B_2 B_3$, and $A_1 B_1 B_2 A_2$ (see Figure 5). Such cells can be easily matched with adjacent ones under the assumption that they all have the same homotopy's center A_0. If an adjacent cell, say cell $A_0' A_1' A_2' A_3' B_0' B_1' B_2' B_3'$, has the face $A_1' A_2' B_2' B_1'$ matching the face $A_0 A_3 B_3 B_0$, then we must assume that S contains the straight segment $[B_0 B_3]$ or $B_1' B_3'$.

forms $|\text{grad}_x^{(r)} u_T|^2 \leq K_{r,r} |\text{grad}_z^{(r)} u_R|^2$, $\text{grad}_x^{(r)} u_T = A_{r,r} \text{grad}_z^{(r)} u_R$ with $A_{1,1} \equiv B \equiv \Pi^*$). $\square$

In the sequel, we make use of the Gram matrix A_G defined by the shift vectors $a_1, \ldots, a_d$. Note that $A_G = \Pi^* \Pi$, that its eigenvalues $\lambda_1 \leq \ldots \leq \lambda_d$ are positive and coincide with those of the matrix $\Pi\Pi^*$ (see Lemma 0.4.6), and that $\lambda_1^{1/2}, \ldots, \lambda_d^{1/2}$ are lengths of semi-axes of the ellipsoid $E \equiv E(T)$ defined by the parameter equation $x = \xi_1 a_1 + \cdots + \xi_d a_d$, where $\xi_1^2 + \cdots + \xi_d^2 = 1$. Hence, $r \leq \lambda_i^{1/2} \leq r'$, $i \in [1, d]$, where the respective $\bar{r} \equiv \bar{r}(T)$ and $\bar{r}' \equiv \bar{r}'(T)$ are radii of balls inscribed in and subscribed around E; this ellipsoid is closely related to the simplex $M(T)$ with vertices

$$M_0 \equiv a_0, \ M_1 \equiv M_0 + a_1, \ M_2 \equiv M_0 + a_2, \ldots, M_d \equiv M_0 + a_d;$$

we shall refer to $M(T)$ as the *associated simplex* [15] with respect to T.

Lemma 8. *For $u \in W_2^1(T)$, let u_R be the function $u_R(z) \equiv u(\Pi z + a_0)$, which is defined on R. Let $m = [m] \geq 1$. Then*

$$\lambda_d^{-1} |\lambda|^{1/2} |u_R|_{1,R}^2 \leq |u|_{1,T}^2 \leq \lambda_1^{-1} |\lambda|^{1/2} |u_R|_{1,R}^2, \ \forall u \in W_2^1(T), \qquad (2.25)$$

$$|u_R|_{m+\gamma,R} \leq K(m,\gamma) |u|_{m+\gamma,T}, \ \forall u \in W_2^{m+\gamma}(T), 0 < \gamma \leq 1, \qquad (2.26)$$

where $K(m,1) \equiv \lambda_d^{(m+1)/2} |\lambda|^{-1/4}$, $K(m,\gamma) \equiv \lambda_d^{(2m+2\gamma+d)/4} |\lambda|^{-1/2}$,

$$|u|_{R,m+\gamma}^2 \equiv \int_R \int_R \frac{\sum_{|\alpha|=m} \frac{m!}{\alpha!} [D^\alpha u(x) - D^\alpha u(y)]^2}{|x-y|^{(d+2\gamma)}} dx dy, \ \gamma \in (0,1). \ (2.27)$$

Proof. We have $|u_T|_{1,T}^2 = |\lambda|^{1/2} (|(\Pi^{-1})^* \nabla_z u|^2, 1)_{0,R}$. Since

$$\lambda((\Pi_1)^* \Pi^{-1}) \in [\lambda_d^{-1}, \lambda_1^{-1}],$$

then (2.25) holds. If $\gamma = 1$, then, by virtue of Lemma 7, we obtain

$$|u_R|_{m+1,r}^2 \leq |\lambda|^{-1/2} (\|\Pi\|^{2(m+1)} |u_T^{(m+1)}|^2, 1)_{0,R}.$$

Since $\|\Pi\| = \lambda_d^{1/2}$, then (2.26) holds for $\gamma = 1$. Let now $0 < \gamma < 1$. For an arbitrary pair of points $x \in R, y \in R$ and their images $x' \equiv \Pi x + a_0, y' \equiv \Pi y + a_0$, we have $|x' - y'| \leq \|\Pi\| |x - y| = \lambda_d^{1/2} |x - y|$. Hence, $|x - y| \geq \lambda_d^{-1/2} |x' - y'|$ and

---

[15]The easily computed eigenvalues of A_G define $|\lambda| \equiv \lambda_1 \lambda_2 \ldots \lambda_d = \det A_G = d! |T|$, $\lambda_d = \|\Pi\|^2$, $\lambda_1^{-1} = \|\Pi^{-1}\|^2$. Note also that $\det A_G = |\det \Pi|^2$, $|\det \Pi| = |T||R|^{-1}$, $|\Pi^{-1}x|^2 \geq \lambda_d^{-1} |x|^2$, $\forall x$, and $|u_R|_{0,R}^2 = |\lambda|^{-1/2} |u|_{0,T}^2$.

$$\frac{1}{|x-y|^{(d+2\gamma)}} \leq \lambda_d^{\frac{d+2\gamma}{2}} \frac{1}{|x'-y'|^{(d+2\gamma)}}. \tag{2.28}$$

Combining (2.27), (2.17) (see Lemma 7), and (2.28) yields the inequality

$$|u|^2_{R,m+\gamma} \leq |\lambda|^{-1}\lambda_d^{1/2}\|\Pi\|^{2m} \int_T \int_T \frac{\sum_{|\alpha|=m} \frac{m!}{\alpha!}[D^\alpha u(x) - D^\alpha u(y)]^2}{|x-y|^{(d+2\gamma)}} dx\,dy,$$

which leads directly to (2.26). $\square$ [16]

2.5. Quasiuniform triangulations; the metric space of simplicial cells. Finally, we introduce the very important notion of a *sequence of quasiuniform triangulations* as an infinite sequence of triangulations $T_h(\hat{\Omega}_h)$ such that there exist constants $\kappa_0 > 0$ and $\kappa_1 > 0$ for which the estimates $\kappa_0 h \leq r(T) \leq r'(T) \leq \kappa_1 h$ hold for every simplex $T \in T_h(\hat{\Omega}_h)$ and all admissible $h > 0$. Here, the respective $2r(T) = 2\lambda_1^{1/2}$ and $2r'(T) = 2\lambda_d^{1/2}$ are diameters of balls inscribed in and circumscribed around the ellipsoid $E(T)$. It is clear that

$$\lambda_i^{1/2} \in [\kappa_0 h, \kappa_1 h] \tag{2.29}$$

for such a sequence of triangulations. It is not very difficult to show that this notion is independent of the ordering of nodes of T, but the constants themselves vary and the use of the best ordering is reasonable as we do below for a more general notion of a *sequence of nondegenerate triangulations*. More precisely, we use this notion if there exists $\kappa > 0$ such that, for every simplex $T \in T_h(\hat{\Omega}_h)$ and all admissible h, we have

$$\rho(T) \equiv \max r'(T)/r(T) \leq \kappa, \tag{2.30}$$

where the maximum is taken with respect to all possible orderings of the vertices of the simplex T. [17]

---

[16] In (2.27), we used coefficients $\frac{m!}{\alpha!}$ (see (0.1.26)) in order to make Lemma 7 applicable and to deal with a norm independent of the Descartes coordinate system. Note that estimates given at least for $m = [m]$ are sharp. Therefore, for $d = 2$, estimates (2.25) and (0.4.15) must coincide (also easily verified using the fact that, for $S_{1,2}$ from (0.4.15), we have $S_{1,2}^2 = |a_1|^2|a_2|^2 - (a_1, a_2)^2$) and, thus, $\lambda_1 = S_{1,2}[\mu - (\mu^2 - 1)^{1/2}]$ and $\lambda_2 = S_{1,2}[\mu + (\mu^2 - 1)^{1/2}]$.) Note that the estimates for a given simplex T coincide with estimates for another simplex if its shift vectors differ from those of T only in their order. The inverse order of vertices is also allowed, but in the general case the estimates depend on the ordering of the vertices.

[17] For $d = 2$, we have $\rho(T) = \mu + (\mu^2 - 1)^{1/2}$ (see (0.4.12)). Evidently, $\rho(T) \geq 1$, with equality when and only when T is a regular simplicial part of a cube. Indeed, $\rho(T) = 1$ implies that $\lambda_1 = \ldots = \lambda_d = (r(T))^2 \equiv c^2$. Thus, $A_G = c^2 I$, and $c^{-1}\Pi$ is an orthonormal matrix corresponding to the change of the Descartes coordinate system, that is, T is a

§ 3. Accuracy estimates for projective-grid methods

The problem of obtaining error estimates for PGMs very often can be reduced to the classical problem of approximation theory in several variables and especially to study of Sobolev spaces approximations (see (1.1.18)). [18]

3.1 Approximation of Sobolev spaces and error estimates for projective-grid methods for polyhedral regions. Recall that to estimate dist$\{u; \hat{G}_h\}$, we may use the inequality

$$\text{dist}_G\{u; \hat{G}_h\} \leq \|\hat{w} - u\|_G, \qquad (3.1)$$

where $\hat{w} \equiv P_a(u)$ is any suitable element of the approximating space $\hat{G}_h$. In this section, we deal with a Hilbert space G, which is a subspace of the Hilbert space $G^{(1)} \equiv W_2^1(\Omega)$ (the case of vector-functions $u \equiv (u_1, \ldots, u_k)$ and the Hilbert space $G \equiv G_1 \times \cdots \times G_k$ is included), and its approximating subspaces $\hat{G}_h \equiv \mathcal{J}^{(m_1)}$ ($m_1 \geq m \geq 1$) (see § 1) associated with triangulations (possibly composite) $T_h^{(1)}(\bar{\Omega})$ and $T_h^{(m_1)}(\bar{\Omega})$; we consider $m_1 \geq m$ to show that PGMs with sufficiently large degrees of polynomials lead to optimal error estimates under assumptions of type (0.5.1) and (0.5.4) and that they retain this property for assumptions with arbitrary $m \in [1, m_1]$.

The simplest analysis can be carried out for the solution $u \in C(\Omega)$ when values of u are defined at nodes (at elementary nodes if the boundary of the region contains slits—recall that then we use several limit values, and by a triangulation of $\bar{\Omega}$ we mean its proper modification in accordance with § 1). This is the case if $u \in W_2^{m+\gamma}(\Omega)$ with $2(m + \gamma) > d$ because

---

regular simplicial part of a cube. This suggests using $f(\rho(T))$ as a measure of deformation that the ideal simplicial cell must undergo to become T, where $f(t)$ is some increasing function for $t \geq 1$ such that $f(1) = 0$ (e.g., we may use $f(t) \equiv t - 1, f(t) \equiv \ln t$, and so on). More precisely, if we agree to identify all regular simplicial parts of cubes, then we may regard the *metric space of simplicial cells* on the basis of the distance $d\{T_1, T_2\} \equiv f(\rho(T_1)) + f(\rho(T_2))$ (it is easy to verify that all axioms of metric space are satisfied). We emphasize again that the case $f(\rho(t)) = 0$ must be considered as optimal and corresponds to T being a regular simplicial part of a cube, and that, from algebraic point of view, $\rho(T) = \|\Pi\|\|\Pi^{-1}\|$ is just the condition number of the corresponding transformation matrix Π (this suggests that a similar distance between bases of a Euclidean space might be useful). It is easy to see that our notions of quasiuniform and nondegenerate triangulations are equivalent to the more traditional ones dealing with balls, associated not with $E(T)$, but directly with T (see [88, 117, 149, 333, 448, 470]); but, in the latter case, the corresponding radii cannot be equal.

[18] There are many relevant investigations (see [18, 117, 394, 448, 470] and references therein). We consider here only questions related to construction of asymptotically optimal methods, paying special attention to general assumptions on the solution of type (0.5.1) and (0.5.4) with admissible fractional $\nu \equiv m - 1 + \gamma$, to regions with slits, to composite triangulations with local refinement, and to relatively accurate estimates of some key constants (see [183, 187, 205]).

this space can be embedded into the space $C(\Omega)$ (see § 0.1). Then in the role of $P_a u$ we can use $\text{int}_{m_1} u$, which denotes the element of $\mathcal{J}^{(m_1)}$ having the same values at the nodes (elementary nodes) A_i (vertices of simplexes from $T_h^{(m_1)}(\bar{\Omega})$) as the function u (values of $\text{int}_{m_1} u$ at the seminodes are obtained through some interpolation of values of u at corresponding nodes). It is clear that when all of these vertices A_i of simplexes from $T_h^{(m_1)}(\bar{\Omega})$ belonging to a simplex $T \in T_h^{(1)}(\bar{\Omega})$ are nodes (e.g., when we use standard (noncomposite) triangulations), then $\text{int}_{m_1} u$ on T is just the Lagrangian polynomial of degree m_1 for u. We consider T and all A_i as images of the reference simplex R and of points $A_i^{(1)}$ of similar interpolation on it under the mapping defined by (1.2). The Hilbert space $W_2^{m+\gamma}(R)$ with $2(m + \gamma) > d$ can be embedded into the space $C(R)$ (see § 0.1). Then we may define the Hilbert space W, with $\|v\|_W \equiv \|v\|_{m+\gamma, R}$, consisting of $v \in W_2^{m+\gamma}(R)$ that vanish at the points $A_i^{(1)}$. We emphasize that, on this subspace of $W_2^{m+\gamma}(R)$, the norm of the space and the indicated norm in W are equivalent (see § 0.1). Using the same argument as in Subsection 0.5.3 based on properties of compact symmetric operators (see also Lemma 1 and § 9.1), we may write $\max\limits_{v \neq 0}\{|v|_{1,R}^2\|v\|_W^{-2}\} \equiv s < \infty$ and $\max\limits_{v \neq 0}\{|v|_{0,R}^2\|v\|_W^{-2}\} \equiv s_0 < \infty$, where $s \equiv s(m_1, m)$ and $_0 s \equiv s_0(m_1, m)$. Hence, for $v \in W_2^{m+\gamma}(R)$ and $\xi \equiv \xi(v) \equiv v - \text{int}_{m_1} v$, we obtain the estimates of the interpolation error: $|\xi|_{1,R} \leq s^{1/2}|v|_{m+\gamma, R}$ and $|\xi|_{0,R} \leq s_0^{1/2}|v|_{m+\gamma, R}$.

Theorem 1. *Let $T \in T_h^{(1)}(\bar{\Omega})$ be the image of R under mapping defined by (2.23). Suppose that $u \in W_2^{m+\gamma}(T)$, with $2(m + \gamma) > d$. Then, for $\zeta \equiv \xi(u)$, we have*

$$|\zeta|_{1,T} \leq q_1|u|_{m+\gamma, T} \text{ and } |\zeta|_{0,T} \leq q_0|u|_{m+\gamma, T},$$

where $q_1 \equiv s^{1/2}\dfrac{\lambda_d^{(m+1)/2}}{\lambda_1^{1/2}}$, $q_0 \equiv s_0^{1/2}\lambda_d^{(m+1)/2}$ if $\gamma = 1$;

$$q_1 \equiv s^{1/2}\frac{\lambda_d^{(2m+2\gamma+d)/4}}{|\lambda|^{1/4}\lambda_1^{1/2}}, \quad q_0 \equiv s_0^{1/2}\frac{\lambda_d^{(2m+2\gamma+d)/4}}{|\lambda|^{1/4}}, \quad \gamma \in (0, 1);$$

if the triangulations $T_h^{(1)}(\bar{\Omega})$ are quasiuniform, then there exist constants K_0 and K_1 such that, for all $T \in T_h^{(1)}(\bar{\Omega})$, $|\zeta|_{1,T} \leq K_1 h^{m+\gamma-1}|u|_{m+\gamma, T}$ and $|\zeta|_{0,T} \leq K_0 h^{m+\gamma}|u|_{m+\gamma, T}$; if $d = 2$ and $\gamma = 1$, then we have $|\zeta|_{1,T} \leq s^{1/2}[\mu + (\mu^2 - 1)^{1/2}]^{1+m/2}S_{1,2}^{m/2}|u|_{m+1, T}$.

Proof. By (2.25), $|\zeta|_{1,T}^2 \leq |\lambda|^{1/2}/\lambda_1|\zeta_R|_{1,R}^2 \leq |\lambda|^{1/2}s/\lambda_1|\zeta_R|_{m+\gamma, R}^2$. Since $|\zeta_R|_{m+\gamma, R}^2 = |u_R|_{m+\gamma, R}^2$, we have $|\zeta|_{1,T} \leq q_1'|u|_{m+\gamma, T}$, where $q_1' \equiv$

$\lambda_1^{-1/2}|\lambda|^{1/4}s^{1/2}K(m,\gamma)$ (see (2.26)). This yields the desired estimate involving q_1. Similarly, we obtain the estimate involving q_0. If triangulations $T_h^{(1)}$ are quasiuniform, then (2.29) holds and, together with the obtained estimates, yields the remaining ones in the formulation of the theorem; the case $d = 2$ is simpler (see the remarks to Lemma 2.8). $\square$ [19]

Theorem 2. For the Hilbert space $G = W_2^1(\bar{\Omega}; \Gamma_0)$, suppose that the solution of operator equation (1.1.1) in G exists, is unique, and satisfies (0.5.4) with $2(m + \gamma) > d$. Suppose that a PGM based on quasiuniform triangulations $T_h^{(1)}(\bar{\Omega})$ and $T_h^{(m_1)}(\bar{\Omega})$, with $m_1 \geq m$, and on the use of the approximating subspace $\hat{G}_h \equiv J_{m_1} \subset G$ is applied. Suppose that each simplex in $T_h^{(1)}(\bar{\Omega})$ belongs to one of the $\bar{\Omega}_s, s = 1, \ldots, p$ (see (0.5.4)), and that error estimate (0.2.4) for this PGM holds. Then

$$\|u - \hat{u}\|_G \leq K h^{m+\gamma-1}[\sum_{s=1}^p |u|_{m+\gamma,\Omega_s}^2]^{1/2}. \tag{3.1}$$

Proof. The solution u on each subdomain $\Omega_s, s \in [1, p]$, is a continuous uniformly bounded function. If we consider a common part of the boundaries of two different subdomains such that it contains no points of Γ, then the trace of $u \in G$ on it is defined. This implies that $u \in C(\Omega)$. Therefore, if for every simplex $T \in T_h^{(1)}(\bar{\Omega}_s)$ we define $\mathrm{int}_{m_1} u \equiv \hat{w}$, then we actually obtain a function $\hat{w} \in \hat{G}$ continuous on Ω and $\mathrm{dist}_G\{u; \hat{G}\} \leq \|u - \hat{w}\|_G$. This together with (0.2.4) and Theorem 1 proves (3.1). $\square$ [20]

3.2. Steklov's and Sobolev's averagings. The above local analysis is not applicable if $2(m + \gamma) \leq d$—this may take place only for $d \geq 3$—and therefore the solution u does not define $\mathrm{int}_{m_1} u$. We shall elaborate on this point at the end of § 5.2, but for now we only note the possibility to use the inequality $\mathrm{dist}_G\{u; \hat{G}\} \leq \|u - v\|_G + \|v - \mathrm{int}_{m_1} v\|_G$, where v is an approximation to u with sufficient smoothness. Very often in the role of such a function we take its *Steklov averaging* (see [129, 394]) defined by

$$Y_\rho u(x) \equiv (2\rho)^{-d}(u(y), 1)_{0,S(x)}, \tag{3.2}$$

where $S(x) \equiv [x_1 - \rho, x_1 + \rho] \times \cdots \times [x_d - \rho, x_d + \rho]$, and u is extended in the proper manner (see the Extension Theorem in § 0.1).

It is known (see [67, 394]) that if, for $u \in G^{(m+\gamma)}$ with $0 < \gamma < 1$, we define $\kappa(m, \gamma) \equiv \gamma$ when $m = 1$ and $\kappa(m, \gamma) = 1$ when $m \geq 2$, then

---

[19] Condition $2(m + \gamma) > d$ is satisfied for all $\gamma > 0$ if $d = 2$ and that for $\gamma = 1$ we may take $q_1 = \|\Pi\|^{m+1}\|\Pi^{-1}\|^{-1}$.

[20] We actually assumed that $T_h^{(1)}(\bar{\Omega})$ yielded a triangulation of each subregion Ω_s.

$Y_\rho u \in G^{(m+\gamma+1)}$,

$$\|u - Y_\rho u\|_{1,\Omega} \le K \rho^{\kappa(m,\gamma)} \|u\|_{m+\gamma,\Omega}, \qquad (3.3)$$

and

$$\|Y_\rho u\|_{2,\mathbf{R}^2} \le K \rho^{\gamma-1} \|u\|_{1+\gamma,\Omega}. \qquad (3.4)$$

Sometimes we shall make use of *Sobolev's averaging* defined by

$$Y'_\rho u(x) \equiv (\kappa \rho^d)^{-1} (\exp\{\rho(|x - y|^2 \rho^{-2} - 1)^{-1}\}, u(y))_{0,S'(x)},$$

where $S'(x) \equiv \{y : |y - x| \le \rho\}$. It is known (see [67, 459]) that for $u \in L_1(\bar{\Omega})$, the function $Y'_\rho u$ has all derivatives. The indicated approach works well if $\Gamma_0 = \emptyset$ and the averaging of the extended solution yields an element of the original Hilbert space G. For example, if we take $\rho \asymp h$, then we can prove (3.1) for $m = 1$ and $d = 3$.

3.3. Error estimates under approximation of the region. In the study of errors associated with approximation of the region the following theorem is of primary importance.

Theorem 3. Let Π_ρ *denote part of* Ω *bounded by two* C^1 *surfaces such that points of one are not farther than* ρ *from the other. Then*

$$|u|_{0,\Pi_\rho} \le K f(\rho) \|u\|_{\gamma,\Omega}, \quad \gamma \in (0,1), \forall u \in G^{(\gamma)}, \qquad (3.5)$$

where $f(\rho) \equiv \rho^{1/2} \ln \rho$ *if* $\gamma = 1/2$ *and* $f(\rho) \equiv \rho^{\alpha(\gamma)}$ *if* $\gamma \ne 1/2$ *and* $\alpha(\gamma) \equiv \min\{\gamma; 1/2\}$; *if additionally* $\rho = O(h^{2m})$ *with* $m = [m] \ge 1$, *then*

$$|u|_{1,\Pi_\rho} \le K h^{m+\gamma-1} \|u\|_{m+\gamma,\Omega}, \quad \forall u \in G^{(m+\gamma)}. \qquad (3.6)$$

Proof. The first assertion of the theorem is the classical Il'in's result (see [67, 283]), and (3.6) is a direct consequence of (3.5) and the assumptions imposed on Π_ρ. □ [21]

Theorem 4. Let conditions of Theorem 2 be satisfied with triangulation $T_h(\bar{\Omega})$ *replaced by triangulation* $T_h(\hat{\Omega})$, *where* $\hat{\Omega} \subset \bar{\Omega}$ *and the boundary* $\hat{\Gamma}$ *of* $\hat{\Omega}$ *differs from* Γ *only by its Pirichlet's part* $\hat{\Gamma}_0$ *whose points are not farther than* $\rho = O(h^{2m})$ *from* Γ_0. *Then estimate (3.1) holds.*

Proof. We construct $\text{int}_{m_1} u \equiv \hat{w}$ on $\hat{\Omega}$ and extend it and $\hat{u}$ to $\Omega \setminus \hat{\Omega}$. Since

$$\|u - \hat{u}\|_{1,\Omega}^2 = \|u - \hat{u}\|_{1,\hat{\Omega}}^2 + \|u\|_{1,\Omega\setminus\hat{\Omega}}^2, \qquad (3.7)$$

---

[21] The general analysis dealing with approximations of the region will be carried out in § 5.1. For now we consider the simplest application of Theorem 3 dealing only with an approximation of the Dirichlet part of the boundary, Γ_0, by $\hat{\Gamma}_0 \in \bar{\Omega}$.

Then (3.1) follows from Theorem 1 and (3.6). $\square$ [22]

3.4. Error estimates for PGMs on composite grids with local refinement. Now, under the same assumptions (0.5.4) as in Theorem 2, we obtain error estimates for PGM based on composite triangulations $T_{c,h}^{(1)}(\bar{\Omega})$ and $T_{c,h}^{(m_1)}(\bar{\Omega})$ with $m_1 \geq m \geq 1$ and the use of approximating subspace $\hat{G}_h \equiv J_{m_1} \subset G$ (see § 1). For each simplex $T \in T_{c,h}^{(1)}(\bar{\Omega})$ with all vertices corresponding to the basis functions, we introduce

$$h(T) \equiv r'(T) \equiv \lambda_d^{1/2} \tag{3.7}$$

(see (2.30)) and the *augmented simplex*

$$T^* \equiv T \cup T^+, \tag{3.8}$$

where T^+ is a union of neighbor simplexes having common seminodes with T (see § 1); note that $T^+ = \emptyset$ if T contains no seminodes. For example, for a triangle T with side $P_0 P_1$ containing a seminode M, its neighbor refined simplexes containing M are depicted in Figure 1 (note that M is a vertex of each).

All nodes A_i belonging to T correspond to the points of the Lagrangian interpolation of degree m_1, but the values of $\hat{v}$ at seminodes are defined through this or similar interpolation for the corresponding faces of smaller dimension (they yield the same values). Assume that the local refinements

---

[22] There are several important results devoted to estimates of accuracy in the norms of such spaces as $L_2(\Omega), W_2^{-m}(\Omega), L_\infty(\Omega), C(\Omega)$, (see [117, 241, 394]). Special attention is paid to *superconvergence* of methods based on regular grids in some specially chosen grid norms (see [319, 254, 334]) provided additional assumptions on the solution are satisfied. The phenomenon of superconvergence relates to the fact that the rate of convergence of grid approximations at some exceptional points of the cells (elements) exceeds the possible global rate. For example, these exceptional points could include the standard nodal points, midpoints of sides of triangular elements, or points of a more complicated nature (Gauss-Legendre, Jacobi, and Lobatto points). Superconvergence of the finite element approximations or its derivatives can also be obtained by means of various post-processing techniques (e.g., averaging), which can sometimes produce an increase of accuracy not only at isolated points, but also in some subregions.

The case of the Dirichlet inhomogeneous boundary condition $u \mid_{\Gamma_0} = g$ frequently presents no problem if the given function g is a trace of a function $\hat{g} \in \hat{G}$. Sometimes, especially when approximation of Γ_0 is desirable, the so-called *boundary penalty method* (see [25, 26, 58, 273]) can be effectively applied in order to convert to natural boundary conditions. For example, if the original problem is to find $u \in G^{(1)}$ such that $u \mid_{\Gamma_0} = g$ and $b(u; v) = l(v), \forall v \in W_2^1(\Omega; \Gamma_0)$ (see (0.1.30)–(0.1.32)), then the boundary penalty method allows one to replace it by the problem of finding $u \in G^1$ such that $b_\varepsilon(u; v) = l_\varepsilon(v), \forall v \in G^{(1)}$, where $\varepsilon > 0$ is a sufficiently small parameter, $b_\varepsilon(u; v) \equiv b(u; v) + 1/\varepsilon(u, v)_{0,\Gamma_0}$, and $l_\varepsilon(v) \equiv l(v) + (\varepsilon)^{-1}(g, v)_{0,\Gamma_0}$. If we denote its solution by u_ε, then estimates of type $\|u - u_\varepsilon\| \leq C\varepsilon \|u\|_{2,\Omega}$ can be proved. Then the real problem is to match ε and the parameter h of the grid (see [58]).

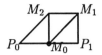

Figure 1. Triangles containing the seminode M_0.

are carried out with the ratio $t = km_1$ and that there are only limited types of geometry for T^*. More precisely, assume that the respective T and A_i are images of the reference simplex R and points $A_i^{(1)}$ of similar interpolation on it under the map defined by (2.23), and that this affine map yields T^* as an image of a reference augmented triangle $R^* \equiv R \cup R^+$, and that there are only several types of R^* (in Figure 1, R^+ is represented). R^* is always a closed domain with a Lipschitz boundary (this is not the case for some types of R^+, e.g., for $m_1 = 2$). Hence, we define the Hilbert space $W_2^{m+\gamma}(R^*)$ in the standard way (see (2.27)). Since $2(m+\gamma) > d$, it can be embedded into the space $C(R^*)$ (see § 0.1) and we may define the Hilbert space $W \equiv W(R^*)$, with $\|v\|_W \equiv \|v\|_{m+\gamma, R^*}$, consisting of $v \in W_2^{m+\gamma}(R^*)$ that vanish at the points $A_i^{(1)} \in R$; on this subspace, the above mentioned norms are equivalent. For a function v continuous on T^*, we denote by $P_a v$ the restriction of $\hat{v} \in \hat{G}_h$ such that its values and values of v at nodes in T^* coincide (on each simplex in T^*, Lagrange interpolation is used for these values and for the corresponding values at seminodes). $P_a v$ for $v \in C(R^*)$ is understood in the same sense.

Lemma 1. For the Hilbert space $W \equiv W(R^*)$ defined above, we have

$$\sup_{v \neq 0} \frac{|v - P_a v|^2_{1,R^*}}{\|v\|_W^2} = \max_{v \neq 0} \frac{|v|^2_{1,R^*}}{\|v\|_W^2} \equiv \bar{s} \equiv \bar{s}(R^*) < \infty \qquad (3.9)$$

and

$$\sup_{v \neq 0} \frac{|v - P_a v|^2_{0,R^*}}{\|v\|_W^2} = \max_{v \neq 0} \frac{|v|^2_{0,R^*}}{\|v\|_W^2} \equiv \bar{s}_0 \equiv \bar{s}_0(R^*) < \infty. \qquad (3.10)$$

Proof. Define the bilinear form

$$b(u; v) \equiv (u - P_a u; v - P_a v)_{1,R^*}, \qquad (3.11)$$

which is evidently bounded and symmetric on W^2. Hence, there exists an operator $A \in \mathcal{L}(W)$ such that $A = A^* \geq 0$ and $(Au, v)_W = b(u; v)$, $\forall u \in W$, $\forall v \in W$ (see § 0.2). Now we prove that A is compact, that is, it maps each bounded set into a precompact one or, equivalently, it maps each weakly convergent sequence into a convergent one (see [292, 323, 341]). Let $\{u^n\}$ be a weakly convergent sequence, that is, there exists $u \in W$ such that $\lim_{n \to \infty} (u_n, v)_W = (u, v)_W$, $\forall v \in W$. Then $\{Au^n\}$ is also a sequence weakly convergent to Au, that is,

$$\lim_{n\to\infty}(Au_n,v)_W = (Au,v)_W, \ \forall v \in W. \tag{3.12}$$

Thus, it suffices to prove that

$$\lim_{n\to\infty}\|Au_n\|_W = \|Au\|_W \tag{3.13}$$

(see [272, 292, 341]). To do this we make use of $Q \equiv I - P_a$, $w^n \equiv Au^n$, $w \equiv Au$ and consider $\{\|w^n\|_W^2\}$. We have $X^n \equiv \|w^n\|_W^2 = (Au^n, Au^n)_W = (Qu^n, Qw^n)_{1,R^*}$. Hence,

$$X^n = (Qu, Qw)_{1,R^*} + (Q(u^n - u), Qw)_{1,R^*} + (Qu^n, QA(u^n - u))_{1,R^*}. \tag{3.14}$$

Recall that the strengthened variants of embedding theorems (see § 0.1) of $W_2^{m+\gamma}(R^*)$ into $W_2^1(R^*)$ and $C(R^*)$ imply that the embedding operator is compact and that $\lim_{n\to\infty}\|u_n - u\|_{1,R^*} = \lim_{n\to\infty}\|u_n - u\|_{C(R^*)} = 0$. From this it follows that

$$\lim_{n\to\infty}\|Q(u_n - u)\|_{1,R^*} = 0. \tag{3.15}$$

Thus, the second term on the right-hand side of (3.14) tends to zero. The third term is equal to

$$X_3 \equiv (Qu, QA(u^n - u))_{1,R^*} + (Q(u^n - u), QA(u^n - u))_{1,R^*}. \tag{3.16}$$

$\{A(u^n - u)\}$ converges weakly in W to 0. Thus, $\lim_{n\to\infty}\|QA(u_n - u)\|_{1,R^*} = 0$ (see (3.15)), and all terms on the right-hand side of (3.16) tend to zero. From (3.14) we obtain (3.13) and the desired property of A. Note that $\bar{s} \equiv \sup_{v\neq 0}\{(Av,v)_W\|v\|_W^{-2}\}$. This, together with classical properties of symmetric compact operators (see [292, 341] and § 9.1), proves (3.9). The proof of (3.10) needs only minor alterations. $\square$ [23]

    Theorem 5. *Let $T \in T_h^{(1)}(\bar{\Omega})$ and T^* be the images of R and of R^*, respectively, under the mapping defined by (2.23). Suppose $u \in W_2^{m+\gamma}(T^*)$, with $2(m + \gamma) > d$. Then, for $\zeta \equiv u - P_a u$, the estimates*

$$|\zeta|_{1,T^*} \leq q_1|u|_{m+\gamma,T^*}, \quad |\zeta|_{0,T^*} \leq q_0|u|_{m+\gamma,T^*}$$

---

[23] Note that $\bar{s} = \sup_{v\neq 0}\{|v - P_a|_{1,R}^2 + |v - P_a|_{1,R+}^2\|v\|_W^{-2}\} \leq \sup_{v\neq 0}[|v - P_a|_{1,R}^2 + |v - P_a|_{1,R+}^2][|v|_{m+\gamma,R}^2 + |v|_{m+\gamma,R+}^2]^{-1}$. This suggests defining estimates for R and R^* separately. For the case depicted in Figure 1 this can be done in a natural way, but in general case, as we mentioned already, this might lead to complicated mathematical problems.

hold, where $q_1 \equiv \lambda_1^{-1/2}|\lambda|^{1/4}\bar{s}^{1/2}K(m,\gamma)$, $q_0 \equiv |\lambda|^{1/4}\bar{s}_0^{1/2}K(m,\gamma)$, with constant $K(m,\gamma)$ from (2.26).

Proof. It suffices to see that, in Lemma 2.8, it is possible to replace T by T^*. We may then use the proof of Theorem 1, with $\text{int}_{m_1} u$ replaced by $P_a u$, and apply Lemma 1. $\square$

Theorem 6. *For $G = W_2^1(\bar{\Omega}; \Gamma_0)$, suppose that the solution of operator equation (1.1.1) in G exists, is unique, and satisfies (0.5.4) with $2(m + \gamma) > d$. Suppose that PGM based on composite triangulations $T_{c,h}^{(1)}(\bar{\Omega})$ and $T_{c,h}^{(m_1)}(\bar{\Omega})$ with $m_1 \geq m$ and on the use of the approximating subspace $\hat{G}_h \equiv J_{m_1} \subset G$ is applied. Suppose that, for all augmented simplexes T^*, only a finite number of reference augmented simplexes R^* is required, that each T^* belongs to one of the $\bar{\Omega}_s, s \in [1,]p$ (see (0.5.4)), and that error estimate (0.2.4) for this PGM holds. Then there exists a constant K such that*

$$\|u - \hat{u}\|_G \leq K[\sum_{T^*}(h(T))^{2(m+\gamma-1)}|u|_{m+\gamma,T^*}^2]^{1/2}. \tag{3.17}$$

Proof. For every simplex T^*, define $P_a u \equiv \hat{w}$. As in the proof of Theorem 2, this yields $\hat{w} \in \hat{G}$. To estimate $\|u - \hat{w}\|_G^2$, we represent it as $\|u - \hat{w}\|_G^2 = \sum_{T^*} \|u - \hat{w}\|_{1,T^*}^2$ and estimate each term via Theorem 5. The assumptions imposed on the grid imply that we have only several possible values of $\bar{s}(R^*)$ and that, for each T^*, the corresponding $\lambda_i, i \in [1,d]$, from Lemma 2.8 and Theorem 5, are such that (2.29) holds with $h(T)$ instead of h and some positive constants κ. This proves (3.17). $\square$

3.5. Increasing the accuracy and a posteriori estimates. The choice of the approximating space $\hat{G}_h$ is of fundamental importance for all features of PGMs. Here we consider briefly only basic ideas for increasing accuracy of the grid method at hand (some aspects will be discussed in Chapter 5 in more detail, especially in connection with the problem of minimizing computational work).

We first mention the possible use of an *additive removal of singularities* suggested by Kantorovich (see [64]) for numerical integration and introduced in [31, 231] for improving the accuracy of FEMs. This works especially well for singularities that have their origin in the geometric irregularities of the boundary (associated with several separate points, like corners and conical points), and it has been thus analyzed very extensively (see [255, 256, 312, 394]). [24]

---

[24]For example, let $d = 2$ and suppose the Dirichlet part of the boundary has an inner corner point $P = [0,0]$. Introducing polar coordinates r, φ, suppose for a sufficiently small $r_0 > 0$ that the sector $S \equiv \{[r,\varphi] : 0 < r < r_0, 0 < \varphi < \alpha\}$ is the common part of Ω and the r_0-neighborhood of the corner point. Then the solution of

Such singularities are not easy to approximate by elements of the given subspace $\hat{G}_h$ associated with a uniform or quasiuniform grid or triangulation. We can take into account the singular part, or at least its leading term, by extending the approximating space to the space $\hat{G}_{h,\chi}$. Indeed, if we include into its basis the required function χ, then the approximation property can be improved dramatically since

$$\text{dist}\{u; \hat{G}_{h,\chi}\} \leq \|u - c\chi - \hat{v}_h\|, \quad \forall \hat{v}_h \in \hat{G}_h,$$

and our concern then can concentrate on approximating function $u - c\chi$ without the singularity. Of course, such a procedure and its modifications dealing with so-called *singular elements* in the neighborhood of a singular point are applicable for problems with several singularities (see [31, 72, 73, 74, 183, 231, 370, 394, 501]), but finding solutions of the resulting systems might become a difficult task because the addition of singular functions to the standard finite element basis destroys the band structure of the matrix and greatly increases its condition number. Therefore, special attention must be paid to construction of effective iterative methods for the resulting systems (see § 1.5 and § 5.1).

The study of singularities in the solutions of elliptic problems is one of the most important mathematical problems (see [255, 256, 312]). For example, it has been shown that for equations with smooth coefficients in the neighborhood of a singular point, the type of the singularity is the same as for the above model problem, and similar results hold for strongly elliptic systems, including the classical elasticity problems.[25] For many problems, especially for three-dimensional regions with edges and surfaces

---

the Laplace's equation has a singularity at this point of type $c\chi$, where c is a constant, $\chi \equiv \chi(r, \varphi) \equiv r^{\pi/\alpha}\zeta(r)\sin(\varphi\pi/\alpha)$, the function $\zeta(r)$ is smooth and monotone on $[0, r_0]$, $\zeta(r) = 1$ if $0 \leq r \leq r_0/3$, and $\zeta(r) = 0$ if $r \geq r_0/3$ (see [256, 312, 394]). Similarly, if P is an inner point of the Neumann part of the boundary and $\pi/2 < \alpha < \pi 3/2$, then $\chi \equiv \chi(r, \varphi) \equiv r^{\pi(2\alpha)^{-1}}\zeta(r)\cos\frac{\varphi\pi}{2\alpha}$; and if P is a boundary point at which the type of boundary condition changes, where the respective Dirichlet and Neumann parts correspond to $\varphi = 0$ and $\varphi = \alpha$, then in the given formula the cosines must be replaced by the sines. Finally, if the straight line segment connecting points $[-1, 0]$ and $[0, 0]$ is a slit in the region, say $\bar{\Omega} \equiv [-1, 1]^2$ (the cracked square), and homogeneous Dirichlet conditions are imposed on it for the harmonic equation, then the leading term in the expansion of the solution in the neighborhood of the point $[0,0]$ takes the form $c\chi$, where c is a constant and $\chi \equiv \chi(r, \varphi) \equiv r^{1/2}\zeta(r)\sin\varphi/2$.

[25] An opposite situation arises when singularities are so numerous that their separate investigation is very difficult. For example, such problems are typical in the theory of elastic and plastic *composites* (materials with highly changing properties corresponding to systems with discontinuous coefficients). In such situations it is very important to consider some averaging of the coefficients and to obtain the corresponding asymptotic error estimates. Such *homogenization procedures* have been the subject of many investigations (see [44, 61]). The use of new models with smooth or even constant coefficients that these studies suggest may radically improve the efficiency of current grid methods.

of discontinuity of coefficients, the study of singularities provides no simple answers. In such cases, it is useful to remember that, for practical purposes, sometimes it is enough to have only approximations of singular functions, perhaps obtained numerically for model problems on very fine grids.

The extremely important questions of obtaining *a posteriori estimates* and *adaptive procedures* for improving the accuracy of the grid methods have been the center of attention of many investigators (see [55, 301, 516]). Suggested algorithms involve the calculation of so-called *a posteriori error estimators* composed of elementwise error indicators; the decision with respect to the refinement of the grid or to the use of more accurate polynomial approximations at each stage is made on the basis of the distribution of these indicators (see [29, 34, 222, 517]). The role of hierarchical basis in working with polynomials of higher degree is very significant and sometimes heuristic approaches of recursive subdivision of the grid are applied (see, e.g., [420]). The most effective refinement procedures must take into consideration the need to solve the resulting discrete systems as is the case for methods based on composite grids with local refinement (see, e.g., [54, 79, 80, 84, 86, 106, 142, 143, 224, 347, 362, 363, 365, 504].) Note that parallelism is most easily exploited when computations are carried out on such structured grids. Of course, if the solution of the problem is smooth enough and the grid is regular, then standard expansions of the error with respect to the mesh size are useful and can lead to more accurate approximations through the well-known *Richardson extrapolation procedure* (see, e.g., [336]). Probably the most general and at the same time simplest procedure of obtaining a posteriori estimates is based on the use of a model operator $B \asymp J$ (see (0.4.3)), $\kappa_0 B \leq J \leq \kappa_1 B, \kappa_0 > 0$ and computation of the residual norm in the Euclidean space $H(B^{-1})$. More precisely, suppose that the vector v is an approximation to the solution obtained on grid h_0 and that, for the much refined grid h, the discretization error is negligible. This implies that instead of $\|\hat{v}_{h_0} - u\|$ we may use $\|\pi_h v - u_h\|_J \equiv \varepsilon_h$, where the vector u_h corresponds to the solution on the refined grid and $\pi_h \in \mathcal{L}(H_{h_0;H_h})$ is a sufficiently simple interpolating operator.

Theorem 7. Let the linear operator L_h satisfy

$$\|L_h^{-1}\|_{H(B^{-1}) \mapsto H(B)} \leq \delta_0^{-1/2}.$$

Suppose that $\zeta_h \equiv L_h(\pi_h v) - f_h$ is the residual for a given vector v. Then

$$\varepsilon_h \leq \frac{\kappa_1^{1/2}}{\delta_0^{1/2}} \|\zeta_h\|_{B^{-1}}. \tag{3.18}$$

Proof. Observe that $L_h(\pi_h v - f_h) = \zeta_h$. Using the condition on L_h, we obtain $\|\pi_h v - u_h\|_B \le \delta_0^{-1/2} \|\zeta_h\|_{B^{-1}}$ and (3.18). □ [26]

§ 4. Model projective-grid operator on parallelepiped grids

4.1. Regular triangulations. There exists a strong connection between PGMs and difference methods in the case of regular triangulations and $d = 2$ (see § 0.2.). Here we generalize these results to $d \ge 3$ but emphasize the significant differences (see [175, 176, 201]). [27]

If diagonals of all cells are parallel, such triangulations are referred to as regular and are denoted by $T_h^{[k]}(\bar{Q})$, $k \in [1, 4]$. Given a triangulation $T_h(\bar{Q})$, we define (see § 1) subspaces $\hat{G} \subset G^{(1)} \equiv W_2^1(Q)$ of functions $\hat{u}(x)$ that are continuous on $\bar{Q}$ and piecewise linear with respect to this triangulation. Denote the standard basis by $\{\hat{\psi}_1(x), \dots, \hat{\psi}_N(x)\}$ and the projective-grid operators by

$$J \equiv J_h \equiv [b(\hat{\psi}_j; \hat{\psi}_i], \tag{4.1}$$

where

$$b(u; v) \equiv (u; v)_{1,Q} + (c, uv)_{0,Q} + (\sigma, uv)_{0,\Gamma_1}. \tag{4.2}$$

For a regular triangulation $T_k^{[k]}(\bar{Q})$, we write $\hat{u}^{[k]}$ and $J^{[k]}$ instead of $\hat{u}$ and J. In the same manner, we define one-dimensional operators J_r and $J_r^{[k]}$ associated with the bilinear form $b(u; v) \equiv (D_r u, D_r v)_{0,Q}$, $r = 1, 2, 3$, and which will enable us to derive the desired representations of J in the simplest case of $c = 0, \sigma = 0$ in (4.2). We make use of the standard difference notation (see § 0.2) $\|h\| \equiv h_1 h_2 h_3$, $u(P_i) \equiv u_i$, $\Delta_r u_i \equiv \bar{\partial}_r \partial_r u_i \equiv h_r^{-2}[u(I_r P_i) - 2u(P_i) + u(I_{-r} P_i)]$ with the shift operators I_r and I_{-r} along the x_r-direction ($r = 1, 2, 3$).

---

[26] In practical application of the theorem, we need to evaluate ζ_h and solve $Bz_h = \zeta_h$ on the refined grid. The computed z_h then determines $(z_h, \zeta_h) = \|z_h\|_b^2 = \|\zeta_h\|_{B^{-1}}^2$. Thus, it is very important to have model operators B spectrally equivalent to J with constants $\kappa \equiv \kappa_1/\kappa_0$ very close to 1 and at the same time admitting relatively easily solved systems. Recall that for estimating δ_0, the adaptation procedure from § 1.3 can be helpful.

[27] We concentrate on the analysis of 3-dimensional problems because they are typical of the more involved multi-dimensional cases. We thus deal with $\bar{Q} \in \mathbf{R}^3$ and a given triangulation $T_h(\bar{Q})$ associated with a given parallelepiped grid with nodes $P_i \equiv [i_1 h_1, i_2 h_2, i_3 h_3]$, where $i \equiv [i_1, i_2, i_3]$ and $h_r > 0$ is the mesh size in the x_r-direction, $r = 1, 2, 3$. Each cell of the grid is partitioned into six congruent simplexes (tetrahedrons) (see § 1); this partition is determined by the given diagonal of the cell and might be parallel to one of the following four directions: $d_1 \equiv [h_1, h_2, h_3], d_2 \equiv [h_1, h_2, -h_3], d_3 \equiv [h_1, -h_2, h_3], d_4 \equiv [-h_1, h_2, h_3]$ (recall that, for $d > 2$, we cannot take arbitrary diagonals in any two neighboring cells). Each of the simplexes has one edge coinciding with the chosen diagonal of the cell and three other edges are parallel to coordinate axes (we refer to these as coordinate edges).

Lemma 1. *Let* $\bar{Q} \equiv [0, l_1] \times [0, l_2] \times [0, l_3], h_r \equiv 1/(N_r + 1)$. *Let the Hilbert space* $G = W_2^1(Q)$ *and its subspace* $\hat{G}_h$ *be associated with the regular triangulation* $T_h^{[1]}(\bar{Q})$. *Suppose that the set* $\gamma_{1,2}$ *consists of edges of* $\bar{Q}$ *parallel to the vector* $[1, 0, 0]$ *and such that they contain either the vertex* $[0, 0, 0]$ *or the vertex* $[l_1, l_2, l_3]$, *and that the set* $\gamma_{1,1}$ *consists of edges of* $\bar{Q}$ *parallel to* $[1, 0, 0]$ *and such that they contain either* $[0, l_2, 0]$ *or* $[0, 0, l_3]$. *Suppose that* $\gamma_1 \equiv \gamma_{1,2} \cup \gamma_{1,1}$ *and* $\gamma_2 \equiv \Gamma \setminus \gamma_1$. *Then*

$$J_1^{[1]} u_i = -\|h\| \alpha_i \Delta_1 u_i, \ if \ 1 \leq i_1 \leq N_1, \tag{4.3}$$

$$J_1^{[1]} u_i = -\|h\| \alpha_i \frac{\partial_1 u_i}{h_1} \ if \ i_1 = 0; \quad J_1^{[1]} u_i \equiv -\|h\| \alpha_i \frac{\bar{\partial}_1 u_i}{h_1} \ if \ i_1 = N_1 + 1,$$

where: $\alpha_i \equiv 1/3$ *if* $P_i \in \gamma_{1,2}$; $\alpha_i \equiv 1/6$ *if* $P_i \in \gamma_{1,1}$; $\alpha_i \equiv 1/2$ *if* $P_i \in \Gamma \setminus (\gamma_{1,2} \cup \gamma_{1,1})$; *and* $\alpha_i = 1$ *if* P_i *is an inner point of* Q.

Proof. As in § 0.2 for a given i, we make use of the support $S_i \subset \bar{Q}$ of the basis function $\hat{\psi}_i$ and of the sets $S_{i,1,1}$ and $S_{i,1,0}$ being unions of simplexes from $T_h^{[1]}(\bar{Q})$ containing line segments $[P_i, I_1 P_i]$ and $[I_{-1} P_i, P_i]$, respectively. Then the same reasoning as in § 0.2 gives the representation

$$J_1^{[1]} u_i = -(\partial_1 u_i, h_1^{-1})_{0, S_{i,1,1}} + (\bar{\partial}_1 u_i, h_1^{-1})_{0, S_{i,1,0}}. \tag{4.4}$$

Hence, what remains is to specify volumes of the sets $S_{i,1,1}$ and $S_{i,1,0}$. For this we notice that every coordinate edge of a simplex from any triangulation $T_h(\bar{Q})$ may belong to m different simplexes, where $m \in [1, 8]$ may assume only the values $1, 2, 3, 6$; moreover, in the case of triangulation $T^{[1]}(\bar{Q})$, each edge belonging to $\gamma_{1,2}$ and $\gamma_{1,1}$ belongs to one or two simplexes, respectively. Hence, (4.3) follows from (4.4). $\square$

In the same way, we can find representation of operators $J_r^{[1]}$ and $J_r^{[k]}$. On the basis of the above lemma, it is not difficult to verify the next three lemmas.

Lemma 2. *Let* $G = W_2^1(Q; \Gamma_0)$, *where* Γ_0 *consists of a number of faces of* $\bar{Q}$. *Suppose that the basis of the subspace* $\hat{G}_h$ *associated with the regular triangulation* $T_h^{[1]}(\bar{Q})$ *contains all* $\hat{\psi}_i$ *with* $P_i \in \bar{Q} \setminus \Gamma_0$. *Then, for all* $P_i \in \bar{Q} \setminus \Gamma_0$, *formula* (4.4) *holds with*

$$u_i = 0 \ if \ P_i \in \Gamma_0. \tag{4.5}$$

Lemma 3. *Let* $G \equiv \overset{o}{W}_2^1(\bar{Q})$. *Let the triangulation* $T_h(\bar{Q})$ *be such that each coordinate edge not belonging to* ∂Q *belongs exactly to six simplexes from* $T_h(\bar{Q})$. *Let* $c = 0, \sigma = 0$ *in the bilinear form* (4.2). *Then the operator* J *from* (4.1) *is such that*

$$J u_i = -\|h\| (\Delta_1 + \Delta_2 + \Delta_3) u_i, \quad P_i \in \bar{Q} \setminus \Gamma_0,$$

with u on Γ_0 *defined by* (4.5).

Lemma 4. *Let* $\bar{Q} \equiv [-1, 1]^3$ *and* $h \equiv (N + 1)^{-1}$. *Suppose the triangulation* $T_h(\bar{Q})$ *is defined by the choice of the diagonal of a cubical cell parallel to the vector* $[\text{sign } x_1, \text{sign } x_2, \text{sign } x_3]$ *where* $[x_1, x_2, x_3]$ *are coordinates of the center of the cell. Let* $G = \overset{\circ}{W_2^1}(Q)$ *and suppose its subspace* $\hat{G}_h$ *is associated with the above triangulation* $T_h(\bar{Q})$. *Let* $c = 0, \sigma = 0$ *in bilinear form* (4.2). *Then the operator* J *from* (4.1) *is such that*

$$Ju_i = -\|h\|[\alpha_{1,i}\Delta_1 u_i + \alpha_{2,i}\Delta_2 u_i + \alpha_{3,i}\Delta_3 u_i, \quad P_i \in Q,$$

where u satisfies (4.5), *and* $\alpha_{r,i} = 4/3$ *if* P_i *belongs to* x_r *coordinate axis and* $\alpha_{r,i} = 1$ *otherwise.*

4.2. Spectral equivalence of projective-grid and difference operators. In this subsection the conditions of Lemma 1 on $\bar{Q}$ and the grid are assumed to hold.

Theorem 1. *Let* $G = W_2^1(Q; \Gamma_0)$ *with* Γ_0 *consisting of a number of faces of* $\bar{Q}$. *Suppose a triangulation* $T_h(\bar{Q})$ *and the regular triangulations* $T_h^{[k]}(\bar{Q})$, $k \in [1, 4]$, *define subspaces* $\hat{G}_h$ *and* $\hat{G}_h^{[k]}$ *of* G *with basis functions* $\hat{\psi}_i(x)$ *and* $\hat{\psi}_i^{[k]}(x)$, *respectively, with* $P_i \in Q_h \equiv \{P_i : P_i \in \bar{Q} \setminus \Gamma_0\}$. *Let the projective-grid operators* J *and* $J^{[k]}$ *be defined by* (4.1), (4.2) *with* $c = 0$ *and* $\sigma = 0$ *in* (4.2) *and with the corresponding sets of the basis functions. Then* $J \asymp J^{[k]}$, $k \in [1, 4]$.

Proof. Denote by $\hat{u}$ and $\hat{u}^{[k]}$ the elements of the subspaces $\hat{G}_h$ and $\hat{G}_h^{[k]}$ generated by the one and the same vector u defined by values of these functions at admissible nodes $P_i \in Q_h$. Then $(Ju, u) = |\hat{u}|_{1,Q}^2$, $(J^{[k]}u, u) = |\hat{u}^{[k]}|_{1,Q}^2$. As in the proof of Lemma 1, we represent $|\hat{u}|_{1,Q}^2$ as a sum of terms of type $\|h\|6^{-1}m_l(\partial_l u)^2$, where l refers to a coordinate edge, $\partial_l u$ refers to the derivative of $\hat{u}$ along this edge, and m_l is the number of simplexes from $T_h(\bar{Q})$ containing this edge. Similar expressions are obtained for $|\hat{u}^{[k]}|_{1,Q}^2$, and comparison of the numbers m_l and corresponding numbers $m_l^{[k]}$ leads to the desired equivalence. □

We also define difference operators

$$\Lambda_r \equiv 4^{-1}(J_r^{[1]} + J_r^{[2]} + J_r^{[3]} + J_r^{[4]}), \quad r = 1, 2, 3, \tag{4.6}$$

$$\Lambda \equiv \Lambda_1 + \Lambda_2 + \Lambda_3, \tag{4.7}$$

structures of which do not depend on the triangulation .

Lemma 5. *Let the conditions of Lemma 2 be satisfied. Then*
$$\Lambda_1^{[1]}u_i = -\|h\|\alpha_i\Delta_1 u_i, \text{ if } 1 \le i_1 \le N_1,$$

$$\Lambda_1^{[1]}u_i = -\|h\|\alpha_i\frac{\partial_1 u_i}{h_1} \text{ if } i_1 = 0; \quad \Lambda_1^{[1]}u_i \equiv -\|h\|\alpha_i\frac{\bar{\partial}_1 u_i}{h_1} \text{ if } i_1 = N_1 + 1,$$

where: $\alpha_i \equiv 1$ *if* P_i *is an inner point of* Q, $\alpha_i \equiv 1/4$ *if* P_i *belongs to an edge of* $\bar{Q}$, *and* $\alpha_i \equiv 1/2$ *otherwise.*

Proof. Since $\Lambda_1 u_i = 1/4(J_1^{[1]}u_1 + J_1^{[2]}u_i + J_1^{[3]}u_1 + J_1^{[4]}u_i)$, then it suffices to apply (4.3) for $J_1^{[1]}u_i$ and similar formulas for $J_1^{[k]}u_i$, with $k = 2, 3, 4$, and to make use of the fact that each fixed edge of an arbitrary parallelepiped cell belonging to $\bar{Q}$ is an edge of exactly two simplexes in some two regular triangulations $T_h^{[k]}(\bar{Q})$ and of only one simplex in the remaining triangulations. $\square$

Lemma 6. *Let conditions of Lemma 2 be satisfied. Let the operator* Λ *be defined by (4.7) and (4.6). Then* $2/3\Lambda \leq J^{[k]} \leq 4/3\Lambda$, $k \in [1,4]$.

Proof. It suffices to make use of (4.7) and the inequalities $2/3\Lambda_r \leq J_r^{[k]} \leq 4/3\Lambda_r, r \in [1,3], k \in [1,4]$, which follow from Lemma 5. $\square$

The integral over a triangle $T \equiv \Delta A_1 A_2 A_3$ of a polynomial of degree of at most 2 is just the area of the triangle (denoted by s(T)) multiplied by the arithmetic mean of the values of the polynomial at the midpoints of the triangle sides (see [117]). Hence,

$$\frac{1}{12} \leq \frac{|\hat{v}|_{0,T}^2}{s(T)[v^2(A_0) + v^2(A_1) + v^2(A_2)]} \leq \frac{1}{3}, \tag{4.8}$$

where $\hat{v}$ is an arbitrary linear function defined on the triangle.

Lemma 7. *Let a planar closed region* $\bar{\Omega}$ *be covered by a finite number of triangles such that each is obtained from a cell of a rectangular grid with mesh sizes* h_1 *and* h_2 *by using one of its diagonals. Let* $\hat{G}$ *be the space functions that are continuous and piecewise linear with respect to the triangulation of* $\bar{\Omega}$ *under consideration. Then*

$$\delta_0 h_1 h_2 \sum_i u_i^2 \leq |\hat{u}|_{0,\Omega}^2 \leq \delta_1 h_1 h_2 \sum_i u_i^2, \quad \forall \hat{u} \in \hat{G}, \tag{4.9}$$

where $\delta_0 = 1/(24)$ *and* $\delta_1 = 4/3$; *if additionally all diagonals of rectangular cells are parallel and* $\hat{u} = 0$ *on* $\partial\Omega$, *then* $\delta_0 = 1/4$ *and* $\delta_1 = 1$.

Proof. It suffices to make use of (4.8) and notice that the number of triangles in the triangulation containing a fixed node may take only values from 1 to 8. $\square$

Theorem 2. *Let* $G = W_2^1(Q; \Gamma_0)$ *with* Γ_0 *consisting of a number of faces of* $\bar{Q}$ *and* $\Gamma_1 \equiv \Gamma \setminus \Gamma_0$. *Let a triangulation* $T_h(\bar{Q})$ *define the subspaces* $\hat{G}_h$ *with basis functions* $\hat{\psi}_i(x)$, *where* $P_i \in Q_h \equiv \{P_i : P_i \in \bar{Q} \setminus \Gamma_0\}$. *Let the projective-grid operator* J *be defined by (4.1), (4.2) with* $c \geq 0$ *constant and* $\sigma \geq 0$ *constant on any fixed face of* Q. *Suppose that the operator* $\Lambda_{c,\sigma}$ *is* *such that $(\Lambda_{c,\sigma} u, u)$ is just*

$$\|u\|_\Lambda^2 + c\|h\| \sum_{P_i \in Q} \alpha_i u_i^2 + \sum_{\Pi_r \subset \Gamma_1} \frac{s(\Pi_r)\sigma(\Pi_r)}{4} \sum_{k=0}^{3} u^2(A_{r,k}), \qquad (4.10)$$

where: $\alpha_i = 1/8$ if P_i is a vertex of $\bar{Q}$, $\alpha_i = 1/4$ if P_i is an inner point of an edge of $\bar{Q}$, $\alpha_i = 1/2$ if P_i is an inner point of a face of $\bar{Q}$, and $\alpha_i = 1$ if P_i is an inner point of Q; and where Π_2 is a rectangular cell of the grid belonging to $\bar{\Gamma}_1$ with vertices $A_{r,k}, 0 \le k \le 3$, $s(\Pi_r)$ is the area of Π_r, and $\sigma(\Pi_r)$ is the value of σ at the center of the cell. Then J and $\Lambda_{c,\sigma}$ are spectrally equivalent operators with bounds independent of c and σ.

Proof. The case with $c = 0, \sigma = 0$ was considered in the proof of Theorem 1. Therefore, it suffices to compare the expression $c|\hat{u}|_{0,\Omega}^2 + (\sigma, \hat{u}^2)_{0,\Gamma_1}$ with $X_1(u) + X_2(u)$, where $X_1(u) \equiv c\|h\| \sum_{P_i \in Q} \alpha_i u_i^2$, and

$$X_2(u) \equiv \sum_{\Pi_r \subset \Gamma_1} \frac{s(\Pi_r)\sigma(\Pi_r)}{4} \sum_{k=0}^{3} u^2(A_{r,k}).$$

The term $(\sigma, \hat{u}^2)_{0,\Gamma_1}$, being a sum of integrals over elementary triangles, is readily bounded by $X_2(u)$ due to (4.8); similarly, $|\hat{u}|_{0,Q}^2$ is a sum of integrals over elementary tetrahedrons in $T_h(\bar{Q})$, and each of these integrals divided by $\|h\|^{-1}$ is bounded by the sum of squares of the values of the function $\hat{u}$ at the vertices of the tetrahedron. [28] □

4.3. The prismatic elements. Let $\bar{\Omega}$ from (1.19) be partitioned into a set of prisms $Z_{k,j}$ (see(1.20)). (We prefer to write i_3 instead of $|j|$ and $[x_1, x_2, x_3] \equiv x$ instead of $[x, y, z]$.) Also, instead of (1.21), we use the equivalent form $\hat{u} = I_3 I_{1,2} u$, where $I_{1,2}$ and I_3 refer to linear interpolation operators with respect to the triangles in $T_h(\bar{\Omega}_2)$ and to the intervals $[h_3 i_3, h_3(i_3 + 1)] \subset [0, l]$. Then the basis functions for the subspace $\hat{G}_h$ can be written in the form $\psi_i(x) \equiv \hat{\psi}_{i_1, i_2}(x_1, x_2)\hat{\psi}_{i_3}(x_3)$.

Lemma 8. *Let $G \equiv \overset{0}{W}_2^1(\Omega)$ and suppose the subspace $\hat{G}_h$ is defined on the basis of the given prismatic partition of $\bar{\Omega}$. Let $b(u; v) \equiv (u, v)_{1,\Omega}$, for all u and v, and the projective-grid operator J be defined by (4.1), (4.2) with the above basis functions, $c = 0$, and $\sigma = 0$. Then*

$$J u_i = -\|h\|(Y_3 \Delta_1 + Y_3 \Delta_2 + Y_{1,2} \Delta_3) u_i, \quad P_i \in Q_h, \qquad (4.11)$$

---

[28] All operators mentioned above are only nonnegative if Γ_0 is an empty set; if Γ_0 contains at least one face of $\bar{Q}$ (or if $c > 0$ or $\sigma > 0$ on a face), then they are positive operators. Also note that many of the given results can be obtained for d-dimensional problems.

where u *satisfies* (4.5), $Y_3 u_i \equiv 1/6[4u_1 + u_{I_1,i_2,i_3-1} + u_{i_1,i_2,i_3-1}]$,

$$Y_{1,2} u_{i_1,i_2} \equiv 1/(12)\{k_{i_1,i_2} u_{i_1,i_2} + \sum u_{j_1,j_2}\},$$

k_{i_1,i_2} *is the number of triangles in* $T_h(\bar{\Omega}_2)$ *containing the point* $[i_1 h_1, i_2 h_2]$ *as a vertex, and summation on the right-hand side of the expression for* $Y_{1,2} u_{i_1,i_2}$ *is carried out with respect to remaining vertices of these triangles.*

Proof. To find the required representation of $(D_r \hat{u}, D_r \hat{\psi}_i)_{0,\Omega} \equiv (J_r u)_i$, we use an approach similar to that used in the proofs of Lemma 0.2.1 and Theorem 0.2.4 (see (0.2.15), (0.2.19), and (0.2.20)). For $r = 1$ and $r = 2$, the reasoning is almost the same and leads to equalities $J_r u_i = -\|h\| Y_3 \Delta_r u_i$, $P_i \in Q_h$. Thus, it remains to prove the desired formula for $(Y_3 u)_i$. If S_{i_1,i_2} is the support of $\hat{\psi}_{i_1,i_2}$, then

$$(D_3 \hat{u}, D_3 \hat{\psi}_i)_{0,\Omega} = -\frac{\|h\|}{h_3}[(I_{1,2}\partial_3 u_i, \hat{\psi}_{i_1,i_2})_{0,S_{i_1,i_2}} - (I_{1,2}\bar{\partial} u_i, \hat{\psi}_{i_1,i_2})_{0,S_{i_1,i_2}}],$$

which implies (4.11). □

Theorem 3. *Let* $\Lambda \in \mathcal{L}^+(H)$ *be defined by*

$$\Lambda u_i = -\|h\|(\Delta_1 + \Delta_2 + \Delta_3)u_i, \quad P_i \in Q_h,$$

and (4.5). *Then* $\Lambda \asymp J$.

Proof. We make use of the inequalities $1/3 I \leq Y_3 \leq I$, notice that $Y_3 \Delta_r = \Delta_r Y_3$, $r = 1, 2$, $Y_{1,2}\Delta_3 = \Delta_3 Y_{1,2}$, then apply Lemma 8. □ [29]

§ 5. Hierarchical bases; estimates of angles between finite element subspaces

5.1. Splittings of finite element subspaces. In what follows, we consider the closure $\bar{Q}$ of a given domain Q in the Euclidean space $\mathbf{R}^d$ that consists of a finite number of d-dimensional simplexes $T_{0,k}$; their collection defines a (possibly composite) triangulation $T^{(0)}(\bar{Q}) \equiv T^{(0)}$. With this triangulation, we associate a spline subspace $\hat{G}^{(0)}$ consisting of functions that are continuous on $\bar{Q}$ and linear on elementary simplexes. Let $Q_h^{(0)}$ be a set of vertices $P_i^{(0)}$ of simplexes $T_{0,k}$ from $T^{(0)}$ such that $P_i^{(0)}$ is in correspondence with the standard basis piecewise linear function $\hat{\psi}_i^{(0)}(x)$ (see § 2.1). Let $\hat{\psi}_i^{(0)}(x), i \in [1, N_0]$, be a basis of $\hat{G}^{(0)}$. It is not important whether this triangulation was obtained as a result of a refinement process.

---

[29] The given results can be readily generalized to $G = W_2^1(\Omega; \Gamma_0)$ and the bilinear form defined by (4.2). The important but very simple case of periodic conditions with respect to $x_3 \equiv z$ can be analyzed in much the same way.

We are concerned only with the single level refinement procedure leading to a new triangulation $T^{(1)}(\bar{Q}) \equiv T^{(1)}$, associated with a new subspace

$$\hat{G}^{(1)} = \hat{G}_1 \oplus \hat{G}_2, \tag{5.1}$$

where $\hat{G}_2 \equiv \hat{G}^{(0)}$. Namely we take a closed subset $\bar{Q}^{(1)} \subset \bar{Q}$, consisting of a number of simplexes from $T^{(0)}$, and a refinement ratio $t_1 > 1$. Next, each simplex from $T^{(0)}$ belonging to $\bar{Q}_1$ (an old cell) is partitioned into a set of t_1^d congruent subsimplexes (new cells).

Figure 1 represents such a 3-dimensional simplex (tetrahedron), which is partitioned into 2^3 smaller congruent simplexes with edges reduced by half; each of the two indicated triangular prisms is composed of three sub-simplexes.

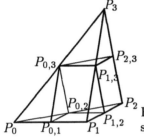

Figure 1. Partition of a three-dimensional simplex.

The old cells not belonging to $\bar{Q}^{(1)}$ and new smaller cells belonging to $\bar{Q}^{(1)}$ define the new triangulation $T^{(1)}$. We take a new wider set $Q_h^{(1)}$ of vertices associated with the standard functions $\hat{\psi}_i^{(1)}(x)$ and obtain $\hat{G}^{(1)} \equiv \{\hat{u} : \hat{u} = \sum u_i \hat{\psi}_i^{(1)}(x), \ P_i^{(1)} \in Q_h^{(1)}\}$. Along with this standard basis for $\hat{G}^{(1)}$, consider the basis consisting of $\bar{\psi}_i^{(1)} \equiv \hat{\psi}_i^{(1)}$ for $P_i^{(1)} \in Q_h^{(1)} \setminus Q_h^{(0)}$ and $\bar{\psi}_i^{(1)} \equiv \hat{\psi}_i^{(0)}$ for $P_i^{(1)} \in Q_h^{(0)}$, assuming that the indices of nodes from $Q_h^{(1)} \setminus Q_h^{(0)}$ to be less than those for nodes from $Q_h^{(0)}$. The indicated choice of the new basis (it is often referred to as an *hierarchical basis*, (see [34, 50, 512, 517])) leads to splitting (5.1) [30] with

$$\hat{G}_2 = \hat{G}^{(0)} \tag{5.2}$$

$$\hat{G}_1 \equiv \{\hat{u} : \hat{u} \in \hat{G}^{(1)}, \hat{u}(P_i^{(0)}) = 0 \text{ for all } P_i^{(0)} \in Q_h^{(0)}\}. \tag{5.3}$$

5.2. Angles between subspaces; local analysis. Consider a finite-dimensional space $\hat{G} \equiv \hat{G}_1 \oplus \hat{G}_2$ with a given semiinner product. Recall

---

[30] In much the same way, hierarchical bases associated with splittings of spline subspaces can be constructed for other choices connected, e.g., with polylinear functions. As we shall see below, such splittings may be regarded as almost orthogonal in some sense, and thus will enable us to use results of § 1.5 in § 3.7.

that the angle α between these subspaces is defined as the minimum $\alpha \geq 0$ satisfying

$$|(\hat{u}_1, \hat{u}_2)| \leq \cos \alpha \|\hat{u}_1\| \|\hat{u}_2\|, \quad \forall \hat{u}_r \in \hat{G}_r, \ r = 1, 2. \tag{5.4}$$

Lemma 1. *Let $\bar{\Omega} \subset \mathcal{R}^d$ be a bounded closed set of the form $\bar{\Omega} = \bar{\Omega}_1 \cup \cdots \cup \bar{\Omega}_m$. Denote the linear spaces of functions that are continuous on $\bar{\Omega}$ and $\bar{\Omega}_i$ by C and C_i, respectively ($i \in [1, m]$). For $u \in C$, the function $u_i \in C_i$ denotes the restriction of u to $\bar{\Omega}_i, i \in [1, m]$. Let the spaces $C_i, i \in [1, m]$ be equipped with semiinner products and seminorms denoted by $(\ ,\ )_i$ and $|\ |_i, i \in [1, m]$, respectively. Consider subspaces U and V of C such that*

$$|(u_i, v_i)| \leq \gamma_i |u_i|_i |v_i|_i, \ i \in [1, m] \tag{5.5}$$

for all $u \in U$ and $v \in V$. Let a semiinner product in the space C be defined by

$$(u, v) \equiv \sum_{i=1}^{d} (u_i, v_i)_i.$$

Then the angle α between U and V is such that

$$\cos \alpha \leq \max_i \gamma_i \equiv \gamma. \tag{5.6}$$

Proof. We have $|(u, v)| \leq \sum_{i=1}^{d} |(u_i, v_i)_i| \leq \sum_{i=1}^{d} \gamma_i |u_i|_i |v_i|_i$ (see (5.5)) and

$$|(u, v)| \leq \gamma [\sum_{i=1}^{d} |u_i|_i^2]^{1/2} [\sum_{i=1}^{d} |v_i|_i^2]^{1/2},$$

whence (5.6) follows. $\square$ [31]

5.3. Estimates of angles between finite element subspaces associated with d-dimensional simplexes.

Theorem 1. *Let all simplexes from triangulations $T^{(l)}(\bar{Q})$ be regular simplicial parts of cubes in $\mathbf{R}^d$ and suppose the finite element subspaces $\hat{G}_2$ and $\hat{G}_1$ in (5.1) are defined by (5.2), (5.3). Suppose also that the semiinner product in $\hat{G}^{(1)}$ is $(u, v) \equiv (u, v)_{1,Q}$. Then the angle α between $\hat{G}_1$ and $\hat{G}_2$ is such that*

$$\cos \alpha \leq [1 - t_1^{1-d}]^{1/2} \equiv \gamma < 1. \tag{5.7}$$

Proof. We have

---

[31] This simple result can be useful for estimating the angle between some spline subspaces by local analysis on cell. The sets $\bar{\Omega}_i$ can be of a very general nature; e.g., they may be manifolds of different dimensions, and they may have common inner points or even coincide. Estimation of angles between various subspaces has been considered by many researchers (see [21, 48, 49, 183, 204, 205, 344, 289]).

$$(\hat{u}_1, \hat{u}_2) = \sum_{T_{0,k} \in T^{(0)}(\bar{Q})} (\hat{u}_1, \hat{u}_2)_{1,T_{0,k}}, \quad \hat{u}_r \in \hat{G}_r, \ r = 1, 2. \qquad (5.8)$$

According to (5.8) and Lemma 1, it suffices to prove that

$$|(\hat{u}_1, \hat{u}_2)_{1,T_{0,k}}| \leq \gamma |\hat{u}_1|_{1,T_{0,k}} |\hat{u}_2|_{1,T_{0,k}}. \qquad (5.9)$$

Note that invariance of $(u, v)_{1,T}$ holds with respect to a change of Descartes coordinates, which we choose so that the edges $P_0P_1, P_1P_2, \ldots, P_{d-1}P_d$ of the simplex $T_{0,k} = [P_0P_1 \ldots P_d]$ are parallel to elements of the new basis associated with coordinates $y_1, \ldots, y_d$. Then

$$(\hat{u}_1, \hat{u}_2)_{1,T_{0,k}} \equiv \int_{T_{0,k}} \left(\frac{\partial \hat{u}_1}{\partial y_1} \frac{\partial \hat{u}_2}{\partial y_1} + \cdots + \frac{\partial \hat{u}_1}{\partial y_d} \frac{\partial \hat{u}_2}{\partial y_d} \right) dy_1 \ldots dy_d.$$

This semiinner product is again a sum of d simpler semiinner products. Therefore, for proving (5.9), it suffices to show that

$$|X_r| \equiv |\int_{T_{0,k}} \frac{\partial \hat{u}_1}{\partial y_r} \frac{\partial \hat{u}_2}{\partial y_r} dy_1 \ldots dy_d| \leq \gamma |\frac{\partial \hat{u}_1}{\partial y_r}|_{0,T_{0,k}} |\frac{\partial \hat{u}_2}{\partial y_r}|_{0,T_{0,k}}. \qquad (5.10)$$

Now, for X_r from (5.10), we use the representation

$$X_r = \sum_{T_{1,k'} \subset T_{0,k}} \int_{T_{1,k'}} \frac{\partial \hat{u}_1}{\partial y_r} \frac{\partial \hat{u}_2}{\partial y_r} dy_1 \ldots dy_d, \qquad (5.11)$$

where $T_{1,k'}$ refers to any of t_1^d subsimplexes from the partition of $T_{0,k}$. Since $\hat{u}_2$ is a linear function on the simplex $T_{0,k}$, its derivative $\frac{\partial \hat{u}_2}{\partial y_r}$ is a constant D_r; similarly, $\frac{\partial \hat{u}_1}{\partial y_r}$ is a constant $d_{r,1,k'}$ on the subsimplex $T_{1,k'}$. If simplex $T_{0,k}$ is a regular simplicial part of a cube $Q_{t_1 h}$, then its volume is $|T_{0,k}| = (t_1 h)^d/(d!)$, and $|T_{1,k'}| = h^d/(d!)$. We have $|\frac{\partial \hat{u}_2}{\partial y_r}|_{0,T_{0,k}}^2 = (2h)^d/(d!)D_r^2$, $|\frac{\partial \hat{u}_1}{\partial y_r}|_{0,T_{1,k'}}^2 = \frac{h^d}{d!}d_{r,1,k'}^2$, and

$$|\frac{\partial \hat{u}_1}{\partial y_r}|_{0,T_{0,k}}^2 = \sum_{T_{1,k'} \subset T_{0,k}} \frac{h^d}{d!} d_{r,l}^2.$$

Since $\hat{u}_1(P_{r-1}) = 0 = \hat{u}_1(P_r)$, the integral over the edge $[P_{r-1}P_r]$ of the function $\frac{\partial \hat{u}_1}{\partial y_r}$ vanishes. Hence, in the sum $X_r = |T_{1,k'}| \sum_{T_{1,k'} \subset T_{0,k}} D_r d_{r,1,k'}$ (see (5.11)), at least a sum of t_1 terms vanishes—such terms correspond to subsimplexes with some edges being parts of the edge $P_{r-1}P_r$ parallel to y_r-axis (in Figure 1, such an edge, in the case of $r = 1$, is $[P_0P_1]$). Hence,

$$|X_r| \leq \frac{h^d}{d!}[(t_1^d - t_1)D_r^2]^{1/2}[\sum_{T_{1,k'} \subset T_{0,k}} d_{r,1,k'}^2]^{1/2}$$

$$\leq (\frac{t_1^d - t_1}{t_1^d})^{1/2}|\frac{\partial \hat{u}_2}{\partial y_r}|_{0,T_{0,k}}|\frac{\partial \hat{u}_1}{\partial y_r}|_{0,T_{0,k}}.$$

Thus, we have proved (5.10), (5.9), and (5.7). □

Theorem 2. Assume the conditions of Theorem 1 with the semiinner product replaced by

$$(u, v) = \sum_{T_{0,k} \in T^{(0)}(\bar{Q})} a(T_{0,k})(u, v)_{1,T_{0,k}}, \tag{5.12}$$

where $a(T_{l,k}) \geq 0$ are constants. Then the statement of Theorem 1 holds.

Proof. As in the proof of Theorem 1 in the case of (5.12), we write $(\hat{u}_1, \hat{u}_2) = \sum_{T_{0,k} \in T^{(0)}(\bar{Q})} a(T_{l,k})(\hat{u}_1, \hat{u}_2)_{1,T_{0,k}}$. In accordance with this relation and Lemma 1, it suffices to prove that

$$|a(T_{l,k})(\hat{u}_1, \hat{u}_2)_{1,T_{0,k}}| \leq \gamma a(T_{l,k})|\hat{u}_1|_{1,T_{0,k}} a(T_{l,k})|\hat{u}_2|_{1,T_{0,k}},$$

which is equivalent to (5.9). Hence, the reasoning from the proof of Theorem 1 applies (see [204, 205]). □

If all simplexes are regular simplicial parts of cubes with edges parallel to the original coordinate axes, then (5.7) remains true for

$$(u, v)_G \equiv \sum_{r=1}^d a_r(T_{l,k})(\frac{\partial u}{\partial x_r}, \frac{\partial v}{\partial x_r})_{0,Q} \tag{5.13}$$

with arbitrary constants $a_r(T_{l,k}) \geq 0$.

5.4. Local numerical estimation of angles between subspaces.
The above local analysis was carried out in accordance with Lemma 1 for separate cells $T_{0,k}$ of the coarser grid and on the basis of estimates (5.9) for angles between local subspaces. Since the dimensions of these subspaces are relatively small, we might evaluate these angles numerically. This is especially reasonable when, on the one hand, explicit estimates are hard to obtain and, on the other hand, only a few types of cells are present. The mathematical basis for such a computation is very simple and was essentially given in § 1.5. More precisely, it is evident that the estimate $S_2(A) \geq s^2 A_{2,2}$, with $s^2 \equiv 1 - \cos^2 \alpha = \sin^2 \alpha$ (see (1.5.17)), cannot be improved. This implies that s^2 is the least positive eigenvalue of the algebraic eigenvalue problem

$$S_2(A)u_2 = \lambda A_{2,2} u_2 \tag{5.14}$$

in the Euclidean space H_2, the dimension of which is just the number of nodes of the coarser (old) cell. All λ are positive if $A > 0$, but if Ker $A \geq 1$, then Ker $S_2(A) =$ Ker $A_{2,2}$, and we can reduce (5.14) to a corresponding eigenvalue problem, with positive matrices, in the Euclidean space H_2^*, which is just the orthogonal complement of Ker $A_{2,2}$ in H_2.

§ 6. Nonconforming finite element methods

6.1. The simplest nonconforming finite element methods. [32] We consider perhaps the simplest but nonetheless important case of methods based on a triangulation $T_h(\bar{\Omega})$ of a closed planar region assuming that the boundary $\Gamma \equiv \partial\Omega$ consists of a finite number of closed line segments (e.g., a polygonal with slits). Also, we deal only with the model elliptic boundary value problem of the form (0.1.16), where $G = W_2^1(\Omega; \Gamma_0)$ and the bilinear form and linear functional are defined by

$$b(u; v) \equiv \sum_{r=1}^{2} (a(x), \frac{\partial u}{\partial x_r} \frac{\partial v}{\partial x_r})_{0,\Omega} \tag{6.1}$$

and

$$l(v) \equiv (g, v)_{0,\Omega}. \tag{6.2}$$

By their additivity, we may rewrite problem (0.1.16), (6.1), and (6.2) in the equivalent form: find $u \in G$ such that

$$b_h(u; v) = l_h(v), \quad \forall v \in G, \tag{6.3}$$

where

$$b_h(u; v) \equiv \sum_{T \in T_h(\bar{\Omega})} b^T(u; v), \quad l_h(v) \equiv \sum_{T \in T_h(\bar{\Omega})} (g; v)_{0,T}, \tag{6.4}$$

and

$$b^T(u; v) \equiv \sum_{r=1}^{2} (a(x), D_r u, D_r v)_{0,T}. \tag{6.5}$$

---

[32] We give an illustration of possible applications of model spectrally equivalent operators (optimal preconditioners) for grid approximations known as *nonconforming finite element methods* (see [91, 117, 127, 193, 475]). They are especially useful for practical solutions of many hydrodynamics and elasticity problems (see Chapters 7 and 8 and [93, 94]). Some of the resulting grid operators have also found applications in multigrid construction of the model operators (see [322]) for standard PGMs.

We introduce finite-dimensional spaces $\hat{G}_h$ of functions $\hat{u}$, defined and linear on each triangle $T \in T_h(\bar{\Omega})$ and such that any two adjacent triangles (their common side belongs to Ω) have the same value at the midpoint of their common side and vanish on sides belonging to Γ_0. Note that these functions, on a common side of two triangles, must be considered as two-valued functions (some authors speak about continuity of these functions at midpoints of sides). It is possible to treat them as elements of $L_2(\Omega)$ (recall that functions in this space need to be defined only for almost all $x \in \Omega$), which on every T are elements of $W_2^1(T)$ and have the same mean values of traces on common sides of the triangles. Thus, the spaces $\hat{G}_h$ do not belong to the original Hilbert space G. Nonetheless, we may consider them as approximations to G, and we may proceed to the discrete problem by replacing the bilinear form and the linear functional in (0.1.16) by their nonconforming finite element approximations b_h and l_h (see, e.g., [91, 117, 127, 475]). This yields the grid problem: find $\hat{u} \in \hat{G}_h$ such that

$$b_h(\hat{u}; \hat{v}) = l_h(\hat{v}), \quad \forall \hat{v} \in \hat{G}_h. \tag{6.6}$$

6.2. The form of grid operators for model regions. We have seen that the algebraic form of the projective method depends critically on the choice of basis for $\hat{G}_h$. The same is true for method (6.6). It is easy to see that functions $\hat{u} \in \hat{G}_h$ are uniquely defined by their values at the midpoints M_i of sides of the triangles $T \in T_h(\bar{\Omega})$ (we denote the set of M_i not belonging to Γ_0 by Ω_h). Actually, we need to define such a function on central triangles like $\triangle M_0 M_1 M_2 \subset T \equiv \triangle P_0 P_1 P_2$ (see Figure 1), then extend them linearly to the whole triangle T.

 Figure 1. Nonconforming element.

With each $M_i \in \Omega_h$, we can then associate a basis function $\hat{\psi}_i(x)$ defined by the following conditions: $\hat{\psi}_i(x) = 1$ if $x = M_i$, $\hat{\psi}_i(x) = 0$ if x coincides with any point from Ω_h different from M_i, and the restriction of $\hat{\psi}_i(x)$ to an arbitrary open triangle from $T_h(\bar{\Omega})$ is a linear function (on the boundary of T, discontinuities are allowed). It is clear that the support, S_i, of each $\hat{\psi}_i$ consists of triangles T containing M_i and that, inside each T, the first partial derivatives of $\hat{\psi}_i$ exist and are constant (nonzero for $T \subset S_i$ and vanishing outside of S_i). Now we make use of the expansion

$$\hat{u} = \sum_{j=1}^{N} u_j \hat{\psi}_j \tag{6.7}$$

and conditions

$$b_h(\hat{u}; \hat{\psi}_i) = l_h(\hat{\psi}_i), \quad i \in [1, N]. \tag{6.8}$$

This gives a system of linear equations $\Lambda' u = f$ for the vector u of components u_i defined from (6.7). More precisely,

$$\Lambda'_{\Omega} \equiv \Lambda'_{\Omega,h} \equiv [b_h(\hat{\psi}_j; \hat{\psi}_i)], \tag{6.9}$$

$$f \equiv [l_h(\hat{\psi}_1), \ldots, l_h(\hat{\psi}_N)]^T. \tag{6.10}$$

It is not difficult to specify the form of the operator Λ'_{Ω} for the model case of $\bar{\Omega} = \bar{Q}$ covered by a number of rectangular cells with mesh sizes $2h_1$ and $2h_2$ (the original triangulation is obtained by drawing diagonals of the cells independently), under assumption that $a(x)$ is constant in each cell. Recall that, in this case, it is convenient to use the vector index $i \equiv [i_1, i_2]$. For simplicity, we restrict ourselves to the following lemma.

Figure 2. Four square cells.

Lemma 1. *Let $a(x) = 1$ and $G \equiv \overset{o}{W}{}^1_2 (Q)$, and suppose that each triangle $T \in T_h(\bar{Q})$ is half of a square of mesh size $2h$. Then the grid operator $\Lambda'_Q \equiv \Lambda'$ takes the form*

$$\Lambda' u_i \equiv - \begin{cases} 2h^2(\Delta_1 + \Delta_2)u_i, & \text{if } M_i \in \omega_{0,h}, \\ 2h^2\Delta_2 u_i, & \text{if } M_i \in \omega_{1,h}, \\ 2h^2\Delta_1 u_i, & \text{if } M_i \in \omega_{2,h}, \end{cases} \tag{6.11}$$

where $\omega_{0,h}$, $\omega_{1,h}$, and $\omega_{2,h}$ denote the sets of points M_i that are centers of square cells, or midpoints of horizontal or vertical sides of the cells, respectively, and $\Delta_r \equiv \bar{\partial}_r \partial_r, r = 1, 2$, refers to the standard difference operator (see Lemma 0.2.1).

   Proof. Points M_i that are centers of square cells of mesh size $2h$ are marked in Figure 2 by $\diamond$; midpoints of horizontal or vertical sides of the cells are marked by $\triangle$ and $\triangleright$, respectively. Observe that each point $M_i \in \omega_{0,h}$ is a vertex corresponding to the right angle exactly in two central triangles for all possible choices of triangulations of independent square cells. Similarly, each point $M_i \in \omega_{1,h}$ or $M_i \in \omega_{2,h}$ corresponds to two central triangles, but it is a vertex lying either on horizontal or on vertical sides of these triangles. Observe also that the derivative $D_r \hat{v}$ on any triangle $T \in T_h(\bar{\Omega})$ is a constant equal either to $\partial_r v_i$ or to $\bar{\partial}_r v_i$, $r = 1, 2$, and for $\hat{v} = \hat{\psi}_i$ is either $-h^{-1}$ or h^{-1}. Thus, for finding $b_h(\hat{u}; \hat{\psi}_i)$, we can apply the same reasoning as in the proof of (0.2.15). $\square$

Theorem 1. Let the conditions of Lemma 1 be satisfied. Then the system $\Lambda' u = f'$ is reduced to the system

$$\Lambda u = g, \tag{6.12}$$

where u is the vector of values of $\hat{u}$ at the midpoints from the set $\omega_{0,h}$, the operator $\Lambda \equiv -h^2(\Delta_1' + \Delta_2')$, and

$$\Delta_r' u_i \equiv [u_{[i_1,i_2]+e_r} + u_{[i_1,i_2]-e_r} - 2u_{[i_1,i_2]}]/h^2.$$

Proof. The equation corresponding to $M_i \in \omega_{0,h}$ has the form

$$-2h^2(\Delta_1 + \Delta_2)u_i = f_i; \tag{6.13}$$

its left-hand side contains the linear combination of the grid function's values at the points with indices

$$[i_1, i_2], [i_1 + 1, i_2], [i_1 - 1, i_2], [i_1, i_2 + 1], \text{ and } [i_1, i_2 - 1]. \tag{6.14}$$

The coefficients are $8, -2, -2, -2$, and -2, respectively. Suppose, e.g., that we eliminate the value $u_{[i_1,i_2+1]}$ from (6.13) using the equation

$$2h^2 \Delta_1 u_{[i_1,i_2+1]} = f_{[i_1,i_2+1]},$$

which is equivalent to

$$2(2u_{[i_1,i_2+1]} - u_{[i_1,i_2+2]} - u_{[i_1,i_2]}) = f_{[i_1,i_2+1]}.$$

This yields a new linear combination of the remaining values with coefficients $7, -2, -2, -2$, respectively, and of the new value $u_{[i_1,i_2+2]}$ with the coefficient 1. Similar reasoning with respect to other indices from (6.14) leads to (6.12). $\square$ [33]

6.3. The spectral equivalence of operators on topologically equivalent triangulations. As in § 0.4, we consider sequences of two topologically equivalent triangulations $T_h(\bar{\Omega})$ and $T_h(\bar{Q})$ (regions might contain slits as mentioned in § 1), such that each triangle from $T_h(\bar{Q})$ is a half

---

[33] In this proof, we obtained a difference operator of the same structure but on another grid and with coefficient determined by a factor of $1/2$ (see [193]).

of a square (see [193]). We define isomorphic nonconforming spaces $\hat{G}_{\Omega,h}$ and $\hat{G}_{Q,h}$ of functions

$$\hat{u}_\Omega(x) = \sum_{i=1}^{N} u_i \hat{\psi}_{\Omega,i}(x) \text{ and } \hat{u}_Q(z) = \sum_{i=1}^{N} u_i \hat{\psi}_{Q,i}(z), \qquad (6.15)$$

where u_i corresponds to the same value of the functions $\hat{u}_\Omega$ and $\hat{u}_Q$ at the equivalent midpoints with index i and $\hat{\psi}_{\Omega,i}, \hat{\psi}_{Q,i}$ are corresponding basis functions (see (6.7)). Let the piecewise constant functions $a(x) > 0$ and $a_Q(z) > 0$ have the same constant values at inner points of the corresponding triangles, and define $\Lambda'_\Omega \equiv \Lambda'_h$ from (6.9) and the corresponding operator $\Lambda'_Q \equiv \Lambda'_{Q,h}$ for the model region Q. For an arbitrary triangle $T \in T_h(\bar{\Omega})$, define $\mu(T)$ by (0.4.12).

Theorem 2. *For each $h \in \{h\}$, assume we are given two topologically equivalent triangulations $T_h(\bar{\Omega})$ and $T_h(\bar{Q})$, with the corresponding grid operators Λ'_Ω and Λ'_Q defined above. Suppose there exists a $\mu < \infty$ such that*

$$\sup \mu(T_h(\bar{\Omega})) \le \mu.$$

Then

$$\delta_{0,\Omega} \Lambda'_Q \le \Lambda'_\Omega \le \delta_{1,\Omega} \Lambda'_Q,$$

where

$$\delta_{1,\Omega} \equiv \mu + (\mu^2 - 1)^{1/2} \equiv \delta_{0,\Omega}^{-1}.$$

Proof. We have

$$(\Lambda'_\Omega u, u) = \sum_{T \in T_h(\bar{\Omega})} (a(x), |\nabla u|^2)_{0,T}$$

and

$$(\Lambda'_Q u, u) = \sum_{T' \in T_h(\bar{Q})} (a_Q(z), |\nabla v|^2)_{0,T'}.$$

Integrals over the corresponding triangles T, T' are estimated in the same manner as in the proof of Theorem 0.4.5. $\square$

Useful generalizations of Theorem 2 are possible (see the end of § 0.4)) that would include grid operators defined on composite triangulations with local refinements.

Chapter 3

Estimates of computational work in solving model grid systems

Chapter 3 is devoted largely to asymptotically optimal iterative methods for solution of elliptic grid systems. We concentrate on several approaches to constructing the optimal and nearly optimal preconditioners mentioned in § 0.3. It is noteworthy that in all of these constructions we observe an interplay between such basic notions as splitting, partitioning, block elimination, factorization, inner iterations, and spectral equivalence. [1] These notions probably form the primary means of dealing with complicated problems, most particularly with problems involving domains of complicated geometry. At the outset, we emphasize that it is impossible to give a universal

---

[1] In § 1 we describe fast direct methods for model grid systems in a d-dimensional parallelepiped and some asymptotically optimal algorithms for solving systems in a rectangle. § 2 is devoted to ADI-methods and their generalizations. Some useful results dealing with splitting of the Euclidean space under consideration are also given there. Factorization of the original operator is the main topic in § 3. Effective inner iterations and so-called two-stage preconditioners are discussed in § 4. Domain decomposition and fictitious domain preconditioners are investigated in § 5 and 6, respectively. § 7 is devoted to multigrid iterative methods, especially to multigrid constructions of asymptotically optimal model operators. While these sections deal with grids and triangulations whose cells are relatively simple, § 8 deals with grids and triangulations that are more general, remaining topologically equivalent to those simpler ones; special attention is paid to the case of nonlinear systems.

prescription for choosing the most effective method for solving concrete problems, but a basic understanding of the nature of these methods is essential for making effective choices for specific applications.

§ 1. Fast direct methods for model grid systems in a rectangle and parallelepiped

For model problems defined on regular grids for rectangles or cubes (parallelepipeds), we consider some remarkable fast direct methods that were mentioned in § 0.3 (see also [46, 119, 159, 164, 281, 421, 447, 474, 502]).

1.1. Separation of variables and the fast discrete Fourier transform. Consider system (0.2.8) with unknown vectors $\mathbf{u} \equiv u \equiv [u_1, \dots, u_N]^T$ ($u \in H \equiv \mathbf{R}^N$) defined at the grid nodes $x_i \equiv [i_1 h_1, \dots, i_d h_d] \in Q_h$ of a regular parallelepiped grid. Here, $u_i \equiv u(x_i)$, $i \equiv [i_1, \dots, i_d]$, Q_h is the set of nodes associated with unknown values of the grid function, and $h \equiv \bar{h} \equiv [h_1, \dots, h_d]$. Any linear operator $A \in \mathcal{L}(H)$ is defined by

$$(Au)_i = \sum_{x_{i+j} \in Q_h} a_{i,j} u_{i+j}, \quad x_i \in Q_h. \tag{1.1}$$

Let $s(A)$ be the set of the shift vectors j such that there exists i in (1.1) corresponding to $a_{i,j} \neq 0$. Denote by lin $\{s(A)\}$ the linear span of vectors in $s(A)$. Then dim lin $s(A) \equiv \dim A$ is called *dimension of the operator A*. If $\dim A = 1$ and e_r is a basis of lin $s(A)$, we refer to A as an operator acting in the x_r-direction. [2]

If $\bar{f}_{i_1} \in H_1^\perp$, $B_1 \bar{u}_{i_1} = \sum_{x_{i_1+j_1} \in \omega_1} a_{i_1,j_1} u_{i_1+j_1}$, $\dim H_1^\perp = N_2 \cdots N_d \equiv M_1$, and $C \in \mathcal{L}(H_1^\perp)$, then we rewrite our system

---

[2] If $\dim A < d$, then under an appropriate ordering of nodes in Q_h it is possible to treat A as a block diagonal matrix and to split the system into a set of independent subsystems. Especially attractive are one-dimensional elliptic grid operators, for they lead to diagonal blocks that are either band matrices with small bandwidth or perturbations of band matrices at several positions (typical for problems with periodic boundary conditions). This implies that such subsystems can be treated either by Gauss elimination or one of its modifications (see § 1.5 and [146, 252], at an obviously optimal asymptotic cost of $W_a(N) \asymp N$. Note that such methods have good stability properties for difference systems like $\Lambda_r u = f$ (see (0.2.21)), or even for systems corresponding to some more general approximations of differential operators like $-D_r(a_r D_r) + b_r D_r + c$ under typical boundary conditions and that are are widely used in practice (see [221, 252, 447]). We emphasize that each of resulting subsystems includes only unknowns at nodes lying on a straight line parallel to the x_r-axis, and that such systems can be solved simultaneously on parallel computers. Below we consider $Q \equiv [a_1, b_1] \times \cdots \times [a_d, b_d]$ and $Q_h \equiv \omega_1 \times \cdots \times \omega_d$, where ω_r is the respective one-dimensional grid on $[a_r, b_r], r \in [1, d]$. We denote the vector defined by values of u at nodes lying in the plane $x_r = i_r h_r$ by $\bar{u}_r(i_r)$, and the corresponding Euclidean space of such vectors by $H_r^\perp$. Then $H = H_r \times H_r^\perp$, where H_r is the Euclidean space of grid functions defined on ω_r and $\dim H_r \equiv N_r$, $r \in [1, d]$.

$$(Bu)_i = f_i, \quad x_i \in Q_h \tag{1.2}$$

in the form $B_1 \bar{u}_{i_1} + C \bar{u}_{i_1} = \bar{f}_{i_1}, \forall x_{i_1} \in \omega_1$. Let $i_0 \equiv [i_2, \ldots, i_d]$, $j_0 \equiv [j_2, \ldots, j_d]$, and $\{\bar{\psi}_{j_0}\}$ be an orthonormal basis of $H_1^\perp$ such that $C\bar{\psi}_{j_0} = \lambda_{j_0}\bar{\psi}_{j_0}$. Then the expansions $\bar{u}_{i_1} = \sum_{j_0} \alpha_{j_0,i_1}\bar{\psi}_{j_0}$, $\bar{f}_{i_1} = \sum_{j_0} \beta_{j_0,i_1}\bar{\psi}_{j_0}$, and the standard method of separation of variables lead to systems

$$\sum_{x_{i_1}+j_1 \in \omega_1} a_{i_1,j_1}\alpha_{j_0,i_1+j_1} + \lambda_{j_0}\alpha_{j_0,i_1} = \beta_{j_0,i_1}, \quad x_{i_1} \in \omega_1. \tag{1.3}$$

If $\psi_{j_0}(i_2 h_2, \ldots, i_d h_d)$ denote the components of the vector $\bar{\psi}_{j_0}$, then

$$\beta_{j_0,i_1} \equiv \sum_{i_0} f_{i_1,i_0}\psi_{j_0}(i_2 h_2, \ldots, i_d h_d). \tag{1.4}$$

Solving (1.2) is reduced to three stages: evaluation of all β_{j_0,i_1} (see (1.4)); evaluation of all α_{j_0,i_1} (we solve M_1 one-dimensional independent systems (1.3) corresponding to different vectors j_0); evaluation of all $\bar{u}_1$. (Note that different indices i_1 can be treated independently at the first and third stages.) [3] Efficient ways to obtain $\bar{\beta}_{i_1}$ when $\bar{\psi}_{j_0}$ are the eigenvectors of the difference Laplace operator are based on the use of the fast discrete Fourier transform (see [41, 120, 286, 416]) under the assumption that

$$N_s = r_1^{\alpha_1} r_2^{\alpha_2} \ldots r_p^{\alpha_p}, \ s \geq 2, \tag{1.5}$$

where r_s is a fixed integer (usually, $r_s = 2$) and $\alpha_k = 1, 2, \ldots$. They lead to estimates of type $W_1 = O(N \ln M_1)$ and $W_3 = O(N \ln M_1)$ for both the first and third stages, so the total computational work in the method under consideration is $W = O(N \ln M_1)$. [4]

The $(d-1)$-dimensional Fourier transform, with respect to variables $x_2, \ldots, x_d$ and for the eigenvectors $\bar{\psi}_{j_0}(x_2, \ldots, x_d) = \psi_{2,j_2}(x_2) \cdots \psi_{d,j_d}(x_d)$, is easily reduced to one-dimensional transforms with respect to the separate variables $x_s, s \geq 2$. Hence, it suffices to describe the basic idea of this

---

[3] The second stage deals with separate one-dimensional grid (difference) systems, which can be solved by the Gauss elimination effectively because of the band structure of their matrices under an appropriate ordering (in the case of (0.2.15), (0.2.16), only diagonal elements and their neighbors can be nonzero). Thus, the computational work for this stage is $W_2 = O(N)$. The first and third stages can be described as evaluations of the matrix-vector products with matrices of a very special form. For example, the first stage actually is a problem of finding N_1 vectors $\bar{\beta}_{i_1} = F\bar{f}_{i_1}$, where $\bar{\beta}_{i_1}$ is a vector with components β_{j_0,i_1} and the rows of the matrix $F \in \mathbf{R}^{M_1 \times M_1}$ are the eigenvectors $\bar{\psi}_{j_0}$ of the matrix $C \in \mathbf{R}^{M_1 \times M_1}$. This implies that the standard matrix-vector product leads to unsatisfactory estimates of computational work $W_1 = O(N_1 M_1^2)$.

[4] The first such estimates were mentioned in the late 1960's in [38, 159, 281].

transform for a function defined on a uniform one-dimensional grid. The simplest case corresponds to a complex valued periodic function. More precisely, let $a > 0, h \equiv a/M, t_k \equiv kh$. Suppose that a complex-valued grid function v is such that $v_k \equiv v(t_k) = v(t_k + a) \; \forall k$ (v is fully defined by the vector $\bar{v} \equiv [v_0, \ldots, v_{M-1}]^T$) and that we are interested in finding the sums

$$F_n(v;h) \equiv \sum_{k=0}^{M-1} v_k e^{-\frac{2\pi n}{a} ikh}, \; n = 0, \ldots, M-1, \tag{1.6}$$

where $i^2 = 1, M = rs$, and $r > 1, s > 1$ are integers. Let the subvectors $v^{(p)} \equiv [v_p, v_{p+r}, \ldots, v_{p+r(s-1)}], \; p \in [0, r-1]$ be defined on coarser grids with the mesh size rh. Note that the indices k and n in (1.6) are uniquely represented in the form $k = k_1 r + k_2, \; k_1 \in [0, s-1], \; k_2 \in [0, r-1], \; n = n_2 s + n_1, \; n_2 \in [0, r-1], \; n_1 \in [0, s-1]$, where k_1, k_2, n_1, and n_2 are integers. We may thus introduce $F_{n_1}(v^{(k_2)}; rh)$ as in (1.6).

Lemma 1. *For* $F_n(v;h)$ *and* $F_{n_1}(v^{(k_2)};rh)$, *we have*

$$F_n(v;h) = \sum_{k_2=0}^{r-1} e^{-\frac{2\pi n k_2 h}{a} i} F_{n_1}(v^{(k_2)}; rh), \; n \in [0, M-1]. \tag{1.7}$$

Proof. Since $F_n(v;h) = \sum_{k_2=0}^{r-1} e^{-\frac{2\pi n k_2 h}{a} i} \sum_{k_1=0}^{s-1} v_{k_1 r + k_2} e^{-\frac{2\pi n}{a} i k_1 rh}$, then we have $F_n(v;h) = \sum_{k_2=0}^{r-1} e^{-\frac{2\pi n k_2 h}{a} i} F_n(v^{(k_2)}; rh)$. Observing that $F_{n_1+s}(v^{(k_2)}; rh) = F_{n_1}(v^{(k_2)}; rh)$ proves (1.7). □

In accordance with (1.7), it is possible to obtain all M values of $F_n(v;h)$ in $2rM$ arithmetic operations (with complex numbers) provided that all similar values $F_{n_1}(v^{(k_2)}; rh)$ on the coarse grid and all coefficients $e^{-\frac{2\pi n k_2 h}{a} i}$ are given. Therefore, when $M = r_1^{\alpha_1} r_2^{\alpha_2} \ldots r_p^{\alpha_p}$ (see (1.5)), we may use this procedure recursively; the obtained multigrid algorithm is called the fast discrete Fourier transform and provides all $F_n(v;h)$ with computational work $W = O(M \ln M)$ (its modifications dealing with real numbers need somewhat more explanation). The problem of finding the sums

$$u_k \equiv \frac{1}{M} \sum_{n=0}^{M-1} w_n e^{\frac{2\pi n}{a} ikh}, \; k = 0, \ldots, M-1, \tag{1.8}$$

(the inverse Fourier transform) can be treated in much the same way. [5]

---

[5] If we rewrite (1.6) in the matrix form $\bar{F} = A\bar{v}$, where $\bar{F} \equiv [F_0(v;h), \ldots, F_{m_1}(v;h)]$, then the above algorithm implies that the matrix A can be represented as a product of $O(\ln M)$ simpler matrices such that each applied to a given vector can be evaluated in $O(M)$ operations. Some particular cases of these algorithms can be associated with the well-known Runge formulas (see, e.g., [64]) used even before the first appearance of

Theorem 1. *Let the conditions of Theorem (2.4.2) be satisfied with nonzero σ only on $S_{1,0} \cup S_{1,1}$. Suppose that N_1 and N_2 have property (1.5) and that system (1.2) with the operator $B \equiv \Lambda_{c,\sigma}$ (see (2.4.10)) has a solution. Then a solution of (1.2) can be found with computational work $W = O(N \ln(N_1 N_2))$.*

Proof. If $S_{3,0} \cup S_{3,1} \subset \Gamma \setminus \Gamma_0$, then we use the even extensions of u and f with respect to x_3 and obtain the grid equations with $|i_3| \leq N_3$. For each node on $S_{3,0}$, we have two (actually identical) equations (see (2.4.10)) of the form $\Lambda_1 u_i + \Lambda_2 u_i + \|h\| c \alpha_{0,i} u_i - \|h\| \alpha_{3,i} h_3^{-1} \partial_3 u_i = f_i$ and $\Lambda_1 u_i + \Lambda_2 u_i + \|h\| c \alpha_{0,i} u_i + \|h\| \alpha_{3,i} h_3^{-1} \bar{\partial}_3 u_i = f_i$, where $i \equiv [i_1, i_2, 0]$, $\|h\| \equiv h_1 h_2 h_3$, $\alpha_{0,i} \equiv \alpha_i$ (see Theorem 2.4.2), and $\alpha_{3,i}$ is $6^{-1}, 3^{-1}, 2^{-1}$ or 1, as was indicated in Lemma 2.4.1. We replace them by their sum, that is, by $2\Lambda_1 u_i + 2\Lambda_2 u_i + 2\|h\| c \alpha_{0,i} u_i - \|h\| \alpha_{3,i} \Delta_3 u_i = f_i$. Each vertex of $S_{3,0}$ is now an inner point of an edge of a new parallelepiped $\bar{Q}_3$ consisting of $\bar{Q}$ and its reflection with respect to the plane $x_3 = 0$. Similarly, nodes on edges of $S_{3,0}$ are now inner points of faces of $\bar{Q}_3$ and inner nodes of the face $S_{3,0}$ become inner points of $\bar{Q}_3$. Hence, the new set of equations (with doubled coefficients for some terms) on $S_{3,0}$ is just the set of equations of type (1.2) constructed for $\bar{Q}_3$. For the obtained system, we repeat the above procedure but now with the plane $x_3 = l_3$ instead of $x_3 = 0$. Since the extended u and f satisfy the grid periodicity conditions $u_{i_1,i_2,i_3+2N_3} = u_i$, $f_{i_1,i_2,i_3+2N_3} = f_i$, we can consider the equations of the new system only for $i_3 \in [0, 2N_3 - 1]$ in combination with the conditions $u_{i_1,i_2,-1} = u_{i1,i_2,2N_3-1}$, $u_{i_1,i_2,2N_3} = u_{i_1,i_2,0}$. Hence, we have reduced system (1.2) to a system of the same type but associated with the periodic conditions $u(x_1, x_2, x_3 + 2l_3) = u(x_1, x_2, x_3)$. Similarly, if $S_{3,0} \subset \Gamma_0$ and $S_{3,1} \subset \Gamma \setminus \Gamma_0$, then we use symmetric extension with respect to the plane containing $S_{3,1}$ and obtain systems associated with the homogeneous Dirichlet boundary conditions which, in turn can be easily reduced to the periodic conditions $u(x_1, x_2, x_3 + 4l_3) = u(x_1, x_2, x_3)$ by using odd extension (note that usually this reduction is not necessary). Hence, all possible variants of the boundary conditions on faces orthogonal to the x_3 direction can be reduced to the periodic case. After that we use the same reduction procedure with respect to the x_2 direction. This yields the system that

---

electronic computers. Applications of the fast discrete Fourier transform to the solution of difference Poisson equation with the homogeneous Dirichlet, Neumann, or periodic boundary conditions are most known (see, e.g., [351, 440, 421, 447]. From a theoretical point of view, the first two types of boundary conditions may be considered as particular cases of the third; we therefore restrict our attention here to how the solution of projective-grid systems from § 2.4 can be reduced to the solution of difference systems with the periodic conditions (see [164, 201]). To this end, let the faces of the parallelepiped Q (see Theorem 2.4.1) lying in the planes $x_r = 0$ and $x_r = l_r$ be denoted by $S_{r,0}$ and $S_{r,1}$, respectively.

can be solved via separation of variables and the use of the fast discrete Fourier transform, since the boundary conditions involving σ can be of importance only in solving one-dimensional grid systems (1.3). Finally, when in (2.4.10) we have $c = 0, \sigma = 0$, and $\Gamma_1 = \Gamma$, then $\Lambda \geq 0$, and it is easy to see that the new obtained system (also with such an operator) is solvable; in solving systems (1.3), we may take the solutions orthogonal to 1. □ [6]

1.2. Partial problems. We first make a rather obvious assertion that is important for understanding the algorithms.

Lemma 2. Suppose we know values of K_1 fixed components of solutions of the systems $A\bar{x}_s = \bar{b}_s, s \in [1, K_2]$. Suppose that $\bar{b} = \sum_{s=1}^{K_2} \beta_s \bar{b}_s$, where all β_s are known. Then we can find the respective components of a solution of the system $A\bar{x} = \bar{b}$ with computational work $W = O(K_1 K_2)$.

Proof. Consider the vector $\bar{x} = \sum_{s=1}^{K_2} \beta_s \bar{x}_s$, where $\bar{x}_s, s \in [1, K_2]$ are the mentioned solutions. It is a solution of the system $A\bar{x} = \bar{b}$, and for obtaining one of its specified components we need only K_2 multiplications and $K_2 - 1$ additions. □

Consider certain particular cases of (1.2) when f is nonzero only on a few grid layers (on several planes for $d = 3$ or straight lines for $d = 2$) and we are interested only in the solution's values on similar layers. [7] The simplest case—when the vectors $\bar{f}_{i_1}$ are nonzero only for several (usually one or two) values of the index i_1 and we seek only similar vectors $\bar{u}_{i_1}$ (usually with the same indices i_1)—was considered in [71, 186, 321] for the iterative domain decomposition method (see § 5). For such a system, we need only know β_{j_0, i_1} for these i_1. Thus, the first stage computational work estimated is only $W_1 = O(M_1 \ln M_1)$. Moreover, if we seek only similar vectors $\bar{u}_{i_1}$, then all corresponding α_{j_0, i_1} in systems (1.3) can be found with computational work $W_2 = O(M_1)$ (see Lemma 2), and the third stage yields the desired $\bar{u}_{i_1}$ with $W_3 = O(M_1 \ln M_1)$. Thus, for systems from Theorem 1, we obtain an algorithm for solving partial problems of the above type with total work

$$W = O(N_2 N_3 \ln(N_2 N_3)). \tag{1.9}$$

A significantly more complicated type corresponds to the case when we are interested, e.g., in values u_i with a fixed index $i_s, s \neq 1$. An ingenious algorithm for this purpose was suggested in [43] for finding ε-approximations to the solution of systems that can be reduced to the grid harmonic equations

---

[6] The described algorithm emphasizes the common mathematical nature of all mentioned variants of the boundary conditions, but, in practical applications, it is reasonable to tackle each variant separately, taking into account the existing symmetry.

[7] Desire to use relevant modifications of the general algorithm for such partial problems is quite natural as was mentioned (e.g., in [164]) for the grid analog of the Schwarz alternating scheme. But probably the first effective algorithms, for the grid Laplace equation, were given by Bakhvalov in the late fifties (see [39] and references therein).

with Dirichlet conditions. [8] If $N_r \asymp N_3$ and $\varepsilon \asymp N_3^{-\alpha}$, $\alpha > 0$, then we obtain the desired ε-approximation with computational work

$$W = O(N_3^2 \ln^2 N_3). \tag{1.10}$$

1.3. Reduction and march methods. Both of the fast direct methods considered here employ some form of block Gaussian elimination. We describe them for solving system (1.2) corresponding to model problem (2.15), (2.16) with $h_1 = h_2 \equiv h$, which we rewrite in the form $C\bar{u}_1 - \bar{u}_2 = \bar{f}_1$, $-\bar{u}_{N_1-1} + C\bar{u}_{N_1} = \bar{f}_{N_1}$, and $-\bar{u}_{i_1-1} + C\bar{u}_{i_1} - \bar{u}_{i_1+1} = \bar{f}_{i_1}$ (for $i_1 \in [2, N_1-1]$), where $C \equiv [c_{i,j}]$ is a matrix such that all $c_{i,i} = 4$, $c_{i,j} = -1$ if $|i-j| = 1$, and vanish otherwise. To describe the class *cyclic reduction* methods (see [119, 440]), we assume that $N_1 = 2^{m+1} - 1$ and consider the above vector equations for $j - 1, j$, and $j + 1$, where $j \equiv i_1$ is an even number. They yield $-\bar{u}_{j-2} + (C^2 - 2I)\bar{u}_j - \bar{u}_{j+2} = C\bar{f}_j + \bar{f}_{j-1} + \bar{f}_{j+1}$. Thus, we can eliminate all $\bar{u}_{i_1}$ with odd index i_1 to obtain a new system involving only $\bar{u}_{i_1}$ with the even i_1. What is important is that the new system has the same properties as (1.2). For its solution, the method of separation of variables could thus be used. But it is more reasonable to apply the reduction procedure recursively, say s times, and to use the method of separation of variables only for the final system with very few unknowns. Such methods lead to estimates (0.3.2) with $r = 1$ or even (0.3.3) for $s \asymp m \asymp |\ln h|$. [9]

The asymptotically optimal estimates (0.3.2) with $r = 0$ are attained by so-called *marching methods* (see [51, 46, 69, 440, 421]). The block elimination that these methods are based can be described in terms of § 1.5 if we renumber our vectors in such a way that $\bar{u}_1 \equiv \bar{v}_{N_1-1}, \bar{u}_1 \equiv \bar{v}_1, \ldots, \bar{u}_{N_1} =$

---

[8] Actually, this scheme deals with grid harmonic functions that vanish on all faces but one, say $S_{1,0}$; therefore, explicit separation of variables formulas for them can be derived that involve the same number terms as there are nodes on $S_{1,0}$. These formulas indicate the rapid convergence to zero of terms corresponding to high frequency oscillations (with respect to one of the variables $x_2, \ldots, x_d$) to zero when $x_1 > 0$ is fixed, and, therefore, such terms may be neglected. The use of relatively small numbers of terms in the representation of an approximate solution for a fixed $x_1 = i_1 h_1$ is the essence of the algorithm and its generalizations suggested in [429, 429, 495]). A possibly simpler and more practical algorithm arises from Lemma 2 (which is especially attractive if we must solve many of systems with the same fixed matrix and the preliminary computational work for obtaining the required components of vectors $\bar{x}_k$ in Lemma 2 is relatively small; similar algorithms have been used with success, e.g., in [119]). It is also possible to use relevant approximate algorithms dealing only with several vectors $\bar{x}_k$ and the relevant approximations of $\bar{b}$ (the choice of the basis vectors $\bar{b}_k$ and a suitable norm for approximating $\bar{b}$ or the corresponding boundary vector deserves special study (see § 5)).

[9] Especially good algorithms are obtained if we use polynomial factorizations (with respect to C) of the resulting matrices. We must pay special attention to the way the right-hand side is calculated at each stage of the reduction because accumulation of the rounding errors is very acute for these methods.

v_{N_1-1} and rewrite our system in the form $\bar{L}\bar{v} = \bar{g}$, where $\bar{v} \equiv [\bar{v}_1, \ldots, \bar{v}_{N_1}]^T$, $\bar{g} \equiv [\bar{f}_2, \ldots, \bar{f}_{N_1}, \bar{f}_1]^T$ and the matrix $\bar{L}$ has the block structure $\bar{L} \equiv [\bar{L}_{i,j}]$, $i \in [1, N_1]$, $j \in [1, N_1]$. It is easy to see that the diagonal blocks are such that $\bar{L}_{1,1} = \cdots = \bar{L}_{N_1-1,N_1-1} = -I$ and $\bar{L}_{N_1,N_1} = 0$ and that all $\bar{L}_{i,i+1} = C, \bar{L}_{i,i+2} = -I$. All remaining blocks vanish with the exception of the two blocks $\bar{L}_{N_1,1} = C$ and $\bar{L}_{N_1,2} = -I$. Consider a new block structure of the same matrix $\bar{L} \equiv A$ of the form (1.5.11), where $A_{2,2} \equiv \bar{L}_{N_1,N_1} = 0$ and the diagonal block $A_{1,1}$ is an upper block triangular matrix with the diagonal blocks being I. Then it is easy to see that systems with the matrix $A_{1,1}$ can be solved in asymptotically optimal computational work as the initial value problems for the grid Poisson equation, but they are exponentially unstable in the sense of growth in roundoff error. The same is true with respect to the method from § 1.5 for solving the original systems with the given block matrix $\bar{L} \equiv A$ (see (1.5.1)), but now it is also very important to use polynomial factorization of the Schur complement $S_2(A)$. Hence, it is natural to appeal to another block structure of the original matrix that would support using the above analog of the well-known shooting procedure for one-dimensional problems only for smaller numbers N_1/k. This is accomplished in a manner similar to the multiple shooting methods when the original segment $[0, l_1]$ is partitioned into k equal parts (in terms of § 1.5, this implies that vectors $\bar{u}_1, \bar{u}_{k+1}, \ldots$ correspond to the vector u_2 in (1.5.1)). [10]

1.4. Fast direct methods for grid systems in a triangle or triangular prism. We will show that grid systems constructed for a triangle or triangular prism can sometimes be reduced to systems for above considered model rectangles or parallelepipeds. Consider, e.g., $Q_\triangle \equiv \triangle A_1 A_3 A_4$, where $A_1 \equiv [0, 0]$, $A_3 \equiv [1, 1]$, and $A_4 \equiv [0, 1]$.

Lemma 3. *Let $h \equiv (N + 1)^{-1}$ be the mesh size of the square grid and $Q_{\triangle,h} \equiv Q_h$ be a set of the nodes that are inner points of the triangle. Then the solution of system* (0.2.15), (0.2.16) *with the above Q_h coincides, for all nodes in Q_h, with the solution of*

$$-h^2(\Delta_1 u_i + \Delta_2 u_i) = g_i, \ i_r = 1, \ldots, N, \ r = 1, 2 \qquad (1.11)$$

$$u_i = 0 \ if \ i_r = 0 \ or \ i_r = N + 1, \ r = 1, 2, \qquad (1.12)$$

where

---

[10] Moreover, it is possible to use a symmetric shooting procedure, where each segment of the length l_1/k is treated in such a way that unstable recursive computations occur only for numbers not greater than $N_1/(2k)$ (see [51, 46, 69, 440]). Model operators associated with these fast direct methods have been used with success for practical problems (see, e.g., [114, 119, 303, 304, 421]).

$$g_i \equiv f_i \ \text{if} \ i_2 > i_1, \ g_i = 0 \ \text{if} \ i_2 = i_1, \ g_i \equiv -f(x_i^*) \ \text{if} \ i_2 < i_1, \qquad (1.13)$$

and x_i^* denotes the node symmetrical to x_i with respect to the side $A_1 A_3$.

Proof. It suffices to observe that system (1.11)–(1.13) constructed for the square has a solution that is odd with respect to the direction of the diagonal $A_2 A_4$. $\square$ [11]

1.5. Basic difference operators on a parallelepiped grid; difference analogs of integration by parts. In the sequel, we will need more detailed information about difference operators associated with d-dimensional parallelepiped grids. We use the same shift operators and the difference operators as for $d = 2$ and $d = 3$ (see § 0.4, 1.4); e.g., $\Lambda_r u_i \equiv -\bar{\partial}_r \partial_r u_i = -h_r^{-2}[I_{-r} u_i - 2u_i + I_r u_i]$, $r \in [1, d]$. We also make use of $\hat{\partial}_r u_i \equiv (2h_r)^{-1}[I_i u_i - I_{-r} u_i]$ and denote any of the differences $\partial_r u_i$, $\bar{\partial}_r u_i$, or $\hat{\partial}_r u_i$ by $\tilde{\partial}_r u_i$, $r \in [1, d]$. Then $\tilde{\partial}^\alpha u_i \equiv \tilde{\partial}_1^{\alpha_1} \times \cdots \times \tilde{\partial}_d^{\alpha_d} u_i$. If a grid function u is defined on a Q_h, then $\tilde{\partial}^\alpha u_i$ is defined only on a subset of Q_h, and we can consider a linear operator $D^\alpha \in \mathcal{L}(H)$ such that $(D^\alpha u)_i = \tilde{\partial}^\alpha u_i$, $\forall x_i \in Q_h$, only if we use an appropriate extension of u. For constructing such an extension, boundary conditions are of primary importance. [12]

For spaces of finite (vanishing outside Q_h) grid functions, the following inner (semi-inner) products and norms (semi-norms) are used: $(\tilde{\partial}^\alpha u', \tilde{\partial}^\alpha v')_0 \equiv \|h\| \sum_{x_i} \tilde{\partial}^\alpha u_i' \tilde{\partial}^\alpha v_i'$, $\|\tilde{\partial}^\alpha u'\|_0 \equiv (\tilde{\partial}^\alpha u', \tilde{\partial}^\alpha u')_0^{1/2}$, where $\|h\| \equiv (h_1^2 + \cdots + h_d^2)^{1/2}$ and $|u'|_0 = (\|h\| \sum_{x_i \in Q_h} |u_i|^2)^{1/2} = \|h\|^{1/2} \|u\|$ is the difference analog of the norm in the space $L_2(Q)$. The reason for using finite grid functions is that difference analogs of integration by parts become extremely simple and take the form

$$(\bar{\partial}^\alpha u', v')_0 = (-1)^{|\alpha|} (u', \partial^\alpha v')_0, \quad (\hat{\partial}^\alpha u', v')_0 = (-1)^{|\alpha|} (u', \hat{\partial}^\alpha v')_0. \quad (1.14)$$

---

[11] Along the same lines we can treat systems associated with projective-grid approximations of the operator $-\Delta$ and the Hilbert space $G \equiv W_2^1(Q_\Delta; \Gamma_0)$, where Γ_0 consists of one or two sides of the triangle. If $A_1 A_3 \subset \Gamma_0$, then we use the odd extension with respect to the direction of the diagonal $A_2 A_4$; otherwise we use the even extension. Also, special attention must be paid to the equations corresponding to the vertices. For example, if $A_1 A_4 \subset \Gamma_0$, then we take $A_1 \in \Gamma_{0.h}$; if $A_1 A_3 \subset \Gamma_0$ and $A_1 A_4 \subset \Gamma \setminus \Gamma_0$, then it is reasonable to take $A_1 \in Q_h$. It is clear that the case $G \equiv W_2^1(Q_\Delta)$ and other combinations of boundary conditions can be considered, as can generalizations to three-dimensional problems involving triangular prisms.

[12] For example, conditions (2.4.5) were used in § 2.4. For similar situations, we can use *finite extensions* of the grid functions u defined on Q_h; we usually denote the corresponding extension by u', so that u' vanishes at all nodes not belonging to Q_h and $u_i' = u_i$, $\forall x_i \in Q_h$. Note that $\tilde{\partial}^\alpha u_i'$ can be nonzero only for the nodes in Q_h and several additional layers of nodes; the number of these layers is not greater than $|\alpha|$.

Indeed, (1.14) follows from

$$(\bar{\partial}_r u', v')_0 = -(u', \partial_r v')_0, \quad r \in [1, d]. \tag{1.15}$$

To prove (1.15), we consider the terms in both parts of (1.15) corresponding to nodes on a straight line parallel to the x_r-axis. Renumbering the nodes on this line, we see that it suffices to prove that $\|h\| \sum_{n=-\infty}^{\infty} \frac{u_n - u_{n-1}}{h_r} v_n = -\|h\| \sum_{n=-\infty}^{\infty} u_n \frac{v_{n+1} - v_n}{h_r}$, which is easily done by comparing the coefficients of $u_n v_n$ and $u_{n-1} v_n$. The above formulas can also be treated as consequences of the well-known Abel transformation in the theory of series, which has the form

$$\sum_{n=1}^{N} (a_n - a_{n-1}) b_n = -\sum_{n=0}^{N-1} a_n (b_{n+1} - b_n) + a_N b_N - a_0 b_0 \tag{1.16}$$

and enables one to use difference analogs of integration by parts for general boundary conditions.

Lemma 4. *Let $h \equiv l(N + 1)^{-l}$ and $P_i \equiv [ih]$ be nodes of the uniform grid on the segment $[0, l]$. Suppose we consider the system of difference equations*

$$-h\Delta_1 u_i + h\alpha_i c u_i = f_i, \quad i = 1, \ldots, N, \tag{1.17}$$

where $\alpha_i = 1$ if $i = 1, \ldots, N$; $\alpha_i = 1/2$ if $i = 0$ or $i = N + 1$. We supplement (1.17) at P_0 by one of the boundary conditions

$$u_0 = 0; \quad or \quad -\partial_1 u_0 + \sigma_0 u_0 = f_0, \tag{1.18}$$

and at P_{N+1} by one of the conditions

$$u_{N+1} = 0; \quad or \quad \bar{\partial}_1 u_{N+1} + \sigma_1 u_{N+1} = f_{N+1}. \tag{1.19}$$

When condition $u_0 = 0$ or $u_{N+1} = 0$ is used, we eliminate the respective variable in the system. Then system (1.17)–(1.19) is an operator equation $\Lambda u = f$ in the standard Euclidean space H of grid functions. Suppose also that $c \geq 0, \sigma_0 \geq 0, \sigma_1 \geq 0$, and that $c + \sigma_0 + \sigma_1 > 0$. Then $\Lambda \in \mathcal{L}^+(H)$ and $sp \Lambda \subset [\kappa_0 h, \kappa_1 h^{-1}]$, where $\kappa_0 > 0$ and $\kappa_1 > 0$ are independent of h.

Proof. The case of $u_0 = 0$ and $u_{N+1} = 0$ was actually considered in § 0.2 (see (2.23)), allowing us to write $\lambda_i(\Lambda) = 4h^{-1} \sin^2 \frac{\pi i h}{2l} + hc$; we see that the assertion of the lemma is true not only for $c \geq 0$, but also for negative c with sufficiently small $|c|$. When one of these conditions is given, we also have a simplified situation. Hence, it suffices to prove Lemma 4 when conditions (1.18) and (1.19) with σ_0 and σ_1 are present and $\dim H = N + 2$. If $(\Lambda u, u) - h \sum_{i=0}^{N+1} \alpha_i c u_i^2 \equiv X$, we have

$$X == -a_0 u_0 + \sigma_0 u_0^2 - \sum_{i=1}^{N} (a_i - a_{i-1}) u_i + a_N u_{N+1} + \sigma_1 u_{N+1}^2, \tag{1.20}$$

where $a_i \equiv \partial_1 u_i, i = 0, \ldots, N$. Next, we make use of (1.16) and obtain

$$-\sum_{i=1}^{N}(a_i - a_{i-1})u_i = h\sum_{i=0}^{N} a_i^2 - a_N u_N + a_0 u_0. \qquad (1.21)$$

Since $a_N u_{N+1} - a_N u_N = h a_N^2$, then (1.20) and (1.21) imply that

$$(\Lambda u, u) = h\sum_{i=0}^{N+1} \alpha_i c u_i^2 + h\sum_{i=0}^{N} a_i^2 + \sigma_0 u_0^2 + \sigma_1 u_{N+1}^2. \qquad (1.22)$$

Observe that $h a_i^2 \leq 2/h(u_i^2 + u_{i+1}^2)$. Hence, there exists κ_1 such that $(\Lambda u, u) \leq \kappa_1/h \sum_{i=0}^{N+1} u_i^2$, $\forall u \in H$. Now if $c > 0$, we may write

$$(\Lambda u, u) \geq \kappa_0 h \sum_{i=0}^{N+1} u_i^2, \ \forall u \in H, \qquad (1.23)$$

where, e.g., $\kappa_0 = c/2$. If $\sigma_0 > 0$, then it suffices to prove existence of a constant $K \neq K(h)$ such that

$$h\sum_{i=0}^{N+1} u_i^2 \leq K[u_0^2 + h\sum_{i=0}^{N} a_i^2], \ \forall u \in H. \qquad (1.24)$$

With this in mind, we use a standard reasoning: for $i \geq 1$, it is easy to see that $u_i^2 = [(u_0 + h\sum_{k=0}^{i-1} a_k]^2 \leq 2[u_0^2 + lh\sum_{k=0}^{i-1} a_k^2]$. Hence, (1.24) holds at least with $K = 2l^2$, and (1.23) holds with $\kappa_0 = K^{-1}\min\{1; \sigma_0\}$. The case $\sigma_1 > 0$ is similar. $\square$

Lemma 5. Let the conditions of Lemma 4 be satisfied and D be a diagonal matrix with positive diagonal elements $d_{i,i}$ such that $0 < k_0 \leq d_{i,i} \leq k_1$. Then the operator $R \equiv D\Lambda \in \mathcal{L}^+(H(D^{-1})$ and $\mathrm{sp}\, R \subset [k_0\kappa_0 h, k_1\kappa_1 h^{-1}]$.

Proof. It suffices to make use of Lemma 4, Lemma 0.3.1, and the obvious fact that $k_0 I \leq D \leq k_1 I$. $\square$ [13]

§ 2. Alternating direction iteration (ADI) methods and splitting operators; additive splitting of the inverse operator

2.1. Basic computational algorithms. [14] We consider here the

---

[13] These lemmas will be used for problems when the associated one-dimensional grid operators differ from the given ones by a factor depending on the grid (see, e.g., § 0.2).

[14] Following [151] we introduce a special class of model grid operators that will enable us to consider separate ADI methods from a unified position; they have become very productive in construction and analysis of such methods, not only for elliptic grid systems, but also for various nonstationary boundary value problems (see, e.g., [152, 157, 158, 162, 164, 168, 171, 227, 378, 519]).

same grids and spaces H with $\dim H \equiv N$ that we did in § 1; $h_r \asymp h \asymp N^{-1/d}, r \in [1, d]$.

We refer to A as a *splitting grid operator* if A can be written in the form

$$A = A_1 A_2 \ldots A_q, \tag{2.1}$$

where $\dim A_r < \dim A$, $r = 1, \ldots, q$ (see § 1). This factorization of A implies that we can reduce solution of a system with A to successive solutions of systems with the operators $A_1, \ldots, A_d$; each A_r under an appropriate ordering of nodes in Q_h is in fact a block diagonal matrix, and a system with A_r splits into a set of independent subsystems (this explains why and in which sense the notion of splitting is used; sometimes under additive splitting of an operator its representation as a sum of operators with prescribed properties is understood). One-dimensional elliptic grid operators $A_r, r \in [1, q]$ are especially attractive (see § 1) because they lead to d one-dimensional systems

$$A_1 u^{1/d} = F, \quad A_r u^{r/d} = u^{(r-1)/d}, \ r \in [2, d], \ u^{d/d} = u. \tag{2.2}$$

Thus, if we have asymptotically optimal algorithms for solving the one-dimensional subsystems on separate grid lines, then we have algorithms of the same type for systems with the splitting operator A. Hence, the use of such operators $A^{(n+1)}$ in iterative methods deserves special attention. In a general setting, we can write these methods in the form

$$A^{(n+1)} u^{n+1} = F^{(n)}(u^n, \ldots, u^0), \ n = 0, 1, \ldots, \tag{2.3}$$

(all $A^{(n+1)}$ are splitting grid operators). They are natural generalizations of alternating direction iteration (ADI) methods (see [405, 138]) suggested for two-dimensional systems of the form

$$Lu \equiv \Lambda_1 u + \Lambda_2 u = f \tag{2.4}$$

with one-dimensional grid operators Λ_1 and Λ_2. The algorithm of passing from u^n to u^{n+1} is defined by

$$\left. \begin{array}{l} (I + \tau_n \Lambda_1) u^{n+1/2} = (I - \tau_n \Lambda_2) u^n + \tau_n f, \\ (I + \tau_n \Lambda_2) u^{n+1} = (I - \tau_n \Lambda_1) u^{n+1/2} + \tau_n f. \end{array} \right\} \tag{2.5}$$

Here we solve one-dimensional systems along the x_1-direction to obtain $u^{n+1/2}$, then we alternate directions to obtain the desired iterate. [15] Elimination of $u^{n+1/2}$ in (2.5) leads to

---

[15] The second equation in (2.5) may be replaced by $(I + \tau_n \Lambda_2) u^{n+1} = 2u^{n+1/2} - u^n + \tau_n \Lambda_2 u^n$.

$$A^{(n)}(u^{n+1} - u^n) = -2\tau_n(Lu^n - f), \quad n = 0, 1, \ldots, \tag{2.6}$$

where $A^{(n)} \equiv (I + \tau_n\Lambda_1)(I + \tau_n\Lambda_2)$. There exist several other algorithms dealing with one-dimensional systems yielding the same basic relation (2.6), e.g., we can use (2.2) for obtaining the correction $u^{n+1} - u^n$. The same applies to more general variants of the method with a positive diagonal matrix D instead of I in (2.5) (see, e.g., [497] and Theorem 3 below). Relations (2.6) suggest their natural generalizations for d-dimensional systems of the form $Lu \equiv \Lambda_1 u + \cdots + \Lambda_d u = f$. They lead to iterative methods of the form

$$A^{(n)}(u^{n+1} - u^n) = -\sigma_n(Lu^n - f), \quad n = 0, 1, \ldots, \tag{2.7}$$

with the splitting operators $A^{(n)} \equiv \prod_{r=1}^{d}(I + \tau_{r,n}\Lambda_r)$, $n = 0, 1, \ldots$ and iteration parameters that must be chosen ($\sigma_n = 2\tau_n$ is typical). We emphasize again that several one-dimensional algorithms yielding (2.7) are possible. One of such schemes was suggested in [138], for $d = 2$ it takes the form
$(I + \tau_n\Lambda_1)u^{n+1/2} = (I - \tau_n\Lambda_2)u_2^n + \tau_n f$,
$(I + \tau_n\Lambda_2)u^{n+1} = 2u^{n+1/2} - u^n + (I - \tau_n\Lambda_2)u^{n+1/2}$.

2.2. Analysis of the commutative case. The strongest results for ADI methods were obtained for the so-called commutative case. More precisely, we assume that grid operators $\Lambda_1, \ldots, \Lambda_d$ are such that

$$\Lambda_r = \Lambda_r^* \geq 0, \quad \Lambda_r\Lambda_l = \Lambda_l\Lambda_r \quad \forall r, \forall l, \tag{2.8}$$

and that

$$L \equiv l(\Lambda) \equiv l(\Lambda_1, \ldots, \Lambda_d) \in \mathcal{L}^+(H). \tag{2.9}$$

Consider method (2.7) of the form

$$\left\{\prod_{r=1}^{d} p_{r,n}(\Lambda_r)\right\}(u^{n+1} - u^n) = -\sigma_n(Lu^n - f), \tag{2.10}$$

where $p_{r,n}(\Lambda_r)$ is an invertible operator-function of Λ_r (e.g., a polynomial of $\Lambda_r, r \in [1, d]$), and σ_n is a number (see, e.g., [155, 162]). Then the reduction error operator Z_M after M iterations is defined by

$$Z_M \equiv Z \equiv \prod_{n=0}^{M-1} \left\{I - \sigma_n[\prod_{r=1}^{d} p_{r,n}(\Lambda_r)]^{-1}L\right\}. \tag{2.11}$$

In accordance with (2.8) and a well-known theorem in linear algebra, there exists an orthonormal basis $\{\psi_i\}$ for H such that

$$\Lambda_r\psi_i = \lambda_{r,i}\psi_i, \quad 0 \leq \lambda_{r,0} \leq \lambda_{r,i} \leq \lambda_r^0, \quad 1 \leq i \leq N. \tag{2.12}$$

We also see that the operator Z_M is symmetric as an element of $\mathcal{L}(H)$. Hence, $\|Z_M\| = \max_i \lambda_i(Z_M)$ and $\|Z_M\| \leq \max_{\lambda \in \Pi} |\rho_M(\lambda)|$, and we arrive at the problem of finding

$$\min_{p_{r,n},\sigma_n} \max_{\lambda \in \Pi} |\rho_M(\lambda)|, \qquad (2.13)$$

where $\rho_M(\lambda) \equiv \prod_{n=0}^{M-1} q_n(\lambda)$, $q_n(\lambda) \equiv 1 - \sigma_n l(\lambda)[\prod_{r=1}^{d} p_{r,n}(\lambda_r)]^{-1}$ and $\lambda \equiv (\lambda_1,\ldots,\lambda_d)$, $\Pi \equiv \{\lambda|\lambda_{r,0} \leq \lambda_r \leq \lambda_r^0, r \in [1,d]\}$.

   Theorem 1. For the iterative method in (2.10), *suppose that conditions* (2.8) *and* (2.9) *are satisfied, where*

$$l(\lambda) \equiv \lambda_1 + \cdots + \lambda_d, \quad p_{r,n}(\lambda_r) \equiv 1 + \tau_n \lambda_r.$$

Suppose also that $\lambda_{r,0}$ *and* λ_r^0 *in* (2.12) *are such that*

$$\lambda_{1,0} + \ldots + \lambda_{d,0} \geq \lambda_0 = Kh^k > 0, \ k > 0, \quad \lambda_1^0 + \ldots + \lambda_d^0 \leq \lambda_0 \asymp 1. \ (2.14)$$

Then there exist $M = O(|\ln h|)$ *and iteration parameters* $\tau_0,\ldots,\tau_{M-1}$ *and* $\sigma_0,\ldots,\sigma_{M-1}$ *such that*

$$\|Z\| \leq \max_{\lambda \in \Pi} |\rho_M(\lambda)| \leq q < 1, \qquad (2.15)$$

where $q \neq q(h)$.

   Proof. For s $r_1 > r_0 > 0$, we define the number M and the iteration parameters $\tau_0,\ldots,\tau_{M-1}$ as follows: $\tau_0 \equiv r_1/\lambda_0$, $\tau_{n+1} \equiv \tau_n r_0/r_1$, $n = 0,\ldots,M-1$, where $M-1$ is the first number leading to the inequality $\tau_{M-1}\lambda_0 \leq r_0$. It is easy to see that $M \asymp |\ln h|$. Next, let $\sigma_n = \gamma\tau_n$, where $\gamma \in (0,2)$. Then $q_n(\lambda) = 1 - \frac{\gamma\tau_n(\lambda_1+\cdots+\lambda_d)}{(1+\tau_n\lambda_1)\cdots(1+\tau_n\lambda_d)}$, where $|q_n(\lambda)| < 1$ for each λ with $\lambda_s > 0, s \in [1,d]$, and $q^* \equiv q^*(r_1,r_2) \equiv \max_{x \in S(r_1;r_2)} \prod_{n=0}^{M-1} q_n^*(x) < 1$,

$$S(r_1;r_2) \equiv \{x : r_1 \leq x_1 + \cdots + x_d \leq r_2, x_r \geq 0, r \in [1,d]\},$$

and $q^*(x) = 1 - \frac{\gamma(x_1+\cdots+x_d)}{(1+x_1)\cdots(1+x_d)}$. Hence, it suffices to prove that, for each $\lambda \in \Pi$, there exists a number n such that $x_n(\lambda) \equiv \tau_n\lambda \in S(r_1;r_2)$ (then we may take $q = q^*$). For this we consider the points $x_n(\lambda), n \in [0, M-1]$. Here $\tau_0(\lambda_1+\cdots+\lambda_d) \geq r_2$ and $\tau_{M-1}(\lambda_1+\cdots+\lambda_d) \leq r_1$. If we assume that there is no mentioned n, then, for some k, we have $\tau_k(\lambda_1 + \cdots + \lambda_d) \geq r_2$ and $\tau_{k+1}(\lambda_1+\cdots+\lambda_d) \leq r_1$ which contradicts the fact that $\tau_{n+1} = \tau_n r_1/r_2$. Hence, the desired n does exist and (2.15) holds. $\square$ [16]

---

[16] It is reasonable to take r_1 and r_2 such that $\max_{x_1+\cdots+x_d=r_1} q^*(x) = \max_{x_1+\cdots+x_d=r_2} q^*(x)$ and $\gamma = 2$ (see [136, 162]; for $d = 2$ such a choice of γ corresponds to method (2.5) and is analyzed below in Subsection 2.3).

Theorem 2. Let the conditions of Theorem 1 be satisfied and an operator $D \in \mathcal{L}^+(H)$ be such that $D\Lambda_r = \Lambda_r D$, $r \in [1, d]$. Then $\|Z\|_D = \|Z\|$.

Proof. We have seen that Z_M is symmetric as an element of $\mathcal{L}(H)$ and that $\|Z_M\| = \max_i \lambda_i(Z_M)$. Hence, it suffices to show that Z_M is symmetric as an element of $\mathcal{L}(H(D))$. With this in mind, we note that each Λ_r, $r \in [1, d]$, is symmetric as an element of $\mathcal{L}(H(D))$. Indeed, $(\Lambda_r u, v)_D = (\Lambda_r u, Dv) = (u, \Lambda_r Dv) = (u, D\Lambda_r v) = (u, \Lambda_r v)_D$. Hence, the desired symmetricity of all the operators in (2.9)–(2.11) holds. $\square$ [17]

2.3. Optimization of iteration parameters for the two-dimensional case. Let $d = 2$, $l(\lambda) \equiv \lambda_1 + \lambda_2$, $p_{r,n}(\lambda_r) \equiv 1 + \tau_{r,n}\lambda_r$, $r = 1, 2$, and $\sigma_n = \tau_{1,n} + \tau_{2,n}$. Then $q_n(\lambda) = \frac{1 - \tau_{2,n}\lambda_1}{1 + \tau_{1,n}\lambda_1} \frac{1 - \tau_{1,n}\lambda_2}{1 + \tau_{2,n}\lambda_2}$ and

$$\rho_M(\lambda) = \prod_{n=0}^{M-1} q_n(\lambda). \tag{2.16}$$

If $\lambda_r \in [\lambda_0, \lambda^0] \equiv [t_0, t_1]$, where $t_0 > 0$, then we have $\tau_{1,n} = \tau_{2,n} \equiv \tau_n$, and the optimal strategy for choosing the parameters $\tau_0, \ldots, \tau_n$ consists in obtaining [18]

$$\min_{\tau_0, \ldots, \tau_{M-1}} \max_{t_0 \le t \le t_1} \left| \frac{\tau_0^{-1} - t}{\tau_0^{-1} + t} \frac{\tau_1^{-1} - t}{\tau_1^{-1} + t} \cdots \frac{\tau_{M-1}^{-1} - t}{\tau_{M-1}^{-1} + t} \right|. \tag{2.17}$$

Lemma 1. If the set of parameters $\tau_0, \ldots, \tau_{M-1}$ is a solution of problem (2.17), then the set $(a\tau_0)^{-1}, \ldots, (a\tau_{M-1})^{-1}$ with $a \equiv t_0 t_1$ is also a solution.

Proof. Under the change of the variable $t = a/t'$, we have $|[\tau_k^{-1} - t]/[\tau_k^{-1} + t]| = |[t'\tau_k a - 1]/[t'\tau_k a + 1]|$. This, together with the fact that $t' \in [t_0, t_1]$ follows from $t \in [t_0, t_1]$, implies that problem (2.17) takes the same form in terms of t'. Hence, the assertion of the lemma is true. $\square$

Lemma 2. Suppose $M = 2k$ and consider problem (2.17) with $M = k$ and $[t_0, t_1]$ replaced by $[t_0^{(k)}, t_1^{(k)}]$, where $t_0^{(k)} = 1$ and $t_1^{(k)} = 2^{-1}[(t_1/t_0)^{1/2} + (t_0/t_1)^{1/2}]$. Let the set $\tau_0^{(k)}, \ldots, \tau_k^{(k)}$ be its solution. Then the solution of (2.17) consists of k pairs of parameters

---

[17] If $L = \Lambda_1 + \cdots + \Lambda_d$, then we may take, e.g., $D = L, D = L^2$; generalizations to the case $L \ge 0$ were indicated in [155, 162]).

[18] Recognized by several researchers in the 1960s, the optimal choice of parameters can be determined by a simple recursive procedure (see [497]) if $M = 2^k$; it is interesting to note that the solution of the general problem (2.17) (it is unique up to parameter order) in terms of elliptic functions was given in 1877 by Zolotarev and was known even earlier to Chebyshev (it was mentioned in [331]). Note also that the case $\lambda_r \in [\lambda_{r,0}, \lambda^{r,0}]$ can be reduced to the given one (see [497]) and that the order of using the best iteration parameters was also investigated from the point of view of numerical stability (see [355]).

$$(t_1 t_2)^{-1/2}[1 - [\frac{1}{(\tau_s^{(k)})^2} - 1]^{1/2}], \ (t_1 t_2)^{-1/2}[1 + [\frac{1}{(\tau_s^{(k)})^2} - 1]^{1/2}], s \in [1, k].$$

$$(2.18)$$

Proof. Let τ and $a\tau$ be the parameters from Lemma 1. Then the function in (2.17) is a product of k functions of the form

$$\left| \frac{\tau^{-1} - t}{\tau^{-1} + t} \frac{a\tau - t}{a\tau + t} \right| = \left| \frac{1 - \frac{2}{a^{1/2}\tau + (a^{1/2}\tau)^{-1}} \frac{ta^{-1/2} + a^{1/2}t^{-1}}{2}}{1 + \frac{2}{a^{1/2}\tau + (a^{1/2}\tau)^{-1}} \frac{ta^{-1/2} + a^{1/2}t^{-1}}{2}} \right|.$$

Next we use the change of variables $2t' = ta^{-1/2} + a^{1/2}t^{-1}$. Since $t' \in [t_0^{(k)}, t_1^{(k)}]$, then $2/[a^{1/2}\tau + (a^{1/2}\tau)^{-1}] \le 1$ is one of the k optimal parameters for the introduced smaller problem. This leads to the recursive formula $2/[a^{1/2}\tau + (a^{1/2}\tau)^{-1}] = \tau_s^{(k)}$. Hence, the optimal τ for problem (2.17) are the roots of the above equations and are given by (2.18). □ [19]

2.4. Estimates of computational work. We will give examples of grid systems for which ADI methods can be effectively applied; we concentrate on systems associated with PGMs like those considered in § 0.2 and 2.4 (see [177]); difference systems, especially corresponding to the Dirichlet conditions, were analyzed in detail in [136, 155, 162, 440, 497]. For $d = 2$ and the notation used in Lemma 0.2.1, consider the system

$$-h_1 h_2(\Delta_1 + \Delta_2)u_i = f_i \text{ if } i_1 \in [1, N_1], \ i_2 \in [0, N_2], \qquad (2.19)$$

$$u_i = 0 \text{ if } i_1 = N_1 + 1 \text{ or } i_2 = N_2 + 1, \qquad (2.20)$$

$$-h_1 \partial_1 u_i - \frac{h_1 h_2}{2}\Delta_2 u_i + \sigma_1 h_2 u_i = f_i \text{ if } i_1 = 0, \ i_2 \in [1, N_2], \qquad (2.21)$$

$$-\frac{h_1 h_2}{2}\Delta_1 u_i - h_2 \partial_2 u_i = f_i \text{ if } i_2 = 0, \ i_1 \in [1, N_2], \qquad (2.22)$$

and

$$-\frac{h_1}{2}\partial_1 u_i - \frac{h_2}{2}\partial_2 u_i + \frac{\sigma_1}{2}h_2 u_i = f_i \text{ if } i_2 = 0, \ i_1 = 0, \qquad (2.23)$$

where $\sigma_1 > 0$. It is easy to see that this system corresponds to the model grid operator mentioned in Subsection 0.2.2 (see also (2.4.10)) and that is does not satisfy commutativity condition (2.8) when considered as an operator equation in the standard Euclidean space H of functions u defined at nodes with $i_1 \in [0, N_1]$ and $i_2 \in [0, N_2]$ (this set of the nodes is denoted by Q_h). Observe that the even reflections of this system with respect to

---

[19] For $M = 1$ the optimal parameter is $\tau_1 = a^{-1/2}$.

x_2 (see Theorem 1.1), together with summation of each pair of equations for nodes with $i_2 = 0$, and similarly for x_1, lead to a system of form (2.4) with the operators satisfying (2.2) and (0.2.22). (This system corresponds to homogeneous Dirichlet conditions on the boundary of the rectangle $Q' \equiv [-l_1, l_1] \times [-l_2, l_2]$ and to relations $\Lambda u_i \equiv -h_1 h_2 (\Delta_1 + \Delta_2) u_i + 2\sigma_1 h_2 u_i$ when $i_1 = 0$; at other inner points we have $\Lambda u_i \equiv -h_1 h_2 (\Delta_1 + \Delta_2) u_i$.) For this system, Theorems 1 and 2 apply with $k = 2$ provided $h_r \asymp h, r = 1, 2$ (see § 0.2 and Lemma 1.4). [20] We multiply equations (2.21) and (2.22) by 2 and equations (2.23) by 4. The obtained system can be written in the form

$$Ru \equiv (R_1 u + R_2 u) = g \equiv Df \qquad (2.24)$$

with the one-dimensional grid operators such that

$$R_2 u_i \equiv -h_1 h_2 \Delta_2 u_i \text{ if } i_2 \in [1, N_2]; \quad R_2 u_i \equiv -2h_1 \partial_2 u_i \text{ if } i_2 = 0, \quad (2.25)$$

$$R_1 = R_{1,0} + \sigma_1 R_{1,1}, \qquad (2.26)$$

$$R_{1,0} u_i \equiv -h_1 h_2 \Delta_1 u_i \ (i_1 \in [1, N_1]); \quad R_{1,0} u_i \equiv -2h_2 \partial_1 u_i, \ i_1 = 0, \quad (2.27)$$

and

$$R_{1,1} u_i \equiv 2h_2 u_i \ (i_1 = 0); \quad R_{1,1} u_i \equiv 0 \ (i_1 \in [1, N_1]). \qquad (2.28)$$

Lemma 3. R_1 *and* R_2 *commute and* $R_s \in \mathcal{L}^+(H(D^{-1}))$, $s = 1, 2$.

Proof. In accordance with the positivity of the original operators as elements of $\mathcal{L}(H)$, we have $R_r \in \mathcal{L}^+(H(D^{-1})), r = 1, 2$ (see Lemma 0.3.1). To verify the commutativity of R_1 and R_2, we compare $X_1 \equiv R_{1,0} R_2 u_i$ and $X_2 \equiv R_2 R_{1,0} u_i$ for all possible types of nodes P_i. For example, the most complicated case is $P_i = [0, 0]$. Then, in accordance with (2.25) and (2.27), $X_1 = 4(-h_1/2\partial_1)(R_2) u_i = 4h_1 h_2 \partial_1 \partial_2 u_{0,0}$. Similarly,

$$X_2 = 4(-h_2/2\partial_2)(R_1) u_i = 4h_1 h_2 \partial_1 \partial_2 u_{0,0} = X_1.$$

---

[20] Note that $\Lambda_r \in \mathcal{L}^+(H'), r = 1, 2$, where H' consists of functions defined at nodes with $|i_r| \in [0, N_r], r = 1, 2$. We thus denote them by Λ_r' and preserve $\Lambda_r, r = 1, 2$, for the system (2.19)–(2.23) as an equation in the standard Euclidean space H; we denote the elements of H' by u'. Note that, for a function u' obtained through the above reflections of the function $u \in H$, we have $\|u'\|^2 = 4\|D_1' D_2' u\|^2$, where D_r' is a diagonal matrix such that $D_r' u_i \equiv u_i$ if $i_r \in [1, N_r]$ and $D_r' u_i \equiv 2^{-1} u_i$ if $i_r = 0$, $r = 1, 2$. Note also that, for the obtained system, its solution u' must be an even function with respect to x_1 and x_2, and that all iterations can be carried out in the subspace of H' of functions with the same symmetry (it is an invariant subspace of the operators Λ_1' and Λ_2'). This indicates that we must transform the original system in such a way that the new operators can be considered as elements of $H(D')$ (in fact, they must be restrictions of Λ_1' and Λ_2' to the above mentioned invariant subspace). A natural way to do this is to multiply original system (2.4) by $D \equiv D_1 D_2$, where $D_r \equiv (D_r')^{-1}, r = 1, 2$ (such a transformation of a certain difference system was used in [9]).

For $P_i = [0, h_2]$, $X_1 = -4h_1/2\partial_1(R_2)u_{0,1}$ and $X_2 = -h_1h_2\Delta_2(-2h_2\partial_1)u_{0,1}$. Thus,

$$X_1 = -2h_2/h_1[-h_1h_2\Delta_2u_{1,1} + h_1h_2\Delta_2u_{0,1}] = -2h_1h_2^2\partial_1\Delta_2u_{0,1}$$

and $X_2 = -2h_1h_2^2\partial_1\Delta_2u_{0,1} = X_1$. For $P_i = [h_1, 0]$, we can use the obtained formulas and write $X_1 = -2h_2h_1^2\partial_2\Delta_1u_{1,0} = X_2$. Other cases are even simpler. Hence, $R_{1,0}R_2 = R_2R_{1,0}$. Next we compare $R_{1,1}R_2u_i$ and $R_2R_{1,1}u_i$. We have $R_{1,1}R_2u_{0,0} = 2h_2(-2h_1\partial_2u_{0,0}) = R_2R_{1,1}u_{0,0}$ and

$$R_{1,1}R_2u_{0,1} = 2h_2(-h_1h_2\Delta_2u_{0,1}) = R_2R_{1,1}u_{0,0}.$$

Also, $R_{1,1}R_2u_i = 0 = R_2R_{1,1}u_i$ if $i_1 \in [1, N_1]$. Thus, $R_1R_2 = R_2R_1$ (see (2.26)). □

Theorem 3. *Suppose that system (2.4), (2.19)–(2.23) is reduced to system (2.24)–(2.28) to which we apply the iterative method*

$$\left.\begin{array}{l}(I + \tau_n R_1)u^{n+1/2} = (I - \tau_n R_2)u^n + \tau_n f, \\ (I + \tau_n R_2)u^{n+1} = (I - \tau_n R_1)u^{n+1/2} + \tau_n f.\end{array}\right\} \qquad (2.29)$$

Then there exist $M = O(|\ln h|)$ and iteration parameters $\tau_0, \ldots, \tau_{M-1}$ such that the error reduction operator Z is symmetric as an element of $\mathcal{L}(H(\Lambda))$ and $\|Z\|_\Lambda \leq q < 1$, where $q \neq q(h)$.

Proof. According to Lemma 3, all operators in (2.29) commute and are elements of $\mathcal{L}^+(H(D'))$. Since Lemma 1.5 applies, then (2.14) holds with $k = 2$. Therefore, Theorem 1 and Lemma 2 yield the error reduction operator Z such that it is symmetric as an element of $\mathcal{L}(H(D')$ and $\|Z\|_{D'} \leq q < 1$. Theorem 2 implies that $Z \in \mathcal{L}(H(D^{-1}R))$ is symmetric. Since $H(D^{-1}R) = H(\Lambda)$, then the desired property of Z follows. □ [21]

Now we consider the case of the three-dimensional system (2.4.10) with the operator $\Lambda \equiv \Lambda_{c,\sigma} \in \mathcal{L}^+(H)$. We multiply it by the diagonal operator $D \equiv h^{-(d-2)}D_1D_2D_3$, where $D_ru_i \equiv u_i$ if $i_r = 1, \ldots, N_r$, $D_ru_i \equiv 2u_i$ if $i_r = 0$ and $S_{r,0} \subset \Gamma \setminus \Gamma_0$, and $D_ru_i \equiv 2u_i$ if $i_r = N_r + 1$ and $S_{r,1} \subset \Gamma \setminus \Gamma_0$ (see Theorem 1.1), $r = 1, 2, 3$. We thus obtain a system of the form

$$Ru \equiv (R_1 + R_2 + R_3)u = g, \qquad (2.30)$$

---

[21] We emphasize that (2.29) can be treated as a method of type (2.5) for the original grid system (it suffices to replace I in (2.5) by D') with the same error reduction operator Z. Note that, in the same way, we can analyze the case where (2.21)–(2.23) is replaced by $-h_1\partial_1u_i - \frac{h_1h_2}{2}\Delta_2u_i + \sigma_1h_2u_i = f_i$ if $i_1 = 0, i_2 \in [1, N_2]$, $-h_1h_2/2\Delta_1u_i - h_2\partial_2u_i + h_2\sigma_2u_i = f_i$ if $i_2 = 0, i_1 \in [1, N_2]$, and $-h_1/2\partial_1u_i - h_2/2\partial_2u_i + \sigma_1/2h_2u_i + \sigma_2/2h_1u_i = f_i$ if $i_2 = 0, i_1 = 0$, with $\sigma_2 > 0$ (positivity conditions can be imposed on other sides of Q and some negative constants σ may be allowed).

where R_r is a grid operator acting along the x_r-direction, $r = 1, 2, 3$, and, e.g.: $R_r u_i \equiv h_1 h_2 h_3 h^{-1}(-\Delta_r + c_r I)u_i$, $i_r \in [1, N_r]$;
$R_r u_i \equiv h_1 h_2 h_3 [-2(hh_r)^{-1}\partial_r u_i + (hh_r)^{-1} 2\sigma_{r,0} u_i]$, $(i_r = 0, S_{r,0} \subset \Gamma \setminus \Gamma_0)$;
$R_r u_i \equiv h_1 h_2 h_3 [2(hh_r)^{-1}\bar\partial_r u_i + (hh_r)^{-1} 2\sigma_{r,1} u_i]$ $(i_r = N_r + 1, S_{r,1} \subset \Gamma \setminus \Gamma_0)$;
and $c_r \geq 0$, $r \in [1, 3]$, $c_1 + c_2 + c_3 = c$. We assume that c_r, $\sigma_{r,0}$, and $\sigma_{r,1}$ are such that Lemmas 1.4 and 1.5 lead to (2.14) with $k = 2$, where all λ refer to the bounds of the eigenvalues of the operators R_1, R_2, R_3, and R.

Theorem 4. *Suppose that, for system (2.30) with the above operator $R \in \mathcal{L}^+(H(D^{-1}))$, we apply iterations of type (2.10). Then there exist $M = O(|\ln h|)$ and parameters $\tau_0, \ldots, \tau_{M-1}$, $\sigma_0, \ldots, \sigma_{M-1}$ such that the error reduction operator Z is symmetric as an element of $\mathcal{L}(H(\Lambda))$ and $\|Z\|_\Lambda \leq q < 1$, where $q \neq q(h)$.*

Proof. The reasoning is much the same as in the proofs of Lemma 3 and Theorem 3 despite the presence of the terms $c_r I$ in any R_r. □ [22]

2.5. Generalized splitting operators. As we have seen, applications of Theorems 1 and 2 were connected with the case of a rectangular or parallelepiped region. It is even possible to show that for a nonrectangular region, such as that depicted in Figure 1, and for the homogeneous Dirichlet conditions, the product of the symmetric one-dimensional grid operators L_r defined by $(L_r u)_i \equiv \Delta_r u_i$, $P_i \in Q_h$, $r = 1, 2$ cannot be symmetric operator (it suffices to compare $L_1 L_2 u_i$ and $L_2 L_1 u_i$ when P_i is the first inner node on the segment $M_0 M_1$ since only one depends on u_{i_1-1,i_2-1}; see [70]).

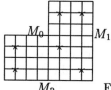

Figure 1. Partition of the region.

Our aim here is show how to generalize the class of splitting operators in order to preserve the desired symmetricity property (see [152]). Below, all nodes not belonging to Q_h are called fictitious; we consider a linear space H' of grid functions u' defined at all nodes of a parallelepiped grid and vanishing at fictitious ones. Define the linear one-dimensional grid operators Λ'_r, A'_r, $r \in [1, d]$, and the splitting operator A'_h by $(\Lambda'_r u')_i \equiv -\Delta_r u'_i$, $(A'_r u')_i \equiv u'_i + \tau_r (\Lambda'_r u')_i \equiv (I' + \tau_r \Lambda'_r)u'_i$, $r \in [1, d]$, and $A'_h \equiv A' \equiv$

---

[22] Z may be regarded as the error reduction operator for the original grid system. Note that generalizations for periodic conditions and the case $\Lambda \geq 0$ are straightforward. Finally, we note that ADI methods are ideally suited for parallel computations (see, e.g., [288]); it seems reasonable to expect that even variants of (2.3) with the splitting operator A^{n+1} containing, say, two-dimensional factors may be useful (e.g., for $d = 4$ and $L = L_1 + L_2 + L_3 + L_4$, we may take $\Lambda_1 \equiv L_1 + L_2, \Lambda_2 \equiv L_3 + L_4$ and apply (2.5) and Lemmas 1 and 2).

$A'_{1,h} \cdots A'_{d,h}$. A grid operator $A \in \mathcal{L}(H)$ is called a *generalized splitting operator* if there exists a splitting operator $A' \in \mathcal{L}(H')$ such that $Au_i = A'u'_i \ \forall x_i \in Q_h$, where $u_i = u'_i, \ \forall x_i \in Q_h$.

Lemma 4. *A generalized splitting operator A is symmetric as an element of $\mathcal{L}(H)$; if $\tau_r \geq 0$, $r \in [1, d]$, then $A \in \mathcal{L}^+(H)$.*

Proof. We have $\|h\|(Au, v) = (A'u', v')_0, \ \forall u, v \in H$, where $u_i = u'_i, \ v_i = v'_i, \ \forall x_i \in Q_h$. For $X \equiv (A'u', v')_0 = ((I' - \tau_1 \Delta_1) \cdots (I' - \tau_d \Delta_d)u', v')_0$, we have $X = (u', v')_0 - \sum_{r=1}^{d} \tau_r(\Delta_r)u', v')_0 + \sum_{r_1 < r_2} \tau_{r_1} \tau_{r_2}(\Delta_{r_1} \Delta_{r_2})u', v')_0 - \cdots + (-1)^d \tau_1 \cdots \tau_d(\Delta_1 \cdots \Delta_d u', v')_0$. Since $\Delta_r = \bar{\partial}_r \partial_r$, then each term on the right-hand side here can be transformed in accordance with (1.14). This yields $-(\Delta_r u', v')_0 = (\partial_r u', \partial_r v')_0, \ldots, (-1)^d(\Delta_1 \cdots \Delta_d u', v')_0 = (\partial_1 \cdots \partial_d u', \partial_1 \cdots \partial_d v')_0$. Hence, X contains the same terms as $(A'v', u')_0 = \|h\|(u, Av)$. This implies that $A^* = A$. It is easy to see that $(Au, u) \geq \|u\|^2 \geq 0$ if $\tau_r \geq 0$, $r \in [1, d]$. $\square$

A generalized splitting operator can not be factored in the form (2.1) with $q = d$, so solving systems with such an operator requires a more complicated algorithm. Here we describe such a scheme for $d = 2$ and some model regions. Consider the grid system defined by the equations

$$A'_1 A'_2 u_i = F_i, \quad x_i \in Q_h, \tag{2.31}$$

and the Dirichlet boundary conditions

$$u_i = \varphi_i, \quad x_i \in \Gamma_h, \tag{2.32}$$

where $A'_r \equiv I' + \tau_r \Lambda'_r$, $\tau_r > 0$, $r = 1, 2$, and the domain $Q \equiv Q$ is depicted in Figure 1. If we eliminate the known boundary values of u in equations (2.31), then we obtain a system with a positive operator (see Lemma 4). Hence, system (2.31), (2.32) has a unique solution u defined on $\bar{Q}_h \equiv Q_h \cup \Gamma_h$. Define $y_i \equiv A'_2 u_i$ where possible. Then (2.31) implies that $A'_1 y_i = F_i$, $x_i \in Q_h$, and the resulting one-dimensional difference systems for almost all horizontal lines can be supplemented with the boundary conditions $y_i = A'_2 \varphi_i$ (we refer to such lines as standard; for Figure 1, only a horizontal line $M_0 M_1$ containing the vertex M_0 of the reentrant corner is an exception and is a continuation of a horizontal segment of the boundary). Thus, we can easily find all y_i at inner nodes of two rectangles (this stage of the algorithm is actually based on decomposition of $\bar{Q}$ into a union of two rectangles by drawing the cutting line $M_0 M_1$ parallel to the x_1-axis); it even suffices to solve partial problems and find y_i only at inner nodes of the vertical line segment $M_2 M_0$. We then obtain all u_i at nodes belonging to the segment $M_2 M_0$ in accordance with the equations $A'_2 u_i = y_i, x_i \in (M_2 M_0)$ and the boundary conditions for u at the nodes M_2 and M_0. We thus reduce the system to two systems with splitting operators associated with two different rectangles obtained by drawing the vertical cutting line $M_2 M_0$, and that the standard procedure of type (2.2) with $q = 2$ applies for each of these rectangles (note that certain y_i were already obtained before unless we made use of partial problems).

Similar algorithms are applicable for certain more complicated regions (see [152, 164, 168]), but there are certain regions (see, e.g., Figure 2) such that, after finding y_i on all standard horizontal lines, it is possible to obtain u_i on no vertical segments (any such segment crosses an exceptional horizontal line). For such regions, a more general algorithm has been suggested (see [152, 164, 186]), which is based on block elimination (see § 1.5) with the vector $u \equiv \mathbf{u} \equiv [\mathbf{u_1}, \mathbf{u_2}]^T$, where $\mathbf{u_2}$ consists of values u_i at several inner nodes (referred to as additional boundary nodes) like those marked by $\triangle$ and ∇ in Figure 2. [23]

Figure 2. A grid region associated with the block elimination.

Lemma 5. Suppose $d = 2$ and that a generalized splitting operator $A \in \mathcal{L}(H)$ is defined by (2.31), (2.32) with $\varphi_i = 0$, $x_i \in \Gamma_h$. Suppose that $\tau_1 = \tau_2 \equiv \tau \asymp h$. Then $A \in \mathcal{L}^+(H)$ and there exist constants $\kappa_0 > 0$ and $\kappa_1 > 0$ such that $h\kappa_0 L \leq A \leq \kappa_1 L$. [24]

Proof. In accordance with (1.14), we have $\|h\|(Au, u) = (A'u', u')_0 = \|u'\|_0^2 + \tau[\|\partial_1 u'\|_0^2 + \|\partial_2 u'\|_0^2] + \tau^2 \|\partial_1 \partial_2 u'\|_0^2$, $\|h\|(Lu, u) = \|\partial_1 u'\|_0^2 + \|\partial_2 u'\|_0^2$, and $\kappa_3 \|u'\|_0^2 \leq \|\partial_r u'\|_0^2 \leq h^{-2} \kappa_4 \|u'\|_0^2$, $\forall u'$, $r = 1, 2$ Hence, $\tau\|h\|(Lu, u) \leq \|h\|(Au, u) \leq (\tau + \kappa_3^{-1} + \tau^2 \kappa_4 h^{-2} 2^{-1})\|h\|(Lu, u)$. These inequalities prove our lemma. $\square$

2.6. Grid methods with generalized splitting operators for nonstationary problems.

Consider the standard time-dependent initial-boundary value problem

---

[23] The latter choice is characterized by the property that these nodes are crossing points of all exceptional horizontal lines and vertical lines containing certain vertical segments of the boundary. The Dirichlet boundary conditions at these nodes together with the known v_i on the corresponding vertical lines enables one to obtain u_i on these lines and, therefore, to reduce system (2.31), (2.32) to separate systems with splitting operators associated with the different rectangles obtained by drawing the corresponding vertical cutting lines (the number of such lines in Figure 2 is two; the above systems correspond to $A_{1,1}\mathbf{u_1} = \mathbf{g_1}$ with block diagonal matrix $A_{1,1}$ if the original system (2.31), (2.32) is rewritten as $Au = F$ (see § 1.5)).

[24] Here and below, κ_s refers only to positive constants independent of h.

(with the initial data at $t = 0$ and the Dirichlet condition at $\Gamma \times [0, T]$) for the parabolic equation

$$D_0 u + L^0 u = f, \ x \in Q \times [0, T]. \tag{2.33}$$

Suppose we discretize this problem as follows: [25]

$$\bar{\partial}_0 u_i^{n+1} + L \frac{u_i^{n+1} + u_i^n}{2} + \frac{\tau^2}{4} \Lambda_1 \Lambda_2 \bar{\partial}_0 u_i^{n+1} = f_i^{n+1/2} \text{ if } P_i \in Q_h, \tag{2.34}$$

$$u_i^n = \varphi_i^n \text{ if } P_i \in \Gamma, \tag{2.35}$$

where $\bar{\partial}_0 u_i^{n+1} \equiv (u_i^{n+1} - u_i^n)/\tau$, $f_i^{n+1/2} \equiv f((n + 1/2)\tau, P_i)$, and

$$u_i^0 = g_i \text{ if } P_i \in Q_h. \tag{2.36}$$

It is easy to see that (2.34) approximates (2.33) with $O(\tau^2 + h^2)$-accuracy (if the solution is smooth enough) and that (2.34) is equivalent to

$$\prod_{r=1}^{2} \left(I - \frac{\tau}{2} \Delta_r \right) \frac{u_i^{n+1} - u_i^n}{\tau} = -(L u_i^n - f_i^{n+1/2}). \tag{2.37}$$

Thus, for a given vector $u^n \equiv \{u_i^n : \forall P_i \in \bar{Q}\}$, we can effectively find the vector u^{n+1} by the algorithms described above.

It is crucial that this scheme exhibits unconditional stability—with no restrictions on τ/h^2—proved by means of energy inequalities; see [153, 168, 171] and references therein. Therefore, this scheme with a generalized splitting operator combines the advantages of explicit and implicit schemes, and can be considered as a natural generalization of the scheme, suggested for a rectangular region in [405], that based on formulas (2.5) with $\tau_n = \tau/2$, provided we set appropriate boundary conditions for intermediate vectors (see [151, 153, 227, 378]). Indeed, suppose we use relations

$$\frac{u_i^{n+1/2} - u_i^n}{\tau/2} - \Delta_1 u_i^{n+1/2} - \Delta_2 u_i^n = f_i^{n+1/2} \text{ if } P_i \in Q_h, \tag{2.38}$$

---

[25] Following [157] we demonstrate that sophisticated difference schemes we obtain may be regarded as standard approximations of somewhat unusual perturbations of the original differential equation. For simplicity, we deal only with two space variables x_1 and x_2. The time variable t is denoted by x_0 and we write $x \equiv [x_0, x_1, x_2]$, $D_s u \equiv \frac{\partial u}{\partial x_s}$, $s = 0, 1, 2$, $-L^0 \equiv D_1^2 + D_2^2$. We use the same region Q and the same grid for the space variables as before, and denote its nodes by P_i; on the time segment $[0, T]$, we use a grid with the mesh size τ and write $u_i^n \equiv u(n\tau, P_i)$.

$$\frac{u_i^{n+1} - u_i^{n+1/2}}{\tau/2} - \Delta_1 u_i^{n+1/2} - \Delta_2 u_i^{n+1} = f_i^{n+1/2} \text{ if } P_i \in Q_h, \qquad (2.39)$$

supplemented by boundary conditions (2.35) and similar conditions

$$u_i^{n+1/2} = \varphi_i^{n+1/2} \text{ if } P_i \in \Gamma, \qquad (2.40)$$

at the intermediate levels. Then, for inner nodes of Q, from (2.38) and (2.39) it follows that $u_i^{n+1} - \tau/2\Delta_2 u_i^{n+1} = 2u_i^{n+1/2} - u_i^n - \tau/2\Delta_2 u_i^n \; P_i \in Q_h$, which also holds on vertical sides of the rectangle Q if the function φ in (2.40) is independent on t. Then (2.38)–(2.40) is equivalent to (2.37). In the general case, we must replace (2.40) by $\varphi_i^{n+1} - \tau/2\Delta_2\varphi_i^{n+1} = 2u_i^{n+1/2} - \varphi_i^n - \tau/2\Delta_2\varphi_i^n$, where P_i is an inner point of a vertical side of the rectangle Q. Hence, the use of schemes of the suggested type enables one to treat intermediate levels from a purely algebraic point of view, based on the factorization (2.1)—for a rectangle—and avoiding approximation and stability issues. [26]

Now we consider the *method of splitting multidimensional scheme* (see [508]) and based on the use of the one-dimensional relations: $\frac{u_i^{n+1/2} - u_i^n}{\tau/2} - \frac{\Delta_1(u_i^{n+1/2} + u_i^n)}{2} = 0, \; \frac{u_i^{n+1} - u_i^{n+1/2}}{\tau/2} - \frac{\Delta_2(u_i^{n+1/2} + u_i^{n+1})}{2} = 0, \; P_i \in Q_h$, instead of (2.38) and (2.39) with $f_i^{n+1/2} = 0$. [27]

---

[26] We stress that both notions for such schemes need a proper choice of norms. In this connection, it is very useful to note that scheme (2.34)–(2.36) is a standard scheme for a similar initial-boundary value problem with (2.33) replaced by $D_0 u + L^0 u + \varepsilon D_1^2 D_2^2 D_0 u = f$, where $\varepsilon = \tau^2$. In the same manner, we can use $D_0 u + L^0 u + \varepsilon D_1^2 D_2^2 u = f$ or $D_0 u + L^0 u + \varepsilon D_1^2 D_2^2 D_0^2 u = f$ (correctness of these problems was analyzed in [157, 158], where it was also shown that their solutions differ from that of the original problem by $O(\varepsilon)$; in [209], boundary conditions of the third type were also analyzed). This approach is fruitful in constructing schemes, with generalized splitting operators, based on projective-grid approximations for L^0 (see [227, 519]). Generalizations for equations $D_0^2 u + L^0 u = f$ and $D_0 L^0 u + L^0 u = f$ and systems of various types are known (see, e.g., [168, 171]; the operator L^0 can be replaced by a general one). It should be noted that, for a rectangular or parallelepiped region, there are many relevant results (see, e.g., [136, 352, 435, 508]).

[27] These relations together with boundary conditions (2.35) and (2.40) lead to (2.34) only for homogeneous Dirichlet conditions (see [171, 168]); for the general case, at nodes having a neighbor on the boundary, additional terms of type $O(\tau^2/h^2)$ may appear and will lead to unsatisfactory approximation. Nevertheless, the simplicity of such schemes and generality of possible extensions made them an object in the 1960s of many investigations (see, e.g., [168, 209, 352, 435, 508]). It was understood (see [209, 435]) that they needed a new notion of *summary approximation* and that they corresponded to the standard schemes for problems based on the replacement of (2.33) by

$$D_0 u - (1 + \omega(t/\delta))D_1^2 u - (1 + \omega(t/\delta))D_2^2 u = 0,$$

where $\omega(\delta t)$ is a periodic function with period δ, e.g., of the form $\omega(t/\delta) \equiv 1$ if $t \in [0, \delta/2)$

2.7. Additive splitting of the inverse operator (additive Schwarz methods). In the early 1960s, ideas of splitting for algebraic systems of equations $Au = f$ with $A \in \mathcal{L}^+(H)$ led to the appearance of the iterations of type $u^{n+1} = u^n - \tau_n[D_0 \bar{B}_0^{-1} + \cdots + D_p \bar{B}_p^{-1}](Au^n - f)$, where $\bar{B}_k, k \in [0,p]$ are some model operators, $D_k, k \in [0,p]$, are diagonal nonnegative matrices such that $D_0 + \cdots + D_p = I$, and $\tau_n = 1$ (see, e.g., [299] and references therein). To indicate possible choices of these operators (matrices) suppose that A has the block form (1.5.11). Then we may take

$$\bar{B}_1 \equiv \begin{bmatrix} A_{1,1} & 0 \\ 0 & I_2 \end{bmatrix}, \ D_1 \equiv \begin{bmatrix} I_1 & 0 \\ 0 & 0 \end{bmatrix}, \text{ and } R_1 \equiv D_1 \bar{B}_1^{-1} = \begin{bmatrix} A_{1,1}^{-1} & 0 \\ 0 & 0 \end{bmatrix}.$$

This implies that $R_1 = (P_1 A P_1)^\dagger$, where P_1 is the orthoprojector of the Euclidean space H onto the relevant subspace H_1. Thus, it is interesting (see, e.g., [384] and references therein) to analyze cases of model operators B such that, for their inverses, we have additive splittings

$$C \equiv R_0 + \cdots R_p, \tag{2.41}$$

where $R_l \equiv (P_l A P_l)^\dagger$ and P_l is the orthoprojector of H onto the relevant subspace $H_l, l \in [0,p]$ for a certain chosen splitting

$$H = H_0 + \cdots + H_p \tag{2.42}$$

of the space H. Below, $\alpha_l(v)$ is the angle between $v \neq 0$ and $H_l, l \in [0,p]$.
 Lemma 6. *Let splitting* (2.42) *and* $\kappa_1 \geq \kappa_0 > 0$ *be such that*

---

and $\omega(t/\delta) \equiv -1$ if $t \in [\delta/2, \delta)$ and $\delta = \tau$. Therefore, we use here an equation with highly oscillating (with respect to t) coefficients as a model for (2.33); this is in contrast to the homogenization procedures mentioned in § 2.3. It was shown in 1962 (see [209]) that the solution $u_\delta(x)$ of the new problem is such that $|u_\delta(x) - u(x)|_{0,Q} = O(\delta)$; moreover, $u_\delta(x) = u(x)$ if $t = 0, \delta, 2\delta, \ldots$; that is, superconvergence holds for special values of t.

For more general problems, a class of *additive schemes* was suggested and investigated to a certain degree (see references in [352, 435]); some theoretical results were and are not especially attractive, but many practical problems were solved on the basis of similar methods. Thus, the notion of splitting became used in a very broad sense (see, e.g., [352]): as a procedure of approximating (partitioning) a complicated problem by simpler ones. Of course, such an approach is so general that it can be found not only in any branch of mathematics, but also in any area of human activity. For example, Descartes estimated great capabilities of such approach for solving scientific problems; Leibnitz, on the contrary, stressed (see, e.g., [413]) that it can be meaningless if the notion of a so-called simpler problem is not clear (almost the same topic was of interest to L. Carrol and W. Gote). On the whole, we should be satisfied that sometimes our analysis (partitioning, simplification, splitting) works; in numerical methods, we can find a lot of examples where the idea of splitting of a space, operator, region, or grid turns out to be fruitful (strictly speaking, even all grid methods are based on partitioning (decomposition) of the region).

$$\kappa_0 \|v\|_A^2 \le \sum_{l=0}^{p} \cos^2 \alpha_l(v) \le \kappa_1 \|v\|_A^2, \ \forall v. \tag{2.43}$$

Then there exists $B \in \mathcal{L}^+(H)$ such that $B^{-1} = C$ and $\kappa_0 B \le A \le \kappa_1 B$.

Proof. In accordance with (2.41), we have

$$(Cu, u) = (R_0 Av, v)_A + \cdots + (R_p Av, v)_A, \tag{2.44}$$

where $v \equiv A^{-1}u$. Now we prove that $R_l A$ is the orthoprojector of the Euclidean space $H(A)$ onto $H_l, l \in [0, p]$. For this we consider an orthonormal basis and the matrices of our operators associated with this basis. We denote these matrices by the same symbols. Next we consider a new orthonormal basis which is a union of orthonormal bases for the Euclidean spaces $H_0, \ldots, H_p$ and such that its first N_l elements define the old basis for a chosen H_l. Then, if we denote the matrices of the operators associated with the new basis by the same symbols but with overbar, then $\bar{A} = T_l^* A T_l$, $\bar{P}_l = T_l^* P_l T_l$, $\bar{R}_l = T_l^* R_l T_l$, where T_l is the orthogonal matrix corresponding to this transformation of the bases $(T_l^* = T_l^{-1})$. It is also known (see, e.g., [7]) that $\bar{R}_l^\dagger = T_l^* R_l^\dagger T_l$. We make use of

$$\bar{A} = \begin{bmatrix} \bar{A}_{1,1} & \bar{A}_{1,2} \\ \bar{A}_{2,1} & \bar{A}_{2,2} \end{bmatrix}; \ \bar{P}_l = \begin{bmatrix} I_1 & 0 \\ 0 & 0 \end{bmatrix}; \ \bar{R}_l = \begin{bmatrix} \bar{A}_{1,1}^{-1} & 0 \\ 0 & 0 \end{bmatrix}.$$

Note that

$$\bar{R}_l \bar{A} = \begin{bmatrix} I_1 & \bar{A}_{l,1}^{-1} \bar{A}_{1,2} \\ 0 & 0 \end{bmatrix} = (\bar{R}_l \bar{A})^2.$$

This implies that $\bar{R}_l \bar{A}$ is a projector. In accordance with this, it is easy to see that $R_l A$ is a projector as well. Indeed, we have $(R_l A)^2 = (T_l \bar{R}_l \bar{A} T_l^*)^2 = T_l \bar{R}_l \bar{A} T_l^* = R_l A$. Since the operator $R_l A$ is symmetric as an element of $\mathcal{L}(H(A))$ and its image is H_l, then $v_l \equiv R_l Av = P_{A,l} v$, where $P_{A,l}$ is the orthoprojector of the Euclidean space $H(A)$ onto the subspace H_l. Therefore, from (2.44) it follows that

$$(Cu, u) = \|P_{A,0} v_0\|_A^2 + \cdots + \|P_{A,p} v_p\|_A^2, \tag{2.45}$$

and $\|v_l\|_A^2 = \cos^2(\alpha(v))\|v\|_A^2$, $l \in [0, p]$. Hence, $\kappa_0(Av, v) \le (Cu, u) \le \kappa_1(Av, v)$ and $\kappa_0(A^{-1}u, u) \le (Cu, u) \le \kappa_1(A^{-1}u, u)$, $\forall u \in H$. Thus, we have $\kappa_0 A^{-1} \le C \le \kappa_1 A^{-1}$. This implies that $C \in \mathcal{L}^+(H)$ and is invertible. Hence, for $B = C^{-1}$ we obtain the desired inequalities. $\square$

We note that $\kappa_1 \le p + 1$; a suitable estimate for κ_0 will be given in Lemma 8; if, instead of $A_{l,l}$, model operators B_l are used such that

$$\sigma_0 B_l \le A_{l,l} \le \sigma_1 B_l, \ l \in [0, p], \tag{2.46}$$

then, for the obtained B, we can show in much the same way that

$$\kappa_0 \sigma_0 B \leq A \leq \kappa_1 \sigma_1 B. \tag{2.47}$$

Often it is convenient to work with splittings of the spline space

$$\hat{G}_h = \hat{G}^{(0)} + \cdots + \hat{G}^{(p)}. \tag{2.48}$$

These spaces are equipped with the inner product $(\hat{u}, \hat{v}) \equiv b(\hat{u}; \hat{u})$ and become the Euclidean space $V, V_0, \ldots, V_p$, so (2.48) can be rewritten as

$$V = V_0 + \cdots + V_p. \tag{2.49}$$

For $\hat{v} \neq 0$, we denote by $\alpha_l(v)$ the angle between $\hat{v}$ and the subspace $V_l, l \in [0, p]$. P_l is now the orthoprojector of the Euclidean space V onto the subspace V_l and $\|P_l \hat{v}\|^2 = \cos^2(\alpha(\hat{v}))\|\hat{v}\|^2$, where $l \in [0, p]$. We now define $P \in \mathcal{L}(V)$ by

$$P \equiv P_0 + \cdots + P_p, \tag{2.50}$$

as was done by several researchers (see [142, 143, 339, 384, 455, 504] and references therein). This leads to a number of optimal or nearly optimal preconditioners (see [78, 144, 347, 363, 365]).

Lemma 7. *Let splitting (2.49) and constants $\kappa_0 > 0$ and $\kappa_1 > 0$ be such that $\kappa_0\|v\|^2 \leq \sum_{l=0}^{p} \cos^2 \alpha_l(\hat{v}) \leq \kappa_1\|v\|^2$, $\forall \hat{v}$. Suppose also that the operator P is defined by (2.50). Then $\kappa_0 I \leq P \leq \kappa_1 I$.*

Proof. We have $(P\hat{v}, \hat{v}) = (P_0\hat{v}, \hat{v}) + \cdots + (P_p\hat{v}, \hat{v})$. Since $P_l^2 = P_l, l \in [0, p]$, then $(P\hat{v}, \hat{v}) = \sum_{l=0}^{p}(P_l\hat{v}, P_l\hat{v}) = \sum_{l=0}^{p}\|P_l\hat{v}\|^2$. This leads to the desired inequalities for P. $\square$

Lemma 8. *Let splitting (2.49) and constant $\kappa_0 > 0$ be such that, for each $\hat{v}$, there exists an expansion*

$$\hat{v} = \hat{v}_0 + \cdots + \hat{v}_p, \ \hat{v}_l \in V_l, \ l \in [0, p] \tag{2.51}$$

such that

$$\kappa_0[\|\hat{v}_0\|^2 + \cdots + \|\hat{v}_p\|^2] \leq \|\hat{v}\|^2. \tag{2.52}$$

Suppose also that the operator P is defined by (2.50). Then $\kappa_0 I \leq P$.

Proof. Since $X \equiv \|\hat{v}\|^2 = \sum_{l=0}^{p}(\hat{v}_l, \hat{v})$, then $X = \sum_{l=0}^{p}(P_l\hat{v}_l, \hat{v}) = \sum_{l=0}^{p}(\hat{v}_l, P_l\hat{v}) \leq [\sum_{l=0}^{p}\|\hat{v}_l\|^2]^{1/2}[\sum_{l=0}^{p}\|P_l\hat{v}\|^2]^{1/2}$. Since $P_l^2 = P_l$, then

$$\|\hat{v}\|^2 \leq [\|\hat{v}_0\|^2 + \cdots + \|\hat{v}_p\|^2]^{1/2}(P\hat{v}, \hat{v})^{1/2},$$

$l \in [0, p]$. This together with (2.52) leads to the desired estimate for P. $\square$ [28]

§ 3. Iterative methods with factored operators

3.1. Basic classes of methods; factored model operators. Consider system (2.4) with the matrix $\Lambda \in \mathcal{L}^+(H)$ represented in the form $\Lambda = I + L_1 + L_2$, where I is its diagonal part and $L_1 = L_2^T$ is its strictly lower triangular part. Then symmetric Gauss-Seidel iteration (see [6]) is given by $(I + L_1)u^{n+1/2} = -(\Lambda u^n - f)$, $(I + L_2)u^{n+1} = -(\Lambda u^{n+1/2} - f)$. Elimination of the intermediate iterate $u^{n+1/2} = (I + L_2)u^{n+1} - L_2 u^n$ leads to the relation $A(u^{n+1} - u^n) = -(\Lambda u^n - f)$, where $A \equiv (I + L_1)(I + L_2) \in \mathcal{L}^+(H)$. Thus, we may consider the symmetric Gauss-Seidel method as our first representative of the class of iterative methods with model factored operator, based on triangular factors. For grid systems defined on the grid Ω_h with N_h nodes, we usually refer to a *factored grid operator* $B \equiv B_h$ if B is represented in the form $B = B_1 B_2$, where B_1 is a strictly lower triangular matrix, $B_2 = B_1^T$, and the number of nonzero elements of each of these matrices is bounded from above by $K N_h$ ($K \neq K(h)$). Such operators appear at SSOR methods ([264, 271]) if we eliminate the intermediate iterate as above for the symmetric Gauss-Seidel method. [29]

3.2. Alternate triangular method. Consider system (2.4) with matrix $\Lambda \in \mathcal{L}^+(H)$ represented in the form $\Lambda = \Lambda_1 + \Lambda_2$, where Λ_1 is its lower triangular part and $\Lambda_2 = \Lambda_1^T$. Moreover, we assume that

$$\Lambda \geq \gamma I, \quad \gamma > 0; \quad \|\Lambda_2 u\|^2 \leq \kappa(\Lambda u, u), \quad \forall u. \tag{3.1}$$

The following simple lemma (see [440]) holds.

---

[28] This lemma can be found in [142, 339, 384]; from its condition, it follows that

$$\sup_{\hat{v}_0, \ldots, \hat{v}_p} \frac{(\hat{v}_0 + \cdots + \hat{v}_p, v)}{\|\hat{v}_0\|^2 + \cdots + \|\hat{v}_p\|^2} \geq \kappa_0^{1/2} \|\hat{v}\|, \quad \forall \hat{v}.$$

Thus, it is connected with correctness of the problem $\min[\|\hat{v}_0\|^2 + \cdots + \|\hat{v}_p\|^2]$ under constraint $\hat{v}_0 + \cdots + \hat{v}_p = \hat{v}$ (such problems will be considered in Chapter 7). A number of efficient iterative methods dealing with splittings of type (2.49) for composite grids with local refinement was suggested in [80, 143, 347, 363, 365, 384, 504].

[29] The first use of factored grid operators as model operators, with no relation to the Gauss-Seidel iterations, was probably suggested by Buleev [102]. He considered them as factored approximations to the original grid operator and used the term "method of the incomplete factorization" (see also [147]). Especially simple constructions of the model factored grid operators were suggested by Samarskii (see [440]). Similar modifications of classical iterative methods are very popular for solving grid systems of very general nature [20, 271, 290, 396, 487], most particularly if these systems are not too large and the aim is to avoid the use of more complicated methods. Since such methods in the best cases require computational work $W = O(h^{-(d+1/2)}|\ln \varepsilon|)$, we only outline some typical results.

Lemma 1. Let conditions (3.1) be satisfied and

$$B \equiv (I + \tau\Lambda_1)(I + \tau\Lambda_2), \tau > 0, \tag{3.2}$$

where $\tau > 0$ is a given iteration parameter. Then

$$\delta_0(\tau)B \leq \Lambda \leq \delta_1(\tau)B, \tag{3.3}$$

$$\delta_0(\tau) \equiv \left(\gamma^{-1} + \tau + \tau^2\kappa\right)^{-1}, \quad \delta_1(\tau) \equiv (2\tau)^{-1} \tag{3.4}$$

and

$$\operatorname{argmin} \frac{\delta_1(\tau)}{\delta_0(\tau)} = \frac{\gamma^{1/2}}{\kappa^{1/2}} \equiv \tau^*. \tag{3.5}$$

Proof. We have

$$(Bu, u) = (u, u) + \tau(\Lambda u, u) + \tau^2\|\Lambda_2 u\|^2. \tag{3.6}$$

Thus, $(Bu, u) \leq (\gamma^{-1} + \tau + \tau^2\kappa)(\Lambda u, u)$, and $\delta_0(\tau)B \leq \Lambda$. On the other hand, $B - 2\tau\Lambda = (I - \tau\Lambda_1)(I - \tau\Lambda_2) \geq 0$ $((I - \tau\Lambda_1)^T = I - \tau\Lambda_2)$. Hence, (3.3) and (3.4) hold. Since $F(\tau) \equiv 2\delta_1(\tau)[\delta_0(\tau)]^{-1} = 1 + [\tau\gamma]^{-1} + \tau\kappa$, then the proof of (3.5) is straightforward. $\square$ [30]

3.3. Incomplete factorization. A wide class of model factored operators can be constructed using various generalizations of incomplete factorization. For example, if we have the standard factorization $\Lambda = A_1 A_2$ associated with Gauss elimination, then the triangular factors have no sparsity property, and the use of such a model operator would result in very large computational work at each iteration. Hence, it is natural (see [358]) to simplify this factorization and to use a model operator

$$A'_h \equiv A' = A'_1 A'_2 \in \mathcal{L}^+(H), \tag{3.7}$$

---

[30] If we consider system (0.2.15), (0.2.16), with $h_1 = h_1 = h$, then the operators Λ_1 and Λ_2 in Lemma 1 (but not in (0.2.15), (0.2.16)) are such that $\Lambda_1 u_i \equiv h(\bar{\partial}_1 u_i + \bar{\partial}_2 u_i)$, $P_i \in Q_h$, and $\Lambda_2 u_i \equiv -h(\partial_1 u_i + \partial_2 u_i)$, $P_i \in Q_h$. This implies that, in (3.1), we may choose the constants $\gamma \equiv h^2 K_0$ and $\kappa \equiv K_1$ such that $K_0 > 0$ and $K_1 > 0$ are independent of h. Thus, $\tau^* \asymp h$ and $\min F(\tau) = O(h^{-1})$, so the use, e.g., of the modified Richardson method leads to the estimate $W = O(h^{-(2+1/2)}|\ln\varepsilon|)$. In much the same way, we obtain $W = O(h^{-(d+1/2)}|\ln\varepsilon|)$ for general d-dimensional approximations. Instead of (3.3), the operator $B \equiv (D + \tau L_1)D^{-1}(D + \tau L_2)$ it is often preferable to use, where $D > 0$ is the diagonal part of Λ and L_1 and L_2 are defined as in the symmetric Gauss-Seidel method.

where $(A_1')^* = A_2'$ retains only a part of nonzero elements of A_1. [31]

3.4. Factorization of the matrix associated with a hierarchical basis. A very interesting use of factored matrices (with non-triangular factors) was given in connection with the use of hierarchical bases and splittings considered in § 2.5 (see [512]). It was found that if we replace the usual nodal bases of finite element spaces by hierarchical bases, then, for $d = 2$, the grid operator of type $-\Delta_h$ corresponds to matrices $\bar{Y}_h$ such that

$$\frac{\delta_0}{p^2} I_h \leq \bar{Y}_h \leq \delta_1 I_h, \tag{3.8}$$

where p refers to the number of successive refinements by a factor of two and positive numbers δ_0 and δ_1 are independent of p (the grid operators Y_h and I_h are nearly spectrally equivalent). More precisely, suppose that we have a sequence of nested triangulations $T^{(0)}(\bar{Q}), \ldots, T^{(p)}(\bar{Q})$ of levels $l \in [0, p]$ and a sequence of nested subspaces $\hat{G}^{(0)}, \ldots, \hat{G}^{(p)}$. Suppose, for $l \in [0, p]$, that

$$\hat{G}^{(l)} \equiv \{\hat{u} : \hat{u} = \sum_{P_i^{(l)} \in Q^{(l)}} u_i \hat{\psi}_i^{(l)}(x)\} \tag{3.9}$$

consists of functions that are continuous on the domain Q, vanishing on Γ_0, and piecewise linear (with respect to the triangulation $T^{(l)}(\bar{Q})$), with the standard nodal basis functions $\hat{\psi}_i^{(l)}(x)$. Then the basis for $\hat{G}^{(p)} \equiv \hat{V}$ defines the standard Gram matrix $-\Delta_h \equiv \Lambda_h \equiv [(\hat{\psi}_j^{(p)}(x), \hat{\psi}_i^{(p)}(x))_{1,Q}]$ and the standard grid system

$$\Lambda_h u_h = f_h \tag{3.10}$$

(an operator equation in the standard Euclidean space H). Suppose that, along with the basis $\{\hat{\psi}_i^{(l+1)}(x)\}$ for $\hat{G}^{(l+1)}, l \in [0, p-1]$, we consider the hierarchical basis $\{\bar{\psi}_i^{(l+1)}(x)\}$, where

$$\bar{\psi}_i^{(l+1)} \equiv \hat{\psi}_i^{(l+1)} \text{ for } P_i^{(l+1)} \in Q_h^{(l+1)} \setminus Q_h^{(l)} \tag{3.11}$$

and

$$\bar{\psi}_i^{(l+1)} = \hat{\psi}_i^{(l)} \text{ for } P_i^{(l+1)} \in Q_h^{(l)}; \tag{3.12}$$

---

[31] If we retain only the terms that fall into the original sparsity structure, then we obtain the operator that is very close to the factored operators given above. On the other hand, if $L_h \asymp \Lambda_h$ and we retain almost all elements, there is reason to expect that $A_h' \asymp L_h$ (neither strategy leads to asymptotically optimal preconditioners but requires a rather significant computational work to obtain the desired operator). Thus, usually all possible fill-in positions are chosen beforehand and numerical tests are carried out to find a satisfactory practical strategy. Applications can be found, e.g., in [20, 271, 290, 396, 487].

we assume that the nodes from $Q^{(l+1)} \setminus Q^{(l)}$ have indices that are less than those of nodes from $Q^{(l)}$ that will be useful for consideration of certain multigrid methods in § 7 (it is the opposite ordering to the one in [512]). The indicated choice of the basis leads to the splitting

$$\hat{G}^{(l+1)} = \hat{G}^{(l)} \oplus \hat{V}^{(l+1)}, \quad l \in [1, p-1], \tag{3.13}$$

where

$$\hat{V}^{(l+1)} \equiv \{\hat{u} : \hat{u} \in \hat{G}^{(l+1)}, \hat{u}(P_i^{(l)}) = 0 \text{ for all } P_i^{(l)} \in Q_h^{(l)}\} \tag{3.14}$$

and the basis for $\hat{V}^{(l+1)}$ is $\{\bar{\psi}_i^{(l+1)} : P_i^{(l+1)} \in Q_h^{(l+1)} \setminus Q_h^{(l)}$. Thus,

$$\hat{G}^{(p)} \equiv \hat{V} \equiv \hat{G}^{(0)} \oplus \hat{V}_1 \oplus \cdots \oplus \hat{V}_p \tag{3.15}$$

and the hierarchical basis for $\hat{V}$ is a union of the indicated bases for $\hat{V}_p, \ldots, \hat{V}_1, \hat{G}_0$. Hence, the expansion of $\hat{u} \in \hat{V}$ with respect to the hierarchical basis leads to the system

$$Y_h \bar{u} = \bar{f}, \tag{3.16}$$

considered as an operator equation in $H \equiv H_0 \times \cdots \times H_p$, where $\bar{u} \equiv [\bar{u}_0, \ldots, \bar{u}_p] \in H$, $\bar{u}_r \in H_r$, $r \in [0, p]$. Under the above ordering, the matrix Y_h takes the form

$$Y_h \equiv \begin{bmatrix} Y_{0,0} & \cdots & Y_{1,p} \\ \vdots & \ddots & \vdots \\ Y_{p,0} & \cdots & Y_{p,p} \end{bmatrix}. \tag{3.17}$$

Here $Y_{p,p}$ corresponds to the approximation on the coarse triangulation. [32]
   Observe that

$$Y_h \equiv E_h^T \Lambda_h E_h, \quad \tilde{f} \equiv E_h^T f_h, \tag{3.18}$$

where E_h is the transition matrix from the nodal basis for $\hat{V}$ to its hierarchical basis. Hence, in order to use the above factorization of Y_h, it suffices to develop fast and stable algorithms needed for change from nodal to hierarchical bases, that is, for obtaining the products $E_h \bar{v}$ and $E_h^T f_h$.

---

[32] Inequalities (3.8) and more general ones for a model operator $D_h \in \mathcal{L}^+(H)$, for two-dimensional problems, were obtained on the basis of the important and well-known inequality $|u_i| \leq K |\ln h|^{1/2} \|\hat{u}\|_{1,Q}$ (which does not hold for $d \geq 3$). When D_h is a block diagonal operator (see (3.17)), with the diagonal blocks $D_{0,0}, \ldots, D_{p-1,p-1}$ being diagonal matrices and $D_{p,p} \equiv Y_{p,p}$, the above choice is effective even for non-quasiuniform triangulations; (3.8) leads to nearly asymptotically optimal convergence of iterative methods with such model operators for systems (3.16). But now it is very important to obtain effective ways for evaluation of the residual $\bar{r}^n \equiv Y_h \bar{u}^n - \bar{f}$ since Y_h is a rather dense and complicated matrix.

Computation of the first product means evaluating a function $\hat{v} \in \hat{V}$, which is given by its coefficients with respect to the hierarchical basis, at the nodes $P_i^{(l)}$. This can be done recursively, beginning with the nodes $P_i^{(0)}$, where the values of $\hat{v}$ are given by the corresponding hierarchical basis coefficients (higher level basis functions vanish at these nodes). If, for $l \leq p-1$, the set of the values $\hat{v}(P_i^{(l)})$ is already known, then we first evaluate the function $\hat{v}^{(l)}$ at the new nodes $P_i^{(l+1)} \in Q_h^{(l+1)} \setminus Q_h^{(l)}$ by virtue of the corresponding interpolation procedure (for our particular case it requires only arithmetical means of certain pairs of the values at neighbor nodes). Then we add the values of $\hat{v}^{(l+1)} - \hat{v}^{(l)}$ at these nodes (they are actually stored in the hierarchical basis coefficient vector $\bar{v}$) to the above obtained interpolated values. This yields the values of $\hat{v}$ at $P_i^{(l+1)} \in Q_h^{(l+1)} \setminus Q_h^{(l)}$ and, thus, the desired values of $\hat{v}$ at all $P_i^{(l+1)} \in Q_h^{(l+1)}$. We emphasize that this algorithm for computing $E_h \bar{v}$ corresponds to a factorization

$$E_h = \bar{E}^{(p)} \dots \bar{E}^{(1)}, \tag{3.19}$$

where $\bar{E}^{(l+1)}$ is a sparse matrix corresponding to the evaluation of the values of $\hat{v}$ at the new nodes $P_i^{(l+1)} \in Q_h^{(l+1)} \setminus Q_h^{(l)}$, where $l \in [0, p-1]$. The matrix of transition from the nodal basis for the space $\hat{G}^{(l+1)}$ to its hierarchical two-level basis is just

$$E^{l+1} \equiv \begin{bmatrix} I_{l,l+1} & E_{1,2}^{l+1} \\ 0 & I_l \end{bmatrix}, \tag{3.20}$$

where the diagonal blocks are the identity matrices; for defining the elements of $(N_l^{(1)} + k)$th column of E^{l+1} (note that $N_l^{(1)} \equiv \dim \hat{V}_l$), it suffices to know the coefficients of the expansion

$$\bar{\psi}_{N_l^{(1)}+k}^{(l+1)} = \sum_{P_i^{(l+1)} \in Q_h^{(l+1)}} d_{N_l^{(1)}+k,i} \hat{\psi}_i^{(l+1)}, \tag{3.21}$$

which are just the values of $\bar{\psi}_{N_l^{(1)}+k}^{(l+1)}(x)$ at the nodes $(k \in [1, N_l])$. This implies that $\bar{E}^{l+1}$ is basically the identity matrix, with the exception of one block, which is just $E^{(l+1)}$, and the desired algorithm for computing $E_h \bar{v}$ is obtained. From (3.19), we see that

$$E_h^T = (\bar{E}^{(1)})^T \dots (\bar{E}^{(1)})^T. \tag{3.22}$$

Hence, computation of $E_h^T f_h$ is also fairly simple. Note that

$$(E^{l+1})^{-1} = \begin{bmatrix} I_{l,l+1} & -E_{1,2}^{l+1} \\ 0 & I_l \end{bmatrix}. \tag{3.23}$$

This yields similar effective factorization procedures for the transformation from hierarchical to nodal bases for $\hat{V}$.

§ 4. Two-stage iterative methods with inner iterations

4.1. Basic computational algorithms. [33] Suppose that, for system

$$\Lambda v = F \tag{4.1}$$

with $\Lambda \in \mathcal{L}^+(H)$, there exists an iterative method, say the $Z(\Lambda)$-method, that is effective in some sense. Suppose that it leads after M iterations to the linear relation

$$z^M = Z z^0, \quad Z \equiv Z_M \in \mathcal{L}(H), \tag{4.2}$$

where, as usual, z^n refers to the error at the nth iteration.

Lemma 1. Suppose that $I - Z$ is invertible. Then the linear operator

$$B \equiv \Lambda(I - Z)^{-1} \tag{4.3}$$

is also invertible and the solution w of system

$$Bw = F \tag{4.4}$$

coincides with the Mth iterate in the $Z(\Lambda)$-method for system (4.1), provided the initial iterate in this method is $v^0 = 0$.

Proof. The operator B is a product of two invertible operators, so B^{-1} exists. Moreover, it has the form $B^{-1} = (I - Z)\Lambda^{-1}$. Using the definition

---

[33] The fruitfulness of the concept of spectrally equivalent operators was shown for the first time by applying *two-stage iterative methods* with inner iterations (see [135, 150] where, on each outer iteration, effective inner iterations like ADI were used to solve model difference systems approximately; actually we need only perform a few inner iterations, so the term approximation is meant here in a very broad sense). The idea has withstood the test of time and was developed in further investigations dealing with difference (see [147, 153, 154, 155, 156, 160, 162, 164, 163, 258, 259, 260, 496, 497]) and finite element (see [165, 166, 170, 172, 174, 175, 176, 177]) systems. Currently, various inner iterations can be found in many variants of multigrid (multilevel) and domain decomposition methods (see § 5 and 7 and references therein). In spite of the appearance of fast direct methods (see § 1), two-stage iterative methods with inner ADI-iterations remain practically very fast, stable, and simply constructed methods that require no assumptions of type (1.5). They are especially useful for grid approximations of fourth-order strongly elliptic systems and for multidimensional problems when algorithms from § 1 are rather complicated and no standard codes are available; they have been used with success for practical solution of some rather difficult linear and nonlinear systems of large order (see, e.g., [210, 214, 421, 467, 497, 510] and § 6.5). They are also very well adapted to parallel computation (see [288]).

of Z, we may write $v^M - v = Z(v^0 - v) = -Zv$. Hence,

$$v = (I - Z)^{-1} v^M. \tag{4.5}$$

Substituting (4.5) into (4.1) leads to the equality $v = w$. $\square$ [34]

Lemma 2. *Suppose that $I - Z$ is invertible and B is defined by (4.3). Suppose we use iterative method (0.3.13). Then the new iterate u^{n+1} in this method is just the Mth iterate in the $Z(\Lambda)$-method for system (4.1) with the vector $F \equiv \Lambda u^n - \tau_n(Lu^n - f)$, provided that the initial iterate in the inner iterations coincides with u^n.*

Proof. We have $v^M - v = Z(u^n - v)$. Hence, $v = (I - Z)^{-1}(v^M - Zu^n)$. Substitution into (4.1) yields $\Lambda(I - Z)^{-1}(v^M - Zu) = \Lambda u^n - \tau_n(Lu^n - f)$ and $\Lambda(I - Z)^{-1}v^M = \Lambda(I - Z)^{-1}(I - Z + Z)u^n - \tau_n(Lu^n - f)$. The latter equality yields $v^M = u^{n+1}$. $\square$

4.2. Spectral equivalence of operators Λ_h and B_h.

Theorem 1. *Let $\Lambda \equiv \Lambda_h \in \mathcal{L}^+(H)$ and suppose $Z \in \mathcal{L}(H(\Lambda))$ is symmetric and*

$$\|Z\|_\Lambda \leq q < 1. \tag{4.6}$$

Let B be defined by (4.3). Then $B \in \mathcal{L}^+(H)$ and

$$(1 - q)B \leq \Lambda \leq (1 + q)B; \tag{4.7}$$

if, additionally, $Z \geq 0$ (as an element of $\mathcal{L}^+(H(\Lambda))$), then

$$(1 - q)B \leq \Lambda \leq B. \tag{4.8}$$

Proof. We start from the representation $B^{-1} = \Lambda^{-1} - Z\Lambda^{-1}$ and prove that $(Z\Lambda^{-1})^* = Z\Lambda^{-1}$. Since Z is symmetric (in the sense of $\mathcal{L}^+(H(\Lambda))$), we have $(Z\Lambda^{-1}u, v) = (Z\Lambda^{-1}u, \Lambda^{-1}v)_\Lambda = (u, Z\Lambda^{-1}v)$. Hence, $(B^{-1})^* = B^{-1}$ and $B^* = B$. We now prove that

$$(1 - q)\Lambda^{-1} \leq B^{-1} \leq (1 + q)\Lambda^{-1}. \tag{4.9}$$

Since $(B^{-1}u, u) = (\Lambda^{-1}u, u) - (Z\Lambda^{-1}u, u)$, then, for $v \equiv \Lambda^{-1}u$, we have

$$|(Z\Lambda^{-1}u, u)| = |(\Lambda Zv, v)| \leq \|Z\|_\Lambda \|v\|_\Lambda^2 \leq q\|u\|_{\Lambda^{-1}}^2. \tag{4.10}$$

---

[34] Solving (4.4) involves inner iterations for (4.1). This explains why we refer to a model operator of form (4.3) as a *two-stage preconditioner*. The algorithm given above is basic and when $B \in \mathcal{L}^+(H)$ enables us to apply effectively all iterative methods considered in § 1.3. Sometimes, another algorithm is applied, as in the following lemma.

Hence, (4.9) follows from the last two relations and leads to (4.7) (see Lemma 0.4.3). To obtain (4.8) (for $Z \geq 0$), it suffices to observe that $(Z\Lambda^{-1}u, u) = (Zv, v)_\Lambda \geq 0$ and to replace (4.10) by

$$0 \leq (Zv, v)_\Lambda \leq q\|v\|_\Lambda^2 = q\|u\|_{\Lambda^{-1}}^2.$$

Then $(1 - q)\Lambda^{-1} \leq B^{-1} \leq \Lambda^{-1}$, and (4.8) holds. $\square$ [35]

4.3. Conditions for relationships of type C^k between operators L_h and B_h. Consider now an operator $L_h \equiv L$ such that L and Λ are connected by any of the relationships $C^0, C^{0,0}, C^{0,1}$, and C^3 from § 1.4.

Lemma 3. Suppose that $L \asymp \Lambda$ and $\delta_0 \Lambda \leq L \leq \delta_1 \Lambda$, $\delta_0 > 0$. Suppose that the operator $B = B^*$ satisfies (4.7) or (4.8) with constant $q \in [0, 1)$ independent of the grid. Then L and B are spectrally equivalent operators as well and

$$\delta_0(1 - q)B \leq L \leq \delta_1(1 + q)B \tag{4.11}$$

(if (4.7) holds) or

$$\delta_0(1 - q)B \leq L \leq \delta_1 B \tag{4.12}$$

(if (4.8) holds).

Proof. The proof is straightforward. $\square$

There is also no problem in analyzing the preservation of the relationships $C^{0,0}$ and $C^{0,1}$ if we observe that, for the case of (4.7),

$$\|w\|_{\Lambda^{-1}}^2 \leq \sigma\|v\|_\Lambda^2 \Rightarrow \|w\|_{B^{-1}}^2 \leq \sigma(1 + q)^2\|v\|_B^2 \tag{4.13}$$

and, for the case of (4.8),

---

[35] Theorem 1 (see [160, 164, 163] and a similar result in [147]) does not assume commutativity of Z and Λ, so it provides the possibility to use modified Richardson methods from § 1.3 with various types of model operators, which in their turn may again use inner iterations (see, e.g., [175]). This observation is especially important in light of some modern multigrid constructions of spectrally equivalent operators (see § 7).

For some investigations of two-stage iterative methods with the model operator B from (4.3) this commutativity is not a restriction (see [154, 155, 258, 496]) due to the simplest form of the region and difference operators under consideration. However, in the first papers devoted to similar methods with inner iterations (see [135, 150]), norms of Z were used in the sense of the original Euclidean space H, and they led to unnecessary complications in the analysis and extra inner iterations.

In section 7, estimates (4.7) and (4.8) will be used extensively (each has its own region of application). However, it is interesting to observe that, if $Z \equiv Z_M$ is the error reduction operator for $M \equiv k$ iterations of the modified Richardson method (see Theorem 1.3.14) with the iteration parameters defined by the set $\{t_i\}$ (see (1.3.21)) or the set $\{t_i^+\}$ (it yields (4.8) with $q \equiv q_k^+$), then $[1 + q_k][1 - q_k]^{-1} = [1 - q_k^+]^{-1}$. Hence, both strategies in choosing iteration parameters in inner Richardson iterations give the same result from the point of view of the ultimate condition number of $B^{-1}\Lambda$.

$$\|w\|_{\Lambda^{-1}}^2 \le \sigma\|v\|_{\Lambda}^2 \;\Rightarrow\; \|v\|_{B^{-1}}^2 \le \sigma\|v\|_B^2. \tag{4.14}$$

Lemma 4. Suppose that L and Λ are connected by relationship C^3 and $\delta_0\Lambda \le L^*\Lambda^{-1}L \le \delta_1\Lambda$, $\delta_0 > 0$. Suppose that the operator $B = B^*$ satisfies (4.7) or (4.8) with constant $q \in [0,1)$ independent of the grid. Then L and B are connected by relationship C^3 as well and, for the case of (4.7),

$$\delta_0(1-q)^2 B \le L^* B^{-1} L \le \delta_1(1+q)^2 B, \tag{4.15}$$

or, for the case of (4.8),

$$\delta_0(1-q)^2 B \le L^* B^{-1} L \le \delta_1 B. \tag{4.16}$$

Proof. Estimates in (4.15) and (4.16) involving δ_1 follow from (4.14). On the other hand, (4.7) or (4.8) implies that $(1-q)\Lambda^{-1} \le B^{-1}$. Hence, $(1-q)\delta_0\Lambda \le (1-q)L^*\Lambda^{-1}L \le L^*B^{-1}L$. This together with the fact that $(1-q)B \le \Lambda$ yields the remaining estimates. $\square$

Lemma 5. Let L and Λ be connected by relationship C^1 and

$$\delta_0\Lambda^2 \le L^* L \le \delta_1\Lambda^2. \tag{4.17}$$

Suppose that the operator $B = B^*$ is defined by (4.3), where

$$Z\Lambda = \Lambda Z, \tag{4.18}$$

$$Z = Z^*, \quad \|Z\| \le q < 1, \tag{4.19}$$

$q \in [0,1)$, and $q \ne q(h)$. Then we have

$$\delta_0(1-q)^2 B^2 \le L^* L \le \delta_1(1+q)^2 B^2; \tag{4.20}$$

if, additionally, $Z \ge 0$ then

$$\delta_0(1-q)^2 B^2 \le L^* L \le \delta_1 B^2. \tag{4.21}$$

Proof. B is a symmetric element of $\mathcal{L}(H)$ as seen from the representation $B^{-1} = \Lambda^{-1} - Z\Lambda^{-1}$ (see the proof of Theorem 1) and the fact that $(Z\Lambda^{-1})^* = \Lambda^{-1}Z^* = \Lambda^{-1}Z = Z\Lambda^{-1}$. Next, we observe that (4.18) and (4.19) imply that $Z \in \mathcal{L}(H(\Lambda))$ is symmetric and, thus, Theorem 1 applies. Hence, we have (4.7), and (4.8) if $Z \ge 0$. Since Λ and B commute, then

$$(1-q)^2 B^2 \le \Lambda^2 \le (1+q)^2 B^2, \tag{4.22}$$

$$(1-q)^2 B^2 \le \Lambda^2 \le B^2, \quad Z \ge 0, \tag{4.23}$$

and (4.17) and (4.22) lead to (4.20). (4.23) yields (4.21). $\square$ [36]

4.4. Nonlinear operators. If we deal with a nonlinear operator L such that problem (1.2.1) with $S = H$ always has a unique solution u and the operators $L\ \Lambda \in \mathcal{L}^+(H)$ are connected by the relationships $C^0(u; r), C^{0,1}(u; r), C^1(u; r), C^2(u; r)$, and $C^3(u; r)$, where $r = \infty$, then there is no significant difference with the above considered linear case. But if $0 < r < \infty$, then we must take into account that our conditions are satisfied on the bounded sets S_r and, therefore, we must consider new balls $S_B(u; r_B)$ with radius r_B chosen so small that $S_B(u; r_B) \subset S_r$. (Similar and more general situations will be investigated in § 8.) We thus confine ourselves to relationship $C^{0,1}(u; r)$.

Theorem 2. Let nonlinear operator L and $\Lambda \in \mathcal{L}^+(H)$ be connected by relationship $C^{0,1}(u; r)$. Suppose that L is such that, at each point $u + z \in S_r \equiv S_\Lambda(u; r)$, for all z', we have $\sigma_0\|z'\|_\Lambda^2 \leq ((L'_{u+z})_s z', z') \leq \sigma_1\|z'\|_\Lambda^2$, $\sigma_0 > 0$, and $\|(L'_{u+z})_a z'\|_{\Lambda^{-1}}^2 \leq \sigma_2\|z'\|_\Lambda^2$. Suppose that the operator $B = B^* > 0$ satisfies (4.7) with constant $q \in [0, 1)$ independent of the grid and $r_B \equiv r(1+q)^{-1/2}$. Then operators L and B are connected by relationship $C^{0,1}(u; r_B)$ and, at each point $u + z \in S_B(u; r_B)$, for all z', we have inequalities $(1 - q)\sigma_0\|z'\|_B^2 \leq ((L'_{u+z})_s z', z') \leq \sigma_1(1+q)\|z'\|_B^2$ and $\|(L'_{u+z})_a z'\|_{B^{-1}}^2 \leq (1+q)^2\sigma_2\|z'\|_B^2$; if (4.8) holds, then $r_B = r$ and $(1+q)$ in the above inequalities may be replaced by 1.

Proof. If $u + z \in S(u; r_B)$, then $\|z\|_\Lambda \leq r_B(1+q)^{1/2} \leq r$, and the conditions of our theorem are satisfied. We can then deal with the given inequalities in much the same way as above. $\square$

4.5. Optimization with respect to inner iterations. Everywhere above, the number $q \in (0, 1)$ was in accordance with the number $M \equiv k$ of inner iterations of the $Z(\Lambda)$-method. The smaller q we wish to obtain, the more inner iterations we perform at each outer iteration. This results in greater computational work at each outer iteration but the number of the required outer iterations [385] becomes definitely smaller by virtue of the better estimates for the constants we used in inequalities like (4.7), (4.13) (they become better for smaller q). These two opposing trends indicate that an optimal value of q and M must exist for fixed desired accuracy. It is also clear that, for complicated L_h or very small h, we should decrease the number of outer iterations and thus use large enough M. A reasonable choice for M might produce a very noticeable effect in practical computations. [37]

---

[36] The case of relationship C^2 is much the same.

[37] It is possible to carry out a theoretical analysis of the problem of minimizing the required computational work, but probably this is not worth the trouble because numerical experiments usually yield the desired M easily enough when we observe the appropriate norm of the residuals. For example, for iterative methods with the convergence in the Euclidean space $H(B)$, we should examine the sequence $\{\|r^n\|_{B^{-1}}\}$ generated by

4.6. Nonnegative operators. Consider now a generalization of the two-stage model operator (see (4.3)) for the case of symmetric and non-negative operator Λ with dim Ker $\Lambda > 0$ and dim Im $\Lambda > 0$ (it will be an essential part of the multigrid construction of asymptotically optimal preconditioners on which we will elaborate in § 7). We use the notation $V_0 \equiv$ Ker Λ, $V_1 \equiv$ Im $\Lambda = V_0^\perp$. The restrictions of Λ to these its invariant spaces are denoted by $\Lambda \mid_{V_0}$ and $\Lambda \mid_{V_1} \equiv \Lambda_1$, respectively (the same applies to restrictions of other operators; e.g., I_0 refers to the restriction of the identity operator). Note that $\Lambda \mid_{V_0} = 0$ and $\Lambda_1 > 0$. By $\Lambda^\dagger$ we denote the pseudoinverse operator for Λ (see § 1.3). Recall that $\Lambda^\dagger f = \Lambda_1^{-1} f_1$, where f_1 is the orthogonal projection of f onto V_1; if $A = A^* \geq 0$ and

$$\sigma_0 A \leq \Lambda \leq \sigma_1 A, \ \sigma_0 > 0, \tag{4.24}$$

then Ker $A = V_0$, Im $A = V_1$, and the operator $D \equiv A^\dagger \Lambda$ is symmetric as an element of $\mathcal{L}(H(\Lambda))$, that is $(\Lambda D u, v) = (\Lambda u, D v)$, it has invariant subspaces V_0 and V_1, $D \mid_{V_0} = 0$, and sp $D \mid_{V_1} 0 \in [\sigma_0, \sigma_1]$ (we agree to use the term Euclidean space for $\mathcal{L}(H(\Lambda))$, keeping in mind that $(\Lambda u, v)$ in fact defines only a semiinner product).

Lemma 7. Let $g \in V_1$. For solving $\Lambda v = g, v \in V_1$, consider the iterations

$$v_{n+1} = v_n - \tau_n A^\dagger (\Lambda v_n - g), \tag{4.25}$$

with the set $\{\tau_n\}$ defined either by the set $\{t_n\}$ or by the set $\{t_n^+\}$ (see Theorem (1.3.14), with constants $\delta_r \equiv \sigma_r, r = 0, 1$ (see (4.24)). Define q_k and q_k^+ as in Theorem (1.3.14). Then the error reduction operator $Z \equiv Z_k \equiv (I - \tau_0 D) \ldots (I - \tau_{k-1} D)$ is symmetric as an element of $\mathcal{L}(H(\Lambda))$ and it has invariant subspaces V_0 and V_1. Moreover, $Z \mid_{V_0} = I_0$ and $\|Z \mid_{V_1} \|_{\Lambda_1} \leq q_k$ for $\{t_n\}$ and $0 \leq Z \mid_{V_1} \leq q_k^+ I_1$ for $\{t_n^+\}$.

Proof. It suffices to combine the statements of Theorem 1.3.14 and Lemma 1.3.5 as was done in proving Theorem 1.3.21. □

Now we are in a position to define B by requiring that it has invariant subspaces V_0 and V_1 and its restrictions to them are

$$B \mid_{V_0} \equiv \Lambda \mid_{V_0} \text{ and } B \mid_{V_1} \equiv \Lambda_1 (I_1 - Z \mid_{V_1})^{-1}. \tag{4.26}$$

---

one outer iteration (for method (1.3.1)) or several (for the modified Richardson method or the modified conjugate gradient method). For relationship C^1, we should deal with $\{\|r^n\|\}$. We note that, in many practical problems associated with elliptic boundary value problems of second order, typical choices for M are 2 or 4 and, for the fourth order problems, such choices are 4 or 8. These choices have been used for strongly nonlinear systems involving $N \approx 5 \times 10^4$ unknowns (see § 6.5). Moreover, a variable number M_n of inner iterations may be used (see [385]).

Theorem 3. *Let the conditions of Lemma 7 be satisfied and* (4.26) *hold. Then $B = B^* \geq 0$ as an element of $\mathcal{L}(H)$, and*

$$(1 - q_k)B \leq \Lambda \leq (1 + q_k)B \text{ for } \{t_n\}, \qquad (4.27)$$

$$(1 - q_k^+)B \leq \Lambda \leq B \text{ for } \{t_n^+\}. \qquad (4.28)$$

Proof. Since $B|_{V_0} = \Lambda|_{V_0}$, it suffices to verify the desired inequalities for B_1 and Λ_1. But $\Lambda_1 = B_1(I_1 - Z|_{V_1})$. By virtue the inequalities for $Z|_{V_1}$ (see Lemma 7) and Theorem 1, we can complete the proof. □ [38]

§ 5. Cutting methods (domain decomposition methods)

5.1. Connection with the block elimination methods; basic computational algorithms. We begin by considering a plane region $\bar{Q}$ such that by, drawing a number of vertical and horizontal cutting lines, it can be partitioned into a union of several basic rectangles (basic blocks, superelements) denoted by $\bar{Q}_r, r \in [1, r_0]$ (actually, we draw several line segments belonging to $\bar{Q}$).

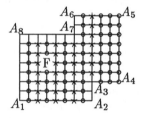

Figure 1. Partition of the region into rectangular blocks.

An example of such a partition of $\bar{Q}$ with Lipschitz boundary is given in Figure 1, where 4 vertical cutting segments are indicated by denoting the nodes belonging to them by $\star$. We stress that domain Q may be like that considered in Subsection 2.1.6 and its boundary Γ may contain slits (this is the case for Figure 2, where double nodes belonging to the vertical slit are marked by □).

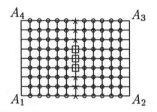

Figure 2. Partition of the region with vertical slit.

---

[38] Iterations (4.25) can be carried out in the space V_1. Lemmas 1 and 2 apply.

Moreover, Q may have a more general boundary, like the Q depicted in Figure 3, and be partitioned in a union of rectangles and triangles.

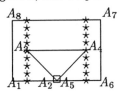

Figure 3. Partition of the region with non-Lipschitz boundary into rectangular and triangular blocks.

Assume now that $\Gamma_0 \subset \partial Q$ consists of several sides of the chosen blocks and that $|\Gamma_0|_{(1)} > 0$. Then for the Hilbert space $G \equiv W_2^1(Q; \Gamma_0)$ we can use the norm $\|u\| \equiv |u|_{1,Q}$. We consider the bilinear form

$$b(u, v) \equiv (a(x)\nabla u, \nabla v)_{0,Q} + (\sigma(x), uv)_{0,\Gamma_1}, \tag{5.1}$$

where $b(u, u) \equiv \|u\|_L^2 \geq \kappa \|u\|^2, \ \forall u \in G, \ \kappa > 0.$ [39] If we now construct the projective-grid system $L_h u = \mathbf{f}$, the resulting operator $L_h \in \mathcal{L}^+(H)$ might be rather complicated (see § 0.2 and § 2). Instead, we may deal with its simplified version $A_h \equiv A \asymp L_h$ (we specify this operator below). We describe *cutting method (domain decomposition method)* for system

$$A\mathbf{u} = \mathbf{f}, \ A \in \mathcal{L}^+(H), \tag{5.2}$$

where $\mathbf{u} \equiv \{u_i : P_i \in Q_h\} \in H$. Next we specify a set of $C_h \subset Q_h$ of (elementary) nodes belonging to the cutting lines segments; we allow some to pass along some parts of Γ_1 in order to decrease the number of segments and to simplify the structure of the resulting systems. [40] The chosen set C_h defines the corresponding splitting of the Euclidean space H. Thus, for each vector $\mathbf{u} \in H$, we write $\mathbf{u} \equiv [\mathbf{u_1}, \mathbf{u_2}]^T$, where $\mathbf{u_1} \equiv \{u_i : P_i \in Q_h \backslash C_h\} \in H_1$ and $\mathbf{u_2} \equiv \{u_i : P_i \in C_h\} \in H_2$. This implies that $H = H_1 \times H_2$, where

$$\dim H \asymp \frac{1}{h^2}, \ \dim H_1 \equiv n_1 \asymp \frac{1}{h^2}, \ \dim H_2 \equiv n_2 \asymp \frac{1}{h}. \tag{5.3}$$

---

[39] It is associated with an original problem, considered as an operator equation in the Hilbert space $G \equiv W_2^1(Q; \Gamma_0)$. In general, the given partition depends not only on the geometry of the domain, but also on the geometry of Γ_0 and the coefficients $a(x)$ and $\sigma(x)$ in (5.1). More precisely, we assume that $a(x)$ on each block Q_r is a positive constant denoted by a_r and that $\sigma(x)$ is a nonnegative constant denoted by $\sigma_{r,l}$ on each side of Q_r belonging to Γ_1 but vanishes on each horizontal side of Q_r belonging to Γ_1. For simplicity, we use the simplest triangulation associated with square grid with the mesh size h (we assume that such a triangulation exists) and define the standard subspace $\hat{G}_h \subset G$ of piecewise linear functions $\hat{u}$. The set Q_h of grid nodes P_i (elementary nodes if $\Gamma_1 \equiv \Gamma \backslash \Gamma_0$ contains parts of the slits (see § 2.2)) corresponds to the set of basis functions $\hat{\psi}_i(x)$ of the space $\hat{G}_h$.

[40] For Figure 1, nodes in C_h are marked by $\star$ and two cutting line segments contain parts of Γ_1.

Now the operator A has the block structure (1.5.1) and we suppose that systems with $A_{1,1}$ are easily solved. Then we can apply the block elimination method (see § 1.5) and work with the system $S_2 u_2 = g_2$, where $S_2 \equiv A/A_{1,1} \equiv S_2(A) \in \mathcal{L}^+(H_2)$ (see (1.5.3) and (1.5.4)). For finding u_2, in contrast to § 1.5, we apply iterative methods of type

$$B_2(\mathbf{u}_2^{n+1} - \mathbf{u}_2^n) = -\tau_n(S_2\mathbf{u}_2^n - \mathbf{g}_2), \tag{5.4}$$

and, most importantly, we do not construct the Schur matrix S_2 in an explicit way. More precisely, given a vector $\mathbf{u}_2^n$, we can obtain the desired residual $\mathbf{r}_2^n \equiv S_2\mathbf{u}_n^n - \mathbf{g}_2$ in four steps in which we evaluate vectors $\mathbf{v}_1^n, \mathbf{w}_1^n$, $A_{2,1}\mathbf{w}_1^n$, and $\mathbf{r}_2^n = S_2\mathbf{u}_2^n - g_2$, where

$$\mathbf{v}_1^n = A_{1,2}\mathbf{u}_2^n, \quad \mathbf{w}_1^n = A_{1,1}^{-1}\mathbf{v}_1^n, \quad \text{and} \quad S_2\mathbf{u}_2^n \equiv A_{2,1}\mathbf{w}_1^n. \tag{5.5}$$

We emphasize that, in solving systems with $A_{1,1}$, we actually deal with each separate rectangle Q_r, and for the corresponding subsystem in Q_r, we may even apply algorithms for solving partial problems (see § 1), requiring computational work

$$W = O(\frac{|\ln h|^k}{h}), \quad k = 1 \text{ or } k = 2, \tag{5.6}$$

where $k = 1$ if only vertical cuttings lines are used.

5.2. Structure of the grid operators. We specify below the structure of L_h and $A_h \equiv A$ necessary for subsequent estimates. We have

$$L_h \equiv [b(\hat{\psi}_j; \hat{\psi}_i)]. \tag{5.7}$$

First, to find $L_h\mathbf{u}_i$, we consider the case when the node $P_0 \equiv P_{0,0}$ is the center of the square $[-h, h] \times [-h, h]$ containing no slits. This node may be a vertex of four cells $\Pi_1 \equiv [0, h] \times [0, h]$, $\Pi_2 \equiv [-h, 0] \times [0, h]$, $\Pi_3 \equiv [-h, 0] \times [-h, 0]$, and $\Pi_4 \equiv [0, h] \times [-h, 0]$. If we find

$$(M\mathbf{u})_{0,0} \equiv b_M(\hat{u}; \hat{\psi}_{0,0}), \tag{5.8}$$

$$b_M(\hat{u}, \hat{v}) \equiv \sum_{r=1}^{4}(a_r\nabla\hat{u}, \nabla\hat{v})_{0,\Pi_r} + X_\sigma, \tag{5.9}$$

$$X_\sigma \equiv \int_0^h \sigma_1\hat{u}\hat{v}dx_1 + \int_0^h \sigma_2\hat{u}\hat{v}dx_2 + \int_{-h}^0 \sigma_3\hat{u}\hat{v}dx_1 + \int_{-h}^0 \sigma_4\hat{u}\hat{v}dx_2, \tag{5.10}$$

then cases when one, two, or three indicated cells belong to the complement of Q present no problems. Also if the square contains slits (they may be only on the coordinate axes), it suffices to consider only the case when P_0 is the end point of the double vertical slit (if P_0 is an elementary node on a certain side of a double slit, we can ignore the remaining side of the slit and make use of formulas for (5.9)). For such a P_0, we must deal with the functions $\hat{u}$ having two limit values $u_{0,1,l}$ and $u_{0,1,r}$ at the double node $[0, h]$ and, hence, we must find

$$X_0 \equiv \sum_{r=1}^{4}(a_r\nabla\hat{u}, \nabla\psi_{0,0})_{0,\Pi_r} + \int_0^h \sigma_{2,l}\hat{u}_l\psi_{0,0}dx_2 + X'_\sigma, \qquad (5.11)$$

where $X'_\sigma \equiv \int_0^h \sigma_{2,r}\hat{u}_r\psi_{0,0}dx_2 + \int_0^h \sigma_2\hat{u}\hat{v}dx_2 + \int_{-h}^0 \sigma_3\hat{u}\hat{v}dx_1 + \int_{-h}^0 \sigma_4\hat{u}\hat{v}dx_2$, $\hat{u}_l$ and $\hat{u}_r$ refer to corresponding limit functions on the left and on the right of the slit. The same approach as in § 0.2 shows that $2(M\mathbf{u})_i$ is just

$$h[-(a_1+a_4)\partial_1 u + (a_2+a_3)\bar{\partial}_1 u - (a_1+a_2)\partial_2 u + (a_3+a_4)\bar{\partial}_2 u]_i + I_\sigma, \quad (5.12)$$

where $I_\sigma \equiv h/3[\sigma_1(2u+I_1u)+\sigma_2(2u+I_2u)+\sigma_3(2u+I_{-1}u)+\sigma_4(2u+I_{-2}u)]_i$ (some σ_k may vanish). Now, for $Y\sigma_i \equiv (\sigma_1+\sigma_2+\sigma_3+\sigma_4)/2$, we may define Au_i by (5.11) if we replace I_σ by $I'_\sigma \equiv Y\sigma_i u_i$. For (5.10), we have $X_0 = \frac{h}{2}[-(a_1+a_4)\partial_1 u + (a_2+a_3)\bar{\partial}_1 u - \frac{a_1}{2}\frac{u_{0,1,r}-u_{0,0}}{h} - \frac{a_2}{2}\frac{u_{0,1,l}-u_{0,0}}{h} + (a_3+a_4)\bar{\partial}_2 u]_i + Z\sigma_i$, $6Z\sigma_i \equiv h[\sigma_{2,l}(u_{0,1,r}+u_{0,0}) + \sigma_{2,r}(u_{0,1,l}+u_{0,0}) + \sigma_2(2u+I_2u)+\sigma_3(2u+I_{-1}u)+\sigma_4(2u+I_{-2}u)]_i$; in this case, we take $Y\sigma_i \equiv (\sigma_1 + (\sigma_{2,l} + \sigma_{2,r}) + \sigma_3 + \sigma_4)/2$ and define Au_i by replacing $Z\sigma_i$ in (5.12) by $\sigma_i u_i$. Thus, the presence of a slit results in the same considerations of separate cells; taking into account the fact that cells Π_1 and Π_2 generate different limit values of a permissible function $\hat{u}$ on two edges of the slit, we can determine the structure of L_h and A_h by assembling the matrices corresponding to different elementary blocks.

Lemma 1. $A_h \in \mathcal{L}^+(H)$ *and*

$$1/3A_h \leq L_h \leq A_h. \qquad (5.13)$$

Proof. It suffices to compare $(A_h\mathbf{u}, \mathbf{u})$ and $(L_h\mathbf{u}, \mathbf{u})$. We have

$$(A_h\mathbf{u}, \mathbf{u}) = \sum_{r=1}^{r_0} a_r|\hat{u}|^2_{1,Q_r} + h\sum_{x_i\in\Gamma_h} Y\sigma u_i^2. \qquad (5.14)$$

For $(L_h\mathbf{u}, \mathbf{u})$ we have a similar expression. Hence, its terms containing the coefficients σ along one of the sides of Q_r have the form indicated in § 0.2; the matrix of the resulting quadratic form is proportional to a tri-diagonal matrix and is easily estimated from above and below. □

5.3. Estimates of the rate of convergence. Here, κ and I_2 refer to certain positive constants and the identity operator in H_2, respectively.

Theorem 1. The above operator A is such that

$$\delta_0 I_2 \leq S_2 \leq \delta_1 I_2, \tag{5.15}$$

where $\delta_0 \equiv h\kappa_0$, $\kappa_0 > 0$, $\delta_1 \asymp 1$.

Proof. Note that

$$(A_{2,2}\mathbf{u}_2, \mathbf{u}_2) = I_0(\mathbf{u}_2) + h \sum_{x_i \in \Gamma_h} Y\sigma_i(\hat{u}_2)_i^2, \tag{5.16}$$

$$I_0(\mathbf{u}_2) \equiv \sum_{r=1}^{r_0} a_r |\hat{w}|_{1,Q_r}^2, \tag{5.17}$$

and $\hat{w}$ takes the same values as $\mathbf{u}_2$ at elementary nodes in C_h and vanishes at other nodes. Hence, $|\hat{w}|_{1,Q_r}^2$ in (5.17) contains only integrals over the triangles having a vertex in C_h and these integrals are easily found. For example, terms in $I_0(\mathbf{w})$ involving $u_2(P)$ for Figures 4 and 5 are $1/2\{(h^2[(a_3+a_4)(\bar{\partial}_2 u_2)^2 + (a_1+a_4)(\partial_1 u_2)^2] + [a_1+a_3+2h/3(\sigma_1+\sigma_2)]u_2^2\}_P$ and $1/2\{(a_2+a_1)[h^2(\partial_2 u_2)^2 + u_2^2] + 2h/3(\sigma_1+\sigma_3)u_2^2]\}_P$, respectively (recall that the points in C_h are marked by $\star$).

Figure 4. Vicinity of a corner node.

Figure 5. Vicinity of a node on a horizontal side.

It is clear that (5.16) yields $A_{2,2} \asymp I_2$. Since $(A\mathbf{u},\mathbf{u}) \geq \kappa_3\|\hat{u}\|^2$ and (0.1.25) holds with $q = 2$, we have

$$(A\mathbf{u}, \mathbf{u}) \geq \kappa_4 |\hat{u}|_{0,C}^2, \tag{5.18}$$

$$(A\mathbf{u}, \mathbf{u}) \geq \kappa_5 h \sum_{x_i \in C_h} u_i^2, \tag{5.19}$$

$$(A\mathbf{u}, \mathbf{u}) \geq \kappa_6 h(A_{2,2}\mathbf{u}_2, \mathbf{u}_2), \quad \forall \mathbf{u} \in H, \tag{5.20}$$

$$\min_{\mathbf{u}_1 \in H_1} (A\mathbf{u}, \mathbf{u}) = (S_2\mathbf{u}_2, \mathbf{u}_2) \geq \kappa_6 h(A_{2,2}\mathbf{u}_2, \mathbf{u}_2), \ \forall \mathbf{u}_2 \in H_2 \qquad (5.21)$$

(see (1.5.5)). Therefore, $S_2 \geq \kappa_6 h A_{2,2}$ from which the lower bound in (5.15) follows. The upper bound is evident. $\square$

Theorem 2. *Suppose that the function* $\sigma(x)$ *in* (5.1) *vanishes and at least one of the sides of each* Q_r *belongs to* Γ_0. *Then*

$$\delta_0' A_{2,2} \leq S_2 \leq A_{2,2}, \qquad (5.22)$$

where $\delta_0' \equiv h\kappa_6'$ *and* κ_6' *is independent of the constants* a_r, $r \in [1, r_0]$.

Proof. It suffices to prove the estimate involving δ_0'. First, we strengthen the estimate for $I_0(\mathbf{u}_2)$ in (5.16). This yields

$$(A_{2,2}\mathbf{u}_2, \mathbf{u}_2) = I_0(\mathbf{u}_2) \leq \kappa_7 h \sum_{r=1}^{r_0} a_r \sum_{P_i \in \partial Q_r \cap C_h} u_i^2. \qquad (5.23)$$

Next, the minimum in (5.21) corresponds to $\mathbf{u}_1 = A_{1,1}^{-1} A_{1,2} \mathbf{u}_2$ (see Lemma 1.5.1). We denote the corresponding element of $\hat{G}_h$ by $\hat{u}$ and write

$$(S_2\mathbf{u}_2, \mathbf{u}_2) = b(\hat{u}; \hat{u}) = \sum_{r=1}^{r_0} a_r |\hat{u}|_{1,Q_r}^2. \qquad (5.24)$$

Now we make use of estimate

$$|\hat{u}|_{1,Q_r}^2 \geq \kappa_r^* |\hat{u}|_{0,\partial Q_r \cap C}^2, \ r \in [1, r_0]. \qquad (5.25)$$

Thus, (5.24) and (5.25) imply that

$$(S_2\mathbf{u}_2, \mathbf{u}_2) \geq \kappa_8 h \sum_{r=1}^{r_0} a_r \sum_{P_i \in \partial Q_r \cap C_h} u_i^2. \qquad (5.26)$$

Combining (5.23) and (5.26) leads to (5.22). $\square$[41]

---

[41] Theorems 1 and 2 imply that in (5.4) we can have either $B_2 = I_2$ or $B_2 = A_{2,2}$; the latter can be replaced by the model operator, which is spectrally equivalent to $A_{2,2}$ and corresponds to a diagonal matrix. As an example of the application of Theorem 2, consider Q depicted in Figure 6 and $\{[A_1 A_6] \cup [A_4 A_5] \cup [A_2 A_3]\} \subset \Gamma_0$. Then $\kappa_5 = [a_1/l_1 + a_2/l_2][3(a_1 + a_2)]^{-1} \in [1/6 \min\{1/l_1; 1/l_2\}, 1/3(1/l_1 + 1/l_2)]$. Thus, the simple and effective estimation of the bounds is provided by elimination of the nodes not belonging to the cutting lines and use of the estimates (5.18) (or (5.25)) for the traces (in the sense of the space $L_2(C)$) of the functions in $W_2^1(Q)$ (or $W_2^1(Q_r)$) suggested in [186, 187] in contrast to an approach used in [321] for the case of the Dirichlet conditions. Currently, more involved traces in the sense of the space $W_2^{1/2}(C)$, are used (see, e.g., [84, 105, 144, 346, 384, 455, 457]) as suggested for the first time by Dryja in [141].

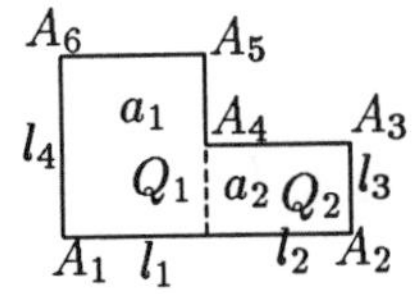

Figure 6. Example of a simple partitioned region.

5.4. Estimates of the required computational work.

Lemma 2. *Let $\sigma(x)$ in (5.1) vanish at all horizontal segments of Γ_1. Suppose also that in iterations (5.4) for (5.2), the operator B_2 is such that solution of each system $B_2 v_2 = g_2$ requires computational work $W(B_2) \leq \kappa_9 N_2 \ln N_2$. Suppose, finally, that all cutting lines are vertical. Then each iteration (5.4) requires computational work $W_2 \leq \kappa_{10} h^{-1} |\ln h|$.*

Proof. It suffices to estimate the computational work in evaluating the vectors indicated in (5.5). Observe that, even for the general case of cutting lines, the matrix $A_{2,1}$ is such that the vector $A_{2,1} \mathbf{w}_1^n$ is completely defined by the components of $\mathbf{w}_1^n$ at nodes next to C_h (we denote this set by C_h'). Also, the matrix $A_{1,2}$ is such that the vector $\mathbf{v}_1^n = A_{1,2} \mathbf{u}_2^n$ vanishes at nodes not belonging to C_h'. Therefore, we need not determine the whole vectors $\mathbf{w}_1^n$ and $\mathbf{v}_1^n$, but only their components corresponding to nodes in C_h'. Thus, the evaluation of $\mathbf{v}_1^n$ is no problem, and only the evaluation of $\mathbf{w}_1^n$ deserves special attention. This evaluation is actually a problem of finding the indicated components of the solution of the system $A_{1,1} \mathbf{w}_1^n = \mathbf{v}_1^n$, and algorithms for solving such partial problems were described in § 1. The case of vertical cutting lines and $\sigma > 0$ only on the vertical sides is very simple and leads to (5.6) with $k = 1$. $\square$ [42]

Now we estimate the effects of approximating the solution of system (5.2). Let $\mathbf{u} \equiv [\mathbf{u}_1, \mathbf{u}_2]^T$, $\mathbf{u}' \equiv [\mathbf{u}'_1, \mathbf{u}'_2]^T$, $\mathbf{z} \equiv \mathbf{u} - \mathbf{u}' \equiv [\mathbf{z}_1, \mathbf{z}_2]^T$.

Lemma 3. Suppose we consider system (5.2) with an operator $A \in \mathcal{L}^+(H)$ having block structure (5.11). Suppose also that the vector $\mathbf{u}_2$ from (1.5.3) is approximated by a vector $\mathbf{u}'_2$ such that $\|\mathbf{z}'_2\|_{A_{2,2}} \leq \varepsilon_2$, where $\mathbf{z}'_2 \equiv \mathbf{u}_2 - \mathbf{u}'_2$. Suppose, finally, that $\mathbf{u}'_1$ is a vector defined by the system

$$A_{1,1} \mathbf{u}'_1 + A_{1,2} \mathbf{u}'_2 = \mathbf{f}_1 + \xi_1, \tag{5.27}$$

---

[42] In this lemma, we can use B_2 such that it becomes a diagonal matrix under a suitable ordering of the nodes; even the choice $B_2 = I_2$ is possible if the constants a_r do not differ too strongly from each other. The choice $B_2 = A_{2,2}$ also leads to relatively simple systems, even for the general case; these systems can be effectively solved by block elimination (see § 1.5) if we partition each vector $\mathbf{u}_2$ as $\mathbf{u}_2 = [\mathbf{u}_{1,2}, \mathbf{u}_{2,2}]^T$, where $\mathbf{u}_{2,2}$ contains all values of $\mathbf{u}_2$ at cross points of cutting lines. For the general case of partial problems, we can obtain (5.6) with $k = 2$ or estimates $W = O(h^{-2})$ (see § 1). Then the choice of the operator B_2 becomes more important, and much attention has been paid to the possibility of using a model operator B_2 such that $S_2 \asymp B_2$ or are nearly spectrally equivalent (see, e.g., [4, 85, 141, 384]).

where $\|\zeta_1\|_{A_{1,1}^{-1}} \leq \varepsilon_1$. *Then*

$$\|z\|_A^2 \leq 2[(\varepsilon_1 + \varepsilon_2)^2 + \varepsilon_2^2]. \tag{5.28}$$

Proof. We have $\|z\|_A^2 \leq 2[\|z_1\|_{A_{1,1}}^2 + \|z_2\|_{A_{2,2}}^2]$, $A_{1,1}z_1 = -A_{1,2}z_2 + \zeta_1$, and $\|z_1\|_{A_{1,1}} \leq \|A_{1,2}z_2\|_{A_{1,1}} + \varepsilon_1 = \|w_1\|_{A_{1,1}} + \varepsilon_1$, where $w_1 \equiv A_{1,1}^{-1}A_{1,2}z_2$ and $(A_{1,1}w_1, w_1) = (A_{1,2}z_2, w_1)$. Hence, $2\|w_1\|_{A_{1,1}}^2 \leq \|z_2\|_{A_{2,2}}^2 + \|w_1\|_{A_{1,1}}^2$, $\|w_1\|_{A_{1,1}} \leq \varepsilon_2$, and $\|z_1\|_{A_{1,1}} \leq \varepsilon_1 + \varepsilon_2$, which yields (5.28). $\square$

Theorem 3. *In the modified Richardson method* (5.4), *let* B_2 *be* I_2 *or* $A_{2,2}$ *so that respective estimates* (5.15) *or* (5.22) *hold. Suppose that* $\|u_2^0 - u_2\| \leq \kappa_{11}$ *and that the number* k *of iterations is such that* $q_k\kappa_{11} \leq \varepsilon\kappa_{12}$ *(see* (1.3.23)*). Let* u_1' *satisfy* (5.27), *where* $\|\zeta_1\|_{A_{1,1}^{-1}} \leq \varepsilon_1 \equiv \kappa_{13}\varepsilon$. *Suppose, finally, that the conditions of Lemma 2 are satisfied. Then* $\|u' - u\|_A \leq \varepsilon[2(\kappa_{12} + \kappa_{13})^2 + 2\kappa_{12}^2]^{1/2}$ *and total computational work is* $W = O(h^{-2} + h^{-3/2}|\ln h \ln \varepsilon|)$.

Proof. The required number of iterations is $k = O(h^{-1/2}|\ln \varepsilon|)$. This, together with Lemma 2, yields the second term in the desired estimate for W. The first term corresponds to the additional work in solving systems with $A_{1,1}$, provided we use the best direct methods considered in § 1. The accuracy is estimated as in (5.28). $\square$ [43]

5.5. Possible generalizations for other types of operators. [44]
For d-dimensional problems, we can obtain estimate

$$W = O(\frac{|\ln h|}{h^d}) + O(\frac{|\ln^\alpha h||\ln \varepsilon|}{h^{d-1/2}}), \tag{5.29}$$

where $\alpha = 1$ if the cutting $(d-1)$-dimensional planes are parallel. Following [186], we consider now the case $L_h = A_h \geq 0$ and $S_2 \geq 0$ which includes, e.g., the case of $\Gamma_0 = \emptyset$ or the periodic conditions.

---

[43] For the general case, we can obtain $W = O(|\ln h|h^{-2} + h^{-3/2}\ln^2 h|\ln \varepsilon|)$. Also, we can use the modified conjugate gradient method.

[44] For composite grids, we use uniform grids on elementary blocks Q_r with mesh size $h_r, r \in [1, r_0]$. Then the main difference from the uniform grid is related to the presence of seminodes; the form of A was indicated in § 0.2; changes in $A_{2,2}$ for the vertices lying on the common boundary of the blocks with the old and refined grids are not significant because (5.16) remains valid. If we want to obtain estimates as in Theorem 1 that are independent of r_0, then we must assume that all refinement ratios are uniformly bounded by t^* and we must replace in (5.15) h by its local analog and I_2 by a suitable diagonal matrix D_2 that is dependent on the grid. Also, in Lemma 2 we must use the estimate $W_2 \leq \kappa_{10} \sum_{r=1}^{r_0} l_r/h_r |\ln l_r/h_r|$, where l_r refers to the length of the vertical side of $Q_r, r \in [1, r_0]$. Generalizations are also possible for problems on two-dimensional manifolds with complex geometry (e.g., a finite number of polygons lying on different planes that have a common side, the surface of a parallelepiped, and so on).

Theorem 4. *Let bilinear form* (5.1) *with* $\Gamma_1 = \Gamma$ *and* $\sigma(x) = 0$ *be considered for all* $u \in G, v \in G$, *where* $G \equiv W_2^1(Q)$. *Suppose that* A_h *is defined by* (5.14). *Then, for all* $\mathbf{u}_2$ *such that* $A_{2,2}\mathbf{u}_2 \perp \mathrm{Ker}\, S_2$, *we have*

$$\kappa_{14}h(A_{2,2}\mathbf{u}_2, \mathbf{u}_2) \leq (S_2\mathbf{u}_2, \mathbf{u}_2) \leq (A_{2,2}\mathbf{u}_2, \mathbf{u}_2). \qquad (5.30)$$

Proof. Let $\mathbf{u} \in H$ be such that $A_{2,2}\mathbf{u}_2 \perp \mathrm{Ker}\, S_2$ and $A_{1,1}\mathbf{u}_1 + A_{1,2}\mathbf{u}_2 = 0$. Then (5.24) holds, and it suffices to prove that $|\hat{u}|_{1,Q}^2 \geq \kappa_{15}\|\hat{u}\|_{1,Q}^2$. Suppose that this is not true and, so there exists a sequence $\{\mathbf{u}^n\}$ such that $\|\hat{u}^n\|_{1,Q}^2 = 1$ and $|\hat{u}^n|_{1,Q}^2 \equiv q_n \leq 1/n$. Then $|\hat{u}^n|_{0,Q}^2 = 1 - q_n$. Now, by virtue of properties of the space $W_2^1(Q)$, we can select a subsequence denoted in the same way and such that it is weakly convergent in $W_2^1(Q)$ and strongly convergent in $L_2(Q)$ and $L_2(C)$ to a function u. This implies that $\|u\|_{1,Q}^2 \leq 1$ and $|u|_{1,Q}^2 = 1$. Hence, $|u|_{1,Q}^2 = 0$ and $u = \mathrm{const}$. Now we rewrite the condition $A_{2,2}\mathbf{u^n}_2 \perp \mathrm{Ker}\, S_2$ for the elements of the indicated subsequence in the form $h(A_{2,2}\mathbf{u^n}_2, 1)_{H_2} = 0$, which is equivalent to $(\hat{u}^n, 1)_{L_2(C)} + g_n(u^n) = 0$. It is easy to verify that $g_n(u^n)$ tends to zero as $n \to \infty$. This implies that our limit function u satisfies $(u, 1)_{L_2(C)} = 0$ and, thus, $u = 0$. This contradicts the established fact $|u|_{1,Q}^2 = 1$. Hence, our supposition was false and (5.30) must hold. $\square$

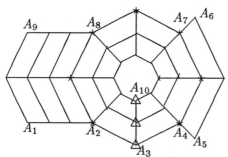

Figure 7. Region with two kinds of grid.

Periodic conditions may arise from using special grids associated in a sense with polar coordinates. For example, the grid in Figure 7 is topologically equivalent to the square grids in Figures 8 and 9. In Figure 8, the points denoted by $\triangle$ and belonging to corresponding vertical lines must be identified (they correspond to the points marked by $\triangle$ in Figure 7), so the planar square grid in Figure 8 is merely a representation of the square grid for the two-dimensional manifold in Figure 9 (in a similar sense other planar square grids in the sequel should be understood; Figure 8 is a development of the lateral surface in Figure 9).

Consider a model operator (see § 0.4) associated with the grid in Figure 7 (8). For solving the resulting systems, we can again apply the above algorithms. For example, if the type of boundary conditions imposed on the outer boundary of the original domain

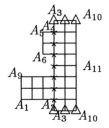

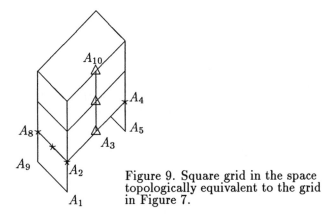

Figure 8. Square grid in the plane equivalent to the grid in Figure 7.

Figure 9. Square grid in the space topologically equivalent to the grid in Figure 7.

does not change, then we can choose the cutting line in Figure 7 corresponding to the vertical line passing through A_2 in Figure 8 and containing A_3 and the nodes marked by $\star$ (of course, use of the above iterative methods is important only when each indicated cell is actually a block containing, say, m^2 cells of the refined grid and m is large). In much the same way, we can deal with a union of several polar-type grids as depicted in Figures 10 and 11.

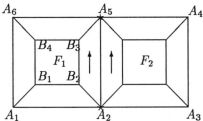

Figure 10. A triply-connected domain with two polar grids.

We emphasize that in Figure 11 the model region is depicted with two double slits that pass along $[A_1 A_2]$ ($[A_3 A_2]$) and $[A_5 A_1]$ ($[A_5 A_3]$); the points denoted by $\square$ are double nodes; each corresponds to two different points in Figure 10). This implies that the grid in Figure 10 is equivalent in a rigorous sense to the square grid in Figure 12.

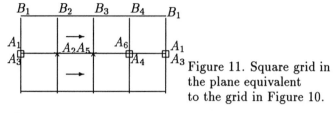

Figure 11. Square grid in
the plane equivalent
to the grid in Figure 10.

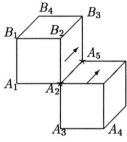

Figure 12. Square grid in the space topologically
equivalent to the grid in Figure 10.

An unconventionally structured grid operator arises from the grid in Figure 13.

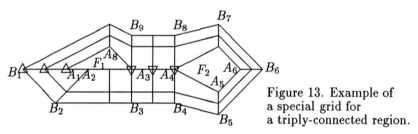

Figure 13. Example of
a special grid for
a triply-connected region.

This grid is topologically equivalent to the square grid in Figure 14, where points denoted by $\triangle$ and lying on the corresponding vertical lines must be identified (they correspond to the points on $[B_1, A_1]$ marked by $\triangle$ in Figure 13).

Moreover, the 6 nodes marked by $\triangledown$ in Figure 14 are actually a double set of the 3 nodes marked by the same symbol in Figure 13, and the equations, e.g., for the two points marked by A_3 in Figure 14 must be identical and include values of the desired solution at the nodes which are neighbors to both points A_3. The grid in Figure 13 is topologically equivalent in the rigorous sense to the square grid in Figure 15 if we use a transformation of the closed polygone $B_3 B_4 A_4 B_8 B_9 A_5$ in Figure 13 into a double square $(B_3 B_4 A_4 A_3$ and $B_5 B_8 A_4 A_3)$ in Figure 15. More precisely, in this double square, all points but those belonging to the side $[A_3 A_4]$ are considered as pairs of points belonging to two different limit planes (such a transformation is used when we fold a sheet of paper along its horizontal axis of symmetry and pass to the limit); all 6 double nodes in Figure 15 are marked by $\square$; each corresponds to two different nodes in Figure 13). If the type of boundary conditions imposed on the outer boundary of the original domain in Figure 13

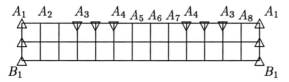

Figure 14. Square grid equivalent to the grid in Figure 13.

does not change, we can choose the horizontal cutting line in Figure 14 corresponding to the the composite cutting line in Figure 14 (it contains the segment A_3A_4, the boundary of the square $A_4A_5A_6A_7$, and the boundary of a regular pentagon; the nodes on this line are marked by $\star$ in Figure 15).

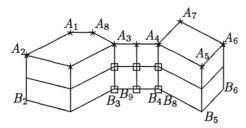

Figure 15. Square grid in the space equivalent to the grid in Figure 13.

Thus, in all of the polar-type grid examples, one of the main features was the use of an effective algorithm for grid systems with the boundary periodic conditions (in contrast to the original boundary conditions). It can even be applied for q-connected domains with $q \geq 3$ if we make use of $q - 2$ double slits like $[A_3A_4]$ in Figure 13 and construct polar-type grids for the resulting doubly-connected domains; we can then ignore these artificial slits. Another and probably better approach is linked with the use of several polar-type grids as in Figure 10.

5.6. Construction of nearly asymptotically optimal model operators; additivity in definition of spaces $V_{1/2}(R)$.

We describe an approach, leading to the construction of the nearly asymptotically optimal model operator. [45] The seminorm $|\hat{u}|_{1,Q_r}$ in (5.24) is replaced by the norm $\|\hat{u}\|_{1,Q_r} \equiv \|\hat{u}\|_{1/2,\partial Q_r,0}$, so that we can find a simpler norm $\|\hat{u}\|_{1/2,\partial Q_r}$,

---

[45] Over the past decade, substantial effort has been spent on construction of model operators B_2 such that S_2 and B_2 are spectrally equivalent or nearly spectrally equivalent (see [79, 86, 130, 132, 144, 236, 302, 345, 384, 455, 457] and references therein). The first such operators were suggested in [141] for the case $G = \overset{o}{W_2^1}(Q)$ and the simple situation considered in Lemma 2. They were based on important results obtained in [10] and devoted to the study of discrete analogs of the space $W_2^{1/2}(\Gamma) \equiv V_{1/2}(\Gamma)$ for rectangular grids (see (0.1.26)). (As indicated in § 0.1, this Hilbert space gives an ultimate answer in characterizing the traces on Γ of functions in $W_2^1(Q)$; a similar fact for grid functions was proven in [10] for several equivalent grid norms, which underscores the significance of such norms from a computational point of view.) In construction of the above B_2, (5.24) serves as a starting point. It implies that, for a given elementary block Q_r, we need to find a suitable seminorm $|\hat{u}|_{1/2,\partial Q_r}$ which is equivalent—uniformly or nearly

that is equivalent—nearly uniformly with respect to the grid—to the norm $\|\hat{u}\|_{1,Q_r}$; that is,

$$\sigma_0\|\hat{u}\|_{1/2,\partial Q_r} \leq \|\hat{u}\|_{1,Q_r} \leq \sigma_1\|\hat{u}\|_{1/2,\partial Q_r}, \quad \forall \hat{u}, \tag{5.31}$$

$$\sigma_0 > 0, \quad \frac{\sigma_1}{\sigma_0} = O(|\ln h|^m), \tag{5.32}$$

and $m \geq 0$ (recall that the function $\hat{u}$ at inner points of the domain Q_r is completely defined by its values at nodes in ∂Q_r, as in (5.24)). In the sequel, we agree to write $\|\hat{u}\|_{1,Q_r} \asymp \|\hat{u}\|_{1/2,\partial Q_r}$ if (5.31) holds with constants $\sigma_0 > 0$ and $\sigma_1 > 0$ independent of h; moreover, we will use this notation of uniform equivalence of norms or seminorms for other pairs; in terms of the corresponding grid operators, this implies their spectral equivalence. But for this case in contrast to [144, 346], we make use of the assumption that all a_r are of the same order. Under this condition, we construct a model operator B_2 such that

$$(B_2\mathbf{u}_2, \mathbf{u}_2) \equiv \sum_{r=1}^{r_0} a_r\|\hat{u}\|_{1/2,\partial Q_r}^2 \tag{5.33}$$

and which is nearly spectrally equivalent to S_2 (see also [455]). Then the key problem to be considered is how to solve the resulting systems (we wish to have algorithms with estimates $W_2 = O(h^{-1}|\ln h|^{m_1}), m_1 \geq 0$). We will return to this problem a little later; for now, we outline possible further developments.

Suppose that we have obtained the above operator B_2. Then, in accordance with the theory of two-stage iterative methods (see § 4), we may also use the operator $D_2 \equiv S_2(I_2 - Z_{2,k})^{-1}$ (see (4.3) with $\Lambda = S_2$), which corresponds to k inner iterations (5.4) and, for an appropriate value of k, may become spectrally equivalent to S_2. Then, as in § 1.5, we can find

$$B \equiv \left[\begin{array}{cc} A_{1,1} & A_{1,2} \\ A_{2,1} & D_2 + A_{2,1}A_{1,1}^{-1}A_{1,2} \end{array} \right]$$

that is a nearly asymptotically optimal model operator; in the simplest case, $D_2 = B_2$.[46]

For $R \equiv [0, 1]$, the norm in the Hilbert space $V_{1/2}(R)$ is defined by

---

uniformly with respect to the grid—to the seminorm $|\hat{u}|_{1,Q_r}$. Such an approach dealing with separate model nonnegative operators on ∂Q_r was suggested in [346] but it needed a special balancing procedure and solver for systems on ∂Q_r.(Convergence of iterations was obtained independent of $a_1, \ldots, a_{r_0}$.)

[46] Such a construction was actually used in [84]; the more difficult case dealing with approximations of $A_{1,1}$ was analyzed in [263].

$$\|u\|_{1/2,R}^2 \equiv |u|_{0,R}^2 + |u|_{1/2,R}^2, \tag{5.34}$$

where

$$|u|_{1/2,R}^2 \equiv \int_R \int_R \frac{|u(x) - u(x')|^2}{|x - x'|^2} dx dx'. \tag{5.35}$$

For $R \equiv [0, l_1]$, $R' \equiv [-l_1', 0]$, the additive definition of the square of the norm in $V_{1/2}(R \cup R')$ (see [506]) leads to (5.34), where

$$|u|_{1/2,R\cup R'}^2 \equiv |u|_{1/2,R}^2 + |u|_{1/2,R'}^2 + F(u), \tag{5.36}$$

$$F(u) \equiv \int_{R_0} \int_{R_0'} \frac{|u(x) - u(x')|^2}{|x - x'|^2} dx dx',$$

and $R_0 \equiv R$ and $R_0' \equiv R'$ or $R_0 = -R_0'$ and $|R_0| = \min\{l/2; l'/2\}$. [47]

Now consider the space $V_{1/2}(R)$, where R is the boundary of a rectangle $Q^{(0)} \equiv S_1 \times S_2 \equiv [M_0 M_1 M_2 M_3]$, where $S_r \equiv [0, l_r]$, $r = 1, 2$, $M_0 \equiv [0, 0]$, $M_1 \equiv [l_1, 0]$, $M_2 \equiv [l_1, l_2]$, and $M_3 \equiv [0, l_2]$. We denote its sides $[M_0 M_1]$, $[M_1 M_2]$, $[M_2 M_3]$, and $[M_3 M_0]$ by R_1, R_2, R_3, and R_4, respectively, and take

$$|u|_{1/2,R}^2 \equiv \sum_{i=1}^4 |u|_{1/2,R_i}^2 + \sum_{i=1}^4 F_{i,i+1}(u), \tag{5.37}$$

$$F_{1,4}(u) \equiv \left| \frac{|u(x_1, 0) - u(0, x_2)|}{|x_1 + x_2|} \right|_{0,S}^2, \quad F_{1,2}(u) \equiv \left| \frac{|u(x_1, 0) - u(l_1, x_2)|}{|l_1 - x_1 + x_2|} \right|_{0,S}^2,$$

$$F_{2,3}(u) \equiv \left| \frac{|u(x_1, l_2) - u(l_1, x_2)|}{|l_1 - x_1 + l_2 - x_2|} \right|_{0,S}^2, \quad F_{3,4}(u) \equiv \left| \frac{|u(x_1, l_2) - u(0, x_2)|}{|x_1 + l_2 - x_2|} \right|_{0,S}^2,$$

and $S \equiv S_1 \times S_2$. The Euclidean space of traces of functions $\hat{u} \equiv u$ with norm defined by (5.37) is denoted by $\hat{V}_{1/2}(R)$. We see that this norm is rather involved. So it is very natural for these finite-dimensional spaces of functions $\hat{u}$ to introduce a simpler (additive):

$$\|\hat{u}\|_{1/2,R,a}^2 \equiv |\hat{u}|_{0,R}^2 + |\hat{u}|_{1/2,R,a}^2, \tag{5.38}$$

---

[47] For functions that are even in a fixed vicinity of x_0, it is possible to take $F(u) \equiv 0$ (this is the case if functions vanish in this vicinity). Note also that, for the segments R and R' lying on crossing straight lines, it is possible to preserve the indicated definition of the norm if x and x' refer to appropriate parameters on the lines.

where $|\hat{u}|_{1/2,R,a}^2 \equiv \sum_{i=1}^4 |\hat{u}|_{1/2,R_i}^2.$ [48] Of course, the key estimates are

$$|\hat{u}|_{1/2,R,a}^2 \leq |\hat{u}|_{1/2,R}^2 \leq \sigma(h)|\hat{u}|_{1/2,R,a}^2, \quad \forall \hat{u}, \qquad (5.39)$$

with the function $\sigma(h)$ of type

$$\sigma(h) = O(|\ln h|^k) \qquad (5.40)$$

(in [455], (5.40) with $k = 2$ was proved but it can probably be improved to $k = 1$). Note that, for the norms in (5.34), we have

$$|\hat{u}|_{0,R} \asymp \|\mathbf{u}\|_D, \qquad (5.41)$$

where $h \equiv 1/(N+1)$, $\mathbf{u} \equiv [u_0, \ldots, u_{N+1}]$, $Du_i \equiv hu_i$, $i \in [1, N]$, $Du_0 \equiv 1/2hu_0$, $Du_{N+1} \equiv 1/2hu_{N+1}$, and

$$|\hat{u}|_{1/2,R}^2 \asymp \sum_{i=0}^{N+1} \sum_{j=0, j\neq i}^{N+1} \frac{(u_i - u_j)^2}{(i-j)^2} \equiv U(\mathbf{u}). \qquad (5.42)$$

In terms of the finite differences $d_1 \equiv u_1 - u_0, \ldots, d_{N+1} \equiv u_{N+1} - u_N$ and their arithmetical means

$$M_{i,1} \equiv d_i, \ M_{i,2} \equiv \frac{d_i + d_{i+1}}{2}, \ldots, M_{1,N} \equiv \frac{d_1 + \cdots + d_{N+1}}{N},$$

we can use the following especially elegant expression:

$$U(\mathbf{u}) = \sum_{i,j} M_{i,j}^2, \qquad (5.43)$$

where the summation is carried over all possible pairs $[i, j]$.

Next, for $F(\hat{u})$ in (5.36), we observe (as in [506]) that

$$F(\hat{u}) \leq \bar{\kappa} \int_{R_0} \frac{|\hat{u}(x) - \hat{u}(-x)|^2}{x} dx.$$

Thus, for finding $\sigma(h)$ in (5.39), it suffices to obtain the estimate

$$\int_{R_0} \frac{|\hat{u}(x)|^2}{x} dx \leq \delta_0(h)|\hat{u}|_{1/2,R_0}^2 \qquad (5.44)$$

---

[48] This approach was developed by Siganevich in [455] and can be generalized relatively easily for d-dimensional problems; several researchers (see [84, 79, 142]) considered similar simplifications, but for special classes of functions vanishing at vertices of elementary blocks; the three-dimensional case was analyzed in [457]; in [384] a more complicated approach leading to the spectral equivalence of S_2 and B_2 for $d = 2$ was suggested.

for $\hat{u}$ such that $u_0 = 0$ (in the sequel, we simply take $R_0 = R$). For such functions, $\int_R \frac{|\hat{u}(x)|^2}{x} dx \asymp \sum_{i=1}^{N+1} \frac{u_i^2}{i} = \sum_{i=1}^{N} iM_{1,i}^2$. Hence, the key problem is to obtain estimate

$$\sum_{i=1}^{N} iM_{1,i}^2 \le \delta(h) \sum_{i,j} M_{i,j}^2, \qquad (5.45)$$

for all numbers $d_1, \ldots, d_{N+1}$. This underscores a connection of (5.44) with the classical Hardy inequality (see [274])

$$\sum_{i=1}^{N} M_{1,i}^2 \le 4 \sum_{i=1}^{N+1} d_i^2. \qquad (5.46)$$

(It is possible to show that $\sum_{i=1}^{N+1} M_{1,i}^2 \le 4 \sum_{i=1}^{N+1} (1 - i^2/(N+2)^2)d_i^2$.) To establish (5.45), we need one more equivalent norm in $\hat{V}_{1/2}(R)$. It makes use of an orthonormal basis for the Euclidean space $H(D)$ composed of eigenvectors of the spectral problem $\Lambda u = \lambda D u$, where $\Lambda u_i \equiv -\Delta u_i, i \in [1, N]$, $\Lambda u_0 \equiv -1/h \partial u_0$, $\Lambda u_{N+1} \equiv 1/h \bar{\partial} u_{N+1}$. The eigenvectors and eigenvalues are: $e_n(x) \equiv 2^{1/2} l^{-1/2} \cos[n\pi x/l]$, $n \in [1, N]$, $e_0(x) \equiv l^{-1/2}$; $e_{N+1}(x) \equiv l^{-1/2} \cos[(N+1)\pi x/l]$, $n \in [1, N]$ and $\lambda_n \equiv 4h^{-2} \sin^2(n\pi h)/(2l)$, $n \in [0, N+1]$ (see, e.g., [437]). Note that $(2n/l)^2 \le \lambda_n \le (\pi n/l)^2$, $n \in [0, N+1]$. Then the desired norm (we refer to it as the spectral one) is defined by

$$\|\hat{u}\|_{1/2,R,s}^2 \equiv \sum_{n=0}^{N+1} a_n^2 (1 + \lambda_n^{1/2}), \qquad (5.47)$$

where the coefficients $a_0, \ldots, a_{N+1}$ correspond to the expansion

$$\mathbf{u} \equiv \sum_{n=0}^{N+1} a_n e_n(x), \qquad (5.48)$$

and $\mathbf{u} \equiv [u_0, \ldots, u_{N+1}]$ is the vector of values of $\hat{u}$ at the nodes.[49]

Lemma 4. *There exists $\bar{\kappa}_1$ such that*

$$\max_{i=0,\ldots,N+1} |u_i| \le \bar{\kappa}_1 \left(\frac{|\hat{u}|_{0,R}}{l^{1/2}} + \ln^{1/2} N |\hat{u}|_{1/2,R,s} \right), \quad \forall \hat{u}. \qquad (5.49)$$

---

[49] Equivalence of the norms and the fact that $|\hat{u}|_{1/2,R,s}^2 \equiv \sum_{n=0}^{N+1} a_n^2 \lambda_n^{1/2} \asymp |\hat{u}|_{1/2,R}^2$ were proved in [10].

Proof. The definition of the basis vectors and (5.48) yield

$$|u_i| \leq \frac{\bar{\kappa}_2}{l^{1/2}} \left(|a_0| + \cdots + |a_{N+1}| \right). \tag{5.50}$$

It is easy to see that $|a_0| \leq \|\mathbf{u}\|_D \leq \bar{\kappa}_3 |\hat{u}|_{0,R}$. We also have $\sum_{n=1}^{N+1} |a_n| \leq [\sum_{n=1}^{N+1} \frac{1}{n}]^{1/2} [\sum_{n=1}^{N+1} a_n^2 n]^{1/2}$. This implies that

$$\sum_{n=1}^{N+1} |a_n| \leq \bar{\kappa}_4 \ln^{1/2} N [\sum_{n=1}^{N+1} a_n^2 \lambda_n^{1/2} l]^{1/2}.$$

Hence, (5.50) leads to (5.49). $\square$

Lemma 5. *There exists $\bar{\kappa}_5$ such that*

$$a_r \|\hat{u}\|_{1/2,R,a}^2 \leq a_r \|\hat{u}\|_{1/2,R}^2 \leq a_r \sigma(h) \|\hat{u}\|_{1/2,R,a}^2, \quad \forall \hat{u}, \tag{5.51}$$

where a_r is an arbitrary positive constant and $\sigma(h) \leq \bar{\kappa}_5 \ln^2 N$.

Proof. We have $\sum_{i=1}^{N} iM_{1,i}^2 = \sum_{i=1}^{N+1} u_i^2/i \leq \max_i |u_i|^2 \sum_{i=1}^{N+1} 1/i$. This and (5.49) yield (5.45) and (5.51). $\square$

If $u_0 = u_{N+1} = 0$ and

$$\mathbf{u} \equiv \sum_{n=1}^{N} a_n e_n(x), \tag{5.52}$$

where $e_n(x) \equiv 2^{1/2} l^{-1/2} \sin[n\pi x/l]$, then (5.47) is replaced by [50]

$$\|\hat{u}\|_{1/2,R,s,0}^2 \equiv \sum_{n=1}^{N} a_n^2 (1 + \lambda_n^{1/2}). \tag{5.53}$$

5.7. Additive splitting of the inverse model operator. To construct a model operator B_2 we define

$$(B_2^{(1)} \mathbf{u}_2, \mathbf{u}_2) \equiv \sum_{r=1}^{r_0} a_r |\hat{u}|_{1/2,\partial Q_r \cap C}^2. \tag{5.54}$$

Note that $\partial Q_r \cap C$ may consist of a number of sides of Q_r (which we denote by $S_{r,1}, S_{r,2}, S_{r,3}, S_{r,4}$ and some of which we allow to be empty).[51] Next we define $B_2^{(2)} \equiv \Lambda_2 \in \mathcal{L}^+(H_2)$ by

---

[50] We emphasize that $\|\hat{u}\|_{1/2,R,s,0}^2 \asymp \|\hat{u}_0\|_{1/2,R'}^2$, where $R' \equiv [-l, 2l]$, $\hat{u}_0(x) = \hat{u}(x)$ if $x \in R$ and $\hat{u}_0$ vanishes at the remaining points of R' (see [10]). Such an approach needs a special procedure to obtain the desired zero values, and it becomes very involved for multidimensional problems.

[51] The equivalence of norms defined by (5.54) and (5.33) is a direct consequence of a similar statement for a separate rectangle which was shown in [10].

$$(\Lambda_2 \mathbf{u}_2, \mathbf{u}_2) \equiv \sum_{r=1}^{r_0} a_r \sum_{k=1}^{4} |\hat{u}|^2_{1/2, S_{r,k,s}}. \tag{5.55}$$

It is easy to see that

$$\bar{\kappa}_6 \Lambda_2 \leq S_2 \leq \bar{\kappa}_7 \ln^2 h \Lambda_2 \tag{5.56}$$

and that $\Lambda_2 \mathbf{u}_2$ is relatively easy to compute for each given vector $\mathbf{u}_2$ (it suffices to apply the Fourier transformation from § 1 on each separate side of each elementary block and assemble the results). [52]

To obtain the desired operator, we use Lemmas 2.6 and 2.8 with the spaces $H_l \equiv H_{2,k,0}$ of vectors that vanish at elementary nodes not belonging to the given side (the spaces $H_{2,k}$ and $H_{2,k,0}$ are isomorphic). More precisely, let R_k denote such a side and let $\mathbf{u}_{2,k} \in H_{2,k}$ and $\mathbf{u}_{2,k,0} \in H_{2,k,0}$ refer to the vector of values of $\mathbf{u}_2$ at nodes belonging to R_k and to its finite extension. Consider $B_{2,k,1} \in \mathcal{L}^+(H_{2,k})$ such that

$$(B_{2,k,1}\mathbf{u}_{2,k}, \mathbf{u}_{2,k}) \equiv \|\hat{u}_{2,k}\|^2_{1/2, R_k, s}, \quad \forall \mathbf{u}_{2,k}, \tag{5.57}$$

where $\hat{u}_{2,k}$ is a standard piecewise linear function on R_k defined by the vector $\mathbf{u}_{2,k}$. We rewrite (5.55) as $(\Lambda_2 \mathbf{u}_2, \mathbf{u}_2) \equiv \sum_k c_k (B_{2,k,1}\mathbf{u}_{2,k}, \mathbf{u}_{2,k})$, where the constant c_k is either a corresponding constant a_r (if R_k is a side of only one Q_r) or $a_r + a_m$ (if R_k is a common side of Q_k and Q_m), and the index k refers to the side under consideration. Let

$$B_2^{-1}\mathbf{u}_2 \equiv \sum_k c_k^{-1} B_{2,k,1}^{-1}\mathbf{u}_{2,k}. \tag{5.58}$$

It is easy to see that the evaluation of $\mathbf{w}_{2,k} \equiv B_{2,k,1}^{-1}\mathbf{u}_{2,k}$ is reduced to finding the expansion of type (5.48) and

$$\mathbf{v}_{2,k} = \sum_{n=0}^{N+1} a_n (1 + \lambda_n^{1/2})^{-1} De_n(x), \tag{5.59}$$

where $N \equiv \dim H_{2,k}$. Thus, $W = O(h^{-1}|\ln h|)$.

Theorem 5. The operator B_2^{-1} defined by (5.58) is the inverse of an operator $B_2 \in \mathcal{L}^+(H_2)$ and is such that

$$\frac{\bar{\kappa}_8}{|\ln h|} B_2 \leq S_2 \leq \bar{\kappa}_9 \ln^2 h B_2. \tag{5.60}$$

Proof. We have $(B_2^{-1}\mathbf{u}_2, \mathbf{u}_2) = \sum_k c_k^{-1}(B_{2,k,1}^{-1}\mathbf{u}_{2,k}, \mathbf{u}_{2,k}) \geq 0$, which vanishes only when all $\mathbf{u}_{2,k} = 0$. Thus, $B_2^{-1} \in \mathcal{L}^+(H_2)$ and $B_2 \in \mathcal{L}^+(H_2)$.

---

[52] Such operators will be used in § 6, but they are useless for iterations (5.4).

For proving (5.60) with $\bar{\kappa}_8$, we will show that there exists a positive constant $\bar{\sigma}_1 \equiv \bar{\kappa}_{10} |\ln h|$ such that, for each $\mathbf{u}_2$, there exists an expansion $\mathbf{u}_{2,k} = \sum_k \mathbf{v}_{2,k,0}$, $\mathbf{v}_{2,k} \in H_{2,k,0}$ satisfying

$$\sum_k \|\mathbf{v}_{2,k}\|^2_{c_k B_{2,k,1}} \leq \bar{\sigma}_1 \|\mathbf{u}_2\|^2_{\Lambda_2}. \tag{5.61}$$

For this we define $\mathbf{v}_{2,k,0}$, for nodes belonging only to one side R_k, as the corresponding values of $\mathbf{u}_2$ and, for the nodes M_i belonging to several different sides, we use a more involved procedure. Let M_i be an end point of sides $R_{k_1}, \ldots, R_{k_m}$, where $m \geq 2$; for $k \in [k_1, k_m]$, we define

$$\mathbf{v}_{2,k,0}(M_i) \equiv \frac{c_k}{c_{k_1} + \cdots + c_{k_m}} \mathbf{u}_2(M_i).$$

For such a $\mathbf{v}_{2,k}$, we have $\mathbf{z}_{2,k} \equiv \mathbf{v}_{2,k} - \mathbf{u}_{2,k}$ and

$$\|\mathbf{z}_{2,k}\|^2_{B_{2,k,1}} \leq \bar{\kappa}_{12} \max_{M_i \in R_k} (\mathbf{u}_2(M_i))^2 \|\hat{g}_{2,k}\|^2_{1/2,R_k},$$

where $\mathbf{w}_{2,k}(M_i) = 1$, which vanishes at other nodes; $\|\hat{g}_{2,k}\|^2_{1/2,R_k} = O(1)$. Hence,

$$\|\mathbf{v}_{2,k}\|^2_{c_k B_{2,k,1}} \leq c_k \bar{\kappa}_{13} |\ln h| \|\mathbf{u}_{2,k}\|^2_{c_k B_{2,k,1}}$$

and (5.61) holds with $\bar{\sigma}_1 \equiv \bar{\kappa}_{14} |\ln h|$. From (5.61) and Lemma 2.8, it follows that

$$\Lambda_2^{-1} \leq \bar{\sigma}_1 \sum_k (P_k \Lambda_2 P_k)^\dagger, \tag{5.62}$$

where P_k denotes the orthoprojector of H_2 onto $H_{2,k,0}$. We see that

$$(P_k \Lambda_2 P_k \mathbf{u}_2, \mathbf{u}_2) = (\Lambda_2 \mathbf{u}_{2,k,0}, \mathbf{u}_{2,k,0}) \equiv (\Lambda_{2,k} \mathbf{u}_{2,k}, \mathbf{u}_{2,k}),$$

where $\Lambda_{2,k} \in \mathcal{L}^+(H_{2,k})$. Moreover, $\Lambda_{2,k} \geq c_k B_{2,k,1}$. If we define $\bar{B}_k \in \mathcal{L}(H_2)$ such that $\bar{B}_k = \bar{B}_k^* = P_k \bar{B}_k P_k \geq 0$, $\bar{B}_k \mathbf{u}_2 = \bar{B}_k \mathbf{u}_{2,k,0}$, and

$$(\bar{B}_k \mathbf{u}_{2,k,0}, \mathbf{u}_{2,k,0}) = (c_k B_{2,k,1} \mathbf{u}_{2,k}, \mathbf{u}_{2,k})$$

then, for $\mathbf{w}_{2,k} \equiv c_k^{-1} B_{2,k,1}^{-1} \mathbf{u}_{2,k}$, its finite extension is $\mathbf{w}_{2,k,0} = (\bar{B}_k)^\dagger \mathbf{u}_2$. Hence, $(\bar{B}_k)^\dagger \leq (P_k \Lambda_2 P_k)^\dagger$ $(P_k \Lambda_2 P_k \geq \bar{B}_k)$ and (5.62) leads to $\Lambda_2^{-1} \leq \bar{\sigma}_1 \sum_k (\bar{B}_k)^\dagger$ and

$$\Lambda_2^{-1} \leq \bar{\sigma}_2 B_2^{-1}, \tag{5.63}$$

where $\bar{\sigma}_2 \equiv \bar{\kappa}_{15} |\ln h|$. (5.63) and (5.56) yield $S_2^{-1} \leq \bar{\sigma}_2 / \bar{\kappa}_6 B_2^{-1}$. This leads to the inequality $\bar{\kappa}_8 B_2 \leq |\ln h| S_2$. To prove the remaining inequality in (5.60), we show that

$$(\bar{B}_k)^\dagger \leq \bar{\sigma}_3(P_k S_2 P_k)^\dagger \leq \bar{\sigma}_3 S_2^{-1}, \tag{5.64}$$

where $\bar{\sigma}_3 \equiv \bar{\kappa}_{17} \ln^2 h$. Let $\hat{u}_{2,k,0}$ be defined by $\mathbf{u}_2 \in H_2$. Then

$$(\bar{B}_k \mathbf{u}_2, \mathbf{u}_2) = c_k(B_{2,k,1} \mathbf{u}_{2,k}, \mathbf{u}_{2,k}) \geq \bar{\kappa}_{17} \|\hat{u}_{2,k,0}\|^2_{1/2,R_k}.$$

Since $(P_k S_2 P_k \mathbf{u}_2, \mathbf{u}_2) \leq q c_k \bar{\kappa}_{18} \|\hat{u}_{2,k,0}\|^2_{1/2,C} \leq q c_k \bar{\kappa}_{19} \ln^2 h \|\hat{u}_{2,k,0}\|^2_{1/2,R_k}$, we see that $P_k S_2 P_k \leq \bar{q} \kappa_{20} \ln^2 h \bar{B}_k$. Hence, (5.64) holds and

$$B^{-1} \leq \bar{\sigma}_4 S_2^{-1}, \tag{5.65}$$

where $\bar{\sigma}_4 \equiv q \bar{\kappa}_{21} \ln^2 h$. This yields (5.60). □

Consider now a d-dimensional parallelepiped $R \equiv S_1 \times \cdots \times S_d$, where $S_r \equiv [0, l_r]$, $r \in [1, d]$. Let $[x_1, \ldots, x_{r-1}, x_{r+1}, \ldots, x_d] \equiv x^{(r)}$ be a set of variables different from $x_r, r \in [1, d]$ and $u(x) \equiv u(x_r; x^{(r)})$. In the definition of the norm $\|u\|_{1/2,R}$ for R, we take

$$|u|^2_{1/2,R} \equiv \sum_{r=1}^{d} \int_{R^{(r)}} U(x^{(r)})^2 dR^{(r)}, \tag{5.66}$$

where

$$U(x^{(r)})^2 \equiv \int_{S_r} \int_{S_r} \frac{|u(x_r; x^{(r)}) - u(x_r'; x^{(r)})|^2}{|x_r - x_r'|^2} dx_r dx_r'$$

as was actually suggested by Gagliardo (see [68, 237]). Then, for $R' \equiv [-l_1', 0] \times S_2 \times \cdots \times S_d$, we may define the square of the norm in $V_{1/2}(R \cup R')$ (as in [506]) in the additive way

$$|u|^2_{1/2,R \cup R'} \equiv |u|^2_{1/2,R} + |u|^2_{1/2,R'} + F(u),$$

(see (5.36)), where $F(u) \equiv \int_{R^{(1)}} \int_{R^{(1)}} F_1(x^{(1)}; u) dQ_1^{(1)} dQ_1^{(1)}$ and

$$F_1(x^{(1)}; u) \equiv \int_{S_1} \int_{S_1'} \frac{|u(x_1; x^{(1)}) - u(x_1'; x^{(1)})|^2}{|x_1 - x_1'|^2} dx_1 dx_1'.$$

The study of the additivity of the square of the norm in $\hat{V}_{1/2}(R \cup R')$ is reduced to the case considered above. Hence, Theorem 5 holds. [53]

---

[53] Important numerical experiments and practical applications of domain decomposition methods can be found, e.g., in [130, 257, 498]. Finally, we mention investigations dealing with the grid analogs of the classical Schwarz method (see [151, 164, 216, 376, 428]) and its additive variants (see § 2 and [142, 143, 144, 504]). The investigations of Poincare and Steklov should also be mentioned as the first attempts to deal with domain decomposition on the differential level; the modern approach to these problems was developed in [313].

§ 6. Fictitious domain iterative methods

6.1. Basic computational algorithms for Neumann and mixed boundary conditions. Consider an elliptic boundary value problem associated with a region Q. Suppose that $\bar{Q}$ can be extended by a regions $\bar{F}_1, \ldots, \bar{F}_p$ to a larger region $\bar{\Pi}$, where $F_r \cap F_m = \emptyset$ if $r \neq m$, and that region Π has relatively simple form so we can solve effectively the relative grid systems for Π. Then it is reasonable to study algorithms for solving the original grid systems for Q that involve solution on Π. In this case, we refer to domains $F_1, \ldots, F_p$ as *fictitious domains* and to Π as the *basic domain* or *extended* domain. [54]

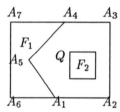

Figure 1. Example of the original region and two fictitious domains.

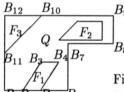

Figure 2. Example of the original region and three fictitious domains.

In the sequel, we always consider the grids for Q and Π consistent in the sense that every cell of the grid for Q is a cell of the grid for Π. The sets Q_h and Π_h consist of elementary nodes that define the corresponding grid systems $L_Q u_Q = f_Q$ and $A_\Pi u_\Pi = g_\Pi$ constructed for Q and Π, respectively. We also make use of the set

---

[54] Examples of such regions with Lipschitz boundaries are given in Figures 1 and 2. There exists a number of different approaches to using the basic domain (see § 2 and [15, 17, 200, 354, 384, 454, 455, 503]). We confine ourselves to the most promising approach where the grid systems for Π serve only as a means of constructing a model grid operator that is spectrally equivalent to the grid operator under consideration (it might be the original grid operator L or an operator Λ connected to L by relationship C^3 or $C^3(u, r)$ (see § 1.4)). Such an approach was used, e.g., in [200, 454]. It is related to the domain decomposition method suggested in [185] and its generalization considered in § 5. We stress that our approach is applicable to cases where the domains Q and Π may have non-Lipschitz boundaries.

$$F_h \equiv \Pi_h \setminus Q_h \equiv F_{1,h} \cup \cdots \cup F_{p,h},$$

where $F_{r,h} \subset \bar{F}_r \setminus \partial Q, r \in [1,p]$. The Euclidean spaces of grid functions defined on Π_h, F_h, and Q_h are denoted by H, H_1, and H_2, respectively, so that $H = H_1 \times H_2$. Note that $\dim H \asymp 1/h^2$, $\dim H_1 \equiv n_1 \asymp 1/h^2$, $\dim H_2 \equiv n_2 \asymp 1/h^2$ if we use uniform or quasiuniform grids. Elements of these spaces are denoted by $\mathbf{u}, \mathbf{u}_1$, and $\mathbf{u}_2$, respectively, so that $\mathbf{u} \equiv [\mathbf{u}_1, \mathbf{u}_2]^T$, $\mathbf{u} \equiv \{u_i : P_i \in \Pi_h\}$, $\mathbf{u}_2 \equiv \{u_i : P_i \in Q_h\}$, $\mathbf{u}_1 \equiv \{u_i : P_i \in \Pi_h \setminus Q_h\}$. The original system $L_Q \mathbf{u}_Q = \mathbf{f}_Q$ is rewritten as

$$L_2(\mathbf{u}_2) = \mathbf{f}_2. \tag{6.1}$$

We represent the model matrix $A_\Pi \equiv A$ in block form (see $(1.5.1)$) and, assuming that $A^* = A > 0$, define

$$S_2 \equiv S_2(A) \equiv A/A_{1,1} \equiv A_{2,2} - A_{2,1} A_{1,1}^{-1} A_{1,2} \in \mathcal{L}(H_2). \tag{6.2}$$

If we show that the operators L_2 and S_2 are connected by relationship C^0 or C^3 (see § 1.4)) then, for system (6.1), we can use the iterative methods from § 1.3 with the model operator $B \equiv S_2 \equiv B_2$, e.g.,

$$B_2(\mathbf{u}_2^{n+1} - \mathbf{u}_2^n) = -\tau_n L_2^* B_2^{-1}(L_2 \mathbf{u}_2^n - \mathbf{f}_2).$$

These iterations can be reduced to those in H of the form

$$A(\mathbf{u}^{n+1} - \mathbf{u}^n) = -\tau_n \begin{bmatrix} 0 & 0 \\ 0 & L_2^* \end{bmatrix} A^{-1} \begin{bmatrix} 0 \\ L_2 \mathbf{u}_2^n - \mathbf{f}_2 \end{bmatrix}, \tag{6.3}$$

where $\mathbf{u}^0 \equiv [0, \mathbf{u}_2^0]^T$. The desired relationships between the operators will be proved to hold only for special types of boundary conditions, including the Neumann, natural, and mixed. Choice of the basic domain Π usually leads to sufficiently simple domains with Lipschitz boundary like a d-dimensional parallelepiped. But there is really no need to restrict ourselves to domains of such type because, as we will see, the choice should be made to account not only for the geometry of the boundary $\partial Q \equiv \Gamma_Q$, but also for the geometry of its part $\Gamma_{Q,0}$ if the original problem is set in the Hilbert space

$$G_Q \equiv W_2^1(Q; \Gamma_{Q,0}). \tag{6.4}$$

Also, this choice should be made on the basis of existing efficient iterative methods for the grid systems $A_\Pi \mathbf{u}_\Pi \equiv A\mathbf{u} = \mathbf{g}_\Pi$ associated with PGMs for a model problem considered as an operator equation in the Hilbert space

$$G_\Pi \equiv W_2^1(\Pi; \Gamma_{\Pi,0}). \tag{6.5}$$

Such methods can be constructed for model domains of various forms, including domains with slits (see § 5 and 7).

6.2. Conditions of spectral equivalence of the model grid operator for the original domain and the Schur complement for the extended domain. Now we consider approximations $\hat{G}_Q$ and $\hat{G}_\Pi$ associated with some triangulations (composite triangulations) of $T_h(\bar{Q})$ and $T_h(\bar{\Pi})$ (see § 2.1) (recall again that each simplex in $T_h(\bar{Q})$ is a simplex in $T_h(\bar{\Pi})$) under the assumption that

$$\Gamma_{Q,0} \subset \Gamma_{\Pi,0}. \tag{6.6}$$

For example, for Q depicted in Figure 1 and $\Gamma_{Q,0} \equiv [A_2, A_3]$, we may take $\Gamma_{\Pi,0} = \Gamma_{Q,0}$; if $\Gamma_{Q,0} \equiv A_1 A_2 A_3 A_4$, we may take $\Gamma_{\Pi,0} = A_6 A_2 A_3 A_4$. If Q is as depicted in Figure 2 and $\Gamma_{Q,0} \equiv [B_1, B_2] \cup [B_2 B_3] \cup [B_3 B_4]$, then we may take $\Gamma_{\Pi,0} = \Gamma_{Q,0}$; thus, we prefer here to use a basic domain with slits. A very instructive example is provided by Figure 3 ($\Gamma_{Q,0} = [A_1 A_{11}]$).

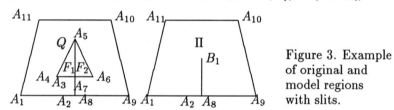

Figure 3. Example of original and model regions with slits.

In this case, we define Π as a rectangle with a double vertical slit with two edges—denoted by $[A_2 A_5]$ and $[A_8 A_5]$—and we take $\Gamma_{\Pi,0} = \Gamma_{Q,0}$ (fictitious domains are $F_1 \equiv \triangle A_3 A_5 A_4$ and $F_2 \equiv \triangle A_7 A_6 A_5$).

For the standard basis functions $\hat{\psi}_i(x)$ (corresponding to elementary nodes in Π_h), we define

$$A \equiv J_\Pi \equiv [(\hat{\psi}_j(x), \hat{\psi}_i(x))]_{1,\Pi}. \tag{6.7}$$

The subset of the same basis functions corresponding to nodes (elementary nodes) in Q_h defines a model operator

$$\Lambda_Q \equiv J_Q \equiv [(\hat{\psi}_j(x), \hat{\psi}_i(x))]_{1,Q} \le A_{2,2}. \tag{6.8}$$

Theorem 1. *For the space $\hat{G}_Q$, suppose there exists a linear extension operator $E \in \mathcal{L}(\hat{G}_Q; \hat{G}_\Pi)$ such that*

$$|\hat{v}|^2_{1,\Pi} \le K_0 |\hat{v}_2|^2_{1,Q}, \quad \forall \hat{v}_2 \in \hat{G}_Q, \tag{6.9}$$

where $\hat{v} \equiv E\hat{v}_2$ and $K_0 \ne K_0(h)$. Suppose S_2 is defined by (6.2). Then

$$\sigma_0 S_2 \le J_Q \le \sigma_1 S_2, \quad \sigma_0 \equiv 1/K_0, \quad \sigma_1 = 1. \tag{6.10}$$

Proof. For any $\mathbf{u}_2 \in H_2$, we have $(J_Q \mathbf{u}_2, \mathbf{u}_2) = |\hat{u}_2|_{1,Q}^2$ and

$$(S_2 \mathbf{u}_2, \mathbf{u}_2) = (A_{2,2} \mathbf{u}_2, \mathbf{u}_2) - (A_{1,1}^{-1} A_{1,2} \mathbf{u}_2, A_{1,2} \mathbf{u}_2) = |\hat{u}|_{1,\Pi}^2,$$

where $\mathbf{u} \equiv p(\mathbf{u}_2)$ (see (1.5.5)) and $\hat{u}(x) = \hat{u}_2(x)$ for $x \in Q$. Thus, $|\hat{u}_2|_{1,Q}^2 \le |\hat{u}|_{1,\Pi}^2$ and the desired inequality involving σ_1 holds. Since $|\hat{u}|_{1,\Pi}^2 \le |E\hat{u}_2|_{1,\Pi}^2$ (see (1.5.5)), then $|\hat{u}|_{1,\Pi}^2 \le K_0 |\hat{u}_2|_{1,Q}^2$, and (6.10) holds. $\square$

Theorem 2. Let the conditions of Theorem 1 be satisfied. Suppose that operators $L_2 \in \mathcal{L}^+(H_2)$ and $B \in \mathcal{L}^+(H)$ satisfy

$$\alpha_{2,0} J_Q \le L_2 \le \alpha_{2,1} J_Q, \quad \alpha_{2,0} > 0, \tag{6.11}$$

$$\alpha_0 B \le A \le \alpha_1 B, \quad \alpha_0 > 0, \tag{6.12}$$

Then $S_2(B) \equiv B_{2,2} - B_{2,1} B_{1,1}^{-1} B_{1,2} \in \mathcal{L}^+(H_2)$ and

$$\alpha_{2,0} \sigma_0 \alpha_0 S_2(B) \le L_2 \le \alpha_{2,1} \sigma_1 \alpha_1 S_2(B), \quad \alpha_{2,0} > 0. \tag{6.13}$$

Proof. Theorem 1 leads to the spectral equivalence of J_2 and S_2 (see (6.10)). Thus, $\alpha_{2,0} \sigma_0 S_2(A) \le L_2 \le \alpha_{2,1} \sigma_1 S_2(A)$, $\alpha_{2,0} > 0$. This, together with (6.12) and (1.5.9), leads to (6.13). $\square$

Theorem 3. Let the conditions of Theorem 1 be satisfied. Suppose that the operators L_2 and J_2 are connected by relationship C^3 and $A \asymp B$. Then the operators L_2 and $S_2(B)$ are connected by the same relationship.

Proof. Theorem 1 leads to the spectral equivalence of J_2 and $S_2(A)$. Thus, $J_2 \asymp S_2(B)$. Next, we apply Lemma 4.4 and conclude that $L_2^*(S_2(B))^{-1} L \asymp S_2(B)$. $\square$ [55]

Theorem 4. Let

$$\Gamma_{Q,0} = \Gamma_{\Pi,0} = \emptyset \tag{6.14}$$

and $L_2 = L_2^*$. Suppose the conditions of Theorem 2 are satisfied. Then $S_2(B) = (S_2(B))^* \ge 0$ and

$$\frac{\alpha_{2,0} \alpha_0}{K_0} S_2 \le L_2 \le \alpha_{2,1} \alpha_1 S_2. \tag{6.15}$$

Proof. It suffices to take into account the fact that $J_2 \ge 0$ and modify the reasoning from the proof of Theorem 2. $\square$

6.3. Grid extension theorems; use of regions with slits.

---

[55] It is easy to obtain generalizations of Theorems 1–3 for the case where, instead of $(u, v)_{1,Q}$, more general inner products are used.

   Theorem 5. *Suppose* $Q, \Phi,$ *and the fictitious domains* $F_1, \ldots, F_p$ *are such that, for all distinct* r *and* $m,$

$$F_r \cap F_m = \emptyset, \bar{F}_r \cap \bar{F}_m \subset \partial Q. \tag{6.16}$$

Suppose that, for each $F_r,$ *there exists a domain* $Q_r \subset Q$ *with Lipschitz boundary* ∂Q_r *such that*

$$\partial F_r \cap \partial Q = \partial Q_r \cap \partial Q, \; r \in [1, p]. \tag{6.17}$$

Let $\Gamma_{Q,0} = \Gamma_{\Pi,0}$ *and the closures of all domains given above be unions of certain d-dimensional simplexes so the given quasiuniform triangulations of* $\bar{\Pi}$ *generate corresponding quasiuniform triangulations of* $\bar{Q}_r$ *and* $\bar{F}_r,$ $r \in [1, p].$ *Then, for the corresponding spaces* $\hat{G}_Q$ *and* $\hat{G}_\Pi,$ *there exists a linear extension operator* $E \in \mathcal{L}(\hat{G}_Q; \hat{G}_\Pi)$ *such that (6.9) holds, where* $\hat{v} \equiv E\hat{v}_2$ *and constant* K_0 *does not depend on* $h.$

   Proof. Under the above assumptions, we construct the desired extension for each pair Q_r and F_r separately, $r \in [1, p].$ For arbitrary $\hat{v}_2 \in \hat{G}_Q,$ consider its restriction $\hat{v}_{2,r} \in \hat{G}_{Q_r}.$ Since Q_r is a domain with Lipschitz boundary, then we can define the extension v_r of $\hat{v}_{2,r}$ to the whole space $\mathbf{R}^d$ in accordance with the classical extension theorems (see § 0.1). We can then construct the desired extension $\hat{v}_r$ in several ways (see, e.g., [15, 454, 448]). We give the simplest proof dealing with the Steklov averaging $w_{r,h}(x) \equiv Y_\rho v_r(x)$ with parameter $\rho \asymp h$ (see (2.3.2)), as was done for $d = 2$ and Neumann boundary conditions in [15].

   The function $w_{r,h}$ is continuous; we use its values at nodes belonging to $\bar{F}_r \setminus \partial Q_r$ to define the desired values of the extension $\hat{v}_r.$ Therefore, we obtain a piecewise linear function $\hat{v}_r$ defined on $\bar{Q}_r \cup \bar{F}_r$ such that $\hat{v}_r(x) = \hat{v}_{2,r}(x)$ for $x \in \bar{Q}_r.$ Now we need only prove that

$$|\hat{v}_r|^2_{1,F_r} \le K_r |\hat{v}_{2,r}|^2_{1,Q_r}, \; \forall \hat{v}_2 \in \hat{G}_Q \tag{6.18}$$

(here and below $K \neq K(h)$). To this end, we introduce $\hat{w}_{r,h}$ as the piecewise linear function defined on $\bar{Q}_r \cup \bar{F}_r$ by values of $w_{r,h}$ at nodes in $\bar{Q}_r \cup \bar{F}_r,$ and we make use of the important inequalities

$$|\hat{w}_{r,h} - w_{r,h}|^2_{1,Q_r \cup F_r} \le K_0^* |w_{r,h}|^2_{2,Q_r \cup F_r} \tag{6.19}$$

and

$$|\hat{w}_{r,h} - v_r|^2_{0,Q_r \cup F_r} \le K_1^* h^2 |\hat{v}_{2,r}|^2_{1,Q_r} \tag{6.20}$$

(we will comment on them shortly). Note that (6.19) and (2.3.4) with $\gamma = 0$ lead to

$$|\hat{w}_{r,h}|^2_{1,F_r} \le K_2^* |v_{2,r}|^2_{1,Q_r}. \tag{6.21}$$

It is important that $\hat{w}_{r,h}(x) = \hat{v}_r(x)$ for all elementary simplexes in $\bar{F}_r$ having no vertices in $\bar{Q}$. Hence, it suffices to prove that

$$\sum_T |\hat{v}_r - \hat{w}_{r,h}|^2_{1,T} \le K'_r |\hat{v}_{2,r}|^2_{1,Q_r}, \tag{6.22}$$

where the summation is carried out with respect to all elementary simplexes in $\bar{F}_r$ having at least a vertex in $\bar{Q}$. Now we consider such a simplex T and note that $|\hat{v}_r - \hat{w}_{r,h}|^2_{1,T} \le \bar{K} h^{d-2} \sum [v_2(P_i) - \hat{w}_{r,h}(P_i)]^2$, where $P_i \in (T \cap \partial Q_r)$. Hence,

$$\sum_T |\hat{v}_r - \hat{w}_{r,h}|^2_{1,T} \le \bar{K}_1 h^{d-2} \sum_{P_i \in (\partial F_r \cap \partial Q_r)} (v_2(P_i) - \hat{w}_{r,h}(P_i))^2 . \tag{6.23}$$

Observe also

$$h^{d-2} \sum_{P_i \in (\partial F_r \cap \partial Q_r)} (v_2(P_i) - \hat{w}_{r,h}(P_i))^2 \le \bar{K}_3 |\hat{w}_{r,h} - v_r|^2_{0,Q_r}. \tag{6.24}$$

Combining (6.23), (6.24), and (6.19), we obtain (6.22), which, together with (6.21), leads to (6.18). For $d \le 3$, (6.19) follows from Theorem 2.3.1 because $w_{r,h} \in W^2_2(Q_r \cup F_r)$ and $2 \times 2 > d$. The same theorem implies that $|\hat{w}_{r,h} - w_{r,h}|^2_{0,Q_r \cup F_r} \le \bar{K}_4^* h^2 |\hat{v}_{2,r}|^2_{1,Q_r}$. This and the well-known estimate

$$|v_r - w_{r,h}|^2_{0,Q_r \cup F_r} \le \bar{K}_5^* h^2 |\hat{v}_{2,r}|^2_{1,Q_r}$$

(see, e.g., [394]) leads to (6.20). It is not also very difficult to prove (6.19) and (6.20) for $d \ge 4$ if the triangulation of $\bar{\Pi}$ is regular. But for quasiuniform triangulations, this becomes rather involved, and refer simply to [448], where from the very beginning a more complicated averaging was used. $\square$

As possible generalizations of Theorem 5 to domains with non-Lipschitz boundaries, we consider the domains depicted in Figures 2–5.

Theorem 6. *Consider Q and the fictitious domains F_1, F_2, F_3 depicted in Figure 2. Let $\Gamma_{Q,0} \equiv [B_1, B_2] \cup [B_2 B_3] \cup [B_3 B_4]$, $\Gamma_{\Pi,0} = \Gamma_{Q,0}$ and quasiuniform triangulations of $\bar{\Pi}$ be used for approximation of the spaces defined in (6.4) and (6.5). Then, for the corresponding spaces $\hat{G}_Q$ and $\hat{G}_\Pi$, there exists a linear extension operator $E \in \mathcal{L}(\hat{G}_Q; \hat{G}_\Pi)$ such that (6.9) holds, where $\hat{v} \equiv E\hat{v}_2$ and the constant K_0 does not depend on h.*

Proof. The corresponding Π is represented in Figure 4. It is easy to see that we need to construct the desired extension only for the domain F_1 (in the role of Q_1, we may simply use Q), and this is done exactly as in the proof of Theorem 5. $\square$

Theorem 7. *Consider Q and the fictitious domains F_1 and F_2 depicted in Figure 3. Let $\Gamma_{Q,0} \equiv [A_1, A_{11}] = \Gamma_{\Pi,0}$ and quasiuniform triangulations of $\bar{\Pi}$ be used for approximation of the spaces defined in (6.4) and (6.5). Then the conclusion of Theorem 6 applies.*

Proof. The corresponding Π is a rectangle with one double vertical slit. In this case, we need to construct the desired extension for the domains F_1 and F_1 (they are open triangles $A_3A_4A_5$ and $A_7A_6A_5$, where A_3 and A_7 are elementary nodes corresponding to a double node). In the roles of Q_1 and Q_2, we may use those parts of Q that are to the left and right, respectively, of the straight line containing $[A_2A_3]$. These extensions are constructed as in the proof of Theorem 5. $\square$

Consider now an instructive example where

$$\Gamma_{Q,0} \neq \Gamma_{\Pi,0} \text{ and } \Gamma_{Q,0} \subset \Gamma_{\Pi,0}. \tag{6.25}$$

Theorem 8. *Let Q and the fictitious domains F_1, F_2, and F_3 be depicted in Figure 2. Suppose that $\Gamma_{Q,0} \equiv [B_1, B_2] \cup [B_5B_6]$ and $\Gamma_{\Pi,0} \equiv [B_1, B_6]$ and that quasiuniform triangulations of $\bar{\Pi}$ are used for approximation of the spaces defined in (6.4) and (6.5). Then the conclusion of Theorem 6 applies.*

Proof. It suffices to construct the desired extension only for the domain F_1 (in the role of Q_1, we may simply use Q). Let $\hat{v}_2 \in \hat{G}_Q$ and $R \equiv B_1B_2B_3B_4B_5B_6$ (R is a broken line). Then

$$\|\hat{v}_2\|_{1/2,R} \leq \bar{K}_6 |\hat{v}_2|_{1,Q}. \tag{6.26}$$

Next, define $v_{2,1}$ on ∂F_1 as the function that coincides with $\hat{v}_2$ on $\partial F_1 \cup \partial Q$ and vanishes on $[B_2B_3]$. By to (6.26) and (5.36), we obtain

$$\|v_{2,1}\|_{1/2,\partial F_1} \leq \bar{K}_7 |\hat{v}_2|_{1,Q}. \tag{6.27}$$

Since F_1 is a domain with Lipschitz boundary, we can define the extension $v_{2,1} \equiv v_1$ of the given trace on ∂F_1 to F_1 such that

$$|v_1|_{1,F_1} \leq \bar{K}_8 |\hat{v}_2|_{1,Q} \tag{6.28}$$

(for this it suffices to solve the harmonic equation with the given Dirichlet conditions—this is the best extension in the sense that it minimizes $|v_1|^2_{1,F_1}$). Moreover, we also extend this function so that it vanishes at the points below $[B_1B_6]$ and its Steklov averaging $w_{1,h}(x) \equiv Y_\rho v_1(x)$ will be such that

$$\sum_{P_i \in [B_2B_5]} (w_{1,h}(P_i) - v_1(P_i))^2 \leq \bar{K}_9 |\hat{v}_2|^2_{1,Q} \tag{6.29}$$

(see (6.24) and (6.20)). We can then construct the desired extension $\hat{v}_1$ as in the proof of Theorem 5, with the only difference that instead of the values $w_{1,h}(P_i)$ for $P_i \in [B_2 B_5]$, we use $v_1(P_i) \equiv 0$. The effect of this is estimated on the basis of (6.29). □

Theorem 9. Consider Q and the fictitious domains F_1, F_2, F_3 depicted in Figure 2. Let $\Gamma_{Q,0} \equiv [B_1, B_2]$ and $\Gamma_{\Pi,0} \equiv [B_1 B_{2,5}]$, where $B_{2,5}$ is the midpoint of the segment $[B_2 B_5]$. Suppose quasiuniform triangulations of $\bar{\Pi}$ are used for approximation of the spaces defined in (6.4) and (6.5). Then the conclusion of Theorem 6 applies.

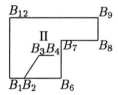

Figure 4. Example of the model region with slits.

Proof. The proof is much the same as the proof of Theorem 8, but now we must modify the boundary conditions on $[B_2 B_5]$ (actually on $[B_{2,5} B_5]$). For this, we use the even (with respect to the point B_5) extension of the function $\hat{v}_2$ considered on $[B_5 B_6]$ (or $[B_5 B_4]$) and take its product with a smooth function $g(x)$ defined on $[B_2 B_5]$ that vanishes on $[B_2 B_{2,5}]$ and equals 1 on $[B'_{2,5} B_5]$, where $B'_{2,5}$ is the midpoint of the segment $[B_{2,5} B_5]$. Then $|g(x)\hat{v}_2(x)|_{1/2,[B_2 B_5]} \le \bar{K}_{10}|\hat{v}_2|_{1,Q}$ (such inequalities are standard and are widely used in context smooth partitions of unity (see [67])). This allows us to apply the solution of the harmonic equation with the given Dirichlet conditions and its appropriate extension to the whole plane. The second significant difference in the proof relates to (6.29), which must now be proved. For $w_{r,h}(x) \equiv Y_\rho v_1(x)$ and $z_i \equiv w_{1,h}(P_i) - v_1(P_i)$, we have

$$\sum_{P_i \in [B_2 B_5]} z_i^2 \le \frac{\bar{K}_{11}}{h}\left(|\hat{w}_{1,h} - w_{1,h}|^2_{0,[B_2 B_5]} + |v_1 - w_{1,h}|^2_{0,[B_2 B_5]}\right). \quad (6.30)$$

Since

$$|v_1 - w_{1,h}|^2_{0,[B_2 B_5]} \le \bar{K}_{12}h|v_1|^2_{1,F_1}, \quad (6.31)$$

then it suffices to prove that

$$|\hat{w}_{1,h} - w_{1,h}|^2_{0,[B_2 B_5]} \le \bar{K}_{13}h|v_1|^2_{1,F_1}. \quad (6.32)$$

As in Theorem 2.3.1, $|\hat{w}_{1,h} - w_{1,h}|^2_{0,[B_2 B_5]} \le \bar{K}_{14}h^2|w_{1,h}|^2_{2,F_1}$, which together with (2.3.4), leads to (6.32). □

We also consider an example when the domain Q may have non-Lipschitz boundary of type different from that considered above. Such a region with no slits is depicted in Figure 5. Π is taken as a rectangle with double slit A_2B_1 and A_5B_1; the points A_2 and A_5 coincide geometrically but represent two different elementary nodes when we use PGMs.

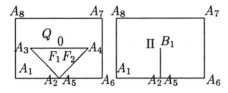

Figure 5. Example of original and model regions with non-Lipschitz boundaries.

Theorem 10. Consider Q and the fictitious domains F_1 and F_2 depicted in Figure 5. Let $\Gamma_{Q,0} \equiv [A_{11}, A_{10}]$ and $\Gamma_{\Pi,0} = \Gamma_{Q,0}$. Suppose that the point B_1 in Π is such that $B_1 = O$ and that quasiuniform triangulations of $\overline{\Pi}$ are used for approximation of the spaces defined in (6.4) and (6.5). Then the conclusion of Theorem 6 applies.

Proof. We construct the desired extensions for the triangles $F_1 \equiv \triangle A_2 A_3 O$ and $F_2 \equiv \triangle A_5 A_3 O$ separately and apply the same reasoning as in the proof of Theorem 5. $\square$ [56]

6.4. Dirichlet boundary conditions and generalizations of the Treftz method. Consider very briefly the case of a Lipschitz boundary $\Gamma_Q = \Gamma_{Q,0}$ that is at a positive distance from Γ_Π. It has been suggested (see [354]) that (6.1) be replaced by

$$L\mathbf{u} \equiv \left[\begin{array}{cc} A'_{1,1} & 0 \\ A'_{2,1} & L_2 \end{array} \right] \left[\begin{array}{c} \mathbf{u}_1 \\ \mathbf{u}_2 \end{array} \right] = \left[\begin{array}{c} 0 \\ \mathbf{f}_2 \end{array} \right] \equiv \mathbf{f}.$$

Under suitable choices of the operators $A'_{1,1}$ and $A'_{2,1}$, the rate of convergence of the iterative method $A(\mathbf{u}^{n+1} - \mathbf{u}^n) = -\tau_n(L\mathbf{u}^n - \mathbf{f})$ was found independent of h. Perhaps, the most logical approach is to construct and analyze similar methods on the basis of elimination of all nodes not belonging to Γ_Q, as was done in § 5. This is typical at the development of the Treftz method. The remaining set of nodes is denoted by C_h. For Q with Lipschitz boundary $\Gamma = \Gamma_0$ and $\Gamma \cap \partial\Pi = \emptyset$, consider uniform rectangular grid and grid system $-h_1 h_2(\Delta_1 u + \Delta_2 u)_i = 0$ for $P_i \in Q_h$, $u_i = \varphi_i$ for $P_i \in \Gamma_h \equiv C_h$.

---

[56] It is also possible to define B_1 in Theorem 10 as the midpoint of $[A_2 O]$ ($[A_5 O]$). Then the extensions for the triangles F_1 and F_2 should generate the same trace on $[B_1 O]$, which vanishes in a prescribed vicinity of B_1. We now make some general remarks about conditions of type (6.6). It is very important that new connected parts of $\Gamma_{\Pi,0} \setminus \Gamma_{Q,0}$ either are continuations of the old ones from $\Gamma_{Q,0}$ or are at a positive distance from $\Gamma_{1,0} \equiv \Gamma_Q \setminus \Gamma_{Q,0}$. Under this assumption, we can obtain useful generalizations even for d-dimensional problems (e.g., if $\overline{\Pi} \equiv [0,3] \times [0,3] \times [0,2]$, $\overline{F} \equiv [1,2] \times [1,2] \times [0,1]$, and $\Gamma_{\Pi,0}$ and $\Gamma_{Q,0}$ correspond to those parts of the boundaries belonging to the plane $x_3 = 0$ (see [454])). Composite grids with an increasing number of local refinements are of special interest (see [455, 448]).

(The general case of a grid Poisson equation can be readily reduced to the one under consideration if we can effectively solve the corresponding grid Poisson equation for the model region II.) We write these systems in the form $A(\mathbf{u}^{(\Pi)}) = \mathbf{f}^{(\Pi)}$ and use the standard block structure of $A \equiv A^{(\Pi)}$ generated by the splitting $\mathbf{u}^{(\Pi)} \equiv \mathbf{u} \equiv [\mathbf{u}_1, \mathbf{u}_2]$, where the vector $\mathbf{u}_1$ consists of values of $\mathbf{u}$ at nodes in C_h. Vectors of traces on C_h of $\mathbf{u}(\Pi)$ satisfying this system with $\mathbf{f}_1 = 0$ can be described as solutions of $S_2 \mathbf{u}_2^n = \mathbf{f}_2$, where $S_2 \equiv S_2(A) \in \mathcal{L}^+(H_2)$. Hence, our goal is to find $\mathbf{u}_2 = \operatorname{argmin} \|\mathbf{u}_2 - \varphi_2\|_{D_2}^2$ using iterations of type $S_2(\mathbf{u}_2^{n+1} - \mathbf{u}_2^n) = -\tau_n D_2(\mathbf{u}_2^n - \varphi_2)$. It is desirable to choose a model operator D_2 that is nearly spectrally equivalent to S_2 and so that the evaluation of $D_2 v_2$ for a given vector v_2 can be done in a nearly asymptotically optimal way. Hence, in the role of D_2 we can use $B_2^{(2)} \equiv \Lambda_2$ defined by (5.55) and (5.56). Then our iterations yield nearly asymptotically optimal algorithms for solving the original grid system as was shown in [455].

§ 7. Multigrid methods; multigrid construction of asymptotically optimal preconditioners

7.1. Basic computational algorithms and classes of multigrid methods. A variety of different iterative methods may be understood under the term *multigrid methods*. Their characteristic feature is the use of a sequence of grids $Q_{h_l} \equiv Q^{(l)}, l = 0, \ldots, p$, with mesh size $h^{(l)} = 2^{p-l} h^{(p)}$, $l \in [0, p]$ (that is, with $h^{(l)} = 2^{-l} h^{(0)}$, $l \in [0, p]$), and a family of corresponding grid problems

$$L_l \mathbf{u}_l = \mathbf{f}_l, \quad \mathbf{u}_l \equiv u_l \in H_l, \quad \mathbf{f}_l \equiv f_l \in H_l, \quad l \in [0, p], \qquad (7.1)$$

used for obtaining an approximation to the solution of the original problem $L_p \mathbf{u}_l = \mathbf{f}_p$, $\mathbf{u}_p \in H_p$, $\mathbf{f}_p \in H_p$, constructed on the finest grid $Q^{(p)} \equiv Q_h$. Here, the parameter l is usually referred to as the level of the grid in the given sequence and H_l is the standard Euclidean space of grid functions defined at nodes $P_i \in Q^{(l)}$, where $l \in [0, p]$.

The simplest variants of such methods, one-way multigrid methods, were already considered in § 1.4 (coarse grid continuations applied to a basic rapidly convergent iterative method). As we saw, they can lead to asymptotically optimal algorithms for solving grid systems provided that $h^{(0)} \asymp 1$ ($p \asymp |\ln h|$). We started there with efficient iterative methods that were already available. One of the most important factors in the acceleration process was the expected closeness of the solutions corresponding to two neighboring levels. But could we not take advantage of this factor for constructing an effective iterative method itself? If so, thus we must now consider using the coarser levels of the grids for supporting each fine grid iteration, so the process becomes two-way.

Such possibilities began to be discussed around 1960 (see [2, 1, 452]). The first effective iterative multigrid method for the Dirichlet problem for difference Poisson's equation on a square was suggested by Fedorenko [228]. Heuristically speaking, the basic idea is to reduce shortwavelength error components with a so-called smoothing procedure and to approximate longwavelength components with a special correction procedure, which is carried out on coarser grids using the residual corresponding to the smoothed approximation. It is very important that, for smoothing, a simple iterative method can be applied: during this step, it is not necessary to decrease error but to reduce its shortwavelength components and to obtain a smooth residual. It was shown that one can obtain an ε-approximation to the solution of the difference system for grid parameter $h \equiv h^{(p)}$ with computational work $W(\varepsilon, h) = O(|\ln \varepsilon|/h^2)$. This result was significantly improved by Bakhvalov (see [37]) by extending this to multigrid methods for more general elliptic problems with variable coefficients. He showed that the estimate holds under a certain correctness property for the difference scheme approximating the Dirichlet problem for a general second-order elliptic equation in a rectangle; this important property which deals with a priori estimates of the difference solution in the norm corresponding to a difference approximation of the space $W_2^2(Q)$, was proved in [156, 164] (see, also, § 6.2 and [266]). Bakhvalov (see [37]) also constructed multigrid algorithms leading to the asymptotically optimal estimate $W(\varepsilon, h) = O(1/h^2)$ provided that $\varepsilon \asymp h^k, k > 0$, and that the solution of the original elliptic boundary value problem is sufficiently smooth in the classical sense. Actually, these algorithms involve a multigrid acceleration of the basic multigrid method of Fedorenko type and are now often referred to as *full multigrid methods*. In [164], the use of multigrid methods for inner iterations was discussed. Then convergence had been proved only in a rather weak space that approximated the space $L_2(Q)$, even for problems with symmetric and positive definite operators. The desirability to modify multigrid constructions in such a way that they would yield asymptotically optimal model operators (spectrally equivalent to the original positive grid operator) was stressed in [162] (recall that modifications of the classical iterative methods based on the use of such operators yield optimal convergence in the energy norm (see § 1.3)).

A very important development was made by Astrakhantsev (see [13]), who obtained optimal estimates for projective-grid approximations of second-order elliptic equations with natural boundary conditions in a polygonal domain. Later investigations (see [89, 267, 270, 360, 364]) drew wide attention and marked the beginning of a rapid development in the theory and application of these methods, which preserved the main features of Fedorenko's method. Although their study is rather complicated, the key questions have now become more or less transparent and standardized: so-called "regularity and approximation assumptions" are used very often (see [89, 267, 270, 349, 360, 364, 350, 361, 362, 367, 499]). Moreover, many researchers understood the significance of weakening these assumptions. The designed *algebraic multigrid methods* turned out very effective for various grid systems (see [83, 87, 89, 133, 267, 270, 291, 350, 361, 362]).

It is likely that, for problems with symmetric and nonnegative operator (precisely such problems are discussed in this section), the most logical multigrid constructions must yield asymptotically optimal model operators with no assumption on smoothness of the solution (it was stressed in [162]).

Currently, several fruitful multigrid approaches to constructing optimal and nearly optimal preconditioners can be suggested to the reader (see, e.g., [24, 50, 48, 76, 78, 203, 205, 322, 512]).

First of all, the importance of an appropriate splitting of the original finite element space into almost orthogonal subspaces should be emphasized (see [49, 344], § 1.5 and § 2.5). Such considerations lead to the so-called *hierarchical basis multigrid methods* (see [50, 48, 512]). For two-dimensional problems, they yield operators $L_p \equiv Y_h \in \mathcal{L}^+(H)$ (see (3.17)) such that $\delta_0 I_h \leq Y_h \leq \delta_1 I_h$, where $\delta_1/\delta_0 = O(p^2)$ (the identity operator is a nearly asymptotically optimal preconditioner).

Asymptotically optimal multigrid preconditioners were constructed in [22, 23, 203, 204, 205, 289] on the basis of similar recursive splittings of the finite element subspace, but in combination with the theory of cooperative operators discussed in § 1.5. We concentrate on them in the final subsections; such operators were obtained even for approximations of d-dimensional problems; the constants δ_0 and δ_1 in (0.3.11), even for problems associated with strongly discontinuous coefficients (see (1.5.12)), are independent of them and very close to 1 (see [203, 204, 205]).

Finally, we remark that surveys of recent developments and of the wide-ranging applicability of multigrid, or multilevel principles, for finite-element systems can be found in [362, 367]; difference systems were considered in [267, 270, 250, 251, 404, 491, 499]). It should also be emphasized that many authors spread confidence in multigrid methods by providing efficient and robust computer codes, some of which are so-called "black box" types (see [133, 270, 362, 367, 499] and references therein).

From this short excursion into the history of multigrid methods, we move to brief consideration of the basic stages of conventional multigrid iteration. We must specify the smoothing and define the operators (intergrid transfers) $I_l^{l+1} \in \mathcal{L}(H_l; H_{l+1})$ and $I_{l+1}^l \in \mathcal{L}(H_{l+1}; H_l)$ that are responsible for the transfer of the data between the spaces H_l and H_{l+1}, where $l \in [0, p-1]$. This can be done done in several ways; often these choices are crucial. But for PGMs their choice is natural. Indeed, suppose that systems (7.1) with $l = p$ correspond to projective problems of type $b_L(\hat{u}_p; \hat{v}_p) = l(\hat{v}_p)$ and that the matrices in (7.1) may be rewritten as

$$L_l \equiv (b_L(\hat{\psi}(l)_j; \hat{\psi}(l)_i)), i \in [1, N_l], j \in [1, N_l], \quad l \in [0, p], \qquad (7.2)$$

where $L_l \in \mathcal{L}^+(H_l), l \in [0, p]$. Consider a nested sequence of the approximating subspaces:

$$\hat{G}_l \equiv \text{lin } \{\hat{\psi}(l)_i\} \subset \hat{G}_{l+1} \subset \hat{G}_p \subset G, \quad l \in [0, p-1] \qquad (7.3)$$

(we write here $\hat{G}_l$ instead of $\hat{G}_{h(l)}, l \in [0, p]$). Then it is natural to choose an interpolation (prolongation) operator I_l^{l+1} with $l \in [0, p-1]$ such that $\hat{u}_l = \widehat{I_l^{l+1} u_l} \in \hat{G}_{l+1}, \forall \hat{u}_l \in \hat{G}_l, l \in [0, p-1]$, and for I_{l+1}^l with $l \in [1, p]$, to take the restriction operator

$$I_{l+1}^l \equiv (I_l^{l+1})^*, l \in [0, p-1] \qquad (7.4)$$

(see [267, 364]). This yields rank $I_l^{l+1} = \dim H_l, l \in [0, p-1]$, and in important representation $L_l = I_{l+1}^l L_{l+1} I_l^{l+1}, l \in [0, p-1]$. Moreover, each Euclidean space H_l with $l \in [1, p]$ is split according to

$$H_l = H_{l,0} \oplus H_{l,0}^\perp, \quad l \in [1, p], \qquad (7.5)$$

where $H_{l,0} \equiv \operatorname{Im} I^l_{l-1}$, $H^\perp_{l,0} \equiv L^{-1}_l \{ \operatorname{Ker} I^{l-1}_l \}$, $l \in [1,p]$, and the orthogonal sum is understood in the sense of the Euclidean space $H_l(L_l)$, $l \in [1,p]$.[57] Next, suppose that we have an iterate $u^n_p \equiv v^0_p$ approximating the solution of the original problem $L_p u_p = f_p$ and that we perform k_1 iterations of a relatively simple iterative method (usually referred to as the **smoother**), e.g., of type

$$v^{k+1}_p = v^k_p - \tau^{(1)}_{p,k}(L_p v^k_p - f_p), \quad k = 0, \ldots, k_1 - 1 \qquad (7.6)$$

(actually the smoother must be available on every grid). Suppose also that the error, though possibly not decreased significantly, becomes significantly smoother in the sense that components in the subspace $H^\perp_{p,0}$ of the error $\tilde{z}^n \equiv \tilde{u}^n_p - u_p$ for the obtained smoothed iterate $\tilde{u}^n_p \equiv v^{k_1}_p$ become sufficiently small.[58] With $L_p \tilde{u}^n_p - f_p \equiv \tilde{r}^n_p$, then $L_p \tilde{z}^n_p = \tilde{r}^n_p$.[59] Since $\tilde{z}^n_p$ is smooth, then it should be possible to approximate it by a coarser grid function z_{p-1}. An appropriate choice is the solution of $L_{p-1} z_{p-1} = g_{p-1} \equiv I^{p-1}_p \tilde{r}^n_p$, where $g_{p-1} \approx \tilde{r}^n_p$. The approximation $\tilde{z}^n_p \equiv I^p_{p-1} L^{-1}_{p-1} g_{p-1}$ can then be regarded as a desirable correction to $\tilde{u}^n_p$, yielding the *coarse grid correction* $u^{n+1}_p \equiv \tilde{u}^n_p - R_p \tilde{r}^n_p$, where

$$R_p \equiv I^p_{p-1} L^{-1}_{p-1} I^{p-1}_p. \qquad (7.7)$$

This describes a single multigrid (two-grid) iteration under assumption that we can solve systems on grid Q^{p-1}. Now we observe that the system on level $p - 1$ has the same form as the fine grid system. Hence, the coarse grid solution can be approximated by a few (say, k_2) inner iterations of the same multigrid scheme (smoothing and coarse grid correction) but now on the coarser level. Carrying this concept recursively to yet coarser levels, we have

$$z^{m+1}_{p-1} = z^m_{p-1} - \tau^{(2)}_{p-1,m} R^{k_2}_{p-1}(L_{p-1} z^m_{p-1} - g_{p-1}), \quad m = 0, \ldots, k_2 - 1, \quad (7.8)$$

where $z^0_{p-1} = 0$, $z^{k_2}_{p-1}$ serves as an approximation to z_{p-1}, and the operator $R^{k_2}_{p-1}$ serves as an approximation to $I^{p-1}_{p-2} L^{-1}_{p-2} I^{p-2}_{p-1}$; $R^{k_2}_{p-1}$ can be described

---

[57] If we could determine that the subspaces $H_{p,0}$ and $H^\perp_{p,0}$ were invariant with respect to L_p, it would have been appropriate to refer to the corresponding error components in these subspaces as smooth and oscillatory, respectively.

[58] Usually this presents no principal problem from a theoretical point of view, but is very essential for practical properties of the methods under consideration; currently, more sophisticated smoothers are often used, so that the concept of smoothing becomes more vague and is gradually replaced by the more general notion of composite methods based on multigrid splittings and different iterations for the corresponding subspaces.

[59] This system differs from the original one only by its right-hand side.

in terms similar to two-stage preconditioners (see § 4) as
$R_{p-1}^{k_2} = I_{p-2}^{p-1}(I_{p-1} - Z_{p-1,k_2})L_{p-2}^{-1}I_{p-1}^{p-2}$, where Z_{p-1,k_2} is the corresponding
error reduction operator in k_2 multigrid iterations on the level $p-1$. Hence,
one complete multigrid iteration on the level p can be defined by $u_p^{n+1} \equiv \tilde{u}_p^n - R_p^{k_2}\tilde{r}_p^n$, but involves the multigrid iterations on the levels $p - 1, \ldots, 1$
(systems on level 0 can be regarded as easily solved by a special direct
method). This allows an algorithmic implementation of the method in a
sufficiently straightforward manner. [60]

**7.2. Simplest estimates of convergence for two-grid meth-
ods.** We confine ourselves to the analysis of two-grid methods, which
deal only with $Q^1 \equiv \Omega_{2h}$ and $Q^2 \equiv \Omega_h$ and where the systems on Ω_{2h}
are solved exactly (estimates for multigrid methods are less optimistic).
We deal with $l = 0, 1$ and the splitting $H_h \equiv H_1 = H_{h,0} \oplus H_{h,0}^\perp$, where
$H_{h,0} \equiv H_{1,0}$, $H_{h,0}^\perp \equiv H_{1,0}^\perp$. The errors at the nth and $(n + 1)$th two-grid
iterations are connected by the relation $z_h^{n+1} = Z_h z_h^n$, where

$$Z_h \equiv (I - R_h L_h)Q_{k_1}, \tag{7.9}$$

$$Q_{k_1} \equiv (I - \tau_0^{(1)}A_h^{-1}L_h)\ldots(I - \tau_{k_1-1}^{(1)}A_h^{-1}L_h), \tag{7.10}$$

$R_h \equiv I_{2h}^h L_{2h}^{-1} I_h^{2h}$, and Q_{k_1} corresponds to the smoothing steps [61]

$$u_h^{k+1} = u_h^k - \tau_k^{(1)}A_h^{-1}(L_h u_h^k - f_h), \quad k = 0, \ldots, k_1 - 1. \tag{7.11}$$

Consider a related "adjoint" two-grid given by

$$z_h^{n+1} = Z_{h,0}z_h^n, \quad Z_{h,0} \equiv Q_{k_1,0}(I - B_h^{-1}L_h), \tag{7.12}$$

where $Z_{h,0} \equiv Q_{k_1,0}(I - R_h L_h)$ and

$$(Q_{k_1})_L' = Q_{k_1,0}; \tag{7.13}$$

---

[60] For $d = 2$, one multigrid iteration requires $O(1/h^2)$ arithmetic operations if $k_2 \leq 3$
(we will present similar but more general estimates a little later); if $k_2 = 1$ or $k_2 = 2$,
then we have the so-called V or W-cycles, respectively. Finally, we note that $H_{p,0}$ is an
invariant subspace with respect to the operator R_p (see (7.7)). For the restriction $R_{p,0}$
of R_p to this subspace (we regard it as the corresponding Euclidean space), it is possible
to show that $R_{p,0} \in \mathcal{L}^+(H_{p,0})$. If $H_{p,0}$ were also an invariant subspace with respect to
the operator L_p, then it could be expected that $\delta_0 B_{p,0} \leq L_{p,0} \leq \delta_1 B_{p,0}$, $\delta_0 \approx 1 \approx \delta_1$,
where $B_{p,0} \equiv R_{p,0}^{-1}$. So it seems reasonable to pay special attention to construction and
study of multigrid methods that are consistent with the idea of spectral equivalence.

[61] Note that (7.11) is of a more general form than (7.6) and that it contains, e.g.,
Gauss-Seidel iterations or modified Richardson methods with factored or block diagonal
model operators.

here, $(Q)'_L$ refers to the adjoint of $Q \in \mathcal{L}(H_h(L_h))$.[62]

Lemma 1. *Let (7.13) be satisfied. Suppose $Q_{k_1,0}$ satisfies*

$$\|Q_{k_1,0}z\|_L^2 \leq \|z_0\|_L^2 + q^2\|z_1\|_L^2, \qquad (7.14)$$

where $z = z_0 + z_1$, $z_0 \in H_{h,0}$, $z_1 \in H_{h,0}^\perp$, $\forall z$. *Then*

$$\|Z_{h,0}\|_L \leq q, \quad \|Z_h\|_L \leq q. \qquad (7.15)$$

Proof. The operator $I - R_h L_h \in \mathcal{L}(H_h(L_h))$ is symmetric. Hence, $(Z_{h,0})'_L = Z_h$, $\|Z_{h,0}\|_L = \|Z_h\|_L$ (see (7.13)). Moreover, $I - R_h L_h$ is the orthoprojector of $H_h(L_h)$ onto $H_{h,0}^\perp$. This and (7.14) lead to (7.15). $\square$ [63]

Suppose that sp $L \subset [c_0 h^2, c_1]$, $c_0 > 0$ (recall that this is the case for PGMs, but for difference methods we usually have sp $L \subset [c_0, c_1/h^2]$) and that $\{y_1, \ldots, y_N\}$ is an orthonormal basis of $H \equiv H_h$ such that $L_h y_i = \lambda y_i$, $i \in [1, N]$. Then, for $0 < \theta < 1$, we define G_θ as the linear span of the vectors y_i with $\lambda_i \in [c_0 h^2, \theta c_1]$ and its orthogonal complement $G_\theta^\perp$ in the sense of the Euclidean space $H_h(L_h)$. If, additionally, we introduce

$$\kappa(\theta, h) \equiv \max_{u \in G_\theta, \|u\|_L = 1} \text{dist}\,\{u; H_{h,0}\} < 1, \qquad (7.16)$$

then, for small enough θ, it is natural to expect that $\kappa(\theta, h) < 1$ because $\dim H_{1,0} \asymp h^{-2}$ and $\dim G_\theta \asymp h^{-1}$. This implies that the angles α between elements of G_θ and the subspace $H_{h,0}$ in the sense of the Euclidean space $H_h(L_h)$ are such that $\cos \alpha \leq \kappa(\theta, h) < 1$. The same estimate for the angles between elements of $G_\theta^\perp$ and the subspace $H_{h,0}^\perp$ holds due to the simple relation $\kappa(\theta, h) \equiv \max_{u \in H_{h,0}^\perp, \|u\|_L = 1} \text{dist}_L\{u; G_\theta^\perp\}$.

Lemma 2. *Suppose that the operator $Q_{k_1,0}$ satisfies*

$$\|Q_{k_1,0}w\|_L^2 \leq \|w_\theta\|_L^2 + q_0^2(\theta, h)\|w_\theta^\perp\|_L^2, \quad \forall w \in H_h, \qquad (7.17)$$

where $w = w_\theta + w_\theta^\perp$, $w_\theta \in G_\theta$, $w_\theta^\perp \in G_\theta^\perp$. *Suppose also that (7.13) and (7.15) are satisfied. Then*

---

[62] If $A_h \in \mathcal{L}^+(H_h)$, then $(Q_{k_1})'_L = Q_{k_1}$ and (7.13) implies that the new method is different from the original only in the order in which the smoothing and correction procedures are used. But if the original two-grid method uses Gauss-Seidel iterations in a prescribed order of unknowns, then the new method uses the same iterations but in the opposite order and after the correction procedure.

[63] Lemma 1 (see [360]) implies that it is possible to attain the desired accuracy ε by the cost of only one of two-grid iteration of either type if the conditions of this lemma are satisfied and $q \leq \varepsilon$. For the two-grid method associated with (7.12), we observe a certain analogy to coarse grid continuation applied to the given smoothing procedure. This analogy becomes even more clear if we make use of Lemma 2 dealing with the subspaces G_θ and $G_\theta^\perp$ defined below (this is consistent with the first variants of multigrid methods).

$$\|Z_{h,0}\|_L = \|Z_h\|_L \leq q_0(\theta, h) + \kappa(\theta, h). \tag{7.18}$$

Proof. Together with the above expansion for w, consider the similar one $z = z_0 + w$ (see Lemma 1). Then $\|Z_{h,0}z\|_L = \|Q_{k_1,0}z_1\|_L \equiv X$, $X \leq \|Q_{k_1,0}w_\theta^\perp\|_L + \|Q_{k_1,0}w_\theta\|_L$, and $\|Z_{h,0}z\|_L \leq q_0(\theta,h)\|w_\theta^\perp\|_L + \kappa(\theta,h)\|z_1\|_L$ (see (7.17)), which leads to (7.18). $\square$ [64]

7.3. Symmetrized multigrid method. Consider now a two-grid method consisting of iteration (7.9) and a subsequent smoothing procedure with error reduction operator $Q_{k_1}^{(s)}$ such that $Q_{k_1}^{(s)} = (Q_{k_1})'_L$ (see (7.13)). Since two-grid error reduction operator $Z_1 \in \mathcal{L}(H_h(L_h))$ is symmetric (shown in the following proof), we use the terms as the *symmetrized two-grid method*.

Lemma 3. *Suppose that the conditions of Lemma 2 are satisfied. Then the error reduction operator Z_M for M iterations of the symmetrized two-grid method is such that*

$$(Z_M)'_L = Z_M, \quad \|Z_M\|_L \leq (q_0(\theta, h) + \kappa(\theta, h))^{2M}.$$

Proof. The error reduction operator Z_1 for one iteration of the method has the form $Z_1 = Q_{k_1}^{(s)}(I - R_h L_h)Q_{k_1}$. Since $I - R_h L_h$ is an orthoprojector, then $(I - R_h L_h)^2 = I - R_h L_h$. Hence, $Z_1 = Z_{h,0}Z_h$. Note that the adjoint of $Z_{h,0} \in \mathcal{L}(H_h(L_h))$ is Z_h. Thus, Z_1 is symmetric as an element of the same space and $\|Z_1\|_L = \|Z_h\|_L^2$. Observe that $\|Z_M\|_L = \|Z_1\|_L^M = \|Z_h\|_L^{2M}$. This together with (7.18) leads to the desired estimate. $\square$ [65]

7.4. Multigrid construction of asymptotically optimal preconditioners. [66] Probably the most fruitful idea in this direction is connected

---

[64] Suppose that (7.6) represents Richardson's method with iteration parameters $\tau_k^{(1)}$ defined by formulas of type (1.3.21) with $\delta_0 \equiv \theta c_1$ and $\delta_1 \equiv c_1$, that is, with $\delta_1/\delta_0 = 1/\theta$. Then (7.17) holds with $q_0(\theta, h) \leq q_0$, where q_0 is an arbitrary fixed positive number and $k_1 \asymp 1$ is sufficiently large. Hence, (7.18) leads to the estimate $\|Z_h\|_L \leq q < 1$ with constant q independent of h provided $\kappa(\theta, h) \leq \kappa(\theta) < 1$ with $\kappa(\theta)$ independent of h. The use of $[\delta_0, \delta_1]$, which contains only a part of the spectrum of L_h, is the main difference between the multigrid algorithm and coarse grid continuation applied to the basic Richardson's iterations.

[65] We emphasize that (7.19) means that the multigrid methods under consideration can be used as effective inner iterations (see § 4). Other results of the same type have been established under other conditions (see [267, 350, 362, 367]).

[66] Several attempts have been made to construct multigrid methods that require no regularity properties of the original elliptic boundary value problem and its grid approximation. We mention iterations $u_p^{n+1} = u_p^n - \gamma_n R_p^{-1}(L_p u_p^n - f_p)$ with a model nonsymmetric operator R_p. The iteration parameter γ_n can be chosen in accordance with the method of minimal residuals (see § 1.3), that is, from the condition $\min_{\gamma_n}\|L_p u_p^{n+1} - f_p\|^2$. To obtain the vector $v_p^{n+1} \equiv R_p r_p^n$ for the residual $r_p^n \equiv r_{(p)} \equiv L_p u_p^n - f_p$, the method uses

with the study of angles between the subspaces in splittings of the original finite element space (see § 2.5) and the use of hierarchical bases (see [49, 344] and, for $d = 1$, [517]). As we have already mentioned it suffices to replace the standard nodal basis by hierarchical basis in order to obtain for $d = 2$ a grid operator that is nearly spectrally equivalent to the identity operator (see [512]). The hierarchical basis multigrid methods developed later (see [50, 48]) also deal with block form (3.17) and can be considered as modifications of the symmetric block Gauss-Seidel method that uses approximations of diagonal blocks constructed in accordance with the theory of two-stage operators (see (4.3)); actually, it is assumed only that systems with the block $Y_{0,0}$ are solved exactly. It is again noteworthy that the matrix Y_h need not be assembled and stored explicitly, and that nearly asymptotically optimal estimates have been obtained even for very nonuniform grids.

Now we concentrate on multigrid constructions of asymptotically optimal preconditioners $B_h \in \mathcal{L}^+(H)$ suggested in [22, 23, 203, 204, 205] and which use hierarchical bases associated with separate two-grid splittings for designing model cooperative operators (see § 1.5). They lead to estimates

$$\sigma_0 B_h \le \Lambda_h \le \sigma_1 B_h, \qquad (7.19)$$

where $\Lambda_h \in \mathcal{L}^+(H)$ is the original grid operator; the positive constants σ_0 and σ_1 are independent of the grid and close to unity even for d-dimensional problems with strongly discontinuous coefficients.

In what follows we suppose that $\bar{Q}$ consists of a finite number of d-dimensional simplexes $T_{0,k}$. In dealing with subspaces of the space $W_2^1(Q; \Gamma_0)$, we also assume that Γ_0 is a part of the boundary of the domain Q and a union of d-dimensional faces of the simplexes $T_{0,k}$ (this means that the domain Q may have a non-Lipschitz boundary). The collection

---

its projections $r_{(l-1)} \equiv I_l^{l-1} r_{(l)}$, $l = p, \ldots, 1$ and the recursive procedure $v_0 \equiv L_0^{-1} r_{(0)}$, $A_l v_l + (L_l - A_l) I_{l-1}^l v_{l-1} = r_{(l)}$, $l = 1, \ldots, p$. Here we can take a sufficiently simple model operator A_l leading to easily solvable systems, e.g., with a diagonal or with factored matrix (see § 3). The last vector v_p defines u_p^{n+1}, and the operator R_p^{-1} is obtained by the recursive procedure $R_l^{-1} = A_l^{-1} \{ I_p - (L_l - A_l) I_{l-1}^l R_{l-1} I_l^{l-1} \}$, $l \ge 1$, where $R_0^{-1} \equiv L_0^{-1}$. If $\gamma_n = 1$, then $\mathrm{Ker}\,(I_p - R_p^{-1} L_p) = H_{p,0}$ and Lemma 1 applies. This method was used with success for solution of many practical three-dimensional elasticity problems (see [62]), its theoretical investigation was not promising for more general situations. It is likely that many variants of multigrid methods have similar features for some concrete problems, but for many their effectiveness becomes doubtful when the role of high oscillatory error components increases its significance and the asymptotic estimates for the rate of convergence become suboptimal. We are thus especially interested in multigrid methods that can be regarded as multigrid constructions of asymptotically optimal or nearly asymptotically optimal preconditioners. Such variants are often referred to as algebraic multigrid methods (see, e.g., [24, 49, 87, 128, 133]).

of simplexes $T_{0,k}$ defines a generalized triangulation $T^{(0)}(\bar{Q}) \equiv T^{(0)}$ (it is a standard triangulation if Q has Lipschitz boundary). All triangulations $T^{(l+1)}(\bar{Q}) \equiv T^{(l+1)}$ of levels $l+1 \in [1,p]$ are obtained by the use of the recurrent refinement of triangulations of the previous level with the corresponding refinement ratio $t_{l+1} > 1, l \in [1,p]$, that is, each simplex $T_l \in T^{(l)}$ is partitioned into a set of t_{l+1}^d congruent subsimplexes $T_{l+1} \in T^{(l+1)}$, with $l \in [0, p-1]$ (see § 2.1). In this way, we obtain a sequence of nested triangulations. We assume that all simplexes in $T(0)$ are regular simplicial parts of cubes with edges parallel to the original coordinate axes. This implies that simplexes in each triangulation $T^{(l)}, l \in [1,p]$ also preserve this property and that now we have $h_{l+1} \equiv h_l/t_{l+1}, l \in [0, p-1]$.

With each triangulation $T^{(l)}$, we associate a standard finite element subspace $\hat{G}^{(l)} \subset G \equiv W_2^1(Q; \Gamma_0)$ consisting of functions on Q that are piecewise linear with respect to this triangulation, $l \in [0,p]$ (see § 2.1) and that vanish on Γ_0 (here, we may include in Γ_0 even some k-dimensional faces of the simplexes in $T(0)$, where $k \in [0, d-2]$, although this makes no sense for the Hilbert space G). Let

$$(u,v) = \sum_{T_{0,k} \in T^{(0)}(\bar{Q})} a(T_{0,k})(u,v)_{1,T_{0,k}}, \qquad (7.20)$$

with arbitrary positive constants $a(T_{0,k})$; the particular case of (7.20) corresponds to $(u,v) = (u,v)_{1,Q}$. Let $Q^{(l)}$ be a set of vertices $P_i^{(l)}$ (recall that elementary vertices or elementary nodes must be used if slits in the region are present (see § 2.1)) of simplexes T_l, which do not belong to Γ_0 and each which are in correspondence with the standard basis functions $\hat{\psi}_i^{(l)}(x)$ on Q that are continuous piecewise linear such that $\hat{\psi}_i^{(l)}(P_i^{(l)}) = 1$ and $\hat{\psi}_i^{(l)} = 0$ at the remaining nodes, and $\hat{\psi}_i^{(l)}(x)$ is linear on each $T_l \in T^{(l)}(Q)$. Then

$$\hat{G}^{(l)} \equiv \{\hat{u} : \hat{u} = \sum_{P_i^{(l)} \in Q^{(l)}} u_i \hat{\psi}_i^{(l)}(x)\}, \quad l \in [0,p], \qquad (7.21)$$

where N_{l+1} is the number of nodes (elementary nodes) in $Q^{(l+1)}$, $N_{l+1} = N_l + N_l^{(1)}$, $\mathbf{R}^{N_{l+1}} \equiv H^{(l+1)} = H_1^{(l+1)} \times H_2^{(l+1)}$, $H_2^{(l+1)} = H^{(l)}$, and $\mathbf{u}_{l+1} = \{u_i\} \in H^{(l+1)}$, $\mathbf{u}_{l+1} = [\mathbf{u}_1^{(l+1)}, \mathbf{u}_2^{(l+1)}]^T$, $\mathbf{u}_s^{(l+1)} \in H_s^{(l+1)}$, $s = 1,2$. Along with the basis $\{\hat{\psi}_i^{(l+1)}(x)\}$ for $\hat{G}^{(l+1)}, l \in [0, p-1]$, we consider the hierarchical basis defined by (3.11) and (3.12) and leading to the splitting

$$\hat{G}^{(l+1)} = \hat{G}_1^{(l+1)} \oplus \hat{G}_2^{(l+1)} \subset G, \quad l \in [1, p-1], \qquad (7.22)$$

$$\hat{G}_2^{(l+1)} = \hat{G}^{(l)}, \ \hat{G}_1^{(l+1)} \equiv \{\hat{u} : \hat{u} \in \hat{G}^{(l+1)}, \hat{u}(P_i^{(l)}) = 0, \forall P_i^{(l)} \in Q^{(l)}\}. \quad (7.23)$$

Then the Gram matrices for these bases take the form

$$\Lambda^{(l+1)} \equiv \begin{bmatrix} \Lambda_{1,1}^{(l+1)} & \Lambda_{1,2}^{(l+1)} \\ \Lambda_{2,1}^{(l+1)} & \Lambda_{2,2}^{(l+1)} \end{bmatrix}, \quad \bar{\Lambda}^{(l+1)} \equiv \begin{bmatrix} \bar{\Lambda}_{1,1}^{(l+1)} & \bar{\Lambda}_{1,2}^{(l+1)} \\ \bar{\Lambda}_{2,1}^{(l+1)} & \bar{\Lambda}_{2,2}^{(l+1)} \end{bmatrix}, \quad (7.24)$$

where $\bar{\Lambda}_{1,1}^{(l+1)} = \Lambda_{1,1}^{(l+1)}$, $\bar{\Lambda}_{2,2}^{(l+1)} = \Lambda^{(l)}$ and the remaining blocks of $\bar{\Lambda}^{(l+1)}$ are also available because the matrix of transition from the nodal basis to the hierarchical two-level one for $\hat{G}^{(l+1)}$ has the form

$$E^{l+1} \equiv \begin{bmatrix} I_{l,l+1} & E_{1,2}^{l+1} \\ 0 & I_l \end{bmatrix},$$

where the diagonal blocks are the identity matrices. For defining the elements of the $(N_l^{(1)} + k)$th column of E^{l+1} ($k \in [1, N_l]$), it suffices to know the coefficients in the expansion (3.21), that is, the values of the function $\bar{\psi}_{N_l^{(1)}+k}^{(l+1)}(x)$ at the nodes $P_i^{(l+1)}$. This implies that $\bar{\Lambda}_{1,2}^{(l+1)} = \Lambda_{1,1}^{(l+1)} E_{1,2}^{l+1}$, $\bar{\Lambda}_{2,1}^{(l+1)} = (\bar{\Lambda}_{1,2}^{(l+1)})^T$. Now using the assumption

$$\cos \alpha_{l+1} \le \gamma, \quad 1 - \gamma^2 \equiv s^2 > 0 \qquad (7.25)$$

(see § 2.5), we apply the theory of cooperative model operators (see § 1.5) with respect to the matrix $\bar{\Lambda}^{(l+1)}, l \in [0, p-1]$. The first step is connected with approximation of the block $\bar{\Lambda}_{1,1}^{(l+1)} = \Lambda_{1,1}^{(l+1)}$.

   Lemma 4. *Let all refinements of the triangulations be such that* $t_{l+1} \le t^*$, $l \in [0, p-1]$. *Then there exists a diagonal matrix* $A_{1,1}^{(l+1)} \in \mathcal{L}(H_l^{(l+1)})$ *and constants* $\sigma_{0,1} \equiv \sigma_{0,1}^{(l+1)} > 0$ *and* $\sigma_{1,1}^{(l+1)} > 0$ *independent on the numbers* $a(T_{0,k})$ *in (7.20), such that*

$$\sigma_{0,1} A_{1,1}^{(l+1)} \le \Lambda_{1,1}^{(l+1)} \le \sigma_{1,1} A_{1,1}^{(l+1)}, \quad l+1 \in [1, p]. \qquad (7.26)$$

   Proof. We have $(\Lambda_{1,1}^{(l+1)} v_{l+1}, v_{l+1}) = \sum_{T_{l,m} \in T^{(l)}(\bar{Q})} a(T_{l,m}) |\hat{v}_{l+1}|_{1,T_{l,m}}^2$, where $a(T_{l,m}) = a(T_{0,k})$ provided $T_{l,m} \subset T_{0,k}, l+1 \in [1, p]$. Next,

$$\kappa_0 (h^{(l+1)})^{d-2} |v^2| \le |\hat{v}_{l+1}|_{1,T_{l,m}}^2 \le \kappa_1 (h^{(l+1)})^{d-2} |v|^2,$$

where $|v|^2 \equiv \sum |\hat{v}_{l+1}(P_i^{(l+1)})|^2$, $P_i^{(l+1)} \in T_{l,m} \cap (Q_h^{(l+1)} \setminus Q_h^{(l)})$, κ_0 and κ_1 are positive numbers defined only by the geometry of the simplexes T_0 and the number $t_{l+1} \le t^*$ (it suffices to map $T_{l,m}$ onto a regular simplicial part of the unique cube). Hence, we choose $A_{1,1}^{(l+1)}$ in such a way that, for

$(A_{1,1}^{(l+1)} \mathbf{v}_{l+1}, \mathbf{v}_{l+1}) \equiv X$, we obtain

$$X = \sum_{T_{l,m} \in T^{(l)}(\bar{Q})} a(T_{l,m}) (h^{(l+1)})^{d-2} \sum_{P_i^{(l+1)} \in T_{l,m} \cap (Q_h^{(l+1)} \setminus Q_h^{(l)})} |\hat{v}_{l+1}(P_i^{(l+1)})|^2.$$

It is easy to see that this choice satisfies the desired conditions. □ [67]

Making use of Lemma 4 for modified Richardson method (7.11) with $A_h = A_{1,1}^{(l+1)}, L_h = \Lambda_{1,1}^{(l+1)}, \sigma_0 = \sigma_{0,1}, \sigma_1 = \sigma_{1,1}, k_1 \equiv k_1^{(l+1)}$, and parameters $\{t_i^+\}$ (see Theorem 1.3.14 with $k \equiv k_1^{(l+1)}$), we obtain

$$B_{1,1}^{(l+1)} \equiv \Lambda_{1,1}^{(l+1)} (I_1 - Z_1^+)^{-1}, \tag{7.27}$$

where $0 \leq Z_1^+ \equiv Z_{1,k_1}^{(l+1)} \leq q_{k_1}^+ I_1$ (in the sense of the Euclidean space $H_1^{(l+1)}(\Lambda_{1,1}^{(l+1)})$) and $q_{k_1}^+ \equiv q_{k_1}^{+,(l+1)}$ is small for large k_1. Our choice of the iteration parameters leading to (7.27) stems from condition (1.5.23), which is important for the application of Theorem 1.5.4 (see (1.5.30) and the proof of Lemma 5 below). Note also that the indicated iterations permit obvious parallelization; a more complicated choice of $A_{1,1}^{(l+1)}$ is possible, especially for square grids, and leads to rapid convergence. These questions probably deserve special attention, but for now we confine ourselves to the indicated results and take the next step in construction of model cooperative operators. We describe it first in matrix form associated with the hierarchical basis for $\hat{G}^{(l+1)}$. Let $l \in [0, p-1]$ and

$$\bar{C}^{(l+1)} \equiv \begin{bmatrix} B_{1,1}^{(l+1)} & \bar{\Lambda}_{1,2}^{(l+1)} \\ \bar{\Lambda}_{2,1}^{(l+1)} & \Lambda^{(l)} \end{bmatrix}, \quad S_2(\bar{C}^{(l+1)}) \equiv \Lambda^{(l)} - \bar{\Lambda}_{2,1}^{(l+1)} (B_{1,1}^{(l+1)})^{-1} \bar{\Lambda}_{1,2}^{(l+1)}.$$

Lemma 5. *Let $B^{(l)} \in \mathcal{L}^+(H^{(l)})$ be such that $\sigma_0^{(l)} B^{(l)} \leq \Lambda^{(l)} \leq \sigma_1^{(l)} B^{(l)}$, where $0 < \sigma_0^{(l)} \leq 1 \leq \sigma_1^{(l)}$. Suppose that $\sin^2 \alpha_{l+1} \geq s_{l+1}^2 > 0$. Then,*

$$s_{l+1}^2 \sigma_0^{(l)} B^{(l)} \leq S_2(\bar{C}^{(l+1)}) \leq \sigma_1^{(l)} B^{(l)}. \tag{7.28}$$

Proof. It is easy to see that $S_2(\bar{C}^{(l+1)}) = S_2(\bar{\Lambda}) + X_2$, where $X_2 \equiv \bar{\Lambda}_{2,1}^{(l+1)} Z_1^+ (\Lambda_{1,1}^{(l+1)})^{-1} \bar{\Lambda}_{1,2}^{(l+1)}$. We have $(X_2 u_2, u_2') = (Z_2^+ v_1, v_1')_{\Lambda_{1,1}^{(l+1)}}$, $u_2 \in H_l, u_2' \in H_l, v_1 \equiv (\Lambda_{1,1}^{(l+1)})^{-1} \bar{\Lambda}_{1,2}^{(l+1)} u_2$, and $v_1' \equiv (\Lambda_{1,1}^{(l+1)})^{-1} \bar{\Lambda}_{1,2}^{(l+1)} u_2'$.

---

[67] The optimal values of κ_0 and κ_1 can be found by solving an eigenvalue problem in $\mathbf{R}^s$, where s is the number of vertices $P_i^{(l+1)} \in T_{l,m} \cap (Q_h^{(l+1)} \setminus Q_h^{(l)})$. For example, if $t_{l+1} = 2$ and $d = 3$, then $\kappa_0 \approx 0.8377$ and $\kappa_1 = 14.691$; for $d = 2$, we have $\kappa_0 = 2 - 2^{1/2}$ and $\kappa_1 = 2 + 2^{1/2}$ (this implies $(\kappa_1/\kappa_0)^{1/2} = 2^{1/2} + 1$).

Hence, $(X_2 u_2, u_2') = (u_2, X_2 u_2')$ and $(X_2 u_2, u_2) \geq 0$. Moreover, $(X_2 u_2, u_2) \leq q_{k_1}^+ \|v_1\|_{\Lambda_{1,1}^{(l+1)}}^2 = q_{k_1}^+ (\bar{\Lambda}_{2,1}^{(l+1)} (\Lambda_{1,1}^{(l+1)})^{-1} \bar{\Lambda}_{1,2}^{(l+1)} u + 2, u_2)$. Hence, $S_2(\bar{\Lambda}^{(l+1)}) \leq S_2(\bar{C}^{(l+1)}) \leq S_2(\bar{\Lambda}^{(l+1)}) + q_{k_1}^+ \bar{\Lambda}_{2,1}^{(l+1)} (\Lambda_{1,1}^{(l+1)})^{-1} \bar{\Lambda}_{1,2}^{(l+1)} \leq \Lambda^{(l)}$. This and (1.5.17) lead to (7.28). □ [68]

Consider now the modified Richardson method

$$B^{(l)}(v_l^{k+1} - v_l^k) = -\tau_{l,k}(S_2(\bar{C}^{(l+1)})v_l^k - g_2^{(l+1)}), \quad k = 0, \ldots, k_2 - 1, \quad (7.29)$$

where $\sigma_0 s_{l+1}^2 \sigma_0^{(l)} \leq 1$, $\sigma_1 = \sigma_1^{(l)} \geq 1$, the iteration parameters are defined by the set $\{t_i\}$ (see (1.3.21)) (with k replaced by $k_2 \equiv k_2^{(l+1)}$), and $v_l^0 \equiv 0$. By Theorem 1.3.14, we have $\|Z_2\|_\Lambda \leq q_{k_2} \equiv q_{k_2}^{(l+1)}$, where $Z_2 \equiv Z_{2,k_2}^{(l+1)}$ is the corresponding error reduction operator. Therefore, in the role of model operator for $S_2(\bar{C}^{(l+1)})$, we may use $S_2(\bar{C}^{(l+1)})(I_2 - Z_2)^{-1}$ (see (4.3)). For small q_{k_2}, this can be considered as an approximation to $S_2(\bar{C}^{(l+1)})$, which suggests defining $\bar{B}^{(l+1)}$ as a model cooperative operator of the form

$$\begin{bmatrix} B_{1,1}^{(l+1)} & \bar{\Lambda}_{1,2}^{(l+1)} \\ \bar{\Lambda}_{2,1}^{(l+1)} & S_2(\bar{C}^{(l+1)})(I_2 - Z_2)^{-1} + \bar{\Lambda}_{2,1}^{(l+1)}(B_{1,1}^{(l+1)})^{-1}\bar{\Lambda}_{1,2}^{(l+1)} \end{bmatrix} \quad (7.30)$$

(we have $S_2(\bar{B}^{(l+1)}) = S_2(\bar{C}^{(l+1)})(I_2 - Z_2)^{-1}$).

We summarize algorithmic aspects of obtaining $\bar{u}^{(l+1)} \equiv (\bar{B}^{(l+1)})^{-1}\tilde{f}$ for a given $\tilde{f} \equiv [\tilde{f}_1, \tilde{f}_2]^T \in H^{(l+1)} = H_1^{(l+1)} \times H_2^{(l+1)}$ with $\tilde{f}_r \in H_r^{(l+1)}$, $r = 1, 2$:

1. We define $g_1^{(l+1)} \equiv (B_{1,1}^{(l+1)})^{-1}\tilde{f}_1^{(l+1)}$ and $g_2^{(l+1)} \equiv \tilde{f}_2^{(l+1)} - \bar{\Lambda}_{2,1}^{(l+1)}g_1^{(l+1)}$.

2. For the obtained $g_2^{(l+1)}$, we perform k_2 iterations (7.29) with $v_l^0 \equiv 0$; the last iterate $v_l^{k_2}$ defines $\bar{u}_2^{(l+1)}$, where $\bar{u}^{(l+1)} \equiv [\bar{u}_1^{(l+1)}, \bar{u}_2^{(l+1)}]^T$.

3. We define $\bar{u}_1^{(l+1)} = g_1^{(l+1)} - B_{1,1}^{(l+1)})^{-1}\bar{\Lambda}_{1,2}^{(l+1)}$.

These three stages are actually those in the block elimination procedure.[69]

---

[68] If instead of the set $\{t_i^+\}$ (see Theorem 1.3.14 with $k \equiv k_1$), we use the standard set $\{t_i\}$, then we can only prove that $((1 + q_{k_1})s_{l+1}^2 - q_{k_1})\sigma_0^{(l)}B^{(l)} \leq S_2(\bar{C}^{(l+1)}) \leq \sigma_1^{(l)}B^{(l)}$.

[69] We emphasize that iterations (7.29) do not need the explicit matrix $S_2(\bar{C}^{(l+1)})$ and that each computation of $S_2(\bar{C}^{(l+1)})v_l^k = (\Lambda^{(l)} - \bar{\Lambda}_{2,1}^{(l+1)}(B_{1,1}^{(l+1)})^{-1}\bar{\Lambda}_{1,2}^{(l+1)})v_l^k$ requires solution of the system with the operator $B_{1,1}^{(l+1)}$ (see (7.27)). Actually, we must solve such systems $k_2^{(l+1)} + 2$ times and the solution of each is reduced to performing $k_1^{(l+1)}$

We concentrate on the study of (7.30), as it leads to better estimates (7.19) (if we use sufficiently large k_1 and k_2, then the constants in (7.19) can be made tending to 1).

Finally, we describe a rather obvious procedure for solving systems with the matrix $B^{(l+1)}$ of the same model operator relative to the standard node basis for the space $\hat{G}^{(l+1)}$. In accordance with (3.18) and (3.20),

$$\bar{B}^{(l+1)} = (E^{(l+1)})^T B^{(l+1)} E^{(l+1)} \tag{7.31}$$

and the solution of $B^{(l+1)} u^{(l+1)} = f^{(l+1)}$ ($f^{(l+1)} \equiv [f_1, f_2]^T \in H^{(l+1)}$, $f_r \in H_r^{(l+1)}$, $r = 1, 2$) is reduced to the solution of $\bar{B}^{(l+1)} \bar{u}^{(l+1)} = \tilde{f} \equiv [\tilde{f}_1, \tilde{f}_2]^T$ ($\tilde{f}_1 \equiv f_1$ and $\tilde{f}_2 \equiv (E_{1,2}^{(l+1)})^T f_1 + f_2$) and to the subsequent transformation $u^{(l+1)} = E^{(l+1)} \bar{u}^{(l+1)}$, which yields $u_1^{(l+1)} = \bar{u}_1^{(l+1)} + E_{1,2}^{(l+1)} \bar{u}_2^{(l+1)}$ and $u_2^{(l+1)} = \bar{u}_2^{(l+1)}$.

7.5. Estimates of spectral equivalence and required computational work. Note that the inequalities $\sigma_0^{(l)} B^{(l)} \leq \Lambda^{(l)} \leq \sigma_1^{(l)} B^{(l)}$ are equivalent to $\sigma_0^{(l)} \bar{B}^{(l)} \leq \bar{\Lambda}^{(l)} \leq \sigma_1^{(l)} \bar{B}^{(l)}$, $l \in [1, p]$ (see (3.18)).

Theorem 1. Let the conditions of Lemma 5 be satisfied and $\sigma^{(l)} \equiv \sigma_1^{(l)} (\sigma_0^{(l)})^{-1}$. *Let* $\bar{\xi}_1^{(l+1)} \equiv q_{k_1}^+ [s_{l+1}^2 (1 - q_{k_1}^+)]^{-1}$. *Then*

$$\sigma_0^{(l+1)} \bar{B}^{(l+1)} \leq \bar{\Lambda}^{(l+1)} \leq \sigma_1^{(l+1)} B^{(l+1)}, \tag{7.32}$$

$$\sigma_0^{(l+1)} \equiv \frac{1 - q_{k_2}}{1 + \bar{\xi}_1} \leq 1, \ \sigma_1^{(l+1)} \equiv 1 + q_{k_2} \geq 1, \tag{7.33}$$

and

$$\sigma^{(l+1)} = f_{l+1}(\sigma^{(l)}). \tag{7.34}$$

Proof. Since $S_2(\bar{B}^{(l+1)}) = S_2(\bar{C}^{(l+1)})(I_2 - Z_2)^{-1}$, then Lemma 5, Theorem 1.3.14, and (4.7) yield $1 - q_{k_2} \kappa_0 S_2(\bar{B}^{(l+1)}) \leq S_2(\bar{C}^{(l+1)}) \leq \kappa_1 S_2(\bar{B}^{(l+1)})$,

---

iterations of type (7.11) with model operator $A_h \equiv A_{1,1}^{(l+1)}$ (see Lemma 4.1). It is also noteworthy that these iterations resemble the smoothing procedure, though only formally because they are carried out only for some components of the iterates. We also emphasize that iterations (7.29) lead to an approximation to $S_2(\bar{C}^{(l+1)})$ in contrast with the correction procedure in conventional multigrid methods, but it is also possible to use the following construction of a cooperative model operator:

$$\tilde{B}^{(l+1)} \equiv \begin{bmatrix} B_{1,1}^{(l+1)} & \bar{\Lambda}_{1,2}^{(l+1)} \\ \bar{\Lambda}_{2,1}^{(l+1)} & \Lambda^{(l)}(I_2 - Z_2)^{-1} + \bar{\Lambda}_{2,1}^{(l+1)}(B_{1,1}^{(l+1)})^{-1}\bar{\Lambda}_{1,2}^{(l+1)} \end{bmatrix}.$$

This is closer to conventional multigrid methods (it was analyzed in [23] under the condition that iterations (7.29) use the parameters defined by the set $\{t_i^+\}$ (see Theorem 1.3.14 with $k \equiv k_1$)).

where $\kappa_0 \equiv 1 - q_{k_2}$ and $\kappa_1 \equiv 1 + q_{k_2}$ (see (1.5.28)). This and Theorem 1.5.4 lead to $\kappa_0/\xi_1 \bar{B}^{(l+1)} \leq \bar{\Lambda}^{(l+1)} \leq \kappa_1 \bar{B}^{(l+1)}$ (see (1.5.30)), where $\xi_1 \equiv \xi_1^{(l+1)} \equiv 1 + \bar{\xi}_1^{(l+1)}$. This yields (7.32) and (7.33). Thus, the proof of (7.34) becomes straightforward because $q_{k_2}^{-1} = T_{k_2}\{[\sigma^{(l)} + s_{l+1}^2][\sigma^{(l)} - s_{l+1}^2]^{-1}\}$. $\square$

Theorem 2. *Let* $\sin^2 \alpha_{l+1} \geq s_{l+1}^2 > 0$, *where* $l \in [0, p - 1]$. *Suppose that the operator* $B \equiv B^{(p)}$ *is defined by the recursive use of (7.30) for* $l \in [0, p - 1]$ *under the assumption that*

$$\sigma_0^{(0)} B^{(0)} \leq \Lambda^{(0)} \leq \sigma_1^{(0)} B^{(0)}, \tag{7.35}$$

where $0 < \sigma_0^{(0)} \leq 1 \leq \sigma_1^{(0)}$. *Then inequalities (7.19) hold for the operators* $L^{(p)} \equiv L$ *and* B, *where*

$$\sigma^{(p)} \equiv \sigma_1/\sigma_0 = F(\sigma^{(0)}) \tag{7.36}$$

$$\text{and } F(t) \equiv f_p(f_{p-1} \cdots f_1(t)) \cdots)), \ t \geq 1. \tag{7.37}$$

Proof. We apply Theorem 1 for $l = 0$. Then $\sigma_0^{(1)} \leq 1 \leq \sigma_1^{(1)}$ and $\sigma^{(1)} = f_1(\sigma^{(0)})$, and Theorem 1 applies by induction. $\square$

Lemma 6. *Equation*

$$t = f_{l+1}(t), \ t \geq 1 \tag{7.38}$$

either has no solution or has only one solution $t_{l+1}^* > 1$ *which exists if and only if there exists* $t \geq \xi_1^{(l+1)}$ *such that*

$$T_{k_2^{(l+1)}}[1 + \frac{2s_{l+1}^2}{t - s_{l+1}^2}] \geq 1 + 2[\frac{t}{\xi_1^{(l+1)}} - 1]^{-1}. \tag{7.39}$$

Proof. It is easy to see that $\frac{f_{l+1}'(t)}{2\xi_1} = \frac{2s^2}{(t-s^2)^2}[T_{k_2}'[1 + \frac{2s^2}{t-s^2} - 1]]^{-2}$, where for notational convenience the indices $l + 1$ are ignored. Since all roots of the polynomial $T_{k_2}(t)$ belong to $[-1, 1]$, the same holds for the roots of its derivatives. Hence, $T_{k_2}'(t) > 0$ and $T_{k_2}''(t) > 0$ for $t \geq 1$, and $f_{l+1}(t)$ is a monotonically increasing function for $t \geq 1$. Likewise, we see that $f_{l+1}''(t) < 0$ if $t > 1$. These properties of $f_{l+1}(t)$ together with the obvious inequality $f_{l+1}(\xi_1^{(l+1)}) > \xi_1^{(l+1)}$ prove Lemma 6. $\square$

Lemma 7. *Let* $T_k(x)$ *be the standard Chebyshev polynomial of degree* k. *Suppose that* $z \geq 0$. *Then*

$$T_k(1 + z) \geq 1 + k^2 z + \frac{k^2(k - 1)}{2} z^2, \ \forall k \geq 1, \tag{7.40}$$

with equality for $k = 1$ *and* $k = 2$.

Proof. It suffices to prove (7.40) for $k \geq 2$. We have $2T_k(1 + z) = [1 + z + (2z + z^2)^{1/2}]^k + [1 + z - (2z + z^2)^{1/2}]^k = 2(1 + z)^k + O(1 + z)^{k-1} + k(k - 1)/2(1 + z)^{k-2}[(2z + z^2)^{2/2} + (2z + z^2)^{2/2}]^k + \cdots$. Hence, $T_k(1 + z) \geq (1 + z)^k + k(k - 1)/2(1 + z)^{k-2}(2z + z^2) \geq 1 + z(k + k(k - 1))$. $\square$

Lemma 8. Suppose that

$$s_{l+1}^2 (k_2^{(l+1)})^2 \equiv s^2 k_2^2 > 1 \tag{7.41}$$

(see (7.35)) and $q_1^+ \approx 0$. Then equation (7.38) has the solution $t_{l+1}^* > 1$.

Proof. If $q_1^+ = 0$, then $\xi^{(l+1)} = 1$. Hence, the change $2/(t - 1) \equiv y$ implies that $1 + \frac{2s^2}{t - s^2} = 1 + \frac{2s^2 y}{2 + \gamma^2 y}$, where $\gamma^2 \equiv 1 - s^2$. Moreover, (7.39) is rewritten in the form $T_k[1 + 2s^2/(2 + \gamma^2 y)] \geq 1 + y$, $y > 0$. This, together with (7.40), implies that it suffices to show existence of a positive solution of the inequality $\frac{2s^2}{2 + \gamma^2 y} k^2 + [\frac{2s^2}{2 + \gamma^2 y}]^2 k(k - 1)^2 \geq 1$, $y > 0$. It is easy to see that this is equivalent to $(2 + \gamma^2 y)^2 - 2s^2 k^2 (2 + \gamma^2 y) - (2s^2)^2 k(k - 1)^2 y \leq 0$, $y > 0$. For $y = 0$, the left-hand side takes a negative value due to (7.41). Hence, there exist positive solutions of the above inequality. $\square$

Note that (7.39) is equivalent to

$$\frac{\xi_1^{(l+1)}}{t} < [T_{k_2^{(l+1)}}[1 + \frac{2s_{l+1}^2}{t - s_{l+1}^2}] - 1][T_{k_2^{(l+1)}}[1 + \frac{2s_{l+1}^2}{t - s_{l+1}^2}] + 1]^{-1}, \tag{7.42}$$

with $t > \xi_1^{(l+1)}$. This form is suitable for computation, especially if we use

$$\xi_1^{(l+1)} = 1 + [4\rho_1^{k_1}][s^2(1 - \rho_1^{k_1})^2]^{-1},$$

where $\rho_1^{k_1} \equiv (\sigma_1^{1/2} - 1)(\sigma_1^{1/2} + 1)^{-1}$, $\sigma_1 \equiv \sigma_{1,1}^{(l+1)}/\sigma_{0,1}^{(l+1)}$, and $k_1 \equiv k_1^{(l+1)}$ (see Lemma 4). It indicates that, given any $\xi_1^{(l+1)} \geq 1$, we can find the desired t for large enough $k_2 \equiv k_2^{(l+1)}$. But, as we will see below, it is reasonable to use only $k_2^{(l+1)}$ such that

$$k_2^{(l+1)} < (t_{l+1})^d. \tag{7.43}$$

This implies that $k_1^{(l+1)}$ must be sufficiently large.

Theorem 3. Suppose that the conditions of Theorem 2 are satisfied and that each equation (7.38) has a solution $t_{l+1}^* > 1$ for $l \in [0, p - 1]$. Then

$$\sigma^{(p)} \leq \max\{\sigma^{(0)}, t_1^*, \ldots, t_p^*\}. \tag{7.44}$$

Proof. Observe that the properties of the function $f_{l+1}(t)$ together with the existence of t_{l+1}^* imply that $f_{l+1}(t) < t$ if $t > t_{l+1}^*$ and $f_{l+1}(t) \geq t$ if

$1 \leq t \leq t^*_{l+1}$. This, (7.36), and (7.37) yield (7.44), the spectral equivalence of $\Lambda^{(p)}$ and $B^{(p)}$. $\square$

Below, K refers to constants independent of l and p.

Theorem 4. *Let the conditions of Theorem 3 be satisfied. Suppose also that* (7.43) *holds for* $l \in [0, p-1]$. *Then computational work for solving a system with the model operator* $B^{(p)}$ *is bounded according to*

$$W^{(p)} \leq K N_p. \tag{7.45}$$

Proof. Observe that we may assume that the $k_1^{(l+1)}$ are uniformly bounded. This yields the estimate $W^{(l)} \leq K_1 N_l + k_2^{(l+1)} W^{(l-1)}$, $l \in [1, p]$. We also may assume that $W^{(0)} \leq K_1 N_0$. Hence, it is easy to see that $W^{(p)} \leq K_1(N_p + k_2^{(p)} N_p + k_2^{(p)} k_2^{(p-1)} N_{p_1} + \cdots + k_2^{(p)} \cdots k_2^{(1)} N_1) \leq K_2 N_p[1 + k_2^{(p)} t_p^{-d} k_2^{(p-1)} t_{p-1}^{-d} + \cdots + k_2^{(p-1)} t_{p-1}^{-d} \cdots k_2^{(1)} t_1^{-d}]$. Since all $k_2^{(l)} t_l^{-d} \leq q^* < 1$, then (7.45) holds. $\square$ [70]

Theorem 5. *Let the conditions of Lemma* (4) *be satisfied and consider the splitting* (7.22). *Suppose that the operators* $\Lambda^{(p)}$ *are defined by* (7.24). *Then the conditions of Theorem 3 are satisfied with* $s^2_{l+1} \equiv t^{1-d}_{l+1}, l \in [0, p-1]$, *and it is possible to choose the numbers* $k_1^{(l)}$ *and* $k_2^{(l)}$, *with* $l \in [0, p-1]$, *in such a way that* $B^{(p)} \equiv B_h$ *becomes an optimal preconditioner.*

Proof. The desired estimates for the angles were obtained in § 2.5 (see (2.5.7)). Hence, conditions (7.41) and (7.43) are satisfied if

$$k_2^{(l+1)} \in [t^{1-d}_{l+1} + 1, (t_{l+1})^d - 1]. \tag{7.46}$$

Thus, Theorem 4 applies. For each $k_2^{(l+1)}$ in (7.46) we define $k_1^{(l+1)}$ which yields small enough $q^+_{k_1}$ and $\bar{\xi}_1^{(l+1)}$ such that each equation (7.38) ($l \in [0, p-1]$) has a solution $t^*_{l+1} > 1$. Thus, Theorem 3 applies. $\square$ [71]

We give now some numerical examples with all refinements ratios equal to 2. For $d = 2$, we have:

	$k_2 = 2$				$k_2 = 3$				
k_1	4	5	6	8	2	3	4	5	6
ξ_1	1.25	1.10	1.04	1.007	3	1.66	1.25	1.10	1.04
σ	1.89	1.44	1.30	1.23	5.77	1.90	1.32	1.14	1.07

Table 1: Dependence of spectral equivalence on the numbers of inner iterations for $d = 2$ and strongly discontinuous coefficients.

For the more difficult case $d = 3$ we have:

---

[70] The result is the same if $k_2^{(l+1)} = (t_{l+1})^d$ (see (7.43)) for several levels; if the number of such levels is not uniformly bounded, then $W^{(p)} \leq K N_p \ln N_p$.

[71] Larger $k_2^{(l+1)}$ yield smaller $k_1^{(l+1)}$ and $\sigma^{(p)}$ (see (7.44)).

k_1	$k_2 = 3$		$k_2 = 4$			$k_2 = 5$	
	6	7	5	6	8	4	6
ξ_1	1.96	1.57	2.68	1.96	1.34	4.1	1.96
σ	9.34	3.28	5.47	2.73	1.54	8.2	2.22

Table 2: Dependence of spectral equivalence on the numbers of inner iterations for $d = 3$ and strongly discontinuous coefficients.

k_1	$k_2 = 6$				$k_2 = 7$			
	6	8	10	12	6	8	12	16
ξ_1	1.96	1.34	1.125	1.047	1.96	1.34	1.05	1.007
σ	2.07	1.37	1.14	1.06	2.01	1.35	1.05	1.009

Table 3: Dependence of spectral equivalence on the numbers of inner iterations for $d = 3$ and strongly discontinuous coefficients.

The constants σ_0 and σ_1 in (7.19) are close to 1 for the constructed model operators B_h, even for problems associated with strongly discontinuous coefficients $a(T_{0,k})$ in (7.20). It is also very important to note that our method can deal effectively even with more general triangulations due to the established fact that the obtained estimates of the angles provide a significant reserve. In other words, if we use triangulations that are not too distorted and the computed angles satisfy the condition $s_{l+1}(t_{l+1}^d - 1) > 1$ (see (7.41)), then the given theorems lead to estimates of the same type. In particular, if $d = 2$ and all $t_{l+1} = 2$, then (see [344]) $s^2 > 1/4$ for arbitrary triangulations. This implies that our multigrid construction of the model operator applies for an arbitrary initial triangulation. We emphasize again that a more complicated choice of $A_{1,1}^{(l+1)}$ may improve practical convergence. For example, if the grid is square and all constants $a(T_{0,k}) = 1$ (see (7.20)), then the operator $\Lambda_{1,1}^{(l+1)}$ has a very simple form and enables natural use of its block structure

$$\Lambda_{1,1}^{(l+1)} \equiv \left[\begin{array}{cc} R_{1,1} & R_{1,2} \\ R_{2,1} & R_{2,2} \end{array} \right].$$

Here, the diagonal blocks $R_{1,1}$ and $R_{2,2}$ correspond to the nodes at the midpoints of sides of the old square cells (with the mesh size $h^{(l)} = 2h^{(l+1)}$) and at the centers of these cells, respectively. Moreover, we have $R_{1,1} = 4I_1$ and it is easy to see that $S_2(\Lambda_{1,1}^{l+1})$ is a diagonally dominant matrix satisfying $2I_2 \leq S_2(\Lambda_{1,1}^{l+1}) \leq 4I_2$. This suggests using a cooperative model operator

$$A_{1,1}^{(l+1)} \equiv \left[\begin{array}{cc} R_{1,1} & R_{1,2} \\ R_{2,1} & 4I_2 + R_{1,2}R_{1,1}^{-1}R_{2,1} \end{array} \right]$$

such that $1/2A_{1,1}^{(l+1)} \leq \Lambda_{1,1}^{(l+1)} \leq A_{1,1}^{(l+1)}$. Then $\sigma^{(l+1)} = 2$ and we obtain the table:

k_1	$k_2 = 2$				$k_2 = 3$				
	2	3	4	5	1	2	3	4	5
ξ_1	1.25	1.04	1.007	1.001	3	1.25	1.04	1.007	1.001
σ	1.89	1.30	1.23	1.21	5.77	1.32	1.07	1.04	1.02

Table 4: Dependence of spectral equivalence on the numbers of inner iterations for $d = 2$ and the model problem.

The problem of selection of $A_{1,1}^{(l+1)}$ becomes especially interesting when t^* in the conditions of Lemma 4 may be large. Such refinements of the triangulations with large t_{l+1} may be attractive from the point of view of enhanced parallelization of the algorithms but this has a negative effect on the angles between the subspaces in our splittings of the finite element space $\hat{G}_h$. It then seems reasonable for construction of $A_{1,1}^{(l+1)}$ to apply the results outlined in § 5. Recall that $\Lambda_{1,1}^{(l+1)}$ is in fact a projective-grid operator for the problem $b(\hat{u}^{(l+1)}, \hat{v}^{(l+1)}) = l(\hat{v}^{(l+1)})$ associated with the additional pointwise Dirichlet conditions $\hat{u}^{(l+1)}(P_i^{(l)}) = 0$, $\forall P_i^{(l)} \in Q_h^{(l)}$. These additional conditions significantly simplify the problem of constructing the model operator $A_{1,1}^{(l+1)}$ spectrally equivalent to $\Lambda_{1,1}^{(l+1)}$ (we may use this notion when there are no limitations on t_{l+1}). If $d = 2$, such conditions permit the use of expansions (5.52) and norm (5.53). Such combinations of multigrid and domain decomposition methods deserve a further study.

7.6. Other types of operators and possible generalizations.

First, we consider generalizations for the case $\Gamma_0 = \emptyset$ (the case of nonnegative operators). Let the semiinner product in the Hilbert space $G \equiv W_2^1(Q)$ be defined by (7.20) and $\Lambda^{(l+1)}$ and $\bar{\Lambda}^{(l+1)}$ be defined by (7.24), where $l \in [0, p-1]$. Note that these operators are symmetric and nonnegative as elements of $\mathcal{L}(H^{(l+1)})$, $l \in [0, p-1]$, and the same property holds for $\Lambda^{(0)}$ as an element of $\mathcal{L}(H^{(0)})$. Our purpose now is to construct model operators $\bar{B}^{(l+1)} \geq 0, l \in [0, p-1]$, which satisfy (7.32) on the basis of assumption (7.35). It is obvious that Lemmas 4 and 5 fully apply, and we need to specify only iterations (7.29) and the sense in which $(I_2 - Z_2)^{-1}$ (see (7.30)) must be understood. We emphasize that Ker $B^{(l)} =$ Ker $S_2(\bar{C}^{(l+1)}) =$ Ker $\Lambda^{(l)}$ (see (7.28)). We denote it by $H_{l+1,0} \equiv V_0$ and make use of the orthogonal (in the Euclidean space $H^{(l+1)}$) splitting $H^{(l+1)} = V_0 \oplus V_1$, where $V_1 = \text{Im } S_2(\bar{C}^{(l+1)}) = \text{Im } B^{(l)} = \text{Im } \Lambda^{(l)}$. Hence, we may apply Lemma 4.7 for the error reduction operator $Z_2 = Z_{2, k_2^{(l+1)}}$ in iterations (7.29) assuming that $g_l \in V_1$ (this is the case if we start with a proper right-hand term (see Lemma 1.5.6)). The subspaces V_0 and V_2 are invariant with respect to this operator, and its restrictions to them are such that $\|Z_2|_{V_1}\|_{S_2(\bar{C}^{(l+1)})} \leq q_{k_2} \equiv q_{k_2}^{(l+1)}$ and $Z_2|_{V_0} = I_{l+1,0}$ (see Theorem 1.3.17). This suggests replacing $(I_2 - Z_2)^{-1}$ (see (7.30)) by an operator Y_2 such that V_0 and V_1 are its invariant subspaces and $Y_2|_{V_0} \equiv Z_2|_{V_0}$, $Y_2|_{V_1} \equiv (I_{l+1,1} - Z_2|_{V_1})^{-1}$ (see (4.8) and Theorem 4.2).

Theorem 6. *Consider the splitting (7.22) with Hilbert space $G \equiv W_2^1(Q)$ and semiinner product defined by (7.20). Let the conditions of Lemma (4) be satisfied and the operators $\Lambda^{(p)}$ be defined by (7.24). Then the assertions of Theorem 4 are valid provided that, in construction of $B^{(p)} \equiv B_h$, instead of (7.30) we use*

$$\bar{B}^{(l+1)} \equiv \left[\begin{array}{cc} B_{1,1}^{(l+1)} & \bar{\Lambda}_{1,2}^{(l+1)} \\ \bar{\Lambda}_{2,1}^{(l+1)} & S_2(\bar{C}^{(l+1)})Y_2 + \bar{\Lambda}_{2,1}^{(l+1)}(B_{1,1}^{(l+1)})^{-1}\bar{\Lambda}_{1,2}^{(l+1)} \end{array} \right]. \quad (7.47)$$

Proof. Theorem 1 holds for the operators defined by (7.47) (see Lemma 1.5.10). We can thus apply Theorems 2–4 under only the additional assumption that, in Theorem 4, we have a system with the right-hand side orthogonal to the kernel of the original operator. □ [72]

A very interesting question relates to the possibility of finding better splittings (closer to orthogonal sums). As an example, consider a square grid and grid refinement ratio 2. As follows from § 2.5, for the standard splitting $\hat{G}_h \equiv \hat{G}^{(l+1)} = \hat{V}_1 \oplus \hat{V}_2$, where $\hat{G}_2^{(l+1)} = \hat{V}_2 = \hat{G}^{(l)} \equiv \hat{V}_{2h}$, (see (7.22)) $\hat{G}_1^{(l+1)} = \hat{V}_1$, and the estimate $\alpha \geq \pi/4$ holds for the angle α between $\hat{V}_1$ and $\hat{V}_2$. Now define $\hat{G}_h = \hat{V}_1 \oplus \hat{V}_2'$, where $\hat{V}_2' \equiv \hat{G}_{2h}'$ is the subspace of $\hat{G}_h$ of the functions $\hat{v}_h \in \hat{G}_h$ such that the following holds: the value of each at the center of an arbitrary cell of the coarse grid is the arithmetic mean of the 4 values of $\hat{v}_h$ at the 4 vertices of the cell; and the value of each at the midpoints of each side of the arbitrary cell of the coarse grid is the arithmetic mean of its values at the 2 vertices of the cell belonging to this side. Then it can be verified that the better estimate $\alpha' \geq \pi/3$ holds for the angle α' between $\hat{V}_1$ and the new subspace $\hat{V}_2'$ provided that each cell of the coarse grid belongs to one of the triangles $T_{0,k}$ (see (7.20)). Using the hierarchical basis for this space $\hat{G}_h$, we obtain a new grid operator

$$\bar{\Lambda}_h' \equiv \left[\begin{array}{cc} \Lambda_{1,1} & \bar{\Lambda}_{1,2}' \\ \bar{\Lambda}_{2,1}' & \bar{\Lambda}_{2h} \end{array} \right],$$

where only one block $\Lambda_{1,1}$ remains the same as for the nodal basis. The operator $\bar{\Lambda}_{2h}$ now becomes more involved and, for obtaining the value of the vector $\bar{\Lambda}_{2h}\mathbf{u}$ at the node P_i, we need to know the 9 values of $\mathbf{u}$ at the vertices of 4 cells (P_i is their common vertex). If we can use direct methods for solving systems with $\bar{\Lambda}_{2h}$, then we can also introduce, e.g., a two-grid

---

[72] For local grid refinement, the analysis of spectral equivalence is the same. But for Theorem 4, it is important that the number of refinements is uniformly bounded. In this case, we may even make use of splitting (2.1.11) and inequalities (1.5.12) for the block diagonal model operator D, with one of the blocks corresponding to the set of new nodes arising in local refinement. Note that our estimates involving $t_l > 2$ serve as a means to reduce the number of the local refinements. For more general cases, the iterative additive Schwarz methods mentioned in § 2 can be applied. Generalizations are possible for problems on two-dimensional manifolds with complex geometry. Such model square grids in space were considered in § 5; as another example, we mention a union of several polygons such that they belong to different planes but have a common side.

cooperative operator of the form

$$\bar{B}_h \equiv \left[\begin{array}{cc} \Lambda_{1,1}(I_1 - Z^+_{1,K_1})^{-1} & \bar{\Lambda}'_{1,2} \\ \Lambda'_{2,1} & \Lambda_{2h} + \bar{\Lambda}'_{2,1} B^{-1}_{1,1} \bar{\Lambda}'_{1,2} \end{array} \right],$$

where $B_{1,1} \equiv \Lambda_{1,1}(I_1 - Z^+_{1,K_1})^{-1}$. We outline briefly some algebraic multi-grid constructions of asymptotically optimal preconditioners which use ideas of splitting in an implicit way (see [76, 322]) and can be linked with the use of operators typical for nonconforming finite element methods (see § 2.6). As an illustration, consider the square grid in (7.48) and all $a(T_{0,k}) = 1$ in (7.20). Let Λ_h be the Gram matrix generated by the standard nodal basis for the space $\hat{G}_h$. Suppose we partition the set Q_h as $Q_h = Q^{(1)} \cup Q^{(2)} \cup Q^{(3)}$, where $Q^{(3)} \equiv Q_{2h}$, $Q^{(1)}$ is the set of the refined grid nodes that are centers of the coarse grid cells, and $Q^{(2)}$ contains the remaining nodes of the re-fined grid (they are midpoints of sides of the refined grid cells). Denote by $\hat{G}^{(k)}$ the linear span of the basis node functions corresponding to the nodes in $Q^{(k)}$, where $k = 1, 2, 3$. Then it is easy to see that the subspaces $\hat{G}^{(1)}$ and $\hat{G}^{(3)}$ are orthogonal, and we can make use of this fact if we weaken the role of the remaining subspace $\hat{G}^{(2)}$. This suggests replacing the grid equations at the nodes in $Q^{(2)}$ by the corresponding equations of the above mentioned nonconforming finite element method (such equations connect only three unknowns at the nodes belonging to a side of a coarse grid cell).[73]

§ 8. Effective iterative methods for general grid systems

8.1. General grid systems for model regions. In the above sec-tions, we considered construction of model symmetric operators $B_h \equiv B > 0$ (or $B \geq 0$) for a given grid symmetric operator Λ_h. On the other hand,

---

[73] To implement it in algebraic form, we partition the vectors $\mathbf{u} \in H$ as $\mathbf{u} \equiv [\mathbf{u}_1, \mathbf{u}_2]^T$, where $\mathbf{u}_1$ and $\mathbf{u}_2$ correspond to the nodes in $Q^{(1)}$ and the remaining nodes, respectively. For the corresponding block operator Λ_h with the blocks $\Lambda_{1,1}, \Lambda_{1,2}, \Lambda_{2,1}$, and $\Lambda_{2,2}$ we define a model cooperative operator

$$\bar{B}_h \equiv \left[\begin{array}{cc} \Lambda_{1,1} & \bar{\Lambda}_{1,2} \\ \Lambda'_{2,1} & R_{2h} + \Lambda_{2,1} \Lambda^{-1}_{1,1} \Lambda_{1,2} \end{array} \right],$$

where R_{2h} is the nonconforming finite element method operator (solution of a system with such an operator is reduced, after elimination of the unknowns at the nodes in $Q^{(2)}$, to the solution of a system with the standard operator $1/2\Lambda_{2h}$, as described in § 2.6). If we want to use this construction recursively, we must replace Λ_{2h} by the two-stage operator $\Lambda_{2h}(I_2 - Z_{2,k_2})^{-1}$ and carry out the analysis of the spectral equivalence in a manner similar to that used above (it is even closer to that used in [22]). The advantage of this approach is a simpler form of the iterations, but at the cost of significantly slower convergence.

the theory of iterative methods with model symmetric operators for a general grid system (0.1.28), (1.2.1) with general grid operator $L_h \equiv L$ was discussed in § 1.3 and 1.4. Thus, we can apply any iterative method from § 1.3 and be sure of its fast convergence provided the operators L_h and B_h are connected by the relationships introduced in § 1.4. We concentrate on relationships C^0 and C^3 and their generalizations for nonlinear operators. For example, if linear L_h and Λ_h are connected by relationship C^3 and we choose any symmetric B_h constructed above in the role of the model operator for Λ_h, then L_h and B_h are connected by the same relationship C^3, and we may apply iterations (1.3.25) or (1.3.37) (note that this is not the case for L_h and Λ_h connected by relationship C^2, where additional properties of B_h are necessary (see Lemma 4.5)).

For nonlinear operators, resulting problems need specification; we discuss this matter for more general situations in Theorems 1 and 2 below. We emphasize that some of our model operators are such that the resulting systems (1.4.2) can be solved only approximately (see, e.g., Lemma 5.3). Then this effect, as well as the effect of rounding errors, can be estimated in accordance with the theorems given in § 1.4. Similar iterative methods can be effectively applied for grid systems associated with grids topologically equivalent to the model grids considered above, as we show it below.

8.2. Use of model regions and corresponding model operators. As already emphasized in § 0.4 and Chapter 2, the use of a model domain Q and a triangulation topologically equivalent to a triangulation of the given domain Ω can yield a model grid operator $\Lambda_{Q,h} \equiv \Lambda_Q$ spectrally equivalent to the original grid operator $\Lambda_{\Omega,h} \equiv \Lambda_\Omega$. This operator Λ_Q can sometimes can be used as an optimal preconditioner B (see § 1) or, in general, can be used for constructing such an optimal preconditioner $B \asymp \Lambda_Q$ (see § 2.4, 3.4–3.7). Now we will show that this B can be used as a model operator for general $L_{\Omega,h} \equiv L_\Omega$.

In approximating the Hilbert space $G \equiv G_\Omega \equiv W_2^1(\Omega; \Gamma_0)$ in the general case, we approximate Ω and Γ_0 by a sequence of $\hat{\Omega}_h \equiv \hat{\Omega}$ and $\hat{\Gamma}_{0,h} \equiv \hat{\Gamma}_0$, respectively. We consider two topologically equivalent triangulations: $T_h(\hat{\Omega})$ of $\hat{\Omega}$ and $T_h(Q)$ of a chosen model domain Q (both triangulations may be composite and the boundaries of the domains may contain slits as indicated in § 2.1 and 6). Let $\Pi_h : Q \mapsto \hat{\Omega}$ be a one-to-one continuous mapping that is affine on each open simplex, that is, on the simplex without its boundary, and suppose that the image of each simplex $T' \in T_h(Q)$ is a corresponding simplex $T \in T_h(\hat{\Omega})$ under the mapping that is a continuous extension of Π_h to the boundary of T' (Π_h can be extended to a piecewise-affine mapping of $\bar{Q}$ onto the closure of $\hat{\Omega}$ when these domains have Lipschitz boundaries; otherwise, we must consider it only on the separate simplexes or on the closed regions on some Riemann's surfaces (see § 2.1); recall also that a vertex of a simplex that belong to the slits must be considered as a set of elementary nodes).

We assume further that $\hat{\Gamma}_0$ consists of a set of $(d-1)$-dimensional faces of the simplexes $T \in T_h(\hat{\Omega})$, and that a set of respective faces of the simplexes $T' \in T_h(Q)$ define $\Gamma_{Q,0}$ so that the spaces $W_2^1(\hat{\Omega}; \hat{\Gamma}_0)$ and $W_2^1(Q; \Gamma_{0,Q}) \equiv G_Q$ are isomorphic (see (1.2.3)). Now define Ω_h and Q_h as sets of nodes (elementary nodes when the boundary is non-Lipschitz (see § 2.1 and 5)) such that they do not belong to $\hat{\Gamma}_0$ and $\Gamma_{0,Q}$, respectively, and each is in correspondence with the respective standard basis piecewise linear function $\hat{\psi}_{\Omega,i}(x)$ or $\hat{\psi}_{Q,i}(z)$. This defines an isomorphism between the finite element spaces $\hat{G}_{\hat{\Omega},h}$ and $\hat{G}_{Q,h}$ of functions $\hat{u}_\Omega(x) = \sum_{i=1}^N u_i \hat{\psi}_{\Omega,i}(x)$ and $\hat{u}_Q(z) = \sum_{i=1}^N u_i \hat{\psi}_{Q,i}(z)$, where u_i correspond to the same values of the functions $\hat{u}_\Omega$ and $\hat{u}_Q$ at the equivalent nodes (elementary nodes) with the index i. Let functions $a(x), a_Q(z)$ have the same constant positive values at inner points of the corresponding simplexes T and T' ($T = \Pi_h\{T'\}$),
$$\Lambda_\Omega \equiv [\textstyle\sum_{r=1}^d (a(x), \frac{\partial \hat{\psi}_{\Omega,i}}{\partial x_r} \frac{\partial \hat{\psi}_{\Omega,i}}{\partial x_r})_{0,\hat{\Omega}}] \text{ and } \Lambda_Q \equiv [\textstyle\sum_{r=1}^d (a_Q(z), \frac{\partial \hat{\psi}_{Q,i}}{\partial z_r} \frac{\partial \hat{\psi}_{Q,i}}{\partial z_r})_{0,Q}].$$

Lemma 1. *Suppose all simplexes T' in the triangulations $T_h(Q)$ are regular simplicial parts of some cubes in $\mathbf{R}^d$ and $h(T')$ denotes the length of an edge of the corresponding cube. For the simplexes T in the triangulations $T_h(\hat{\Omega})$, suppose that conditions (2.2.29) are satisfied, where h refers to $h(T')$ and constants κ_0' and κ_1' are independent of the triangulation. Then the grid operators Λ_Ω and Λ_Q are spectrally equivalent with estimates independent of the functions $a(x)$.*

Proof. The indicated expansions for $\hat{u}_\Omega(x)$ and $\hat{u}_Q(z)$ lead to $(\Lambda_\Omega u, u) = (a(x), |\nabla u|^2)_{0,\hat{\Omega}}$, $(\Lambda_Q u, u) = (a_Q(z), |\nabla v|^2)_{0,Q}$, where $|\nabla u|^2 \equiv \sum_{r=1}^d [\frac{\partial \hat{u}_\Omega}{\partial x_r}]^2$, and $|\nabla v|^2 \equiv \sum_{r=1}^d [\frac{\partial \hat{u}_Q}{\partial z_r}]^2$. Integrals over $\hat{\Omega}$ and Q are standard sums of integrals over all possible simplexes T and T', and it suffices to obtain the desired inequalities for $|\hat{u}_\Omega|_{1,T}^2$ and $|\hat{u}_Q|_{1,T'}^2$. This can be easily done due to (1.2.5), (1.2.25), and (1.2.29). $\square$ [74]

The complicated form of Λ_Q sometimes forces its replacement by a new model operator B (see § 2.4, 4–7). We may assume that
$$\delta_{0,\Omega} B \leq \Lambda_\Omega \leq \delta_{1,\Omega} B, \quad \delta_{0,\Omega} > 0 \tag{8.1}$$
implies spectral equivalence of Λ_Ω and B or its weakened variant.

---

[74] Constants in the spectral equivalence relation for such model operators $B = \Lambda_Q$, even in the case of strongly varying coefficients, are independent of these coefficients and are completely defined by the original sequence of triangulations. Lemma 1 remains true even when $\hat{\Gamma}_0$ contains some faces of smaller dimensions, but this case is meaningless for the Sobolev spaces $G^{(1)}$. Note also that, when $\hat{\Gamma}_0 = \emptyset$, we have $\Lambda_\Omega = \Lambda_\Omega^* \geq 0$. If $|\hat{\Gamma}_0|_{d-1} > 0$, then $\Lambda_\Omega \in \mathcal{L}^+(H)$.

8.3. Applicability of effective iterative methods. We restrict ourselves to the case of a nonlinear system $L(u) = f$ and iterative methods (1.3.1) and (1.3.12) where the operators L_h and B_h are connected by relationships $C^0(u; r)$ or $C^3(u; r)$ (see § 1.4). We can thus obtain rates of convergence independent of the grid. For simplicity, we apply slightly stronger conditions on L.

Theorem 1. *Let* $\Lambda \equiv \Lambda_\Omega \in \mathcal{L}^+(H)$ *and* $B \in \mathcal{L}^+(H)$ *be such that* (8.1) *holds. Suppose that u is a solution of* (1.2.1) *and that* $(L(u+z) - L(u), z) \geq \sigma_0 \|z\|_\Lambda^2$, $\sigma_0 > 0$, $\forall z \in S_\Lambda(r)$ *and* $\|L(u+z) - L(u)\|_{\Lambda^{-1}}^2 \leq \sigma_1 \|z\|_\Lambda^2$. *Then, for all $z \in S_B(r')$ with* $r' \equiv r\delta_{1,\Omega}^{-1/2}$, *inequalities* (1.3.2) *and* (1.3.3) *hold, where* $\delta_0(t) \equiv \sigma_0 \delta_{0,\Omega}$ *and* $\delta_1(t) \equiv \sigma_1 \delta_{1,\Omega}^2$.

Proof. For all z with $\|z\|_B \leq r'$, we have $\|z\|_\Lambda \leq r$ (see (8.1)). Hence, for such z, the inequalities involving σ_0 and σ_1 hold. Together with (8.1), they lead to (1.3.2) and (1.3.3) with the above indicated constants. □

Theorem 2. *Let the operators* $\Lambda \equiv \Lambda_\Omega \in \mathcal{L}^+(H)$ *and* $B \in \mathcal{L}^+(H)$ *be such that* (8.1) *holds. For a solution u of* (1.2.1), *suppose that L is continuously differentiable on $S \equiv S_\Lambda(u; r)$ and satisfies*

$$\sigma_0 \|v\|_\Lambda^2 \leq \|L'_{u+z} v\|_{\Lambda^{-1}}^2 \leq \delta_1 \|v\|_\Lambda^2, \quad \sigma_0 > 0, \forall u + z \in S, \forall v, \tag{8.2}$$

and

$$\|(L'_{u+\theta z} - L'_{u+z})v\|_{\Lambda^{-1}}^2 \leq \delta_2(\|z\|_\Lambda)\|v\|_\Lambda^2, \quad \forall \theta \in (0,1), \forall v, \tag{8.3}$$

where $\delta_2(t)$ is nondecreasing on $[0,r]$ and such that $\lim_{t \to 0} \delta_2(t) = 0$. *Let* $r' \in (0, r\delta_{1,\Omega}^{-1/2})$ *satisfy* $\delta_2 \equiv \delta_{1,\Omega}^4 \sigma_1 \sigma_2 (r'\delta_{1,\Omega}^{1/2}) < \delta_0^2$, *where* $\delta_0 \equiv \delta_{0,\Omega} \sigma_0$, $\delta_1 \equiv \delta_{1,\Omega} \sigma_1$. *Then, for all $u + z \in S$, inequalities* (1.3.16) *and* (1.3.17) *hold, with $B_1 = B_2 = B$ and constants δ_0, δ_1, and δ_2 above specified. Moreover, method* (1.3.12), *for $u^0 \in S$ and the iteration parameter from* (1.3.13), *converges with estimate* (1.3.15), *where $D = B$.*

Proof. Note again that $\|z\|_B \leq r'$ implies $\|z\|_\Lambda \leq r$ (see (8.1)). Hence, for such z, (8.2) and (8.1) yield (1.3.16). From (8.3) and (8.1), it follows that $\|(L'_{u+\theta z} - L'_{u+z})v\|_{B^{-1}}^2 \leq \delta_{1,\Omega}^2 \sigma_2(r'\delta_{1,\Omega}^{1/2})(\|v\|_B^2 = \delta_2/\delta_1$. This yields (1.3.17) and, thus, Theorem 1.3.11 applies. □

8.4. Use of linearization of nonlinear operators. Since, for a complicated nonlinear operator L, the problem of evaluating a residual $r^n \equiv L(u^n) - f$ requires considerable computational work, it is important to reduce the required number of iterations for obtaining the desired accuracy, even when the rate of convergence is independent of the grid. As we emphasized already (see § 1.4), a considerable effect can be achieved via

various continuation procedures. Here we consider certain iterative methods based on linearization of the given operator and approximate solution,
via some effective inner iterations, of the obtained linearized systems. [75]

Generally speaking, we can linearize $L(u) - L(v)$ in a variety of ways if
we replace it by $A_v(u - v)$, where A_v is a linear operator defined by a given
$v \in S$. We start by considering iterations

$$A_{u^n}(u^{n+1} - u^n) = -L(u^n) + f \equiv g^n. \tag{8.4}$$

For a continuously differentiable operator L, it is possible to use the standard choice $A_v = L'_v$ (see [41, 292]), where L'_v is the Jacobian matrix. We
thus obtain the classical Newton-Raphson method

$$L'_{u^n}(u^{n+1} - u^n) = -L(u^n) + f, \tag{8.4'}$$

and its analysis is well known (see [292]).

To perform one iteration (8.4), we need to solve system $Av = g^n$, where
$v = u^{n+1} - u^n$. If we use inner iterations for obtaining an approximation to
this v, then we can analyze such a two-stage method in the way suggested in
§ 4, and we can replace (8.4) by $A(I - Z^{(n)})^{-1}(u^{n+1} - u^n) = g^n$, where $Z^{(n)}$
refers to the error reduction operator in solving (8.4). Similarly, instead of
(8.4'), we obtain

$$L'_{u^n}(I - Z^{(n)})^{-1}(u^{n+1} - u^n) = -L(u^n) + f, \tag{8.5}$$

with the corresponding operator $Z^{(n)}$.

We emphasize that the number of inner iterations may depend on n (we
denote it by k_n) and that they provide the possibility to use the adaptation
procedure of iteration parameters outlined in § 1.3. What we want to
investigate is the use of the method (1.3.25), which takes the form

$$B(v^{m+1} - v^m) = -\tau_m(L'_{u^n})^* B^{-1}(L'_{u^n} v^m + L(u^n) - f) \tag{8.6}$$

where $m = 0, \ldots, k_n - 1$, and

---

[75] Of course, the selection of such inner iterations (taking into account that only a few
of them is really required) is of fundamental importance (as demonstrated already in
the theory of two-stage iterative methods (see § 4) for linear operators). On the other
hand, it is well known that, for continuously differentiable operators, very promising
linearizations can be constructed by the Newton-Raphson method since they lead to
quadratic convergence when the linearized systems are solved exactly (see [41, 292]).
Considering such methods, we concentrate on approximate solution of linearized systems under the general and natural condition of uniform boundedness of the inverse to
the operators obtained through the chosen linearization (see [201, 206]). Simpler cases
were analyzed in [175, 217, 219, 510].

$$\|Z^{(n)}\|_B \le q_n < 1. \tag{8.7}$$

Theorem 3. *Let u be a solution of $L(u) = f$. Let $B \in \mathcal{L}^+(H)$ and $S \equiv S_B(u; r)$. Suppose that L is continuously differentiable on S such that*

$$\|L'_v\|_{H(B) \mapsto H(B^{-1})} \le \delta_1^{1/2}, \ \ \|(L'_v)^{-1}\|_{H(B^{-1}) \mapsto H(B)} \le \delta_0^{-1/2}, \ \forall v \in S, \tag{8.8}$$

$$and \ \|L'_v - L'_w\|_{H(B) \mapsto H(B^{-1})} \le l\|v - w\|_B, \ \ \ \forall v \in S, \ \ \forall w \in S. \tag{8.9}$$

Suppose also that $u^n \in S$ and the error reduction operator $Z^{(n)}$ for k_n inner iterations (8.6) satisfies (8.7). Then the rate of convergence of method (8.5) is estimated in accordance with

$$\|z^{n+1}\|_B \le \frac{1}{\delta_0^{1/2}} \left(\frac{l}{2} \|z^n\|_B^2 + \xi^n \right), \tag{8.10}$$

where $\|z^n\|_B \le r$ and

$$\xi^n \equiv q_n \delta_1^{1/2} (1 - q_n)^{-1} \|u^{n+1} - u^n\|_B.$$

Proof. We have $L'_{u^n} z^{n+1} = g_0 - g_1$, $g_0 \equiv L(u) - L(u^n) - L'_{u^n} z^n$, and $g_1 \equiv L'_{u^n} [(I - Z^{(n)})^{-1} - I](u^{n+1} - u^n)$ (see (8.5)). Hence,

$$\|z^{n+1}\|_B \le \frac{1}{\delta_0^{1/2}} \left(\|g_0\|_{B^{-1}} + \|g_1\|_{B^{-1}} \right). \tag{8.11}$$

Note that (1.2.16) implies that

$$\|g_0\|_{B^{-1}} \le \frac{l}{2} \|z^n\|_B^2$$

and that (8.8) and (8.9) yield the estimate $\|g_1\|_{B^{-1}} \le \xi^n$. Hence, (8.10) follows from (8.11). $\square$ [76]

Theorem 4. *Let the conditions of Theorem 3 with respect to the operators L and B be satisfied. Suppose r and all q_n in (8.7) are small so that $q_n \delta_1^{1/2} [(1 - q_n) \delta_0^{1/2}]^{-1} \le q < 1$ and*

$$1/(1 - q)[rl/2\delta_0^{-1/2} + q] \le \rho < 1. \tag{8.12}$$

Then method (8.5) with $u^0 \in S$ converges with the rate estimated in accordance with

$$\|z^k\|_B \le \rho^k \|z^0\|_B \le \frac{\rho^k}{\delta_0^{1/2}} \|r^0\|_{B^{-1}}. \tag{8.13}$$

---

[76] Note that ξ^n can be computed in the process of iterations and that we obtain the standard recursive estimates for the classical Newton-Raphson method if $\xi^n = 0$.

Proof. If $\|z^n\|_B \leq r$, then (8.11) implies

$$\|z^{n+1}\|_B \leq lr\|z^n\|_B \delta_0^{-1/2} + q(\|z^{n+1}\|_B + \|z^n\|_B).$$

This and (8.12) yield $\|z^{n+1}\|_B \leq \rho\|z^n\|_B$. Thus, all iterates in (8.5) belong to S and (8.13) holds. $\square$

Constants δ_0 and δ_1 can be obtained from the adaptation procedure for iterations (8.6) if our assumption on the invertibility of L'_v holds. Constant l from (8.9) is more difficult to estimate. In practical cases, it is possible to obtain an approximation to l by analyzing values of $\|L(u^p) - f\|_{B^{-1}}$ with $p = n$ and $p = n - 1$. If we have (8.8) and (8.9) for $S \equiv S_B(u^0; r')$ and small enough $\|L(u^0) - f\|_{B^{-1}}$, then Theorem 1.2.11 yields the existence of a solution of (1.2.1). This implies that instead of conditions (8.8) and (8.9) associated with $S \equiv S_B(u; r)$, we may formulate them for $S_B(u^0; 2r)$. As for (8.12), we see that lr and q_n must be sufficiently small. Usually it is possible to obtain a reasonable choice of k_n by numerical experiments.

8.5. Use of inner iterations with the nonnegative error reduction operator. In the above analysis, we considered inner iterations from the Theorem 1.3.14 based on the use of the set (1.3.21) for the iteration parameters. From a theoretical point of view, such a choice of the parameters is much better than the choice leading to the nonnegative error reduction operator $Z^{(n),+} \in \mathcal{L}(H(A))$. Nonetheless, in some practical problems, the latter choice might useful, especially if we want to avoid systems with almost degenerate matrices. Indeed, Theorem 4.1 with $\Lambda \equiv A$ implies that $B^{(n)} \equiv A(I - Z^{(n),+})^{-1}$ is a symmetric positive operator and that $A \leq B^{(n)} \leq 1/(1 - q^+)A$ (see (4.8)). Hence, in the case of $A_{u^n} > 0$ with the smallest eigenvalue being rather small, we deal with the operators

$$B^{(n)} \equiv A_{u^n}(I - Z^{(n),+})^{-1} \geq A_{u^n}.$$

This might be useful in tracing solution branches in the vicinity of bifurcation points when the Jacobian matrices necessarily degenerate. Note that this topic is very important in many applications (see [8, 100, 245, 300, 467]), but is beyond the scope of our discussion.

Chapter 4

Construction of topologically equivalent grids

§ 1. Basic approaches to constructing grids

1.1. General ideas. We return now to the problem of mapping of a given domain Ω into a simpler model domain Q and concentrate on numerical construction of topologically equivalent triangulations (or grids). These problems are of utmost importance for many branches of modern computational mathematics. For example, the generation of computational grids is very often a critical element for the numerical study of fluid dynamics problems (see [63, 304, 478]); the review [478] contains a list of several hundreds of papers dealing with the fluid dynamic problems and with applications of grid generation procedures. Recall that the desired mapping is defined by the transformation (2.1.2) of the original variables $x \equiv [x_1, \ldots, x_d]$, associated with points of Ω, into the variables z, associated with points of Q (we can even use one-to-one correspondence between closures of domains provided that they have Lipschitz boundaries; but for general cases it is necessary to define appropriate limit boundary points; see Subsection 2.1.6). Given this transformation, we can choose a simply structured grid for Q and define its analog for Ω. [1]

---

[1] Cells of the grid for Ω may have complicated geometry; e.g., if $d = 2$, then the analog of a rectangular grid for Q may contain curvilinear cells and we would in fact use curvilinear coordinates for Ω. This a serious drawback, so it is natural to replace such cells by quadrilateral ones. Sometimes this can be done if we find analogs of four

Before discussing ways to transform the geometry of the given region, we should pay special attention to the goal of such transformation. First, the necessity to maintain coordinate lines (or surfaces for $d = 3$) coincident with the boundaries must be mentioned. This is especially important for nonstationary problems with moving boundaries when standard difference approximations are applicable to general configurations without the need of special procedures at the boundaries. Second, a significant simplification in coding, most particularly with regard to boundary conditions, can be attained in constructing the grid operator. Third, a concentration of lines in regions of high gradients is desirable and can increase accuracy of the method. Fourth, for stationary problems or nonstationary problems approximated by implicit schemes, the use of grids of the mentioned type can result in the very favorable possibility to apply effective methods for solving the arising systems. This in combination with a posteriori estimates can even lead to self-adjusting procedures for solving the given boundary problem when a user needs only specify this problem and the desired accuracy (usually not very high); in this respect, see, e.g., [301, 516]. Finally, an important aim is to construct asymptotically optimal algorithms for solving the arising grid systems (see Chapter 3). The goal is to develop asymptotically optimal algorithms for solving a given elliptic boundary value problem, allowing practical treatment of very large systems and with very high accuracy (see [177, 178, 180]).

As we have already emphasized, this is of utmost importance to us. To realize it, we make use of geometric-algebraic procedures for generation of topologically equivalent triangulations for regions with complicated geometry. They are based on partitioning $\bar{\Omega}$ into a union of a finite number of blocks (see § 3, where the most important mathematical problems dealing with the existence of the desired partition will be considered) for which of these blocks, the desired quasiuniform triangulations are constructed separately in accordance with some relatively simple algebraic algorithms (see § 2). Since our approximations are based on PGMs, there is no need for smooth coordinate lines, and triangulations of a very general nature can be used. Moreover, the choice of blocks and composite triangulations can be made consistent with expected regions of strong changes in the solution, and the desired spectral equivalence of grid operators (see § 0.4 and Lemma

---

vertices of a cell for Q and connect them by straight line segments that are chords of arcs corresponding to sides of the given cell for Q. Unfortunately, this simple procedure of obtaining a quadrilateral grid for Ω does not work because the given segments may intersect; we then have to employ a new grid for Q with the hope that it will produce better results. This feature is typical for all completely automated algorithms for finding the desired set of nodes for Ω, in contrast to geometric-algebraic methods we consider.

2.2.8) can be established as well.[2]

1.2. Conformal and quasiconformal mappings for $d = 2$. Conformal mappings using elementary transformations in the complex plane have long been used to generate orthogonal coordinate systems about special boundary curves that are contours of the mapping. They preserve angles and yield orthogonal coordinate systems (we refer to them and their corresponding grids using the term conformal). Moreover, for the important case of the Laplace equation these transformations lead to the same equation in the new variables. [3]

First note that the Cauchy–Riemann conditions for the desired conformal mapping lead to the Laplace equations

$$\Delta z_k(x) = 0, \quad k = 1, 2, \tag{1.1}$$

$$\Delta x_k(z) = 0, \quad k = 1, 2. \tag{1.2}$$

Since they are lacking boundary Dirichlet conditions, it was suggested in the beginning of the 60's to make use of minimization of the functional

$$\Phi \equiv ([l^{1/2}\frac{\partial x_1}{\partial z_1} - l^{-1/2}\frac{\partial x_2}{\partial z_2}]^2 + [l^{-1/2}\frac{\partial x_1}{\partial z_2} + l^{1/2}\frac{\partial x_2}{\partial z_1}]^2, 1)_{0,S}, \tag{1.3}$$

where the functions $x_1(z)$ and $x_2(z)$ are defined on the unit square S, they define a one-to-one correspondence between the sides of the square and their chosen analogs on $\partial\Omega$, and they are such that (1.3) makes sense. The minimization here is carried out with respect to these functions and the number l (similar variational settings were studied by Riemann). Note that (1.3) can be replaced by

$$\Phi_1 \equiv ((\frac{\partial x_1}{\partial z_1})^2 + (\frac{\partial x_2}{\partial z_1})^2 + \frac{1}{l}[(\frac{\partial x_1}{\partial z_2})^2 + (\frac{\partial x_2}{\partial z_2})^2], 1)_{0,S} \tag{1.4}$$

because these functions differ by the constant term $2|\Omega|$. Now (1.4) can be approximated by a simple difference scheme associated with a square grid

---

[2] Creation of the grid can be approached by several ways (see [117, 177, 178, 180, 304, 478, 518]), some of which and several of these are summarized below.

[3] In classical works, a simply connected region Ω is usually mapped onto a disc; but for the purpose of generating conformal grids, it is natural to study its mapping onto a rectangle Q in the transformed plane. Denote the Descartes coordinates for Ω by $x \equiv [x_1, x_2]$ and the corresponding curvilinear coordinates for Ω (they may be also regarded as the Descartes coordinates for Q) by $z \equiv [z_1, z_2]$; if complex variables are preferred then we write $X \equiv x_1 + ix_2 \in \Omega$ and $Z \equiv z_1 + iz_2 \in Q$. There are several ways to construct approximately the desired conformal mapping and these are useful in practice (see [478] and references therein).

and minimized by iterative methods involving some iterations with respect to l and inner iterations for solving the arising grid systems (see § 3.4).) [4]

Methods of generating general orthogonal or nearly orthogonal grids have several advantages over conformal mappings: distribution of the nodes along the boundary can be very arbitrary; it is easier to control the behavior of the grid near the boundary; and orthogonal coordinate systems may be generated from hyperbolic as well as from elliptic systems.

Many numerical procedures have been designed to produce coordinates that are nearly orthogonal. For example, the use of such coordinates in combination with difference methods and effective iterative methods (see § 3.4 and 6.5) was successful for practical solution of some shell theory problems associated with various types of tires in the 1960s (more complicated nonlinear problems of aviation tires and even special tires for "Buran" were solved in the 1980s). Actually, these coordinate systems were approximate natural coordinate systems on the surface of the shell (tire).

We also mention some very general procedures based on the use of *quasiconformal mappings*, which lead to variational problems more general than (1.4) with the functional of the form

$$\Phi_1 \equiv (a, (\frac{\partial x_1}{\partial z_1})^2 + (\frac{\partial x_2}{\partial z_1})^2)_{0,S} - 2(b, \frac{\partial x_1}{\partial z_1}\frac{\partial x_1}{\partial z_2} + \frac{\partial x_2}{\partial z_1}\frac{\partial x_2}{\partial z_2})_{0,S}$$

$$+(c, (\frac{\partial x_1}{\partial z_2})^2 + (\frac{\partial x_2}{\partial z_2})^2)_{0,S}, \qquad (1.5)$$

where a, b, and c are functions of z_1 and z_2 of a particular type and are to be chosen as part of the minimization procedure (see [63, 478] and references therein). Such procedures can be applied separately for blocks in a given partition of the region because they can start from the prescribed sets of

---

[4] Often, approximations of the classical formulas can be used. For example, the well-known Schwarz–Christoffel formula gives a conformal mapping of a region with polygonal boundary onto the upper half plane and can be used as the basis for a numerical procedure. Approximate mappings of this type have been constructed for a wide range of bounded or unbounded polygonal regions. In similar role, transformation Fourier series methods can be used as well. The above procedures have aimed primarily at regions of a particular type. For regions with arbitrary curved boundaries, integral equation methods are popular. For example, we can make use of the representation for a harmonic function as a single layer logarithmic potential of unknown source density and obtain an integral equation on the boundary with respect to this density function. Of course, the choice of numerical method for this equation is of utmost importance. Conformal mappings of regions of connectivity greater than two have received some attention. In general, conformal mappings do not allow an arbitrary specification of boundary nodes. Even when we specify boundary points corresponding to the vertices of the rectangle, such a map exists only for a special ratio of sides lengths and this number (e.g., l in (1.4)) must be also determined.

nodes on all interfaces of these blocks (sometimes, additional procedures are used to smooth the obtained coordinate lines). The coordinate lines that such an approach produces are generally not orthogonal.

1.3. Use of elliptic generating systems. For generation of nonorthogonal coordinate systems and corresponding grids, several types of elliptic generating systems can be mentioned, including a certain class of quasilinear equations. For example, the basic equations of the form

$$\Delta z_k(x) = F_k(z)|\nabla z_k(x)|^2, \quad k \in [1, d], \tag{1.6}$$

are very popular for construction of three-dimensional grids (see [478] and references therein); they are frequently applied separately for blocks in a given partition of the region, and special attention is paid to the choice of two-dimensional grids on interfaces of these blocks. [5]

§ 2. Algebraic methods for constructing quasiuniform triangulations of special type for standard blocks

2.1. Triangulations of type $T(\triangle; p)$ for a standard quasitriangle. For two-dimensional manifolds, the most natural choice of a block is a triangle with two straight sides and at most one curvilinear side (see § 2.2). Consider the unit square $\square \equiv A_1A_2A_3A_4$, with $A_1 \equiv [0, 0], A_2 \equiv [1, 0], A_3 \equiv [1, 1]$, and $A_4 \equiv [0, 1]$, covered by the square grid with the mesh size $h \equiv 1/p$, $p \geq 2$. Suppose each cell is subdivided into two triangles by drawing the diagonal parallel to A_1A_3 (from lower left to upper right). Then we obtain triangulations of the square and of the triangle $\triangle A_1A_3A_4$, which we denote by $T^{[1]}(\square; p)$ and $T^{[1]}(\triangle; p)$, respectively. Note that $T^{[1]}[\triangle; p]$ is topologically equivalent to the triangulation of a triangle obtained by its refinement with ratio p (see § 2.1). Triangulations of a polygon that are topologically equivalent to the above triangulations for $\square A_1A_2A_3A_4$ or $\triangle A_1A_3A_4$ are referred to as being of type $T^{[1]}(\square; p)$ and $T^{[1]}(\triangle; p)$, respectively (see Figure 1). If different cells may be triangulated by drawing nonparallel diagonals, then we speak about more general triangulations of type $T(\square; p)$ and $T(\triangle; p)$, respectively.

Now consider a standard quasitriangle $T' \equiv A_1A_4A_3$ of order 1 (see Figure 2.2.1)), which we also denote by G' together with some triangulations

---

[5]It is important that the control functions $F_k(z)$, $k \in [1, d]$, have prescribed signs in order for the maximum principle like one for the harmonic functions to be valid. Then similar equations can be written for the functions $x_k(z)$, $k \in [1, d]$. The resulting Dirichlet problems associated with separate blocks can be approximated by difference schemes associated with cubical grids. For solving the resulting nonlinear systems, some effective inner iterations (see § 3.4) may be applied. Additional procedures may be used to smooth the obtained surfaces and lines.

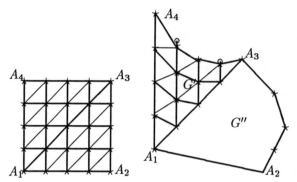

Figure 1. Model triangulation.   Figure 2. Standard quasisquare.

as shown in Figure 2, where the smooth concave arc $\widetilde{A_4 A_3}$ (see (2.1)) is approximated by a broken line. In the Descartes coordinate system $[y_1, y_2]$ used for Figure 2.2.1, assume that $A_1 = [0, 0]$, $A_4 = [0, f(a)]$, $A_3 = [a, f(a)]$, where $a > 0$, $f(a) > 0$, $\angle A_4 A_1 A_3 \equiv \varphi \in (0, \pi)$, and the arc $A_4 A_3 \equiv \Gamma_3$ is defined by the equation $y_2 = f(y_1)$, $0 \le y_1 \le a$, $f \in C^2$. We rewrite the equations of the straight lines containing the line segments $[A_1 A_3]$, $[A_4 A_3']$, and $[A_4 A_3]$ as $y_2 = \lambda_0 y_1$, $y_2 - f(a) = \lambda_1 (y_1 - a)$, and $y_2 - f(a) = \lambda_2 (y_1 - a)$, respectively. We also assume that either

$$f''(y_1) \ge 0 \tag{2.1}$$

or

$$f''(y_1) \le 0. \tag{2.2}$$

Everywhere below in this section, $K_1 \equiv 1/2 \max_{y_1} |f''(y_1)|$, $l \equiv a/p$, and for $p \to \infty$ there exist positive constants κ_0 and κ_1 such that $\kappa_0 \le l/h \le \kappa_1$; the symbols κ and K will be used only for positive and nonnegative constants. A standard quasitriangle of order 1 for which (2.1) or (2.2) holds is referred to as *standard quasitriangle*.

In the construction of approximating polygons $\hat{G}_h'$ (approximating G') and its triangulations, we distinguish the two cases $\hat{G}_h' \subset G'$ and $G' \subset \hat{G}_h'$. The first is desirable when the original problem is set in the Hilbert space $W_2^1(\Omega; \Gamma_0)$ and $\Gamma_3 \subset \Gamma_0$ (see the convergence theorems in § 2.3). The second is preferable when $\Gamma_3 \subset \Gamma \setminus \Gamma_0$. For the case $\hat{G}_h' \subset G$ and (2.1), we use the additional condition

$$K_1 l < \lambda_1 - \lambda_0. \tag{2.3}$$

Theorem 1. *Consider a standard quasitriangle G' and suppose that (2.3) is satisfied. Then it is possible to construct a sequence of polygons $\hat{G}' \equiv \hat{G}_n'$*

obtained from G' by replacing the arc $A_3A_4 \equiv \tilde{\Gamma}_3$ by the broken line $\hat{\Gamma}_{3,h}$ connecting the points A_3 and A_4 and such that: either $\hat{G}'_h \subset G'$ or $G' \subset \hat{G}'_h$; between Γ_3 and $\hat{\Gamma}_{3,h}$ there exists a one-to-one piecewise differentiable correspondence

$$\hat{y} = y + \xi(y) \quad y \in \Gamma_3, \ \hat{y} \in \hat{\Gamma}_{3,h}, \tag{2.4}$$

where $\xi(y) \equiv [\xi_1(y), \xi_2(y)]$ and

$$|\xi(y)| \le K_2 h^2; \quad |\frac{\partial \xi_r}{\partial y_l}| \le K_3 h, \ r = 1, 2; \ l = 1, 2; \tag{2.5}$$

for each polygon $\hat{G}'_h$, $\exists \, T_h(\hat{G}'_h)$ of type $T^{[1]}(\triangle; p)$ for which

$$\kappa_2 \le \frac{l(\triangle)}{h} \le \kappa_3, \tag{2.6}$$

$$\kappa_4 \le \frac{s(\triangle)}{h^2} \le \kappa_5, \tag{2.7}$$

and

$$\mu(T_h(\hat{G}'_h)) \le \mu < \infty, \tag{2.8}$$

where $l(\triangle)$ and $s(\triangle)$ denote the length of an arbitrary side and the area of arbitrary triangle in $T_h(\hat{G}'_h)$, respectively, and $\mu(T_h(\hat{G}'_h))$ is defined by (0.4.13).

Proof. First we describe the algorithm for constructing $\hat{G}'_h$ and its triangulation assuming that $p = h^{-1}$ is given. Let the points $A_{r,r}$ and $A_{0,r}$, with $0 \le r \le p$ and $A_{0,0} \equiv A_1$, $A_{p,p} \equiv A_3$, and $A_{0,p} \equiv A_4$, partition the sides $[A_1A_3]$ and $[A_1A_4]$, respectively, into p equal line segments. Through the points $A_{r,r}$ we draw vertical straight lines (parallel to $[A_1A_4]$) and choose points $A_{r,p} = [lr, \tilde{f}(lr)]$ such that

$$|\tilde{f}(lr) - f(lr)| \le K_4 l^2, \tag{2.9}$$

and $\tilde{f}(y_1) \le f(y_1)$ if we want $\hat{\Gamma}_{3,h} \subset G'$ and $\tilde{f}(y_1) \ge f(y_1)$ otherwise (the function $\tilde{f}(y_1)$ will be specified below; in Figure 2 we deal with the smooth concave arc $\widetilde{A_4A_3}$ and denote its points not belonging to the graph of the function $y_2 = \tilde{f}(y_1)$ by $\circ$, in contrast to their common points marked by $\star$). This yields the polygon $[A_1 A_{0,p} A_{1,p} \ldots A_{p,p}] \equiv \hat{G}'_h$ and the broken line $[A_{0,p}, A_{1,p}, \ldots, A_{p,p}]$, with $A_{0,p} = A_4$ and $A_{p,p} = A_3$, which is just $\hat{\Gamma}_{3,h}$.

Next, we partition vertical line segments $[A_{r,r} A_{r,p}]$ into a union of $p - r$ equal small segments by choosing points $A_{r,q}$, with $r \le q \le p$, and connect all the obtained pairs of the points $A_{r-1,q}$ and $A_{r,q}$ by line segments; the same is applied with respect to all pairs of the points $A_{r,q}$ and $A_{r+1,q+1}$. This yields a triangulation of $\hat{G}'_h$ of type $T^{[1]}(\triangle; p)$ (see Figures 1 and 2).

It is now clear that we may take $\tilde{f}(y_1) = f(y)$ and $K_4 = 0$ in (2.9), if (2.2) holds and we want to have $\hat{\Gamma}_{3,h} \subset G'$. The same is true if (2.1) holds and we want to have $G' \subset \hat{G}'_h$.

The two remaining cases are more complicated but similar to each other. Consider, e.g., the case of (2.1) and $\hat{\Gamma}_{3,h} \subset G'$. If p is even, we may choose the points $A_{r,p}$ as: points belonging to Γ_3 if r is even; the lowest points of intersections of the vertical line $y_1 = rl$ with the tangents to Γ_3 at previously located points $A_{r-1,p}$ and $A_{r+1,p}$ on Γ_3 if r is odd. We then have $K_4 \leq K_1$, and $\tilde{f}(lr) > f(a) + \lambda_1(lr - a)$ (see (2.3)). If $p = 2k + 1$, then we may apply the above algorithm for $r \leq 3$, with $\tilde{f}(rl) = f(rl)$ if $r = 3, 5, \ldots, p$. Remaining values $\tilde{f}(l)$ and $\tilde{f}(2l)$ can be defined by using the tangents at points with $y_1 = 0$ or $y_1 = 3l$. This yields $K_4 \leq 2K_1$. Since $2K_1 l^2 < (\lambda_1 - \lambda_0)2l$, then we again obtain the desired polygon $\hat{G}'_h \subset G'$.

The case of (2.2) and $G' \subset \hat{G}'_h$ is quite similar but even simpler because now the highest points of intersections of the lines $y_1 = rl$ with tangents to Γ_3 at previously located points on Γ_3 correspond to the points $A_{r,p}$ (for even p and odd r). We see now that $xi_1(y) = 0$ in (2.4) and that, on each $(rl, (r+1)l)$, $\xi_2(y) = a_r + t_r y_1 - f(y_1)$. Hence, $K_2 \leq 4K_1 \kappa_1^2$. The straight line with the equation $y_2 = a_r + t_r y_1$ is determined by two points, one of which is always on Γ_3 and the second of which is a vertically shifted point on a tangent (with a shift Y) such that $|Y| \leq K_1 l^2$. Thus, it is not difficult to show that either $|t_r - f'(y_1)| \leq |f'(\xi_r) - f'(y_1)| + lK_1$, with some ξ_r such that $|\xi_r - y_1| \leq l$, or $|t_r - f'(y_1)| \leq |f'(\xi_r) - f'(y_1)| + l/2K_1$, with some ξ_r such that $|\xi_r - Y_1| \leq 2l$. Hence, we can take $K_3 = 9/2K_1 \kappa_1$ (in the general case) and $K_3 = 2K_1 \kappa_1$ (for the simplest two cases).

We verify now that the suggested triangulations of $\hat{G}'_h$ satisfy conditions (2.6)–(2.8). We begin by estimating the lengths of the vertical segments $[A_{r,q} A_{r,q+1}]$. Since $A_{r,p} = [rl, \tilde{f}(rl)]$, we have $|A_{r,q} A_{r,q+1}| = [\tilde{f}(rl) - \lambda_0 rl]/(p - r)$. Moreover, since $\lambda_0(a - y_1) - \lambda_1(a - y_1) \leq f(y_1) - \lambda_0 y_1 \leq \lambda_0(a - y_1) - \lambda_2(a - y_1)$, we have

$$(\lambda_0 - \lambda_1)l - K_1' \frac{l^2}{p - r} \leq |A_{r,q} A_{r,q+1}| \leq (\lambda_0 - \lambda_2)l + K_1'' \frac{l^2}{p - r}, \quad (2.11)$$

with some $K_1' \geq 0$, $K_1'' \geq 0$ such that $K_1' K_1'' = 0$ and $K_1' \leq K_1$, $K_1'' \leq K_1$. Hence, we may write

$$\kappa_6 \leq \frac{|A_{r,q} A_{r,q+1}|}{l} \leq \kappa_7 \quad (2.12)$$

(see (2.3)) and

$$\lambda_0 - \lambda_1 \leq \frac{b_r}{l} \leq \lambda_0 - \lambda_2, \tag{2.13}$$

where $b_r \equiv (f(rl) - \lambda_0 rl)(p-r)^{-1}$. We now estimate now lengths of vectors $\overrightarrow{A_{r-1,q} A_{r,q}} = [l, y]$, where

$$y \equiv \lambda_0 l + \frac{q-r}{p-r} \tilde{g}(rl) - \frac{q-r+1}{p-r+1} \tilde{g}((r-1)l), \tag{2.14}$$

where $\tilde{g}(y_1) \equiv \tilde{f}(y_1) - \lambda_0 y_1$. To make our estimates more precise, we introduce for fixed r a linear function $\hat{f}_r$ that interpolates values of f at $y_1 = lr$ and $y_1 = a$. Then y in (2.14) takes the form

$$y = \lambda_0 l - b_r + \frac{q-r+1}{p-r+1} \left(\hat{f}_r((r-1)l) - f((r-1)l) \right) +$$

$$\frac{q-r}{p-r} \left(\tilde{f}(rl) - f(rl) \right) - \frac{q-r+1}{p-r+1} \left(\tilde{f}((r-1)l) - f((r-1)l) \right). \tag{2.15}$$

Observe that $|\hat{f}_r((r-1)l) - f((r-1)l)| \leq lV_r$, where V_r is the variation of $f'(y_1)$ on $[rl, a]$. Hence, $|y| \leq |\lambda_0 l - b_r| + lV_r + K_1 l^2$,

$$l \leq |A_{r-1,q} A_{r,q}| \leq l[1 + (\lambda_0 - \frac{b_r}{l} + V_r + K_1 l^2)^2]^{1/2} \leq \kappa_8 l, \tag{2.16}$$

$$\sin \angle A_{r-1,q-1} A_{r-1,q} A_{r,q} \geq \frac{1}{\kappa_8}. \tag{2.17}$$

From (2.12), (2.16), and (2.17), it follows that

$$2\mu(\triangle A_{r-1,q-1} A_{r-1,q} A_{r,q}) \leq \kappa_8 \left(\frac{\kappa_8}{\kappa_6} + \kappa_7 \right). \tag{2.18}$$

Similarly, the estimate for $\mu(\triangle A_{r-1,q} A_{r,q} A_{r,q+1})$ is obtained. Therefore, (2.8) holds. Now from (2.8), (2.12), (2.16), and (2.17), it is not difficult to deduce that (2.6) and (2.7) hold. $\square$

For $G' = \triangle A_1 A_3 A_4$, the above estimate gives

$$\mu = \frac{|A_2 A_4|/|A_4 A_3| + |A_4 A_3|/|A_1 A_4|}{2 \sin \angle A_1 A_4 A_3} \tag{2.19}$$

and it is sharp. The proof provides an algorithm for constructing the desired triangulations; our attention to the requirements $\hat{G}'_h \subset G'$ or $G' \subset \hat{G}'_h$ is motivated by the simplicity of the convergence analysis of PGMs. For practical purposes, it is possible to ignore conditions (2.1) and (2.2) and to apply the simplest case of the algorithm with $K_1 = K_4 = 0$. Then its connection with the linear homotopy considered in § 2.2 becomes especially pronounced.

Lemma 1. *Let a standard quasisquare* $G \equiv A_1A_2A_3A_4$ *be partitioned by its diagonal* A_1A_3 *into two triangles* $G' \equiv \triangle A_1A_3A_4$ *and* $G'' \equiv \triangle A_1A_2A_3$, *with* $0 < \angle A_3A_1A_4 < \pi$, $0 < \angle A_2A_1A_3 < \pi$. *Suppose that, for a prescribed* $p \geq 2$, *each of these triangles is partitioned into a set of* p^2 *equal subtriangles. Then the resulting* $2p^2$ *triangles defines a quasiuniform triangulation* $T_h(G)$ *of type* $T(\square; p)$ *and* $\mu(T_h(G)) = \max\{\mu(\triangle A_1A_4A_3); \mu(\triangle A_1A_2A_3),\}$ *(see (0.4.13)).*

Proof. It suffices to observe that the triangulation of G' used in Theorem 1 for $G' = \triangle A_1A_4A_3$ is exactly the same as in Lemma 1 and corresponds to the standard refinement procedure with ratio p. $\square$

We will occasionally need more general triangulations of a triangle.

Lemma 2. *Let, for a triangle* $G' \equiv \triangle A_1A_3A_4$ *and a given* $p \geq 2$, *the points* $A_{0,0}, A_{0,1}, \ldots, A_{0,p}$ *on* $[A_1A_4]$ *and the points* $A_{0,p}, A_{1,p}, \ldots, A_{p,p}$ *on* $[A_4A_3]$ *be defined as in the proof of Theorem 1. Suppose that the points* $A_{0,0}, A_{1,1}, \ldots, A_{p,p}$ *partition the side* A_1A_3 *into a union of* p *arbitrary subsegments. Then there exists a triangulation of* G' *of type* $T(\triangle; p)$ *such that the set of vertices on its boundary coincides with the prescribed set of points* $A_{r,q}$.

Proof. Connect the points $A_{r,r}$ and $A_{r,p}$, for each $r \in [1, p-1]$, by the line segments $[A_{r,r}A_{r,p}]$ (in general, they are not vertical) and partition each such segment by the points $A_{r,r}, A_{r,r+1}, \ldots, A_{r,p}$ into a union of r equal subsegments, $r \in [1, p-1]$. Next, connect all possible pairs of points $A_{r-1,q}$ and $A_{r,q}$ and pairs of points $A_{r,q}$ and $A_{r+1,q+1}$ by line segments. The obtained partition of G' is a desired triangulation. $\square$ [6]

2.2. Triangulations of type $T(\square; p)$ and $T(\square; p_1, p_2)$ for a standard quasisquare. A standard quasisquare $G \equiv A_1A_2A_3A_4$ (see Figure 2) of order 1 that can be partitioned by the straight line segment A_1A_3 into a union of two standard quasitriangles G' and G'' is called a *standard quasisquare*. Its triangulation of type $T^{[1]}(\square; p)$ is defined as a union of triangulations of type $T(\triangle; p)$ for G' and G''. More general triangulations of type $T(\square; p_1, p_2)$, and their particular cases of type $T(\square; p)$ corresponding to $p_1 = p_2 = p$, are defined below.

Suppose that we cover the rectangle $[0, p_1] \times [0, p_2]$, with p_1 and p_2 integers, by the square grid with mesh size 1 and that we partition each cell of the grid into two triangles by drawing one of its diagonals. Then each trian-

---

[6] We will make use of this Lemma only for fixed p when we wish to construct only the coarse triangulation.

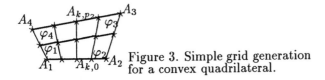

Figure 3. Simple grid generation for a convex quadrilateral.

gulation of a polygon topologically equivalent to the obtained triangulation is called a *triangulation of type* $T(\Box; p_1, p_2)$. A simple enough algorithm for constructing such a triangulation can be suggested for a convex quadrilateral G (see Figure 3). To describe it, we denote the inner angle with vertex A_r by φ_r, $r \in [1,4]$, and the lengths of sides $A_1 A_2, A_2 A_3, A_3 A_4$, and $A_4 A_1$ by l_1, l_2, l_3, and l_4, respectively; next, we partition sides $A_1 A_2$ and $A_4 A_3$ by the points $A_{k,0}$ and A_{k,p_2}, with $k \in [0, p_1]$ and $A_{0,0} = A_1$, $A_{p_1,0} = A_2$, $A_{0,p_2} = A_4$, $A_{p_1,p_2} = A_3$, into p_1 equal segments; further, we connect the points $A_{k,0}$ and A_{k,p_2}, with $k \in [1, p_1 - 1]$, by line segments and partition each segment $[A_{k,0} A_{k,p_2}]$, with $k \in [0, p_1]$, by points $A_{k,q}$, with $k \in [0, p_1]$ and $q \in [0, p_2]$, into p_2 equal subsegments. If now we connect all possible pairs of points $A_{k,q}$ and $A_{k+1,q}$ by line segments, then we obtain a partition of G into a union of $p_1 p_2$ convex quadrilateral cells denoted by $G_{k,q} \equiv [A_{k,q} A_{k+1,q} A_{k+1,q+1} A_{k,q+1}]$, $k \in [0, p_1 - 1]$, $q \in [0, p_2 - 1]$. In what follows we refer to a partition of Ω topologically equivalent to the given one as being of type $P(\Box; p_1; p_2)$. Finally, if in each cell $G_{k,q}$ we draw one of its diagonals, we obtain the desired triangulation of G.

Theorem 2. *Suppose integers p_1 and p_2 tend to infinity and h tends to 0 in such a way that*

$$\kappa_9 \le h p_1 \le \kappa_{10}, \ 0 < \kappa_{11} \le h p_2 \le \kappa_{12}. \tag{2.20}$$

Then, for the above triangulations of G, inequalities (2.6)–(2.8) hold.

Proof. We make use of a Descartes coordinate system $[x, y]$, with $A_1 = [0, 0]$ and the abscissa axis along $A_1 A_2$. Without loss of generality, we may take

$$\psi \equiv \varphi_1 + \varphi_2 - \pi \ge 0; \tag{2.21}$$

otherwise, we can consider the reflection of G with respect to the vertical axis. It is easy to verify that

$$A_{k,0} \vec{A}_{k,p_2} = [X, Y], \tag{2.22}$$

where $X \equiv l_4 \cos \varphi_1 + t(l_3 \cos \psi - l_1)$, $Y \equiv l_4 \sin \varphi_1 + t l_3 \sin \psi$, with $t \equiv k/p_1$, and that

$$|X| \le \max\{l_4 |\cos \varphi_1|; l_2 |\cos \varphi_1|\}. \tag{2.23}$$

Taking, into account (2.20), for $Z \equiv |\cos \angle A_1 A_{k,0} A_{k,p_2}|$, we obtain

$$Z \leq [1 + \sin^2 \varphi_1 (\max\{|\cos \varphi_1|; l_2/l_4 |\cos \varphi_2|\})^2]^{-1/2}. \qquad (2.24)$$

In much the same way, we estimate inner angles of the quadrilateral $G_k \equiv A_{k,0} A_{k+1,0} A_{k+1,p_2} A_{k,p_2}$ which correspond to vertices on the side $A_4 A_3$. Then, for $p_1 \to \infty$ and all inner angles α of each quadrilateral G_k, it is possible to find such an angle α_0, independent of p_1, that $0 < \alpha_0 \leq \alpha \leq \pi - \alpha_0$. We can take, e.g., $\alpha_0 = \arccos[1 + \kappa_{13}]^{1/2}$, where $\kappa_{13} \equiv [\min_r \sin \varphi_r]^2 [\max\{l_2/l_4; l_4/l_2\} \max_r |\cos \varphi_r|]^{-2}$. With respect to the sides of G_k, we have: $|A_{k,0} A_{k+1,0}| = l_1/p_1$, $|A_{k,p_2} A_{k+1,p_2}| = l_3/p_1$, and $l_4 \sin \varphi_1 \leq |A_{k,0} A_{k,N_2}| \leq \max\{l_2 \sin \varphi_4; l_4 \sin \varphi_3\}(\sin \alpha_0)^{-1}$. Now considering the partition of G_k into a union of p_2 convex quadrilaterals $G_{k,q}$, with $q \in [1, p_2]$, for all inner angles β of quadrilaterals $G_{k,q}$, it is possible to find an angle β_0, independent of p_1 and p_2, and such that

$$0 < \beta_0 \leq \beta \leq \pi - \beta_0. \qquad (2.25)$$

With respect to the sides of $G_{k,q}$, we have

$$\frac{l_4 \sin \varphi_1}{p_2} \leq |A_{k,q} A_{k,q+1}| \leq \frac{\max\{l_2 \sin \varphi_4; l_4 \sin \varphi_3\}}{p_2 \sin \alpha_0} \qquad (2.26)$$

and

$$\frac{\min\{l_1; l_3\} \sin \alpha_0}{p_1} \leq |A_{k,q} A_{k,q+1}| \leq \frac{\max\{l_1; l_3\}}{p_1 \sin \beta_0}. \qquad (2.27)$$

From (2.25)–(2.27) we can easily deduce all of the desired inequalities. □

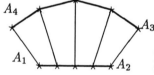

Figure 4. Simple grid generation for a curvilinear quadrilateral.

For a parallelogram G, we have

$$\mu = [p_2 l_1 (p_1 l_2)^{-1} + p_1 l_2 (p_2 l_1)^{-1}][2 \sin \varphi_1]^{-1}.$$

The above algorithm is very natural and has been used for automatization of grid generation in many papers (see, e.g., [478, 518]). Its generalizations for a standard quasisquare $G \equiv A_1 A_2 A_3 A_4$ like the one depicted in Figure 4 that has only one curvilinear side, are straightforward (in Figure 4, a smooth convex arc $\widehat{A_4 A_3}$ (see (2.2)) is approximated by a set of its chords).

We also emphasize the usefulness of a similar approach for more general two-dimensional blocks associated, not with a plane, but with a smooth two-dimensional surface in the space.

§ 3. Algebraic-geometric methods

3.1. Combination of geometric and algebraic methods. As we have emphasized several times, for a given region $\bar{\Omega}$ with complicated geometry, construction of grids that are topologically equivalent to those of a significantly simpler model region $\bar{Q}$ can be carried out effectively by combining geometric and algebraic methods. The former ones partition (cut) $\bar{\Omega}$ into the union of a finite number of standard blocks (supercells, superelements) $\bar{S}_{\Omega,k}, k \in [1, k_0]$ (different domains S_i and S_j do not have common points). The latter ones deal with very simple information about the blocks and their grids and they construct grids that are topologically equivalent to grids for the corresponding model blocks $S_{Q,k}$ of $\bar{Q}$ separately and automatically (see [117, 177, 178, 180, 379, 390, 394, 478, 518]).

For obtaining quasiuniform triangulations for $\bar{\Omega}$, we require that these grids coincide on common parts of the boundaries of two neighboring blocks (*grid matching*). Very often we use quasiuniform grids only within each block, having in mind composite grids with local refinement for $\bar{\Omega}$. In this case, we need a rather *weak matching of grids* (one must be a subset of the other) or no matching at all. Thus the above grid generation is especially well suited for methods based on composite grids and triangulations. We stress that, not only the geometry of the given region must influence the initial partition of $\bar{\Omega}$, but also of prime importance are questions concerning the resulting grid approximations and the applicability of existing effective algorithms for solving the associated grid systems. Since these questions are connected directly with aspects of the original elliptic boundary value problem (such as types of boundary conditions, lines (surfaces) of discontinuity of coefficients, and so on), these aspects must be taken into account at the start. Also, the structure of data for the given and unknown functions depends strongly on the grid's structure, and factors such as the experience and intuition of the mathematician and engineer must be considered. We may therefore say that the geometric part of the grid generation approach outlined above is the most intellectual for complicated regions and problems. [7]

---

[7] Not surprisingly, some mathematical problems concerning the theory of such partitions turned out to be rather difficult even for $d = 2$ (see [180]) and have remained unsolved for $d = 3$. But, for many practical problems dealing with complicated but fixed regions, quite reasonable initial partitions are often naturally obtained, and can even be improved on the basis of certain a posteriori estimates (see § 2.3), possibly to composite

3.2. Cutting of the region into several standard quasisquares and standard quasitriangles. For two-dimensional manifolds, the most natural choice of a block is a standard quasitriangle or standard quasisquare from § 2 corresponding to a model triangle or square (rectangle). Such a choice can be especially effective for relatively simple partitions of the original region into a union of standard quasisquares and quasitriangles, which together with the algebraic algorithms from § 2 enable one to construct triangulations topologically equivalent to those of a model region $\bar{Q}$. Consider a simply connected bounded domain Ω having Lipschitz boundary Γ partitioned by points A_1, A_2, A_3, and A_4 (numbered counter-clockwise) into a set of four consecutive arcs

$$\Gamma^{(1)} \equiv A_1\tilde{A}_2, \Gamma^{(2)} \equiv A_2\tilde{A}_3, \Gamma^{(3)} \equiv A_3\tilde{A}_4 \text{ and } \Gamma^{(4)} \equiv A_4\tilde{A}_1. \tag{3.1}$$

Such a closed region, with the above partitioned boundary, is called a *curvilinear quadrilateral* and is denoted by $[\Omega; A_1, A_2, A_3, A_4] \equiv \Omega_A$.

A partition of a curvilinear quasiquadrilateral Ω_A is *of type* $\bar{\mathcal{P}}(\Box; m_1, m_2)$ if it is topologically equivalent to the partition of the rectangle $[0, m_1] \times [0, m_2]$ by the square grid with the mesh size 1 and, in this topological correspondence, the points $A_s, s = 1, 2, 3, 4$ and the vertices of the rectangle correspond to each other.

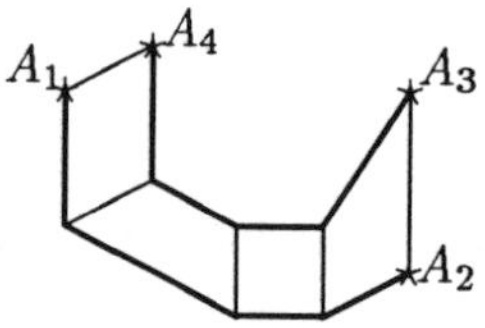

Figure 1. Partition of type $\bar{\mathcal{P}}(\Box; 4, 1)$.

Figures 1 and 2 correspond to the simplest partitions of this type with the natural choices of vertices of $\bar{\Omega}$ in the role of the points A_1, A_2, A_3, A_4 (they correspond to the vertices of the rectangle Q denoted by the same symbols). Here and below, the cutting lines in $\bar{\Omega}$ corresponding to horizontal and vertical lines in Q are depicted by thick and thin lines, respectively, and positions of the points A_1, A_2, A_3, A_4 are indicated by $\star$. Also, the straight line segments in all our figures can be considered as chords of smooth arcs, so these figures can be related to regions with curvilinear boundaries. Figures 3–5 illustrate complicated choices of the points A_1, A_2, A_3, and A_4; here we deal with simpler regions. Remember that the choice of these points can be motivated by aspects of the elliptic boundary value problem under consideration.

---

grids with local refinement. It seems that, even in this respect, the use of structured grids can be advantageous.

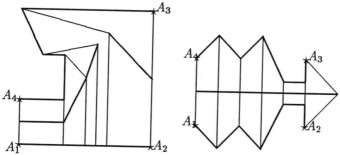

Figure 2. Partitions of type $\bar{\mathcal{P}}(\Box; 5, 2)$.

If all $\Gamma^{(s)}$ are broken lines, then Ω_A is called a *polygonal quadrilateral* (see Figures 1–5). If all $\Gamma^{(s)}$ are straight line segments, then Ω_A is called a *quadrilateral*; $\bar{\Omega}$ here may be a triangle (see Figure 5).

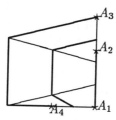

Figure 3. Partition of type $\mathcal{P}(\Box; 3, 2)$ for a quadrilateral.

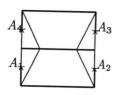

Figure 4. Partition of type $\mathcal{P}(\Box; 3, 2)$ for a quadrilateral.

For a region with slits, the notion of a generalized curvilinear quasi-quadrilateral Ω_A can sometimes be introduced as the limit of certain sequence of curvilinear quasiquadrilaterals. For example, Figure 6 represents such a region Ω_A and its *partition of type* $\mathcal{P}(\Box; m_1, m_2)$ (it differs from a partition of type $\bar{\mathcal{P}}(\Box; m_1, m_2)$ in that the topological equivalence holds for all inner points of region Ω_A and the rectangle Q, and a limit correspondence in the sense of Subsection 2.1.6 between the boundary points is required). (The end points of the double slit are marked by $\triangledown$ and $\triangle$.)

Consider now a doubly-connected region Ω with boundary Γ consisting of two disjoint closed curves $\Gamma^{[1]}$ and $\Gamma^{[2]}$ that are Lipschitz boundaries of bounded simply connected domains $\Omega^{[1]}$ and $\Omega^{[2]}$ with $\bar{\Omega}^{[2]} \subset \Omega^{[1]}$ (that is,

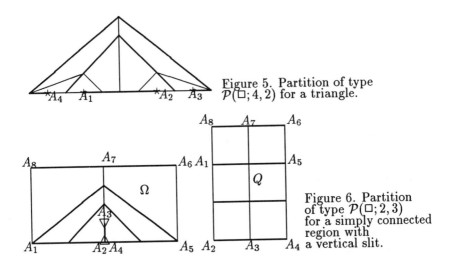

Figure 5. Partition of type $\mathcal{P}(\square; 4, 2)$ for a triangle.

Figure 6. Partition of type $\mathcal{P}(\square; 2, 3)$ for a simply connected region with a vertical slit.

$\Omega = \Omega^{[1]} \setminus \bar{\Omega}^{[1]}$). The region $\bar{\Omega}$ is topologically equivalent to a ring or to a rectangle, say Q^0, obtained by unrolling the side surface of a cylinder, that is, the vertical sides of Q^0 are identified.

For definiteness, let $Q^0 \equiv [0, m_1] \times [0, m_2]$ with integers m_1 and m_2, which we assume is partitioned by a square grid with the mesh size 1. Let each cell of the grid be partitioned into two equal triangles. Then each partition of $\bar{\Omega}$ into a set of $m_1 m_2$ standard quasisquares that is topologically equivalent to the above grid is called *of type* $\bar{\mathcal{P}}(Q^0; m_1, m_2)$ (see Figure 7). If we draw one of the diagonals in each cell, we obtain a class of triangulations of *type* $\bar{T}(Q^{(0)}; m_1, m_2)$.

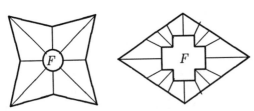

Figure 7. Partitions of type $\bar{\mathcal{P}}(Q^0; m_1, m_2)$ for singly-connected regions.

Similarly, we introduce partitions and triangulations of type $\mathcal{P}(Q^0; m_1, m_2)$, $T(Q^0); m_1, m_2)$, and $T(Q^{(0)}; m_1, m_2)$, respectively, for regions with slits. In place of a simply-connected region in the role of the model region, we can select a more complicated single-connected region with Lipschitz boundary (see, e.g., Figure 8; the topologically equivalent partition for the model region is typical for domain decomposition methods and is of the form like in Figure 3.5.1).

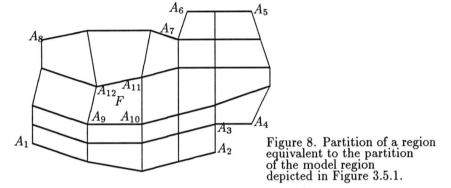

Figure 8. Partition of a region equivalent to the partition of the model region depicted in Figure 3.5.1.

We also emphasize that certain model regions with slits, like the region depicted in Figure 9, can be very attractive.

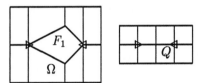

Figure 9. Partition of a region equivalent to the partition of the model region with a slit.

When we wish to deal with model rectangular or even square blocks, we cannot always find such a model region in the plane and must instead consider two-dimensional manifolds in space (see Figures 3.5.7–3.5.15 and 10, where • is related to the end points of double slit in the horizontal plane and ⋆ is related to the end points (A_6 and A_3) of the double vertical slits in the vertical plane).

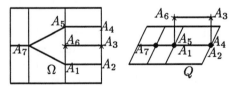

Figure 10. A special partition of a simply-connected region equivalent to a square grid on a model two-dimensional manifold in the space.

Along the same lines, we can also treat multiconnected regions (see, e.g., Figures 11 and 12).

3.3. Geometric problems dealing with special partitions of two-dimensional regions into standard blocks. Now we consider briefly the most complicated geometric problems (see [177, 178, 179, 180]) concerning the possibility to construct triangulations of the desired simple type for fairly general two-dimensional regions $\bar{\Omega}$.

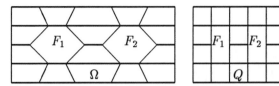

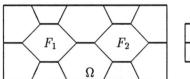

Figure 11. Equivalent partitions of doubly-connected regions.

Figure 12. Partition of doubly-connected region equivalent to a partition of a model region with two slits.

Theorem 1. *For an arbitrary polygonal quadrilateral Ω_A, it is possible to find a partition of type $\bar{\mathcal{P}}(\square; m_1, m_2)$ with some integers m_1 and m_2.*

The proof of this theorem can be found in [180]. Unfortunately, it is not simple and the algorithm for constructing such a partition is based on cutting Ω_A into a set of simpler parts. For example, if the partition of the model region $\bar{Q} \equiv \square$ has axes of symmetry, it is natural to find cutting lines in Ω_A corresponding to these axes. Examples given in Figures 2–6 indicate that, very often, such successive simplifications of the task do lead to the simplest cases dealing with quadrilaterals.

Especially instructive are Figures 4–6, which indicate that a straight line segment at each stage of these simplifications in the role of the cutting line may be used; in Figure 5, such segments cut off simpler parts of the region. The validity of such segment line cuttings is trivial for convex Ω. For the general case, our next theorem 2 about domains with so-called colored boundaries is crucial.

Let the boundary of a domain Ω be a closed Jordan curve Γ and suppose that Γ is partitioned by points A_1, A_2, A_3, and A_4 in the same way as for a standard quasisquare Ω_A (see (3.1)). Suppose that the inner points of $\Gamma^{(1)}$ and $\Gamma^{(3)}$ are colored black and those of $\Gamma^{(2)}$ and Γ_4 are colored red. Then the domain Ω is called a *domain with colored boundary*.

Theorem 2. *For each domain Ω with colored boundary there exists a pair of points B_1 and B_2 of the same color belonging to different parts of the boundary such that inner points of the straight line segment $[B_1 B_2]$ are inner points of the domain Ω (see [177]).* [8]

---

[8] For any fixed region $\bar{\Omega}$ arising in practice, the task of finding a partition indicated in Theorem 1 is much simpler than for the general case, although it often needs (see, e.g.,

Theorem 3. *For an arbitrary polygonal quadrilateral Ω_A, it is possible to construct quasiuniform triangulations of type $\bar{T}(\square; pm_1, pm_2)$ with some integers m_1 and m_2 and any integer p.*

Proof. Theorem 1 implies that there exists a partition of Ω_A of type $\bar{P}(\square; m_1, m_2)$. Each of the polygonal quadrilaterals in this partition can be subdivided into two triangles. Next, to each of these triangles the refinement procedure can be applied with ratio p. This yields a triangulation of type $\bar{T}(\square; p)$ for each polygonal quadrilateral in $\bar{P}(\square; m_1, m_2)$. Since each side of the above polygonal quadrilateral is partitioned into p equal intervals, the union of these triangulations yields the desired triangulation of Ω_A. $\square$

Theorem 4. *Let the boundary Γ of a simply connected region Ω consist of a finite number of arcs γ_r meeting at the interior angles β_s such that $0 < \beta_0 \leq \beta_s \leq \pi - \beta_0$, $\forall s$. Suppose that each arc is twice continuously differentiable and is either convex or concave with respect to Ω. Then, for an arbitrary choice of the points $A_1, A_2, A_3,$ and A_4 on Γ defining a curvilinear quadrilateral Ω_A, partitions of Ω_A of type $\bar{P}(\square; m_1, m_2)$ can be constructed which yield a set of $m_1 m_2$ standard quasisquares, with curvilinear sides being arcs of Γ.*

Proof. We outline the basic stages of obtaining the desired partitions. First, consider the set of points $B_i, i \in [1, n+1]$ (enumerated in the same direction as points $A_1, A_2, A_3,$ and A_4), such that it contains all points $A_1, A_2, A_3,$ and A_4 and all points of intersections of neighboring arcs γ_r and γ_{r+1}. Assume also that $B_1 = A_1 = B_{n+1}$ and that all arcs $\overparen{B_i B_{i+1}}$ have sufficiently small variations of tangent direction (e.g., the tangent direction at B_{i+1} is obtained by rotation of the tangent at B_i through angle α_i, with $|\alpha_i| \leq \alpha^*$). In the next stage, the inner points $C_i, i \in [0, n+1]$, are obtained; $C_0 = C_{n+1}$ and each C_i is a shifted point B_i. More precisely, if B_i is a point such that the tangent to Γ at B_i exists, then we take the shift in the direction of the inner normal to Γ at B_i. If B_i is an angle point of Γ, then the shift is taken in the direction of the bisector of the corresponding inner angle. The lengths $d_i \equiv |B_i C_i|$ must be sufficiently small to ensure that $C_i \in \Omega$. On the other hand, if arc $B_i B_{i+1}$ is concave with respect to our domain, then d_i and d_{i+1} must be sufficiently large to ensure that line segments $[C_i C_{i+1}] \subset \Omega$. To understand why this is possible, assume that there exist positive constants κ, independent of i, such that $\kappa_0 \leq n|B_i \vec{B}_{i+1}| \leq \kappa_1$, $\kappa_2 \leq n d_i \leq \kappa_3$, $\forall i$. Then, between the points of $B_i \tilde{B}_{i+1}$ and the chord $[B_i B_{i+1}]$, we can establish a correspondence of type (2.4) with $|\xi(y)| = O(n^{-2})$. This implies that $[C_i C_{i+1}] \subset \Omega$ for $d_i \asymp 1/n \asymp d_{i+1}$. So the second stage yields a polygonal quadrilateral

---

Figures 5 and 6).

$[C_1C_2 \ldots C_n] \subset \Omega$ such that line segments $[B_iC_i], [C_iC_{i+1}], [C_{i+1}B_{i+1}]$, and arc $\widetilde{B_iB_{i+1}}$ define a standard quasisquare, $i \in [1, n]$. The boundary of $[C_1 \ldots C_n]$ is naturally subdivided by the four points that are the shifted points A_1, A_2, A_3, and A_4 into the set of four broken lines $\Gamma_{C,1}, \Gamma_{C,2}, \Gamma_{C,3}$, and $\Gamma_{C,4}$ (we denote them by D_1, D_2, D_3, and D_4).

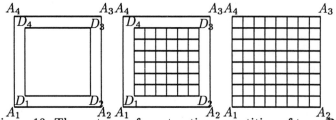

Figure 13. Three stages of constructing a partition of type $\mathcal{P}(Q; m_1, m_2)$, depicted for a model region.

This and the two further stages of constructing the desired partition are illustrated in Figure 13, where their topological analogs for the model region are shown. In the third stage, for the polygonal $[C_1C_2 \ldots C_n]$ with prescribed partition of its boundary, we construct a partition $\mathcal{P}(\Omega)$ of type $\mathcal{P}(\square; m_1', m_2')$ (see Theorem 1) into $m_1'm_2'$ of quadrilaterals. Finally, we continue $\Gamma_{C,i}, i \in [1, 4]$ and the cutting lines in the obtained partition and draw some additional broken lines in standard quasisquares in $\mathcal{P}(\Omega)$ neighboring Γ to obtain the desired partition of Ω_A. $\square$

Theorem 5. *Suppose that a standard quasiquadrilateral Ω_A satisfies the conditions of Theorem 4. Then there exist quasiuniform triangulations of type $\bar{T}(\square; pm_1, pm_2)$ for all large enough integers p.*

Proof. We may take the partition of Ω_A of type $\bar{\mathcal{P}}(\square; m_1, m_2)$ from Theorem 4 into m_1m_2 standard quasisquares G. For each we obtain quasiuniform triangulations of type $\bar{T}(\square; p)$ on the basis of Theorem 2.1. If for each G its approximation $\hat{G}_h$ is such that $\hat{G}_h \subset G$, then we can easily unite all resulting triangulations to obtain a quasiuniform triangulation of Ω_A of type $\bar{T}(\square; pm_1, pm_2)$ for all integers p. However, if there are $\hat{G}_h$ such that $Y \equiv \hat{G}_h \setminus G \neq \emptyset$, then we must demand that different Y have no common points. This and (3.2) imply that p must be sufficiently large. $\square$ [9]

Theorem 6. *Consider a doubly-connected region $\bar{\Omega}$ with Lipschitz boundary consisting of two closed curves such that each of them satisfies the con-*

---

[9] We emphasize again that, for concrete regions Ω_A it makes sense to obtain the desired partitions on the basis of successive simplifications of the task by cutting off simpler parts of Ω_A corresponding to vertical or horizontal strips of squares in the model partition of the rectangle. It is also useful to find cutting lines corresponding to possible axes of symmetry in the model partition. Theorem 4 can also be a very useful tool in dealing with multiconnected regions.

ditions of Theorem 4. Then there exists a partition of type $\bar{\mathcal{P}}(Q^0; m_1, m_2)$ into $m_1 m_2$ standard quasisquares.

Proof. It is not difficult to prove that it is possible to find points $A_1 A_2$ on $\Gamma^{[2]}$ and points $A_3 A_4$ on $\Gamma^{[1]}$ such that the segments $[A_1 A_4]$ and $[A_2 A_3]$ cut out of $\bar{\Omega}$ a standard quasisquare $(A_1 A_2 A_3 A_4)$ (its curvilinear sides $A_1 \tilde{} A_2$ and $A_4 \tilde{} A_3$ belong to $\Gamma^{[2]}$ and $\Gamma^{[1]}$, respectively). Then the closure of the remaining part of $\bar{\Omega}$ may be treated as a standard quadrilateral Ω_A from Theorem 4, and a partition of type $\bar{\mathcal{P}}(\square; m_1', m_2')$ can be constructed for it. It then suffices to partition the quasisquare $(A_1 A_2 A_3 A_4)$ in accordance with already prescribed partitions of line segments $A_1 A_4$ and $A_3 A_4$ $\square$.

Theorem 6 indicates that, for $\bar{\Omega}$ depicted in Figures 7 and 8, partitions of type $\bar{T}(Q^0; m_1, m_2)$ can be found. But we emphasize again that the extra simplicity of the model grid may require a rather complicated grid generation procedure for $\bar{\Omega}$. Thus, for example, the partition shown in Figure 8 is more natural because it yields nearly square blocks, for which estimates (2.8) hold with sufficiently small μ. Similar examples of partitions were given in § 3.5.

We now briefly consider problems concerning partition of the region Ω with non-Lipschitz boundary (see Subsection 2.1.6)). Fortunately, in constructing a partition of $\bar{\Omega}$ with several slits their presence may be ignored if Γ_s consists of a finite number of line segments and we require that the desired partition of $\bar{\Omega}$ must be such that each of these segments is a union of several sides of a few standard quasitriangles in the partition. Only while constructing quasiuniform grids for these separate blocks we must take into account that some of their sides correspond to some parts of the slits, so grids may not match the grids on the opposite sides of double slits. Even if we prefer matching grids on double slits, it is essential to remember that nodes on the slits are actually multinodes, as explained in Subsection 2.1.6, and may correspond to several basis functions. Instead of partitions of a planar $\bar{\Omega}$, we can construct those of closed regions on certain two-dimensional manifolds (including Riemann's surfaces) and make use of Theorem 4 for parts of the region. For example, instead of triangulations of a sphere considered in [59], it is possible to construct triangulations that are topologically equivalent to those of the surface of the unit cube, or even of a nonstandard planar region obtained as the limit of a parallelepiped's surface as its height tends to zero (to obtain this triangulation, it suffices to cut the sphere into two hemispheres and for each to make use of triangulation of type $\bar{T}(\square; p)$). For such initial partitions, we can also take advantage of composite triangulations with local refinement combined with effective iterative methods.

§ 4. Generalizations for multidimensional blocks

4.1. Refinement of an n-dimensional simplex. We return to the problem of partitioning a given n-dimensional simplex $S \equiv [P_0, P_1, \ldots, P_n]$ into m^d congruent subsimplexes (elementary simplexes). This was considered partially in § 2.1, where existence of such a partition for $n > 3$ was just an assumption. (We prefer now to use integers $n \geq 2$ and $m \geq 2$, instead of d and t, because our proofs are based on induction. [10]

In what follows, we deal with n-dimensional simplexes, which are regular simplicial parts of cubes in $\mathbf{R}^n$ (see § 2.1). More precisely, we consider a cube $Q_m \equiv \{x : 0 \leq x_i \leq m, i \in [1, n], \}$, where $x_1, \ldots, x_n$ refer to the Descartes coordinates of a point $x \equiv [x_1, \ldots, x_n] \in Q_m$. We regard the direction $[1, 1, \ldots, 1]$ as one of its chosen diagonal and we take its vertices

$$P_0 \equiv 0, P_1 \equiv P_0 + me_1, P_2 \equiv P_1 + me_2, \ldots, P_n \equiv P_{n-1} + me_n \quad (4.1)$$

as vertices of our n-dimensional simplex

$$S \equiv [P_0, P_1, \ldots, P_n] \quad (4.2)$$

(Q_m contains $n!$ such different congruent simplexes and $|S| = m^n/(n!)$).

Lemma 1. *Point $[x_1, \ldots, x_n]$ belongs to the simplex S if and only if*

$$m \geq x_1 \geq x_2 \geq \cdots \geq x_n \geq 0. \quad (4.3)$$

---

[10] For $n = 2$ and a triangle S, this problem is trivial because we can apply an obvious induction with respect to m. It becomes more involved for $n = 3$ and a tetrahedron S, respectively, but remains within an ability to illustrate it geometrically and, e.g., for $m = 2$ and $P_{i,j}$ referring to the midpoints of the edges $[P_i P_j]$, $i \in [0,3]$, $j \in [0,3]$, $j \neq i$, we obtain the following list of the elementary simplexes:

n	st	shifts	simplex	n	st	shifts	simplex
1	P_0	[1,2,3]	$P_0 P_{0,1} P_{0,2} P_{0,3}$	5	$P_{0,2}$	[1,2,3]	$P_{0,2} P_{1,2} P_2 P_{2,3}$
2	$P_{0,1}$	[1,2,3]	$P_{0,1} P_1 P_{1,2} P_{1,3}$	6	$P_{0,2}$	[1,3,2]	$P_{0,2} P_{1,2} P_{1,3} P_{2,3}$
3	$P_{0,1}$	[2,1,3]	$P_{0,1} P_{0,2} P_{1,2} P_{1,3}$	7	$P_{0,2}$	[3,1,2]	$P_{0,2} P_{0,3} P_{1,3} P_{2,3}$
4	$P_{0,1}$	[2,3,1]	$P_{0,1} P_{0,2} P_{0,3} P_{1,3}$	8	$P_{0,3}$	[1,2,3]	$P_{0,3} P_{1,3} P_{2,3} P_3$

Table 1: Elementary simplexes in the partition of the tetrahedron $S \equiv [P_0 P_1 P_2 P_3]$.

Here st refers to the starting point and these 8 simplexes are provided so that the basic idea of identifying an elementary simplex with a chain of shifts (see § 2.1) is emphasized, and thus prepares us for similar partitions in the general case. Different partitions of a three-dimensional simplex can be obtained under different orderings of its vertices, in contrast with the case $n = 2$. For example, under the ordering associated with the sequence $P_0 P_2 P_1 P_3$, we obtain simplex $P_{0,2} P_{0,1} P_{0,3} P_{2,3}$ as an elementary one that is not present in the given list.

Proof. It is well known that $x \in S$ if and only if there exist nonnegative constants $\alpha_0, \ldots, \alpha_n$ such that $\alpha_0 + \cdots + \alpha_n = 1$ and $x = \alpha_0 P_0 + \cdots + \alpha_n P_n$. Hence, $x_1 = m(\alpha_1 + \cdots + \alpha_n)$, $x_2 = m(\alpha_2 + \cdots + \alpha_n), \ldots, x_n = m\alpha_n$, and, therefore, conditions (4.3) do characterize points of S. □

Next, consider the cubic grid for the cube Q_m with mesh size 1 and a regular triangulation of Q defined by the chosen grid and a common direction of the cell's diagonals. Recall that each cubic cell is partitioned into a union of $n!$ elementary simplexes

$$S_{el} \equiv [A_0, A_1, \ldots, A_n], \tag{4.4}$$

where

$$A_0 \equiv [x_{0,1}, \ldots, x_{0,n}], \quad A_1 \equiv A_0 + e_{j_1}, \ldots, A_n \equiv A_{n-1} + e_{j_n}, \tag{4.5}$$

$j_1, \ldots, j_n$ are different integers and each $j_r \in [1, n]$ (Q_m contains $m^n n!$ different elementary simplexes and $|S_{el}| = 1/(n!)$). In other words, the chain (sequence) $A_0, A_1, \ldots, A_n$ (see (4.4)) is defined by the starting point A_0 and the shift vector

$$\vec{j} \equiv [j_1, \ldots, j_n]. \tag{4.6}$$

Each A_0 in (4.5) is such that its coordinates are integers in $[0, m-1]$. Now our main task is to show that our simplex S contains exactly m^n elementary simplexes S_{el} (see (4.4)). If $n = 3$, the situation is again fairly simple and we can use induction with respect to m: if the simplex (tetrahedron) $S_{m-1} \equiv S \cap \{x : x_3 \geq 1\}$ is partitioned into a union of $(m-1)^3$ elementary simplexes S_{el}, then the remaining lower part of S can be represented as the union of two triangular prisms and a tetrahedron (it suffices to draw the planes $x_2 = 1$ and $x_1 = 1$ as shown in Figure 1 for $m = 3$; see also Figure 2.1.3). It is easy then to see that one of the prisms ($M_0 M_1 P_2 M_0' M_1' M_2'$ in Figure 1) contains $3(m-1)^2$ elementary simplexes and another ($B_1 M_0 M_0' P_1 M_1 M_1'$ in Figure 1) contains $3(m-1)$ simplexes. Hence, S contains $(m-1)^3 + 3(m-1)^2 + 3(m-1) + 1 = m^3$ elementary simplexes.

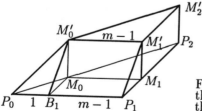

Figure 1. Partition of the lower part of the tetrahedron.

For $n \geq 4$, we replace such geometric proofs by combinatorial ones. With this in mind, we emphasize that A_0 in (4.5) can be a starting point for an elementary simplex $S_{el} \subset S$ if and only if its coordinates $x_{0,1}, \ldots, x_{0,n}$ are integers such that $m - 1 \geq x_{0,1} \geq \cdots \geq x_{0,n} \geq 0$. Moreover, if we partition the vector A_0 into subvectors with equal coordinates (different for different subvectors), then it is easy to see that the shift j_1 (more precisely, the vector e_{j_1}) must correspond to one of the first coordinates of these subvectors; the same rule applies to the shift j_k (see (4.4)), with respect to the partitioned vector A_{k-1}, with exception of coordinates that have been already increased (in particular, the coordinates equal to m).

For example, if $n = 4, m = 2$, and $A_0 \equiv [1,1,0,0]$, then e_{j_1} is either e_1 or e_3. In accordance with this, for the case $n = 4, m = 2$, we obtain the following list of the different elementary simplexes belonging to $S \equiv [P_0 P_1 P_2 P_3 P_4]$:

n	A_0	shifts	A_1	A_2	A_3	A_4
1	[0,0,0,0]	[1,2,3,4]	[1,0,0,0]	[1,1,0,0]	[1,1,1,0]	[1,1,1,1]
2	[1,0,0,0]	[1,2,3,4]	[2,0,0,0]	[2,1,0,0]	[2,1,1,0]	[2,1,1,1]
3	[1,0,0,0]	[2,1,3,4]	[1,1,0,0]	[2,1,0,0]	[2,1,1,0]	[2,1,1,1]
4	[1,0,0,0]	[2,3,1,4]	[1,1,0,0]	[1,1,1,0]	[2,1,1,0]	[2,1,1,1]
5	[1,0,0,0]	[2,3,4,1]	[1,1,0,0]	[1,1,1,0]	[1,1,1,1]	[2,1,1,1]
6	[1,1,0,0]	[1,2,3,4]	[2,1,0,0]	[2,2,0,0]	[2,2,1,0]	[2,2,1,1]
7	[1,1,0,0]	[1,3,2,4]	[2,1,0,0]	[2,1,1,0]	[2,2,1,0]	[2,2,1,1]
8	[1,1,0,0]	[1,3,4,2]	[2,1,0,0]	[2,1,1,0]	[2,1,1,1]	[2,2,1,1]
9	[1,1,0,0]	[3,1,2,4]	[1,1,1,0]	[2,1,1,0]	[2,2,1,1]	[2,2,1,1]
10	[1,1,0,0]	[3,1,4,2]	[1,1,1,0]	[2,1,1,0]	[2,2,1,1]	[2,2,1,1]
11	[1,1,0,0]	[3,4,1,2]	[1,1,1,0]	[1,1,1,1]	[2,1,1,1]	[2,2,1,1]
12	[1,1,1,0]	[1,2,3,4]	[2,1,1,0]	[2,2,1,0]	[2,2,2,0]	[2,2,2,1]
13	[1,1,1,0]	[1,2,4,3]	[2,1,1,0]	[2,2,1,0]	[2,2,1,1]	[2,2,2,1]
14	[1,1,1,0]	[1,4,2,3]	[2,1,1,0]	[2,1,1,1]	[2,2,1,1]	[2,2,2,1]
15	[1,1,1,0]	[4,1,2,3]	[1,1,1,1]	[2,1,1,1]	[2,2,1,1]	[2,2,2,1]
16	[1,1,1,1]	[1,2,3,4]	[2,1,1,1]	[2,2,1,1]	[2,2,2,1]	[2,2,2,2]

Table 2: List of the elementary simplexes $A_0 A_1 A_2 A_3 A_4$ in the partition of S.

Theorem 1. *For an arbitrary $n \geq 2$, there exists a partition of the simplex S with $m = 2$ into a union of 2^n elementary simplexes.*

Proof. In (4.4) and (4.5), the starting point can be either $A_0 \equiv A_{0,0} \equiv [0, \ldots, 0]$ or such that its first k coordinates ($1 \leq k \leq n$) equal 1, and the remaining $n - k$ coordinates vanish (we denote such points by $A_{k,0}$). It then suffices to show that

$$N_n(A_{k,0}) = C_n^k, \qquad (4.7)$$

where $N_n(A_{k,0})$ is the number of possible chains (shift vectors $\vec{j}$ (see (4.6))) starting with $A_{k,0}$ ($0 \leq k \leq n$) and

$$C_n^k \equiv 1 \text{ if } k(n-k) = 0 \text{ and } C_n^k \equiv \frac{n!}{k!(n-k)!} \text{ otherwise.}$$

Indeed, if (4.7) holds, then the general number of possible chains (elementary simplexes belonging to S) is $1 + C_n^1 + C_n^2 + \cdots + 1 = 2^n$. It is easy to see that (4.7) holds if $k = 0$ or $k = n$, so we need consider only $1 \leq k \leq n-1$ assuming that (4.7), holds for $n-1$. For our

$$A_{k,0} = [\underbrace{1, \ldots, 1}_{k}, \underbrace{0, \ldots, 0}_{n-k}]$$

we have only two possible first shifts $j_1 = 1$ or $j_1 = k+1$. They yield the points A_2 of the form

$$[2, \underbrace{1, \ldots, 1}_{k-1}, \underbrace{0, \ldots, 0}_{n-k}], \quad [\underbrace{1, \ldots, 1}_{k+1}, \underbrace{0, \ldots, 0}_{n-k-1}].$$

Now it is easy to see that all subsequent shifts correspond to shifts for $(n-1)$-dimensional problems, with starting points obtained from our two points A_2 by deleting the first and $(k+1)$th coordinates. This implies that

$$N_n(A_{k,0} = N_{n-1}^k + N_{n-1}^{k-1} = C_{n-1}^k + C_{n-1}^{k-1} = C_n^k. \tag{4.8}$$

Therefore, (4.8) leads to (4.7).[11] $\square$

Theorem 2. For arbitrary $n \geq 2$ and $m \geq 2$, there exists a partition of the simplex S into a union of m^n elementary simplexes.

Proof. Let $m = 2$. Then the proof (as the proof of Theorem 1 for $d = 2$) is based on the induction with respect to m. For $m \geq 3$ at hand we denote by

$$N^{(n)} \equiv N^{(n)}(m \geq x_1 \geq \cdots \geq x_n \geq 0)$$

the number of elementary simplexes in S. Here and below, we indicate the conditions specifying the given closed domain as the argument of N. We partition S by drawing the plane $x_n = 1$. Then $N^{(n)} = N_0^{(n)} + \bar{N}_1^{(n)}$, where

$$N_0^{(n)} \equiv N_0^{(n)}(m \geq x_1 \geq x_2 \geq \cdots \geq x_n \geq 1)$$

and $\bar{N}_1^{(n)} \equiv \bar{N}_1^{(n)}(m \geq x_1 \geq x_2 \geq \cdots \geq x_n \geq 0, x_n \leq 1)$. Next, we partition the region associated with $\bar{N}_1^{(n)}$ by drawing the plane $x_{n-1} = 1$. Then

$$\bar{N}_1^{(n)} = N_1^{(n)} + \bar{N}_1^{(n)}(m \geq x_1 \geq x_2 \geq \cdots \geq x_n \geq 0, \ x_n \leq 1, x_{n-1} \leq 1),$$

---

[11] For $n = 3$ we thus obtained a combinatorial proof of an already established assertion.

where

$$N_1^{(n)} \equiv \bar{N}_1^{(n)}(m \geq x_1 \geq x_2 \geq \cdots \geq x_{n-1} \geq 1,\ 0 \leq x_n \leq 1).$$

We repeat this procedure linked with the planes $x_{n-2} = 1, \ldots, x_1 = 1$ and obtain

$$N^{(n)} = \sum_{k=0}^{n} N_k^{(n)}, \tag{4.9}$$

where, for $k \in [0, n]$, $N_k^{(n)}$ denotes

$$N_k^{(n)}(m \geq x_1 \geq x_2 \geq \cdots \geq x_{n-k} \geq 1,\ \underbrace{0 \leq x_{n-k+1} \leq 1, \ldots, 0 \leq x_n \leq 1}_{k}),$$

and $N_n^{(n)} = N_n^{(n)}(1 \geq x_1 \geq x_2 \geq \cdots \geq x_n \geq 0 = 1$. Now, if we could show that

$$N_k^{(n)} = C_n^k (m-1)^{n-k}, \quad k \in [0, n], \tag{4.10}$$

then (4.9) and (4.10) would imply that $N^{(n)} = \sum_{k=0}^{n} C_n^k (m-1)^{n-k} = m^n$, and Theorem 2 would follow. We therefore concentrate on proving (4.10) under the assumption that it holds for smaller dimensions and for smaller m in the n-dimensional case under consideration. We further restrict ourselves to the case $1 \leq k \leq n-1$ because

$$N_0^{(n)} = N_0^{(n)}(m \geq x_1 \geq x_2 \geq \cdots \geq x_n \geq 1)$$

is just $(m-1)^n$ by the induction hypothesis. For the region defined by the conditions

$$m \geq x_1 \geq x_2 \geq \cdots \geq x_{n-k} \geq 1,\ \underbrace{0 \leq x_{n-k+1} \leq 1, \ldots, 0 \leq x_n \leq 1}_{k},$$

we see that k variables $(x_{n-k+1}, \ldots, x_n)$ play a special role; their shifts (see (4.6)) may correspond to an arbitrary set of k different integers in $[1, n]$ and must always be uniquely defined by this set. The number of such sets is C_n^k. Moreover, these shifts have no influence on the possible shifts with respect to the first $n - k$ coordinates. Thus, given the above special shifts, we may simply delete the last k coordinates and deal only with the $(n-k)$-dimensional problem for finding possible chains (see (4.6)) for the simplex in $\mathbf{R}^{n-k}$ defined by

$$m \geq x_1 \geq x_2 \geq \cdots \geq x_{n-k} \geq 1.$$

The number of them is $(m-1)^{n-k}$ and, therefore, (4.10) holds. $\square$ [12]

4.2. Triangulations of type $T(S;p)$ for a standard quasisimplex.
In the sequel, we return to use our standard notations d and p. For d-dimensional regions, the most natural choice of the block T is probably a simplex with only one possible curvilinear face (see § 2.2). But, in contrast to the case $d = 2$, we meet complications even in the definition of T as a standard quasisimplex: the assumption that its curvilinear face $\tilde{\Gamma}$ is either concave or convex with respect to T (see conditions (2.1) and (2.2) for $d = 2$) is rather restrictive, though it may be useful for many types of regions. Nevertheless, if we use this assumption, then we can obtain algorithms for approximating $\tilde{\Gamma}$ by a piecewise linear surface $\hat{\Gamma}_h$ (with the additional requirement that either $\hat{\Gamma}_h \subset T$ or does not have common points with the open domain defined by T), and for constructing triangulations of the corresponding polyhedron $\hat{T}$ (it approximates T), as natural generalizations of the algorithms for $d = 2$. Moreover, they lead to triangulations of $\hat{T}$ that are topologically equivalent to the obtained in Theorem 2 with $h^{-1} \asymp m = p \geq 2$ (we refer to them as triangulations of type $T(S;p)$). For more general regions and $d = 3$, we can define a standard quasisimplex as either of order 1 (see Figure 2.2.4) or of type $[\vec{l}, m]$ (see Figure 2.2.3). If we ignore the previous condition that either $\hat{T}_h \subset T$ or $T \subset \hat{T}_h$, then the vertices for elementary simplexes in $\hat{T}_h$ can be defined as images of the vertices for elementary simplexes in S (see Theorem 2). For p large enough, we obtain quasiuniform triangulations of type $T(S;p)$, that is, there exist constants $\kappa > 0$, independent of p, such that

$$\kappa_0 \leq \frac{l}{h} \leq \kappa_1, \ \kappa_2 \leq \frac{s}{h^2} \leq \kappa_3, \ \kappa_4 \leq \frac{v}{h^3} \leq \kappa_5, \tag{4.11}$$

where l, s, and v refer to the length of an arbitrary edge, the area of an arbitrary face, and the volume of an arbitrary elementary simplex in the constructed triangulation of type $T(S;p)$, respectively. These inequalities, together with Lemma 2.2.8, imply that

$$\kappa_{0,\lambda} \leq \frac{\lambda_i^{1/2}(A)}{h} \leq \kappa_{1,\lambda}, \tag{4.12}$$

where $\lambda_i(A)$ (see (2.2.29)) refers to an arbitrary eigenvalue of the Gram matrix A defined by the vectors a_1, a_2, a_3 in the elementary simplex under consideration (see (2.2.23), (2.2.24)). Similar generalizations hold for $d \geq 4$.

4.3. Triangulations of type $T(C;p)$ for a standard quasicube.
Consider the d-dimensional unit cube

---

[12] For $d = 3$, the given proof is equivalent to the geometric one, which was based on the partition of the lower part of S (see Figure 1).

$$C \equiv Q \equiv [0,1]^d \tag{4.13}$$

and a cubic grid with mesh size $h = 1/p, p \geq 2$. If we use a regular partition of each cubic cell (defined, say, by its diagonal parallel to $[1, \ldots, 1]$) into a union of $d!$ elementary simplexes, then we obtain a regular triangulation of C denoted by $T(C; p)$. Let the points of C be denoted by $z \equiv [z_1, \ldots, z_d]$ and consider the image $\hat{\Omega}$ of C under the mapping

$$x = \Pi(z) \tag{4.14}$$

(see (2.1.2)) (continuous on C and even on $\mathbf{R}^d$) such that every elementary simplex in $T(C; p)$ is mapped into a simplex in a triangulation of $\hat{\Omega}$. Then the obtained triangulation $T_h(\hat{\Omega})$ is referred to as being *of type* $\bar{T}(C; p)$. In this case, $\hat{\Omega}$ is a polyhedron. However, if we use (4.14) under the more general assumption that Π and Π^{-1} are continuous only at inner points of C and $\hat{\Omega}$ (see Subsection 2.1.6), then the boundary of $\hat{\Omega}$ may contain slits. In this case, we refer to our triangulations as being *of type* $T(C; p)$.

Among triangulations of type $\bar{T}(C; p)$, we can choose the two most remarkable cases when $\hat{\Omega}$ is a standard quasicube (see Figure 2.2.5 for $d = 3$) consisting of either $d!$ standard quasisimplexes of type $[\vec{l}, m]$ or $d!$ standard quasisimplexes of order m (see Subsection 2.2.3) mapped into the corresponding $d!$ simplexes in C by a linear or central homotopy (such a standard quasicube may have only d curvilinear faces corresponding to the faces of C belonging to the planes $x_1 = 1, \ldots, x_d = 1$). Again, if we ignore the condition that either $\hat{T}_h \subset T$ or $T \subset \hat{T}_h$, then the vertices for elementary simplexes in $\hat{\Omega}$ can be defined as images of the vertices for elementary simplexes in S (see Theorem 2), and they define a quasiuniform triangulation of type $\bar{T}(C; p)$.

Finally, we consider a parallelepiped $P \equiv [0, a_1] \times \cdots \times [0, a_d]$ and a parallelepiped grid with mesh sizes $h_k \equiv a_k/p_k, k \in [1, d]$. A regular partition of each cell (defined by, say, its diagonal parallel to $[a_1, \ldots, a_d]$) into a union of $d!$ elementary simplexes yields a regular triangulation of P denoted by $T(P; p_1, \ldots, p_d)$. This suggests we define triangulations of type $\bar{T}(P; p_1, \ldots, p_d)$ and $T(P; p_1, \ldots, p_d)$ in the same manner as before for $P = C$. For particular examples of $\hat{\Omega}$, grid generation schemes were analyzed in Theorem 2.2.

4.4. Triangulations of type $T(Pr; p, p')$ for a standard simplicial quasiprism. Now consider a d-dimensional simplicial prism

$$Pr \equiv S^{(d-1)} \times [0, a] \equiv [P_0 \ldots P_{d-1} P_0' \ldots P_{d-1}'], \tag{4.15}$$

where $P_0 \equiv 0, P_1 \equiv P_0 + pe_1, \ldots, P_{d-1} \equiv P_{d-2} + pe_{d-1}$, $P_0' \equiv P_0 + p'e_d, P_1' \equiv P_1 + p'e_1, \ldots, P_{d-1}' \equiv P_{d-1} + p'e_d = [p, p, \ldots, p, p']$, and $S^{(d-1)} \equiv [P_0 \ldots P_{d-1}]$ is the basic $(d-1)$-dimensional simplex partitioned into a union of p^{d-1} elementary $(d-1)$-dimensional simplexes in accordance with Theorem 2. Denote these partitions by $S_k^{(d-1)}, k \in [1, p^{d-1}]$, and define the simplicial prisms

$$Pr \equiv Z_{k,j} \equiv S_k^{(d-1)} \times [j-1 \le z_d \le j], \ j \in [0, p'-1],$$

where the points of Pr are $z \equiv [z_1, \ldots, z_d]$.

Lemma 2. *Each of the prisms* $Z_{k,j}, k \in [0, p^{d-1}], \ j \in [0, p'-1],$ *can be partitioned into a union of d elementary simplexes S_{el} (see (4.4)), and the union of these $dp^{d-1}p'$ elementary simplexes defines a triangulation* $T(Pr; p, p')$ *of the prism Pr.*

Proof. Consider the d-dimensional parallelepiped $Q \equiv [0, p] \times \cdots \times [0, p] \times [0, p']$ partitioned by the cubic grid with mesh size 1. Consider next its regular triangulation, defined by this grid and the direction $[1, \ldots, 1]$, into a union of $p^{d-1}pd!$ elementary simplexes. The cubic cell containing the prism $Z_{k,j}$ is then the union of precisely $d!$ elementary simplexes defined by $d!$ shift vectors $\vec{j}$ (see (4.6)). The shift direction e_d may correspond to an arbitrary coordinate of $\vec{j}$. Deleting these coordinates in the vectors $\vec{j}$, we obtain d sets of new vectors $\vec{j}^{(d-1)} \in \mathbf{R}^{d-1}$, each containing exactly $(d-1)!$ equal vectors. It is easy to see that a union of $(d-1)!$ elementary simplexes associated with $\vec{j}$ having the same vector $\vec{j}^{(d-1)}$ defines an elementary simplicial prism and that one of them is just our prism $Z_{k,j}$. $\square$

It is natural to define *triangulations of type* $\bar{T}(Pr; p, p')$ *and* $T(Pr; p, p')$ in the same manner as for triangulations of type $\bar{T}(C; p)$ and $T(C; p)$. Such triangulations are important because we can describe a d-dimensional standard block for which they can be constructed algebraically. The basic idea is as follows. Consider barycentric coordinates $\lambda_0, \ldots, \lambda_{d-1}$ for two given $(d-1)$-dimensional simplexes $T \equiv [M_0 \ldots M_{d-1}]$ and $T' \equiv [M_0' \ldots M_{d-1}']$ belonging to two different $(d-1)$-dimensional planes in the Euclidean space $\mathbf{R}^d$. Suppose that neither has a common point with the above mentioned plane for another simplex (see, e.g., Figure 2 for $d = 3$).

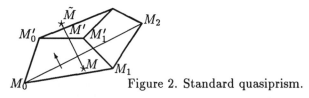

Figure 2. Standard quasiprism.

Lemma 3. *Suppose that the points $M \in T$ and $M' \in T'$ have the same barycentric coordinates $\lambda_0, \ldots, \lambda_{d-1}$ for given $(d-1)$-dimensional simplexes T and T' in the Euclidean space $\mathbf{R}^d$. Then, for $\vec{a}_i \equiv M_i M_i'$, we have*

$$\vec{MM'} = \sum_{i=0}^{d-1} \lambda_i \vec{a}_i. \tag{4.16}$$

Proof. If $O \equiv [0, \ldots, 0]$, then

$$\vec{OM} = \sum_{i=0}^{d-1} \lambda_i \vec{OM}_i, \quad \vec{OM'} = \sum_{i=0}^{d-1} \lambda_i \vec{OM'}_i,$$

and

$$\vec{MM'} = \vec{OM'} - \vec{OM} = \sum_{i=0}^{d-1} \lambda_i \left(\vec{OM'}_i - \vec{OM}_i \right),$$

which yields (4.16). □ [13]

For each point $M \in T$ with the barycentric coordinates $\lambda_0, \ldots, \lambda_{d-1}$ with respect to T, define the vector

$$\vec{a}(M) \equiv \sum_{i=0}^{d-1} \lambda_i \vec{a}_i.$$

Then standard simplicial quasiprism $\hat{Pr} \equiv [M_0 \ldots M_{d-1} M'_0 \ldots M'_{d-1}]$ is defined as

$$\hat{Pr} \equiv \{ F : \vec{OF} = \vec{OM} + t \frac{\vec{a}(M)}{|\vec{a}(M)|}, \quad \forall M \in T, \ \forall t \in [0,1] \}. \tag{4.17}$$

Finally, a general definition of the standard simplicial quasiprism is

$$\tilde{Pr} \equiv \{ F : \vec{OF} = \vec{OM} + t \frac{\vec{a}(M)}{|\vec{a}(M)|}, \quad \forall M \in T, \ \forall t \in [0, f(M)], \tag{4.18}$$

where $f(M)$ is a smooth positive function such that

$$f(M_i) = 1, \ i \in [0, d-1],$$

---

[13] The simplest variant of Lemma 3 for $d = 2$ is the well-known generalization of the classic property of the trapezoid middleline. We also note that this lemma holds for arbitrary T and T' (e.g., they may intersect), but our assumptions are essential to the definition of a *standard simplicial quasiprism*, which in the simplest case is the locus of points belonging to all possible straight line segments $[MM']$ connecting the points $M \in T$ and $M' \in T'$, with the same barycentric coordinates with respect to our $(d-1)$-dimensional simplexes.

(see Figure 2). $\tilde{P}r$ can be obtained from $\hat{P}r$ from linear homotopy with changing direction $\vec{a}(M)$ (see Subsection 2.2.3). More precisely, this mapping is from a point $F \equiv z$ in (4.17) (with coordinates $[z_1, \ldots, z_d]$ defined by M and t) to the point $\tilde{F} \equiv x$ in (4.18) (with the coordinates $[x_1, \ldots, x_d]$ defined by the same M and $\tilde{t}$) such that

$$\tilde{t} = tf(M).$$

This enables one, for a given triangulation of $\hat{P}r$ of type $\bar{T}(Pr; p, p')$, to find images of the vertices of the d-dimensional simplexes in this triangulation, and to construct the desired simplexes in the triangulation of an approximate $\tilde{P}r$. Also, algorithms for finding triangulations of $\hat{P}r$ of type $\bar{T}(Pr; p, p')$ are not especially involved and, for $d = 3$, can be significantly simplified if the faces $M_0 M_1 M_1' M_0'$, $M_1 M_2 M_2' M_1'$, and $M_2 M_0 M_0' M_2'$ are planar standard blocks as considered in Subsection 2.2. For $d = 2$ such an algorithm was analyzed in Theorem 2.2. We also mention papers [66, 333] dealing with certain grids for d-dimensional regions.

4.5. Partitions of multidimensional regions. First note that, for $d = 3$, the geometric results presented in § 3 find an easy application for 3-dimensional regions $\bar{\Omega} \equiv \bar{\Omega}_2 \times [0, l]$ (see 2.1.19) or for more general regions partitioned into a union of blocks of such form. If we choose a triangulation of $\bar{\Omega}$ of type $T(Q_2)$ for some model region Q_2 on the plane (more generally, on a two-dimensional manifold), then on the basis of this triangulation we can obtain a partition of $\bar{\Omega}$ into a union of elementary prisms $Z_{k,j}$ (see (2.1.20)). Hence, we obtain special prismatic grids, which we call grids of type $T(Q_2) \times \omega_3$, where ω_3 refers to a chosen uniform grid on $[0, l]$. The model region can be of any desirable form as discussed in § 3 and may have a non-Lipschitz boundary. In the role of $T(Q_2)$ we can have $T(\square; N_1, N_2)$. It is clear that, for the domain Ω_2 with slits, similar planar slits are present in Ω. Recall that composite grids with local refinement are allowed. For many concrete three-dimensional regions, their partitions into unions of standard quasisimplexes, quasicubes, quasiparallelepipeds, and quasiprisms are fairly straightforward, even if a significant topological simplicity of such a partition is required. For example, if for an ellipsoid region $\bar{\Omega}$, we can draw three planes passing through its chosen inner point (e.g, the center of the ellipsoid) and partition it into 8 standard quasicubes in the sense of § 2.2.2. For the ellipsoid the planes are just canonical coordinate planes. Note that this partition is a particular case of a grid that is topologically equivalent to the cubic grid with mesh size 1 for

$$P \equiv [0, m_1] \times [0, m_2] \times [0, m_3],$$

where m_1, m_2, and m_3 are integers (we refer to such grids as being *of type*

$\bar{\mathcal{P}}(m_1, m_2, m_3)$. For each of these standard quasicubes, the algorithm based on central homotopy (see (2.2.21)) generates triangulations of type $\bar{T}(C; p)$ and, thus, triangulations of $\bar{\Omega}$, which we call of type $\bar{T}(C_0; 2p)$ (recall that, in Lemma 2.4.4, the simplest representative of such a triangulation was considered with $p = N + 1$). Of course, such an algorithm associated with the central homotopy is also useful for nonconvex starlike regions (e.g., for the ellipsoid without some standard quasisimplexes in the above indicated 8 standard quasicubes). It can also be applied to d-dimensional problems.

In § 2.2.2, we discussed certain difficulties that arise in partitioning of three-dimensional regions into unions of standard blocks, in the context of matching of neighbor blocks. We were forced to assume that some curvilinear faces contained common straight line segments. If such assumptions are unsuitable, then even for concrete regions we must combine partition with an approximating procedure for the surface. In this respect, for relatively complex three-dimensional geometries, the use of standard simplicial quasiprisms seems promising, especially when they are constructed from Ω by four cutting planes. On the whole, for $d \geq 3$, there is no geometric theory like that presented in § 3 for $d = 2$. In fact, no analog of Theorem 3.3 for a polyhedron and 6-faced cells is known (each face of a cell is a quadrilateral or even a union of two triangles belonging to different planes). It seems reasonable to expect that these difficult geometric problems will draw appropriate attention in the future because they are directly related to domain decomposition methods and attempts to design effective parallel algorithms (see § 3.5 and 3.6). In certain practical situations where grids of general type are used, the above mentioned problems become of smaller importance and various approaches to constructing grids are suggested (see, e.g., [304, 343, 478] and references therein).

Chapter 5

Asymptotic minimization of computational work in solving second-order elliptic equations and systems

This chapter may be regarded as culmination of the preceding chapters. It provides construction of asymptotically optimal algorithms for a wide class of correct elliptic boundary value problems associated with bounded multidimensional domains.[1]

§ 1. Basic boundary value problems for elliptic equations associated with positive definite quadratic forms

1.1. Modified projective-grid methods for two-dimensional problems. We consider families of pairs of topologically equivalent triangulations: $T_h(\bar{\Omega})$ of the original region Ω (or $T_h(\hat{\Omega})$, where $\hat{\Omega} \equiv \hat{\Omega}_h$ is an approximation of $\bar{\Omega}$) and $T_h(\bar{Q})$ of a chosen model region Q. We assume

---

[1] Its main theoretical outcome is the proof of the strengthened variant of the Kolmogorov-Bakhvalov hypothesis (see § 0.5). Our exposition is based on the results obtained in [165, 166, 176, 177, 178, 179, 180, 181, 182, 183, 186, 187, 200, 201, 205]; many relevant references were given in Chapters 0–4.

that the above equivalence is defined by a piecewise affine transformation

$$z = Z(x), \quad Z \equiv Z_h, \tag{1.1}$$

where $x \equiv [x_1, x_2] \in \hat{\Omega}_h$, $z \equiv [z_1, z_2] \in \bar{Q}$. Composite triangulations with local refinements are allowed, but everywhere we assume that the number of such refinements is bounded, that is, all triangulations are quasiuniform. In choosing the form of the model region, one must keep in mind what ultimate grid approximations and grid systems will be obtained and which of the methods from Chapter 3 can be effectively applied for their solution. Consider an elliptic boundary value problem that is equivalent to a linear operator equation in the Hilbert space $W_2^1(\Omega; \Gamma_0)$, and suppose we wish the iterative methods from § 3.5 for the associated grid systems. Suppose also that the boundary Γ of this bounded $(q+1)$-connected domain Ω (the case $q = 0$ corresponds to a simply connected domain) consists of closed curves $\Gamma^{(r)}, r \in [0, q]$, where: the domain with the boundary $\Gamma^{(0)}$ contains all other curves; each $\Gamma^{(r)}$ is a piecewise smooth curve that does not intersect any other $\Gamma^{(j)}j$; and $\Gamma^{(r)}$ may only contain corner points with angles from $(0, 2\pi)$. Suppose finally that $|\Gamma_0|_{(1)} > 0$ and that $\Gamma_0 = \bar{\Gamma}_0$ consists of a finite number of simple (connected) arcs. Then, in accordance with the theorems from § 4.3, we can draw broken lines to partition $\bar{\Omega}$ into the union of blocks $\bar{\Omega}_1, \ldots, \bar{\Omega}_p$, each a standard quasisquare (see § 4.2) for which construction of quasiuniform triangulations of type $\bar{T}(\square; p)$ or $\bar{T}(Q; p_1; p_2)$ are possible (see Theorem 4.2.2). Such blocks correspond to rectangles in a partition of $\bar{Q}$. Now denote by $\hat{\Gamma}$ the boundary of $\hat{\Omega}$ and by $\hat{\Gamma}_0$ and $\hat{\Gamma}_1$ its parts corresponding to Γ_0 and $\Gamma_1 \equiv \Gamma \setminus \Gamma_0$, respectively. Their images defined by map (1.1) are denoted by $\Gamma_{Q,0}$ and $\Gamma_{Q,1}$. We assume that each side of the above rectangles that belongs to Γ is a part of either $\Gamma_{Q,0}$ or $\Gamma_{Q,1}$. [2]

Let Ω_h be a set of vertices of the triangles, which do not belong to $\hat{\Gamma}_0$, and let each vertex (node) $P_i \in \Omega_h$ be in correspondence with the standard continuous piecewise linear function $\hat{\psi}_i(x)$. The above triangulation was described for our original problem defined by (0.1.16), where $b(u; v) \equiv b(\Omega; u; v)$ (see (0.1.30)) and, e.g.,

$$l(v) = (g, v)_{0,\Omega} + (f_\Gamma, v)_{\Gamma_1} \tag{1.2}$$

(the basic Hilbert space here is $G = W_2^1(\Omega; \Gamma_0)$ with $\|u\|_G = |u|_{1,\Omega}$). For solving of the problem, we assume the representation

---

[2] An example of such partition is depicted in Figure 4.3.8 for $q = 1$ and Γ_0 containing arcs $A_1 A_8 A_7 A_6$ and $A_2 A_3$; the model region consisting of 6 rectangles is represented in Figure 3.5.1. The cutting lines for this Q are 4 vertical lines, so the algorithms from § 3.1 for partial grid problems are applicable. In Figure 3.5.1, only those nodes that do not belong to $\Gamma_{Q,0}$ are marked and $\diamond$ correspond to nodes on the cuttings lines.

$$u(x) = \sum_{k=1}^{n_2} c_k \chi_k(x) + u_0(x), \quad \|u_0(x)\|_{W_2^{m+\gamma}(\Omega)} \le K^*, \quad 0 < \gamma \le 1, \quad (1.3)$$

where $m = 1$, $\chi_k(x), k \in [1, n_2]$, are known singular functions describing the asymptotic behavior of u in neighborhoods of certain singular points (see § 2.3), and the coefficients $c_k, k \in [1, n_2]$, are unknown. The simplest case with $n_2 = 0$ in (1.3) corresponds to the absence of the singular part of u; we then deal with a compact set in G defined by (0.5.1). The general case of (1.3) relates to an unbounded set in G (which might be called swimming compact set); but, as we show, the asymptotically optimal algorithms for this type of problem are characterized by the same estimates of computational work.

To simplify the study, we assume that the support Π_k of the singular function $\chi_k(x)$ corresponding to the singular point M_k is a polygonal domain $\hat{\Pi}_k \subset \Omega$ (with a possible slit) such that M_k lies on its boundary and the closure of $\hat{\Pi}_k$ is the union of the triangles in $T_h(\Omega)$, $k \in [1, n_2]$. This implies that the parts of the boundary Γ in certain neighborhoods of the singular points must consist of several straight line segments. We will make use of e $\Omega_{0,h} \equiv \Omega \setminus \hat{\Omega}$ and $\Omega_{1,h} \equiv \hat{\Omega} \setminus \Omega$, under the assumption that

$$\left.\begin{array}{l} \hat{\Gamma}_0 \subset \bar{\Omega}, \quad \hat{\Gamma}_1 \cap \Omega = \emptyset, \\ |\Omega_{0,h}| = O(h^2), \quad |\Omega_{1,h}| = O(h^2). \end{array}\right\} \quad (1.4)$$

If Γ_1 consists of several straight line segments, then

$$\hat{\Gamma}_1 = \Gamma_1, \quad \hat{\Omega} \subset \Omega, \Omega_{1,h} = \emptyset; \quad (1.5)$$

if Γ_1 contains some curvilinear arcs, then we assume that there exists domain Ω' such that: $\Omega \subset \Omega'$; $\hat{\Omega} \subset \Omega'$; the solution u and the coefficients of bilinear form can be extended to Ω' so that

$$\|u_0\|_{m+\gamma,\Omega'} \le K \|u_0\|_{m+\gamma,\Omega}; \quad (1.6)$$

and the sequence of forms $b(\hat{\Omega}; u; v)$ defined on $(\tilde{G}_h)^2 \equiv (W_2^1(\hat{\Omega}; \hat{\Gamma}_0))^2$ satisfies conditions (0.1.13) and (0.1.14) uniformly in h (in defining these forms, we replace integrals over Ω in (0.1.30) by those over $\hat{\Omega}$ and the integral $(\sigma, uv)_{\Gamma_1}$ by $(\sigma(x - \xi(x)), uv)_{\hat{\Gamma}_1}$, where the function $\xi(x)$ maps Γ_1 onto $\hat{\Gamma}_1$ (see (4.2.4) and (4.2.5)).

We reassert that all the nodes in Ω_h correspond to the standard basis of continuous piecewise linear functions $\hat{\psi}_i(x), i = 1, \ldots, n_1$. We account for the singularities by extending the approximating space to the space

$\hat{G}_{h,\chi} \equiv \hat{V}$; it was shown in § 2.3 that this can dramatically improve the accuracy of PGM radically. We include these functions $\chi_k, k \in [1, n_2]$, in the basis for the space $\hat{V}$ and denote them by $\hat{\psi}_r$ with $r = n_1+1, \dots, n_1+n_2$. Our basic PGM can be written in the form

$$b(\hat{\Omega}; \hat{u}; \hat{v}) = l(\hat{v}), \quad \forall \hat{v} \in \hat{V} \qquad (1.7)$$

(see (0.2.5)) and it yields correct grid problems (see Theorem 0.2.3); $z \equiv u - \hat{u}$ corresponds to its error, $\|z\| \equiv |z|_{1,\Omega}$, and $\|z\|_h \equiv |z|_{1,\hat{\Omega}}$. [3]

Theorem 1. *Suppose that condition (1.3) for the solution of the original problem is satisfied. Suppose also that the assumptions made above with respect to the bilinear forms, the singular functions, the geometry of the domain, and the structure of PGM are satisfied. Then,*

$$\|z\|_h \leq K_0 h^\gamma \|u_0\|_{1+\gamma,\Omega}, \quad \|z\| \leq K_0' h^\gamma \|u_0\|_{1+\gamma,\Omega}. \qquad (1.8)$$

Proof. If (1.5) holds, we may apply Theorem 1.1.2, leading to

$$\|z\|_h \leq K_1 \mathrm{dist}_{W_2^1(\hat{\Omega})}\{u; \hat{V}\} \leq K_2 \|u_0 - \mathrm{int}_1 u_0\|_h. \qquad (1.9)$$

Together with (2.3.1) and (2.3.6), this yields (1.8).

If $\hat{\Gamma}_1$ is not a subset of Γ_1, then the term $b(\hat{\Omega}; u; v)$ in (1.7) calls for additional study. We outline the most significant steps (see [177, 394]). First observe that

$$b(\hat{\Omega}; z; z) = \zeta_1 + \zeta_\Omega + \zeta_\Gamma, \qquad (1.10)$$

where $\zeta_1 \equiv b(\hat{\Omega}; z; u-\hat{v})$, $\zeta_\Omega \equiv b^{(0)}(\Omega_{1,h}; u; \hat{v}-\hat{u})$ (see (0.1.30) and (0.1.31)), and $\zeta_\Gamma \equiv (\sigma(x - \xi(x)), \hat{u}(x)(\hat{v}(x) - \hat{u}(x)))_{\hat{\Gamma}_1} - (\sigma(x), u(x)(\hat{v}(x) - \hat{u}(x)))_{\Gamma_1}$, $\forall \hat{v} \in \hat{V}$. If $\hat{v} \equiv c_1 \chi_1(x) + \dots + c_{n_2} \chi_{n_2}(x) + \mathrm{int}_1 u_0$ (see (1.3)), then

$$\|u - \hat{v}\|_h \leq K_3 h^\gamma |u_0|_{1+\gamma,\Omega}, \qquad (1.11)$$

$$\delta_0 \|z\|_h^2 \leq K_4 h^\gamma |u_0|_{1+\gamma,\Omega} + |\zeta_\Omega| + |\zeta_\Gamma| \qquad (1.12)$$

(see (1.10) and (0.1.14)). But $|u|_{0,\Omega_{1,h}} \leq K_5 h |u|_h$ (see (2.3.5)) and

$$|\zeta_\Omega| \leq K_6 \|u\|_{1,\Omega_{1,h}} \|\hat{v} - \hat{u}\|_{1,Q_{1,k}} \leq K_7 \|u\|_{1,\Omega_{1,h}} \{\|z\|_h + \|u - \hat{v}\|_h\}. \qquad (1.13)$$

---

[3] We can define $\hat{v} = 0$ on $\Omega_{0,h}$ so that $\hat{\Omega}$ on the left-hand side of (1.7) can be replaced by Ω and that the right-hand side of (1.7) assumes $\hat{v}$ is defined on Ω (here we do not consider approximation of the functional l because it is a much simpler task in comparison with the study of the perturbed bilinear form).

Further, we introduce $w(x) \equiv u(x)(\hat{v}(x) - \hat{u}(x))$ and make use of the representation $\zeta_\Gamma = \zeta^{(1)} + \zeta^{(2)}$, where $\zeta^{(1)} \equiv \int_{\Gamma_1} \sigma(x)w(x + \xi(x))(d\hat{s} - ds)$ and $\zeta^{(2)} \equiv \int_{\Gamma_1} \sigma(x)[w(x + \xi(x)) - w(x)]ds$. It is easy to see that $|\zeta^{(1)}| \leq K_8 h \|u\|_{L_2(\hat{\Gamma}_1)} \|\hat{v} - \hat{u}\|_{L_2(\hat{\Gamma}_1)} \leq K_9 h \|u\|_h \|\hat{v} - \hat{u}\|_h$, and a similar estimate holds for $|\zeta^{(1)}|$. This yields

$$|\zeta_\Gamma| \leq K_{10} h \|u\|_h \|\hat{v} - \hat{u}\|_h, \tag{1.14}$$

$$\|u\|_{1,\Omega_{1,h} \cup \Omega_{0,h}} \leq K_{11} h^\gamma \|u_0\|_{1+\gamma,\Omega} \tag{1.15}$$

(see (1.6) and (2.3.5)). Hence, (1.12)–(1.15) lead to (1.8). $\square$ [4]

In approximating Γ_1, our assumption $\Gamma_1 \cap \Omega = \emptyset$ is not necessary. For example, if $\hat{\Gamma}$ consists of chords of Γ, our additional considerations are connected with the elementary triangles $T \equiv A_1 A_2 A_3$ ($A_1 \in \Omega, A_k \in \bar{\Gamma}_1, k = 2, 3,$) approximating curvilinear triangles $\tilde{T}$ with the arc $\widehat{A_2 A_3}$ instead of the chord $[A_2 A_3]$ and such that $\tilde{T} \setminus T \equiv T_+ \neq \emptyset$ (of course, $|T_+| = O(h^3)$). It is easy to see that now, instead of the estimate $|\hat{v} - \hat{v}|_{1,T} \leq K_1 h^\gamma |u|_{1+\gamma,T}$ (see Theorem 2.3.1), we need a similar estimate with T replaced by T', where $\tilde{T} \subset T'$ and $\hat{v}$ is linear on T'. Without loss of generality, we may assume, e.g., that $T' = \triangle A_1 A_2' A_3'$, where $\vec{A_1 A_k'} = 2 \vec{A_1 A_k}, k = 2, 3$. We emphasize that the function $\hat{v}$, which is linear on T', is now defined by the same values of u at the vertices of T, so we deal with a generalized interpolation. Nonetheless, it is easy to see that asymptotic estimates given in Theorem 2.3.1 remain true with new constants s_0 and s_1 defined for a reference triangle R^* (T' is its image) containing the triangle $\mathcal{R}$ from this theorem (it is easy to indicate the constants q_1 and q_0). Thus, Theorem 1 can be generalized for PGMs associated with the simpler triangulations.

1.2. Asymptotically optimal iterative methods. Problem (1.7) is reduced by standard means (see (0.2.2) and (0.2.8)) to the grid system

$$Lu = f, \quad u \in H, \ f \in H, \tag{1.16}$$

in the Euclidean space $H \equiv H_h \equiv \mathbf{R}^N$, with $N = n_1 + n_2$; the vector $u \equiv [u_1, \ldots, u_N]^T$ is defined by the coefficients in the expansion of $\hat{u} \in \hat{V}$ with respect to the basis of $\hat{V}$. Under the given assumptions on the bilinear form, Theorem 0.4.2 applies with $J \equiv \Lambda \equiv \Lambda_\Omega$ (see (0.4.3)) where $(u, v)_G \equiv (u, v)_{1,\hat{\Omega}}$. This implies that the grid operators L and Λ are connected by the relationship $C^{0,0}$ (even C^0 provided $L = L^*$) (see § 1.4).

---

[4] Theorem 1 can be generalized for the case $\Gamma_0 = \emptyset$ and $\|u\|_G \equiv \|u\|_{W_2^1(\Omega)}$ in a straightforward manner.

Theorem 2. Suppose that the conditions of Theorem 1 are satisfied, with $m = 1$ and $n_2 = 0$ (see (1.3)). Suppose also that we use PGM (1.7) with $h^\gamma \asymp \varepsilon$. Then there exists an optimal model operator $B = B^* > 0$ such that $M = O(|\ln \varepsilon|)$ iterations of method (1.3.1) lead to the desired $O(\varepsilon)$-approximations to the solutions of (1.7) and the original elliptic problem with computational work

$$W(\varepsilon) = O(\varepsilon^{-2/\gamma}|\ln \varepsilon|). \tag{1.17}$$

Proof. Since $n_2 = 0$, we may make use of the fact that $\Lambda \asymp \Lambda_Q$ (see (0.4.15)). The operators Λ_Q associated with the chosen model region Q were studied in detail in Chapter 3, and various constructions of model operators B spectrally equivalent to Λ_Q can be used in iterations (1.3.1). The operators L and B are connected by relationship $C^{0,0}$ (see Subsection 3.4.3). Thus, reason convergence of iterations (1.3.1) are independent of the grid, and we can obtain the estimate $\|\mathbf{u}^M - \mathbf{u}\|_B \le K_{12}\varepsilon$ for $M = O(|\ln \varepsilon|)$. This implies that $\|\mathbf{u}^M - \mathbf{u}\|_\Lambda \le K_{13}\varepsilon$ and $|\hat{w} - \hat{u}|_{1,\hat{\Omega}} \le K_{13}\varepsilon$, where $\hat{w} \equiv \hat{\mathbf{u}}^M$ is the standard piecewise linear extension of e iterate $\mathbf{u}^M$ (see (1.4.2)). In other words, we have obtained an $O(\varepsilon)$-approximation to the solution of (1.7); the computational work is estimated as $O(|\ln \varepsilon|/h^2) = O(\varepsilon^{-2/\gamma}|\ln \varepsilon|)$ (see (1.17)). Next, in accordance with (1.8), we see that $\|\hat{w} - u\|_h \le K_{14}\varepsilon$. Moreover, if we define $\hat{w}$ on $\Omega_{0,h}$ as vanishing, then $\|\hat{w} - u\|_{1,\Omega} \le K_{15}\varepsilon$ (see (1.8)), which completes the proof. $\square$ [5]

Now consider the more complicated case where the basis for $\hat{G}$ contains singular functions (see (1.3)); their numbers correspond to $n_1 + 1, \ldots, n_1 + n_2$. Then, in the role of grid operators $L \equiv L_{\Omega,\chi}$, $\Lambda_{\Omega,\chi}$, and $\Lambda_{Q,\chi}$, we obtain

$$L \equiv \begin{bmatrix} L_{1,1} & L_{1,2} \\ L_{2,1} & L_{2,2} \end{bmatrix}, \quad \Lambda_{\Omega,\chi} \equiv \begin{bmatrix} \Lambda_{1,1}^{(\Omega)} & \Lambda_{1,2}^{(\Omega)} \\ \Lambda_{2,1}^{(\Omega)} & \Lambda_{2,2}^{(\Omega)} \end{bmatrix}, \quad \Lambda_{Q,\chi} \equiv \begin{bmatrix} \Lambda_{1,1}^{(Q)} & \Lambda_{1,2}^{(Q)} \\ \Lambda_{2,1}^{(Q)} & \Lambda_{2,2}^{(Q)} \end{bmatrix},$$

where: $L_{1,1}$ and $\Lambda_{1,1}^{(\Omega)}$ are just the operators L and Λ_Ω considered above for $n_2 = 0$; $\Lambda_{Q,\chi}$ is the corresponding Gram matrix for the basis functions $\psi_i^{(Q)}(z)$, which are images of elements in the basis for $\hat{G}$ under mapping (1.1); and $\Lambda_{1,1}^{(Q)}$ is the operator Λ_Q considered above for $n_2 = 0$.

---

[5] Method (1.3.1) can be replaced by more sophisticated and practical iterations based on relationship C^3 (we will consider them in § 2). Of course, for $L = L^* > 0$, we can use modifications of Richardson iterations or conjugate gradients as was indicated in § 1.3; such choices actually make the constants in our estimates much smaller, and thus lead to more practically efficient algorithms. This remark applies to similar situations considered below. However, for many nonlinear problems (see § 5), it is difficult to apply methods that are essentially different from iterations (1.3.1) unless we use linearization as described in § 3.8.

Lemma 1. $\Lambda_{\Omega,\chi}$ *and* $\Lambda_{Q,\chi}$ *are spectrally equivalent operators.*

Proof. For $u^{(Q)} \equiv u^{(Q)}(z) = u_1 \psi_1^{(Q)}(z) + \ldots + u_N \psi_N^{(Q)}(z)$, we have

$$(\Lambda_{\Omega,\chi}\mathbf{u}, \mathbf{u}) = (\|\hat{u}\|_h)^2, \quad (\Lambda_{Q,\chi}\mathbf{u}, \mathbf{u}) = |u^{(Q)}|_{1,Q}^2.$$

The integrals over Ω and Q are sums of integrals over the corresponding triangles. We can easily estimate each such integral in accordance with general Lemma 2.2.1 (if functions on the triangles are piecewise linear, then even Theorem 0.4.4 applies). □

Theorem 3. *Suppose that the conditions of Theorem 1 are satisfied, with* $m = 1$ *and* $n_2 \geq 1$ *(see* (1.3)*). Suppose also that we use PGM* (1.7) *with* $h^\gamma \asymp \varepsilon$. *Then there exists an optimal model operator* $B = B^* > 0$ *such that* $M = O(|\ln \varepsilon|)$ *iterations of method* (1.3.1) *lead to the desired* $O(\varepsilon)$-*approximations to the solutions of* (1.7) *and with computational work as in Theorem 2.*

Proof. The operators L and $B = \Lambda_{Q,\chi}$ are connected by relationship $C^{0,0}$, and the convergence of iterations (1.3.1) is analyzed as in the proof of Theorem 2. We therefore need to specify only the algorithm for implementing a single iteration under consideration. First note that the number n_2 is independent of the grid. This implies that the dense blocks $L_{1,2}, L_{2,1}$, and $L_{2,2}$ contain only $O(N)$ elements and that the matrix-vector product $L\mathbf{u}^n$ requires computational work estimated as $O(N)$. Second, for solving of systems with our model operator B, we can apply block elimination (see § 1.5). The Schur matrix $S_2(B) \equiv B_{2,2} - B_{2,1}B_{1,1}^{-1}B_{1,2}$ (and even its inverse) can be formed at the cost of solving n_2 systems with operator $\Lambda_Q = B_{1,1}$, where Theorem 3.5.3 applies. We can solve them with such a high accuracy that it is possible to speak about obtaining the exact solution at a cost of $O(N)$ operations. In fact, however, approximate solutions of these systems can be taken into account as indicated in Theorem 1.4.1 and Lemma 3.5.3: it suffices to take $\varepsilon_2 = \varepsilon_{B^{-1}} = O(\varepsilon)$ so that our basic iterative algorithm is the perturbed iteration (1.3.1) analyzed in Theorem 1.4.1. If we obtain $(S_2(B))^{-1}$ in a preprocessing stage, then each iteration requires solving two additional systems with Λ_Q, so the total computational work is estimated as in Theorem 2. □ [6]

1.3. The choice of model operators and regions. As was seen from our construction of model operators, the basic idea was to preserve their spectral equivalence with the operator $J \equiv \Lambda \equiv \Lambda_\Omega$ (see (0.4.3)) such that $(\mathbf{u}, \mathbf{u})_J = (\hat{u}, \hat{u})_G$. Recall that $G = W_2^1(\Omega; \Gamma_0)$, $\|u\|_G = |u|_{1,\Omega}$, and

---

[6] In practical implementation of such algorithms (in the presence of rounding errors), it is sometimes reasonable to take $\varepsilon_2 = O(\varepsilon^m)$ and $m > 1$, and to carry out computations involving $S_2(B)$ with double precision (the condition number of this matrix is actually estimated in (1.5.20)).

J is a grid operator approximating the identity operator in $\mathcal{L}(G)$. This corresponds to the simplest elliptic operator $-\Delta$ with Dirichlet and Neumann boundary conditions on Γ_0 and Γ_1, respectively, if we deal with the classical setting (see § 0.1) (practically, all sections in Chapter 3 provide constructions of B spectrally equivalent to our J). In terms of the classical setting our original grid operator L is an approximation of a general elliptic operator with variable coefficients and with Dirichlet and natural boundary conditions on Γ_0 and Γ_1, respectively, where the last conditions have the form $\sum_{r,l=1}^{2} a_{r,l}(x)\cos\alpha_r(x)D_l u + \sigma(x)u = f_\Gamma(x)$, $x \in \Gamma_1$ ($\alpha_r(x)$ denotes the angle between the x_r-axis and the outward normal to the boundary at the point $x \in \Gamma_1$, $r = 1, 2$; see (0.1.30), (0.1.31)). This implies that, in passing to J, we may simplify the structure of the original elliptic operator (in particular, we omit the term $(\sigma(x)u, v)_{0,\Gamma_1}$ in our bilinear form), but we must pay special attention to preserving the types (Dirichlet or natural) of boundary conditions. [7]

Suppose we approximate the problem $-u''(x) = f(x)$, $0 < x < 1$, $-u'(0) + \sigma u(0) = 0$, $u'(1) + \sigma u(1) = 0$, by the scheme $-1/h^2(u_{i-1} - 2u_i + u_{i+1}) = f_i$, $i \in [1, N]$, $-1/h(u_1 - u_0) + \sigma u_0 = 0$, $1/h(u_{N+1} - u_N) + \sigma u_{N+1} = 0$, where $\sigma \geq 0$ and $h \equiv 1/(N+1)$. We consider it as an operator equation in the Euclidean space H_0 of vectors $\mathbf{u} \equiv [u_1, \ldots, u_N]^T$ with inner product $(\mathbf{u}, \mathbf{v}) \equiv h(\mathbf{u}, \mathbf{v})_{\mathbf{R}^N}$ (see Subsection 3.1.5). For this we eliminate u_{N+1} in accordance with the prescribed boundary conditions $u_0 = (1 + h\sigma)^{-1}u_1$ and $u_{N+1} = (1 + h\sigma)^{-1}u_N$. Then our operator equation in H_0 takes the form $\Lambda_\sigma \mathbf{u} = \mathbf{f}$, where $(\Lambda_\sigma \mathbf{u}, \mathbf{u}) = h\sum_{i=0}^{N}(\frac{u_{i+1}-u_i}{h})^2 + \sigma u_0^2 + \sigma u_{N+1}^2$, $u_0 = (1 + h\sigma)^{-1}u_1$, $u_{N+1} = (1 + h\sigma)^{-1}u_N$ (see (3.1.16)). Now $(\Lambda_\sigma \mathbf{u}, \mathbf{u}) = 1/h[u_1^2 + (u_2 - u_1)^2 + \cdots + (u_N - u_{N-1})^2 + u_N^2] - 1/h(1+h\sigma)^{-1}(u_0^2 + u_N^2)$. If we define the model operator B associated with the conditions $u_0 = 0 = u_{N+1}$, then $(B\mathbf{u}, \mathbf{u}) = 1/h[u_1^2 + (u_2 - u_1)^2 + \cdots + (u_N - u_{N-1})^2 + u_N^2]$ (see (3.1.15)). Thus, $(\Lambda_\sigma \mathbf{u}, \mathbf{u}) = (B\mathbf{u}, \mathbf{u}) - \frac{1}{h}\frac{1}{1+h\sigma}(u_0^2 + u_N^2)$, and the inequality $\Lambda_\sigma \leq B$ cannot be improved. But the inequality $\delta_0 B \leq \Lambda_\sigma$ with nonnegative δ_0 holds only when $\delta_0 \leq h\sigma/(1 + h\sigma)$. Hence, Λ_σ and B cannot be spectrally equivalent operators. [8]

---

[7] The importance of this principle was emphasized probably for the first time in [162, 164] in contrast to several suggestions to use simplified operators with Dirichlet boundary conditions on the whole boundary. We describe a simple but enlightening example given in [162], which deals with the one-dimensional case and difference operators (see also Subsection 0.2.4).

[8] Along the same lines, it is possible to study the case of boundary conditions with $\sigma < 0$ (we assume then that $1 + \sigma h > 0$) or when, say, the condition $-u'(0) + \sigma u(0) = 0$ is replaced by $u(0) = 0$. Note that it is easy to apply the indicated results to projective-grid operators and to the standard Euclidean space $H = \mathbf{R}^N$. It should be mentioned that the same conclusion about importance of preserving types of boundary conditions was obtained recently in [357] as a result of general theoretical analysis of limit properties of

It is very important from a practical point of view that we can also construct the model operators B on the basis of their spectral equivalence not to J, but to a more complicated operator. Suppose that the original bilinear form is $b(\Omega; u; v) = (a_1, \nabla u \nabla v)_{0,\Omega_1} + \ldots + (a_p, \nabla u \nabla v)_{0,\Omega_p}$ with variable coefficients $a_1(x), \ldots, a_p(x)$ such that each is almost constant on the corresponding subregion. Then it is reasonable to choose the operator Λ_Ω associated with the bilinear form

$$b^{(0)}(\Omega; u; v) \equiv a_1'(u, v)_{1,\Omega_1} + \ldots + a_p'(u, v)_{1,\Omega_p},$$

where a_r' denotes a middle (averaged) value of the function $a_r(x)$ on Ω_r, $r \in [1, p]$. The reason for this is better constants in the estimates of spectral equivalence of Λ_Ω and L (they may be made independent of

$$\delta_0 \equiv \min_{k \in [1,p]} \inf_{x \in \Omega_k} a(x) \text{ and } \delta_1 \equiv \max_{k \in [1,p]} \sup_{x \in \Omega_k} a(x),$$

in contrast to the constants in the inequalities connecting L and J). Moreover, it is also possible to include in $b^{(0)}(\Omega; u; v)$ the term $(\sigma'(x), uv)_{\Gamma_1}$, where $\sigma'(x)$ becomes a constant on each part of Γ_1 in a chosen partition of Γ_1. Frequently, it is convenient to use model regions and forms

$$b(Q; u; v) \equiv \sum_{r=1}^{p} (a_r', \nabla_z u \nabla_z v)_{0,Q_r} + (\sigma', uv)_{\Gamma_1(Q)}, \qquad (1.18)$$

similar to those indicated above (model operators spectrally equivalent to such Λ_Q can be found in § 3.1, 3.2., 3.4, 3.5). We recall that, in § 3.5, we required that $\sigma'(x)$ vanish on the horizontal sides of the blocks $Q_k, k \in [1, p]$, belonging to Γ_1. However, this was not the case for Sections 3.1, 3.2, and 3.4, although difference simplifications of our operators were used there. The indicated operators were even constructed for $b(Q; u; v)$ involving the term $(c', uv)_{0,Q}$ with a piecewise constant function c' (see also (2.4.10)). Thus, several asymptotically optimal and practically effective model operators associated with the use of model regions of various types are currently available. In choosing the model region Q, we must pay special attention not only to the possibility to obtain the desired model operator $B \asymp \Lambda_Q$, but also to the possibility to obtain topologically equivalent triangulations with a relatively small value μ (see (0.4.14)), which implies that geometries of Ω and Q must be sufficiently alike. [9]

---

grid operators.

[9] For example, if we wish to apply iterative methods with model operators from § 3.7, then we suppose, in accordance with theorems from § 4.3, that by drawing some broken lines we partition $\bar{\Omega}$ into a set of blocks such that each is a standard quasitriangle

1.4. Analysis of multigrid acceleration of the basic iterative algorithm. As we have seen, the use of asymptotically optimal model operators in iterative methods may lead to algorithms (e.g., (1.3.1) or (0.3.13)) for finding ε-approximations to the solutions of grid systems that yield computational work estimates of type (0.3.15) with $r = 0$. Recall that relations $\varepsilon \asymp h^\gamma$ and $N \asymp h^{-2}$. Here we eliminate the multiplier $\ln \varepsilon$ in these estimates and obtain optimal estimates

$$W(\varepsilon) = O(\varepsilon^{-2/\gamma}) \qquad (1.19)$$

for the general case of curvilinear boundaries ([181, 187]; a simpler case was considered in § 1.4; for relevant results see [515]). The multigrid acceleration procedure for this purpose deals with $p + 1$ grid systems

$$L_l u_l = f_l, \quad u_l \in H_l, \quad f_l \in H_l, \quad l \in [0, p],$$

associated with triangulations $T^{(l)}(\hat{\Omega}_{h^{(l)}})$ of levels $l \in [0, p]$ and with subspaces $\hat{G}_{h(l)} \equiv \hat{V}_l$ consisting of the piecewise linear and singular functions considered above, $l \in [0, p]$. Note that the grid parameter here is $h^{(l)} = 2^{p-l} h^{(p)}, l \in [0, p]$, $h^{(0)} \asymp 1$, $h^{(p)} = h$, $h^\gamma \asymp \varepsilon$, and that the vector $\mathbf{v_l} \equiv v_l \in H_l$ is defined by the coefficients in the expansion of $\hat{v}_l \in \hat{V}_l$ with respect to the basis of $\hat{V}_l$, $l \in [0, p]$. We note also that our triangulations

---

of order 1 (see Figure 2.2.1), that is, a triangle with two straight sides and only one possibly curvilinear side. Then this partition is topologically equivalent to some initial triangulation $T^{(l)}(\bar{Q})$; all triangulations $T^{(l)}(\bar{Q}) \equiv T^{(l)}$ of levels $l \in [1, p]$ are obtained by recurrent refinement of the triangulations of the previous level with corresponding refinement ratio 2; that is, each triangle in $T^{(l)}$ is partitioned into a set of 4 subtriangles in $T^{(l+1)}$ by drawing its midlines, where $l \in [0, p-1]$; and the ultimate triangulation $T^{(p)}(\bar{Q})$, with $p \asymp |\ln h|$, will be topologically equivalent to $T_h(\hat{\Omega}_h)$, where $\hat{\Omega}_h$ is obtained as a union of approximations to standard quasitriangles (see Theorem 4.1.1) based on the replacement of the arc by a broken line. But each such partition of $\bar{\Omega}$ leads to its own value of μ and may differ greatly in practical convergence of asymptotically optimal iterative methods.

It should also be mentioned that sometimes we agree to deal with rather unusual grids, though they are composed of sufficiently simple blocks (see Figures 2.1.4, 2.1.5, 3.5.7–3.5.15) to take advantage of this simplicity for parallel computing. We thus see that certain concrete problems definitely need further investigation in order to determine the most suitable method, and that only numerical experiments can confirm the best choice. The situation here is typical for all nontrivial mathematical recomendations: they apply in a straightforward manner for relatively simplified model problems, but for significantly complicated problems they need additional information, investigation, and practical experiments; in the latter cases, sometimes we are able to obtain sufficiently good or satisfactory answers, but sometimes we can only understand that the problem at hand is too hard for our theoretical analysis. Even when we need to carry out significant numerical experiments, iterative methods with convergence independent of the grid provide additional advantages because all experiments can be restricted to coarse grids. Applications of similar algorithms can be found in [288, 396, 398, 406, 410, 421].

$T^{(l)}(\hat{\Omega}_{h^{(l)}}) \equiv T^{(l)}(\Omega)$ on level l are constructed in accordance with Theorem 4.2.1 for separate standard quasitriangles, and that they employ partitions of each standard side (line segment) into, say, $M_0 2^l$ equal parts. Moreover, these triangulations satisfy the following condition: $\hat{\Pi}$ is a union of triangles in $T^{(l)}(\Omega)$ and each such triangle T_l is a union of four triangles $T_{l+1} \in T^{(l+1)}(\Omega)$, where $l \in [0, p-1]$.

The operators $I_l^{l+1} \in \mathcal{L}(H_l; H_{l+1})$ and $p_l \in \mathcal{L}(H_l; \hat{V}_l)$ are used to provide initial guesses in the basic iterations on the level $l+1$ through their outcome on the coarser level. In these constructions, we use the above expansions of $\hat{v}_l \in \hat{V}_l$. The singular functions in the bases are assumed to be the same, but other functions depend on l and $T^{(l)}(\hat{\Omega}_{h^{(l)}})$. In accordance with this, we define these operators in the standard manner on the basis of the equality $p_l v_l = \hat{v}_l \in \hat{V}_l$, $l \in [0, p]$, but now we represent $\hat{v}_l$ in the form $\hat{v}_l = \tilde{v}_l + \hat{v}_{l,0}$, where $\tilde{v}_l$ is a linear combination of the singular functions defined by the corresponding components of the vector v_l and $\hat{v}_{l,0}$ is a piecewise linear function with respect to the triangulation under consideration (we will refer to $\tilde{v}_l$ and $\hat{v}_{l,0}$ as the respective singular and standard parts of $\hat{v}_l$). To define $I_l^{l+1} \in \mathcal{L}(H_l; H_{l+1})$, we introduce the triangulation $T^{(l,2)}(\hat{\Omega}_{h^{(l)}})$, which is obtained by drawing midlines in each triangle in $T^{(l)}(\hat{\Omega}_{h^{(l)}})$; the obtained subtriangles are denoted by $T_l^{(2)}$. We emphasize that $T^{(l,2)}(\hat{\Omega}_{h^{(l)}})$ and $T^{(l+1)}(\hat{\Omega}_{h^{(l+1)}})$ are topologically equivalent triangulations, and we denote the correspondence between their triangles by $T_l^{(2)} \sim T_{l+1}$.

Now, given a vector $v_l \in H_l$, we define $I_l^{l+1} v_l \equiv w_{l+1} \in H_{l+1}$ so that the components of vectors v_l and w_{l+1} corresponding to the same singular basis functions coincide and the remaining components of w_{l+1} that correspond to the values of the standard part of $\hat{w}_l \in \hat{V}_{l+1}$ at vertices of triangles T_{l+1} are just values of the standard part $\hat{v}_{l,0}$ of $\hat{v}_l$ at corresponding vertices of triangles $T_l^{(2)} \sim T_{l+1}$. Observe that these values at common vertices of triangles $T_l^{(2)}$ and T_l coincide with values of $\hat{v}_{l,0}$ there and that those at vertices of $T_l^{(2)}$, which are midpoints of sides of triangles T_l, are just the arithmetical mean of the values of $\hat{v}_{l,0}$ at the corresponding two vertices.

If Γ consists of broken lines and all the standard quasitriangles in the partition of $\bar{\Omega}$ are triangles, then

$$\hat{w}_{l+1} \equiv p_{l+1} I_l^{l+1} v_l \equiv (I_l^{l+1} v_l)_h = \hat{v}_l = p_l v_l, \qquad (1.20)$$

and we may apply the analysis from § 1.4. The case of curvilinear boundary is more involved and needs special consideration since we only have

$$\hat{w}_{l+1}(x) = \hat{v}_l(x), \quad x \in \hat{\Pi}. \qquad (1.21)$$

This is due to the assumption about the supports of singular functions. Note that it is possible to preserve (1.21) even for curvilinear parts of the boundary close to the points of singularity: instead of Theorem 4.2.1, we use its simplified variant involving curvilinear triangles with curvilinear sides on Γ and replace the corresponding piecewise linear functions by the more complicated ones indicated in Lemma 2.2.2.

To carry out the above mentioned analysis, we formulate and prove three important lemmas. First recall that Theorem 4.2.1 implies the existence of positive constants κ independent of the grid such that

$$\left. \begin{array}{c} \kappa_0 \leq l_\Delta h_l^{-1} \leq \kappa_1, \ \kappa_2 \leq s_\Delta h_l^{-2} \leq \kappa_3, \ \sin \phi_\Delta \geq \kappa_4, \\ r(l, l+1) \leq \kappa_5 h_l^2 \equiv z_l, \end{array} \right\} \tag{1.22}$$

where l_Δ is the length of an arbitrary side of a triangle $T_l \in T(l)(\hat{\Omega}_{h^{(l)}})$, φ_Δ and s_Δ refer to an arbitrary angle and the area of T_l, respectively, $r(l, l+1)$ denotes the maximal distance between the corresponding vertices of the triangles $T_l^{(2)}$ and T_{l+1}, where $T_l^{(2)} \sim T_{l+1}$ (below we denote their respective vertices $T_l^{(2)}$ by A_1, A_2, and A_3 and, correspondingly, by A_1', A_2', and A_3'), and $h_l \equiv h^{(l)}, l \in [0, p]$. If we use the coordinate system $y \equiv [y_1, y_2]$ from this theorem, then the corresponding vertices A_k and A_k' are such that

$$A_k = [y_{k,1}, y_{k,2}], \ A_k' = [y_{k,1} y_{k,2} + c_k], \ k = 1, 2, 3,$$

where the vertical shift c_k is such that $|c_k| \leq z_k = \kappa_5 h_l^2$ (see (1.22)). Different combinations of signs of $c_k, k = 1, 2, 3$, yield 8 possible cases, three of which are illustrated 3 in Figure 1.

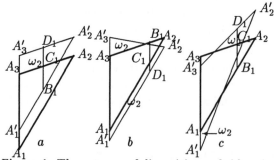

Figure 1. Three types of dispositions of sides $A_3 A_2$ and $A_3' A_2'$.

For $l \in [0, p-1]$, define the sets

$$\omega_3(T_{l+1}) \equiv T_{l+1} \setminus [\omega_1(T_{l+1} \cup \omega_2(T_{l+1})], \tag{1.23}$$

where $\omega_1(T_{l+1}) \equiv T_{l+1} \cap T_l^{(2)}$, $\omega_2(T_{l+1}) \equiv \omega_2^+ \cup \omega_2^-$, and

$$\omega_2^+ \equiv \omega_2^+(T_{l+1}) \equiv \{y : y \in T_{l+1} \setminus T_l^{(2)}, \ [y_1, y_2 - z_l] \in \omega_1(T_{l+1})\},$$

$$\omega_2^- \equiv \omega_2^-(T_{l+1}) \equiv \{[y_1, y_2)] : (y_1, y_2) \in T_{l+1} \setminus T_l^{(2)}, \ [y_1, y_2 + z_l] \in \omega_1(T_{l+1})\}$$

(in Figure 1, the sets denoted by ω_2 are parts of $\omega_2(T_{l+1})$ and their config-
uration is defined by the condition $|C_1 B_1| = z_l, |C_1 D_1| \leq z_l$).

Lemma 2. *If h_0 is small enough, then*

$$|\omega_3(T_{l+1})| \leq \kappa^{(0)} h_l^4, \quad \forall T_{l+1} \in T^{(l+1)}(\hat{\Omega}_{h^{(l+1)}}), \quad l \in [0, p-1]. \quad (1.24)$$

Proof. It suffices to consider only the case when $z_l < l_\Delta / 2$. There are
four different configurations between sides $A_3 A_2$ and $A_3' A_2'$, but only the
three depicted in Figure 1 need be considered. Define the set ω_3^+ of points
of T_{l+1} above the line $A_3 A_2$; note that $\omega_3^+ = \emptyset$ for Figure 1$_b$, and for Figures
1$_a$ and 1$_c$ we have $|\omega_3^+| \leq |\triangle B_1 D_1 A_2'| \leq z_l/2 |y_1 - y_1'|$, where $B_1 \equiv [y_1, y_2]$
and $A_2' \equiv [y_1', y_2']$ (we emphasize that $|C_1 D_1| \leq z_l$). This, together with the
condition involving κ_4, yields the desired inequality for $|\omega_3^+|$. In a similar
manner, we estimate $|\omega_3^-|$, where $\omega_3^- \equiv \omega_3(T_{l+1}) \setminus \omega_3^+$. $\square$

Let $\Omega^l \equiv \hat{\Omega}_{h^{(l)}}$, $\hat{D}_{l+1} \equiv \Omega^{l+1} \setminus \hat{\Pi}$, $|\hat{v}_l|_{1,l} \equiv |\hat{v}_l|_{1,\Omega^l}$,
$\hat{w}_{l+1} \equiv p_{l+1} I_l^{l+1} v_l \in \hat{V}_{l+1}$, and $\kappa \neq \kappa(h)$.

Lemma 3. *If h_0 is small enough and $l \in [0, p-1]$, then*

$$\sum_{T_{l+1} \in T^{(l+1)}(\Omega^{l+1})} |\hat{w}_{l+1} - \hat{v}_l|_{1, \omega_1(T_{l+1})}^2 \leq \kappa_6 h_l^2 |\hat{v}_l|_{1,l}^2, \quad \forall \hat{v}_l \in \hat{V}_l. \quad (1.25)$$

Proof. It suffices to consider only $T_{l+1} \subset \hat{D}_{l+1}$ (see (1.20)). Denote by
v the linear extension of $\hat{v}_l$ from the set $\omega_1(T_{l+1})$ to the triangle T_{l+1}. Then
at its vertices we have $|\hat{w}_{l+1} - v| \leq z_l |\frac{\partial \hat{v}_l}{\partial y_2}|$. Hence, for $y \in \omega_1(T_{l+1})$, we
obtain $|\nabla(\hat{w}_{l+1} - \hat{v}_l)|^2 \leq \kappa^{(1)} h_l^2 |\nabla \hat{v}_l|^2$, which leads to (1.25). $\square$

Below we make use of the notation $\omega_2^+ - z_l \equiv \{y : [y_1, y_2 + z_l] \in \omega_2^+\}$.

Lemma 4. *Assume that (1.3) and (1.6) hold with $m = 1$. Let h_0 be
small enough and assume that $|\hat{v}_l|_{1,l} \leq K^{(0)}$ for all $l \in [0, p-1]$ and that
$z \equiv \hat{w}_{l+1} - u = p_{l+1} I_l^{l+1} v_l - u$. Then*

$$|z|_{1,l+1} \leq \kappa_7 |\hat{v}_l - u|_{1,l} + \kappa_8 h_l^\gamma. \quad (1.26)$$

Proof. We have $\hat{D}_{l+1} = Q_1 \cup Q_2 \cup Q_2' \cup Q_3$, where the sets Q_1, Q_2, Q_2',
and Q_3 consist of the sets $\omega_1(T_{l+1})$, $\omega_2^+(T_{l+1})$, $\omega_2^-(T_{l+1})$, and $\omega_3(T_{l+1})$,
respectively, with $T_{l+1} \subset \hat{D}_{l+1}$ (see (1.23) and (1.21)). Then

$$|z|^2_{1,l+1} = |z|^2_{1,\hat{\Pi}} + |z|^2_{1,Q_1} + |z|^2_{1,Q_2} + |z|^2_{1,Q'_2} + |z|^2_{1,Q_3}. \qquad (1.27)$$

Now we may write

$$|z|^2_{1,Q_2} \leq 2(A_1 + A_2), \qquad (1.28)$$

where $A_1 \equiv \sum_{T_{l+1}} |\hat{w}_{l+1} - Y^{(l)}u|^2_{1,\omega_2^+}$, $A_2 \equiv \sum_{T_{l+1}} |Y^{(l)}u - u|^2_{1,\omega_2^+}$, $Y^{(l)}u \equiv Y_{h_l}u$ is the Steklov's averaging of u with $\rho \asymp h_l$ (see § 2.3), and $T_{l+1} \subset \hat{D}_{l+1}$. For these triangles, we have $u = u_0$ and $Y^{(l)}u = Y^{(l)}u_0$. Theorem 2.3.1 implies that

$$A_2 \leq |Y^{(l)}u_0 - u_0|^2_{1,\hat{D}_{l+1}} \leq \kappa_9 h_l^{2,\gamma}|u_0|^2_{1+\gamma,\Omega}. \qquad (1.29)$$

Moreover, for the same triangles, $X \equiv |\hat{w}_{l+1} - Y^{(l)}u_0|^2_{1,\omega_2^+}$ is estimated as

$$X \leq 2|\hat{w}_{l+1} - Y^{(l)}u_0|^2_{1,\omega_2^+ - z_l} + 2|Y^{(l)}[u(y_1, y_2) - u(y_1, y_2 - z_l)]|^2_{1,\omega_2^+}. \text{ Hence,}$$

$$X \leq 2|\hat{w}_{l+1} - Y^{(l)}u_0|^2_{1,\omega_1(T_{l+1})} + \kappa_{10}z_l \|Y^{(l)}u_0\|^2_{2,T'_{l+1}}, \qquad (1.30)$$

where T'_{l+1} is a union of several triangles from $T^{(l+1)}(\Omega)$ that have common points with T_{l+1}. According to (1.28)–(1.30) and (2.3.4), we have

$$|z|^2_{1,Q_2} \leq \kappa_9 h_l^{2\gamma}|u_0|^2_{1+\gamma,\Omega} + 2|z|^2_{1,Q_1} + \kappa_{11}h_l^2 \leq 2|z|^2_{1,Q_1} + \kappa_{12}h_l^2. \qquad (1.31)$$

A similar estimate holds for $|z|^2_{1,Q'_2}$. For estimating $|z|^2_{1,Q_3}$, (see (1.27)) we introduce the sets

$$\omega_4 \equiv \omega_4(T_{l+1}) \equiv \omega_3(T_{l+1}) \cap \Omega^l, \quad \omega_5 \equiv \omega_3(T_{l+1}) \setminus \omega_4.$$

Note that all $\omega_5 \equiv \omega_5(T_{l+1})$ belong to a boundary strip of width $O(h_l^2)$. On the basis of Lemma 2, we obtain the inequalities: $|z|^2_{1,\omega_4} \leq |\hat{v}_l - u|^2_{1,\omega_4} + \kappa_{1,3}h_l^2$, $|z|^2_{1,\omega_5} \leq 2|u|^2_{1,\omega_5} + \kappa_{1,4}h_l^2$, and $\sum_{T_{l+1}} |u|^2_{1,\omega_5} \leq \kappa_{15}h_l^{2\beta(\gamma)}|u|^2_{1+\gamma,\Omega}$, where $\beta(\gamma) = 1$ if $\gamma > 1/2$ and $\beta(\gamma) = \gamma$ if $\gamma \leq 1/2$ (see (2.3.5) with $\rho = h^2$). Thus,

$$|z|^2_{1,Q_3} \leq |v_l - u|^2_{1,Q_3} + \kappa_{16}h_l^{2\gamma}. \qquad (1.32)$$

From (1.27), (1.31), and (1.32) we obtain $|z|^2_{1,l+1} \leq \kappa_{17}|z|^2_{1,Q_1} + \kappa_{18}|\hat{v} - u|^2_{1,\Omega^l \setminus Q_1} + \kappa_{19}h_l^{2\gamma}$, and

$$|z|^2_{1,Q_1} \leq 2(|\hat{v}_l - u|^2_{1,Q_1} + |\hat{w}_{l+1} - \hat{v}_l|^2_{1,Q_1}) \leq 2|\hat{v} - u|^2_{1,Q_1} + 2\kappa_6 h_l^2(K^{(0)})^2$$

(see (1.25)). This yields (1.26). □ [10]

---

[10] If $\gamma \in (0, 1/2)$, then a simplification of the given proof is possible.

Theorem 4. *Suppose that the conditions of Theorem 3 and Lemma 4 are satisfied. Suppose also, for the solutions $\hat{u}_l$ of (1.7) associated with triangulations $T^{(l)}(\Omega)$ and the subspaces $\hat{V}_l$, that $\|\hat{u}_l - u\|_{1,\Omega^l} \leq K_0 h^\gamma$ and $\|\hat{u}_l - u\| \leq K_0' h^\gamma$ for all $l \in [0,p]$, where $h^p \equiv h \asymp \varepsilon^{1/\gamma}$. Suppose finally that, on each level $l \in [0,p]$, the basic iterative method is the same as in Theorem 3, and that we perform t iterations, with the initial iterate taken from level $l - 1$ (for $l \geq 1$) or defined by other means as in Theorem 1.4.4. Then it is possible to choose the number $t \asymp 1$ of iterations on each level in such a way that the final iteration on level p yields the desired $O(\varepsilon)$-approximation to the solution of the original problem, and the computational work in this algorithm is estimated as in (1.19).*

Proof. Since

$$\alpha_0 |\hat{v}_l|_{1,l} \leq \|v_l\|_{B_l} \leq \alpha_1 |\hat{v}_l|_{1,l}, \ \alpha_0 > 0, \tag{1.33}$$

then we may say that t iterations (1.3.1) on level l with initial iterate w_l and ultimate iterate v_l lead to the estimate

$$|\hat{v}_l - \hat{u}_l|_{1,\Omega^l} \leq q |\hat{w}_l - \hat{u}_l|_{1,\Omega^l}, \ q < 1. \tag{1.34}$$

For $l = 0$, we assume only (as in Theorem 1.4.4) that $|\hat{w}_0 - \hat{u}_0|_{1,\Omega^0} \leq \bar{K}_0$; very often it is possible to assume even that $\bar{K}_0 = 0$, and then no basic iterations on the level $l = 0$ are necessary. For $l \geq 1$, we define the initial iterate by $w_l \equiv I_{l-1}^l v_{l-1}$, where the vector v_l is taken from the lower level. Then, on the basis of (1.34) and (1.26), we see that

$$\xi_l \equiv |u - \hat{v}_l|_{1,l} \leq |u - \hat{u}_l|_{1,l} + q |\hat{w}_l - \hat{u}_l|_{1,l} \leq q \kappa_7 \xi_{l-1} + \kappa h_l^\gamma. \tag{1.35}$$

Thus, the choice of t from the condition $\rho \equiv q \kappa_7 < 1$ leads to the situation considered in Theorem 1.4.4, and we can obtain v_p such that $|\hat{v}_p - u|_{1,\Omega^p} = O(\varepsilon)$ for certain $t \asymp 1$. This implies (see Theorem 1.4.4) that (1.19) holds. Now we take $\hat{v}_p = 0$ on $\Omega_{0,h}$ (see (1.4)) and estimate $|u|_{1,\Omega_{0,h}}$ in accordance with (2.3.6). This implies that our $\hat{v}_p$ is an $O(\varepsilon)$-approximation to u in the sense of the original Hilbert space G. □

1.5. Proof of the strengthened variant of the Kolmogorov-Bakhvalov hypothesis. Theorem 4 implies that, for a class of elliptic boundary value problems associated with two-dimensional domains, not only the strengthened variant of the Kolmogorov-Bakhvalov hypothesis with $n_2 = 0$ in (1.3) holds, but so does its generalization with $n_2 \geq 1$ in (1.3).

We emphasize that estimate (1.19) was obtained for all $\gamma \in (0,1]$ and that similar estimates for $n_2 = 0, \gamma = 1$ are typical of the best theoretical results for multigrid methods (see [13, 270, 367, 368]). Note also that the geometry of our domain with Lipschitz boundary is very general. Now we show that the condition for u_0 in (1.3) can be replaced by

$$u_0 \in G, \quad \|u_0\|_{m+\gamma,\Omega_r} \le K_r^*, \quad r \in [1,p'], \tag{1.36}$$

which is of type (0.5.4), where $\Omega_r, r \in [1,p']$, are standard quasitriangles in our partition of $\bar{\Omega}$. This is of special importance for problems with discontinuous coefficients along interfaces of the blocks as well as for problems associated with domains with non-Lipschitz boundaries.

Theorem 5. Suppose we apply the algorithm described in Theorem 4 under condition (1.36) with $m = 1$ on u_0 in (1.3). Then this algorithm yields the desired $O(\varepsilon)$-approximation to the solution of the original elliptic boundary value problem with asymptotically optimal work estimate (1.19).

Proof. To generalize Theorem 1, we replace (1.8) by

$$\|z\|_h \le K_0 h^\gamma \|u_0\|_{1+\gamma,\Omega}^*, \quad \|z\| \le K_0' h^\gamma \|u_0\|_{1+\gamma,\Omega}^*, \tag{1.37}$$

where

$$\|u_0\|_{1+\gamma,\Omega}^* \equiv [\sum_{r=1}^{p} \|u_0\|_{1+\gamma,\Omega_r}^2]^{1/2}. \tag{1.38}$$

Then, in the proof of Theorem 1, we replace $\|u_0\|_{1+\gamma,\Omega}$ by $\|u_0\|_{1+\gamma,\Omega}^*$ and, instead of (1.6), we use similar extensions from our blocks Ω_r to $\Omega_r', r \in [1,p']$ (it is possible to assume that their interfaces belong to our domain Ω). It is important to note that the extended function $u_0(x)$ will be continuous at all nodes and thus $\mathrm{int}_1 u_0$ may be used as in the proof of Theorem 1. Indeed, $2(1+\gamma) > 2$, so we can treat u_0 as a function that is continuous on the closure of each $\Omega_r', r \in [1,p']$. On their interfaces (line segments), the traces must coincide because $u_0 \in G$ and has a unique trace on each interface C_q (e.g., in the sense of $L_2(C_q)$). It is quite clear that, in deriving (1.14), we may work with separate parts of Γ_1 belonging to one of the blocks. Next, Theorems 2 and 3 need only the above changes in the conditions. Finally, in the proof of modified Lemma 4, each term on the right-hand side of (1.27) can be represented as the sum of terms related to each separate block for which estimates given in the proof of Lemma 4 apply and yield (1.26). Then the proof of Theorem 4 applies. $\square$

Now we consider briefly changes in our proofs needed for for elliptic boundary value problems associated with domains with non-Lipschitz boundaries (e.g., Figure 2.1.6). Assume that we use blocks of the same type as above and that we allow slits corresponding to several line segments that are interfaces of our standard blocks; more precisely, each such slit is assumed to be a double slit corresponding to two different limit segments, each of which belongs either to Γ_0 or Γ_1. Then, in the construction of PGMs, it is important to remember that Ω_h is now a set of elementary nodes (see § 2.1)). Model asymptotically optimal operators for such domains and problems were considered in § 3.5–3.7 (when no singular basis functions are used), and the case when (1.36) holds in combination with the use of singular functions is much the same as in Theorem 5.

We therefore see that the strengthened variant of the Kolmogorov-Bakhvalov hypothesis has actually been proved even for a class of regions with the non-Lipschitz boundaries. It is clear (see § 2.2, 2.3, 4.2) that slits corresponding to smooth arcs can in fact be treated in much the same way if we agree to use generalized triangulations containing triangles with one curvilinear side (belonging to one of the slits). Such generalized triangulations can be also useful if we deal with the general case where $\hat{\Pi}$ may have curvilinear boundary.

1.6. Solutions with more smoothness. Here we consider the case of conditions (1.3) or (1.36) with $m = [m] \geq 2$. For simplicity, assume that $n_2 = 0$ and that all standard quasitriangles in the partition of $\bar{\Omega}$ are triangles (otherwise our analysis can be easily modified). We start by considering triangulations $T_h(\bar{\Omega})$ and $T_h^{(m)}(\bar{\Omega})$ (see Subsection 2.1.3), for which we define the respective spline spaces $\mathcal{J}^{(m)}$ and $\mathcal{J}^{(1)}$ ($\hat{u}^{(m)} \in \mathcal{J}^{(m)}$ inside each triangle $T \in T_h(\bar{\Omega})$ and $\hat{u} \equiv \hat{u}^{(1)} \in \mathcal{J}^{(1)}$ inside each triangle $T_m \in T_h^{(m)}(\bar{\Omega})$ are Lagrange polynomials of degree m and 1, respectively; we write here inside (instead of on) because slits in the domain are allowed). The standard bases for these spaces associated with the set of elementary nodes belonging to $\bar{\Omega}$ define the grid operators $\Lambda_\Omega^{(m)}$ and $\Lambda_\Omega^{(1)}$ associated with the same bilinear form (see (0.4.10)) with the piecewise constant coefficient $a(x)$. In other words, we consider two different grid operators defined on the same grid, one of which corresponds to a higher order approximation of the problem at hand and has a more complicated structure.

A natural question arises: are they spectrally equivalent operators? For difference operators, this was shown in [153] (see also § 6.1) to be the case. For projective-grid operators, similar results were obtained in [181, 314], and recently new interesting investigations (see [33, 28]) have appeared dealing with construction of model operators for large p. We concentrate here (see [181]) on the spectral equivalence of $\Lambda_\Omega^{(m)}$ and $\Lambda_Q^{(1)}$, where the last grid operator is constructed for a triangulation $T_h^{(m)}(\bar{Q})$ in accordance with (0.4.11), ($T_h^{(m)}(\bar{\Omega})$ and $T_h^{(m)}(\bar{Q})$ are topologically equivalent triangulations; the same holds for $T_h^{(1)}(\bar{\Omega})$ and $T_h^{(1)}(\bar{Q})$).

Theorem 6. *Let $T_h^{(1)}(\bar{Q})$ contain only rectangular isosceles triangles with the area $h^2/2$. Then $\Lambda_\Omega^{(m)}$ and $\Lambda_Q^{(1)}$ are spectrally equivalent operators.*

Proof. It suffices to prove that

$$\delta_0 |\hat{u}(z)|^2_{1,Q} \leq |\hat{u}^{(m)}(x)|^2_{1,\Omega} \leq \delta_1 |\hat{u}(z)|^2_{1,Q}, \quad \forall \mathbf{u} \in H, \delta_0 > 0, \qquad (1.39)$$

where $\hat{u}^{(m)}(x) \in J^{(m)} \subset G$ and $\hat{u}(z)$ is the piecewise linear extension on $T_h^{(m)}(Q)$ of the vector $\mathbf{u}$.

Let $T \equiv \triangle P_0 P_1 P_2 \in T_h^{(1)}(\Omega)$ and $T' \equiv \triangle P_0' P_1' P_2' \in T_h^{(1)}(Q)$ correspond to each other in the mapping (1.1), where $\angle P_0' P_1' P_2' = \pi/2$ and $\angle P_0 P_1 P_2 = \alpha$, $|P_0 P_1| = m l_1$, $|P_0 P_2| = m l_2$, $|P_0' P_1'| = |P_0' P_2'| = mh$. Now taking into account the fact that

$$|\sin \alpha| \geq \kappa_0 > 0, \ l \equiv \frac{l_1}{l_2} \in [\kappa_2^{-1}, \kappa_1], \ \kappa_2 > 0, \qquad (1.40)$$

we will prove inequalities

$$\delta_0 Z \leq X \leq \delta_1 Z, \quad \delta_0 > 0 \qquad (1.41)$$

(for arbitrary values of $\mathbf{u}$ at the nodes in T'), where $Z \equiv |\hat{u}(z)|^2_{1,T'}$, $X \equiv |\hat{u}^{(m)}(x)|^2_{1,T}$. We may take $T' \equiv \{[z_1, z_2] : 0 \leq z_k \leq mh, k = 1, 2, \ z_1 + z_2 \leq mh\}$ and denote the value of $\mathbf{u}$ at $[i_1 h, i_2 h]$ by u_{i_1, i_2}. Let also $u_{i_1+1, i_2} - u_{i_1, i_2} \equiv \xi_{i_1, i_2} \equiv d_1 u_{i_1, i_2}$, $u_{i_1, i_2+1} - u_{i_1, i_2} \equiv \eta_{i_1, i_2} \equiv d_2 u_{i_1, i_2}$ ($d_r = h \partial_r, r = 1, 2$). Then

$$Z = \frac{1}{2} \sum_{i_1=0}^{m-1} \xi_{i_1, 0}^2 + \frac{1}{2} \sum_{i_2=0}^{m-1} \eta_{0, i_2}^2 + \sum_{i_2=1}^{m-1} \sum_{i_1=0}^{m-i_2-1} \xi_{i_1, i_2}^2 + \sum_{i_1=1}^{m-1} \sum_{i_2=0}^{m-i_1-1} \eta_{i_1, i_2}^2.$$

Now using affine coordinates $t' \equiv [t_1', t_2'] \equiv [t_1/l_1]t_2/l_2] \equiv t$ in T such that $T = \{t' : 0 \leq t_1' \leq l_1, \ 0 \leq t_2' \leq l_2, \ t_1'/l_1 + t_2'/l_2 \leq 1\}$ and the polynomials $P_{i_r}(t_r) \equiv t_r(t_r - 1) \ldots (t_r - i_r + 1)$, we verify that

$$X = \sin \alpha \int_0^m dt_1 \int_0^{m-t_1} \left(\frac{f_1^2(t)}{l} + \frac{l}{\sin^2 \alpha} \left(f_2(t) - \frac{f_1(t) \cos \alpha}{l} \right)^2 \right) dt_2,$$

where $f_1(t) \equiv \sum_{k=1}^m \sum_{i_1+i_2=k, i_1 \geq 1} \frac{1}{i_1! i_2!} P_{i_2}(t_2) \frac{\partial}{\partial t_1} P_1(t_1) d_1^{i_1} d_2^{i_2} u_{0,0}$ and $f_2(t)$ is obtained by replacing i_1 and t_1 by i_2 and t_2, respectively (see (2.1.16) and the proof of Theorem 0.4.5). We emphasize that $X \geq 0$ is independent of h and, if $X = 0$, then $|\hat{u}^{(m)}(x)|^2_{1,T} = 0$, and all our $d_1^{i_1} d_2^{i_2} u_{0,0} = 0$ if $i_1 + i_2 > 0$. This implies that if we take $u_{0,0} = 0$ and represent X and Z as quadratic forms of remaining values u_{i_1, i_2} (they number $(m+2)(m+1)/2 - 1$), then we have two positive definite quadratic forms associated with matrices A_Ω and

A_Q. Finding the minimal and maximal eigenvalues of problem $A_\Omega = \lambda A_Q$ (they depend only on m, l, α), we obtain (1.41) with $\delta_0 \equiv \delta_0(T) > 0$ and $\delta_1 \equiv \delta_1(T)$. We have only several types of T and thus (1.39) holds. $\square$ [11]

Theorem 7. Suppose that the elliptic boundary value problem in Theorem 1 is associated with a domain such that the standard quasitriangles in our partition of $\bar{\Omega}$ are triangles. Suppose that its solution $u = u_0$ satisfies condition (1.36) with $m = [m] \geq 2$ and $\gamma \in (0, 1]$. Suppose also that we use the PGM associated with triangulations $T^{(m)}(\bar{\Omega})$ and subspaces $\mathcal{J}^{(m)}$ as in Theorem 6. Then it is possible to construct a computational algorithm that yields an $O(\varepsilon)$-approximation with asymptotically optimal work estimate

$$W(\varepsilon) = O(\varepsilon^{-2/\nu}), \qquad (1.42)$$

where $\nu \equiv m - 1 + \gamma$.

---

[11] For the general case, we have $\delta_0 = \min_{l,\alpha} \delta_0(T) > 0$, $\delta_1 = \max_{l,\alpha} \delta_1(T)$, where l and α satisfy conditions (1.48) for our triangulations. Simpler estimates can be obtained if we use the space of variables ξ_{i_1,i_2} and η_{i_1,i_2} (they number to $m(m+1)$) and represent each $d_1^{i_1} d_2^{i_2} u_{0,0}$ in the expression for $f_1(t)$ by a linear combination of the variables ξ_{i_1,i_2} (it is unique) and the same $d_1^{i_1} d_2^{i_2} u_{0,0}$ in the expression for $f_2(y)$ by a linear combination of η_{i_1,i_2}. We illustrate this approach for $m = 2$ using the notation

$$a_1 \equiv \xi_{0,0}, \; a_2 \equiv \xi_{1,0}, \; a_3 \equiv \xi_{0,1}, \quad b_1 \equiv \eta_{0,0}, \; b_2 \equiv \eta_{0,1}, \; b_3 \equiv \eta_{1,0}.$$

Then $Z = 1/2(a_1^2 + a_2^2) + a_3^2 + 1/2(b_1^2 + b_2^2) + b_3^2$,

$$\hat{u}^{(2)}(t') = u_{0,0} + \frac{t_1'}{l_1} d_1 u_{0,0} + \frac{t_1'(t_1'-l_1)}{2l_1^2} d_1^2 u_{0,0} + \frac{t_2'}{l_2} d_2 u_{0,0} + \frac{t_2'(t_2'-l_2)}{2l_2^2} d_2^2 u_{0,0} + \frac{t_1' t_2'}{l_1 l_2} d_1 d_2 u_{0,0},$$

$\frac{\partial \hat{u}^{(2)}(t)}{\partial t_1'} = 1/l_1[d_1 u_{0,0} + (t_1 - 1/2)d_1^2 u_{0,0} + t_2 d_1 d_2 u_{0,0}]$, $f_1(t) = a_1 + (t_1 - 1/2)(a_2 - a_1) + t_2(a_3 - a_1)$, and $f_2(t) = b_1 + (t_2 - 1/2)(b_2 - b_1) + t_1(b_3 - b_1)$. Thus, $f_1(t) = 0$ implies that $a_1 = a_2 = a_3 = 0$ and $f_2(t) = 0$ implies that $b_1 = b_2 = b_3 = 0$. Hence, we may treat Z and X as positive quadratic forms associated with matrices A_Z and A_X. Observe that A_Z is a diagonal matrix with elements $1/2$ and 1, and if we use new variables $a_1' \equiv a_1 2^{-1/2}$, $a_2' \equiv a_2 2^{-1/2}$, $a_3' \equiv a_3$, $b_1' \equiv b_1 2^{-1/2}$, $b_2' \equiv b_2 2^{-1/2}$, $b_3' \equiv b_3$, then the corresponding matrices are $A_Z' = I$ and A_X'. It can be verified in a straightforward manner that $A_X' \equiv A = A^T$ has the block form in (1.5.1), where $A_{1,1} \equiv B, A_{2,2} \equiv C, A_{1,2} \equiv D$, and

$$B = \frac{1}{3l \sin \alpha} \begin{bmatrix} 3 & -1 & 0 \\ -1 & 3 & 0 \\ 0 & 0 & 4 \end{bmatrix}, C = l^2 B, \quad D = -\frac{\cos \alpha}{6 \sin \alpha} \begin{bmatrix} 6 & -2 & 0 \\ -2 & -2 & 2^{1/2}4 \\ 0 & 2^{1/2}4 & 4 \end{bmatrix}.$$

In particular, $\delta_0 = 2/3$ and $\delta_1 = 4/3$ for $l = 1, \alpha = \pi/2$. This implies that practical convergence of iterative methods with $B = \Lambda_Q^{(1)}$ or $B \times \Lambda_Q^{(1)}$ may be very fast, at least for relatively small m and L close to $\Lambda_\Omega^{(m)}$. The above operators are nonnegative, but if instead of the spline spaces $\mathcal{J}^{(m)}$ we consider their subspaces associated with approximations of the Hilbert space $G = W_2^1(\Omega; \Gamma_0)$ and the smaller sets Ω_h of the nodes, then our operators become positive and Theorem 6 applies (the constants may only change for the better).

Proof. As in Theorem 5, we replace (1.8) by

$$\|z\| \le K_0^* h^\nu \|u\|_{m+\gamma,\Omega}^*, \tag{1.43}$$

where $\|u\|_{m+\gamma,\Omega}^* \equiv [\sum_{r=1}^p \|u_0\|_{m+\gamma,\Omega_r}^2]^{1/2}$. It is clear that (1.43) follows from Theorem 2.3.2 (see the proof of Theorem 5) and yields the estimates

$$\|\hat{u}_l - u\|_{1,\Omega} \le K_0^* h^\nu, \quad l \in [0, p],$$

which are important for application of Theorem 1.4.4 (in it, γ was used instead of our ν; see (1.4.21). Now $h^p \equiv h \asymp \varepsilon^{1/\nu}$. The basic iterative method can be defined as in the above theorems, but with model operator $B \asymp \Lambda_Q^{(1)}$. Theorem 1.4.4 then applies and yields the desired optimal work estimates. $\square$

Theorem 7 implies that the strengthened variant of the Kolmogorov-Bakhvalov hypothesis holds for problems whose solutions satisfy condition (1.36) with $m = [m] \ge 2$ and $\gamma \in (0, 1]$. [12]

§ 2. General linear elliptic equations and multidimensional regions

2.1. Iterative methods for symmetrized grid systems. For the grid systems considered in § 1 and especially for the more general case involving an invertible operator L such that inequalities (0.3.19) hold, it is reasonable to apply modified classical iterative methods for the symmetrized system (0.3.17), with $A = B^{-1}L^*B^{-1}L$. [13] We emphasize that $A \in \mathcal{L}^+(H(B))$, and its spectrum is localized in accordance with Lemma 1.3.2 ((0.3.18) holds with constants δ_0, δ_1 independent of h). We show below that conditions (0.3.19) may be regarded as a consequence of the correctness of the original elliptic boundary value problem (this was established in [164] for difference methods and in [166] for PGMs and FEMs).

2.2. General case of invertible elliptic operator. We start by considering the same planar regions and the Hilbert space $G = W_2^1(\Omega; \Gamma_0)$ as we did in § 1, but now with the bilinear form

$$b(u; v) \equiv b(\Omega; u; v) \equiv b_1(u; v) + b_0(u; v), \tag{2.1}$$

---

[12] The same optimal estimates hold for PGM associated with triangulations $T^{(m_1)}(\bar{\Omega})$ and subspaces $\mathcal{J}^{(m_1)}$ provided $m_1 \ge m$ (the basic estimate of accuracy follows from Theorem 2.3.2). This implies that such PGMs have an adaptation property with respect to possible better smoothness of the solutions, but usually it is reasonable to use $m_1 = m$ and avoid grid operators with complicated structure.

[13] Relevant results and numerical experiments can be found in [60, 218].

where

$$b_1(u;v) \equiv \sum_{r,l=1}^{d} (a_{r,l}, D_l u D_r v)_{0,\Omega} + \frac{1}{2} \sum_{r=1}^{d} ((b_r - b_r'), v D_r u + u D_r v)_{0,\Omega}$$

$$+ (c + c_0, uv)_{0,\Omega} + (\sigma, uv)_{0,\Gamma_1}, \tag{2.2}$$

$$b_0(u;v) \equiv \frac{1}{2} \sum_{r=1}^{d} ((b_r + b_r'), v D_r u - u D_r v)_{0,\Omega} - (c_0, uv)_{0,\Omega}, \tag{2.3}$$

$d = 2$, and the constant c_0 in $b_1(u;v)$ and $b_0(u;v)$ is introduced to ensure that $b_1(u;v)$ exhibits a certain positivity property. More precisely,
we assume below that all given functions are such that there exist operators L, L_1, and L_0 in $\mathcal{L}(G)$ defined by the relations $b(u;v) = (Lu,v)_G$,
$b_1(u;v) = (L_1u,v)_G$, $b_0(u;v) = (L_0u,v)_G$, for all u and v in G (e.g., this is
the case if the above functions are bounded); we require further that the
conditions of Theorem 0.1.4 are satisfied for L_1, that is,

$$L_{1,s} \equiv \frac{L_1 + L_1^*}{2} \in \mathcal{L}^+(G) \tag{2.4}$$

(e.g., it suffices to take conditions (0.1.32) and sufficiently large c_0). L_1 is
symmetric when $a_{1,2} = a_{2,1}$. It should be emphasized that L_0 is a compact
operator (see, e.g., [326]). We establish a stronger property here that will
also be used in Chapter 9.

Lemma 1. *Suppose that a Hilbert space G_1 is compactly embedded in a
Hilbert space G_0 and that a bilinear form $b_0(u;v)$ defined on $G_1 \times G_1$ is such
that $|b_0(u;v)| \leq \delta[|u|_1|v|_0 + |u|_0|v|_1]$, $\forall u \in G_1$, $\forall v \in G_1$, where $|u|_1$ and
$|v|_0$ denote the norms in G_1 and G_0. Then there exists a unique compact
operator $L_0 \in \mathcal{L}(G)$ such that $b_0(u;v) = (L_0u, v)_{G_1}$, $\forall u \in G_1$, $\forall v \in G_1$.*

Proof. Our bilinear form is bounded, and existence of a unique associated operator $L_0 \in \mathcal{L}(G)$ associated with our bilinear form is a consequence
of the well-known Riesz Theorem (see Theorem 0.1.4). It therefore suffices
to prove that L_0 is compact. Suppose we have a bounded sequence $\{u_n\}$
(in G_1). Then $\{L_0 u_n\}$ is also bounded, and compactness of the embedding
operator implies existence of a subsequence $\{u_{n_k} \equiv v_k\}$ such that $\{v_k\}$ and
$\{L_0 v_k\}$ are convergent in the Hilbert space G_0. Hence, they are fundamental sequences and we prove now that $\{L_0 v_k\}$ is a fundamental sequence in
the Hilbert space G_1 as well (which is equivalent to the desired convergence
in G_1). Indeed, for $z \equiv v_{k+p} - v_k$, we have

$$|L_0 z|_1^2 = b_0(z; L_0 z)| \leq \delta (|z|_1 |L_0 z|_0 + |z|_0 |L_0 z|_1)$$

and $|L_0 z|_1^2 = b_0(z; L_0 z)| \leq \delta'[|L_0 z|_0 + |z|_0]$ with some constant δ'. This implies that $\{L_0 v_k\}$ is a fundamental sequence in the Hilbert space G_1. $\square$

Consider PGMs of the same type as in § 1. Along with the subspaces $\hat{G} \subset W_2^1(\hat{\Omega}; \hat{\Gamma}_0)$, we make use of the subspaces $\tilde{G} \subset G$ of functions $\tilde{u}$ that vanish on $\Omega_{0,h}$ (see (1.4)) and coincide with the corresponding elements $\hat{u} \in \hat{G}$ on $\hat{\Omega} \cap \Omega$. We emphasize that the functions $\hat{u} \in \hat{G}$ and $\tilde{u} \in \tilde{G}$ correspond to the same vector $\mathbf{u} \in H$. In the same way, along with the operators $L_\Omega \equiv L_h$ and $\Lambda_\Omega \equiv \Lambda$ defined in § 1, we introduce the operators $\tilde{L}_\Omega \equiv \tilde{L}$ and $\tilde{\Lambda}_\Omega \equiv \tilde{\Lambda} \equiv J$ in accordance with the following relations: $(\tilde{L}\mathbf{u}, \mathbf{v}) = b(\tilde{u}, \tilde{v})$, $(J\mathbf{u}, \mathbf{v}) = (\tilde{u}, \tilde{v})_{1,\Omega}$, $\forall \mathbf{u} \in H, \forall \mathbf{v} \in H$. In the sequel, $K > 0$ and $K \neq K(h)$.

Lemma 2. *Suppose that the operator* $L \in \mathcal{L}(G)$ *associated with the original elliptic boundary value problem is invertible and satisfies (2.4). Then there exists* $h_0 > 0$ *such that, for all* $h < h_0$, *the projective operator* $\tilde{L}$ *is invertible and*

$$\|(\tilde{L})^{-1}\|_{H(J^{-1}) \mapsto H(J)} \leq \tilde{K}_0, \quad \|\tilde{L}\|_{H(J) \mapsto H(J^{-1})} \leq \tilde{K}_1. \qquad (2.5)$$

Proof. The first inequality follows from Theorem 0.4.3, which applies here because $\tilde{L} = \tilde{L}_1 + \tilde{L}_0$ and $\tilde{L}_0$ corresponds to projective approximation of the compact operator L_0 (see Lemma 2). The second inequality is a consequence of the boundedness of L (see (0.4.8)). $\square$ [14]

Lemma 3. *If* $h \leq h_0'$, *where* h_0' *is small enough, then* $\Lambda \asymp J$.

Proof. We have $J = J^*$ and $(J\mathbf{u}, \mathbf{u}) = |\hat{u}|_{1,\Omega \cup \hat{\Omega}}^2$, $(\Lambda\mathbf{u}, \mathbf{u}) = |\hat{u}|_{1,\hat{\Omega}}^2$. These integrals are sums of integrals over the triangles $T \in T_h(\hat{\Omega})$ and their parts $\tilde{T} \equiv T \cap \Omega$. It is easy to see that the conditions of Lemma 3 guarantee existence of a positive constant $\kappa^{(1)} < 1$ such that $|\tilde{T}| \geq \kappa^{(1)}|T|$. Hence, $\kappa^{(1)}\Lambda \leq J \leq \Lambda$. $\square$

Lemma 4. *Let the conditions of Lemma 3 be satisfied. Then*

$$\|R\|_{H(\Lambda) \mapsto H(\Lambda^{-1})} \leq K_2 h, \qquad (2.6)$$

where

$$R \equiv L_\Omega - \tilde{L}. \qquad (2.7)$$

Proof. We have $|(R\mathbf{u}, \mathbf{v})| = |b(\hat{\Omega}; \hat{u}, \hat{v}) - b(\hat{u}, \hat{v})| \equiv X$ or, using the notation of § 1,

$$X \leq |(R\mathbf{u}, \mathbf{v})| = |b^{(0)}(\Omega_{1,h}; \hat{u}, \hat{v})| + |(\sigma(x - \xi(x), \hat{u}\hat{v})))_{0,\hat{\Gamma}_1} - (\sigma(x, \hat{u}\hat{v})))_{0,\Gamma_1}|.$$

---

[14] Lemma 2 implies that $\tilde{L}$ and J are connected by relationship C^3 (see § 1.4).

Hence, $|(R\mathbf{u}, \mathbf{v})| \leq K_3 h |\hat{u}|_{1,\hat{\Omega}} |\hat{v}|_{1,\hat{\Omega}} = K_3 h \|\mathbf{u}\|_\Lambda \|\mathbf{v}\|_\Lambda$ (see (1.13) and (1.14)), which with the choice $\mathbf{v} = \Lambda^{-1} R\mathbf{u}$ leads to the desired inequality. $\square$

Theorem 1. *Suppose that the operator $L \in \mathcal{L}(G)$ associated with the original elliptic boundary value problem is invertible and satisfies (2.4). Suppose also that $h_0' \tilde{K}_0 K_2 < 1$ (see Lemma 2) and that $\Lambda \asymp B \equiv B_h$. Then, for $h < \min\{h_0, h_0'\}$, the operators L_Ω and B are connected by relationship C^3.*

Proof. We apply Lemma 2 and conclude that the operators $\tilde{L}$ and J are connected by relationship C^3. Next, we regard $L_\Omega = \tilde{L} - R = \tilde{L}(I - (\tilde{L})^{-1} R)$ as a small perturbation of $\tilde{L}$ provided $K_0 K_2 h < 1$ (see (2.5)–(2.7)), so the operators L_Ω and J are also connected by relationship C^3. Finally, the same relationship connects the operators L_Ω and Λ (see Lemmas 2 and 3.4.4) and L_Ω and B. $\square$

Theorem 2. *Suppose that the conditions of Theorem 1 are satisfied. Suppose also that, for the solution of the original problem, assumptions (1.3), (1.6) with $m = 1$ hold. Then, for the PGM under consideration, estimates of accuracy (1.8) hold.*

Proof. If our domain and the boundary conditions are such that (1.5) holds, then we may apply Theorem 1.1.3 or even 1.1.5 (for the latter case, we may take $A = L_1$) to conclude that (1.8) follows from (1.1.18). For the general case, it is necessary to combine the proofs of Theorems 1.1.3 and 1.1. Suppose that the vectors $\mathbf{u} \in H, \mathbf{w} \in H$ correspond to the solution of our projective problem (1.7) and best approximation in $\hat{G}$ to the solution $u \in G$ of the original problem, respectively. Then $|u - \hat{w}|_{1,\hat{\Omega}} \leq K_4 k^{(0)} h^\gamma$ and $|\hat{w}|_{1,\Omega_{1,h}} \leq K_5 K^{(0)}$ (see (1.4), (1,9)). Thus, in the same way as in (1.1.20), we obtain $L_\Omega(\mathbf{u} - \mathbf{w}) = \zeta_1 + \zeta_2$, where

$$(\zeta_1, \mathbf{v}) = b(\Omega; \hat{w}; \hat{v}) - b(\hat{\Omega}; \hat{w}; \hat{v}), \quad (\zeta_2, \mathbf{v}) = -b(\Omega; \hat{w} - u; \hat{v}).$$

This yields the estimates $|b(\Omega_{1,h}; \hat{w}; \hat{v})| \leq K_6 |\hat{w}|_{1,\Omega_{1,h}} |\hat{v}|_{1,\hat{\Omega}}$, $\|\zeta_1\|_{\Lambda^{-1}} \leq K_7 h |\hat{u}|_{1,\Omega}$, $\|\zeta_2\|_{\Lambda^{-1}} \leq K_8 |\hat{w} - \hat{u}|_{1,\hat{\Omega}} \leq K_9 K^{(0)} h^\gamma$, and $|\hat{w} - \hat{u}|_{1,\hat{\Omega}} = \|\mathbf{u} - \mathbf{w}\|_\Lambda \leq K_{11} \|\zeta\|_{\Lambda^{-1}} \leq K_{11} K^{(0)} h^\gamma$, which leads to (1.8). $\square$

Theorem 3. *Suppose that the conditions of Theorem 2 are satisfied. Then, for PGM (1.7), an iterative method of type (1.3.25) yields the desired ε-approximation to the solution of the original elliptic boundary value problem with computational work $W(\varepsilon) = O(\varepsilon^{-d/\gamma}) |\ln \varepsilon|$; multigrid acceleration of this basic iterative algorithm improves the work estimate to (1.19).*

Proof. Theorem 1 and Lemma 1.3.2 guarantee effectiveness of modified iterative methods (1.3.25) with optimal preconditioner B. So proof of the first part of Theorem 3 is similar to that of Theorem 1.3. Proof of the second part is similar to that of Theorem 1.4 as we now show. We again use the same sequence of triangulations $T_{h^{(l)}}^{(l)}(\Omega^l)$ and subspaces $\hat{V}_l$. The important difference is that we must now consider level $l = 0$ with the sufficiently refined triangulation in order to be able to apply Theorem 1 to all grids under consideration and to be sure that the required estimates for $\|\hat{u}_l - u\|_{1,\Omega^L}$ and $\|\hat{u}_l - u\|$ hold for all $l \in [0, p]$. The choice of the number $t \asymp 1$ of iterations (1.3.25) on each level and estimation of computational work are essentially the same as in the proof of Theorem 1.4. $\square$ [15]

2.3. Generalizations for multidimensional problems. Our above construction of asymptotically optimal algorithms involves three basic subproblems: construction of asymptotically optimal PGMs, construction of asymptotically optimal preconditioners leading to asymptotically optimal iterative methods, and application of coarse grid continuation. For multidimensional cases, these subproblems retain their fundamental significance, but yield additional theoretical and practical obstacles.

We have already underscored the significance of some geometric problems associated with the generation of simplex grids whose corresponding PGMs have asymptotically optimal error estimates of type (1.8) and yield grid systems for which effective iterative algorithms from Chapter 3 can be applied with estimates of type (0.3.8) and (0.5.3) (see § 4.4 and Lemma 2.2.8). If we confine ourself to multidimensional regions whose desired partitions on standard blocks are known, then our results relevant to construction of triangulations of desired type and model operators apply. Hence, the second basic subproblem has already been considered for certain types of regions, and the original partition of $\bar{\Omega}$ into standard blocks should be chosen in agreement with our construction of model operators B in Chapter 3. Asymptotically optimal model operators B for $d \geq 3$ have been suggested only recently (see § 3.7 and 3.6), although nearly asymptotically optimal preconditioners for simple regions have been known for many years (see § 3.4 and 3.5). Moreover, all necessary inequalities connecting grid operators (e.g., those indicated in Theorem 1) are obtained in much the same way as for $d = 2$. The case of prismatic grids should be mentioned. We now focus on the first basic subproblem dealing with estimates (1.8), (1.37), and (1.43) for the PGMs under consideration; we concentrate on the case $\bar{\Omega} = \hat{\Omega}_h$ (the effect of approximating the region can be analyzed as for $d = 2$ on the basis of Theorem 2.3.3); fundamental inequalities (2.5) apply to our new operators. If we consider condition (1.3) or its generalizations involving the use of the norms $\|u_0\|^*_{m+\gamma,\Omega}$ (see (1.38) and (1.43)), then our two-dimensional analysis can be applied in a straightforward manner provided $2(m + \gamma) > d$ and local interpolants as in Theorems 1, 5, and 6 can be constructed. Such an analysis yields estimates of type

---

[15] Theorem 3 implies that, for many correct problems associated with two-dimensional domains, the strengthened variant of the Kolmogorov-Bakhvalov hypothesis holds. Generalization of Theorem 1.6 to the case $m > 1$ is straightforward.

$$\|z\|_h \leq K_0 \left[\sum_{T \in T_h(\bar{\Omega})}^{p} h(T)^{2\nu} \|u_0\|_{m+\gamma,T}^2 \right]^{1/2}, \qquad (2.8)$$

where $\nu \equiv m - 1 + \gamma$, $h(T) \equiv r(T)$ is defined in § 2.2 (see (2.2.30); $r(T)$ serves as a typical linear size of our simplex T), and we use a sequence of nondegenerate triangulations (possibly composite) instead of quasiuniform ones (see also Theorems 2.3.5 and 2.3.6). This remark is essential only for highly nonuniform grids; but, for our purposes even estimates (1.8), (1.37), and (1.43) are quite good, and we define h by the relation $h^\nu \asymp \varepsilon$ (in this case, we obtain grid systems with $N \asymp 1/h^d \asymp \varepsilon^{-d/nu}$). For $2(1+\gamma) > 3$, modifications of Theorems 1.4 and 1.5 hold for certain regions like balls (e.g., if $\hat{\Omega} \subset \bar{\Omega}$ and its triangulations, topologically equivalent to those in Lemma 2.4.4, are constructed as in § 4.4). Thus, for certain d-dimensional problems we can obtain asymptotically optimal algorithms with estimates

$$W(\varepsilon) = O(\varepsilon^{-d/nu}), \qquad (2.9)$$

and it is very important to preserve these estimates when

$$2(m + \gamma) \leq d. \qquad (2.10)$$

Now the solution is not defined at the nodes, and instead of local interpolants some more general projectors P_a must be used (see (2.3.1)). Nonetheless, the most applied important case $d = 3$ can be analyzed in much the same way if we make use of the Steklov averaging $Y_\rho u_0(x) \equiv w$ (see (2.3.2), where u_0 is extended as in § 2.3). (We use new indices for K.)

Lemma 5. Let Ω be a bounded three-dimensional domain with Lipschitz boundary such that $\bar{\Omega}$ is a union of several simplexes defining an initial triangulation $T^{(0)}(\bar{\Omega})$. Suppose that quasiuniform triangulations $T_h(\bar{\Omega})$ are constructed as in § 2.1 through consecutive global refinements and possibly through additional local refinements with a bounded number of levels. Suppose that Γ_0 consists of a finite number of faces of the simplexes in $T^{(0)}(\bar{\Omega})$ and that the spline subspaces $\hat{G}_h \subset W_2^1(\Omega; \Gamma) \equiv G$ are associated with our triangulations $T_h(\bar{\Omega})$. Suppose, finally, that

$$u \in W_2^1(\Omega; \Gamma_0) \cap W_2^{1+\gamma}, \qquad (2.11)$$

where $0 < \gamma \leq 1/2$. Then there exist functions $\hat{u}_h \in \hat{G}_h$ such that

$$\|\hat{u} - u\|_G \leq K_0 h^\gamma |\ln h|^\alpha \|u\|_{\Omega, 1+\gamma}, \qquad (2.12)$$

where $\alpha \equiv 0$ if $\gamma < 1/2$ (or $\gamma = 1/2$ and $|\Gamma_0|_{(2)} = 0$) and $\alpha \equiv 1$ if $\gamma = 1/2$ and $|\Gamma_0|_{(2)} > 0$.

Proof. The case $|\Gamma_0|_{(2)} = 0$ where $G = W_2^1(\Omega)$ is the simplest. We then take $\hat{u} \equiv \text{int}_1 Y_\rho u(x)$ (see (2.3.2), where $\rho \asymp h$. Hence, $\|\hat{u}_h - u\|_G \leq \|u - w\|_G + \|w - \hat{u}_h\|_G$, where $w \equiv Y_\rho u(x)$ and $w \in W_2^2(\Omega)$ (see (2.3.4)). Thus,

$$\|w - u\|_G \leq K_1 \rho^\gamma \|u\|_{1+\gamma,\Omega} \tag{2.13}$$

(see (2.3.3)) and $\|\hat{u}_h - w\|_G \leq K_2 h \|w\|_{2,\Omega}$ (our above reasoning applies), which together with (2.3.4) yields

$$\|\hat{u}_h - w\|_G \leq K_4 h^\gamma \|u\|_{1+\gamma,\Omega}. \tag{2.14}$$

Hence, (2.13) and (2.14) lead to (2.12) with $\alpha = 0$.

Consider now the case $|\Gamma_0|_{(2)} > 0$ where we need to approximate the Dirichlet conditions on Γ_0. Defining w as above, let $\hat{w} \equiv \text{int}_1 w$, which is piecewise linear but does not satisfy the Dirichlet conditions on Γ_0. We therefore replace it by $\hat{u}_h \in \hat{G}_h$ so that their values at the nodes differ only for nodes belonging to Γ_0. Clearly, $\|\hat{u}_h - u\|_G \leq \|u - w\|_G + \|w - \hat{w}\|_G + \|\hat{w} - \hat{u}_h\|_G$, so it suffices to prove that

$$\|\hat{w} - \hat{u}_h\|_G \leq K_5 h^\gamma |\ln h|^\alpha \|u\|_{\Omega,1+\gamma}. \tag{2.15}$$

This is equivalent to

$$\sum_T \|\hat{u}_h - \hat{w}\|_{1,T}^2 \leq K_5^2 h^\gamma |\ln h|^\alpha \|u\|_{\Omega,1+\gamma}, \tag{2.16}$$

where the summation is carried out with respect to all $T \in T_h(\bar{\Omega})$ having at least one vertex belonging to Γ_0. Now consider such a T and note that $\|\hat{u}_h - \hat{w}\|_{1,T}^2 \leq K_6 h^{d-2} \sum_{P_i \in \Gamma_0} (w(P_i))^2$, where

$$|w(P_i)| = |Y_\rho u \mid_{P_i}| \leq \frac{K_7}{h^d} (|u|, 1)_{0,S(P_i)} \leq \frac{K_8}{h^{d/2}} |u|_{0,S(P_i)}$$

and $S(P_i)$ is a cube with center at P_i, which is used for the definition of the Steklov averaging $Y_\rho u(x)$ (see (2.3.2)). Thus, (2.16) implies that

$$\|\hat{u}_h - \hat{w}\|_{1,T}^2 \leq \frac{K_9}{h^2} \sum_T |u|_{0,S(P_i)}^2. \tag{2.17}$$

Letting $B(P_i) \equiv \{x : |x - P_i| \leq d^{1/2}\rho\}$ (this ball contains $S(P_i)$), we make use of the important inequality

$$|u|_{0,S(P_i)}^2 \leq K_* h^2 |u|_{1,B(P_i)}^2, \tag{2.18}$$

which is a modification of (0.1.9) that we prove in Lemma 6. From (2.17) and (2.18), it follows that

$$\sum_T \|\hat{u}_h - \hat{w}\|_{1,T}^2 \leq K_{10}\|u\|_{1,Q_\rho}^2, \tag{2.19}$$

where $Q_\rho \equiv \{x : \exists P_i \in \Gamma_0, x \in B(P_i)\}$. We assumed that $\rho \asymp h$ and that the extended u is such that $\|u\|_{1,Q_\rho} \leq K_{11}\|u\|_{1,\Omega}$.
Since $\|u\|_{1,Q_\rho} \leq K_{12}f(\rho)\|u\|_{1+\gamma,\Omega}$ (see (2.3.5)), then (2.12) holds. $\square$

Lemma 6. *Let the conditions of Lemma 5 be satisfied. Then there exists a constant K_* such that (2.18) holds.*

Proof. We define $\lambda(S(P_i)) \equiv \max |u|_{0,S(P_i)}^2 |u|_{1,S(P_i)}^{-2}$ and
$\bar{\lambda}(B(P_i)) \equiv \max |u|_{0,S(P_i)}^2 |u|_{1,S(P_i)}^{-2}$, where $u \in W_2^1(S(P_i); \Gamma_{[i]})$, $u \neq 0$, and $\Gamma_{[i]} \equiv S(P_i) \cap \Gamma_0$. Note that $\Gamma_{[i]}$ here is not a part of the boundary of our cube $S(P_i)$, but $W_2^1(S(P_i); \Gamma_{[i]})$ and $W_2^1(B(P_i); \Gamma_{[i]})$ are nonetheless Hilbert spaces. Note also that $\lambda(S(P_i))$ and $\bar{\lambda}(B(P_i))$ correspond to maximal eigenvalues of certain symmetric compact operators in these spaces (see Lemma 1 and § 9.1). Of course, $\lambda(S(P_i)) \leq \bar{\lambda}(B(P_i)) < \infty$. The mappings $x - P_i = \rho z$ define the reference cube C, ball B, and corresponding Hilbert spaces $G_{[i]} \equiv W_2^1(B(P); \Gamma_{[i]}^*)$, where $\Gamma_{[i]}^*$ is the image of $\Gamma_{[i]}^*$ containing several $(d-1)$-dimensional simplexes (one of their vertices is the point 0). Then $\bar{\lambda}(B(P_i)) = \rho^2 \bar{\lambda}(G_{[i]}))$, where

$$\bar{\lambda}(G_{[i]}) \equiv \max_{u \in G_{[i]}, u \neq 0} \frac{|u|_{0,C}^2}{|u|_{1,B}^2}. \tag{2.20}$$

Thus, it suffices to find K_* such that

$$\bar{\lambda}(G_{[i]}) \leq K_* \tag{2.21}$$

for all Hilbert spaces $G_{[i]}$. With this in mind, we define a small enough reference $(d-1)$-dimensional simplex $\Gamma^* \equiv [0A_1 \ldots A_{d-1}]$ such that it is congruent to $(d-1)$-dimensional simplexes in $\Gamma_{[i]}^*$ for all i under consideration (possible because of the quasiuniformity of our triangulations). Then, for the Hilbert space $G_* \equiv W_2^1(B; \Gamma^*)$, we have

$$\bar{\lambda}(G_{[i]}) \leq \max_{u \in G_*, u \neq 0} \frac{|u|_{0,S}^2}{|u|_{1,B}^2},$$

and we can take the right-hand side of this inequality in the role of K_* in (2.21). $\square$ [16]

---

[16]Lemmas 6 and 5 hold for more general quasiuniform triangulations. It is even possible to replace $B(P_i)$ by $S(P_i)$ on the right-hand side of (2.18) if we have only a finite

Lemma 7. Let $S \equiv [P_0 \ldots P_d]$ be a regular simplicial part of the cube $C \equiv [0, m]^d$ partitioned in accordance with Theorem 4.4.1 into m^d elementary simplexes $(m \geq 1)$. Denoting the set of their vertices by $\omega \equiv \{A_i\}$, suppose it is used in constructing the Lagrange polynomial $int_m w$ of degree m for a continuous function w on S and that

$$Q \equiv Q(S) \equiv \cup_{A_i \in \omega} C(A_i), \quad C(A_i) \equiv \{x : x - A_i \in [-1, 1]^d\}. \quad (2.22)$$

Suppose also that $u \in W_2^{m+\gamma}(Q)$, with $2(m + \gamma) \leq d$ and that $P_a u \equiv int_m Y_1 u$. Suppose, finally, that $m = 1$. Then there exists constants K_1^* and K_0^* such that, for $\zeta \equiv P_a u - u$, we have

$$\max_{u \in W_2^{1+\gamma}(Q), u \neq 0} \frac{|\zeta|_{1,S}^2}{|u|_{m+\gamma,Q}^2} = K_1^*, \qquad \max_{u \in W_2^{1+\gamma}(Q), u \neq 0} \frac{|\zeta|_{0,S}^2}{|u|_{m+\gamma,Q}^2} = K_0^*. \quad (2.23)$$

Proof. First observe that $S \subset Q$, Q has Lipschitz boundary, and that $W(Q) \equiv \{v : v \in W_2^{m+\gamma}(Q), \ Y_1 v \mid_{A_i} = 0, \ \forall A_i \in \omega\}$ is a subspace in the Hilbert space $W_2^{1+\gamma}(Q)$. Second, it is easy to see that the Steklov averaging Y_ρ preserves linear functions (it suffices to verify this property for the one-dimensional case and the polynomial t^k, $k = 0, 1$; this is not the case for $k \geq 2$). Third, in accordance with the Equivalent Norm Theorem (see § 0.1), we can treat $W(Q)$ as a Hilbert space with inner product $(u, v) \equiv (u, v)_{m+\gamma,Q}$. Finally, we make use of Lemma 1 and conclude that

$$\max_{u \in W(Q), \neq 0} \frac{|u|_{1,S}^2}{|u|_{m+\gamma,Q}^2} \equiv \bar{K}_1 < \infty, \qquad \max_{u \in W(Q), u \neq 0} \frac{|u|_{0,S}^2}{|u|_{m+\gamma,Q}^2} \equiv \bar{K}_0 < \infty,$$

which is a natural generalization of Theorem 2.3.1. Hence, (2.23) holds with $K_1^* = \bar{K}_1$ and $K_0^* = \bar{K}_0$. □ [17]

---

number of different configurations of $\Gamma_{[i]}$. Lemma 5 implies that, under conditions (1.3) with $m = 1$, for solutions of the indicated three-dimensional elliptic boundary value problems, the PGM under consideration leads to asymptotically optimal estimates of accuracy (1.8) for $\gamma \neq 1/2$ and slightly degraded estimates for $\gamma = 1/2$. Therefore, we conclude that, for many correct elliptic boundary value problems associated with three-dimensional domains, the strengthened variant of the Kolmogorov-Bakhvalov hypothesis is true. It is not difficult to prove (1.8) for $d \geq 4$ if $m = 1, \Gamma_0 = \emptyset$, and all our triangulations are regular. We only sketch a possible proof based on modifications of the theorems in § 2.3.

[17] Lemma 7 implies that (1.8) holds for $d \geq 4$ if $\Gamma_0 = \emptyset$ and all our triangulations are regular. Moreover, it is clear how to generalize these results to PGMs associated with certain composite triangulations, provided they are obtained from regular triangulations

A more involved approach for construction of $P_a u$ was suggested recently in [448]. It leads to desired estimates (2.23) (with $m \geq 1$) and, thus, to (1.8) at least for $\gamma = 1$ (it is likely that these results can be generalized for $0 < \gamma < 1$ and composite triangulations). We briefly describe only the construction of $P_a u$ for $u \in W_2^{1+\gamma}(\Omega)$ assuming that we have a standard nodal basis $\{\hat{\psi}_i, A_i \in \Omega_h\}$. For each node $A_i \in \Omega_h$, we choose a $(d-1)$-dimensional simplex $S_{d-1}(A_i)$ such that A_i is a vertex of $S_{d-1}(A_i)$ and $S_{d-1}(A_i)$ is a $(d-1)$-dimensional face of one of our simplexes in $T_h(\Omega)$. We then consider traces on $S_{d-1}(A_i)$ of our basis functions associated with vertices of $S_{d-1}(A_i)$. These d linear functions form a basis of a d-dimensional subspace W in the Hilbert space $V_i \equiv L_2(S_{d-1}(A_i))$. Numbering these functions in a convenient way, we obtain, say, functions $\psi_1, \ldots, \psi_d$ with ψ_1 corresponding to the trace of $\hat{\psi}_i$. Next, define the V_i-dual basis for W consisting of linear functions $f_1, \ldots, f_d$ such that $(\psi_r, f_r)_{V_i} = 1$ and $(\psi_r, f_l)_{V_i} = 0$ if $r \neq l$, $r \in [1, d]$, $l \in [1, d]$. Then, for $f_1 \equiv f_{A_i}$, we use

$$P_a u \equiv \sum_{A_i \in \Omega_h} (u, f_{A_i})_{V_i}, \qquad (2.24)$$

where u refers to the trace of the given function $u \in W_2^{1+\gamma}(\Omega)$. It is important to note that, for local conditions of type (1.36) and estimates (1.37), it suffices to chose $S_{d-1}(A_i)$ for A_i belonging to the interface of two standard blocks as a part of the interface. Thus, for certain correct elliptic boundary value problems associated with d-dimensional domains the strengthened variant of the Kolmogorov-Bakhvalov hypothesis holds.

2.4. Iterative methods with orthogonalization. Consider briefly applications of iterative methods of type (1.3.27) for grid problems with $L = L^*$ (these methods were suggested in [187]). Suppose we approximate the problem from Theorem 1 with

$$a_{1,2} = a_{2,1}, \ b_r = b'_r, \ L_1 = L_1^* \equiv A, \ L_0 = L_0^* = -c_0 M$$

on the basis of PGM (1.7). To simplify the study, we assume that $l(\hat{v}) = (\hat{f}, \hat{v})_{0,\hat{\Omega}}$. Then the resulting grid system takes the form

$$L_\Omega \mathbf{u} \equiv (A - c_0 M)\mathbf{u} = M\mathbf{f}, \qquad (2.25)$$

where $(M\mathbf{u}, \mathbf{v}) = (\hat{u}, \hat{v})_{0,\hat{\Omega}}$ and $M \in \mathcal{L}^+(H)$. In the Euclidean space $H' \equiv H(M)$, we replace (2.25) by the equivalent operator equation

and contain only simplexes with a finite number of geometric forms (then we can obtain generalizations of Theorems 2.3.5 and 2.3.6, but now with a reference augmented simplex $R^* \equiv S^*$ and augmented region $Q(R^*)$ (see (2.22))). The above natural approach is problematic when $|\Gamma_0|_{(d-1)} > 0$ or when we consider general quasiuniform triangulations so that general affine transformations of our simplexes are necessary.

$$L'\mathbf{u} \equiv (A' - c_0 I)\mathbf{u} = \mathbf{f}, \qquad (2.26)$$

where $A' \equiv M^{-1}A \in \mathcal{L}^+(H')$. Suppose that we know several of its left eigenvalues $\lambda_1 \leq \lambda_2 \leq \ldots \lambda_k$, with $\lambda_i \equiv \lambda_{i,h}$ enumerated in increasing order. Let their corresponding eigenvectors $\mathbf{y}_1, \ldots, \mathbf{y}_k$ be an orthogonal basis (in the sense of the Euclidean space H') of the subspace $U_k \equiv \operatorname{lin}\{\mathbf{y}_1, \ldots, \mathbf{y}_k\}$. Let P denote the orthoprojector of H' onto U_k and $P^\perp \equiv I - P$. It is easy to verify that $MP^\perp = P^\perp M$ and $LP^\perp = P^\perp L$, and that iterative method (1.3.26) with $B' \equiv M^{-1}B \in \mathcal{L}^+(H')$ can be rewritten in the form $B'(\mathbf{v}^{n+1} - \mathbf{v}^n) = -\tau_n P^\perp(L'P^\perp\mathbf{v}^n - \mathbf{f})$ or

$$B(\mathbf{v}^{n+1} - \mathbf{v}^n) = -\tau_n P^\perp(LP^\perp\mathbf{v}^n - M\mathbf{f}). \qquad (2.27)$$

Theorem 1.3.16 applies provided that inequalities (1.3.26) hold for all $\mathbf{v} \in V \equiv \{\mathbf{v} : (\mathbf{y_r}, \mathbf{v})_B = 0, \ r \in [1, k]\}$ (these inequalities are equivalent to those connecting L' and B' in the sense of H'). We may assume that

$$L_\Omega \geq \nu_0 B - c_0 M, \ \nu_0 > 0, \ c_0 \geq 0 \qquad (2.28)$$

and that $\lambda_{i,h}$ and $\hat{y}_i$, $i \in [1, k]$, converge to the corresponding eigenvalues and eigenfunctions of the differential eigenvalue problem $b(u; v) = \lambda(u, v)_{0,\Omega}$, $\forall v \in G$ (see Section 9) as h tends to zero. If the $(k + 1)$th eigenvalue of the differential problem, say λ_{k+1}, is positive, then it is possible to prove that

$$(L_\Omega\mathbf{v}, \mathbf{v}) \geq \sigma_0\|\mathbf{v}\|_M^2, \ \forall\mathbf{v} \in U_k^\perp, \ \sigma_0 > 0, \qquad (2.29)$$

if $h < h_0$ and h_0 is small enough. The proof is analogous to the proof of Theorem 3.3; the assumption that (2.29) does not hold and the compactness principle contradicts the fact that $\lambda_{k+1} > 0$. From (2.29), it follows that $(L_\Omega\mathbf{v}, \mathbf{v}) \geq \nu_0[1 + c_0/\sigma_0]^{-1}\|\mathbf{v}\|_B^2$, $\forall\mathbf{v} \in U_k^\perp$ which leads to (1.3.26).

2.5. Singular elliptic operators. Consider now, as a typical example, the Neumann problem for an elliptic equation. For the basic Hilbert space $G' \equiv W_2^1(\Omega)$, we consider a particular case of bilinear form (0.1.30): $b(u; v) \equiv \sum_{r,l=1}^d (a_{rl}, D_l u D_r v)_{0,\Omega}$, where the matrix $A \equiv [a_{r,l}]$ is symmetric and satisfies (0.1.32). This problem is equivalent to the operator equation $Lu = f$ in $G' \equiv V$, with $L = L^* \geq 0$ and $\operatorname{Ker} L = \operatorname{lin}\{1\} = \operatorname{Ker}\hat{L}$, where $\hat{L}$ is our standard grid operator associated with the spline space $\hat{V}$. To apply theorems dealing with convergence of PGMs for correct operator equation we can reduce our problem to a correct operator equation, not G' but in the subspace $G \equiv V \setminus 1 \equiv \{v : v \in G', (v, 1)_{0,\Omega} = 0\}$. The same applies for our projective problem if we replace $\hat{V}$ by $\hat{G} = \hat{V} \cup G$. This reduction requires specification of estimates for $\text{dist}_G\{u; \hat{G}\}$, which can be done in accordance with the next general but simple theorem (see [197]).

Theorem 4. Let V be a Hilbert space and its subspace G satisfy

$$\varphi_i(u) \equiv (u, g_i), \quad i = 1, \ldots, k, \tag{2.30}$$

where $g_1, \ldots, g_k$ are linearly independent elements of G'. Suppose that $\{\hat{G}'_h\}$ is a sequence of subspaces that approximates G' (see (0.2.12)), and define the sequence of subspaces $\{\hat{G}_h\}$ by $\hat{G}_h \equiv \hat{G}'_h \cap G$. Then there exist numbers $h_0 > 0$ and $K > 0$ such that

$$\text{dist}_G\{u; \hat{G}\} \le K \text{dist}_V\{u; \hat{V}_h\}, \quad \forall u \in G, \ h < h_0 > 0. \tag{2.31}$$

Proof. Let $P \equiv \hat{P}_h$ be the orthoprojector of V onto $\hat{V}_h$. Take $\hat{u} \equiv Pu, \hat{g}_i \equiv Pg_i, i \in [1, k]$, and construct $\hat{u}_g \equiv u - \sum_{i=0}^{k} c_i \hat{g}_i$ in such a way that $(\hat{u}_g, g_i) = 0$, $i \in [1, k]$. This system, obtained with respect to the unknown vector $\vec{c} \equiv [c_1, \ldots, c_k]^T$, will be definite for small enough h because its matrix tends to the Gram matrix of the functions $g_i, i \in [1, k]$, as $h \to 0$. The right-hand side of the ith equation is $(\hat{u}, g_i) = (\hat{u} - u, g_i) = O(\text{dist}_V\{u; \hat{V}_h\})$. Hence, $\|\vec{c}\| = O(\text{dist}_V\{u; \hat{V}_h\})$, $X \equiv \|c_1\hat{g}_1 + \cdots + c_k\hat{g}_k\| \le K\text{dist}_V\{u; \hat{V}\}$, and $\text{dist}_V\{u; \hat{G}\} \le X + \text{dist}_V\{u; \hat{V}\}$, which completes the proof. $\square$

The above theorem implies that $\{\hat{G}_h\}$ approximates G. We return now to the algebraic side of the problem. Of crucial importance is again the choice of the basis and the Euclidean space H. If we decide to deal with the subspace $\hat{G}_h$, then the natural choice of basis is $\{\bar{\psi}_i \equiv \hat{\psi}_i - \alpha_i\}$, where all α_i are constants. This yields dense Gram matrices. In order to return to standard grid operators, it is natural to use a projective formulation, not in $\hat{G}$, but in the space $\hat{V}$ (with the standard nodal basis) and look for $\hat{u} \in \hat{V}$ such that

$$\left. \begin{array}{c} b(\hat{u}; \hat{v}) = l(\hat{v}), \ \forall \hat{v} \in \hat{V}, \\ (\hat{u}, 1)_{0,\Omega} = 0. \end{array} \right\} \tag{2.32}$$

We then obtain the algebraic problem

$$L_\Omega \mathbf{u} = \mathbf{f}, \mathbf{u} \in S, \tag{2.33}$$

where $S \equiv \{\mathbf{u} : (\mathbf{u}, \bar{1})_M = 0\}$, $\bar{1} \equiv [1, \ldots, 1]^T$, and $(\mathbf{f}, \bar{1}) = 0$ in accordance with the condition $l(1) = 0$; this guarantees correctness of the original problem and its grid analog (2.32). Now $L_\Omega = L_\Omega^* \ge 0$ and if $B = B^* \ge 0$ is such that $\kappa_{0,Q} B \le L_\Omega \le \delta_1 Z$, $\kappa_{0,Q} > 0$, (see Chapter 3) then Ker $B = $ Ker L_Ω and Lemma 1.3.5 and Theorem 1.3.17 apply. This shows that the case $L = L^* \geq 0$ is essentially the same as the case $L \in \mathcal{L}^+(H)$.

§3. Strongly elliptic systems and elasticity problems

3.1. Construction of PGMs and model grid operators. Boundary value problems for systems of equations contain k unknown functions $u_r(x) \equiv u_r$, $r \in [1, k]$, where $x \equiv [x_1, \ldots, x_d] \in \Omega$ and has the same sense as in § 2.1. For second order strongly elliptic systems, we therefore deal with the unknown vector-function

$$\bar{u} \equiv u \equiv [u_1, \ldots, u_k] \in G \subset (G^{(1)})^k. \tag{3.1}$$

In the role of the space $G^{(m)}, m \geq 0$, we use the Hilbert space with the norm

$$\|u\|_{m,\Omega} \equiv [\sum_{r=1}^{k} \|u_r\|_{m,\Omega}^2] 1/2. \tag{3.2}$$

In the role of G, we use the Hilbert space with the norm $\| \ \|_1$ or

$$\|u\|_G \equiv |u|_{1,\Omega} \equiv [\sum_{r=1}^{k} |u_r|_{1,\Omega}^2]^{1/2}; \tag{3.3}$$

we also write $(u, v)_0$ instead of $(u, v)_{0,\Omega}$. A given problem is written in standard form (0.1.16) or (0.1.28) (we use a variational formulation (0.1.4) when it is possible). We call $L \in \mathcal{L}(G)$ a *strongly elliptic operator* if there exist $\nu > 0$ and $c_0 \geq 0$ such that

$$(Lv, v) \equiv b(v; v) \geq \nu |v|_{1,\Omega}^2 - c_0 |v|_0^2, \ \forall v \in G. \tag{3.4}$$

If $c_0 = 0$, then the correctness of problem (0.1.16) follows from Theorem 0.1.4; otherwise, further investigations are necessary (see, e.g., Lemma 1.1.1 or the elasticity problems below).

Considering construction of asymptotically optimal algorithms for these problems, we must distinguish two situations. The first is characterized by

$$G = G_1 \times \cdots \times G_k, \tag{3.5}$$

with $G_r \equiv W_2^1(\Omega; \Gamma_{0,r})$. We then can take

$$\hat{G} = \hat{G}_1 \times \cdots \times \hat{G}_k, \tag{3.6}$$

with $\dim \hat{G} = \dim \hat{G}_1 + \cdots + \dim \hat{G}_k$. If $\{\hat{\psi}_{r,i_r}\}$ is a known basis for the space $\hat{G}_r$, $r \in [1, k]$, then we can work with the basis for the space $\hat{G}$ consisting of the vector-functions

$$\hat{\Psi}_{i_r}^{(r)} \equiv [\underbrace{0, \ldots,}_{r-1} \hat{\psi}_{r,i_r}, \underbrace{0, \ldots, 0}_{k-r}]^T, \ r \in [1, k]. \tag{3.7}$$

This implies that, in the construction of PGMs and study of their convergence, no new difficulties are introduced. Also, this implies that standard expansions of $\hat{u}$ lead to systems $L_h \mathbf{u} = \mathbf{f}$ that can be considered as operator equations in the Euclidean space $\mathbf{H} \equiv H \equiv H_1 \times \cdots \times H_k$, where $\mathbf{u} \in H, \mathbf{u}_r \in H_r, \ r \in [1, k]$. Under suitable ordering of the unknowns, the matrix L_h takes the block form $L_h \equiv [L_{i,j}], \ i \in [1, k], j \in [1, k]$. Hence it is reasonable to construct model block diagonal operators

$$B\mathbf{u} \equiv [B_1 \mathbf{u}_1, \ldots, B_k \mathbf{u}_k]^T, \tag{3.8}$$

with $B_r, \ r \in [1, k]$ indicated in § 1 (for difference methods such an approach was suggested in [153, 154] and it was extended later for PGMs in [166, 170]). [18]

3.2. Boundary value elasticity problems with positive definite operators. Consider two- and three-dimensional elasticity problems in coordinates $x \equiv [x_1, \ldots, x_d]$ ($d = 2$, or $d = 3$) for the *displacement vector* $u \equiv \vec{u} \equiv [u_1, \ldots, u_d]^T \equiv [u_1(x), \ldots, u_d(x)]^T, \ x \in \Omega$, where Ω is a region in $\mathbf{R}^d$ corresponding to an elastic body. Then the components $\varepsilon_{i,j}(u) = \varepsilon_{j,i}(u)$ of the *strain tensor* and $\sigma_{i,k}(u) = \sigma_{k,i}(u)$ of *stress tensor* are defined in linear elasticity by the relations

$$\varepsilon_{i,j}(u) \equiv \frac{D_j u_i + D_i u_j}{2}, \quad \sigma_{i,k} \equiv \sum_{l,m} c_{i,k,l,m} \varepsilon_{l,m}(u), \tag{3.9}$$

where $i \in [1, d]$, $j \in [1, d]$, $k \in [1, d]$. The piecewise smooth functions $c_{i,k,l,m} \equiv c_{i,k,l,m}(x)$ are such that

$$\mu_0 \varepsilon^2(v) \leq \sum_{l,m,i,k=1}^{d} c_{i,k,l,m} \varepsilon_{i,k}(v) \varepsilon_{l,m}(v) \leq \mu_1 \varepsilon^2(v), \ \forall v \in G, \tag{3.10}$$

---

[18] Iterative methods with such model operators (see, e.g., (0.3.13), (1.3.19), and (1.3.37)) are very well suited for parallel computation, but recent investigations (see § 4) have yielded some model operators not having the block diagonal structure. Also, there is no difficulty with the coarse grid continuation procedure compared to that given in § 1.

The second and more complicated situation is characterized by the lack of factorizations (3.5) and (3.6). Here we have additional problems, not only in the construction and investigation of PGMs, but also in the construction of model operators. Nonetheless, for some problems, the additional analysis is based on that described above (see [195, 197]). We will return to this subject later in § 4 and Chapter 8.

with $\mu_0 > 0$, $\varepsilon^2(v) \equiv \sum_{i,k=1}^{d} \varepsilon_{i,k}^2(v)$ and

$$c_{i,k,l,m} = c_{k,i,l,m} = c_{l,m,i,k}. \tag{3.11}$$

So far, we have not specified what $\vec{u}$ are permissible. To formulate the basic boundary value elasticity problems as correct variational problems in a Hilbert space G (see § 0.1), we introduce $W \equiv W_2^1(\Omega; \Gamma_0) \equiv G_r$, $r \in [1, d]$, with $|\Gamma_0|_{(d-1)} > 0$, as the permissible space for all components u_r ($r \in [1, d]$) of $\vec{u}$ and define the Hilbert space $G \equiv \vec{V} \equiv W^d$ and the inner product $(\vec{u}, \vec{v}) \equiv (\vec{u}, \vec{v})_{1,\Omega} \equiv \sum_{r=1}^{d}(u, v)_{1,\Omega}$. Note that Γ_0 corresponds to the part of the boundary that is subject to clamping conditions, which are similar to the homogeneous Dirichlet conditions from Lemma 0.1.1. Next, define the linear functional

$$l(v) \equiv (\vec{F}, v)_0 + (\vec{F}_\Gamma, v)_{0,\Gamma_1}, \tag{3.12}$$

where $\vec{F} \in (L_2(\Omega))^d$ and $\vec{F}_\Gamma \in (L_2(\Gamma_1))^d$ correspond to the given forces distributed over Ω and Γ_1, respectively. We also define the energy functional

$$\Phi(v) \equiv \frac{1}{2}I(v) - l(v), \tag{3.13}$$

where $I(v) \equiv I_\Omega(v) \equiv \sum_{i,k=1}^{d}(\sigma_{i,k}(v), \varepsilon_{i,k}(v))_0$ (or $I(v) = \sum_{l,m,i,k=1}^{d}(c_{i,k,l,m}, \varepsilon_{i,k}(v)\varepsilon_{l,m}(v))_0$) and

$$\mu_0(\varepsilon^2(v), 1)_0 \leq I(v) \leq \mu_1(\varepsilon^2(v), 1)_0 \leq \bar{\mu}_2|v|_{1,\Omega}^2, \quad \forall v \in \vec{V}, \tag{3.14}$$

(see (3.10)). In the simplest case of isotropic elastic material, we have

$$I(v) \equiv \lambda|\operatorname{div} v|_0^2 + 2\mu\left(\sum_{i,k=1}^{d} \varepsilon_{i,k}^2(v), 1\right)_0, \tag{3.15}$$

where $\operatorname{div} v = \sum_{s=1}^{d} \varepsilon_{s,s}(v) = \sum_{s=1}^{d} D_s v_s$, with $\lambda > 0$ and $\mu > 0$ the *Lame parameters*. An indispensable tool in the study of the correctness of elasticity problems is the *Korn-Friedrichs inequality* (see [124, 279]):

$$(\varepsilon^2(v), 1)_0 \geq \mu_2\|v\|_G^2, \quad \mu_2 > 0, \quad \forall v \in G, \tag{3.16}$$

which holds even for $G \equiv (W_2^1(\Omega))^d$, that is, for $\Gamma_0 = \emptyset$ (for the simplest case, we give a proof of (3.16) below). Since we consider the case when $|\Gamma_0|_{(d-1)} > 0$ and (0.1.9) holds, then we have $l \in G^*$ and $I(v) \geq \mu_0\mu_2\|v\|^2$. Hence, theorems from § 0.1 apply for (0.1.4), with $\Phi(v)$ from (3.13). The variational problem for (3.13) is of the form (0.1.16) or (0.1.28), with $L \in \mathcal{L}^+(G)$ and

$$b(u;v) \equiv \sum_{i,k,l,m=1}^{d} (c_{i,k,l,m}\varepsilon_{l,m}(u), \varepsilon_{i,k}(v))_0 \equiv \sum_{r,l=1}^{d} b_{\Omega,r,l}(u_l;v_r), \quad (3.17)$$

$r \in [1,d], l \in [1,d]$, and $(u,v)_0 \equiv (u,v)_{0,\Omega}$. For (3.15) with $d = 2$, we have
$I(v) = (2\nu+\lambda)[|D_1v_1|_0^2+|D_2v_2|_0^2]+2\mu[|D_2v_1|_0^2+|D_1v_2|_0^2]+2\lambda(D_1v_1,D_2v_2)_0$
$+2\mu(D_2v_1,D_1v_2)_0, b_{\Omega,1,1}(u_1;v_1) = (2\nu+\lambda)(D_1u_1,D_1v_1)_0+\nu(D_2u_1,D_2v_1)_0,$
$b_{\Omega,1,2}(u_2;v_1) = \lambda(D_1v_1,D_2u_2)_0 + \nu(D_2v_1,D_1u_2)_0, b_{\Omega,2,1}(u_1;v_2) =$
$\lambda(D_1u_1,D_2v_2)_0+\nu(D_2u_1,D_1v_2)_0, b_{\Omega,2,2}(u_2;v_2) = (2\nu+\lambda)(D_2u_2,D_2v_2)_0+$
$\nu(D_1u_2,D_1v_2)_0$. This splitting of $b(u;v)$ yields a block structure of the resulting grid system.

PGMs for elasticity problems are much the same as those considered in § 1 and 2: we use the same $\hat{\Omega}, \hat{\Gamma}, \hat{\Gamma}_0, \hat{\Gamma}_1$, and a quasiuniform triangulation (possibly composite) $T_h(\hat{\Omega})$. We assume again that the interface between Γ_0 and Γ_1 consists of vertices of the triangles T in $T_h(\hat{\Omega})$ if $d = 2$, or of edges of the tetrahedrons T in $T_h(\hat{\Omega})$ if $d = 3$, and that $\exists \kappa_0 \leq 0$ such that

$$|(T \setminus \Omega)| \leq \kappa_0 h|T|, \ \forall T \in T_h(\hat{\Omega}), \quad (3.18)$$

(all coefficients in the bilinear form are properly extended to $\hat{\Omega}$). The set Ω_h of nodes P_i is associated with the same spline space as in § 1 ($d = 2$) or 2 ($d = 3$) (it is a subspace of $W_2^1(\hat{\Omega}; \hat{\Gamma}_0)$ of, say, piecewise linear functions), but it is denoted by $\hat{W}$ now because we approximate our original Hilbert space $G \equiv \vec{V}$ by $\hat{G} \equiv (\hat{W})^d$. [19]

Next, we consider construction of asymptotically optimal iterative methods for solving the resulting grid systems. Since we deal with the Euclidean space $\hat{G} \equiv (\hat{W})^d$, then each node $P_i \in \Omega_h$ is associated with d basis vector-functions of form (3.7). We always number them in such a way that the vector-functions with index (r) in (3.7) precede those with index $(r + 1), r \in [1, d - 1]$. Then, if $\hat{\psi}_i \in \hat{W}$ corresponds to node P_i (numbered in any suitable way) and

---

[19] If (1.5) holds, then $\|\hat{u}_h - u\|_G \leq \bar{\nu}_1^{1/2} \nu_0^{-1/2} \mu_2^{-1/2} |\hat{u}_h - u|_{1,\Omega}$, (see Theorem 0.2.2) which yields estimates of type (1.8) and (1.37). The singular basis functions can be used along the same lines as in § 1 and 2 (important results concerning the study of singularities in elasticity problems can be found in [312]). If (1.5) does not hold, then the convergence of our PGMs should be analyzed separately as in § 1 and 2; this can be done in a straightforward manner. Thus, we conclude that problems dealing with accuracy of PGMs for elasticity problems are essentially the same as that considered in § 1 and 2.

$$L_{r,l} \equiv [b_{\hat{\Omega},r,l}(\hat{\psi}_j; \hat{\psi}_i)], \qquad (3.19)$$

where $b_{\hat{\Omega},r,l}(u; v) \equiv b_{r,l}(\hat{\Omega}; u; v)$, $r \in [1, d]$, $l \in [1, d]$, then (3.19) defines a block in the matrix, associated with (3.17). As a typical example, we take $d = 3$ and the grid system

$$L_{\Omega}\mathbf{u} = \mathbf{f}, \qquad (3.20)$$

where $L_{\Omega} = [L_{i,j}]$, $i \in [1, 3], j \in [1, 3]$, $\mathbf{u} = [\mathbf{u}_1, \mathbf{u}_2, \mathbf{u}_3]^T$, $\mathbf{u}_s \in H_{(1)}, \mathbf{f}_s \in H_{(1)}, s = 1, 2, 3$, and $\mathbf{f} \equiv [\mathbf{f}_1, \mathbf{f}_2, \mathbf{f}_3]^T$. We consider (3.20) as an operator equation in the Euclidean space $\vec{H} \equiv (H_{(1)})^3 \equiv H$, where $H_{(1)}$ corresponds to the Euclidean space associated with $\hat{W}$ (denoted by H in § 1 and 2). It is easy to see that $L_{\Omega} \equiv L_h \in \mathcal{L}^+(H)$. We introduce block diagonal operators

$$\Lambda_{\Omega} \equiv D(\Lambda; 3), \qquad (3.21)$$

where $\Lambda \equiv [(\hat{\psi}_j; \hat{\psi}_i)_{1,\Omega}]$ (see (3.19)).

Theorem 1. *There exists $h_0 > 0$ such that $L_{\Omega} \asymp \Lambda$ when $|h| \leq h_0$.*

Proof. For arbitrary vector $\mathbf{u}_s \equiv [\mathbf{u}_1, \mathbf{u}_2, \mathbf{u}_3]^T \in \vec{H}$, define corresponding $\hat{u} \equiv [\hat{u}_1, \hat{u}_2, \hat{u}_3] \in \vec{V}$. Then $(L_{\Omega}\mathbf{u}, \mathbf{u}) = I_{\hat{\Omega}}(\hat{u})$, and it suffices to prove existence of constants $\delta_1^{(1)} > 0$ and $\delta_0^{(1)} > 0$ such that

$$\delta_0^{(1)}|\hat{u}|_{1,\hat{\Omega}}^2 \leq I_{\hat{\Omega}}(\hat{u}) \leq \delta_1^{(1)}|\hat{u}|_{1,\hat{\Omega}}^2, \ \forall \hat{u} \in \hat{G}. \qquad (3.22)$$

Actually, we need only prove the inequality involving $\delta_0^{(1)}$, which is a direct consequence of (3.14) and (3.16) if (1.5) holds. For the general case, we have $I_{\hat{\Omega}}(\hat{u}) \geq \mu_0(\varepsilon^2(\hat{u}), 1)_0 \geq \mu_0\mu_2|\hat{u}|_{1,\Omega}$, $\forall \hat{u} \in \hat{G}$, and using the properties of the triangulations (see (3.18)), we conclude that

$$I_{\hat{\Omega}}(\hat{u}) \geq \mu_0\mu_1(1 - \kappa_0 h)|\hat{u}|_{1,\Omega}^2, \ \forall \hat{u} \in \hat{G}. \qquad (3.23)$$

This completes the proof. $\square$ [20]

Lemma 1. *Define the Hilbert space $G = (W)^d$, where $W \equiv \overset{o}{W}_2^1(\Omega)$, and the quadratic functional I by (3.15). Then*

$$\mu\Lambda \leq L_{\Omega} \leq (\lambda + 2\mu)\Lambda. \qquad (3.24)$$

Proof. We consider only the more difficult case $d = 3$. As in the proof of Theorem 1, we deal with $I_{\hat{\Omega}}(v) \equiv X(v)$, where $v \equiv \hat{u}$,

$$X(v) = X_1(v) + X_2(v), \qquad (3.25)$$

---

[20] This proof from [175, 182] is based only on (3.16), not on similar inequalities for $\hat{\Omega}_h$.

$$X_1(v) \equiv \lambda |\text{div } \mathbf{v}|_0^2 + 2\mu \left(|D_1 v_1|^2 + |D_2 v_2|^2 + |D_3 v_3|^2, 1 \right)$$
$$+\mu \left(|D_2 v_1|^2 + |D_1 v_2|^2 + |D_3 v_1|^2 + |D_1 v_3|^2 + |D_3 v_2|^2 + |D_2 v_3|^2, 1 \right),$$
$$(3.26)$$

$$X_2(v) \equiv 2\mu \left(D_2 v_1 D_1 v_2 + D_3 v_1 D_1 v_3 + D_3 v_2 D_2 v_3, 1 \right), \qquad (3.27)$$

and inner product is in the sense of $L_2(\Omega)$. We may extend v as a vanishing vector-function outside of domain Ω and replace index $0, \Omega$ by $0, C$, where C is a cube containing $\bar{\Omega}$. Moreover, we approximate v by a sequence $\{w^{(n)}\}$, where $w^{(n)} \equiv Y'_{\rho_n} v$ is the Sobolev averaging of v (see § 2.3), $\rho_n \equiv \rho_0/n$, $n = 1, 2, \ldots$, and ρ_0 is small so each $w^{(n)}$ and its derivatives vanish on the boundary of C. Then $X_2(v) = \lim_{n \to \infty} W^n$, where

$$W^n \equiv 2\mu[(D_2 w_1, D_1 w_2)_{0,C} + (D_3 w_1, D_1 w_3)_{0,C} + (D_3 w_2, D_2 w_3)_{0,C}] \quad (3.28)$$

(we omit the index (n) in the components of $w^{(n)}$). We also have
$W^n = 2\mu[(D_1 w_1, D_2 w_2)_{0,C} + (D_1 w_1, D_3 w_3)_{0,C} + (D_2 w_2, D_3 w_3)_{0,C}]$. This, together with (3.25)–(3.27), yields

$$X(w^{(n)}) = (\lambda + \mu)|\text{div } w^{(n)}|^2 + \mu|w^{(n)}|_{1,C}^2 \geq \mu|w^{(n)}|_{1,C}^2. \qquad (3.29)$$

Passing to the limit now yields

$$X(v) \geq \mu|v|_{1,C}^2 = \mu|v|_{1,\Omega}^2. \qquad (3.30)$$

To obtain the estimate from above for $X(v)$, we write

$$X(v) = \bar{X}_1(v) + \bar{X}_2(v) + \bar{X}_3(v), \qquad (3.31)$$

$\bar{X}_1(v) \equiv (\lambda + 2\mu)[|D_1 v_1|_0^2 + |D_2 v_2|_0^2 + |D_3 v_3|_0^2]$,
$\bar{X}_2(v) \equiv 2\lambda[(D_1 v_1, D_2 v_2)_0 + (D_1 v_1, D_3 v_3)_0 + (D_2 v_2, D_3 v_3)_0]$,
$\bar{X}_3(v) \equiv \mu[|D_2 v_1|_0^2 + |D_1 v_2|_0^2 + |D_3 v_1|_0^2 + |D_1 v_3|_0^2 + |D_3 v_2|_0^2 + |D_2 v_3|_0^2] + 2\mu[(D_2 v_1, D_1 v_2)_0 + (D_3 v_1, D_1 v_3)_0 + (D_3 v_2, D_2 v_3)_0]$. It is easy to see that
$\bar{X}_3(v) \leq 2\mu[|D_2 v_1|_0^2 + |D_1 v_2|_0^2 + |D_3 v_1|_0^2 + |D_1 v_3|_0^2 + |D_3 v_2|_0^2 + |D_2 v_3|_0^2]$.
Next, as for $X_2(v)$, we have $\bar{X}_2(v) = \lim_{n \to \infty} \bar{X}'_2(w^{(n)})$, where
$\bar{X}'_2(w^{(n)}) \equiv 2\lambda[(D_1 v_1, D_2 v_2)_{0,C} + (D_1 v_1, D_3 v_3)_{0,C} + (D_2 v_2, D_3 v_3)_{0,C}] = 2\lambda[(D_2 w_1, D_1 w_2)_{0,C} + (D_3 w_1, D_1 w_3)_{0,C} + (D_3 w_2, D_2 w_3)_{0,C}] \leq \lambda[|D_2 w_1|_{0,C}^2 + |D_1 w_2|_{0,C}^2 + |D_3 w_1|_{0,C}^2 + |D_1 w_3|_{0,C}^2 + |D_3 w_2|_{0,C}^2 + |D_2 w_3|_{0,C}^2]$.
Hence, (3.31) yields the desired inequality. $\square$

This proof is based on [162, 164]. The estimates

$$I(v) \geq \mu|v|_{1,\Omega}^2, \qquad (3.32)$$

$$I(v) \leq (d\lambda + 2\mu)|v|^2_{1,\Omega} \tag{3.33}$$

are well known (see [124, 418]), with (3.33) holding even without Dirichlet conditions because it is based on the simple inequality

$$|\text{div } v| \leq d\left(|D_1 v_1|^2_0 + |D_2 v_2|^2_0 + |D_3 v_3|^2_0\right). \tag{3.34}$$

Theorem 2. *Let the conditions of Theorem 1 be satisfied. Suppose that* $B_Q \in \mathcal{L}^+(H_{(1)})$ *and* $\sigma_0 B_Q \leq \Lambda_\Omega \leq \sigma_1 B_Q$, $\sigma_0 > 0$. *Suppose also that, for* L_Ω *in* (3.30), *we choose a model block diagonal operator*

$$B \equiv D(B_Q; 3). \tag{3.35}$$

Then $L_\Omega \asymp B$; *if* B_Q *is an asymptotically optimal model operator for* Λ_Ω, *then* B *is an asymptotically optimal model operator for* L_Ω.

Proof. It is easy to see that $\delta_0^{(1)}\sigma_0 B \leq L_\Omega \leq \delta_1^{(1)}\sigma_1 B$ (see (3.22)). If we also know that it is possible to obtain solutions of systems with B_Q in asymptotically optimal computational work $W_Q \leq KN$ (N is the number of nodes in Ω_h), then computational work for solving a system with B is $W \leq 3W_Q$. $\square$

Q in B_Q refers to a model region chosen for Ω, and such operators were considered in Chapter 3. Thus, asymptotically optimal iterative methods for (3.20) can be constructed in accordance with the theory in § 1.3.[21] We can thus obtain asymptotically optimal algorithms for elasticity problems as in § 1 and 2. The practical effectiveness of the algorithms suggested here depends crucially on the character of the given coefficients and regions. For example, as seen from Lemma 1, convergence can be slow for small ratio μ/λ (we also meet large numbers in (1.8)). For such situations, the problem of constructing asymptotically optimal algorithms with estimates of computational work independent of a parameter (e.g., λ in Lemma 1) becomes very important; we will study such problems in Chapter 7. Moreover, we will construct model operators B that do not have form (3.35) in § 4.

3.3. Elasticity problems with nonnegative operators. We restrict ourselves to the study of two important examples. The first is our second basic boundary value problem in elasticity, where surface traction is prescribed; the Hilbert space is $G \equiv (W_2^1(\Omega))^d$ (for definiteness, we consider $d = 3$); and the problem is given by (0.1.28) in G with $L = L^* \equiv L_{el} \geq 0$ and dim Ker $L_{el} \geq 1$ (see [279, 418]). More precisely, if $I(\vec{v}) = (L_{el}\vec{v}, \vec{v})$ (see (3.14)), then

---

[21] Such methods for solving practical elasticity problems have been applied with success at Moscow State University for many years (starting in the sixties for difference approximations and B_Q from § 3.4); for more recent work see, e.g., [62, 450].

Ker $L_{el} = \{\vec{v} : \vec{v} = \vec{a} + \vec{b} \times \vec{x}\} = \lin\{\vec{e}_1, \ldots, \vec{e}_6\}$, where $\vec{a} \in \mathbf{R}^3$ and $\vec{b} \in \mathbf{R}^3$ are vectors independent of $x \equiv [x_1, x_2, x_3]$, and $\vec{e}_1 \equiv [1, 0, 0]^T$, $\vec{e}_2 \equiv [0, 1, 0]^T$, $\vec{e}_3 \equiv [0, 0, 1]^T$, $\vec{e}_4 \equiv [0, -x_3, x_2]^T$, $\vec{e}_5 \equiv [x_3, 0, -x_1]^T$, $\vec{e}_6 \equiv [-x_2, x_1, 0]^T$. There exist $\mu_3 > 0$ and $\mu_4 > 0$ such that

$$(\varepsilon^2(v), 1)_0 \geq \mu_3 \|v\|_G^2 - \mu_4 |v|_0^2, \quad \forall v \in G, \tag{3.36}$$

(see [279, 418]); (3.16) holds for some $\mu_2 > 0$ if we replace G by $V \equiv \{v : v \in G \text{ and } \varphi_k(v) = 0, \ k \in [1, 6]\}$, where $\varphi_k(v) \in G^*$, $k \in [1, 6]$ are such that $V \cap \text{Ker } L_{el} = \emptyset$. For example, the most natural choice gives $V = (\text{Ker } L_{el})^\perp$ when $\varphi_k(v) = (\vec{e}_k, v)_G$, $k \in [1, 6]$; then, for $f \in V$ (that is, when $l(\vec{e}_k) = 0$, $k \in [1, 6]$), we can reduce our problem to a correct operator equation in the Hilbert space V (the restriction of L_{el} to this subspace is a positive definite operator).

We apply PGM of the same type as before (necessary specifications were given in Subsection 2.5), yielding systems (3.20) with $L_\Omega = L_\Omega^* \equiv L_h \geq 0$ in the Euclidean space H. We also define $B = B^* \geq 0$ as in Theorem 2 (see (3.35)) but now instead of general $B_Q \asymp \Lambda_\Omega$ we take more specific operators. For $d = 2$ we might take $B_Q = \Lambda_Q \geq 0$ (if Q is a rectangle then for systems with B_Q we may apply fast direct algorithms (see Theorems 0.3.4 and 3.1.1)). We will thus be able to make use of the relation

$$(B\mathbf{u}, \mathbf{v}) = (\hat{u}, \hat{v})_{1,Q}. \tag{3.37}$$

For $d = 3$ and a parallelepiped Q, we define $B_Q = \Lambda$ by (2.4.6), which has the structure of a difference operator and

$$(B\mathbf{u}, \mathbf{v}) = \frac{1}{4} \sum_{k=1}^{4} (\hat{u}^{[k]}, \hat{v}^{[k]})_{1,Q}. \tag{3.38}$$

Here, $\hat{u}^{[k]}$ denotes the corresponding element of the vector spline subspace $\hat{G}_h^{[k]}(Q) \subset (W_2^1(Q))^3$ associated with the regular triangulation $T_h^{[k]}(\bar{Q})$, $k = 1, 2, 3, 4$ (see Lemma 2.4.6)). For systems with B_Q, we may apply fast direct algorithms (see Theorem 3.1.1). Our analysis is based on application of Lemma 1.3.6; the operator $M \in \mathcal{L}^+(H)$ is defined by

$$(M\mathbf{u}, \mathbf{v}) \equiv (\hat{u}, \hat{v})_0. \tag{3.39}$$

We therefore need the representation Ker L_h = Ker $B \oplus Q_0$, where the three-dimensional subspaces Ker B and Q_0 are orthogonal in the sense of the Euclidean space $H(M)$ (as seen from (3.39), the corresponding spline subspaces must be orthogonal in the sense of $(L_2(\Omega))^3$). Since $\vec{e}_1, \vec{e}_2, \vec{e}_3$ comprise an orthogonal (in the sense of $(L_2(\Omega))^3$) basis for Ker B, as a basis for Q_0 we choose $\vec{e}_4', \vec{e}_5', \vec{e}_6'$, where $\vec{e}_k' \equiv \vec{e}_k' - c_{k,1}\vec{e}_1 - c_{k,2}\vec{e}_2 - c_{k,3}\vec{e}_3$, $k \in [4,6]$. We also agree to write $\vec{e}_k'$ instead of $\tilde{e}_k$ for $k \in [1,3]$.

If the vector $\mathbf{a}_r \in H$ consists of the coefficients of the expansion of $\vec{e}_k' \in \hat{G}$ with respect to the basis for the spline space $\hat{G}$ ($k \in [1,6]$), then Ker $B = \text{lin}\{\mathbf{a}_1, \mathbf{a}_2, \mathbf{a}_3\}$, and Ker $L_h = \text{Ker } B \oplus Q_0$, where $Q_0 \equiv \text{lin}\{\mathbf{a}_4, \mathbf{a}_5, \mathbf{a}_6\}$ (see Lemma 1.3.6). In accordance with the conditions of Lemma 1.3.6, we define the subspace

$$S_1 \equiv \{\mathbf{v} : (M\mathbf{v}, \mathbf{a}_k) = 0, \ (k \in [1,3]; \varphi_{k,h}(\mathbf{v}) = 0 \ (k \in [4,6], \}, \qquad (3.40)$$

where $\varphi_{k,h}(\mathbf{v}) \equiv (B\mathbf{v}, \mathbf{a}_k, v)$, $k \in [4,6]$.

Theorem 3. Let the topologically equivalent triangulations $T_h(\Omega)$ and $T_h(Q)$ be constructed by applying the refinement algorithm described in § 4.4 for simplexes $T_s \equiv T$ and $T_{Q,r} \equiv T_Q$ in initial triangulations of the original and model regions. Suppose that $h \le h_0$, where h_0 is small enough. Then there exist positive constants δ_0 and δ_1, independent of h, such that inequalities (0.3.9) hold for all $\mathbf{v} \in S_1$.

Proof. It suffices to prove existence of $\delta_{0,1} > 0$, independent of h, such that $q_h(\hat{v}) \equiv (\varepsilon^2(\hat{v}), 1)_0 \ge \delta_{0,1}\|\hat{v}\|_G^2$, $\forall \hat{v} \in \hat{G}$. Suppose this is not the case and consider sequences $\{h\}$ and $\{v_h\}$ such that $v_h \in S_h(B)$, $\|\hat{v}_h\|_G = 1$, and $\lim_{h\to 0} q_h(\hat{v}_h) = 0$. Then $\exists \ v \in G$ to which a subsequence $\{\hat{v}_h\}$ converges weakly in G and strongly in $(L_2(\Omega))^3$. From (3.36), it follows that $|v|_0 > 0$. We prove that $v \in \text{Ker } L_{el}$. Indeed, define the symmetric and nonnegative operator $L_{el,1} \in \mathcal{L}(G)$ by $(L_{el,1}\vec{v}, \vec{v}) = (\varepsilon^2(\vec{v}), 1)_0$, $\forall \vec{v} \in G$. Then, for fixed $\vec{w} \in G$, we have $(L_{el,1}\hat{v}_h, \vec{w}) \le (q_h(\hat{v}_h))^{1/2}(\varepsilon^2(\vec{w}), 1)_0^{1/2} \to 0$. Since $(L_{el,1}\vec{v}, \vec{w}) = \lim_{h\to 0}(L_{el,1}\hat{v}_h, \vec{w})$, then $(L_{el,1}\vec{v}, \vec{w}) = 0$, $v \in \text{Ker } L_{el,1} = \text{Ker } L_{el}$, and $\vec{v} = \sum_{k=1}^6 \alpha_k \vec{e}_k'$. Note that all $\hat{v}_h$ correspond to $\mathbf{v} \in S_1$ (see (3.40)). Then $(\vec{v}, \vec{e}_k')_0 = \lim_{h\to 0}(\hat{v}_h, \vec{e}_k')_0 = 0$ for $k = 1, 2, 3$ (see (3.40) and (3.39)). Thus, $\alpha_k = 0$ if $k \in [1,3]$, and $\vec{v}$ is determined by α_r, $r \in [4,6]$, which vanish as we show below. With this in mind, we analyze the remaining three conditions

$$(B\mathbf{v}, \mathbf{a}_r) = \frac{1}{4}\left[\sum_{k=1}^4 (\hat{v}^{[k]}, \hat{g}_{r,h}^{[k]})_{1,Q}\right], \qquad (3.41)$$

where $r \in [4,6]$ and $\hat{g}_{r,h}^{[k]} \in \hat{G}_h^{[k]}(Q) \subset (W_2^1(Q))^3$ is defined by the vector $\mathbf{a}_r$, $k \in [1,4]$, $r \in [4,6]$, and the same notation is used for $\hat{v}^{[k]}$, $k \in [1,4]$, and $\mathbf{v}$. Denote by $\vec{g}_r^Q$ the piecewise linear vector-functions (with respect to the initial triangulation of Q) that correspond (see (2.2.4)) to the linear (on Ω) functions $\vec{e}_r'$, $r \in [4,6]$. For each simplex T_Q in the initial triangulation of Q, let $T_Q^{[k]}$ denote its subset containing all elementary simplexes in the regular triangulation $T_h^{[k]}(\bar{Q})$, $k \in [1,4]$, that have no common points with the boundary of T_Q. It is then easy to see that $|T_Q \backslash T_Q^{[k]}| = O(h)$ and that on $T_Q^{[k]}$ the functions $\hat{g}_{r,h}^{[k]}$ with different indices coincide with the linear vector-function $\vec{g}_r^Q$, $r \in [4,6]$. Thus, $\hat{g}_{r,h}^{[k]} \to \vec{g}_r^Q$ as $h \to 0$, $r \in [4,6]$, $k \in [1,4]$, in the sense of $(W_2^1(Q))^3$. Without loss of generality, we may treat the bounded sequence $\{\hat{v}_h^{[k]}\}$ as weakly convergent (as $h \to 0$) in $(W_2^1(Q))^3$ and strongly convergent in $(L_2(Q))^3$ to functions $\vec{v}^{Q,[k]}$, $k \in [1,]4$. But it can be proved that $|\hat{v}_h^{[k]} - \hat{v}_h^{[k']}|_{0,Q}^2 \leq Kh^2$ for all admissible k and k'. Thus,

$$\lim_{h \to 0} \vec{v}^{Q,[k]} = \vec{v}^Q, \quad k \in [1,4], \tag{3.42}$$

which corresponds to the limit $\vec{v} \in G$. We thus have, $\vec{v}^Q = \sum_{k=4}^6 \alpha_k \vec{g}_k^Q$, and conditions (3.41) imply that $(\vec{v}^Q, \vec{g}_r^Q)_{1,Q} = 0$, $r \in [4,6]$, and that $|\vec{v}^Q|_{1,Q} = 0$. Since $\vec{g}_r^Q$ with $r \in [1,]6$ are linearly independent, we thus conclude that $\vec{v}^Q = 0$, which contradicts the fact $|\vec{v}|_0 > 0$. □ [22]

3.4. Domain symmetry. We emphasized in Subsection 2.1.7 that use of symmetry of the solution in asymptotically optimal algorithms can increase their practical effectiveness. The same holds for elasticity problems, but now some additional comments are necessary. We restrict ourselves to symmetry with respect to the plane $x_1 = 0$ and problems dealing the energy functional defined by (3.13), (3.15), and (3.12). More precisely, we assume that $d = 3$, that Ω and Γ_0 are symmetric with respect to the plane $x_1 = 0$, and that the vectors of the given forces satisfy either

---

[22] Theorem 3 can be generalized to include curvilinear boundaries (see [182]). Of special importance is the second example dealing with periodic conditions imposed with respect to all d variables. Such problems are essential for homogenization procedures for problems with periodically structured media (see [44, 61]). For simplicity, we consider briefly the case $d = 2$. We may thus take $\bar{\Omega} \equiv [0, l_1] \times [0, l_2]$. The Hilbert space G is the completion in the sense of $(W_2^1(\Omega))^2$ of vector-functions $\vec{v} \equiv [v_1, v_2]$ that are smooth and periodic with respect to x_r (with period $l_r, r = 1, 2$). It is not difficult to show that $I(v)$ from (3.15) can now be rewritten as $I(v) \equiv \mu|v|_{1,\Omega}^2 + (\lambda + \mu)|\text{div } v|_0^2$. For PGMs based on triangulations of type $\bar{T}(Q^0; N_1, N_2)$ (see § 4.3) and spaces of type $J^{(m)}$, we may effectively apply model operators B that are spectrally equivalent to $\Lambda \equiv \Lambda_h \equiv D(-\Delta_h; 2)$, where $\Delta_h \equiv h_1 h_2(\Delta_1 + \Delta_2)$ (see (0.2.15)). For $m = 1$, we even have estimates $\mu\Lambda_h \leq L_h \leq (\lambda + 2\mu)\Lambda_h$. Then Ker L_h = Ker Λ = Ker B (they correspond to constant vector-functions), and the conditions of Lemma 1.3.5 are satisfied. In the same way, we can consider problems connected with a combination of the second type and periodic conditions for one (if $d = 2$) or two (if $d = 3$) variables. Of course, the corresponding kernels become even simpler. In case of PGMs based on the use of rectangular grids and piecewise bilinear basis functions, we can use the same operators, but now $1/3\mu\Lambda_h \leq L_h \leq (\lambda + 2\mu)\Lambda_h$ (see Theorem 2.4.3).

$$\vec{F}(-x_1, x_2, x_3) = \vec{F}(x_1, x_2, x_3), \ \vec{F}_\Gamma(-x_1, x_2, x_3) = \vec{F}_\Gamma(x_1, x_2, x_3), \quad (3.43)$$

or

$$\vec{F}(-x_1, x_2, x_3) = -\vec{F}(x_1, x_2, x_3), \ \vec{F}_\Gamma(-x_1, x_2, x_3) = -\vec{F}_\Gamma(x_1, x_2, x_3).$$
$$(3.44)$$

For (3.43), we define the symmetry operator $S \in \mathcal{L}(G)$ by

$$S\vec{v}(x) \equiv [-v_1(-x_1, x_2, x_3), v_2(-x_1, x_2, x_3), v_3(-x_1, x_2, x_3)] \qquad (3.45)$$

and the corresponding symmetric subspace

$$G_{(s)} \equiv \{\vec{v} : \vec{v} \in G, S\vec{v} = \vec{v}\}, \qquad (3.46)$$

(see (2.1.28)) which is considered as a new Hilbert space (with old inner product). It is easy to verify that

$$\operatorname{div}(S\vec{v}) = \operatorname{div}\vec{v}, \ \varepsilon_{s,s}(S\vec{v}) = \varepsilon_{s,s}(\vec{v}), \ s = 1, 2, 3,$$

$$\varepsilon_{1,2}(S\vec{v}) = -\varepsilon_{1,2}(\vec{v}), \quad \varepsilon_{1,3}(S\vec{v}) = -\varepsilon_{1,3}(\vec{v}), \quad \varepsilon_{2,3}(S\vec{v}) = \varepsilon_{2,3}(\vec{v})$$

(see (3.9)). Hence, $\Phi(S(\vec{v})) = \Phi(\vec{v})$ and

$$\vec{u} \in G_{(s)} \qquad (3.47)$$

(see (2.1.29) and Theorem 2.1.1), where $\vec{u}$ is the unique solution of the problem under consideration. For (3.44), we define
$S\vec{u}(x) \equiv [v_1(-x_1, x_2, x_3), -v_2(-x_1, x_2, x_3), -v_3(-x_1, x_2, x_3)]$. Now
$\operatorname{div}(S\vec{v}) = -\operatorname{div}\vec{v}, \ \varepsilon_{s,s}(S\vec{v}) = -\varepsilon_{s,s}(\vec{v}), \ s \in [1, 3], \ \varepsilon_{1,2}(S\vec{v}) = \varepsilon_{1,2}(\vec{v}),$
$\varepsilon_{1,3}(S\vec{v}) = \varepsilon_{1,3}(\vec{v}), \ \varepsilon_{2,3}(S\vec{v}) = -\varepsilon_{2,3}(\vec{v})$, and we again obtain (3.47). [23]

---

[23] Both symmetries imply that the components of $\vec{u}$ are either odd or even functions with respect to x_1. If we use PGMs associated with grids that are symmetric with respect to our plane x_1, then for the corresponding solutions $\hat{u}$ we obtain the same symmetry property, and we can deal directly with the subspaces $\hat{G}_{(s),h}$ of functions in $\hat{G}_h$ satisfying condition (3.47). So, to obtain the desired asymptotically optimal algorithms, we must choose model operators B_Q such that solutions of the corresponding systems are odd or even grid functions with respect to x_1, where the right-hand sides of the systems are of the same type. This is the case for almost all operators constructed in Chapter 3 if, in their construction, we preserve the basic symmetry property (e.g., cuttings surfaces in § 3.5 or regions Π in § 3.6 must be symmetric with respect to the plane x_1). Clearly, then, if we take an initial iterate with the desired symmetry property, it will be retained through all iterations and they need be carried out only in the corresponding subspace. Coarse grid continuation can be also implemented for our symmetric spline subspaces in a straightforward manner. In the same fashion, we can deal with symmetries with respect to several coordinate planes (in [159], such an approach was used for shell problems).

3.5. Cylindrical coordinates. Consider now a region Ω that is obtained by rotation of a two-dimensional region Ω_2 around an axis (say, the z axis; we assume it belongs to the plane containing Ω_2; see, e.g., Figure 0.2.3, with the horizontal line corresponding to the axis of the rotation). Then we can use cylindrical coordinates $[z, r, \varphi] \equiv [x_1, x_2, x_3] \equiv x$, where $[r, \varphi]$ are polar coordinates in the plane of Ω_2. If we confine ourselves to regions such that

$$0 < r_0 \le r \le r_1, \ \forall x \in \bar{\Omega}, \tag{3.48}$$

then we can easily convert the original elasticity problems to variational ones that are equivalent to operator equations in a new Hilbert space G of vector-functions of the vector variable x. Also, in constructing PGMs, we may use grids of special type, including composite ones as we did before. For example, practical problems dealing with composites and the above mentioned regions were solved with success on prism grids (in the space of x) of type $\bar{\mathcal{P}}(\Box; N_1, N_2) \times \omega_3$ (see § 4.3) based on the use of simple model operators that are spectrally equivalent to the original grid operator (we give the necessary specifications below).

Consider the symmetric elasticity problem, where one of three displacement functions (with respect to x_3) vanishes. We can thus formulate the problem in terms of functions $u \equiv u_1$ and $w \equiv u_2$ that describe displacements in the x_1 and x_2 directions, respectively. Letting $G \equiv (W_2^1(\Omega_2; \Gamma_0))^2$, $\vec{u} \equiv [u_1, u_2]$, and $\varepsilon_s(\vec{u}) \equiv \varepsilon_s, s \in [1, 4]$, we then define $\varepsilon_1 \equiv D_1 u$, $\varepsilon_2 \equiv D_2 w$, $\varepsilon_3 \equiv w/x_2$, and $\varepsilon_4 \equiv (D_2 u + D_1 w)/2$. Elastic properties of the concrete media under consideration are defined by a matrix $A = A^T \equiv [\alpha_{i,j}] \in \mathbf{R}^{3 \times 3}$, with sp $A \subset [\lambda_0, \lambda_1], \lambda_0 > 0$, and a constant $\gamma_4 > 0$, which yield the quadratic functional (see (3.13))

$$I_{\Omega_2}(\vec{u}) \equiv \sum_{i,k=1}^{3} (\alpha_{i,k} x_2, \varepsilon_i \varepsilon_k)_{0,\Omega_2} + (\gamma_4 x_2, \varepsilon_4^2)_{0,\Omega_2}. \tag{3.49}$$

Theorem 4. There exist $\mu_3 \equiv \mu_3(\Omega_2, \lambda_0, \gamma_4, r_0, r_1) > 0$ *and* $\mu_4 \equiv \mu_4(\Omega_2, \lambda_1, \gamma_4, r_0, r_1) > 0$ *such that*

$$\mu_3 \left(|\vec{u}|^2_{1,\Omega_2} + |u_2|^2_{0,\Omega_2} \right) \le I_{\Omega_2}(\vec{u}) \le \mu_4 (|\vec{u}|^2_{0,\Omega_2} + |u_2|^2_{0,\Omega_2}), \ \forall \vec{u} \in G. \tag{3.50}$$

Proof. From (3.49) and our assumptions, it follows that
$I_{\Omega_2}(\vec{u}) \ge (r, \lambda_0[(D_1 u)^2 + (D_2 w)^2 + w^2/r^2])_{0,\Omega_2} + \gamma_4(r, (D_2 u + D_1 w)^2)_{0,\Omega_2}$
$\ge r_0 \lambda_0 [|D_1 u|^2_{0,\Omega_2} + |D_2 w|^2_{0,\Omega_2}] + r_0 \gamma_4 |D_2 u + D_1 w|^2_{0,\Omega_2} + \lambda_0/r_1 |w|^2_{0,\Omega_2}$. We now apply the Korn-Friedrichs inequality (see (3.16)) and write
$I_{\Omega_2}(\vec{u}) \ge r_0 \mu_2 \min\{\lambda_0; \gamma_4\} |\vec{u}|^2_{1,\Omega_2} + \frac{\lambda_0}{r_1} |w|^2_{0,\Omega_2}$. Hence, we may take

$\mu_3 \equiv \min\{\mu_2 \min\{\lambda_0; \gamma_4\}; \lambda_0/r_1\}$. In the same manner, we obtain
$I_{\Omega_2}(\vec{u}) \leq (r, \lambda_1[(D_1 u)^2 + (D_2 w)^2 + w^2 r^{-2}])_{0,\Omega_2} + \gamma_4(r, (D_2 u + D_1 w)^2)_{0,\Omega_2}$
$\leq r_1 \lambda_1 [|D_1 u|^2_{0,\Omega_2} + |D_2 w|^2_{0,\Omega_2}] + \frac{\lambda_1}{r_0} |w|^2_{0,\Omega_2} + 2\gamma_4 r_1 (|D_2 u|^2 + |D_1 w|^2_{0,\Omega_2})$. Thus,
$\mu_4 = \max\{r_1 \max\{\lambda_1; 2\gamma_4\}; \lambda_1/r_0\}$. $\square$ [24]

Now define the symmetric grid operators in $H \equiv H^2_{(1)}$ by

$$(L_h \mathbf{u}, \mathbf{u}) = I_{\Omega_2}(\vec{u}), \quad (\Lambda_h \mathbf{u}, \mathbf{u}) = |\hat{u}|^2_{1,\Omega_2} + |\hat{w}|^2_{1,\Omega_2}, \quad \forall \vec{u} \in \hat{G}, \qquad (3.51)$$

where $\vec{u} \equiv [\hat{u}_1, \hat{u}_2] \equiv [\hat{u}, \hat{w}] \in \hat{G}$ corresponds to to $\mathbf{u} \equiv [\mathbf{u}_1, \mathbf{u}_2] \in H$.

Theorem 5. If $|\Gamma_0|_{(1)} > 0$, then $L_h \asymp \Lambda_h$.

Proof. It suffices to combine (3.51), Theorem 4, and an inequality of type (0.1.9). $\square$ [25]

It makes sense, instead of Λ_h, to use the operator $\Lambda_{d,h}$ obtained by replacing the block $\Lambda_{2,h}$ by $\Lambda_{d,h} \equiv \Lambda_{2,h} + d I_2$, where $d > 0$ is a constant. Then it is easy to verify our next theorem.

Theorem 6. Let $G \equiv (W^1_2(\Omega_2; \Gamma_0))^2$. Then $L_h \asymp \Lambda_{d,h}$. [26]

§ 4. Multigrid construction of asymptotically optimal preconditioners for two-dimensional elasticity problems

4.1. Original splitting of the finite element space. [27] Let

$$\vec{V} \equiv (W^1_2(Q; \Gamma_0))^2, \qquad (4.1)$$

---

[24] Theorem 4 implies that $I_{\Omega_2}(\vec{u}) = 0 \Leftrightarrow \vec{u} = [c_1, 0]$, where c_1 is a constant. Hence, if we consider the Hilbert space $G \equiv (W^1_2(\Omega_2; \Gamma_0))^2$, with $\Gamma_0 \subset \partial\Omega_2$, and define the symmetric operator L by $(L\vec{u}, \vec{u}) = I_{\Omega_2}(\vec{u})$, $\forall \vec{u} \in G$, then either Ker $L = 0$ (if $|\Gamma_0|_{(1)} > 0$) or dim Ker $L = 1$ and its basis is $[1^*, 0]$, where 1^* denotes the function of value 1 at all points. Hence, we may use the subspaces $\hat{G}$ and PGMs of the same type as before (e.g., we always assume that Γ_0 is approximated by $\hat{\Gamma}_0 \subset \bar{\Omega}_2$).

[25] The block diagonal operator $\Lambda_h \equiv \Lambda_{\Omega,h}$ with the diagonal blocks $\Lambda_{1,h}$ and $\Lambda_{2,h}$ may be replaced by a model operator $B \asymp \Lambda_h$ of the same structure.

[26] This theorem applies to the case where $\Gamma_0 = \emptyset$ and $L \geq 0$ as well. Estimates (3.50) can be improved in two ways. One is connected with large γ_4 if integration by parts as in Lemma 1 is applicable. Then, in μ_4, it is possible to replace $2\gamma_4$ by $q\gamma_4$ for some $q \in (1, 2)$. The second and more radical improvement for regions with large r_1/r_0 (see (3.48)) is connected with the use of a model operator $B \asymp \Lambda_h$ such that, for all $\mathbf{u} \in H$, $(\Lambda_h \mathbf{u}, \mathbf{u}) = \sum_{s=1}^{p} r^*_s |\hat{u}|^2_{1,\Omega_{2,s}}$, where $\bar{\Omega}_{2,s}, s \in [1, p]$, define a partition of $\bar{\Omega}$ and r^*_s refers to an averaged value of r over $\bar{\Omega}_{2,s}, s \in [1, p]$ (see, e.g., the operators B in § 3.7, where the constants defining the spectral equivalence of Λ_h and B were close to 1).

[27] We consider here only the multigrid construction of asymptotically optimal preconditioners that were suggested in [204] for grid approximations of basic boundary value problems in the theory of elasticity. For relevant results see [76, 289, 291, 367] and references therein. We restrict ourselves to the consideration of two-dimensional elasticity problems in regions Q with boundaries specified below.

$$\vec{u} \equiv [u^{(1)}, u^{(2)}] \in \vec{V}, \tag{4.2}$$

$$b_Q(\vec{u}; \vec{v}) \equiv \bar{\beta}[(u_{x_1}^{(1)}, v_{x_1}^{(1)})_{0,Q} + (u_{x_2}^{(2)}, v_{x_2}^{(2)})_{0,Q}]$$

$$+\beta[(u_{x_2}^{(1)}, v_{x_2}^{(1)})_{0,Q} + (u_{x_1}^{(2)}, v_{x_1}^{(2)})_{0,Q}]+$$

$$\beta[(u_{x_2}^{(1)}, v_{x_1}^{(2)})_{0,Q} + (u_{x_1}^{(2)}, v_{x_2}^{(1)})_{0,Q} + (u_{x_1}^{(1)}, v_{x_2}^{(2)})_{0,Q} + (u_{x_2}^{(2)}, v_{x_1}^{(1)})_{0,Q}], \tag{4.3}$$

where $\beta \equiv \mu/\lambda$, $\mu > 0$ and $\lambda > 0$ are the Lame parameters (see § 3) and $\bar{\beta} \equiv 1 + 2\beta$. Introducing the symmetric matrix

$$A_4 \equiv [a_{i,j}], \ i \in [1,4], j \in [1,4], \tag{4.4}$$

with the nonzero elements $a_{1,1} = \bar{\beta} = a_{4,4}$, $a_{2,2} = \beta = a_{3,3} = a_{2,3} = a_{3,2}$, $a_{1,4} = 1 = a_{4,1}$, and the vectors $X(x) \equiv [u_{x_1}^{(1)}(x), u_{x_2}^{(1)}(x), u_{x_1}^{(2)}(x), u_{x_2}^{(2)}(x)]^T$ and $Y(x) \equiv [v_{x_1}^{(1)}(x), v_{x_2}^{(1)}(x), v_{x_1}^{(2)}(x), v_{x_2}^{(2)}(x)]^T$. Then, $b_Q(\vec{u}; \vec{v}) = ((A_4 X(x), Y(x))_{\mathbf{R}^4}, 1)_{0,Q}$ and $b_Q(\vec{v}; \vec{v}) = \lambda I(\vec{v})$ (see (3.15)). [28]

In accordance with (3.16), we may also use the energy Hilbert space $\vec{G}$ with the inner product

$$(\vec{u}; \vec{v}) \equiv b_Q(\vec{u}; \vec{v}). \tag{4.5}$$

We use the same nested generalized triangulations $T^{(0)}, \ldots, T^{(p)}$ as in Subsection 3.7.3, with the difference now that we consider only a refinement ratio 2 and deal with isosceles rectangular triangles. With each triangulation $T^{(l)}$, we also associate the same spline subspace $\hat{G}^{(l)} \subset G = W_2^1(Q; \Gamma_0)$ as in § 3.7, the set $Q^{(l)}$, and the set of the standard basis nodal functions $\hat{\psi}_i^{(l)}(x)$. Along with the basis $\{\hat{\psi}_i^{(l+1)}(x)\}$ for $\hat{G}^{(l+1)}$, $l \in [0, p-1]$, we consider the hierarchical basis (see § 3.7), leading to splitting (3.7.22). Along with this splitting for $\hat{G}^{(l+1)}$, we consider

$$\vec{G}^{(l+1)} = \vec{G}_1^{(l+1)} \oplus \vec{G}_2^{(l+1)} \subset \vec{V}, \quad l \in [0, p-1], \tag{4.6}$$

where the components of the vector-functions

$$\vec{u}^{(l+1)} \equiv [\hat{u}^{(1,l+1)}, \hat{u}^{(2,l+1)}] \in \vec{G}^{(l+1)} \tag{4.7}$$

---

[28] A_4 is a nonnegative matrix of the rank 3. In what follows, we suppose that the boundary of Q consists of several line segments on straight lines that form angles $0, \pi/4, \pi/2$ with the horizontal coordinate line, so we can make use of triangulations consisting of isosceles rectangular triangles and obtain necessary estimates like those in § 2.5. But we do not assume that ∂Q is Lipschitz, and we allow ts part Γ_0 to include whole or partial double slits (see § 2.1).

belong to the spaces $\hat{G}_1^{(l+1)}$ and $\hat{G}_2^{(l+1)}$, respectively. We emphasize that $\vec{G}_2^{(l+1)} = \vec{G}^{(l)}$ and that the components of $\vec{u}^{(l+1)} \in \vec{G}_1^{(l+1)}$ vanish at the vertices of triangles $T_k \in T^{(l)}$ (see Figure 1 below). We note also that the Gram matrices for the two indicated bases for the space $\vec{G}_1^{(l+1)}$ take the standard block form

$$L^{(l+1)} \equiv \begin{bmatrix} L_{1,1}^{(l+1)} & L_{1,2}^{(l+1)} \\ L_{2,1}^{(l+1)} & L_{2,2}^{(l+1)} \end{bmatrix}, \quad \bar{L}^{(l+1)} \equiv \begin{bmatrix} \bar{L}_{1,1}^{(l+1)} & \bar{L}_{1,2}^{(l+1)} \\ \bar{L}_{2,1}^{(l+1)} & L_{2,2}^{(l)} \end{bmatrix}. \quad (4.8)$$

4.2. Estimates for the angle between the subspaces.

Lemma 1. For an arbitrary $T_k \in T^{(l)}$ with $|T_k| \equiv 2h_{l+1,k}^2$ and $\vec{u} \equiv [\hat{u}^{(1,l+1)}, \hat{u}^{(2,l+1)}] \in \vec{G}_1^{(l+1)}$ with $l \in [0, p-1]$, we have

$$b_{T_k}(\vec{u}; \vec{u}) \geq \frac{2}{3} h_{l+1}^2 (A_4 Y, Y)_{\mathbf{R}^4}, \quad (4.9)$$

where the vector

$$Y \equiv Y^k \equiv [d_1^{(1)}, d_2^{(1)}, d_1^{(2)}, d_2^{(2)}]^T \in \mathbf{R}^4 \quad (4.10)$$

refers to the differences of functions $\hat{u}^{(s,l+1)} \in \hat{G}_1^{(l+1)}$, $s = 1, 2$ along any leg of T_k.

Proof. Without loss of generality, we consider the triangle T_k depicted in Figure 1.

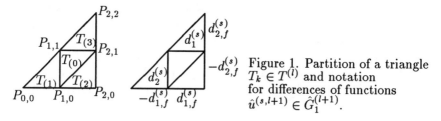

Figure 1. Partition of a triangle $T_k \in T^{(l)}$ and notation for differences of functions $\hat{u}^{(s,l+1)} \in \hat{G}_1^{(l+1)}$.

We have $|P_{0,0}P_{2,0}| = |P_{2,0}P_{2,2}| = h_{l,k} = 2h_{l+1,k}2$. Besides, $d_{1,f}^{(s)} \equiv -1/h_{l+1,k}\hat{u}^{(s,l+1)}(P_{1,0})$, $d_{2,f}^{(s)} \equiv -1/h_{l+1,k}\hat{u}^{(s,l+1)}(P_{2,1})$, $s = 1, 2$,

$$b_{T_k}(\vec{u}; \vec{u}) - I_0 = \sum_{i=1}^{3} I_i, \quad (4.11)$$

$$I_i \equiv b_{T_{(i)}}(\vec{u}; \vec{u}), \quad i \in [0, 3], \quad (4.12)$$

and $T_{(i)}, i \in [0, 3]$, refer to the triangles in $T^{(l+1)}$ belonging to T_k (see Figure 1). Inside each $T_{(i)}, i \in [0, 3]$, derivatives of the functions $\hat{u}^{(1,l+1)}$

and $\hat{u}^{(2,l+1)}$ are constants, for which we use the notation indicated in Figure 1 (though there are 8 such constants, only 6 of them correspond to degrees of freedom). In the sequel, we drop index k. Now $I_0 = h_{l+1}^2/2\{\bar{\beta}[(d_1^{(1)})^2 +$ $(d_2^{(2)})^2] + \beta[(d_2^{(1)})^2 + (d_1^{(2)})^2 + d_2^{(2)}d_2^{(1)} + 2d_1^{(2)}d_2^{(1)}] + 2d_1^{(1)}d_2^{(2)}\}$, which implies that $b_{T_0}(\vec{u}_1;\vec{u}_1) = h_{l+1}^2/2(A_4Y^k, Y^k)$,

$I_1 = h_{l+1}^2/2\{\bar{\beta}[(d_{1,f}^{(1)})^2 + (d_2^{(2)})^2] + \beta[(d_2^{(1)})^2 + (d_{1,f}^{(2)})^2 - 2d_{1,f}^{(2)}d_2^{(1)}] - 2d_{1,f}^{(1)}d_2^{(2)}\}$,

$I_2 = h_{l+1}^2/2\{\bar{\beta}[(d_{1,f}^{(1)})^2 + (d_2^{(2)})^2] + \beta[(d_{2,f}^{(1)})^2 + (d_{1,f}^{(2)})^2 - 2d_{1,f}^{(2)}d_{2,f}^{(1)}] - 2d_{1,f}^{(1)}d_{2,f}^{(2)}\}$,

$I_3 = h_{l+1}^2/2\{\bar{\beta}[(d_1^{(1)})^2 + (d_{2,f}^{(2)})^2] + \beta[(d_{2,f}^{(1)})^2 + (d_1^{(2)})^2 + 2d_1^{(2)}d_{2,f}^{(1)}] + 2d_1^{(1)}d_{2,f}^{(2)}\}$.

Hence, (4.11) and (4.12) yield the formula

$$b_{T_k}(\vec{u};\vec{u}) - I_0 = \frac{h_{l+1}^2}{2}\left\{\bar{\beta}[(d_1^{(1)})^2 + (d_2^{(2)})^2 + 2(d_{1,f}^{(1)})^2 + 2(d_{2,f}^{(2)})^2]\right.$$

$$+\beta[(d_2^{(1)})^2 + (d_1^{(2)})^2 + 2(d_{2,f}^{(1)})^2 + 2(d_{1,f}^{(2)})^2] + 2\beta[d_1^{(2)}d_{2,f}^{(1)}$$

$$\left. - d_2^{(1)}d_{1,f}^{(2)} - d_{2,f}^{(1)}d_{1,f}^{(2)}] + 2[d_1^{(1)}d_{2,f}^{(2)} - -d_{1,f}^{(1)}d_2^{(2)} - d_{1,f}^{(1)}d_{2,f}^{(2)}]\right\}. \qquad (4.13)$$

We now denote our 8 constants in accordance with Figure 2.

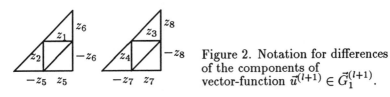

Figure 2. Notation for differences of the components of vector-function $\vec{u}^{(l+1)} \in \vec{G}_1^{(l+1)}$.

In other words, we make use of the vector

$$Z \equiv [z_1, z_2, z_3, z_4, z_5, z_6, z_7, z_8]^T \in \mathbf{R}^8, \qquad (4.14)$$

where $z_1 \equiv y_1$, $z_2 \equiv y_2$, $z_3 \equiv y_3$, $z_4 \equiv y_4$, $z_5 \equiv d_{1,f}^{(1)}$, $z_6 \equiv d_{2,f}^{(1)}$, $z_7 \equiv d_{1,f}^{(2)}$, and $z_8 \equiv d_{2,f}^{(2)}$. Then, introducing the matrix

$$A_8 \equiv \begin{bmatrix} \bar{\beta} & 0 & 0 & 0 & 0 & 0 & 0 & 1 \\ 0 & \beta & 0 & 0 & 0 & 0 & -\beta & 0 \\ 0 & 0 & \beta & 0 & 0 & \beta & 0 & 0 \\ 0 & 0 & 0 & \bar{\beta} & -1 & 0 & 0 & 0 \\ 0 & 0 & 0 & -1 & 2\bar{\beta} & 0 & 0 & -1 \\ 0 & 0 & \beta & 0 & 0 & 2\beta & -\beta & 0 \\ 0 & -\beta & 0 & 0 & 0 & -\beta & 2\beta & 0 \\ 1 & 0 & 0 & 0 & -1 & 0 & 0 & 2\bar{\beta} \end{bmatrix}, \qquad (4.15)$$

we see that

$$b_{T_k}(\vec{u}; \vec{u}) = \frac{h_{l+1}^2}{2} \left((A_4 Y^k, Y^k) + (A_8 Z, Z) \right). \tag{4.16}$$

To obtain (4.9), we need to estimate $(A_8 Z, Z)$ from below in terms of $(A_4 Y, Y)$ (see (4.10) and (4.14)). First, we estimate $(A_8 Z, Z)$ in accordance with Lemma 1.5.1 and the block representation

$$A_8 = \begin{bmatrix} A_4' & C \\ C^T & B_4 \end{bmatrix}, \quad B_4 \in \mathbf{R}^{4 \times 4}.$$

This leads to

$$(A_8 Z, Z) \geq ((A_4' - C B_4^{-1} C^T) Y, Y). \tag{4.17}$$

To obtain the Schur complement $A_4' - C B_4^{-1} C^T \equiv D_4$, we simply eliminate unknowns z_5, z_6, z_7, and z_8 in the system $A_8 Z = 0$. To this end, observe that $(A_8 Z, Z) = F_1(z_2, z_3, z_6, z_7) + F_1(z_1, z_4, z_5, z_8)$, where the constituent simpler quadratic forms are associated with the respective matrices

$$E_4 \equiv \begin{bmatrix} \beta & 0 & 0 & -\beta \\ 0 & \beta & \beta & 0 \\ 0 & \beta & 2\beta & -\beta \\ -\beta & 0 & -\beta & 2\beta \end{bmatrix}, \quad E_4' \equiv \begin{bmatrix} \bar\beta & 0 & 0 & 1 \\ 0 & \bar\beta & -1 & 0 \\ 0 & -1 & 2\bar\beta & -1 \\ 1 & 0 & -1 & 2\bar\beta \end{bmatrix}.$$

Thus, we can carry out the necessary elimination for the matrices E_4 and E_4' separately, making use of their block forms

$$E_4 \equiv \begin{bmatrix} E_{1,1} & E_{1,2} \\ E_{1,2}^T & E_{2,2} \end{bmatrix}, \quad E_4' \equiv \begin{bmatrix} E_{1,1}' & E_{1,2}' \\ (E_{1,2}')^T & E_{2,2}' \end{bmatrix},$$

where

$$E_{1,1}' \equiv \begin{bmatrix} \bar\beta & 0 \\ 0 & \bar\beta \end{bmatrix}, \quad E_{1,2}' \equiv \begin{bmatrix} 0 & -1 \\ -1 & 0 \end{bmatrix}, \quad E_{2,2}' \equiv \begin{bmatrix} 2\bar\beta & -1 \\ -1 & 2\bar\beta \end{bmatrix}.$$

From the sixth and seventh equations in the system $A_8 Z = 0$ (corresponding to the third and fourth rows in E_4), we have $z_6 = 1/3(z_2 - 2z_3)$ and $z_7 = 1/3(2z_2 - z_3)$. Substitution of these expressions in the second and third equations in the system shows that

$$E_{1,1} - E_{1,2} E_{2,2}^{-1} E_{2,1} = \frac{\beta}{3} \begin{bmatrix} 1 & 1 \\ 1 & 1 \end{bmatrix}. \tag{4.18}$$

From the fifth and eighth equations in the system $A_8 Z = 0$ (corresponding to the third and fourth rows in E'_4), we obtain $z_5 = \frac{1}{4\bar{\beta}^2-1}(-z_1 + 2\bar{\beta}z_4)$, $z_8 = \frac{1}{4\bar{\beta}^2-1}(-2\bar{\beta}z_1 + z_4)$, which yields

$$E'_{1,1} - E'_{1,2}(E'_{2,2})^{-1}E'_{2,1} = \begin{bmatrix} \bar{\beta}(1-2p) & p \\ p & 2\bar{\beta}(1-2p) \end{bmatrix}, \qquad (4.19)$$

where $p \equiv (4\bar{\beta}^2 - 1)^{-1} < 1/3$. Hence, (4.18) and (4.19) imply that

$$A'_4 - CB_4^{-1}C^T \equiv D_4 = \begin{bmatrix} \bar{\beta}(1-2p) & 0 & 0 & p \\ 0 & 1/3\beta & 1/3\beta & 0 \\ 0 & 1/3\beta & 1/3\beta & 0 \\ p & 0 & 0 & \bar{\beta}(1-2p) \end{bmatrix}. \qquad (4.20)$$

Thus, we have $(D_4 Y, Y)_{\mathbf{R}^4} - 1/3(A_4 Y, Y)_{\mathbf{R}^4} = F(y_2, y_3)$, where the simpler quadratic form $F(y_2, y_3)$ is associated with the matrix

$$\begin{bmatrix} \bar{\beta}(1-2p-1/3) & p-1/3 \\ p-1/3 & 2\bar{\beta}(1-2p-1/3) \end{bmatrix}$$

and is positive because $0 < p < 1/3$. Hence, $D_4 \geq \frac{1}{3}A_4$, and

$$(A_8 Z, Z) \geq \frac{1}{3}(A_4 Y, Y). \qquad (4.21)$$

It is easy to see that (4.9) follows from (4.16) and (4.21). $\square$

Theorem 1. *The angle α between the subspaces $\vec{G}_2^{(l+1)} = \vec{G}^{(l)}$ and $\vec{G}_1^{(l+1)}$ is bounded below by $\pi/6$, $l \in [0, p-1]$.*

Proof. Let $T_k \in T^{(l)}$, $\vec{v}_2 \equiv [\hat{v}^{(1,l)}, \hat{v}^{(2,l)}] \in \vec{G}^{(l)}$, $\vec{u}_1 \equiv [\hat{u}_1^{(1,l+1)}, \hat{u}_1^{(2,l+1)}] \in \vec{G}_1^{(l+1)}$, and $b_{T_k}(\vec{v}_2; \vec{u}_1) \equiv W^k$. Then, due to Lemma 2.5.1 (see (2.5.5)), it suffices to show that

$$|W^k| \leq \frac{3^{1/2}}{2}\|\vec{v}_2\|\|\vec{u}_1\|. \qquad (4.22)$$

Observe that

$$W^k = \sum_{i=0}^{3} b_{T_{(i)}}(\vec{v}_2; \vec{u}_1) \qquad (4.23)$$

(see Figure 1). Define the constants $D_r^{(s)} \equiv \frac{\partial \hat{v}^{s,l}}{\partial x_r}$ and $d_r^{(s)}, d_{r,f}^{(s)}$, $r = 1, 2$, $s = 1, 2$, as in Lemma 1 (see Figure 1). Then we can represent $b_{T_{(i)}}(\vec{v}_2; \vec{u}_1)$ on the right-hand side of (4.23) in much the same form as was used in the proof of Lemma 1 for $I_i, 1 \in [0,3]$. But it is important to observe that we need define only the terms in $b_{T_{(i)}}(\vec{v}_2; \vec{u}_1)$ containing the differences $d_r^{(s)}$, because the terms with $d_{r,f}^{(s)}$ occur in pairs in our triangles $T_{(i)}$ with opposite signs, so they cancel. This implies that

$$W^k = h_{l+1,k}^2 \left\{ \bar{\beta} \left(D_1^{(1)} d_1^{(1)} + D_2^{(1)} d_2^{(2)} \right) + \beta \left(D_2^{(1)} d_2^{(1)} + D_1^{(2)} d_1^{(2)} \right) + \right.$$

$$\left. \beta \left(D_1^{(2)} d_2^{(1)} + D_2^{(1)} d_1^{(2)} \right) + D_2^{(2)} d_1^{(1)} + D_1^{(1)} d_2^{(2)} \right\}. \qquad (4.24)$$

Moreover, (4.24) yields the important formula

$$b_{T_k}(\vec{v}_2; \vec{u}_1) = h_{l+1,k}^2 (A_4 X^k, Y^k)_{\mathbf{R}^4}, \qquad (4.25)$$

with $X^k \equiv X \equiv [D_1^{(1)}, D_2^{(1)}, D_1^{(2)}, D_2^{(2)}]^T$, $Y^k \equiv Y \equiv [d_1^{(1)}, d_2^{(1)}, d_1^{(2)}, d_2^{(2)}]^T$, and the matrix A_4 defined by (4.4). From (4.25), it follows that

$$h_{l+1,k}^2 |(A_4 X, Y)| \le \left(h_{l+1,k}^2 (A_4 X, X) \right)^{1/2} \left(h_{l+1,k}^2 (A_4 Y, Y) \right)^{1/2}, \qquad (4.26)$$

$$\left(h_{l+1,k}^2 (A_4 Y, Y) \right)^{1/2} \le \frac{3^{1/2}}{2^{1/2}} \|\vec{u}_1\|, \qquad (4.27)$$

and

$$\|\vec{v}_2\|^2 = 2 h_{l+1,k}^2 (A_4 X, X). \qquad (4.28)$$

Therefore, (4.25)–(4.28) yield (4.22). □

4.3. Multigrid construction of spectrally equivalent operators.
Now we return to the block matrices $L^{(l+1)}$ and $\bar{L}^{(l+1)}$ (see (4.8)) regarded as operators in $\mathcal{L}(\vec{H}^{(l+1)})$, where the Euclidean space $\vec{H}^{(l+1)} = \vec{H}_1^{(l+1)} \times \vec{H}^{(l)}$. We emphasize that each of these spaces is actually a Descartes square of the Euclidean spaces associated with grid functions of type $\hat{u}^{(l+1,s)}$, $s = 1, 2$; e.g., $\vec{H}_1^{(l+1)} = H_1^{(l+1)} \times H^{(l)}$. In accordance with the theory given in § 3.7, we can approximate the block $\bar{L}_{1,1}^{(l+1)} = L_{1,1}^{(l+1)}$ on the basis of the following lemma.

Lemma 2. *There exists a diagonal matrix $A_{1,1}^{(l+1)} \in \mathcal{L}(\vec{H}_1^{(l+1)})$ and constants $\sigma_{0,1} > 0$ and $\sigma_{1,1} > 0$, independent of l, such that*

$$\sigma_{0,1} A_{1,1}^{(l+1)} \le L_{1,1}^{(l+1)} \le \sigma_{1,1} A_{1,1}^{(l+1)}, \quad l+1 \in [1,p]. \qquad (4.29)$$

Proof. Inequalities (4.29) and (3.7.26) are closely related, and their proofs are very similar. For $\mathbf{u}_{l+1} \in \vec{H}_1^{(l+1)}$ and corresponding $\vec{u}_{l+1} \in \vec{G}_1^{(l+1)}$, we have

$$(L_{1,1}^{(l+1)}\mathbf{u}_{l+1}, \mathbf{u}_{l+1}) = \sum_{T_k \in T^{(l)}} b_{T_k}(\vec{u}; \vec{u}). \tag{4.30}$$

We express $\vec{u}$ and and the differences used in the proof of Lemma 1 as functions of six parameters, which are just the values of the functions $\hat{u}^{(s,l+1)}$ at the vertices of the triangle $T_{(0)}$ (we denote the corresponding vector by $U \equiv [u_1, \ldots, u_6]^T \in \mathbf{R}^6$). Then we may write

$$b_{T_k}(\vec{u}; \vec{u}) = h_{l+1}^2 (A_6 U, U)_{\mathbf{R}^6}, \tag{4.31}$$

where A_6 is a symmetric matrix dependent only on β and positive if the quadratic form $b_{T_k}(\vec{u}; \vec{u})$ is positive. To prove positiveness of b_{T_k}, we assume that $b_{T_k}(\vec{u}; \vec{u}) = 0$. Then, on each triangle $T_{(i)} \subset T_k$, we have $D_1 \hat{u}^{l+1,1} = 0$, $D_2 \hat{u}^{l+1,2} = 0$, and $D_2 \hat{u}^{l+1,1} + D_1 \hat{u}^{l+1,2} = 0$. The first two of these relations are satisfied only when all values of $\hat{u}^{(l+1,1)}$ and $\hat{u}^{(l+1,2)}$ at the vertices of the triangles $T_{(i)}$ (see Figure 1 above), vanish with the possible exception $\hat{u}^{(l+1,1)}(P_{1,1}) = \hat{u}^{(l+1,1)}(P_{2,1}) \equiv a$ and $\hat{u}^{(l+1,2)}(P_{1,0}) = \hat{u}^{(l+1,2)}(P_{1,1}) \equiv b$. The third relation requires that $a+b = 0$ (on $T_{(1)}$) and $a-b = 0$ (on $T_{(0)}$). Hence, the quadratic form b_{T_k} and the matrix A_6 are positive. Denote the respective minimal and maximal positive eigenvalues of A_6 by κ_0 and κ_1. Suppose we choose

$$A_{1,1}^{(l+1)} \equiv h_{l+1}^2 D(D_1; 2), \tag{4.32}$$

where D_1 is a diagonal matrix such that, for each $u \in H_1^{(l+1)}$, we have $D_1 u \mid_{M_i} = n_i u(M_i)$ and n_i refers to the number of the triangles $T_k \in T^{(l)}$ containing the node M_i ($n_i = 1, 2$). Then it is easy to see that this choice satisfies (4.29) with $\sigma_{0,1} = \kappa_0$ and $\sigma_{1,1} = \kappa_1$. $\square$

We can now apply the construction of the model cooperative operators $\bar{B}^{(l+1)}$ and $B^{(l+1)}$ from § 3.7.

Theorem 2. *Suppose that the model cooperative operators $\bar{B}^{(l+1)}$ for the operators defined in (4.8) are constructed in accordance with (3.7.30), with Λ replaced by L. Then there exist $k_2 \leq 3$ and k_1, independent of the level l, such that the pairs $\bar{L}^{(p)}, \bar{B}^{(p)}$ and $L^{(p)}, B^{(p)}$ are spectrally equivalent operators and $B^{(p)}$ is an asymptotically optimal preconditioner for $L^{(p)}$.*

Proof. The inequalities $\sigma_0^{(l+1)} \bar{B}^{(l+1)} \leq \bar{L}^{(l+1)} \leq \sigma_1^{(l+1)} B^{(l+1)}$ are obtained in the same way as in the proof of Theorem 3.7.1. Due to Theorem 1, we have $s_{l+1}^2 \leq s^2 \equiv 1/4$. This implies that equation (3.7.38) has a solution t^* if (3.7.41) is satisfied, which is the case if $k_2 = 3$ and q_1^+ (defined by k_1 and Lemma 2) is small enough (see Lemma 3.7.8). Then, due to Theorem 3.7.3, we have

$$\sigma^{(p)} \equiv \sigma_1^{(p)}/\sigma_0^{(p)} \le \max\{\sigma^{(0)}, t^*\}. \tag{4.33}$$

This yields the desired spectral equivalences. The required computational work is estimated as in (3.7.45). □ [29]

§5. Quasilinear elliptic problems

5.1. Weakly nonlinear monotone operators. Consider problems dealing with the mildly nonlinear equation

$$-\sum_{r=1}^{d} D_r a_r(x, \zeta(u)) + a_0(x, \zeta(u)). \tag{5.1}$$

Here

$$\zeta(u) \equiv [D_0 u, D_1 u, \dots, D_d u] \equiv [\zeta_0, \zeta_1, \dots, \zeta_d] \tag{5.2}$$

(that is, $\zeta_0 \equiv u$ and $\zeta_r \equiv D_r u \equiv \frac{\partial u}{\partial x_r}$, $r \ge 1$) and $a_r(x, \xi)$, defined for almost all $x \in \bar{\Omega}$ and all $\xi \in \mathbf{R}^{d+1}$, are bounded piecewise continuous function such that

$$|A_r| \equiv |a_r(x, \zeta + \xi) - a_r(x, \zeta)| \le K \left(|\xi_0| + \dots + |\xi_d|\right), \quad r \in [0, d], \tag{5.3}$$

$$A_0 + \dots + A_d \ge \nu_0 \left(\xi_1^2 + \dots + \xi_d^2\right) - c_0 \xi_0^2, \tag{5.4}$$

where $A_r \equiv a_r(x, \zeta + \xi) - a_r(x, \zeta), r \in [0, d]$, and

$$\nu_0 > 0, \quad c_0 \ge 0, \quad \nu_0 - c_0 \gamma^2 \equiv \mu_0 > 0 \tag{5.5}$$

(see (0.1.9)). Here, K is used only for nonnegative constants. For a precise formulation of our problem, assume we deal with a bounded d-dimensional region Ω and a part Γ_0 of its boundary Γ of the same type as in § 1–4 $(|\Gamma_0|_{(d-1)} > 0)$. We make use of the Hilbert space

$$G \equiv W_2^1(\Omega; \Gamma_0), \tag{5.6}$$

---

[29] We note that a more complicated choice of $A_{1,1}^{(l+1)}$ may result in more faster convergence of the ultimate multigrid iterative method. We emphasize also that it is easy to apply the obtained results for certain other boundary value elasticity problems dealing with subspaces of $\vec{V}$ (see (4.1)). For example, this is the case when on the part Γ_3 of Γ we deal with the condition $(\vec{u}, \vec{n})_{\mathbf{R}^2} = 0$.

with norm $\|u\| \equiv |u|_{1,\Omega}$, and define a quasibilinear form by

$$b(u; v) \equiv b(\Omega; u; v) \equiv \sum_{r=0}^{d}(a_r(x, Du), D_r v)_0. \tag{5.7}$$

Thus, we can deal with our problem in the form of (1.1.8), that is, we seek $u \in G$ such that $b(u; v) = l(v)$, $\forall v \in G$. Below we give sufficient conditions for its correctness (we can also use (0.1.28) if the nonlinear operator L is defined by the relation $b(u, v) = (L(u), v), \forall u, \forall v)$.

Lemma 1. *Let, for quasibilinear form (5.7), conditions (5.3)–(5.5) be satisfied. Then this form is continuous and, for all u, v, and w, inequalities (1.1.12)–(1.1.14) hold with $\delta_0 = \sigma_0 = \mu_0$ and $\sigma_1 = K(d + 1)(1 + \gamma^2)$.*

Proof. We have $b(u + z; z) - b(u; z) \geq \nu_0\|z\|^2 - c_0|z|_0^2$ (see (5.4)), which, with (0.1.9) and (5.5), lead to (1.1.13), where $v = u + z$. Since (5.3) implies that $b(0; v) = 0, \forall v$, then (1.1.12) follows from (1.1.13). Next, (5.3) implies that, for $r \in [0, d]$, we have (see (0.1.9))

$$|(a_r(x, \zeta(v)) - a_r(x, \zeta(u)), D_r w)_0| \leq K(d + 1)^{1/2}$$

$$\times [|v - u|_0^2 + |D_1(v - u)|_{0,\Omega}^2 + \ldots + |D_d(v - u)|_{0,\Omega}^2]^{1/2}|D_r w|_{0,\Omega}$$

$$\leq K(d + 1)^{1/2}(1 + \gamma^2)^{1/2}|v - u|_{1,\Omega}|D_r w|_0,$$

$$|b(v; w) - b(u; w)| \leq K(d + 1)^{1/2}(\gamma^2 + 1)^{1/2}|v - u|_{1,\Omega}$$

$$\times [|w|_0 + |D_1 w|_0 + \cdots + |D_d w|_0],$$

which leads to (1.1.14). $\square$ [30]

Theorem 1. *Suppose that $a_r(x, \zeta), r \in [0, d]$, for almost all $x \in \bar{\Omega}$ and all $\zeta \in \mathbf{R}^{d+1}$, are continuously differentiable functions with respect to ζ. Suppose also that all first derivatives are uniformly bounded and satisfy the condition*

$$\sum_{r=0}^{d}\sum_{l=0}^{d}\frac{\partial a_r(x, \zeta)}{\partial \zeta_l}\xi_r\xi_l \geq \nu_0\left(\xi_1^2 + \ldots + \xi_d^2\right) - c_0\xi_0^2,$$

and that (5.5) holds. Then, for the quasibilinear form (5.7), estimates (1.1.13) and (1.1.14) hold with $\sigma_0 = \mu_0$ and positive σ_1.

Proof. In the sequel, we use $v \equiv u + z$. First transform the left-hand side of (1.1.13) in accordance with (1.2.17), where $w = z$. Then

$$b(u + z; z) - b(u; z) = \sum_{r=0}^{d}\sum_{l=0}^{d}(\frac{\partial a_r(x, \zeta(u + \theta z))}{\partial \zeta_l}D_l z, D_r z)_{0,\Omega}.$$

---

[30] More detailed conditions than (5.3) can improve the estimate for σ_1.

Hence, we obtain the same estimates for $b(u + z; z) - b(u; z)$ from below as in the proof of Lemma 1. To prove (1.1.14), we also transform the left-hand side of (1.1.14) in accordance with (1.2.17):

$$|b(v; w) - b(u; w)| = |\sum_{r=0}^{d} \sum_{l=0}^{d} (\frac{\partial a_r(x, \zeta(u + \theta z))}{\partial \zeta_l} D_l z, D_r w)_{0,\Omega}| \equiv F.$$

Hence, $F \leq K_1 |\sum_{r=0}^{d} \sum_{l=0}^{d} (|D_l z|, |D_r w|)_{0,\Omega} \leq K_2 \|z\| \|w\|$. □ [31]

Using the same triangulations and subspaces $\hat{G}_h$ as in § 1 and 2, we obtain the nonlinear problems $b_{\hat{\Omega}}(\hat{u}; \hat{v}) = l(\hat{v})$, $\forall \hat{v} \in \hat{G}_h$, where $b_{\hat{\Omega}}(u; v) \equiv \sum_{r=0}^{d} (a_r(x, Du), D_r v)_{0,\hat{\Omega}}$; all given functions are extended to $\hat{\Omega}$ so that their indicated properties are preserved. We use the same Euclidean space H and operators Λ_Ω and $B \equiv B_Q \asymp \Lambda_\Omega$ as in § 1 and 2. Then our PGM leads to a nonlinear system $L_h(\mathbf{u}) = \mathbf{f}$ in H, and for the nonlinear operator $L_h \equiv L_\Omega \equiv L$, with $(L_h(\mathbf{u}))_i \equiv b_{\hat{\Omega}}(\hat{u}; \hat{\psi}_i)$, $i \in [1, N]$, inequalities (1.3.2) and (1.3.3) hold for all $u \equiv \mathbf{u}, z \equiv \mathbf{z}$ with the constants $\delta_0 > 0$ and δ_1 (Lemma 1 or Theorem 1 implies that L_h and Λ_Ω are connected by relationship $C_1(\mathbf{u}; \infty)$, which together with Subsection 3.4.4 yields the desired estimates). Thus, correctness of the grid problems and convergence of the modified method of the simple iteration follow from Theorems 1.2.5 and 1.3.1. [32]

Theorem 2. Suppose that we approximate the function $a_s(x, \zeta(\hat{v}))$ by a simpler function $a_s^*(x)$ such that

$$|a_s(x, \zeta(\hat{v})) - a_s^*(x)| \leq K^* \varepsilon^*, \quad \forall \mathbf{v} \in S_r \equiv S_B(\mathbf{u}, r), \quad s \in [0, d] \quad (5.8)$$

(see Theorem 1.4.1). Let $\mathbf{e} \equiv \{e_i\}$, $e_i \equiv \sum_{r=0}^{d} \sum_T E_{T,r,i}(v)$, where $E_{T,r,i}(v) \equiv (a_r(x, \zeta(\hat{v})) - a_r^*(x), D_r \hat{\psi}_i)_{0,T}$. Then there exists a constant K_1^* such that $\|\mathbf{e}\|_{B^{-1}} \leq K_1^* e^*/h$.

Proof. The proof is similar to that of Theorem 1.2.3. If $\Lambda \mathbf{w} = \mathbf{e}$, then $(\Lambda \mathbf{w}, \mathbf{w}) = \|\mathbf{e}\|_{\Lambda^{-1}}^2 = (\hat{w}, \hat{e})_{1,\hat{\Omega}}$. Since $|E_{T,r,i}(v)| \leq K_2^* \varepsilon^*/h$, then $|\hat{e}|_{1,\hat{\Omega}} \leq K_3^*/h$ and $|\hat{w}|_{1,\Omega}^2 \leq K_4^* \varepsilon^*/h |\hat{w}|_{1,\Omega}$. Thus, we estimate $|\hat{w}|_{1,\Omega} = \|\mathbf{e}\|_\Lambda$ and $\|\mathbf{e}\|_{B^{-1}}$ (see Lemma 0.4.3). □ [33]

---

[31] These simple results were used in [154, 155, 162]; similar results can be found in [145, 238, 244, 315, 337, 426, 442, 514].

[32] In practical applications deal with approximate evaluations of $L_h \mathbf{u}^n$, for which Theorem 1.4.1 applies if we specify the algorithm a_1 for obtaining ε_L-approximation in $H(B^{-1})$ to vector the $L_h(v)$, where $v \equiv \mathbf{u}^n$. It is easy to see that actually we approximate separate terms $I_{T,s,i}(v) \equiv (a_s(x, \zeta(\hat{v})), D_s \hat{\psi}_i)_{0,T}$, where T is a simplex in $T_h(\hat{\Omega}_h)$ and index i corresponds to its vertex (see § 0.2, 0.5), $s \in [0, d]$.

[33] Theorem 2 allows us to take $\varepsilon_L \asymp \varepsilon^* \asymp \varepsilon$ in the application of Theorem 1.4.1.

Theorem 3. *Suppose $d = 2$ and the conditions of either Lemma 1 or Theorem 1 are satisfied. Suppose also that, for the solution of our problem, conditions (1.3) and (1.6) are satisfied. Then on the basis of our PGM and method (1.3.1) in combination with its multigrid acceleration, we obtain an algorithm leading to the desired ϵ-approximation with computational work*

$$W(\epsilon) = O(\epsilon^{-2/\gamma}t(\epsilon)). \tag{5.9}$$

Proof. In analyzing PGMs for nonlinear problems, our purpose is to obtain the same accuracy estimates as for linear problems. Again, we have two situations. The first is when (1.5) holds, allowing us to apply Theorems 1.1.2 and 1.1.5 directly. In the second situation, when (1.5) does not hold, we need to modify the proof of Theorem 1.1. This is more involved matter, but all stages of this proof can nonetheless be modified in a straightforward manner. Hence, estimates of accuracy of PGMs have the old form. In analyzing asymptotically optimal algorithms, we emphasize that our basic iterative method (1.3.1) works well for an arbitrary initial iterate. In Theorem 1.4.1, we can then take $r > \|\mathbf{u}^0 - \mathbf{u}\|$. As for application of the coarse grid continuation, we again have (1.25) and thus we can easily modify the proofs of Theorems 1.4 and 1.5. $\square$ [34]

5.2. Nonlinearity of bounded power. So far we have considered only the case of mildly nonlinear problems. Here we consider the more general and difficult case of nonlinearity characterized by the assumption that our quasibilinear form $b(u; v)$ is defined on G^2 (the corresponding nonlinear operator L is defined on G) and, for each ball $S_G(r)$, there exists a constant $\sigma_1 \equiv \sigma_1(r)$ such that inequality (1.1.14) holds for all $w \in G, u \in S_G(r), v \in S_G(r)$ (in contrast with Lemma 1, we now have $\lim_{r \to \infty} \sigma_1(r) = \infty$). Again, our most important condition (1.13) (associated with monotonicity of L) will be satisfied only for a certain ball in G. In verifying these conditions, of fundamental importance is the generalized Hölder inequality

$$(|u_1 \cdots u_n|, 1) \leq |u_1|_{0,q_1} \cdots |u_n|_{0,q_n} \tag{5.10}$$

Of course, to meet condition (5.8), we must take into account the computational work required by such an approximation. In general, we may estimate it as $W^*(\epsilon, h) = O(t(\epsilon)/h^d)$ (e.g., if $a_s^*(x)$ is an interpolant on T to our function $a_s(x, \zeta(\hat{v}))$, $s \in [0, d]$, then we need to evaluate it with prescribed tolerance at several points in T). For many smooth functions, $t(\epsilon) \leq \kappa |\ln \epsilon|^{1/2}$ (see [41]) and, for piecewise polynomial functions, $t(\epsilon) = O(1)$.

[34] Algorithms of this type have been analyzed in [52, 116, 167, 169, 183, 426, 434, 514]; important applications can be found in [8, 100, 214, 245, 217, 219, 284, 380].

Under the conditions of Theorem 1, algorithms based on linearization are also possible (see § 3.8 and [206]). We note also that the presence of the term $b_{\Gamma_1}(u; v) \equiv (a_{\Gamma_1}(u), v)_{0,\Gamma_1}$ in our quasibilinear form demands only simple additional considerations.

(see [387]) where $q_r \geq 1, r \in [1, n], 1/q_1 + \cdots + 1/q_n = 1$, and

$$|u|_{0,q} \equiv \|u\|_{L_q(\Omega)}, \quad q \geq 1. \tag{5.11}$$

The case $q = \infty$ is allowed, for which then $|u|_{0,\infty} \equiv$ vrae max $|u(x)|$ (for a continuous function u, $|u|_{0,\infty} \equiv \max_{x \in \bar{\Omega}} |u(x)|$).

We consider L as a perturbation of a linear operator R from (1.2.4), denoted by $\mathcal{R}$, and write

$$b(u; z) = b_1(u; z) + \nu b_2(u; z), \quad \nu \geq 0, \tag{5.12}$$

where $b_1(u; v) = (\mathcal{R}u, v), \forall u, \forall v$. We assume that

$$b_1(z; z) \geq \sigma_2 \|z\|^2, \quad \sigma_2 > 0, \quad |b_1(u; z)| \leq \sigma_3 \|u\| \|z\|, \tag{5.13}$$

$$b_2(v; v - u) - b_2(u; v - u) \geq -\sigma_4(\|u\|; \|v\|) \|v - u\|^2, \tag{5.14}$$

and

$$|b_2(v; z) - b_2(u; z)| \leq \sigma_5(\|u\|; \|v\|) \|v - u\| \|z\|, \tag{5.15}$$

where u, v, and z refer to arbitrary elements in the Hilbert space G; nonnegative functions σ_4 and σ_5 are nondecreasing with respect to both variables $(\sigma_4 \leq \sigma_5)$.

Lemma 2. *Let conditions (5.12)–(5.15) be satisfied. Then, for all $u \in S \equiv S(R)$ and $v \in S$, inequalities (1.1.13) and (1.1.14) hold with constants*

$$\sigma_0 = \sigma_2 - \nu \sigma_4(R; R), \quad \sigma_1 = \sigma_3 + \nu \sigma_5(R; R). \tag{5.16}$$

Proof. The proof, based on the additive representation of $b(v; v - u) - b(u; v - u)$ and $b(v; w) - b(u; w)$, is straightforward. $\square$ [35]

In the sequel, we often encounter functions of the form

$$v(x) = v_0(x) \prod_{r=1}^{k+k'} |v_r(x)|^{s_r}, \quad s_r > 0, r \in [1, k + k'], \tag{5.17}$$

and

$$w(x) = \prod_{n=0}^{k} |w_n(x)|^{s_n}, \quad s_n > 0, n \in [0,]k, \tag{5.18}$$

where $v_0(x) \in L_\infty(\Omega), v_r(x) \in L_{p_r}(\Omega), 1 \leq p_r < \infty, r \in [1, k + k']$, and

$$w_n(x) \in L_{p_n}(\Omega), 1 \leq p_n < \infty, \ n \in [0, k].$$

---

[35] Clearly, $\sigma_0 > 0$ when either ν is small for a fixed radius R, or R is small for a fixed ν and $\lim_{R \to 0} \sigma_4(R; R) = 0$ (as is usually the case).

Lemma 3. *Consider the function* $v(x)$ *in* (5.17). *Suppose that* $\rho \equiv s_1/p_1 + \cdots + s_k/p_k < 1$ *and that* $v_r(x) \in L_p(\Omega)$ *for arbitrary* $p > 1$, $r > k$. *Suppose that*

$$\nu \equiv s_{k+1} + \cdots + s_{k+k'} \geq 1 - \rho \tag{5.19}$$

and $1/p_0 \equiv (1 - \rho)/\nu$. *Then*

$$|v(x)|_{0,1} \leq |v_0(x)|_{0,\infty} \prod_{r=1}^{k} |v_r(x)|_{0,p_r}^{s_r} \prod_{r=k+1}^{k+k'} |v_r(x)|_{0,p_0}^{s_r}. \tag{5.20}$$

Proof. The proof is based on (5.10). For $r \in [1, k]$, we choose $q_r = p_r/s_r$. Then, for $r \in [k + 1, k + k']$, we see that $1/q_1 + \cdots + 1/q_{k+k'} = 1$. This justifies our use of (5.10) and leads to (5.20). $\square$

To satisfy condition (5.19), we may introduce an additional factor 1 in the right-hand side of (5.17) (e.g., we may take $v_{k+k'}(x) = 1$).

Lemma 4. *Consider the function* $w(x)$ *defined by* (5.18). *Suppose that* $\rho \equiv s_0/p_0 + \cdots + s_k/p_k \leq 1$ *and that* $v_r(x) \in L_p(\Omega)$ *for* $p > 1$ *if* $r > k$. *Then*

$$|w(x)|_{0,1} \leq \prod_{n=0}^{k} |w_n(x)|_{0,p_n}^{s_n} |\Omega|^{1-\rho}. \tag{5.21}$$

Proof. The proof is similar to the proof of Lemma 2 if we introduce an additional factor $w_{n+1}(x) = 1$ in the right-hand side of (5.18). $\square$

Lemma 5. *Let* $d > 2, p_0 \equiv 2d/(d - 2)$, *and* $p_l \equiv 2$, $l \in [1, d]$. *Suppose that the functions* $a_r^{(2)}(x, \zeta)$, *for almost all* $x \in \bar{\Omega}$ *and all* $\zeta \in \mathbf{R}^{d+1}$ *and* $\xi \in \mathbf{R}^{d+1}$, *satisfy*

$$|a_r^{(2)}(x, \zeta + \xi) - a_r^{(2)}(x, \zeta)| \leq \sum_{\alpha=1}^{m_r} g_{r,\alpha}(x) \prod_{l=0}^{d} \left(|\zeta_l|^{s_{r,\alpha,l}} |\xi_l|^{t_{r,\alpha,l}}\right), \tag{5.22}$$

$$g_{r,\alpha}(x) \in L_{p_{r,\alpha}}(\Omega), \quad \frac{1}{p_{r,\alpha}} + \sum_{l=0}^{d} \frac{s_{r,\alpha,l} + t_{r,\alpha,l}}{p_l} + \frac{1}{p_r} \leq 1, \tag{5.23}$$

where $r \in [0, d]$ *and* $l \in [0, d]$. *Then, for the quasibilinear form*

$$b_2(u; v) \equiv \sum_{r=0}^{d} (a_r^{(2)}(x, \zeta(u)), D_r v)_0, \tag{5.24}$$

we have

$$X \equiv |b_2(v; z) - b_2(u; z)| \leq K_3 \sum_{r,\alpha} \|u\|^{s_{r,\alpha}} \|v - u\|^{t_{r,\alpha}} \|z\|, \qquad (5.25)$$

where $s_{r,\alpha} \equiv s_{r,\alpha,0} + \cdots + s_{r,\alpha,d}$ *and* $t_{r,\alpha} \equiv t_{r,\alpha,0} + \cdots + t_{r,\alpha,d}$.
  Proof. From (5.24) and (5.22), it follows that

$$X \leq K \sum_{n=0}^{d} \sum_{\alpha=0}^{m_r} (g_{r,\alpha} \prod_{l=0}^{d} [|\zeta_l(u)|^{s_{r,\alpha,l}} |\xi_l(v - u)|^{t_{r,\alpha,l}}, D_r z)_0. \qquad (5.26)$$

Each term on the right-hand side of (5.26) is an integral of a product of $2d + 2$ factors and is estimated from above in accordance with Lemma 4. For this, it is important to take advantage of condition (5.23) and the embedding theorem of our space G into $L_{p_0}(\Omega)$. The indicated estimation yields (5.25). □ [36]
  Lemma 6. *Let* $d > 2$ *and suppose that* $a_r^{(2)}(x, \zeta)$, *for almost all* $x \in \bar{\Omega}$ *and all* $\zeta \in \mathbf{R}^{d+1}$, *are continuously differentiable functions with respect to* ζ *that satisfy*

$$\left| \frac{\partial a_r^{(2)}(x, \zeta)}{\partial \zeta_l} \right| \leq \sum_{\alpha=1}^{m_r} g_{r,\alpha}(x) \prod_{l=0}^{d} |\zeta_l|^{s_{r,\alpha,l}}, \qquad (5.27)$$

where $s_{r,\alpha,l} \geq 0$,

$$g_{r,\alpha}(x) \in L_{p_{r,\alpha}}(\Omega), \quad 1/p_{r,\alpha} + 1/p_r + \sum_{l=0}^{d} s_{r,\alpha,l}/p_l \leq 1, \qquad (5.28)$$

$r \in [0, d], l \in [0, d]$. *Let*

$$\sigma_5 \equiv K_4 \sum_{r,\alpha} [\max\{\|u\|; \|v\|\}]^{s_{r,\alpha}}.$$

Then, for $b_2(u; v)$ *(see (5.24)), estimate (5.15) holds.*
  Proof. To prove (5.15) we transform its left-hand side as in the proof of Theorem 1, which, in accordance with (5.27), yields

$$|b_2(v; z) - b_2(u; z)| = |\sum_{r=0}^{d} \sum_{l=0}^{d} \left(\frac{\partial a_r(x, \zeta(u + \theta z))}{\partial \zeta_l} D_l z, D_r z \right)_{0, \Omega}|.$$

---

[36] We emphasize that (5.25) with $t_{r,\alpha} \geq 1$ yields (5.15), where $\sigma_5 = K \sum_{r,\alpha} \|u\|^{s_{r,\alpha}} \|v - u\|^{t_{r,\alpha}-1}$.

Together with (5.27) and Lemma 4, thus implies that

$$|b_2(v; z) - b_2(u; z)| \leq K_5 \sum_{r,\alpha} \|u + \theta z\|^{s_{r,\alpha}} \|v - u\| \|z\|.$$

Note that $\|u + \theta z\| \leq \max\{\|u\|; \|v\|\}$. Therefore, (5.15) holds. $\square$

We now we pay special attention to the case $d = 2$, where the embedding theorem of our space G into arbitrary $L_p(\Omega)$ holds. We may thus choose p_0 as a very large number, and (5.23) and (5.28) with $1/p_0 \equiv 0$ should be written as strict inequalities (it is then possible to find an appropriate $p_0 < \infty$). For example, if

$$a_0^{(2)}(x, \zeta(u)) = \sum_{l=1}^{d} g_l u^{s_l} D_l u + g_0 u^{s_0}, \qquad (5.29)$$

then for $d = 2$ we may take $g_l \in L_\infty(\Omega)$ and $s_l = 0, 1, \dots$ for $l = 0, 1, 2$. For $d = 3$ in (5.29), we may take $g_l \in L_\infty(\Omega), s_0 = 3$, and $s_l = 1, l = 1, 2, 3$. Lemmas 5 and 6 yield (5.15) and sufficient conditions for the fundamental inequalities (1.1.13) and (1.1.14) (actually, (1.1.12)–(1.1.14)) for all u and v in a certain ball $S \equiv S_G(R)$.

For PGMs when (1.5) holds, our spline spaces $\hat{G}_h$ are subspaces of the Hilbert space G. Moreover, under the standard condition $\|l\|/\delta_0 \equiv R_0 < R$ on the functional l (see [337, 489] and Theorem 1.2.1), we conclude that the original problem and its projective approximation, respectively, have unique solutions u and $\hat{u}$ in S (even in $S_G(R_0)$) and Theorem 1.2.1 applies. This leads to estimates of accuracy of PGMs of the same type as for linear problems. Further, if for each Euclidean space H we consider the ball $S_h \equiv S_{\Lambda_\Omega}(u; R - R_0)$ ($u \equiv \mathbf{u} \in H$ corresponds to $\hat{u}$), then the operators L_Ω and Λ_Ω are connected by relationship $C^0(u, R - R_0)$ (inequalities (1.3.2) and (1.3.3) with $B = \Lambda_\Omega$ hold for all z with $\|z\|_{\Lambda_\Omega} \leq R - R_0$). This, together with Subsection 3.4.4, implies that, for any model $B_Q \equiv B \asymp \Lambda_\Omega$, the operators L_Ω and B are connected by relationship $C^0(u, r)$ as well (r is defined by $R - R_0$ as indicated in § 3.4). Thus, method (1.3.1) is asymptotically optimal provided $u^0 \in S_B(u, r)$. This additional restriction is severe compared to that of the iterative algorithms considered above. [37] However, for some types of nonlinear problems (leading, e.g., to small R_0 with respect to R), we may take $u^0 = 0$. Nearly asymptotically optimal algorithms with computational work $W(\epsilon, h) = O(t(\varepsilon)|\ln \varepsilon|/\varepsilon^{d/\nu})$ can then be constructed as for linear problems. We can improve them by applying the coarse grid continuation, which is straightforward if R_0 and h_0 are so

---

[37] In practical problems, various continuation methods with respect to a parameter are very useful for obtaining u^0 (see [64, 100, 214, 244, 245, 467]).

small that the functions $\hat{w}_{l+1}$ remain in the appropriate ball on the level $l+1, l \in [0, p-1]$ (see Lemma 1.3 and (1.25)). Hence, even for such nonlinear problems, we obtain asymptotically optimal algorithms if $t(\varepsilon) = O(1)$. [38]

5.3. Antisymmetric quadratic nonlinearity. We consider now a remarkable case where the nonlinear perturbation of the linear operator A (see (5.12)) is of the form

$$b_2(u; z) \equiv \sum_{l=1}^{d} (g_l, uzD_lu - uzD_lz)_0, \qquad (5.30)$$

or

$$b_2(u; z) \equiv \sum_{l=1}^{d} (uD_lu, z)_0, \qquad (5.31)$$

and satisfies $b_2(z; z) = 0$, $\forall z$. (For (5.31), we assume that $G \equiv \overset{o}{W_2^1}(\Omega)$; a similar property is typical for the Navier-Stokes system.) This case yields simplifications in our analysis. First, simple localizations of the solutions of the original problem and those of its projective approximations are possible (see Theorem 1.2.1), so we may assume that $\|u\| \leq r_1, \|\hat{u}\| \leq r_1$. Second, we can simplify estimates (5.14) and (5.15) (see [159]).

Lemma 7. *Let $d > 2, p_0 \equiv 2d/(d-2)$, and $g_l \in L_{q_l}, l \in [1, d]$ (see (5.30)), where $s_l \equiv 1/q_l + 1/2 + 2/p_0 \leq 1$. Let $b_2(u; v)$ be defined either by (5.30) or (5.31). Then there exist constants K^* and K_1^* such that*

$$|b_2(v; z) - b_2(u; z)| \leq K^* \|u\| \|z\|^2, \quad \forall u, \forall v = u + z \qquad (5.32)$$

$$|b_2(v; z) - b_2(u; z)| \leq K_1^* (\max\{\|u\|; \|v\|\})^2 \|z\|, \quad \forall u, \forall v, \forall z. \qquad (5.33)$$

Proof. We have $X \equiv b_2(v; z) - b_2(u; z) = \sum_{l=1}^{d} (g_l, z^2 D_l u + uz D_l z)_0$ (see (5.30)). We apply Lemma 4, e.g.,

$$|(g_l; uzD_lz)_{0,\Omega}| \leq |g_l|_{0,q_l} |u|_{0,p_0} |z|_{0,p_0} |D_lz|_{0,2} |\Omega|^{1-s_l}.$$

---

[38] The case when (1.5) does not hold needs special modifications of our proofs. The main difference is that in applying Lemmas 5 and 6 we now deal with a family of regions $\hat{\Omega}_h$. But the embedding operators mapping the space $W_2^1(\hat{\Omega}_h)$ into $L_p(\hat{\Omega}_h)$ may be regarded as uniformly bounded with respect to h by virtue of well known results for Sobolev spaces (see, e.g., [3, 148]). This yields the necessary basic inequalities (1.1.12)–(1.1.14) that hold uniformly with respect to all $\hat{G}_h$. A similar problem is connected with estimating the term $|b(\Omega_{1,h}; u; \hat{v})|$ (in the linear case, it corresponds to estimating $|\zeta_\Omega|$ (see (1.1.10) and (1.1.13)) for the region $\Omega_{1,h}$ belonging to $O(h^2)$-boundary strip. All other modifications are fairly transparent.

Since $|v|_{0,p_0} \leq K(p_0)\|v\|$, then (5.32) holds. The case of (5.31) and (5.33) is simpler. $\square$

The case $d = 2$ enables us to use $1 < p_0 < \infty$. If, e.g.,

$$b_2(u; z) \equiv \sum_{l=1}^{2}(a_l u^{k_l} D_l u, z)_0, \qquad (5.34)$$

where $G \equiv \overset{0}{W_2^1}(\Omega)$, a_l is a constant, and k_l is a positive integer $(l = 1, 2)$, then, on the right-hand sides of (5.32) and (5.33), $\|u\|$ and $(\max\{\|u\|; \|v\|\})^2$ must be replaced by $\sum_{l=1}^{2}|a_l|\|u\|^{k_l-1}$ and $\sum_{l=1}^{2}|a_l|(\max\{\|u\|; \|v\|\})^{k_l}$, respectively. We consider only PGMs associated with $\hat{\Omega}_h \subset \bar{\Omega}$ (see (1.5)).

Theorem 4. Let the original quasibilinear form be defined by (5.12), where $b_1(u; v)$ and $b_2(u; v)$ satisfy (5.13) and the conditions of Lemma 7. Suppose that $\sigma_2 - \nu K^* r_1 \equiv \sigma_0 > 0$. Then the arising nonlinear grid operator $L_h \equiv L_\Omega$ and the model operator $B_h \equiv B_Q \in \mathcal{L}^+(H)$ are connected by relationship $C^{0,0}(\mathbf{u}, r)$ with any $r < \infty$ and any $\mathbf{u}$ such that $\|\hat{u}\| \leq r_1$. Moreover, if conditions (1.3) or (1.6) are satisfied, then the conclusion of Theorem 3 remains true with $t(\epsilon) = O(1)$ for (5.31).

Proof. In accordance with Lemma 7, the assumption on $\nu K^* r_1$ ensures the desired crucial property (1.1.13) in any ball $S_G(u; r)$ with $\|u\| \leq r_1$, so it holds in $S_G(\hat{u}; r_1)$. It is easy to that (1.1.14) presents no problems (see (5.15)). Thus, Theorem 1.1.2 applies and we have the same error estimates for our PGMs as in the linear case. Since $b(\hat{u}; \hat{v}) = (L_h(\mathbf{u}, \mathbf{v}))$, it is a simple matter to show that the grid operators L_h and Λ_Ω are connected by relationship $C^{0,0}(\mathbf{u}, r')$ with arbitrary $r' < \infty$. Thus, L_h and B_h are also connected by relationship $C^{0,0}(\mathbf{u}, r)$ with arbitrary r. For (5.31), the integrals $(D_l \hat{u}^n \hat{u}^n, \hat{\psi})_{0,T}$ can easily be found. $\square$ [39]

5.4. Nonlinear perturbation of linear invertible operator. Here, we consider $L \equiv \mathcal{R} + \mathcal{P}$ as a perturbation of a linear invertible operator $\mathcal{R}$, that is, for (5.12), instead of (5.13), we have only $\|\mathcal{R}\| < \infty, \|\mathcal{R}^{-1}\| < \infty$. Suppose that $\mathcal{P}$ can be considered as a small operator in the sense that (5.15) holds with small enough $\nu\sigma_5$ for all $u \in S_G(u_0; r_0)$, $v \in S_G(u_0; r_0)$ and all z, where $\mathcal{R}u_0 = f$. Then we can apply Theorems 1.1.4, 1.2.2, 1.2.3, 1.2.6, and 1.2.7. Also, we can apply iterations of type (1.3.18).

We specify these results to the case of Lemma 6 and differentiable operator $\mathcal{P}$, where the implicit function theorem applies (see Subsection 1.2.4) and the operators $L'_v = \mathcal{R} + \mathcal{P}'_v$ are invertible and continuous in v for v in

---

[39] Note that (5.31) can be replaced by (5.34). Note also that global convergence of our iterations holds.

the vicinity S of the solution u; $\mathcal{P}'_w$ corresponds to the form

$$\nu \sum_{r=0}^{d} \sum_{j=0}^{d} \left(\frac{\partial a_r^{(2)}(x, \zeta(\hat{w}))}{\partial \zeta_j} D_j \hat{u}, D_r \hat{v}\right)_0.$$

Considering the projective operators $\hat{L}$ ($\hat{L}(\hat{v}) \equiv PL(\hat{v})$ (see Theorem 0.4.4)) and their derivatives $\hat{L}'_{\hat{u}} = PL'_{\hat{u}}$, we may assume further that

$$\|(\hat{L}'_{\hat{v}})^{-1}\| \leq K_0, \quad \|\hat{L}'_{\hat{v}}\| \leq K_1, \quad \forall \hat{v} \in S_G(\hat{w}, r_1), \tag{5.35}$$

where $\hat{w} \equiv Pu \to u$ and r_1 is small enough. This implies invertibility of the *Jacobian matrices* $L'_{h,\mathbf{v}} \equiv S$ and $\|S^{-1}\|_{H(J-1) \mapsto H(J)} \leq K_0$, $\|S\|_{H(J) \mapsto H(J-1)} \leq K_1$, with $J \equiv \Lambda_\Omega$ (see (0.4.7), (0.4.8), and (1.2.18) in Theorem 1.2.11). Moreover, it is not difficult to show (under additional assumptions on the functions $a_r^{(2)}(x, \zeta), r \in [0, d]$, as in [206]) that there exists a constant $\bar{K}$ such that, for all $\hat{v}$ and $\hat{v}'$ in $S_G(\hat{w}, r_2)$,

$$\|L'_{h,\mathbf{v}} - L'_{h,\mathbf{v}'}\|_{H(J) \mapsto H(J-1)} \leq \bar{K}\|\mathbf{v} - \mathbf{v}'\|_J, \tag{5.36}$$

where $r_2 \leq r_1$. We thus obtain (1.2.19) and by Theorem 1.2.11 we conclude that correctness of the grid method holds at least for $S_{F_h} = S_{\Lambda_\Omega}(L_h\mathbf{w}; r_f)$ with small enough r_f. Further, Theorem 1.1.4 applies and we have the same error estimates for our PGMs as in the linear case. Now we need to assume small enough h_0 and $\hat{u} \in S_G(\hat{w}, r_2)$. [40]

5.5. Variational inequalities. In conclusion, we briefly consider an interesting variational problem that generalizes (0.1.4):

$$\min_{v \in S} \Phi(v), \quad \Phi(v) \equiv b(v; v) - 2l(v), \tag{5.37}$$

where S is a nonempty, closed, and convex set in the Hilbert space $G \equiv W_2^1(\Omega, \Gamma_1)$ and the bilinear form satisfies the conditions of Theorem 0.1.3 (see [35, 220, 244, 246, 280, 337]). [41]

It is especially easy to analyze correctness of (5.37) if we rewrite $\Phi(v)$ as in the proof of Theorem 0.1.3 and conclude that $\|u-f\|_L \leq \|v-f\|_L$, $\forall v \in S$ (recall that $(u, v)_L \equiv b(u; v)$ and $l(v) = (f, v)_L$). Considering the triangle in

---

[40] For the arising nonlinear grid systems, conditions (5.35) and (5.36) imply that the conditions of Theorem 1.3.11 are satisfied with $S_r \equiv S_B(\mathbf{u}, r), r = r(r_2)$, and the chosen model operator $B \equiv B_h \equiv B_Q \in \mathcal{L}^+(H)$. Hence, our basic iterative algorithm uses (1.3.12). Analysis of coarse grid continuation presents no additional problems.

[41] Construction of asymptotically optimal algorithms remains an open problem for this case. However, some of the above constructions turned out fairly effective for certain practical problems (see [425, 211] and § 6.5, where certain contact problems for shells are considered), and they provide the best asymptotic estimates of computational work among the known classes of algorithms.

$G(L)$ with vertices f, u, and v it is also easy to see that the angle α between $f - u$ and $v - u$ satisfy $\alpha \geq \pi/2$. This implies that $(v - u, f - u)_L \leq 0$, $\forall v \in S$. Hence, we obtain the problem of finding $u \in S$ such that

$$b(u; v - u) - l(v - u) \geq 0, \quad \forall v \in S, \tag{5.38}$$

which is an example of a *variational inequality* (see [246]). Projective approximations for (5.38) with spline subspaces $\hat{G}_h$ take the form

$$b(\hat{u}; \hat{v} - \hat{u}) - l(\hat{v} - \hat{u}) \geq 0, \quad \forall \hat{v} \in \hat{S} \equiv \hat{G}_h \cup S, \tag{5.39}$$

and have been analyzed by many (see [246, 280] and references therein). We consider only problem (5.37) with

$$S \equiv \{u : u \in G \text{ and } u(x) \geq 0, \text{ for almost all } x \in \Omega\}, \tag{5.40}$$

and replace its variational approximation in $\hat{S}$ (or (5.39)) by the nonlinear problem of finding $\hat{u} \in \hat{G}_h$ such that

$$b(\hat{u}; \hat{z}) + \frac{((\hat{u})_-(x), \hat{z})_{0,\Omega}}{\alpha} = l(\hat{z}), \quad \forall \hat{z} \in \hat{G}, \tag{5.41}$$

where $u_-(x) \equiv u(x)$ if $u(x) < 0$ and $u^-(x) \equiv 0$ if $u(x) \geq 0$, and $\alpha > 0$ is a small penalty parameter. We rewrite (5.41) (in terms of H) as

$$L_{h,\alpha}(\mathbf{u}) = \mathbf{f}, \tag{5.42}$$

where $L_{h,\alpha} \equiv A + 1/\alpha \mathcal{P}_h(\mathbf{u}) = \mathbf{f}$ (A here corresponds to the linear operator L_Ω in § 1 and $A = \Lambda_\Omega \equiv \Lambda$ if $b(u, v) = (u, v)_{1,\Omega}$).

Theorem 5. *The nonlinear grid operator $L_h \equiv L_{h,\alpha}$ and the model operator $B_h \equiv B_Q \in \mathcal{L}^+(H) \times \Lambda_\Omega$ are connected by relationship $C^{0,0}(\mathbf{u}, r)$ with arbitrary $r < \infty$ and $\mathbf{u}$. Moreover, the solution of (5.42) is unique and its ε-approximation in $H(\Lambda_\Omega)$ can be found by the modified method of the simple iteration (1.3.1) with computational work $W(\varepsilon, \alpha) = O(|\ln \varepsilon|/\alpha h^2)$.*

Proof. We first show that

$$(L_h(\mathbf{v}) + \mathbf{z}) - L_h(\mathbf{v}), \mathbf{z}) \geq \delta_0 \|\mathbf{z}\|_\Lambda^2, \quad \forall \mathbf{v}, \forall \mathbf{z} \tag{5.43}$$

(see (0.1.14) and (1.3.2)). To prove (5.43), it suffices to show that

$$(u_- - v_-, u - v)_0 \geq 0, \quad \forall u \in \hat{G}_h, \forall v \in \hat{G}_h. \tag{5.44}$$

For any $\hat{u} \equiv u \in \hat{G}_h$ and all $x \in \Omega$, we use the representation $u(x) = u_+(x) + u_-(x)$ ($u_+(x) \geq 0$ is the nonnegative part of $u(x)$) and write $(u_-(x) - v_-(x))(u(x) - v(x)) = (u_-(x) - v_-(x))^2 + X$, where

$$X \equiv (u_-(x) - v_-(x))(u_+(x) - v_+(x)) = -u_-(x)v_+(x) - v_-(x)u_+(x) \geq 0.$$

Thus, (5.44), (5.43), and (1.3.2) hold. Next we show that

$$\|L_h(\mathbf{v}) + \mathbf{z}) - L_h(\mathbf{v}\|_{\Lambda^{-1}} \le \delta_1 \|\mathbf{z}\|_{\Lambda} + \frac{\gamma^2}{\alpha} \|\hat{z}\|_0, \quad \forall \mathbf{v}, \forall \mathbf{z}, \qquad (5.45)$$

where δ_1 and γ are defined by (0.1.13) and (0.1.9). It suffices to show (see Lemma 0.4.5 and Theorem 0.4.1) that

$$Y \equiv |(\hat{v} + \hat{z})_- - \hat{v}_-(x), \hat{w})_0| \le \|\hat{z}\|_0 \|\hat{w}\|_0, \quad \forall \hat{v}, \forall \hat{z}, \forall \hat{w},$$

which follows from the inequality $|(\hat{v} + \hat{z})_- - \hat{v}_-(x)| \le |\hat{z}|$, (0.1.6), and (0.1.9). (5.45) implies that L_h is continuous and Theorem 1.2.1 applies with $r = \infty$. Moreover, (5.43) and (5.45) yield the desired relationship of our operators. Thus, Theorem 1.3.1 applies. The number of the required iterations is estimated as $O(|\ln \varepsilon|/\alpha)$ if $\tau \asymp \alpha$. $\square$

Various strategies for choosing α were considered in [425, 211].

Chapter 6

Estimates of computational work of optimal type for difference methods

Difference methods are usually understood (see [232, 248, 378, 437, 439]) as particular cases of grid methods that, at some stages, approximate derivatives by corresponding difference quotients (on regular grids, they are proportional the corresponding finite differences). As stressed in § 1.1, convergence is usually established as a consequence of correctness and approximation. Appropriate choice of the corresponding norm spaces U_h and F_h is of paramount importance in the study of these properties for difference schemes: it may even enable convergence and corresponding error estimates that are very similar to those for PGMs associated with the same grids or triangulations (see, e.g., [164, 424, 439]). We will concentrate in § 1 on similar results, but for relatively simple regions and grids because of the additional difficulties that difference methods introduce. Special attention (see, e.g., [304, 439] and references therein) is paid to the study of various modifications of difference methods that can be applied on general grids, but the results relevant to asymptotically optimal algorithms are much weaker from those in Chapter 5. However, when we confine ourselves to simple regions and grids, then the asymptotically optimal algorithms are even simpler than those for PGMs, and they were obtained (e.g., for elasticity problems) even earlier. They also found more practical applications then, although the trend is reversing. For a class of simple elliptic boundary value problems, we show that the strengthened variant of the Kolmogorov–Bakhvalov hypothesis can be proved for difference methods. We will also pay special attention to the case when U_h is a difference analog of the space $G^{(2)}$ (important in theory of multigrid methods for linear problems and enables one to use Theorems 6–8 when the theory for nonlinear systems given in § 5.5 does not apply). We will discuss briefly certain difference approximations for fourth-order elliptic equations and systems, and indicate certain important classes of practical problems (mainly in theory of shells) that were solved by asymptotically optimal iterative methods.

§ 1. Estimates of computational work for second-order elliptic equations and systems on model regions

1.1. The first boundary value problem for the equation with variable coefficients in a d-dimensional model region. We confine ourselves to the simplest case of the difference approximation of the first boundary value problem (see (0.1.16) with $G \equiv \overset{\circ}{W}^1_2 (Q)$ and $l(v) \equiv (f, v)_{0,Q}$) in a domain Q such that $\bar{Q}$ is a union of several parallelepipeds. We use parallelepiped grids with mesh sizes $h_r \asymp h, r \in [1, d]$, and nodes $P_i \equiv [i_1 h_1, \ldots, i_d h_d]$ ($i \equiv [i_1, \ldots, i_d]$ is a vector index), and we assume that all vertices of Q are nodes of the given grids.

Let $Q_h \equiv \{P_i : i_r \in Q\}$. Then the grid system associated with the difference approximation takes the form

$$(L_h \mathbf{u})_i \equiv Lu_i \equiv -\frac{1}{2} \sum_{r=1}^{d} \sum_{l=1}^{d} \left(\bar{\partial}_r(a_{r,l}\partial_l u + \partial_r(a_{r,l}\bar{\partial}_l u)\right)_i$$

$$+ \sum_{r=1}^{d} \left(b_r \hat{\partial}_r u + \hat{\partial}_r(b'_r u)\right)_i + c_i u_i = f_i, \quad \forall P_i \in Q_h, \tag{1.1}$$

$$u_i = 0, \text{ if } P_i \notin Q_h, \tag{1.2}$$

and $u_i \equiv u(P_i)$ denotes the value of the grid function at the node P_i. The simplest case of our difference operator is

$$\Lambda u_i \equiv -(\Delta_1 + \cdots + \Delta_d)u_i, \quad P_i \in Q_h. \tag{1.3}$$

1.2. Conditions for correctness of the difference problem. For difference operators, we prefer to define the basic Euclidean space H differently that we did for PGMs. More precisely, we use the inner product

$$(\mathbf{u}, \mathbf{v})_H \equiv \|h\| \sum_{P_i \in Q_h} u_i v_i, \tag{1.4}$$

with $\|h\| \equiv h_1 \times \cdots \times h_d$, which means that the norm in H is just a difference analog of the norm in $L_2(Q)$. For a grid function $\mathbf{u} \in H$ defined on a set Q_h, it is convenient to use its finite extension u' (see § 3.1.3). If

$$(u', v')_0 \equiv \|h\| \sum_{P_i} u_i v_i, \tag{1.5}$$

where the summation is carried out with respect to all nodes (see Subsection 3.1.5), then the difference analogs of integration by parts take the simple forms defined by (3.1.14). If we eliminate in (1.1) all $u_i = 0$ for which $P_i \notin Q_h$, then we obtain an operator equation $L_h \mathbf{u} = \mathbf{f}$ in the Euclidean space H. We sometimes omit the dependence on h and write $L \equiv L_h$, e.g.,

$$(L\mathbf{v}, \mathbf{v}) = \sum_{r=1}^{d} \|\partial_r v'\|_0^2 = \sum_{r=1}^{d} \|\bar{\partial}_r v'\|_0^2, \quad \forall \mathbf{v} \in H$$

(see (3.1.14)). It is then easy to prove that

$$\|v\|_0 \leq \gamma \|v\|_\Lambda, \quad \forall v \in H, \tag{1.6}$$

with $\gamma \neq \gamma(h)$ (for small enough h, γ differs slightly from the γ defined by (0.1.9), but the proof is complicated by the need to use difference approximations of the eigenvalues of $-\Delta$ (see, e.g., [166])).

Lemma 1. *Suppose that the conditions of Lemma 0.1.2 are satisfied with constant γ defined by (1.6). Then*

$$(L\mathbf{v}, \mathbf{v}) \geq \delta_0 \|\mathbf{v}\|_\Lambda^2, \quad \forall \mathbf{v} \in H, \tag{1.7}$$

and

$$\|L\mathbf{v}\|_{\Lambda^{-1}} \leq \delta_1 \|v\|_\Lambda, \quad \forall \mathbf{v} \in H, \tag{1.8}$$

where $\delta_0 \equiv \nu - \nu'$ and $\delta_1 > 0$ is a constant independent of the grid.
   Proof. It is easy to verify that

$$(L\mathbf{u}, \mathbf{v}) = \Phi^+(u'; v') + \Phi^-(u'; v'), \tag{1.9}$$

where $2\Phi^+(u'; v') \equiv X$ and $2\Phi^-(u'; v') \equiv Y$ are defined by

$$X \equiv \sum_{r=1}^{d}\sum_{l=1}^{d}(a_{r,l},\partial_l u'\partial_r v')_0 + \sum_{r=1}^{d}[(b_r\partial_r u', v')_0 - (b'_r u', \partial_r v)_0] + (c, u'v')_0$$

$$(1.10)$$

and

$$Y \equiv \sum_{r=1}^{d}\sum_{l=1}^{d}(a_{r,l},\bar{\partial}_l u'\bar{\partial}_r v')_0 + \sum_{r=1}^{d}[(b_r\bar{\partial}_r u', v')_0 - (b'_r u', \bar{\partial}_r v)_0] + (c, u'v')_0.$$

$$(1.11)$$

The same estimation as in the proof of Lemma 0.1.2, but with γ from (1.6), yields $2\Phi^+(u', v') \le \delta_0\|\mathbf{u}\|_\Lambda^2$ and $2\Phi^-(u', v') \le \delta_0\|\mathbf{u}\|_\Lambda^2$, $\forall \mathbf{v} \in H$. This and (1.9)–(1.11) lead to (1.7). Next, for $|\Phi^+(u'; v')|$ and $|\Phi^-(u'; v')|$, we apply the same estimation as in the proof of Lemma 0.1.1. This yields $|\Phi^+(u'; v')| \le \frac{\delta_1}{2}\|\mathbf{u}\|_\Lambda\|\mathbf{v}\|_\Lambda$. Hence,

$$|(L\mathbf{u}, \mathbf{v})| \le \delta_1\|\mathbf{u}\|_\Lambda\|\mathbf{v}\|_\Lambda, \quad \forall \mathbf{u} \in H, \quad \forall \mathbf{v} \in H. \qquad (1.12)$$

It remains to observe that (1.12) and Lemma 0.4.5 yield (1.8). □

Lemma 1 implies that inequalities

$$\|L^{-1}\|_{H(\Lambda^{-1})\mapsto H(\Lambda)} \le K_0, \quad \|L\|_{H(\Lambda)\mapsto H(\Lambda^{-1})} \le K_1 \qquad (1.13)$$

hold with constants K_0 and K_1 independent of h (here, the symbols K and κ are used only to denote nonnegative constants independent of the grid). We can even prove a more general result (see [164, 166] and Theorem 1 below), for which we need the following lemma. (Note that $|\bar{v}|_{0,Q} = \|v'\|_0$.)

Lemma 2. *For each* $\mathbf{v} \in H$, *let* $\bar{v}$ *be a piecewise constant function that equals* v'_i *inside each cell* $\Pi_i \equiv [i_1h_1, (i_1+1)h_1] \times \cdots \times [i_dh_d, (i_d+1)h_d]$, *where* v' *is the finite extension of* $\mathbf{v}$. *Suppose we partition each* Π_i *into a union of congruent* $d!$ *simplexes (see § 2.1) such that each has a common vertex* $P_i \equiv [i_1h_1, \ldots, i_dh_d]$. *Suppose also that this regular triangulation of* $\bar{Q}$, *for a given* v', *defines a standard piecewise linear function* $\hat{v} \in G$. *Then*

$$|\hat{v} - \bar{v}|_{0,Q} \le K_2 h\|\mathbf{v}\|_\Lambda, \quad \forall \mathbf{v} \in H. \qquad (1.14)$$

Proof. It suffices to show that $|\hat{v} - \bar{v}|_{0,T}^2 \le K_2^2 h^2|\hat{v}|_{1,T}^2$, where T is an arbitrary simplex in our triangulation $T_h(\bar{Q})$. One of several possible ways to do this is to rewrite this inequality in the form

$$|\hat{z}|_{0,T}^2 \le K_2^2 h^2|\hat{z}|_{1,T}^2,$$

where $\hat{z} \equiv \hat{v} - \bar{v}$, and observe that $\hat{z}(P_i) = 0$. We then apply almost the same reasoning as in Subsections 0.5.3 and 2.3.1 to establish this inequality (with $h = 1$) for the reference simplex that is a regular simplicial part of the unit cube (we deal with a finite-dimensional eigenvalue problem). We then make use of the h-transformation of this simplex and observe that

$$|\hat{v}|_{1,Q} = \|\mathbf{v}\|_\Lambda, \quad \forall \mathbf{v} \in H, \tag{1.15}$$

(Λ and H here differ from those in Chapter 5). $\square$

Theorem 1. *Suppose that, in our original boundary value problem, all coefficients $a_{r,l}, b_r, b'_r, c$, and f are continuous on $\bar{Q}$. Let $\bar{Q}$ be partitioned into a union of parallelepipeds $\bar{Q}_1, \ldots, \bar{Q}_p$. Suppose that, on each domain $Q_s, s \in [1,p]$, these coefficients have uniformly bounded first derivatives and that conditions (0.1.32) with respect to the coefficients $a_{r,l}$ are satisfied. Suppose also that the operator $L \in \mathcal{L}(G)$, defined by bilinear form (0.1.30), is invertible and that we approximate it by the difference operators $L_h \in \mathcal{L}(H)$ (see (1.1) and (1.2)) with $h \le h_0$, where h_0 is small enough. Finally, suppose that all vertices of the parallelepipeds $\bar{Q}_1, \ldots, \bar{Q}_p$ are nodes of the grids under consideration. Then the difference operator L_h is also invertible and inequalities (1.13) hold with constants K_0 and K_1 independent of h.*

Proof. Observe that $(L_h \mathbf{v}, \mathbf{v}) \ge \delta \|\mathbf{v}\|_\Lambda^2 - \kappa_0 \|\mathbf{v}\|^2$, $\forall \mathbf{v} \in H$, where $\delta \equiv \nu/2$ and $\kappa_0 > 0$ (see (0.1.32)). Thus, it suffices to prove only (1.13) with K_0. Suppose that there exists no such constant. Then there exist a sequence $\{h^{(n)}\}$ with $h^{(n)} \to 0$ and an associated sequence $\mathbf{u}^{(n)}$ such that

$$\|\mathbf{u}^{(n)}\|_\Lambda = 1, \quad \|L_h \mathbf{u}^{(n)}\|_{\Lambda^{-1}} \to 0. \tag{1.16}$$

Then from (1.16) we see that

$$|(L_h \mathbf{u}^{(n)}, \mathbf{u}^{(n)})| \le \|L_h \mathbf{u}^{(n)}\|_{\Lambda^{-1}} \|\mathbf{u}^{(n)}\|_\Lambda \to 0. \tag{1.17}$$

From (1.16) and (1.17), it follows that

$$\|\mathbf{u}^{(n)}\| \ge \kappa_1 > 0. \tag{1.18}$$

Next, for our grid functions $\mathbf{u}^{(n)}$, we define functions $\bar{u}^{(n)}$ and $\hat{u}^{(n)}$ as indicated in Lemma 2. Then $|\hat{u}^{(n)}|_{1,Q} = 1$ and $|\hat{u}^{(n)}|_{0,Q} \ge \kappa_2 > 0$. Hence, we can choose a subsequence (for convenience, we identify it with the original sequence) such that it converges (weakly in our Hilbert space G and strongly in the space $L_2(Q)$) to a function $u \in G$. It is obvious that $|u|_{0,Q} \ge \kappa_2$. Now, for smooth functions $v(x) \in C_0^\infty(Q)$, we define the vectors $\mathbf{v}$ and the corresponding functions $\bar{v}$ and $\hat{v}$. Then we have

$$(a_{r,l}, \partial_l((u^{(n)})' \partial_r v')_0 = \sum_{T \in T_h(\bar{Q})} (\bar{a}_{r,l}(I_T x), D_l \hat{u}^{(n)} D_r \hat{v})_{0,T},$$

where $a_{r,l}(I_T x)$ refers to a possible shift to a neighbor simplex in the parallelepiped cell. Since inside each cell Π_i our coefficients have uniformly bounded first derivatives, it is easy to prove that

$$\lim_{n \to \infty} (a_{r,l}, \partial_l ((u^{(n)})' \partial_r v')_0 = (a_{r,l}, D_l u D_r v)_{0,Q}.$$

In the same way, we treat all the remaining terms in the expression for $(L_h \mathbf{u}^{(n)}, \mathbf{v})$ (see (1.9)–(1.11)). Hence, $\lim_{n \to \infty} (L_h \mathbf{u}^{(n)}, \mathbf{v}) = b(u; v)$ and $b(u; v) = 0$. Since this relation holds for arbitrary smooth v, then $u \in \mathrm{Ker}\ L$ and $\mathrm{Ker}\ L \neq 0$, which contradicts invertibility of L. $\square$

1.3. Error estimates. Here, we deal only with $d = 2$ and $d = 3$, so $u \in V \equiv G^{(2)}$ implies that $u \in C(\bar{Q})$.

Theorem 2. Let the conditions of Theorem 1 be satisfied and suppose that all coefficients $a_{r,l}, b_r, b'_r, c$, and f have continuous and uniformly bounded first derivatives. Suppose also that the solution u of our problem belongs to $G^{(2)}$. Then, for the piecewise linear extension $\hat{u}_h$ of the difference solution $\mathbf{u}_h$, we have the error estimate

$$|\hat{u}_h - u|_{1,Q} \leq K_3 h. \tag{1.19}$$

Proof. Let $\mathbf{w} \in H$ be the vector whose components are values of the solution of the original problem at nodes $P_i \in Q_h$. (Note that $\|\mathbf{w}\|_\Lambda \leq K_4$.) Define the vector ζ_h by its components $\zeta_i \equiv L_h \mathbf{w}_i - f_i = L_h \mathbf{w}_i - L_h \mathbf{u}_i$, $\forall P_i \in Q_h$. We then show that $\zeta_h \to 0$ as $h \to 0$. Indeed,

$$(\zeta_h, \mathbf{v}) = (L_h \mathbf{w}, \mathbf{v}) - (\mathbf{f}, \mathbf{v}), \quad \forall \mathbf{v} \in H. \tag{1.20}$$

Next, from (1.20) and (1.9)–(1.11), it follows that

$$(\zeta_h, \mathbf{v}) = \Phi^+(w'; v') + \Phi^-(w'; v') - (\bar{f}, \bar{v})_{0,Q}. \tag{1.21}$$

We compare each term on the right-hand side of (1.21) with the corresponding term in the expression for $b(u; \hat{v}) - (f, \hat{v})_{0,Q}$. For example, for $X_{r,l} \equiv (a_{r,l}, \partial_l w' \partial_r v')_0 - (a_{r,l}, D_l u D_r v)_{0,T}$, we can show in a straightforward manner that $|X_{r,l}| \leq \kappa_{r,l} h \|\mathbf{v}\|_\Lambda$ (it suffices to deal with separate simplexes in the indicated regular triangulation of $\bar{Q}$ and to take into account the assumptions on the given coefficients). For $X_f \equiv (\bar{f}, \bar{v})_{0,Q} - (f, \hat{v})_{0,Q}$, we have $|X_f| \leq |(\bar{f} - f, \bar{v})_{0,Q}| + |(f, \bar{v} - \hat{v})_{0,Q}|$ and we can make use of (1.14). These simple estimates imply that

$$|(\zeta_h, \mathbf{v})| \leq K_5 h \|\mathbf{v}\|_\Lambda. \tag{1.22}$$

Since (1.22) holds for each $\mathbf{v} \in H$, then $\|\zeta_h\|_{\Lambda^{-1}} \leq K_5 h$ (see Lemma 0.4.5). For $\mathbf{z} \equiv \mathbf{u} - \mathbf{w}$, we have $L_h \mathbf{z} = \zeta_h$, and (1.13) implies that $\|\mathbf{z}\|_\Lambda \leq K_0 K_5 h$

and $|\hat{w} - \hat{u}_h|_{1,Q} \le K_0 K_5 h$ (in accordance with (1.15)). But the relation $\hat{w} = \mathrm{int}_1 u$ and Theorem 1.3.1 now imply that (1.19) holds. $\square$ [1]

1.4. Asymptotically optimal algorithms. Here, $\varepsilon > 0$ is a prescribed tolerance.

Theorem 3. Let the conditions of Theorem 2 be satisfied. Then there exists an optimal model operator $B = B^* > 0 \asymp \Lambda$ such that $M = O(|\ln \varepsilon|)$ iterations of the modified iterative methods considered in § 1.3 lead to vectors $\mathbf{u}^M \equiv \mathbf{v}$ such that $|\hat{v} - u|_{1,Q} \le \varepsilon$, with computational work $W(\varepsilon) = O(\varepsilon^{-d}|\ln \varepsilon|)$; multigrid acceleration of these basic iterative algorithms improves the work estimate to $W(\varepsilon) = O(\varepsilon^{-d})$.

Proof. Our difference operator Λ is the same as the operator Λ_Q in Chapter 5 except for the factor $\|h\|$. Thus, determining an optimal model operator $B \asymp \Lambda$ is no problem. Under general conditions (1.13), we may use, f. e., iterations (1.3.25) (iterations (0.3.13) are applicable for $L_h = L_h^* \ge 0$). We choose M in such a way that $|\mathbf{v} - \mathbf{u}|_\Lambda \le \varepsilon/2$ and $|\hat{v} - \hat{u}|_{1,Q} \le \varepsilon/2$ (see (1.15)). If $h \asymp \varepsilon$ is small enough that (1.19) yields $|\hat{u}_h - u|_{1,Q} \le \varepsilon/2$, then $\hat{v}$ is the desired ε-approximation to u; the required computational work is $W(\varepsilon) = O(\varepsilon^{-d}|\ln \varepsilon|)$. Finally, multigrid improvement to the estimate $W(\varepsilon) = O(\varepsilon^{-d})$ is achieved by applying Theorem 1.4.4. $\square$ [2]

1.5. Strongly elliptic systems and boundary value problems in the theory of elasticity. We show that the strengthened variant of the Kolmogorov-Bakhvalov hypothesis can be proved on the basis of difference methods for elasticity problems associated with the Hilbert space $G \equiv (\overset{\circ}{W}{}_2^1\,(Q))^3$ and Q as above. Here, $\bar{u}_h \equiv [\mathbf{u}_1, \mathbf{u}_2, \mathbf{u}_3]^T \in H$, where $H \equiv H_{(1)}^3$ and $H_{(1)}$ corresponds to the old H in Subsection 1.2 and we use the same finite extensions of our functions (when we want to stress that we deal with such an extension, we write u'_k instead of $\mathbf{u}_k$, $k = 1, 2, 3$).

In lieu of (5.3.9), we use

$$\epsilon_{l,s}^+(\bar{u}_h) \equiv \frac{\partial_s u'_l + \partial_l u'_s}{2}, \quad \epsilon_{l,s}^-(\bar{u}_h) \equiv \frac{\bar{\partial}_s u'_l + \bar{\partial}_l u'_s}{2}, \tag{1.23}$$

---

[1] The conditions in Theorem 2 on the coefficients may be weakened, and even approximation of the boundary conditions can be analyzed in much the same way (see [164] where also the case of arbitrary d was investigated under the assumption that $u \in C^2(\bar{Q})$). First results of the above type for difference methods (under less general conditions) were obtained in [424] by a more complicated analysis.

[2] We emphasize again that modified gradient methods can be used as well. Theorem 3 implies that, for a class of simple elliptic boundary value problems, the strengthened variant of the Kolmogorov-Bakhvalov hypothesis is established on the basis of difference methods that can actually be regarded as small perturbations of PGMs.

$$\sigma_{r,r}^+(\bar{u}_h) \equiv (\lambda + 2\mu)\epsilon_{r,r}^+(\bar{u}_h) + \sum_{l \neq r} \epsilon_{r,r}^+(\bar{u}_h), \quad r = 1, 2, 3, \qquad (1.24)$$

$$\sigma_{r,r}^-(\bar{u}_h) \equiv (\lambda + 2\mu)\epsilon_{r,r}^-(\bar{u}_h) + \sum_{l \neq r} \epsilon_{r,r}^-(\bar{u}_h), \quad r = 1, 2, 3, \qquad (1.25)$$

$$\sigma_{r,l}^+(\bar{u}_h) \equiv 2\mu\epsilon_{r,l}^+(\bar{u}_h), \quad \sigma_{r,l}^-(\bar{u}_h) \equiv 2\mu\epsilon_{r,l}^-(\bar{u}_h), \quad r \neq l. \qquad (1.26)$$

We also make use of the vectors

$$\left.\begin{array}{l} \sigma_s^+(\bar{u}_h) \equiv [\sigma_{1,s}^+(\bar{u}_h), \sigma_{2,s}^+(\bar{u}_h), \sigma_{3,s}^+(\bar{u}_h)]^T, \\ \sigma_s^-(\bar{u}_h) \equiv [\sigma_{1,s}^-(\bar{u}_h), \sigma_{2,s}^-(\bar{u}_h), \sigma_{3,s}^-(\bar{u}_h)]^T, \end{array}\right\} \qquad (1.27)$$

$s = 1, 2, 3$. We then can use difference approximations of type

$$(L_h \bar{u}_h)_i \equiv -\frac{1}{2} \sum_{s=1}^{3} \left(\bar{\partial}_s \sigma_s^+(\bar{u}_h) + \partial_s \sigma_s^-(\bar{u}_h) \right)_i = (\bar{f}_h)_i, \quad P_i \in Q_h, \qquad (1.28)$$

where $(\bar{f}_h)_i \equiv [\mathbf{F}_1, \mathbf{F}_2, \mathbf{F}_3]_i^T$ and (1.2) holds with $\bar{u}_h$ in place of u. If we number the unknowns in such a way that the values of $(\mathbf{u}_k)_i$ have smaller numbers than values of $(\mathbf{u}_l)_j$ for $k < l$, then the corresponding difference system can be considered as the operator equation

$$L_h \bar{u}_h = \bar{f}_h \qquad (1.29)$$

in the Euclidean space H. Here, L_h corresponds to the block matrix

$$L_h \equiv [L_{r,l}], \quad r = 1, 2, 3, \; l = 1, 2, 3. \qquad (1.30)$$

The difference operator $\bar{\Lambda}_h \equiv \bar{\Lambda}$ is taken as the block diagonal matrix with the diagonal blocks

$$\Lambda_{r,r} = \Lambda, \quad r = 1, 2, 3, \qquad (1.31)$$

where Λ is defined by (1.3) and (1.2).

   Theorem 4. *L_h and $\bar{\Lambda}_h$ are spectrally equivalent operators; moreover,*

$$\mu\bar{\Lambda}_h \leq L_h \leq (\lambda + 2\mu)\bar{\Lambda}_h. \qquad (1.32)$$

   Proof. It is easy to verify that

$$(L_h \bar{u}_h, \bar{v}_h) = \Phi^+(\bar{u}_h; \bar{v}_h) + \Phi^-(\bar{u}_h; \bar{v}_h), \tag{1.33}$$

where $\Phi^+(\bar{u}_h; \bar{v}_h) \equiv X$ and $\Phi^-(\bar{u}_h; \bar{v}_h) \equiv Y$ are defined by

$$X \equiv \lambda\left(\sum_{s=1}^{3} \epsilon^+_{s,s}(\bar{u}_h), \sum_{s=1}^{3} \epsilon^+_{s,s}(\bar{v}_h)\right)_0$$

$$+2\mu\left(\sum_{s=1}^{3}\sum_{l=1}^{3} \epsilon^+_{s,l}(\bar{u}_h), \sum_{s=1}^{3} \epsilon^+_{s,l}(\bar{v}_h)\right)_0, \tag{1.34}$$

$$Y \equiv \lambda\left(\sum_{s=1}^{3} \epsilon^-_{s,s}(\bar{u}_h), \sum_{s=1}^{3} \epsilon^-_{s,s}(\bar{v}_h)\right)_0$$

$$+2\mu\left(\sum_{s=1}^{3}\sum_{l=1}^{3} \epsilon^-_{s,l}(\bar{u}_h), \sum_{s=1}^{3} \epsilon^-_{s,l}(\bar{v}_h)\right)_0. \tag{1.35}$$

Hence, L_h is a symmetric operator. Moreover, we may write

$$(L_h \bar{u}_h, \bar{u}_h) = \mu \sum_{s=1}^{3}\sum_{l=1}^{3} \|\partial_s u'_l\|^2_0 + (\lambda+\mu)\left[\sum_{s=1}^{3}[\|\partial_s u'_s\|^2_0 + \|\bar{\partial}_s u'_s\|^2_0] + R\right], \tag{1.36}$$

$$R \equiv \sum_{r<l}^{3} \left[(\partial_r u'_r, \partial_l u'_l)_0 + (\bar{\partial}_r u'_r, \bar{\partial}_l u'_l)_0\right]. \tag{1.37}$$

This and (1.36) imply that $L_h \geq \mu\bar{\Lambda}_h$. To obtain the upper bound, we transform the terms on the right-hand side of (1.37) via (3.1.14). Then we have $|(\partial_r u'_r, \partial_l u'_l)_0| = |(\bar{\partial}_l u'_r, \bar{\partial}_r u'_l)_0| \leq 1/2[\|\partial_l u'_r\|^2_0 + \|\partial_r u'_l\|^2_0]$, $|(\bar{\partial}_r u'_r, \bar{\partial}_l u'_l)_0| = |(\partial_l u'_r, \partial_r u'_l)_0| \leq 1/2[\|\partial_l u'_r\|^2_0 + \|\partial_r u'_l\|^2_0]$. Therefore, we have $L_h \leq (\lambda + 2\mu)\bar{\Lambda}_h$, and (1.39) holds (it was shown in [162, 164]). $\square$

Theorem 5. *Let the solution $\vec{u} \in (\overset{\circ}{W}^1_2(Q))^3 \cap (W^2_2(Q))^3$. Then there exists an optimal model operator $\bar{B} \asymp \bar{\Lambda}$ (see (1.31)) such that $M = O(|\ln \varepsilon|)$ iterations of the modified Richardson method (0.1.13) lead to vectors $\bar{u}^M \equiv v$ such that $\|\hat{v} - \vec{u}\|_G \leq \varepsilon$ with computational work $W(\varepsilon) = O(\varepsilon^{-3}|\ln \varepsilon|)$; multigrid acceleration of these basic iterative algorithms improves the work estimate to $W(\varepsilon) = O(\varepsilon^{-3})$.*

Proof. Theorem 4 implies that $L_h \asymp \bar{\Lambda}_h$. We can easily determine an optimal model operator $\bar{B}_h \asymp \bar{\Lambda}_h$ by taking it as a block diagonal matrix corresponding to $\bar{\Lambda}_h$ (it suffices to define the diagonal blocks B of $\bar{B}_h$ as optimal preconditioners for Λ). Thus, convergence of (0.3.13) presents no problem. The only difficulty that deserves a special consideration is the generalization of Theorem 2, but this is straightforward. □ [3]

1.6. Difference schemes with higher order approximation. We indicate briefly (following [153]) how the model operators B_h constructed above can be used for difference schemes with higher order approximation on the same grid. This approach was later generalized for PGMs (see Subsection 5.1.6). For the simplest difference Poisson equation $(\Delta_1 + \Delta_2)u_i = f_i$ on a square grid then, we have

$$(\Delta_1 + \Delta_2)w_i - \frac{h^2}{12}(D_1^4 + D_2^4)w = f_i + O(h^4).$$

If w and f are smooth enough, we can improve the approximation by way of the equation

$$-L_{h,4}u_i \equiv (\Delta_1 + \Delta_2)u_i + \frac{h^2}{6}(\Delta_1\Delta_2)u_i = f_i + \frac{h^2}{12}(\Delta_1 + \Delta_2)f_i$$

(here we use identity $D_1^4w + D_2^4w + 2D_1^2D_2^2w = (D_1^2 + D_2^2)f$). Thus, $-L_{h,4}w_i = f_i + h^2/12(\Delta_1 + \Delta_2)f_i + O(h^4)$, but now our new operator $L_{h,4} \equiv \Lambda - [h^2/6]\Lambda_{1,2}$ has a more involved structure. Nonetheless, it is easy to verify that $L_{h,4} = L_{h,4}^*$ and that $2/3\Lambda \leq L_{h,4} \leq \Lambda$. Indeed, $(L_{1,2}\mathbf{u}, \mathbf{u}) = |\partial_1\partial_2 u'|_0^2$ (see (3.1.14)) and

$$|\partial_1\partial_2 u'|_0^2 \leq \frac{2}{h^2}\left(|\partial_1 u'|_0^2 + |\partial_2 u'|_0^2\right) = \frac{2}{h^2} = (\Lambda\mathbf{u}, \mathbf{u}).$$

Thus, $L_{h,4} \asymp \Lambda$. In [153] even a more general differential equation $(D_1^2 + 2a_{1,2}D_1D_2 + D_2^2)w = f$ with $|a_{1,2}| < 2/3$, was considered (see also [435]).

Finally, we stress the role of averaging in construction of difference schemes on general grids. Note that $\partial u(x_0) = 1/h(u'(x_0 + \xi), 1)_{0,\Pi}$, where $\Pi \equiv [x_0, x_0 + h]$. It is therefore reasonable for the case of, say, a triangular or quadrilateral cell Π to apply the approximation $\partial u(x_0) \approx \frac{1}{|\Pi|}(D_1u, 1)_{0,\Pi} = \frac{1}{|\Pi|}\oint_{\partial\Pi} vdx_2 \approx \sum_{P_i} \alpha_i u(P_i)$, where $x_0 \in \Pi$ and the sum is taken with respect to the vertices P_i of Π (we obtain the coefficients α by applying the trapezoid formula for the integral over each side of Π). [4]

---

[3] Nearly asymptotically optimal iterative methods with $\bar{B}_h \asymp \Lambda_h$ have been successfully in practice for many years (first practical applications for $d = 2$ and rectangular regions date back to the beginning of the sixties (see, e.g., [161, 210, 211]), where nearly orthogonal grids on the shell surface of type $\bar{P}(Q^0; m_1, m_2)$ were also used).

[4] These simple ideas in combination with the standard variational approach (when, say, $((D_1u)^2, 1)_{0,\Pi}$ is approximated by $|\Pi|(\sum_{P_i} \alpha_i u(P_i))^2$) were used widely, especially for practical applications. Current attention is paid to three-dimensional problems (see,

§ 2. A difference analog of $W_2^2(Q)$ for a d-dimensional parallelepiped.

2.1. Basic inequalities. [5]
Lemma 1. The operator Λ is such that

$$\|\Lambda \mathbf{v}\|^2 = \sum_{r=1}^{d} \|v'_{\bar{r},r}\|^2_{0,Q_h} + 2\sum_{r\neq l}^{d} \|v'_{r,l}\|^2_0, \quad \forall \mathbf{v} \in H. \tag{2.1}$$

Proof. For P_i belonging to the boundary of our parallelepiped $\bar{Q}$, we have $\bar{\partial}_l \partial_l \bar{\partial}_r \partial_r v' P_i = 0$ provided $l \neq r$. Hence, we may write

$$(\Lambda \mathbf{v}, \mathbf{v}) = X_1 + X_2, \tag{2.2}$$

$$X_1 \equiv \sum_{r=1}^{d} \|v'_{\bar{r},r}\|^2_{0,Q_h}, \quad X_2 \equiv 2\sum_{r\neq l}^{d} (v'_{\bar{r},r}, v'_{\bar{l},l})_0, \quad \forall \mathbf{v} \in H.$$

It is easy to verify that $(v'_{\bar{r},r}, v'_{\bar{l},l})_0 = (v'_{r,l}, v'_{r,l})_0$ (see (3.1.14)), which with (2.2) yields (2.1). □

In the study of the properties of $L \equiv L_h$ (see (1.1) and (1.2)) and the operator $A \equiv A_h$ defined by

$$Au_i \equiv \frac{1}{2} \sum_{s=1}^{d} \sum_{l=1}^{d} \left((I_s a^{s,l}) u'_{\bar{s},l} + (I s a^{s,l}) u'_{s,\bar{l}} \right)_i, \quad P_i \in Q_h \tag{2.3}$$

and (1.2), product rules for differences play an important role. We write them in symmetric form in terms of the operators

$$M_l u_i \equiv \frac{1}{2}(u'(P_i) + u'(I_l P_i)), \quad M_{-l} u_i \equiv \frac{1}{2}(u'(P_i) + u'(I_{-l} P_i)), \quad l \in [1, d]$$

(arithmetic average of neighbor nodes). Here we may write u instead of u' as the finite extension of a grid function $\mathbf{u} \in H$.

e.g., [304, 450]) and grids of type $\bar{P}(C; m_1, m_2, m_3)$, which allow use above of the model operators B_h constructed on parallelepiped grid. Theoretical aspects of relevant approximations on general grids were considered, e.g., in [53, 106, 268, 275, 366, 335].

[5] We confine ourselves to the simplest case of difference approximation of the first boundary value problem in a parallelepiped $\bar{Q} \equiv [0, l_1] \times \cdots \times [0, l_d]$ (see (1.1) and (1.2)). Our exposition is based on [156, 164]; there are many relevant results (see, e.g., [139, 266] and references therein). We assume that $h_r \equiv l_r (N_r + 1)^{-1}, r \in [1, d]$, and $Q_h \equiv \{P_i : P_i \in Q.\}$ Here we write v'_r and v'_r instead of the respective $\partial_r v'$ and $\bar{\partial}_r v', r \in [1, d]$; old coefficients $a_{s,l}$ will be denoted by $a^{s,l}, s \in [1, d], l \in [1, d]$; the simplest difference operator Λ is the same as in § 1.

Lemma 2. *For arbitrary grid functions u and a, we have*

$$(au)_l = (M_l a)u_l + a_l(M_l u), \quad l \in [1, d], \tag{2.4}$$

$$(au)_{\bar{l}} = (M_{-l} a)u_{\bar{l}} + a_{\bar{l}}(M_{-l} u), \quad l \in [1, d], \tag{2.5}$$

$$au_l = (au)_l - a_l(M_l u) - \frac{h_l}{2} a_l u_l, \quad l \in [1, d], \tag{2.6}$$

and

$$au_{\bar{l}} = (au)_{\bar{l}} - a_{\bar{l}}(M_{-l} u) + \frac{h_l}{2} a_{\bar{l}} u_{\bar{l}}, \quad l \in [1, d]. \tag{2.7}$$

Proof. It is easy to rewrite (2.4) and (2.5) in the respective forms

$$a_{n+1}u_{n+1} - a_n u_n = \frac{a_{n+1} + a_n}{2}(u_{n+1} - u_n) + (a_{n+1} - a_n)\frac{u_{n+1} + u_n}{2},$$

$$a_n u_n - a_{n-1}u_{n-1} = \frac{a_{n-1} + a_n}{2}(u_n - u_{n-1}) + (a_n - a_{n-1})\frac{u_{n-1} + u_n}{2},$$

where indices $n-1, n$ and $n+1$ correspond to the repective nodes $I_l P_i, P_i,$ and $I_l P_i$. Hence, (2.4) and (2.5) hold. We similarly rewrite (2.6) and (2.7) in the respective forms

$$a_n(u_{n+1} - u_n) = a_{n+1}u_{n+1} - a_n u_n - (a_{n+1} - a_n)[\frac{u_{n+1} - u_n}{2} + \frac{u_{n+1} + u_n}{2}],$$

$$a_n(u_n - u_{n-1}) = a_n u_n - a_{n-1}u_{n-1} + (a_n - a_{n-1})[\frac{u_n - u_{n-1}}{2} - \frac{u_{n-1} + u_n}{2}],$$

which are easily verified. $\square$

Everywhere below, the constants $\nu, \kappa, \sigma, \delta,$ and K are nonnegative and independent of the grid. We may stress sometimes the use of the inner product in H by writing $(u, v)_H$ (see § 1). Note that, for $D \in \mathcal{L}^+(H)$ and arbitrary $\epsilon > 0$, we have

$$\|v\|_D^2 \leq \frac{\epsilon}{2}\|Dv\|^2 + \frac{1}{2\epsilon}\|v\|^2, \quad \forall v \in \mathbf{H}. \tag{2.8}$$

Theorem 1. *Let constants $\nu_0 > 0, \nu_1 > 0,$ and $\kappa \geq 0$ be such that*

$$\nu_0|\xi|^2 \le \sum_{s=1}^{d} \sum_{l=1}^{d} a^{s,l} \xi_s \xi_l \le \nu_1 |\xi|^2, \quad \forall \xi \equiv [\xi_1, \ldots, \xi_d], \tag{2.9}$$

$$\max_{s,l} \left\{ |D_s a^{s,l}|_{C(\bar{Q})} \right\} \le \kappa, \quad \max_{k,s \ne l} \left\{ |D_k a^{s,l}|_{C(\bar{Q})} \right\} \le \kappa. \tag{2.10}$$

Then, for h_0 small enough, there exist constants $\sigma_0 > 0, \sigma_1 > 0$, and $\kappa_0 \ge 0$, independent of $h \le h_0$ and such that

$$\sigma_0 \|\Lambda \mathbf{v}\|^2 - \kappa_0 \|\mathbf{v}\|^2 \le (A\mathbf{v}, \Lambda \mathbf{v}) \le \sigma_1 \|\Lambda \mathbf{v}\|^2, \quad \forall \mathbf{v} \in H; \tag{2.11}$$

if we also have $\kappa = 0$, then $\kappa_0 = 0$.

Proof. We have $(A\mathbf{v}, \Lambda \mathbf{v}) = J_1 + \cdots + J_d$, where, for $r \in [1, d]$,

$$2J_r \equiv \sum_{s=1}^{d} \sum_{l=1}^{d} (M_s a^{s,l} v_{\bar{s},l}, v_{\bar{r},r})_H + \sum_{s=1}^{d} \sum_{l=1}^{d} (M_s a^{s,l} v_{s,\bar{l}}, v_{\bar{r},r})_H. \tag{2.12}$$

We proceed now to transform each term on the right-hand side of (2.12). First consider the case $s \ne r$ and $l \ne r$. In accordance with (2.6), we have $X_1 \equiv (M_{-s} a^{s,l} v_{\bar{s},l}, v_{\bar{r},r})_H = ((M_{-s} a^{s,l} v_{\bar{s}})_l, v_{\bar{r},r})_H + R_1, R_1 \equiv R_1(s, l, r)$,

$$R_1 \equiv -h_l/2((M_{-s} a^{s,l})_l v_{\bar{s},l}, v_{\bar{r},r})_H - (M_{-s} a^{s,l} M_l v_{\bar{s}})_l, v_{\bar{r},r})_H.$$

Since $s \ne r$ and $l \ne r$, then $((M_{-s} a^{s,l} v_{\bar{s}})_l, v_{\bar{r},r})_H = ((M_{-s} a^{s,l} v_{\bar{s}})_l, v_{\bar{r},r})_0$ (which holds for an arbitrary extension of our coefficients to the nodes $P_i \notin \bar{Q}$). Hence (see (3.1.14)), $((M_{-s} a^{s,l} v_{\bar{s}})_l, v_{\bar{r},r})_0 = ((M_{-s} a^{s,l} v_{\bar{s}})_r, v_{\bar{l},r})_0$ and $X_2 \equiv ((M_{-s} a^{s,l} v_{\bar{s}})_r, v_{\bar{l},r})_0 = (M_r M_{-s} a^{s,l} v_{\bar{s},r}, M_{-l} v_{\bar{l},r})_0 + R_2$, where $R_2 \equiv R_2(s, l, r) \equiv (M_{-s} a_r^{s,l} M_r v_{\bar{s}}, v_{\bar{l},r})_0$. Finally,

$$X_1 = (a^{s,l} v_{\bar{l},r}, v_{\bar{s},r})_0 + R_1 + R_2 + R_3,$$

where $R_3 \equiv R_3(s, l, r) \equiv ((M_r M_{-s} a^{s,l} - a^{s,l}) v_{\bar{s},r}, v_{\bar{l},r})_0$. Similarly, we have $\bar{X}_1 \equiv (M_s a^{s,l} v_{\bar{l},s}, v_{\bar{r},r})_H = (a^{s,l} v_{\bar{l},l}, v_{\bar{r},s})_0 + \bar{R}_1 + \bar{R}_2 + \bar{R}_3$. For the remaining terms with $s = r$ or $l = r$ on the right-hand side of (2.12), we have: $(M_{-s} a^{s,l} v_{\bar{s},l}, v_{\bar{r},r})_H = (a^{s,l} v_{\bar{s},l}, v_{\bar{r},r})_H + R_4$ and $(M_s a^{s,l} v_{\bar{l},s}, v_{\bar{r},r})_H = (a^{s,l} v_{\bar{l},s}, v_{\bar{r},r})_H + \bar{R}_4$, where $R_4 \equiv R_4(s, l, r) \equiv \frac{h_s}{2}((a_{\bar{s}}^{s,l} v_{\bar{s},l}, v_{\bar{r},r})_H$ and $\bar{R}_4 \equiv \bar{R}_4(s, l, r) \equiv \frac{h_s}{2}((a_s^{s,l} v_{\bar{l},s}, v_{\bar{r},r})_H$. Therefore,

$$J_r = \frac{1}{2} \sum_{s=1}^{d} \sum_{l=1}^{d} (a^{s,l} v_{\bar{l},r}, v_{\bar{s},r})' + (a^{s,l} v_{\bar{r},l}, v_{\bar{r},s})' + Y_r. \tag{2.13}$$

Here, $(a^{s,l}v_{\bar{l},r}, v_{\bar{s},r})' \equiv (a^{s,l}v_{\bar{l},r}, v_{\bar{s},r})_0$ and $(a^{s,l}v_{\bar{r},l}, v_{\bar{r},s})' \equiv (a^{s,l}v_{\bar{r},l}, v_{\bar{r},s})_0$ provided $s \neq r, l \neq r$; they refer to the inner products in H otherwise, and $Y_r \equiv \sum_{s \neq r} \sum_{l \neq r} [\sum_{k=1}^{3} R_k(s,l,r) + \sum_{k=1}^{3} \bar{R}_k(s,l,r)] + \sum_{s \neq r} (R_4(s,l,r) + \bar{R}_k(s,l,r)) + \sum_{l \neq r} (R_4(s,l,r) + \bar{R}_k(s,l,r)) + R_4(r,r,r) + \bar{R}_4(r,r,r)), r \in [1,d]$. Next, condition (2.10) shows that, for $X \equiv \sum_{r=1}^{d} |Y_r|$, we have

$$X \leq (\kappa_1 \epsilon + \kappa_2 h) \|\Lambda \mathbf{v}\|^2 + \kappa_3 \max\{1; (\epsilon)^{-1}\} \|\mathbf{v}\|_{\Lambda}^2,$$

which with (2.8) for $D \equiv \Lambda$ leads to

$$\sum_{r=1}^{d} |Y_r| \leq (\kappa_4 \epsilon + \kappa_2 h) \|\Lambda \mathbf{v}\|^2 + \kappa_5 \max\{1; (\epsilon)^{-1}\} \|\mathbf{v}\|^2. \tag{2.14}$$

But

$$2(J_r - Y_r) = Z_r + \bar{Z}_r, \tag{2.15}$$

where $Z_r \equiv \sum_{s=1}^{d} \sum_{l=1}^{d} (a^{s,l}v_{\bar{l},r}, v_{\bar{s},r})'$ and $\bar{Z}_r \equiv (a^{s,l}v_{\bar{r},l}, v_{\bar{r},s})$. Now rewrite Z_r in the form $Z_r = Z_{r,1} + Z_{r,2}$, where

$$Z_{r,1} = \|h\| \sum_{P_i \in Q_h} \sum_{s=1}^{d} \sum_{l=1}^{d} a^{s,l}(P_i) v_{\bar{l},r}(P_i) v_{\bar{s},r}(P_i), \tag{2.16}$$

$$Z_{r,2} = \|h\| \sum_{P_i \notin Q_h} \sum_{s \neq r} \sum_{l \neq r} a^{s,l}(P_i) v_{\bar{l},r}(P_i) v_{\bar{s},r}(P_i). \tag{2.17}$$

Note that both right-hand sides in (2.16) and (2.17) can be estimated from above and below in accordance with (2.9) because nonzero terms correspond to $P_i \in \bar{Q}$. For example,

$$\nu_0 \sum_{s \neq r} (v_{\bar{s},r}(P_i))^2 \sum_{s \neq r} \sum_{l \neq r} a^{s,l}(P_i) v_{\bar{l},r}(P_i) v_{\bar{s},r}(P_i) \leq \nu_1 \sum_{s \neq r} (v_{\bar{s},r}(P_i))^2.$$

This and a similar procedure for $Z_{r,2}$ enables us to write

$$\nu_0 F_r \leq J_r - Y_r \leq \nu_1 F_r \quad \forall \mathbf{v} \in H, \quad r \in [1,d], \tag{2.18}$$

and

$$F_r \equiv \sum_{r=1}^{d} \|v_{\bar{r},r}\|_H^2 + 1/2 \sum_{r \neq l} [\|v_{r,\bar{s}}\|_0^2 + \|v_{s,\bar{r}}\|_0^2],$$

$r \in [1, d]$. Note that (2.17) leads (see (2.1)) to

$$\nu_0 \|\Lambda \mathbf{v}\|^2 \leq \sum_{r=1}^{d} (J_r - Y_r) \leq \nu_1 \|\Lambda \mathbf{v}\|^2, \quad \forall \mathbf{v} \in H, \quad r \in [1, d]. \tag{2.19}$$

Now, from (2.19) and (2.14), we conclude that, for all $\mathbf{v} \in H$,

$$(A\mathbf{v}, \Lambda \mathbf{v}) \geq (\nu_0 - (\kappa_4 \epsilon + \kappa_2 h)) \|\mathbf{v}\|_\Lambda^2 - \kappa_5 \max\{1; (\epsilon)^{-1}\} \|\mathbf{v}\|^2. \tag{2.20}$$

Hence, we can find $\sigma_0 > 0$ such that estimate (2.11) from below holds if $\nu_0 - \kappa_2 h > 0$ and we choose $\epsilon > 0$ small enough. The estimate from above is simpler and follows for any of h. It remains to observe that, under the additional condition of our theorem, we can take $\kappa_i = 0, i \in [1, 5]$. $\square$ [6]

2.2. Correctness conditions.

Theorem 2. *Suppose that conditions of Theorem 1 are satisfied and the functions $b_r, b_r', D_r b_r'$, and $c - c_0$ (see (1.1)) are uniformly bounded, where c_0 is a constant. Then, for h_0 small enough, there exist constants $\bar{\sigma}_0 > 0, \bar{\sigma}_1 > 0$, and $\bar{\kappa}_0 \geq 0$ independent of $h \leq h_0$ such that*

$$\bar{\sigma}_0 \|\Lambda \mathbf{v}\|^2 - \bar{\kappa}_0 \|\mathbf{v}\|^2 \leq (L_h \mathbf{v}, \Lambda \mathbf{v}) \leq \bar{\sigma}_1 \|\Lambda \mathbf{v}\|^2, \quad \forall \mathbf{v} \in H. \tag{2.21}$$

Proof. We have $L - c_0 I = A + A_1$, where the operator A_1 contains no differences higher than first order. This and (2.11) yields

$$((L_h - c_0 I)\mathbf{v}, \Lambda \mathbf{v}) \geq (\sigma_0 - \bar{\kappa}_2 \epsilon) \|\Lambda \mathbf{v}\|^2$$

$$- (\bar{\kappa}_0 + \bar{\kappa}_3 \max\{1; (\epsilon)^{-1}\}) \|\mathbf{v}\|^2, \tag{2.22}$$

$\forall \mathbf{v} \in H$. If $\bar{\kappa}_2 = 0$, then we take $\bar{\sigma}_0 = \sigma_0$. If $\bar{\kappa}_2 > 0$, then, for $\sigma_0 - \bar{\kappa}_2 \epsilon = \sigma_0/2$, we have $\bar{\sigma}_0 = \sigma_0/2$. For such ϵ, (2.22) leads to $((L_h - c_0 I)\mathbf{v}, \Lambda \mathbf{v}) \geq \bar{\sigma}_0 \|\Lambda \mathbf{v}\|^2 - \bar{\kappa}_4 \|\mathbf{v}\|^2, \forall \mathbf{v} \in H$. Therefore,

$$(L_h \mathbf{v}, \Lambda \mathbf{v}) \geq \bar{\sigma}_0 \|\Lambda \mathbf{v}\|^2 - (\bar{\kappa}_4 - c_0) \|\mathbf{v}\|^2, \quad \forall \mathbf{v} \in H \tag{2.23}$$

and, in (2.21), $\bar{\kappa}_0 \equiv \bar{\kappa}_4 - c_0$. The estimate involving σ_1 is much simpler. $\square$

Theorem 3. *Suppose that conditions of Theorem 2 are satisfied and that c_0 is a large enough constant. Then there exists a constant $\delta_0 > 0$ independent of $h \leq h_0$ such that*

$$\delta_0 \|\Lambda \mathbf{v}\|^2 \leq (L_h \mathbf{v}, \Lambda \mathbf{v}), \quad \forall \mathbf{v} \in H, \tag{2.24}$$

---

[6] Condition (2.10) can be replaced by corresponding Lipschitz conditions.

and

$$\|L_h^{-1}\|_{H \mapsto H(\Lambda^2)} \leq \frac{1}{\delta}. \tag{2.25}$$

Proof. It suffices to take $c_0 = \bar{\kappa}_4$ in (2.23) to obtain $\delta_0 = \bar{\sigma}_0$ in (2.24). Next, from (2.24), we obtain $\delta_0 \|\Lambda \mathbf{v}\|^2 \leq \|L_h \mathbf{v}\| \|\Lambda \mathbf{v})\|$, $\forall \mathbf{v} \in H$, which, after the change $L_h \mathbf{v} = \mathbf{w}$, leads to (2.25). $\square$ [7]

Theorem 4. *Suppose that conditions of Theorem 2 are satisfied and, for $h \leq h_0$, inequality (1.13) with K_0 holds. Then (2.25) holds.* [8]

Proof. From (2.21), it follows that

$$\bar{\sigma}_0 \|\Lambda \mathbf{v}\|^2 \leq \|L_h \mathbf{v}\| \|\Lambda \mathbf{v}\| + \bar{\kappa}_0 \|\mathbf{v}\|_\Lambda \|\mathbf{v}\|_{\Lambda^{-1}}, \quad \forall \mathbf{v} \in H. \tag{2.26}$$

Observe that

$$\|\mathbf{v}\|_\Lambda \|\mathbf{v}\|_{\Lambda^{-1}} \leq K_0 \|L_h \mathbf{v}\|_{\Lambda^{-1}} \|\mathbf{v}\|_{\Lambda^{-1}}, \forall \mathbf{v} \in H,$$

which with (1.6) yields $\|\mathbf{v}\|_\Lambda \|\mathbf{v}\|_{\Lambda^{-1}} \leq K_0 \gamma^4 \|L_h \mathbf{v}\| \|\Lambda \mathbf{v}\|$, $\forall \mathbf{v} \in H$. This and (2.26) in turn imply that

$$\|\Lambda \mathbf{v}\| \leq \frac{1}{\delta_0} \|L_h \mathbf{v}\| \|\Lambda \mathbf{v}\|, \quad \forall \mathbf{v} \in H, \tag{2.27}$$

where $1/\delta_0 = [1 + \bar{\kappa}_0 K_0 \gamma^4][\bar{\sigma}_0]^{-1}$. Thus, (2.25) holds. $\square$

2.3. Asymptotically optimal algorithms for grid systems.

Theorem 5. *Suppose that the solution u of the original differential problem is in $C^3(\bar{Q})$. Suppose also that (2.25) holds for $h \leq h_0$. Then $\|\Lambda(\mathbf{u} - \mathbf{w})\| \leq \bar{K}h$, where $\mathbf{u}$ and $\mathbf{w}$, respectively, denote the solution of the difference system and the vector of values $u(P_i)$ at the nodes $P_i \in Q$.*

Proof. It suffices to apply (2.25) because $L_h \mathbf{z}_h = \zeta_h$, where $\mathbf{z}_h \equiv \mathbf{u} - \mathbf{w}$ and $\|\zeta_h\| = O(h)$. $\square$

Theorem 6. *Let the conditions of Theorem 2 be satisfied and, for $h \leq h_0$, inequality (1.13) with K_0 holds. Then the operators L_h and Λ_h are connected by the relationship C^1.*

Proof. It suffices to combine inequality (2.27) and the fact that $\|L_h \mathbf{v}\|^2 \leq \delta_1 \|\Lambda \mathbf{v}\|^2$, $\forall \mathbf{v} \in H$. $\square$

Theorems 6 and 1.3.14 imply that iterations (1.3.19), with $B_1 = I$ and $B_2 = B$, where B is defined by Λ as in Lemma 3.4.5, apply. However, it seems preferable to make use of the fact that L_h and Λ_h are connected by relationship C^3 as well (see Theorem 1.1) and apply iterations of type

---

[7] It is possible to obtain (2.24) and (2.25) even when $c_0 - \bar{\kappa}_4 < 0$ provided $|c_0 - \bar{\kappa}_4| < \bar{\sigma}_0 \gamma^4$ (see (1.6)). Moreover, (2.25) was established under much more general conditions in [164].

[8] We emphasize that (1.13) was proved in Theorem 1.1.

(1.3.25). Finally, we note that estimates of type (2.25) can be also obtained for PGMs associated with regular triangulations of $\bar{Q}$ (the resulting grid operators are very close to the difference operators (see § 2.4)).

§3. Difference methods for quasilinear elliptic problems

3.1. Difference embedding theorems. The case of mildly nonlinear difference problems differs only slightly from the linear case considered in Sections 1 and 2. But if we want to deal with difference approximations of nonlinear problems like those considered in § 5.5, then our investigations of correctness and applicability of effective iterative methods depend crucially on the ability to use difference analogs of the classical Sobolev embedding theorems (see [18, 164, 169, 458]). [9]

We define

$$|u'|_{L_p(s,P_j)} \equiv [h_1 \cdots h_s \sum_{P_i \in \Omega_h(s,P_j)} |u_i|^p]^{1/p} \tag{3.1}$$

and [10]

$$|u'|_{m,p} \equiv [\sum_{|\alpha|=m} |\partial^\alpha u'|_{0,p}^p]^{1/p}. \tag{3.2}$$

Theorem 1. Suppose $d < p$. Then

$$\|u'\|_C \leq K' (|u'|_{0,1} + |u'|_{1,p}), \quad \forall u'. \tag{3.3}$$

Suppose that $d \geq p, s > d - p$, and $q \leq q^* \equiv sp[d - p]^{-1}$. Then

$$\|u'\|_{L_q(s,P_j)} \leq K'' (|u'|_{0,1} + |u'|_{1,p}), \quad \forall u'; \tag{3.4}$$

if $d = p$, then $q \geq 1$ in (3.4) may be any number.

Proof. It suffices to prove (3.3) and (3.4) under the assumption that $|u'|_{0,1} + |u'|_{1,p} = 1$. For each such u', we define its respective piecewise linear and piecewise constant extensions $\hat{u}$ and $\bar{u}$ as in § 1. Then

---

[9] First such theorems were proved by Sobolev in [458] and were connected with the use of polylinear functions (see § 2.1) inside cells of a parallelepiped grid. We simplify and specify these using regular triangulations of these cells and associated spaces of piecewise linear functions. For a bounded domain $\Omega \subset \mathbf{R}^d$ with Lipschitz boundary, we use parallelepiped grids as in § 1, but now define Ω_h as the set of the nodes $P_i \in \Omega$. For each grid function $\mathbf{u} \in H$ we use its finite extension u' (see (1.2)) and introduce a family of norms. For this we consider a node P_i and let $\Omega_h(s, P_j)$ denote the set of nodes having the $d - s$ last coordinates the same as P_i (they belong to a $(d - s)$-dimensional plane $\mathcal{P}(s, P_j)$ and, if $s = 0$, then $\Omega_h(0, P_j)$ is merely the set of all nodes).

[10] Note that $|u|_{0,p} = \|u'\|_{L_p(\Omega_h)} = |u'|_{L_p(d,P_j)}$. Here we let K and κ denote only nonnegative constants independent of the grid.

$$\|u'\|_C = \|\hat{u}\|_C, \quad |\partial_r u'|_{0,p} = \|D_r \hat{u}\|_{L_p(\Omega)}, \quad \forall u'. \tag{3.5}$$

Moreover,

$$0 < \kappa_0^p \|u'\|_{L_p(s,P_j)}^p \leq \|\hat{u}\|_{L_p(s,P_j)}^p \leq \kappa_1^p \|u'\|_{L_p(s,P_j)}^p.$$

Indeed, it suffices to prove that

$$\kappa_0^p \|\bar{u}\|_{L_p(T)}^p \leq \|\hat{u}\|_{L_p(T)}^p \leq \kappa_1^p \|u'\|_{L_p(T)}^p, \tag{3.6}$$

where T refers to any $(d - s)$-dimensional simplex belonging to the plane $\mathcal{P}(s, P_j)$ and is a face of a simplex in our regular triangulation. Inequalities (3.6) are equivalent to similar inequalities on the reference simplex, which are easy to verify. In accordance with (3.5) and (3.6), we obtain (3.3) and (3.4) as consequences of the corresponding theorems for the space $W_2^1(\Omega)$ and the fact that

$$\kappa_2(|u'|_{0,1} + |u'|_{1,p}) \leq \|u'\|_{1,p} \leq \kappa_3(|u'|_{0,1} + |u'|_{1,p}), \quad \forall u', \tag{3.7}$$

which follows from the equivalence of the respective norms in $W_2^1(\Omega)$. $\square$ [11]

Theorem 2. Suppose $d < mp$. Then

$$\|u'\|_C \leq K_0 \|u\|_{m,p}, \quad \forall u'. \tag{3.8}$$

Suppose also that $d \geq s > d - mp \geq 0$ and $q \leq q^* \equiv sp[d - mp]^{-1}$. Then

$$\|u'\|_{L_q(s,P_j)} \leq K_1 \|u'\|_{m,p}, \quad \forall u'. \tag{3.9}$$

Proof. The case $d < p$ enables us to use (3.3) and (3.7) recursively m times. For $D \geq p$, we use (3.4) with $s = d$. For $|\alpha| = m - 1$ and $1/q^{(m-1)} \equiv 1/p - 1/d$, we see that

$$|\partial^\alpha u'|_{0,q^{(m-1)}}| \leq \bar{K}_0 (\|u'\|_{0,1} + |\partial^\alpha u'|_{1,p}) \leq \bar{K}_1 \|u'\|_{m,p}, \quad \forall u'. \tag{3.10}$$

If $q^{(m-1)} > d$, then we may again apply (3.4) with $p = q^{(m-1)}$. If $q^{(m-1)} \leq d$, then, for $|\alpha| = m - 2$ and

$$1/q^{(m-2)} \equiv 1/q^{(m-1)} - 1/d = 1/p - 2/d,$$

we make use of (3.4) and obtain

---

[11] Note that $K' = K'(p, d, \Omega)$ and $K'' = K''(q, p, s, d, \Omega)$.

$$|\partial^\alpha u'|_{0,q^{(m-2)}} \le \bar{K}_2 \|u'\|_{m,p}, \quad \forall u'. \tag{3.11}$$

Repeating of this procedure yields

$$|\partial_r u'|_{0,q^{(1)}} \le \bar{K}_{m-1} \|u'\|_{m,p}, \quad \forall u', \tag{3.12}$$

where $1/q^{(1)} \equiv 1/p - (m-1)/d$. (Note that this procedure yields $q^{(1)} > d$ if $d < mp$ and $q^* = sq^{(1)}(d - q^{(1)})^{-1} \ge 1$ if $d \ge mp$.) Hence, we can apply Theorem 1 with $p = q^{(1)}$ and obtain the desired inequalities. $\square$ [12]

3.2. Applicability of effective iterative methods. The study of nonlinear difference schemes in accordance with Theorem 2 is very much like the study of PGMs (see § 5.5) even in case of bounded nonlinearity. An important point is that some desired inequalities for difference operators can be considered as consequences of the corresponding differential inequalities for reasonable difference approximations and small enough h. The constants in these inequalities are only slightly worse than in the differential case (e.g., if the symmetric positive definite operators L and Λ satisfy the relation $L \ge \delta_0 \Lambda, \delta_0 > 0$, then we cannot prove that this relation remains true for their difference approximations even when they are symmetric positive operators; but we can prove (see, e.g. [166]) that it holds with a positive constant $\delta_0' < \delta_0$ for certain approximations provided h is small enough). In this respect, we emphasize importance of difference schemes obtained as discretizations of problems of type (0.1.4) or (0.1.16) (see, e.g., [150, 164, 166, 324, 329, 342, 439]). Actually, such constructions follow the ideas used in [123] and especially in the papers of Ladyzenskaya in the fifties devoted to the study of convergence of difference methods .

We confine ourself to the study of bounded nonlinearities in the space $W_2^2(\Omega)$ (they can be much stronger than those in the space $W_2^1(\Omega)$ (see § 5). Detailed examples of nonlinear difference systems for which our theory applies can be found in [163, 206]; see also Subsection 5.3.

Here we consider the vector $\zeta \equiv \{\zeta_\alpha\}$ corresponding to possible derivatives of order not greater than 2 ($\zeta \in \mathbf{R}^6$ if $d = 2$), and we use $\zeta(u) \equiv \{\zeta_\alpha(u)\}$ and the following difference approximations: $\zeta_\alpha(u) \equiv u$ if $|\alpha| = 0$, $\zeta_\alpha(u) \equiv \hat{\partial}^\alpha u$ if $|\alpha| = 1$, $\zeta_\alpha(u) \equiv 1/2[\bar{\partial}_r \partial_l + \partial_r \bar{\partial}_l] u$ if $\alpha = e_r + e_l$.

Lemma 1. *Let $d = 2$ and consider a nonlinear difference operator $\mathcal{P}(u) \equiv a(x, \zeta(u))$, where the function $a(x, \zeta)$, for almost all $x \in \bar{\Omega}$ and all $\zeta \in \mathbf{R}^6$, is continuously differentiable with respect to ζ. Suppose that all the corresponding first derivatives are such that*

---

[12] Similar results can be also obtained when u is defined on Ω_h and $\partial^\alpha u$ on the appropriate subset (see § 2). Certain estimates depending on h are also useful. For example, if $d = 2$, then $|u|_C \le \kappa_4 |\ln h|^{1/2} \|u\|_{1,2}$ (see, e.g., [437]).

$$\left|\frac{\partial a_r(x,\zeta)}{\partial \zeta_\beta}\right| \le \sum_{n=1}^{t} K_{\beta,n} \prod_{|\alpha|\le 2} |\zeta_\alpha|^{s_{\beta,\alpha,n}}, \tag{3.13}$$

where

$$\frac{2}{p_\beta} + \sum_{|\alpha|\le 2}^{d} \frac{2s_{\beta,\alpha,n}}{p_\alpha} \le 1, \tag{3.14}$$

$p_\alpha \equiv \infty$ *if* $|\alpha| = 0$, $1 \le p_\alpha < \infty$ *if* $|\alpha| = 1$, *and* $p_\alpha = 2$ *if* $|\alpha| = 2$. *Then,*
for all u and u' in $S_r \equiv S_{\Lambda^2}(r)$, we have

$$\|\mathcal{P}(u) - \mathcal{P}(u')\| \le K(r)\|\Lambda(u-u')\|, \tag{3.15}$$

where the function $K(r) \to 0$ as $r \to 0$.

Proof. By (1.2.15), we have $\|\mathcal{P}(u) - \mathcal{P}(u')\| \le \|\mathcal{P}'_v \Lambda^{-1}\|\|\Lambda(u-u')\|$,
where $v \equiv u + \theta(u' - u) \in S_r$. Hence, it suffices to estimate

$$\|\mathcal{P}'_v \Lambda^{-1}\| = \max_{z \ne 0} \|\mathcal{P}'_v z\|\|\Lambda z\|^{-1}.$$

Conditions (3.13) and (3.14) yield

$$\|\mathcal{P}'_v z\| \le \sum_{|\beta|\le 2} \sum_{n=1}^{t} K_{\beta,n} Q_{\beta,n}, \quad Q_{\beta,n} \equiv [\prod_{|\alpha|\le 2} |\zeta_\alpha(v)|^{2s_{\beta,\alpha,n}}, |\zeta_\beta(z)|^2]^{1/2}.$$

Next, we apply (5.5.21) and obtain

$$Q_{\beta,n} \le |\zeta_\beta(z)|_{0,2,p_\beta} \prod_{|\alpha|\le 2} |\zeta_\alpha(v)|^{s_{\beta,\alpha,n}}.$$

Now noting that

$$|\zeta_\alpha(v)|_{0,p_\alpha}| \le \kappa\|\Lambda v\|, \tag{3.16}$$

we obtain (3.15). $\square$ [13]

---

[13] Similar estimates hold for $d \ge 3$, but p_0 and p_1 must be chosen in accordance
with (3.16). Lemma 1 implies that the perturbation $\mu\mathcal{P}$ of a linear operator A_h, for
which (2.25) holds with $L_h = A_h$, may be considered as small if $f(r) \equiv |\mu|K(r)$ is
small enough, where $|\mu(\mathcal{P}(u) - \mathcal{P}(v), \Lambda z)| \le K(r)|\mu|\|\Lambda(u-v)\|\|\Lambda z\|$, $\forall u \in S_{\Lambda^2(r)}, \forall v \in S_{\Lambda^2(r)}$. Moreover, if the operators ΛA_h and Λ^2 are connected by relationship $C^{0,0}$, then
Theorems 1.2.4 and 1.2.8 can be applied provided $f(r)$ is small enough. Localization
of the solution ($\|\Lambda u\| \le r_0 < r$) leads to the fact that ΛL_h and Λ^2 are connected by
relationship $C^0 u, r - r_0$. Then it is also possible to find a model operator B such that
BL_h and B^2 are connected by the relationship $C^0(u, r - r_0)$, so we can apply iterations
(1.3.1). Similarly, if A_h and B are connected by relationship C^1 then iterations (1.3.10)
apply under natural additional conditions.

3.3 Geometrical nonlinear problems in the theory of elasticity and shells. It is easy to see that Lemma 1 applies to difference approximations of the differential operator $\mathcal{P}_0(u) \equiv g_1 u^{k_1} D_r D_l u + g_2 u^{k_2} (D_r u)^{k_3}$, where k_1 and k_2 are nonnegative integers and g_1 and g_2 are bounded functions. Significantly stronger and more difficult types of nonlinearities correspond to so-called geometrical nonlinear problems in the theory of elasticity, where instead of (5.3.9), say for $d = 2$, we have the following relations:

$$2\epsilon_{i,j}(u) \equiv D_j u + D_i u_j + D_i u_1 D_j u_1 + D_i u_2 D_j u_2, \quad i = 1, 2, j = 1, 2. \quad (3.17)$$

(In this case, $\mathcal{P}(u)$ contains additional terms like $\mu(D_k u)^3 D_k D_l u, k = 1, 2, l = 1, 2$.) Such problems cannot be regarded as having a bounded power of nonlinearity but iterative methods can still be effective. We indicate here only a special class of shell problems related to the design of various types of tires (see also § 5) with the above strong nonlinearity that were solved in sixties and seventies (the results are partly published in [210, 211]; for later results (which deal with more complicated mathematical models), see § 5). The use of orthogonal coordinates on the shell surface led to the region $\Omega = Q = [-l_1, l_1] \times [-l_2, l_2]$ and boundary conditions

$$\vec{u}\,|_{|x_1|=l_1} = 0, \quad \vec{u}(x_1, x_2 + 2l_2) = u(x_1, x_2), \quad (3.18)$$

where $\vec{u} \equiv [u_1(x), u_2(x), u_3(x)]$ is the standard displacement vector and periodic conditions imply that we deal actually with functions defined for all x_2. For the simplest linear problem, the original Hilbert space $G \equiv \vec{G}$ is the completion of the indicated vector field in the sense of $(W_2^1(Q))^3$, and correctness of the variational problem of type (1.4) was proved in [160]. Moreover, for arising difference operators L_h, it was proved that they are spectrally equivalent to the model block diagonal operator B of the form

$$B\mathbf{u} \equiv [\Lambda_1(I - Z_{1,k})^{-1}\mathbf{u}_1, \Lambda_2(I - Z_{2,k})^{-1}\mathbf{u}_2, \Lambda_3(I - Z_{3,k})^{-1}\mathbf{u}_3]^T, \quad (3.19)$$

where $\Lambda_r \equiv -\Delta_1 - \Delta_2 + c_r I, c_r \geq 0$, k is the number of inner ADI iterations for the systems $\Lambda_r u_r = f_r$, and $Z_{r,k}$ is the corresponding reduction operator, $r = 1, 2, 3$ (see § 3.2 and 3.4). [14]

---

[14] Typical grids corresponded to the number of the unknowns $N \in [10^4, 5 \times 10^4]$, and very often symmetry of the solution with respect to x_1 and x_2 (as indicated in § 5.3 for elasticity problems) was exploited. The number k of inner iterations was $k = 2$ or $k = 4$. For linear problems, the modified Richardson method was used and the necessary constants were defined by numerical experiments for problems on a coarse grid (see Subsection 1.3.7). For nonlinear ones, the modified method of the simple iteration was used only for sufficiently good initial iterates, obtained through solution of either the linearized problem or a similar nonlinear problem with neighboring values of a parameter (usually related to intensity of load). Continuation procedures were essential for practical solution of these highly nonlinear problems.

§ 4. Boundary value problems in a rectangle for fourth-order elliptic equations and systems

4.1. The first boundary value problem for equations with variable coefficients. We confine ourselves[15] to the simplest case of difference approximations of the first boundary value problem in a rectangle $Q \equiv [0, l_1] \times [0, l_2]$ for the equation

$$Lu \equiv \sum_{|\alpha| \leq 2} \sum_{|\beta| \leq 2} (-1)^\alpha D^\alpha \left(a^{\alpha,\beta} D^\beta u \right) = f(x),$$

where all $a^{\alpha,\beta}$ are piecewise continuous functions of $x \equiv [x_1, x_2]$ such that

$$\sum_{|\alpha| \leq 2} \sum_{|\beta| \leq 2} \xi^\alpha \xi^\beta \geq \nu_0 \left((\xi^{[2,0]})^2 + (\xi^{[0,2]})^2 \right), \quad \nu_0 > 0, \qquad (4.1)$$

$(\alpha \equiv [\alpha_1, \alpha_2], |\alpha| \equiv \alpha_1 + \alpha_2)$. Note that our problem can be formulated as an operator equation in the Hilbert space $G \equiv \overset{o}{W}_2^2 (Q)$.[16]

The grid system associated with the simplest difference approximation takes the form

$$(L_h \mathbf{u})_i \equiv Lu_i \equiv \sum_{|\alpha| \leq 2} \sum_{|\beta| \leq 2} (-1)^\alpha \bar{\partial}^\alpha \left(a^{\alpha,\beta} \partial^\beta u \right)_i = f_i, \quad \forall P_i \in Q_h, \qquad (4.2)$$

where $u_i = 0$ if P_i does not belong to Q_h (see (1.2)) (this is a simple but nonetheless instructive example from the point of view of selection of asymptotically optimal model operators B). If in (4.2) we eliminate values u_i with x_i not belonging to Q_h, then we obtain an operator equation $L_h \mathbf{u} = \mathbf{f}$ in the Euclidean space H with inner product $(\mathbf{u}, \mathbf{v}) \equiv h_1 h_2 \sum_{i_1=2}^{N_1-1} \sum_{i_2=2}^{N_2-1} u_i v_i$. Note that

$$(L_h \mathbf{u}, \mathbf{v}) = \sum_{|\alpha| \leq 2} \sum_{|\beta| \leq 2} \left(a^{\alpha,\beta}, \partial^\beta u \partial^\alpha v \right)_0 \qquad (4.3)$$

(see (1.2) and (3.1.14)). We also consider a much simpler operator $\Lambda \in \mathcal{L}^+(H)$ defined by $\Lambda u_i \equiv (\Delta_1^2 + \Delta_2^2) u_i$, $P_i \in Q_h$, so that

$$(\Lambda \mathbf{u}, \mathbf{u}) = \|\Delta_1 u'\|_0^2 + \|\Delta_2 u'\|_0^2. \qquad (4.4)$$

---

[15] The general results related to construction of asymptotically optimal algorithms will be given in Chapter 8.

[16] We consider rectangular grids with mesh sizes $h_r \equiv l_r/(N_r + 1)$, $r = 1, 2$, and nodes $P_i \equiv [i_1 h_1, i_2 h_2]$ $(i \equiv [i_1, i_2])$, and $u_i \equiv u(P_i)$ denotes a value of the grid function at the node. We also use $Q_h \equiv \{P_i : i_r \in [2, N_r - 1], r = 1, 2\}$.

Lemma 1. *There exist constants γ_0 and γ_1 independent of the grid such that*

$$\|\mathbf{u}\| \le \gamma_0 \|u\|_\Lambda, \quad \|u'\|_{1,2} \le \gamma_1 \|u\|_\Lambda, \quad \forall \mathbf{u} \in H; \tag{4.5}$$

moreover,

$$\|\mathbf{u}\|_\Lambda^2 \le \|u'\|_2^2 \equiv \sum_{|\alpha| \le 2} \|u'\|_0^2 \le 2\|\mathbf{u}\|_\Lambda^2. \tag{4.6}$$

Proof. First note that γ_0^{-2} and γ_1^{-2} can be expressed via the slightly diminished smallest eigenvalues of the problems $\Delta^2 u = \lambda u$ and $\Delta^2 u = -\lambda\Delta u$ in $G \equiv \overset{\circ}{W}_2^2(Q)$ provided h is small enough. But we prefer to give their explicit bounds for all h. For this, we observe that

$$\|\mathbf{u}\|_\Lambda^2 = \|\Delta_1 u'\|_0^2 + \|\Delta_2 u'\|_0^2, \tag{4.7}$$

where u' is the finite extension of $\mathbf{u}$ (see (1.2) and (3.1.14)) and, e.g., $\Delta_1 u_i'$ vanishes when either $i_1 < 1$, $i_1 > N_1$, $i_2 \le 1$, or $i_2 \ge N_2$. This suggests we use the new Euclidean space H_+ of grid functions defined as in § 2 (for $1 \le i_r \le N_r, r = 1, 2$) and the operator $-\Delta_+ \in \mathcal{L}(H_+)$ defined by the relation $-\Delta_+ \mathbf{u}_i \equiv -(\Delta_1 + \Delta_2)u_i'$ for $1 \le i_r \le N_r, r = 1, 2$. (We agree here not to distinguish elements H and H_+ if they correspond to the same finite extension u'.) Next we make use of the following orthonormal basis for H_+:

$$e_j(x) \equiv e_{j_1,j_2}(x_1, x_2) \equiv (2/l_1)^{1/2} \sin[\pi j_1 x_1 l_1^{-1}](2/l_2)^{1/2} \sin[\pi j_2 x_2 l_1^{-1}]),$$

where $j_r \in [1, N_r]$, $x_r \equiv i_r h_r, i_r \in [1, N_r]$, and $r = 1, 2$ (see (0.2.23)). Recall that $\Delta_r e_j(x) = -\lambda_{r,j} e_r(x)$, where $\lambda_{r,j} \equiv \frac{4}{h_r^2} \sin^2 \frac{\pi j_r h_r}{2l_r} \ge \frac{4}{l_r^2}, r = 1, 2$, $(\sin(\pi t)/2 \ge t$ for $t \in [0, 1])$. This enables us to write

$$\mathbf{u} = \sum_j c_j e_j(x) \in H_+, \quad \|\mathbf{u}\|_\Lambda^2 = \|\Delta_+ \mathbf{u}\|^2 = sum_{r=1}^2 \sum_j c_j^2 \lambda_{r,j}^2. \tag{4.8}$$

In (4.5), we can thus take the constants defined by $\gamma_0^2 = 16(l_1^{-4} + l_2^{-4})$ and $\gamma_1^2 = 4(l_1^{-2} + l_2^{-2})$. For (4.6), it is easy to verify that

$$2\|\partial_1 \partial_2 u'\|_0^2 = 2(\Delta_1 u', \Delta_2 u')_0 = \sum_j c_j^2 \lambda_{1,j} \lambda_{2,j} \le (\Lambda \mathbf{u}, \mathbf{u})) \tag{4.9}$$

(see (4.8) and (4.7)). $\square$

We consider now the operator L_h defined by (4.3) and (1.2).

Lemma 2. *There exist constants $\gamma_{0,0} > 0$, $\kappa_0 \leq 0$, and $\kappa_1 \leq 0$ independent of the grid such that*

$$(L_h\mathbf{u}, \mathbf{u}) \geq \gamma_{0,0}\|\mathbf{u}\|_\Lambda^2 - \kappa_1|u'|_{1,2}^2 - \kappa_0|u'|_0^2. \qquad (4.10)$$

Proof. The proof is straightforward (see the proof of Lemma 0.1.2) and is based on (4.3) and (4.1). For estimating, say, of the typical term $X \equiv (a^{\alpha,\beta}, \partial^\beta u' \partial^\alpha u')_0$, with $|\alpha| = 2$ and $|\beta| \leq 1$, we use the inequality $X \geq -|a^{\alpha,\beta}|_C(\epsilon|\partial^\alpha u'|_0^2 + |\partial^\beta u'|_0^2/\epsilon)$, where $\epsilon > 0$ can be chosen so small that ν_0 is larger than the sum of all $\epsilon|a^{\alpha,\beta}|_C$ with $|\alpha| = 2$ and $|\beta| \leq 1$. $\square$

Theorem 1. *Suppose, for all α and β in (4.2), that $a^{\alpha,\beta} = a^{\beta,\alpha}$. Suppose also that $\delta_0 \equiv \gamma_{0,0} - \kappa_0\gamma_0^2 - \kappa_1\gamma_1^2 > 0$. Then, $L_h \asymp \Lambda_h$.*

Proof. Since $|(\partial^\alpha(a\partial^\beta u', \partial^\alpha u')_0)| \leq K\left(\|\partial^\alpha u'\|_0^2 + \|\partial^\beta u'\|_0^2\right)$, then we have $(L_h\mathbf{u}, \mathbf{u}) \leq \delta_1(\Lambda\mathbf{u}, \mathbf{u})$, $\forall \mathbf{u} \in H$, with a constant $\delta_1 \neq \delta_1(h)$. On the other hand, $(L_h\mathbf{u}, \mathbf{u}) \geq \delta_0(\Lambda\mathbf{u}, \mathbf{u})$, $\mathbf{u} \in H$, is a direct consequence of (4.10), (4.5), and the conditions of our theorem. $\square$ [17]

Theorem 2. *Suppose that the conditions of Theorem 1 are satisfied. Consider the model operator*

$$B_h \equiv \Lambda_h(I - Z_k)^{-1}, \qquad (4.11)$$

where Z_k is the error reduction operator for k iterations of the ADI method (see (3.2.5), (3.2.15), and (3.2.17)). Suppose also that $\|Z_k\|_\Lambda \leq q < 1$, where $q \neq q(h)$. Then, $L_h \asymp B_h$.

Proof. First, observe that $\Lambda = \Lambda_1 + \Lambda_2$ and that (3.2.8) and (3.2.9) hold. Moreover, in analyzing ADI iterations, we may apply (3.2.17) with $\lambda_0 \asymp 1$ and $\lambda^0 \asymp h^{-4}$. Thus, Theorems 3.2.1 and 3.2.2 apply. Together with Theorems 3.4.1, they lead to the desired spectral equivalence. $\square$ [18]

It is easy to study the $O(h^2)$-approximations

$$D^\alpha(aD^\beta u) \approx \frac{1}{2}\left(\bar{\partial}^\alpha(a\partial^\beta u) + \partial^\alpha(a\bar{\partial}^\beta u)\right). \qquad (4.12)$$

The case of periodic conditions is extremely simple; other more practical approximations of the boundary conditions are considered in § 5. The nonlinear equation $\Delta^2 u + g(u) = f$, where $g(u)$ is continuously differentiable and $g'(u) \geq 0$ or $g'(u) \geq -c^2$ with small c^2, can serve as the simplest example for which the derivative $L'_{h,w} \asymp B$ and iterations (1.3.1) can

---

[17] For the classic equation $\Delta^2 u = f$, the constants of spectral equivalence are to 1 and 2; generalizations for d-dimensional problems were given in [150].

[18] Theorem 2 implies that we can find ϵ-approximations to the solution of systems with L_h by modified iterative methods using the model operator B, with computational work $W(\epsilon, h) = O(|\ln \epsilon||\ln h|/h^2)$ (see [154, 163]).

be applied. The case when $L'_{h,w}$ has several negative eigenvalues $\lambda(L'_{h,w})$ ($|\lambda(L'_{h,w})| \geq d > 0$) and iterations (1.3.12) or (3.8.6) are applied was considered in [206].

4.2. Strongly elliptic systems. We outline possible generalizations to the case of strongly elliptic systems (in the sense of Nierenberg), where we deal with an operator equation in the Hilbert space

$$G \equiv \overset{\circ}{W}{}_2^{m_1}(Q) \times \cdots \times \overset{\circ}{W}{}_2^{m_k}(Q) \qquad (4.13)$$

associated with the system

$$\sum_{l=1}^{k} \sum_{|\alpha_r| \leq m_r} (-1)^{\alpha_r} D^{\alpha_r} \sum_{|\beta_l| \leq m_l} a^{r,l,\alpha_r,\beta_l} D^{\beta_l} u(x) = f_r(x), x \in Q, \qquad (4.14)$$

of k equations, where $m_r \geq 1, r \in [1, k]$, and there exists $\nu_0 > 0$ such that, $b_0(\bar{u}; \bar{u}) \equiv \sum_{r=1}^{k} \sum_{l=1}^{k} \sum_{|\alpha_r|=m_r} \sum_{|\beta_l|=m_l} (a^{r,l,\alpha_r,\beta_l}, D^{\beta_l} u_l D^{\alpha_r} u_r)_{0,Q} \geq \nu_0 \sum_{r=1}^{k} |u_r|_{m_k,Q}^2, \forall \bar{u} \equiv [u_1, \ldots, u_k] \in G$. We approximate (4.14) by

$$[\sum_{l=1}^{k} \sum_{|\alpha_r| \leq m_r} (-1)^{\alpha_r} \bar{\partial}^{\alpha_r} \sum_{|\beta_l| \leq m_l} a^{r,l,\alpha_r,\beta_l} \partial^{\beta_l} u]_i - f_{r,i}, \ P_i \in Q_{r,h}, \qquad (4.15)$$

where $Q_{r,h} \equiv \{P_i : i_s \in [m_r, N_r + 1 - m_r], s = 1, 2,\}$ and $r \in [1, k]$. Then, for the resulting difference operator $\bar{L}_h \equiv L$ in the Euclidean space $\bar{H} \equiv H \equiv H_{(1)} \times \times H_{(k)}$ ($H_{(r)}$ corresponds to grid functions defined on $Q_{r,h}$); block diagonal operators $\bar{\Lambda}$ were suggested in [153, 163] such that

$$\bar{\Lambda}\bar{u} \equiv [\Lambda_{(1)}\mathbf{u}_1, \ldots, \Lambda_{(k)}\mathbf{u}_k]^T$$

and $(\Lambda_{(r)}\mathbf{u}_r)_i \equiv (-1)^{m_r}(\Delta_1^{m_r} + \Delta_2^{m_r})u'_r |_i, \ P_i \in Q_{r,h}, \ r \in [1, k].$ [19]

§ 5. Linear and nonlinear problems of plates and shells theory

5.1. Linear problems for elastic plates. Problems (4.1), (1.2) are related to bending of elastic clamped plates with boundary conditions

$$u \mid_\Gamma = \frac{\partial u}{\partial n} \mid_\Gamma = 0 \qquad (5.1)$$

---

[19]For each of the operators $\Lambda_{(r)} = \Lambda_{(r,1)} + \Lambda_{(r,2)}, r \in [1, k]$, the theorems given in § 4.2 hold. We can thus construct the model block diagonal operator $\bar{B} \equiv B$ such that $\bar{B}\bar{u} \equiv [B_1\mathbf{u}_1, \ldots, B_k\mathbf{u}_k]^T$ and $B_r \asymp \Lambda_{(r)}$. Under appropriate conditions on the coefficients, it is possible to prove that $\bar{L}_h \asymp \bar{B}$. More general conditions lead to relationships $C^{0,0}$ or C^3. Finally, we remark that similar model operators can be used for certain nonlinear problems (see [163, 206]).

(n is the outward normal on Γ). Instead of (1.2), so-called *staggered bound-ary conditions* of higher order approximation are used in practice (they are especially suitable for approximations (4.12)). To describe them, we define Q_h and H as in § 2 ($Q_h \equiv \{P_i : 1 \leq i_1 \leq N_1, 1 \leq i_2 \leq N_2\}$) and con-sider only the simplest case of governing equation $\Delta^2 u = f$. At the nodes $P_i \in Q_h$, we use the difference equations

$$Lu_i \equiv (\Delta_1 + \Delta_2)^2 u_i, \quad P_i \in Q_h,$$

where necessary values u_i for $P_i \notin Q_h$ are defined through the difference boundary conditions

$$u_i = 0, \text{ if } P_i \in \Gamma, \quad \hat{\partial}_l u_i = 0 \text{ if } P_i \in \Gamma_r^+, l \neq r, r = 1, 2, \qquad (5.2)$$

where $\Gamma_r^+ \equiv \{P_i : i_r = 0 \text{ or } i_r = N_r, 1 \leq i_l \leq N_l\}, l \neq r, r = 1, 2, \ l = 1, 2$ ((5.2), e.g. for $i = [0, i_2]$ with $i_2 \in [1, N_2]$, implies that $u(0, i_2 h_2) = 0$ and $u(-h_1, i_2 h_2) = u(h_1, i_2 h_2)$). We define thus the difference operators $L_h \equiv L \in \mathcal{L}(H)$ and $\Lambda \in \mathcal{L}(H)$ by the above formula for Lu_i and (4.4), re-spectively, under the assumption that $\mathbf{u} \in H_1$ is extended to the additional nodes through the difference boundary conditions (5.2). We compare them with similar operators $L^{(0)} \in \mathcal{L}(H)$ and $\Lambda^{(0)} \in \mathcal{L}(H)$, which require the finite extentions of $\mathbf{u} \in H$ (a slightly different Q_h was used in § 4).

Lemma 1. The above operator Λ belongs to $\mathcal{L}^+(H)$ and satisfies (4.5).

Proof. The operator Λ is clearly symmetric and

$$h_1 h_2 \sum_{i_1=1}^{N_1} \sum_{i_2=1}^{N_2} v_i \Delta_1^2 u_i = (\Delta_1 u', \Delta_1 v')_0 + \frac{h_2}{h_1^3} \sum_{i_2=1}^{N_2} (u_{1,i_2} v_{1,i_2} + u_{N_1,i_2} v_{N_1,i_2}).$$

This and the analogous representation for $h_1 h_2 \sum_{i_1=1}^{N_1} \sum_{i_2=1}^{N_2} v_i \Delta_2^2 u_i$ imply that $\Lambda \geq \Lambda^{(0)}$. □

We also define one-dimensional difference operators $\Lambda_r \in \mathcal{L}(H_1)$ and $\Lambda_r^{(0)} \in \mathcal{L}(H_1), r = 1, 2$, by the same relations $\Lambda_r u_i \equiv \Delta_r^2 u_i, \ P_i \in Q_h$, and $\Lambda_r^{(0)} u_i \equiv \Delta_r^2 u_i, \ P_i \in Q_h$, but under different extensions of $\mathbf{u} \in H$.

Lemma 2. The operators Λ_1 and Λ_2 commute, $\Lambda_r \in \mathcal{L}^+(H)$, $r = 1, 2$, sp $\Lambda_r \subset [\Lambda_0, \lambda^0]$, $0 < \lambda_0 \asymp 1$, and $\lambda^0 \asymp h^{-4}$.

Proof. We have $\Lambda_r = \Lambda_r^{(0)} + h_r^{-4} D_r$, where: $D_r u_i \equiv 0$ if $P_i \in Q_h$ and $i_r \in [2, N_r - 1]$; $D_r u_i \equiv u_i$ if $P_i \in Q_h$ and $i_r = 0$ or $i_r = N_r$; and $r = 1, 2$ (see the proof of Lemma 1). Thus, the symmetry of Λ_r is obvious and $\Lambda_r \in \mathcal{L}^+(H), r = 1, 2$. Moreover, $\Lambda_r \geq \Lambda_r^{(0)}$, and we can easily localize its spectrum, $r = 1, 2$. Next, it is easy to see that $D_1 \Lambda_2^{(0)} = \Lambda_2^{(0)} D_1$ and $D_2 \Lambda_1^{(0)} = \Lambda_1^{(0)} D_2$, which yields the desired property $\Lambda_1 \Lambda_2 = \Lambda_2 \Lambda_1$. $\square$ [20]

Theorem 1. *The operators L and B are spectrally equivalent.*

Proof. We have $L = L^{(0)} + \sum_{r=1}^{2} \frac{1}{h_r^4} D_r$ and $\Lambda = \Lambda^{(0)} + \sum_{r=1}^{2} \frac{1}{h_r^4} D_r$, which with (4.6) lead to $\Lambda \leq L \leq 2\Lambda$ and, hence, to the desired equivalence of L and B. $\square$

We consider now very briefly other types of the boundary conditions, assuming for simplicity that they are posed only on the left side of Q. The hinge support of the plate implies that $u \mid_{x_1=0} = D_1^2 u \mid_{x_1=0} = 0$. We then deal with approximations

$$u_{0,i_2} = 0 \text{ if } i_2 \in [0, N_2+1], \quad u_{-1,i_2} = -u_{1,i_2} \text{ if } i_2 \in [1, N_2]. \tag{5.3}$$

For so-called free boundary conditions $D_1^2 u \mid_{x_1=0} = D_1^3 u \mid_{x_1=0} = 0$, their difference approximations [21] are defined by

$$\Delta_1 u_{0,i_2} = 0 \text{ if } i_2 \in [1, N_2], \quad \partial_1 \Delta_1 u_{0,i_2} = 0 \text{ if } i_2 \in [1, N_2]. \tag{5.4}$$

We show now that similar algorithms can be applied for certain stiffened plates. As an example, we consider only one stiffener (bar) along the straight line $x_1 = a_1 = i_1^* h_1$. We then deal with minimization of $\Phi(v) \equiv \|v\|_G^2 - 2(f, v)_{0,Q}$, where

$$\|v\|_G^2 \equiv |v|_{2,Q}^2 + \kappa \int_0^{l_2} [D_2^2 u(a_1, x_2)]^2 dx_2$$

and the Hilbert space G is the completion of the preHilbert space of functions in C_0^∞ having the inner product

$$(u, v) \equiv (u, v)_{2,Q} + \kappa \int_0^{l_2} [D_2^2 u(a_1, x_2)][D_2^2 v(a_1, x_2)] dx_2.$$

Difference approximation for this problem lead to the operator equation

---

[20]Lemma 1 implies that ADI-methods (3.2.5) for system $\Lambda v = f$ can be effectively applied. Thus, in accordance with Theorem 3.2.1, we choose a model operator $B \equiv \Lambda(I - Z_k)^{-1}$, where Z_k is the error reduction operator for k iterations of the ADI method and $\Lambda \asymp B$ (see Theorem 3.4.1).

[21]They bring changes in the structure of our new L_h, Λ, and B. It is easy to verify that Theorem 1 holds for these new difference schemes and that the corresponding iterative methods yield the same computational work estimate as in § 4. More cumbersome are variants with conditions (5.3) and (5.4) prescribed on neighboring sides of Q. Nonetheless, all of these variants lead to very practical algorithms (typical calculations in the seventies were on 80×80 grids and can be found in [214]).

$$(L + \kappa L_{2,1})\mathbf{u} = \mathbf{f}, \tag{5.5}$$

where $\Lambda_{2,1}u_i \equiv 0$ if $i_1 \neq i_1^*$ and $\Lambda_{2,1}u_i \equiv 1/h_1(\Delta_2^2 u_i)$ if $i_1 = i_1^*$. Let

$$\Lambda_{st} = \Lambda_1 + \Lambda_{2,st}, \tag{5.6}$$

with the one-dimensional difference operator $\Lambda_{2,1}$ defined by $\Lambda_{2,st}u_i \equiv \Delta_2^2 u_i + \kappa L_{2,1}u_i$ if $P_i \in Q_h$, and conditions (5.2).

Theorem 2. *The operator* $\Lambda_{st} \equiv \Lambda$ *in* (5.6) *satisfies the conditions of Theorem 3.2.1.*

Proof. It is easy to see that one-dimensional difference operators Λ_1 and $\Lambda_{2,st}$ belong to $\mathcal{L}^+(H)$ and commute. To localize sp $\Lambda_{2,st}$, we write $(\Lambda_{2,st}\mathbf{u}, \mathbf{u}) = |\Delta_2 u'|_0^2 + \frac{\kappa h_1 h_2}{h_1}\sum_{i_2}(\Delta_2 u_{i_1^*,i_2})^2$. We thus have sp $(\Lambda_{2,st}) \subset [\lambda_0, \lambda^0]$, $0 < \lambda^0 = O(1)$, $\lambda^0 = O(h^{-4} + \kappa h^{-5})$. $\square$ [22]

5.2. Linear problems of shells theory. Consider now more complicated problems in the theory of cylindrical and slanting shells associated with strongly elliptic systems. We show, following [162, 214], that they can be treated in much the same manner. For example, for the displacement vector $\vec{u} \equiv [u, v, w]$, consider the equilibrium equations

$$L\vec{u} \equiv \begin{bmatrix} L_{1,1} & L_{1,2} & L_{1,3} \\ L_{2,1} & L_{2,2} & L_{2,3} \\ L_{3,1} & L_{3,2} & L_{3,3} \end{bmatrix} \begin{bmatrix} u \\ v \\ w \end{bmatrix} = \begin{bmatrix} f_1 \\ f_2 \\ f_3 \end{bmatrix}, \tag{5.7}$$

where: $L_{1,1}u \equiv -(bu_x)_x - \nu(bu_y)_y$; $L_{1,2}v \equiv -\mu(bv_x)_x - \nu(bv_x)_y$; $L_{1,3}w \equiv \frac{\kappa_2\mu(bw)_x}{4}$; $L_{2,1}u \equiv -\mu(bu_x)_y - \nu(bu_y)_x$; $L_{2,2}v \equiv -\nu(bv_x)_x - (bv_y)_y$; $L_{2,3}w \equiv \frac{\kappa_2(bw)_y}{4}$; $L_{3,1}u \equiv b\frac{\kappa_2\mu u_x}{4}$; $L_{3,2}v \equiv -\frac{b\kappa_2 v_y}{4}$; $L_{3,3}w \equiv -1/12[(dw_{x,x})_{x,x} + \mu(dw_{x,x})_{y,y} + \mu(dw_{y,y})_{x,x} + 2(1-2\mu)(dw_{x,y})_{x,y} + (dw_{y,y})_{y,y}] + b\kappa_2^2 w/16$; $\mu > 0$ is the Poisson's ratio; κ_2 is the curvature of the cylindrical shell; $\nu \equiv (1-\mu)/2 > 0$, b and d are positive functions determined by the breadth of the shell; and the coordinates $[x, y]$ are used instead of $[x_1, x_2]$. Boundary conditions correspond to clamped edges, that is,

$$u\mid_\Gamma = v\mid_\Gamma = w\mid_\Gamma = \frac{\partial w}{\partial n}\mid_\Gamma = 0, \tag{5.8}$$

or they are imposed on vertical sides of Q and are supplemented by periodic conditions with respect to x. [23] The difference approximations

---

[22] We can choose a model operator $B \equiv B_{st} \equiv \Lambda_{st}(I - Z_k)^{-1} \asymp \Lambda_{st}$ as in Theorem 1 (Z_k is the error reduction operator for k iterations of the ADI method for the system $\Lambda_{st}\mathbf{v} = \mathbf{g}$). The computational work is estimated as in § 4.

[23] The first variant of the boundary conditions implies that our problem can be formulated as an operator equation in the Hilbert space $G \equiv (\overset{0}{W_2^1}(Q))^2 \times \overset{0}{W_2^2}(Q)$.

are straightforward (see Subsections 4.2 and 1). The elimination of nodes not belonging to the domain Q (on the basis of the chosen boundary difference conditions) yields a system of grid equations $L_h \bar{u} = \bar{f}$, where $\bar{u} \equiv [\mathbf{u}_1, \mathbf{u}_2, \mathbf{u}_3]^T \in H \equiv H_{(1)} \times H_{(2)} \times H_{(3)}$. Then corresponding difference operator takes the block form

$$\vec{L}_h \equiv [L_{r,l}], \quad r = 1, 2, 3, l = 1, 2, 3.$$

We consider also block diagonal operators $\vec{\Lambda}$ and $\vec{B} \equiv B_h$ with the diagonal blocks

$$\Lambda_{r,r} \equiv \Lambda_r, \quad B_{r,r} \equiv B_r, \quad r = 1, 2, 3, \tag{5.9}$$

where: $\Lambda_1 u_i \equiv -b_0(\Delta_1 + \nu\Delta_2)u_i$, $\Lambda_2 v_i \equiv -b_0(\nu\Delta_1 + \Delta_2)u_i$, $\Lambda_3 w_i \equiv d_0(\Delta_1^2 + \Delta_2^2)w_i/12 + b_0\kappa_2^2 w_i/16$, $b_0 > 0$, and $d_0 > 0$; the operators B_1 and B_2 are spectrally equivalent to the operator $-\Delta_h$, and the operator B_3 is as in § 4 or Theorem 1. [24]

5.3. Nonlinear problems of shell theory; von Karman type systems. Large deflections for a shell with two curvatures κ_1 and κ_2 are often described by the system

$$\frac{1 - \mu^2}{12}\Delta^2 u_1 - \frac{\kappa_1 D_2^2 u_2 + \kappa_2 D_1^2 u_2}{4} - [u_1; u_2] = f_1, \tag{5.10}$$

$$2\Delta^2 u_2 + \frac{\kappa_1 D_2^2 u_1 + \kappa_2 D_1^2 u_1}{4} + [u_1; u_1] = f_2, \tag{5.11}$$

where $u_1 \equiv w, u_2 \equiv \Phi$, and

$$[u_1; u_2] \equiv D_1^2 u_1 D_2^2 u_2 + D_2^2 u_1 D_1^2 u_2 - 2(D_1 D_2 u_1)(D_1 D_2 u_2). \tag{5.12}$$

As an example, consider the clamped boundary conditions

$$w \mid_\Gamma = \frac{\partial w}{\partial n} \mid_\Gamma = \Phi \mid_\Gamma = \frac{\partial \Phi}{\partial n} \mid_\Gamma = 0. \tag{5.13}$$

This implies that we can formulate our problem in the Hilbert space $G \equiv (\overset{o}{W_2^2}(Q))^2$. We approximate the linear terms as in § 4 (see (4.12) and (5.2)) and, for the nonlinear terms, we use

---

[24] In [162] it was proved that $\vec{L}_h \asymp \vec{B}$. For this and more general problems that deal with two different curvatures (dependent on x and y), iterative methods with given B are practically effective; according to [214], methods with split operators B required typically 20–30 times more of computing time.

$$[u_1; u_2]_h \equiv \Delta_1 u_1 \Delta_2 u_2 + \Delta_2 u_1 \Delta_1 u_2 - \bar{\partial}_1 \partial_2 u_1 \bar{\partial}_1 \partial_2 u_2 - \partial_1 \bar{\partial}_2 u_1 \partial_1 \bar{\partial}_2 u_2. \quad (5.14)$$

Now $H \equiv (H_{(1)})^2$, where $H_{(1)}$ denotes the space H in Lemma 1, and we obtain the nonlinear problem

$$L_h \mathbf{u} \equiv L\mathbf{u} = A\mathbf{u} + \mathcal{P}\mathbf{u} = \mathbf{f}, \quad (5.15)$$

where $A \in \mathcal{L}(H)$ and $\mathcal{P}\mathbf{u} \equiv [-[u_1; u_2]_h, [u_1; u_1]_h]^T$.

Lemma 3. The operator P possesses antisymmetry property (1.2.5).

Proof. It is easy to verify that $2[u; v]_h = \bar{\partial}_1(\Delta_2 u \partial_1 v - \partial_1 \partial_2 u \partial_v) + \bar{\partial}_2(\Delta_1 u \partial_2 v - \bar{\partial}_1 \partial_2 u \bar{\partial}_2 v) + \partial_1(\Delta_2 u \bar{\partial}_1 v - \bar{\partial}_1 \partial_2 u \partial_2 v) + \partial_2(\Delta_1 u \bar{\partial}_2 v - \bar{\partial}_1 \partial_2 u \partial_1 v)$. Thus, for arbitrary $u, v,$ and w, we have $([u; v]_h, w)_0 = ([u; w]_h, v)_0$. This implies that $([v; v]_h, v)_0 = ([v; v]_h, v)_0$ and, hence, $(\mathcal{P}\mathbf{u}, \mathbf{u}) = 0$, $\forall \mathbf{u}$ (see (1.2.5) and [173, 337]). □

Now define $\bar{\Lambda} \in \mathcal{L}^+(H)$ by

$$\bar{\Lambda} \mathbf{u}_i \equiv [(\Delta_1 + \Delta)\mathbf{u}_1, (\Delta_1 + \Delta)\mathbf{u}_2]_i^T, \quad \forall P_i \in Q_h, \quad (5.16)$$

and the indicated generalizations of (5.2).

Lemma 4. Suppose that, for the linear operator A, there exists a constant $\delta_0 > 0$ independent of h such that $(A\mathbf{u}, \mathbf{u}) \geq \delta_0 \|\mathbf{u}\|^2$, $\forall \mathbf{u} \in H$. Then (5.15) has a solution and all of its solutions are such that

$$\|\mathbf{u}\|_{\bar{\Lambda}} \leq 1/\delta_0 \|\mathbf{f}\|_{\bar{\Lambda}^{-1}} \equiv r_0. \quad (5.17)$$

Proof. It suffices to combine Lemma 4 and Theorem 1.2.1. □ [25]

Below, K is used only for constants independent of h.

Lemma 5. There exist constants K and K' such that, for all $\mathbf{u}, \mathbf{v}, \mathbf{z} \in H$,

$$|(\mathcal{P}\mathbf{v}, \mathbf{z}) - (\mathcal{P}\mathbf{u}, \mathbf{z})| \leq K\|\mathbf{u} - bf v\|_{\bar{\Lambda}} \|\mathbf{z}\|_{\bar{\Lambda}} \max\{\|\mathbf{u}\|_{\bar{\Lambda}}; \|\mathbf{v}\|_{\bar{\Lambda}}\} \quad (5.18)$$

and

$$|(\mathcal{P}\mathbf{v}, \mathbf{z}) - (\mathcal{P}\mathbf{u}, \mathbf{z})| \leq K\|\mathbf{z}\|_{\bar{\Lambda}}^2 \|\mathbf{u}\|_{\bar{\Lambda}}, \quad \forall \mathbf{u} \quad \mathbf{v} \equiv \mathbf{u} + \mathbf{z}. \quad (5.19)$$

Proof. We observe that $([u'_1; u'_2]_h, z'_1)_0 - ([v'_1; v'_2]_h, z'_1)_0 = ([u'_1 - v'_1; u'_2]_h, z'_1)_0 + ([v'_1; u'_2 - v'_2]_h, z'_1)_0$. Each term on the right-hand side can be estimated in a uniform manner, e.g.,

$$|([u'_1 - v'_1; u'_2]_h, z'_1)_0| \leq \|z'\|_C |u'_1 - v'_1|_2 |u'_2|_2$$

(see (4.6)). Note that $\|z'\|_C \leq K_2 |z'_1|_2 \leq K_3 \|\mathbf{z_1}\|_{\bar{\Lambda}}$. From this we obtain the first desired inequality. For the second, it suffices to note that, in

---

[25] The conditions of Lemma 4 are satisfied provided κ_1 and κ_2 in (5.10), (5.11) are small enough.

accordance with Lemma 3, we may deal only with the terms containing two factors that depend on $\mathbf{z}$. $\square$

In the role of the model operator $\bar{B}$, we choose a block-diagonal operator with the diagonal blocks $B_1 = B_2 \equiv \Lambda(I - Z_k)^{-1}$ (see Theorem 1).

Theorem 3. Let the conditions of Lemma 4 be satisfied and $\delta_0 - K'r_0 \equiv \bar{\delta}_0 > 0$. Then the solution of (5.15) is unique and iterative method (1.3.1) with $B = \bar{B}$ yields an ϵ-approximation to the solution in the sense of the Euclidean space $H(\bar{\Lambda})$ with computational work estimate as in Theorem 2.

Proof. We have $(L(\mathbf{u} + \mathbf{z}) - L(\mathbf{u}), \mathbf{z}) \geq (\delta_0 - K'\|\mathbf{u}\|_{\bar{\Lambda}})\|\mathbf{z}\|_{\bar{\Lambda}}^2$, $\forall \mathbf{u}, \mathbf{z}$. If (5.17) holds, then we have

$$(L(\mathbf{u} + \mathbf{z}) - L(\mathbf{u}), \mathbf{z}) \geq \bar{\delta}_0 \|\mathbf{z}\|_{\bar{\Lambda}}^2, \quad \forall \mathbf{z}, \tag{5.20}$$

and the uniqueness of the solution is obvious. Denote this solution by $\mathbf{u} \equiv u$. Next, in accordance with (5.16), the operators L and Λ are connected by relationship $C^0(u; r)$ for arbitrary $r > 0$. The same is true with respect to the operators L and B (see Theorem 3.8.1). Thus, convergence of iterations (1.3.1) with $u^0 \in S_B(u, r)$ is independent of h provided the iteration parameter is chosen as in Theorem 1.3.1. $\square$ [26]

5.4. Nonlinear shell problems written in displacements. Instead of problem (5.10)–(5.13), another formulation is frequently used. It involves three unknown functions $u_1 \equiv u, u_2 \equiv v$, and $u_3 \equiv w$, and is a generalization of the linear problem defined by (5.7), (5.8). We specify only the type of the new nonlinear terms in (5.7) when we replace Lu by $\tilde{L}(u) \equiv Lu + \mathcal{P}(u)$, where $\mathcal{P}(\vec{u}) \equiv [\mathcal{P}_1(\vec{u}), \mathcal{P}_2(\vec{u}), \mathcal{P}_3(\vec{u})]^T$ and $\vec{u} \equiv [u_1, u_2, u_3]^T \equiv [u, v, w]^T$. The typical terms, for $\mathcal{P}_1(\vec{u})$ and $\mathcal{P}_2(\vec{u})$, have the form $D_k(D_r u_3 D_l u_3)$, and in $\mathcal{P}_3(\vec{u})$ a term of the form $(D_r u_1 + D_l u_2 + D_l u_3 D_2 u_3) D_1 D_2 u_3$ appears.

We sketch now the proof that resulting difference nonlinearities are of bounded type in the Euclidean space $H(\vec{\Lambda})$ (see (5.9)). Let, e.g., $X_1 \equiv |(\bar{\partial}_k(\partial_r u_3' \partial_l u_3'), z_1')_0 - (\bar{\partial}_k(\partial_r v_3' \partial_l v_3'), z_1')_0|$, $X_2 \equiv |(\partial_1 u_3' \partial_2 v_3' \partial_l \partial_2 w_3', z_3')_0|$. We transform X_1 in accordance with (3.1.14). This and elementary inequalities yield $X_1 \leq |(\partial_r(u_3' - v_3')\partial_l u_3', \partial_k z_1')_0| + |(\partial_r v_3' \partial_l(u_3' - v_3'), \partial_k z_1')_0|$. Next, we apply (5.5.10) (see also (3.2)) to obtain $X_1 \leq \|\partial_k z_1'\|_{0,2}[\|\partial_r(u_3' - v_3')\|_{0,4}\|\partial_l u_3'\|_{0,4} + \|\partial_r v_3'\|_{0,4}\|\partial_l(u_3' - v_3')\|_{0,4}]$. This together with (3.9) yields the ultimate estimate

$$X_1 \leq K_4 \|\mathbf{u} - \mathbf{v}\|_{\bar{\Lambda}} \max\{\|\mathbf{u}\|_{\bar{\Lambda}}; \|\mathbf{v}\|_{\bar{\Lambda}}\}\|\mathbf{z}\|_{\bar{\Lambda}}.$$

In the same way, we obtain $X_2 \leq \|z_3'\|_C \|\partial_1 p_2 w_3'\|_{0,2}\|\partial_l u_3'\|_{0,4}\|\partial_2 v_3'\|_{0,4} \leq K_5 \|(\max\{\|\mathbf{u}\|_{\bar{\Lambda}}; \|\mathbf{v}\|_{\bar{\Lambda}}\})^2 \|\mathbf{u} - \mathbf{v}\|_{\bar{\Lambda}}\|\mathbf{z}\|_{\bar{\Lambda}}$. Hence, for arbitrary $\bar{u}, \bar{v}$, and $\bar{z}$, we obtain the inequality

---

[26] It is possible to apply iterations considered in § 3.8.

$$|(\mathcal{P}(\bar{v}), \bar{z}) - (\mathcal{P}(\bar{u}), \bar{z})| \le K_6(g + g^2)\|\bar{v} - \bar{u}\|_{\bar{\Lambda}}\|\bar{z}\|_{\bar{\Lambda}},$$

where $g \equiv \max\{\|\bar{u}\|_{\bar{\Lambda}}; \|\bar{v}\|_{\bar{\Lambda}}\}$. Thus, theorems from § 1.1 apply and lead to localization of the solution of type $\|\bar{u}\|_{\bar{\Lambda}} \le K_7\|\bar{f}\|_{\bar{\Lambda}-1} \equiv r_0$. If this r_0 is small enough, then correctness of our difference problems in the corresponding ball S_{r_0} can be proved (see § 1.2), and effective iterative methods of type (1.3.1) can be constructed. [27]

5.5. Two-dimensional flow of viscous incompressible fluids. The above mentioned problem in terms of the stream function $\psi \equiv u^*$ and vorticity $w \equiv -\Delta u^*$ is frequently reduced to the equation $\nu\Delta^2 u^* + \frac{\partial(u^*;w)}{\partial(x_1;x_2)} = 0$ under the boundary conditions $u \mid_{\Gamma} = g,\ \frac{\partial u}{\partial n}\mid_{\Gamma} = g'$, where

$$\frac{\partial(u;w)}{\partial(x_1;x_2)} \equiv [u;w] \equiv D_1 u D_2 w \psi - D_2 u D_1 w.$$

If $g \in W_2^2(Q)$ satisfies our boundary conditions, then the change $u = u^* - g$ yields the problem (1.1.8) with $G \equiv \overset{o}{W_2^2}(Q)$,

$$b(u;v) \equiv \nu(u,v)_{2,Q} + (\Delta u, D_1(vD_2 u) - D_2(vD_1 u))_{0,Q}$$

$$+(\Delta g, D_1(vD_2 u) - D_2(vD_1 u))_{0,Q} + (\Delta u, D_1(gD_2 u) - D_2(vD_1 g))_{0,Q}$$

($l(v)$ in (1.1.8) is easy to specify). The nonlinear term is such that $(\mathcal{P}(u); v) \equiv (\Delta u, D_1(vD_2 u) - D_2(vD_1 u))_{0,Q}$ and $(\mathcal{P}(u); u) = 0, \forall u \in G$. It is important that its difference analog satisfy (1.2.5). [28]

---

[27] In the role of the model operator B, a block-diagonal operator can be used as in the linear case (the operators B_1 and B_2 are spectrally equivalent to the operator $-\Delta_h$, and the operator B_3 is of type considered in Theorem 1). At the present time, more complicated problems are being solved that deal with plasticity problems for multilayer shells (see, e.g., [467]). Similar algorithms were constructed and applied for geometric nonlinear problems associated with certain net shells and tires of various structure in the Computer Center of Research Institute of Tire Industry in Moscow (the first results can be found in [211]). In the eighties, they were applied for designing certain aviation tires, including those for Buran (the Russian analog of Shuttle). It should be emphasized that these problems take into account contact of the tires with the landing surface (leading to the restriction $u_3 \ge 0$), and requiring penalty methods as indicated in § 5.5.

[28] For example, if we approximate $[u; \Delta u]$ by $\mathcal{P}u_i \equiv 1/2\{[\partial_2(\Delta_h u \bar{\partial}_1 u) + \bar{\partial}_2(\Delta_h u \partial_1 u)] - [\partial_1(\Delta_h u \bar{\partial}_2 u) + \bar{\partial}_1(\Delta_h u \partial_2 u)]\}_i$ then (1.2.5) holds (see, e.g., [159, 298] and references therein). This implies that Lemmas 4, 5 and Theorem 3 are generalized for our simpler case. Analogous results were also obtained for some problems of meteorology, dealing, e.g., with the system $\nu_1 D_1^2 \Delta u_1 + \nu_2 D_2^2 \Delta u_1 - c[u_1; \Delta u_1] + b_1 D_1 u_2 = 0$, $\nu_1 D_1^2 u_2 - \nu_2 D_2^2 u_2 + c[u_1; u_2] + b_2 D_1 u_1 = 0,\ \nu_1 > 0, \nu_2 > 0$ (see [298] and references therein). Numerical experiments with the given iterative methods show promise.

Chapter 7

Minimization of computational work for systems of Stokes and Navier-Stokes type

This chapter is devoted to asymptotically optimal algorithms for solving boundary value problems associated with systems of Stokes or Navier-Stokes type. Such equations are of fundamental importance in the theory of elasticity and shells, hydrodynamics, meteorology, magnetohydrodynamics, and other fields of science (see [42, 244, 262, 476]). The most significant feature of the problems we study here is that they involve divergence-free vector fields (defined for $x \in \Omega \subset \mathbf{R}^d$), i.e., that div $\vec{u} = 0$. If we introduce subspaces of the Hilbert space $G \equiv (W_2^1(\Omega))^d$ such that their elements $\vec{u}$ meet this incompressibility condition, it then possible to consider original problems as elliptic boundary value problems associated with this new Hilbert space (see [323, 337, 475]) and apply well-known approaches for their study.

An alternative setting (important for construction of asymptotically optimal algorithms) is connected with the use of an additional function p (corresponding to the pressure in hydrodynamics), which plays the role of the Lagrangian multiplier for minimization problems with linear constraint div $\vec{u} = 0$. Such minimization problems are then reduced to finding saddle points of the Lagrangian function. Correctness of these and more general problems was shown in the seventies (see § 1). [1]

§ 1. Saddle-point problems and saddle operators

1.1. Lagrangian function and saddle operators.

Hereafter, G_1 and G_2 are Hilbert spaces, and the original problem is formulated in the Hilbert space $G \equiv G_1 \times G_2$ as the operator equation

$$Lu \equiv \begin{bmatrix} L_{1,1} & L_{1,2} \\ L_{2,1} & -L_{2,2} \end{bmatrix} \begin{bmatrix} u_1 \\ u_2 \end{bmatrix} = \begin{bmatrix} f_1 \\ f_2 \end{bmatrix}, \tag{1.1}$$

where

$$L_{i,j} \in \mathcal{L}(G_j; G_i), \ L_{i,j}^* = L_{j,i}, \ L_{1,1} \in \mathcal{L}^+(G_1), \ L_{2,2} \geq 0 \tag{1.2}$$

and

$$-S_2(L) \equiv L_{2,2} + L_{2,1}L_{1,1}^{-1}L_{1,2} \geq \sigma^2 I_2, \ \sigma > 0. \tag{1.3}$$

The operator $L \in \mathcal{L}(G)$ is called a *saddle operator* if (1.2) and (1.3) hold. If, additionally,

$$L_{2,1}L_{1,1}^{-1}L_{1,2} \geq \sigma^2 I_2, \ \sigma > 0, \tag{1.4}$$

then L is called a *strongly saddle operator*.

For this operator problem, we introduce the *Lagrangian function*

$$\Phi(v_1; v_2) \equiv (L_{1,1}v_1, v_1) - (L_{2,2}v_2, v_2) + 2(L_{2,1}v_1, v_2) - 2(f_1, v_1) - 2(f_2, v_2). \tag{1.5}$$

A point $u \in G$ is called a *saddle point* of $\Phi(v_1; v_2)$ if for all $v_i \in G_i$, $i = 1, 2$, we have $\Phi(u_1; v_2) \leq \Phi(u_1; u_2) \leq \Phi(v_1; u_2)$.

Lemma 1. Problem (1.1) with saddle operator L has a unique solution and is equivalent to the problem of finding a saddle point of $\Phi(v_1; v_2)$.

---

[1] Numerical methods, especially mixed finite element methods, for problems of Stokes or Navier-Stokes types have been studied by many (see, e.g., [244, 261, 411, 475, 476] and references therein). But questions related to the construction of asymptotically optimal algorithms are especially difficult and significant progress in this direction is fairly recent (see [191, 198, 201, 205] and for results with multigrid methods, see [57, 269, 368, 411, 499] and references therein). We pay special attention here to the difficult case of regions with non-Lipschitz boundary; we construct asymptotically optimal algorithms for a wide class of regions when their approximations associated with composite grids with local refinement must be taken into account. Our PGMs (see [191]) are based on piecewise linear approximations for the velocity components and are especially well suited for use with model operators developed in preceding chapters. Moreover, we consider rather general boundary conditions analogous to those in Subsection 5.3.2 and, for problems with the restriction div $\vec{u} - \alpha p = 0$ involving parameter $\alpha \in [0, \alpha_0]$, we obtain estimates of computational work independent of α.

Proof. As in § 1.5, we replace (1.1) by the equivalent system

$$S_2(L)u_2 = f_2 - L_{2,1}L_{1,1}^{-1}f_1 \equiv g_2, \quad L_{1,1}u_1 + L_{1,2}u_2 = f_1. \tag{1.6}$$

Note that $-S_2(L) \in \mathcal{L}^+(G_2)$. Thus, (1.1) is uniquely solvable, and we denote its solution by u. Now, using the evident equality $2(L_{1,2}u_2, v_1) = 2(f_1, v_1) - 2(L_{1,1}u_1, v_1)$, we find that, for all v_1 in G_1,

$$\Phi(v_1; u_2) = \|v_1 - u_1\|_{L_{1,1}}^2 - \|u_1\|_{L_{1,1}}^2 - (L_{2,2}u_2; u_2) - 2(f_2; v_2). \tag{1.7}$$

Hence, $\Phi(u_1; u_2) \le \Phi(v_1; u_2)$, $\forall v_1 \in G_1$. Analogously, we see that $\Phi(u_1; v_2) \le \Phi(u_1; u_2)$, $\forall v_2 \in G_2$, which completes the proof. □ [2]

Lemma 2. *Condition* (1.4) *with positive* σ *is equivalent to each of the following conditions:*

$$\|L_{1,2}v_2\|_{L_{1,1}^{-1}} \ge \sigma\|v_2\|, \ \forall v_2 \in G_2, \tag{1.8}$$

$$\sup_{v_1 \ne 0} \frac{(L_{2,1}v_1, v_2)}{\|v_1\|_{L_{1,1}^{-1}}} \ge \sigma\|v_2\|, \ \forall v_2 \in G_2, \tag{1.9}$$

$$\text{Im } L_{2,1} = G_2. \tag{1.10}$$

Moreover, each of these conditions imply that Im $L_{1,2}$ *is a subspace of* G_1 *and* G_1 *is an orthogonal sum*

$$G_1 = \text{Ker } L_{2,1} \oplus \text{Im } L_{1,2}. \tag{1.11}$$

Proof. (1.4) is equivalent to $(L_{2,1}L_{1,1}^{-1}L_{1,2}v_2, v_2) \ge \sigma^2\|v_2\|^2$, $\forall v_2 \in G_2$, $\sigma > 0$, which by (1.2) can be rewritten in the form $(L_{1,1}^{-1}L_{1,2}v_2, L_{1,2}v_2) \ge \sigma^2\|v_2\|^2$, $\forall v_2 \in G_2$, $\sigma > 0$. Thus, (1.4) is equivalent to (1.8). For the left-hand side of (1.9) we apply Lemma 0.4.5 and conclude that (1.9) and (1.8) are equivalent. Finally, if (1.9) and (1.4) hold, then (1.6) with $L_{2,2} = 0, f_1 = 0$, has a solution for each $f_2 = v_2 \in G_2$. Hence, (1.10) holds. Conversely, for each $v_2 \in G_2$, suppose there exists $u_1 \in G_1$ such that

---

[2] A particular and very important case of the above problem, corresponding to a *constrained minimization problem*, is defined by $L_{2,2} = 0$. Then, from (1.7), it follows that u_1 corresponds to the solution of the variational problem $u_1 = \text{argmin } \Phi_1(v_1)$, where the energy functional is defined by $\Phi_1(v_1) \equiv (L_{1,1}v_1, v_1) - 2(f_1; v_1)$ (see § 0.1), and minimization is carried out with respect to v_1 such that $L_{2,1}v_1 = f_2$ (see § 0.1 and [97, 233, 323, 475]). Indeed, if $[u_1, u_2]$ is a solution of (1.1) then $\Phi_1(v_1) = (L_{1,1}v_1, v_1) - 2(L_{1,1}u_1, v_1) - 2(L_{1,2}u_2, v_1)$ and, in accordance with the fact that $L_{2,1}v_1 = f_2$, we see that $\Phi_1(v_1) = \|v_1 - u_1\|_{L_{1,1}}^2 - \|u_1\|_{L_{1,1}}^2 - 2(u_2, f_2)$. Hence, $u_1 = \text{argmin } \Phi_1(v_1)$.

$L_{2,1}u_1 = v_2$. Consider the Hilbert space $G_1(L_{1,1})$ (see § 0.1) and represent it as $G_1(L_{1,1}) = \text{Ker } L_{2,1} \oplus W_1$, where $W_1 \equiv (\text{Ker } L_{2,1})^{\perp}$. It is easy to see that W_1 is a Hilbert space and that $L_{2,1}$ is a one-to-one mapping of W_1 onto G_2. Then, by virtue of the classical Banach Theorem (see, e.g., [292, 341]), there exists an inverse bounded mapping (denote it by $[L_{2,1}]^{-1}$). Thus, if $u_1 = [L_{2,1}]^{-1}v_2$, then $\|u_1\|_{L_{1,1}} \leq \|[L_{2,1}]^{-1}\|_{G_2 \mapsto W_1}\|v_2\|$ and $1/\sigma = \|[L_{2,1}]^{-1}\|_{G_2 \mapsto W_1}$ (see (1.9)). We now prove (1.11). First observe that each fundamental sequence $\{L_{1,2}v_2^n\}$ in G_1 corresponds to the fundamental sequence $\{v_2^n\}$ in G_2 (see (1.8)). This implies that $\lim_{n \to \infty} L_{1,2}v_2^n = L_{1,2}v_2$, where $v_2 = \lim_{n \to \infty} v_2^n$. Hence, Im $L_{1,2}$ is a subspace of the Hilbert space G_1. It is easy to see that Ker $L_{2,1}$ is a subspace of G_1 and that its orthogonal complement $(\text{Ker } L_{2,1})^{\perp}$ contains Im $L_{1,2}$. Suppose now that $(\text{Ker } L_{2,1})^{\perp}$ contains $u_1 \neq 0$ orthogonal to Im $L_{1,2}$. Then $L_{2,1}u_1 \equiv v_2 \neq 0$ and $0 < (L_{2,1}u_1, v_2) = (u_1, L_{1,2}v_2) = 0$. Hence, (1.11) holds. $\square$ [3]

1.2. Correctness of problems with strongly saddle operators. We study problem (1.1) in a somewhat more general setting. This enables us to deal also with problems depending on the parameter

$$\alpha \in [0, \alpha_0), \quad \alpha_0 > 0, \tag{1.12}$$

and with more general operators $L_{1,1}$ (important examples will be given in Subsection 6). First, we replace (1.1) by

$$Lu \equiv \begin{bmatrix} L_{1,1} & L_{1,2} \\ L_{2,1} & -\alpha L_{2,2} \end{bmatrix} \begin{bmatrix} u_1 \\ u_2 \end{bmatrix} = \begin{bmatrix} f_1 \\ f_2 \end{bmatrix}. \tag{1.13}$$

Second, we replace (1.2) by the equivalent but more detailed conditions

$$\left.\begin{array}{c} L_{i,j} \in \mathcal{L}(G_j; G_i), \ L_{i,j}^* = L_{j,i}, i \in [1,2], j \in [1,2], \\ \gamma_0 I_1 \leq L_{1,1} \leq \gamma_1 I_1, \ \gamma_0 > 0, \gamma_2 I_2 \leq L_{2,2} \leq \gamma_3 I_2, \ \gamma_2 \geq 0, \\ \|L_{2,1}\| \equiv \|L_{2,1}\|_{G_1 \mapsto G_2} \leq \kappa_1. \end{array}\right\} \tag{1.14}$$

Finally, instead of (1.4) (or (1.9)), we make use of

$$\sup_{v_1 \neq 0} \frac{|b_{2,1}(v_1; v_2)|}{\|v_1\|} \geq \sigma_0 \|v_2\|, \ \forall v_2 \in G_2, \tag{1.15}$$

---

[3]It is easy to see that we may replace $(L_{2,1}v_1, v_2)$ in (1.9) by $|(L_{2,1}v_1, v_2)|$, and that the constant $\sigma > 0$ remains the same for all Descartes coordinates and under the change of variables $x = hy$. We emphasize that the facts $L_{2,1}^* = L_{1,2}$ and (1.11) for a Euclidean space G_1 are immediate. But for a Hilbert space and bounded $L_{1,2}$, (1.11) is rare and it implies that $L_{2,1}$ is a *normally solvable operator* (true for $L_{2,1} = L_{1,2}^*$ as well). We also note that the assumption that Im $L_{1,2}$ is a subspace of G_1 implies that inequality (1.8) involving a positive constant σ holds for all v_2 orthogonal to Ker $L_{1,2}$.

with positive constant σ_0 (which, of course, is equivalent to (1.4) if $L_{1,1} \in \mathcal{L}^+(G_1)$).

Theorem 1. *Problem* (1.13) *under conditions* (1.14), (1.15) *is correctly posed and the components of its solution have the a priori estimates*

$$\|u_1\| \leq \frac{1}{\gamma_0}(\|f_1\| + \kappa_1 t_2) \quad \|u_2\| \leq \frac{\gamma_1}{\sigma_0^2}\left(\|f_2\| + \frac{\kappa_1}{\gamma_0}\|f_1\|\right), \qquad (1.16)$$

with $t_2 \equiv \gamma_1/\sigma_0^2[\|f_2\| + \kappa_1/\gamma_0\|f_1\|]$ *(independent of* $\alpha \leq 0$*).*

Proof. As in the proof of Lemma 1, we obtain

$$S_2(L)u_2 = \equiv -(\alpha L_{2,2} + L_{2,1}L_{1,1}^{-1}L_{1,2})u_2 = f_2 - L_{2,1}L_{1,1}^{-1}f_1 \equiv g_2. \quad (1.17)$$

Since $(L_{1,1}^{-1}L_{1,2}v_2, L_{1,2}v_2) \geq 1/\gamma_1\|L_{1,2}v_2\|^2 \geq \sigma_0^2/\gamma_1\|v_2\|^2$, then $(\alpha\gamma_2 + \sigma_0^2/\gamma_1)I_2 \leq -S_2$. Hence, (1.17) implies that $\|u_2\| \leq \gamma_1/\sigma_0^2\|g_2\|$. Since $\|g_2\| \leq \|f_2\| + \|L_{2,1}L_{1,1}^{-1}f_1\| \leq \|f_2\| + \kappa_1/\gamma_0\|f_1\|$, then we obtain the desired estimate for $\|u_2\|$. To complete the proof, it suffices to note that $\|u_1\| \leq 1/\gamma_0(\|f_1\| + \kappa_1\|u_2\|_2)$. $\square$

Lemma 3. *Let* (1.14), (1.15) *be satisfied. Then* $S_2(L)$ *is such that*

$$(\alpha\gamma_2 + \sigma_0^2/\gamma_1)I_2 \leq -S_2 \leq (\alpha\gamma_3 + \kappa_1^2/\gamma_0)I_2. \quad (1.18)$$

Proof. We may use the proof of Theorem 1 and observe also that

$$(L_{1,1}^{-1}L_{1,2}v_2, L_{1,2}v_2) \leq \frac{1}{\gamma_0}\|L_{1,2}v_2\|^2 \leq \frac{\kappa_1^2}{\gamma_0}\|v_2\|^2.$$

Therefore, inequalities (1.18) hold. $\square$

Theorem 2. *The operator* L *defined by* (1.13)–(1.15) *is invertible and*

$$\|L^{-1}\| \leq K, \qquad (1.19)$$

$$K \equiv \max\{\frac{1}{\gamma_0}(\|f_1\| + \kappa_1 t_2); \frac{\gamma_1}{\sigma_0^2}(\|f_2\| + \frac{\kappa_1}{\gamma_0}\|f_1\|)\}, \qquad (1.20)$$

with $t_2 \equiv \gamma_1/\sigma_0^2[\|f_2\| + \kappa_1/\gamma_0\|f_1\|]$.

1.3. Correctness of more general problems. We consider now problems of type (1.13)–(1.15) involving nonsymmetric operators $L_{1,1}$ and $L_{2,2}$ (their symmetric parts are $L_{1,1,s}$ and $L_{2,2,s}$, respectively). We carry out the analysis of correctness under conditions that reduce to (1.13), (1.14) for symmetric operators.

Lemma 4. *Let* G *be a Hilbert or Euclidean space and* $A \in \mathcal{L}(G)$. *Let* $A_s \equiv 1/2(A + A^*) \geq \alpha_0 I$, *with* $\alpha_0 > 0$. *Then* A *is invertible and* $(A^{-1}u, u) \geq \alpha_0/\|A\|^2\|u\|^2, \forall u \in G$.

Proof. We have $(Av, v) = (A_s v, v) \geq \alpha_0 \|v\|^2$, $\forall v \in G$. Hence, A is invertible. Replacing v by $A^{-1}u$, then $(A^{-1}u, u) \geq \alpha_0 \|A^{-1}u\|^2$, $\forall u \in G$. Observe that $u = AA^{-1}u$ and $\|u\| \leq \|A\|\|A^{-1}\|$, $\forall u \in G$. Combining those two inequalities leads to the desired one. $\square$

Theorem 3. Under conditions

$$\left.\begin{array}{c} L_{1,2}^* = L_{2,1}, \\ \gamma_0 I_1 \leq L_{1,1,s}, \ \gamma_0 > 0, \ \|L_{1,1}\| \leq \gamma_1, \\ \gamma_2 I_2 \leq L_{2,2}, \ \gamma_2 \geq 0, \ \|L_{2,2}\| \leq \gamma_3, \\ \|L_{2,1}\| \equiv \|L_{2,1}\|_{G_1 \mapsto G_2} \leq \kappa_1, \end{array}\right\} \quad (1.21)$$

and (1.15), *problem* (1.13) *is correctly posed and the components of its solution satisfy the a priori estimates*

$$\|u_1\| \leq \frac{1}{\gamma_0}(\|f_1\| + \kappa_1 t_2), \quad \|u_2\| \leq \frac{1}{\gamma_4}[\|f_2\| + \frac{\kappa_1}{\gamma_0}\|f_1\|], \quad (1.22)$$

with $\gamma_4 \equiv \alpha\gamma_2 + \gamma_0\sigma_0^2/\gamma_1^2$ *and* $t_2 \equiv \gamma_1/\sigma_0^2[\|f_2\| + \kappa_1/\gamma_0\|f_1\|]$, *independent of* $\alpha \geq 0$.

Proof. We start again from (1.17) rewritten in the form $-S_2(L)u_2 = -g_2$. Now, on the basis of Lemma 4, we have

$$(L_{1,1}^{-1}L_{1,2}v_2, L_{1,2}v_2) \geq \frac{\gamma_0}{\gamma_1^2}\|L_{1,2}v_2\|^2 \geq \frac{\gamma_0\sigma_0^2}{\gamma_1^2}\|v_2\|^2,$$

and $\gamma_4\|u_2\|^2 \leq -(S_2 u_2, u_2)$, with $\gamma_4 \equiv \alpha\gamma_2 + \gamma_0\sigma_0^2/\gamma_1^2$. Hence, (1.17) implies that $\gamma_4\|u_2\| \leq \|g_2\|\|u_2\|$. Since $\|g_2\| \leq \|f_2\| + \kappa_1/\gamma_0\|f_1\|$, then the desired estimate for $\|u_2\|$ holds. To complete the proof we combine this and the estimate $\|u_1\| \leq 1/\gamma_0[\|f_1\| + \kappa_1\|u_2\|_2]$. $\square$ [4]

Lemma 5. The solution of the problem (1.13), (1.15), (1.21) *with* $f_2 = 0$ *is such that* $\|u_1\| \leq \frac{1}{\gamma_0}\|f_1\|$ *and* $\|u_2\| \leq (1 + \gamma_1)/\sigma_0\|f_1\|$.

Proof. We have $(L_{1,1}u_1, u_1) + \alpha(L_{2,2}u_2, u_2) = (f_1, u_1)$ (see (1.13)). The standard approach yields $\gamma_0\|u_1\| \leq \|f_1\|\|u_1\|$. We, thus, apply the representation $L_{2,1}u_2 = f_1 - L_{1,1}u_1$, condition (1.8) (with $L_{1,1}$ and σ replaced by I and σ_0), and the evident inequality for $\|L_{1,1}u_1\|$. This yields the estimate for $\|u_1\|$. $\square$

Theorem 3 implies that (1.19) must be valid for some K. For construction of projective methods (see § 2), it will be convenient to rewrite the problem under consideration using bilinear forms $b_{r,l}$, defined and bounded on $G_l \times G_r$ and connected with the operators $L_{r,l}$ by the standard equalities $b_{r,l}(v_l; v_r) = (L_{r,l}v_l, v_r), r \in [1, 2], l \in [1, 2]$. Then (1.13) is just the problem of finding $u \in G$ such that

---

[4] For $f_2 = 0$, there is an another way of estimating the solution.

$$b_{1,1}(u_1; v_1) + b_{1,2}(u_2; v_1) = l_1(v_1), \ \forall v_1, \ \left.\right\} \tag{1.23}$$
$$b_{2,1}(u_1; v_2) - \alpha b_{2,2}(u_2; v_2) = l_2(v_2), \ \forall v_2. \ \left.\right\}$$

Conditions (1.21) can be rewritten in the form

$$b_{1,2}(v_2; v_1) = b_{2,1}(v_1; v_2), \ |b_{2,1}(v_1; v_2)| \leq \kappa_1 \|v_1\| \|v_2\|, \ \left.\right\}$$
$$\gamma_0 \|v_1\|^2 \leq b_{1,1}(v_1; v_1), \ \gamma_0 > 0, \ |b_{1,1}(v_1; w_1)| \leq \gamma_1 \|v_1\| \|w_1\|, \ \left.\right\} \tag{1.24}$$
$$\gamma_2 \|v_2\|^2 \leq b_{2,2}(v_2; v_2), \ \gamma_2 \geq 0, \ |b_{2,2}(v_2; w_2)||| \leq \gamma_3 \|v_2\| \|w_2\|, \ \left.\right\}$$

where v_r and w_r refer to arbitrary elements of G_r, $r = 1, 2$.

1.4. Perturbation of the parameter. Here, we study dependence of the solution of (1.13) on the parameter α (see [191, 195]).

Theorem 4. Suppose conditions of Theorem 3 are satisfied and denote the operators L in problems (1.13)–(1.15) and their solutions by L_α and $u(\alpha) \equiv [u_1(\alpha), u_2(\alpha)]$, respectively. Then there exists K' such that

$$\|u(\alpha) - u(\alpha')\| \leq K'|\alpha - \alpha'|, \quad \forall \alpha \geq 0, \ \alpha' \geq 0. \tag{1.25}$$

Proof. For $z \equiv u(\alpha) - u(\alpha')$, we have

$$L_\alpha z = [0, -(\alpha - \alpha')L_{2,2}u_2(\alpha')]^T. \tag{1.26}$$

Hence, applying estimates from Theorem 3, we obtain (1.25). □ [5]

1.5. Variational problems with large parameters; the penalty method. We prove the following simple but important theorem (see, [337]).

Theorem 5. Under conditions $L_{1,1} \in \mathcal{L}^+(G_1)$, $L_{2,2} \in \mathcal{L}^+(G_2)$, $L_{2,1}^* = L_{1,2}$, and $\alpha > 0$, problem (1.13) is correctly posed and the first component of its solution coincides with the solution of problem

$$u_1 = argmin \ \Phi_\alpha(v_1), \tag{1.27}$$

where

$$\Phi_\alpha(v_1) \equiv (L_{1,1}v_1, v_1) + \frac{1}{\alpha}(L_{1,2}L_{2,2}^{-1}L_{2,1}v_1, v_1) - 2(g_1, v_1), \tag{1.28}$$

$g_1 \equiv f_1 + \frac{1}{\alpha}L_{2,2}^{-1}f_2$; *and conversely, variational problem (1.27), (1.28) is correct and its solution is just the first component of that for (1.13)–(1.15).*

Proof. Since $S_2 \geq \alpha L_{2,2} > 0$, the correctness of (1.13) follows from Theorem 1. Now we make use of the converse elimination, that is, of u_2 from (1.13). This gives $u_2 = 1/\alpha L_{2,2}^{-1}(-f_2 + L_{2,1}u_1)$ and

---

[5] Similar sharp estimates for Stokes and Navier-Stokes systems were given in [462].

$S_{1,\alpha}u_1 \equiv L_{1,1} + 1/\alpha L_{1,2}L_{2,2}^{-1}L_{2,1}u_1 = g_1$, with $S_{1,\alpha} \in \mathcal{L}^+(G_1)$. Hence, Theorem 0.1.3 applies and leads to the correctness of our problem. □ [6]

1.6. Examples of problems from hydrodynamics and elasticity.
Here, we use notation somewhat different from that used in § 5.3. For d-dimensional problems, with $d = 2$ or $d = 3$, and $x \in \Omega \subset \mathbf{R}^d$, we use $\vec{u} \equiv u_1 \equiv [u_{1,1}, u_{1,2}]^T$ if $d = 2$ and $\vec{u} \equiv u_1 \equiv [u_{1,1}, u_{1,2}, u_{1,3}]^T$ if $d = 3$. For the Stokes system

$$-\Delta\vec{u} + \text{grad } p = \vec{f'}, \left.\begin{array}{c} \\ \\ \end{array}\right\} \qquad (1.29)$$
$$\text{div } \vec{u} = 0,$$

involving an unknown vector-function $\vec{u}$ (velocity vector) and function p (pressure), we consider homogeneous Dirichlet conditions for $\vec{u}$. From this classical setting of the problem, we pass to its modern formulation of type (1.1) in a Hilbert space $G = G_1 \times G_2$, where

$$G_1 \equiv (\overset{o}{W_2^1}(\Omega))^d, \ G_2 \equiv G_2' \setminus 1, \qquad (1.30)$$

$G_2' \equiv L_2(\Omega)$, $G_2' \setminus 1 \equiv \{p : p \in L_2(\Omega) \text{ and } (p, 1)_{0,\Omega} = 0\}$ (in other words G_2 is just $\text{Ker } \varphi_0$, where the linear functional φ_0 is defined on G_2' by $\varphi_0 \equiv (p, 1)_{0,\Omega}$, and $\|u_1\|_{G_1} \equiv |\vec{u}_1|_{1,\Omega}$. Then, in (1.1), we have $L_{1,1} = I_1$, $L_{2,2} = 0$, $(L_{2,1}u_1, v_2) = (\text{div } u_1, v_2)_{0,\Omega} = (L_{1,2}v_2, u_1)$, $(f_1, v_1) = (\vec{f'}, v_1)_{0,\Omega}$, $f_2 = 0$, and u_i and v_i are arbitrary elements of $G_i, i = 1, 2$. Inequality (1.9) takes the well-known form

$$\sup_{\vec{u} \in G_1} \frac{(\text{div } \vec{u}, p)_{0,\Omega}}{\|\vec{u}\|_{G_1}} \geq \sigma|p|_{0,\Omega}, \ \sigma > 0, \ \forall p \in G_2. \qquad (1.31)$$

---

[6]This theorem does not give estimates independent of α, and it cannot lead to estimates of type (1.25). Nonetheless, some weaker estimates, even just the convergence of $u(\alpha)$ to $u(0)$ were obtained (see [298, 475]) as justification of the idea to approximate problems (1.13) with $\alpha = 0$ by problems of type (1.27)–(1.29), that is, of the basic idea of the *penalty method* (or the *regularization method*) (see [244, 337]). Unfortunately, the method for hydrodynamics problems needs very small α (see [244]) even when (1.25) holds. This complicates the construction of model grid operator and does not lead to asymptotically optimal algorithms, although such methods were used with success for important practical hydrodynamics problems [244]. Some practical results were also obtained for shell problems in the mid 60's based on the possibility to use not too small a parameter α (this sometimes even led to an improvement in accuracy of the mathematical model) (see § 6.5 and [161, 210] and references therein) and optimal preconditioners. Thanks to an understanding of the role of (1.15) and its grid analogs in the theory of PGMs and iterative processes (we shall discuss the subject in detail in § 2–6), it now seems reasonable to regard problems (1.13) as basic and, instead of problems (1.27), (1.28) and more general ones involving a large parameter $1/\alpha$, to work with corresponding problems (1.13) (see, e.g., [94, 191, 309, 486]). We give examples of such elasticity problems below.

It was proved for regions with sufficiently smooth boundaries (see, e.g., [27, 127] and references therein) and with Lipschitz boundaries (see, e.g., [310, 325] and references therein). [7]

From (1.31) and Theorem 1, the next useful result is easily deduced.

Theorem 6. *Let Ω have Lipschitz piecewise smooth boundary Γ and suppose that vector-function $\vec{g}$ is prescribed on Γ such that $(\vec{g}, \vec{n})_{0,\Gamma} = 0$ and each of its component $g_s \in W_2^{1/2}(\Gamma), s \in [1, d]$. Then there exists a solenoidal vector field $\vec{v} \in (W_2^1(\Omega))^3$ and a constant K_{sol} such that $\vec{v} \mid_\Gamma = \vec{g} \mid_\Gamma$ and $\|\vec{v}\|_{W_2^1(\Omega)} \leq K_{sol} \|\vec{g}\|_{1/2,\Gamma}$.*

Proof. We first use the classical result (see § 0.1 and [3, 67]) and construct $\vec{w} \in (W_2^1(\Omega))^3$ such that $\vec{w} \mid_\Gamma = \vec{g} \mid_\Gamma$. We then define $\vec{u}$ as the solution of problem (1.13), with G_1, G_2 defined by (1.30), $L_{1,1} = I_1$, $L_{2,2} = 0$, $(L_{2,1}u_1, v_2) = (\operatorname{div} u_1, v_2)_{0,\Omega}$, $f_1 = 0$, and $f_2 = \operatorname{div} \vec{w} \in G_2$. It is then easy to see that $\vec{v} = \vec{w} - \vec{u}$. □ [8]

In conclusion, we give important examples of the reduction of elasticity problems, involving certain large parameters, to problems (1.13)–(1.15). We start by considering the variational problem $\vec{u} = \arg \min_{\vec{v} \in G_1} \Phi(v)$, where the energy functional Φ is defined by (5.3.13), (5.3.12), and (5.3.15). We split now the large Lame parameter λ as $\lambda = \lambda_0 + (\lambda - \lambda_0)$, with a relatively small $\lambda_0 > 0$ such that $0 < \lambda - \lambda_0 \equiv 1/\alpha$. Now denoting div $\vec{v}$ by αp, we may write $2\Phi(\vec{v}) = 2\Phi(\vec{v}; p)$ with

---

[7]In the definition of $L_{2,1}$ and $L_{1,2} = L_{2,1}^*$, it is possible to use arbitrary elements of G_2' because $(\operatorname{div} \vec{u}, 1)_{0,\Omega} = (\vec{u}, \vec{n})_{0,\Gamma} = 0$, $\forall \vec{u} \in G_1$, where n denotes the outward normal to the boundary and $(\vec{u}, \vec{n})_{0,\Gamma}$ corresponds to the flow of the vector field $\vec{u}$ through Γ. But this is not the case for more general spaces G_1. The classical variational approach to the study of (1.29) was suggested by Hopf and developed in many respects by Ladyzenskaya (see [313, 323]). This analysis is based on the use of the Hilbert space $V \equiv \{\vec{u} : \vec{u} \in G_1 \text{ and } \operatorname{div}\vec{u} = 0\}$ with inner product $(\vec{u}, \vec{v})_{1,\Omega}$. The first component of the solution of (1.1) is then just $\vec{u} = \operatorname{argmin}\Phi(\vec{v})$, where $\Phi(\vec{v}) \equiv \|\vec{v}\|_{G_1}^2 - 2(\vec{f}', \vec{v})_{0,\Omega}$ and the minimum is taken over the Hilbert space V. We may thus apply theorems from § 0.1, but the real problem was to study properties of the divergence operator and show that V is a completion of the corresponding space of smooth *solenoidal vector fields*, that is of fields $\vec{u}$ with div $\vec{u} = 0$ (see [323]). It was precisely such study of divergence in [323] that indicated a way to (1.31) and to problems of type (1.1). There are also relevant results of Magenes, Mikhlin, Necas (see, e.g., [27, 375] and references therein). It should also be mentioned that the first results dealing with eigenvalues of the problem grad div $\vec{u} = \lambda\Delta\vec{u}$ were obtained almost a hundred years ago in [121]. Another but quite obvious motive for working with (1.1) is the significant difficulties that arise in constructing subspaces of V. It would not have been right to say that the reasonable choice of subspaces of G is easy (see § 2 and 3), but it can be done in a number of ways. Such PGMs are widely applied in practice (see, e.g., [99, 230, 233, 242, 244, 262, 411]), and some do lead to asymptotically optimal algorithms for problems of Stokes type (see § 3 and 4).

[8]Theorem 6 also yields a corresponding extension theorem for a solenoidal vector field $\vec{v} \in (W_2^1(\Omega))^3$ to a solenoidal vector field $\vec{v} \in (W_2^1(\Omega'))^3$, where $\bar{\Omega} \subset \Omega'$.

$$2\Phi(\vec{v};p) \equiv \lambda_0 |\text{div } v|^2_{0,\Omega} + 2\mu(\sum_{i,k=1}^{d} \epsilon^2_{i,k}(v), 1)_{0,\Omega} + \frac{1}{\alpha}|p|^2_{0,\Omega_2} - 2l(\vec{v}). \quad (1.32)$$

This implies that we may treat the original problem as a variational one under the constraint

$$\text{div } \vec{u} - \alpha p = 0 \quad (1.33)$$

(it corresponds to $L_{2,1}u_1 - \alpha L_{2,2}u_2 = 0$ in (1.13)). We emphasize that here $G_1 \equiv (W^1_2(\Omega_2;\Gamma_0))^d$ and, for $\Gamma_0 = \Gamma$, we may use the pair of spaces G_1 and G_2 defined by (1.30). But if

$$|\Gamma \setminus \Gamma_0|_{(d-1)} > 0, \quad (1.34)$$

then, as we show in § 3, condition (1.15) holds for the choice

$$G_1 \equiv (W^1_2(\Omega))^d, \quad G_2 \equiv G'_2 = L_2(\Omega), \quad (1.35)$$

which simplifies the construction of PGMs and optimal algorithms. [9]

§ 2. Projective methods for problems with saddle operators

2.1. General scheme. Below we consider a sequence of finite-dimensional subspaces $\hat{G}_h \equiv \hat{G}_{1,h} \times \hat{G}_{2,h} \in G$ approximating the original Hilbert space G (G_r is approximated by the sequence $\hat{G}_{r,h} \equiv \hat{G}_r, r = 1,2$). We approximate (1.23) by the problems of finding $\hat{u} \equiv [\hat{u}_1, \hat{u}_2]^T \hat{G}_h$ such that

$$b_{1,1}(\hat{u}_1; \hat{v}_1) + b_{1,2}(\hat{u}_2; \hat{v}_1) = l_1(\hat{v}_1), \ \forall \hat{v}_1 \in \hat{G}_1, \quad (2.1)$$

$$b_{2,1}(\hat{u}_1; \hat{v}_2) - \alpha b_{2,2}(\hat{u}_2; \hat{v}_2) = l_2(\hat{v}_2), \ \forall \hat{v}_2 \in \hat{G}_2. \quad (2.2)$$

We also rewrite (1.23) in the equivalent form

$$\hat{L}\hat{u} \equiv \begin{bmatrix} \hat{L}_{1,1} & \hat{L}_{1,2} \\ \hat{L}_{2,1} & -\alpha\hat{L}_{2,2} \end{bmatrix} \begin{bmatrix} \hat{u}_1 \\ \hat{u}_2 \end{bmatrix} = \begin{bmatrix} \hat{f}_1 \\ \hat{f}_2 \end{bmatrix}, \quad (2.3)$$

---

[9] Along the same lines, for the problem associated with (5.3.49) and large γ_4, we introduce $\Phi_1(\vec{v};p) \equiv \sum_{i,k=1}^{3}(\alpha_{i,k}x_2, \varepsilon_i\varepsilon_k)_{0,\Omega_2} + \lambda_0(x_2, (D_2v_1 + D_1v_2)^2)_{0,\Omega_2} + 1/\alpha(x_2,p^2)_{0,\Omega_2} - 2l(\vec{v})$, with $1/\alpha \equiv \gamma_4 - \lambda_0 > 0$, which we minimize over $\vec{v} \in G$ such that $D_2v_1 + D_1v_2 - \alpha p = 0$. Under the reverse numbering of space variables, we can rewrite this constraint in standard form (1.33).

with $\hat{L}_{r,l} \in \mathcal{L}(\hat{G}_l, \hat{G}_r), r = 1, 2, l = 1, 2$ (see (1.13) and (0.2.6)). As an analog of (1.15), we take

$$\sup_{\hat{v}_1 \in \hat{G}_1, \hat{v}_1 \neq 0} \frac{b_{2,1}(\hat{v}_1; \hat{v}_2)}{\|\hat{v}_1\|} \geq \hat{\sigma}_0 \|\hat{v}_2\|, \ \forall \hat{v}_2 \in \hat{G}_2, \ \hat{\sigma}_0 > 0, \tag{2.4}$$

where $\hat{\sigma}_0$ is independent of h. [10] From the definition of $\hat{\sigma}_0$ in (2.4), the next simple lemma follows.

Lemma 1. *Suppose that, for a sequence of subspaces* $\hat{G}_h = \hat{G}_{1,h} \times \hat{G}_{2,h}$, *property (2.4) holds. Suppose also that a new sequence of subspaces* $\hat{G}'_h \equiv \hat{G}'_{1,h} \times \hat{G}'_{2,h}$ *is such that* $\hat{G} \subset G'_h$ *and* $\hat{G}'_{2,h} \subset \hat{G}_{2,h}$. *Then, for this new sequence, (2.4) also holds with the same constant* $\hat{\sigma}_0$.

Theorem 1. *Under conditions (1.24), (2.4), problems (2.3), are correctly posed. Moreover, there exists a constant* $K_0 \equiv K_0(\gamma_0, \gamma_1, \kappa_1, \hat{\sigma}_0)$, *independent of* h *and* $\alpha \geq 0$, *such that*

$$\|\hat{L}^{-1}\| \leq K_0. \tag{2.5}$$

Proof. This is just a repeat of the proof of Theorem 1.1, with obvious simplifications because now we are not interested in an explicit representation of K_0. $\square$

2.2. Error estimates. To study convergence of projective method (2.1), (2.2), we denote its error by $z \equiv [z_1, z_2]^T \equiv [\hat{u}_1 - u_1, \hat{u}_2 - u_2]^T$ (for relevant results, see [36, 98, 127, 242, 417, 443]).

Theorem 2. *Let the conditions of Theorem 1 for projective problems (2.3) and condition (1.12) for the parameter* α *be satisfied and* $b_{2,2}(v_2; w_2) = b_{2,2}(w_2; v_2), \ \forall v_2 \in G_2, \ \forall w_2 \in G_2$. *Let*

$$K_1 \equiv \frac{1}{2\gamma_0} \left((1 + \frac{\kappa_1}{\sigma_0})\gamma_1 \rho_1 + \kappa_1 \rho_2 \right), \ K_2 \equiv \frac{\kappa_1}{\gamma_0} \left(1 + \frac{\kappa_1}{\sigma_0} \right) \rho_1 \rho_2 + \frac{\alpha_0 \gamma_3 \rho_2^2}{4\gamma_0},$$

$\rho_1 \equiv dist_{G_1}\{u_1; \hat{G}_1\}$ *and* $\rho_2 \equiv dist_{G_2}\{u_2; \hat{G}_2\}$. *Then we have the estimates*

$$\|z_1\| \leq K_1 + [(K_1^2 + K_2]^{1/2}, \tag{2.6}$$

---

[10] This independence from the dimensions of the approximating subspaces $\hat{G}_h$ plays a fundamental role in the investigation of PGMs and in the construction of effective algorithms for solving resulting systems (see, e.g., [27, 97, 99, 117, 126, 127, 191, 233] and references therein, where problems (2.3), with $\alpha = 0$ and certain symmetric $L_{1,1}$, were analyzed); our more general analysis is based on the results of [191] and differs from the mentioned ones in many respects even for $\alpha = 0$. The whole of § 3 will be devoted to the choice of $\hat{G}_{1,h}$ and $\hat{G}_{2,h}$ for which (2.4) holds (known as *inf-sup* or *Ladyzenskaya-Babushka-Brezzi condition.*

$$\|z_2\| \le (1 + \frac{\kappa_1}{\sigma_0})\rho_2 + \frac{\gamma_1}{\hat{\sigma}_0}\|z_1\|. \tag{2.7}$$

Proof. In (1.23), let $v = \hat{v}$. Then

$$b_{1,1}(u_1; \hat{v}_1) + b_{1,2}(u_2; \hat{v}_1) = l_1(\hat{v}_1), \ \forall \hat{v}_1 \in \hat{G}_1. \tag{2.8}$$

Combining (2.1) and (2.8), we see that

$$b_{1,1}(z_1; \hat{v}_1) = -b_{2,1}(\hat{v}_1; z_2), \ \forall \hat{v}_1 \in \hat{G}_1. \tag{2.9}$$

Analogously,

$$b_{2,1}(z_1; \hat{v}_2) - \alpha b_{2,2}(z_2; \hat{v}_2) = 0, \ \forall \hat{v}_2 \in \hat{G}_2. \tag{2.10}$$

These basic equalities will be used several times in our proof. We estimate $\|z_2\|$ starting from the evident inequality $\|z_2\| \le \|\hat{v}_2 - u_2\| + \|\hat{u}_2 - v_2\|$. For the second term on the right-hand side, we apply condition (2.4). This gives $\|\hat{u}_2 - v_2\| \le 1/\hat{\sigma}_0 \sup_{\hat{v}_1}\{|b_{2,1}(\hat{v}_1; \hat{u}_2 - \hat{v}_2)|\|\hat{v}_1\|^{-1}\}$. Since $\hat{u}_2 - \hat{v}_2 = u_2 - \hat{v}_2 + \hat{u}_2 - u_2$, we then have

$$\|\hat{u}_2 - v_2\| \le \frac{1}{\hat{\sigma}_0} \sup_{\hat{v}_1} \frac{|b_{2,1}(\hat{v}_1; \hat{u}_2 - \hat{v}_2)|}{\|\hat{v}_1\|} + \frac{1}{\hat{\sigma}_0} \sup_{\hat{v}_1} \frac{|b_{2,1}(\hat{v}_1; z_2)|}{\|\hat{v}_1\|}.$$

Hence,

$$\|z_2\| \le \|\hat{v}_2 - u_2\| + \frac{1}{\hat{\sigma}_0} \left(\kappa_1\|\hat{u}_2 - \hat{v}_2\| + \frac{1}{\hat{\sigma}_0} \sup_{\hat{v}_1} \frac{|b_{2,1}(\hat{v}_1; z_2)|}{\|\hat{v}_1\|} \right).$$

To finish with $\|z_2\|$, we observe that $b_{2,1}(\hat{v}_1; z_2)$ can be replaced in accordance with (2.9). Thus,

$$\sup_{\hat{v}_1} \frac{|b_{2,1}(\hat{v}_1; z_2)|}{\|\hat{v}_1\|} = \sup_{\hat{v}_1} \frac{|b_{1,1}(\hat{z}_1; \hat{v}_1)|}{\|\hat{v}_1\|} \le \gamma_1\|z_1\|.$$

Therefore,

$$\|z_2\| \le \|\hat{v}_2 - u_2\| + \frac{1}{\hat{\sigma}_0} \left(\kappa_1\|\hat{u}_2 - \hat{v}_2\| + \gamma_1\|z_1\| \right) \tag{2.11}$$

and, due to the freedom in choosing $\hat{v}_2$, we obtain the inequality $\|z_2\| \le \rho_2 + 1/\hat{\sigma}_0[\kappa_1\rho_2 + \gamma_1\|z_1\|]$, which is equivalent to the desired estimate (2.7). What remains is to estimate $\|z_1\|$. Since $\gamma_0\|z_1\|^2 \le b_{1,1}(z_1; z_1)$, then we transform its right-hand side as

$$b_{1,1}(z_1; z_1) = b_{1,1}(z_1; \hat{v}_1 - u_1) + b_{1,1}(z_1; \hat{u}_1 - \hat{v}_1).$$

Replacing the second term on the right-hand side, again using (2.9), we have $b_{1,1}(z_1; z_1) = b_{1,1}(z_1; \hat{v}_1 - u_1) - b_{2,1}(\hat{u}_1 - \hat{v}_1; z_2)$ and

$$I_0 \equiv \gamma_0 \|z_1\|^2 \leq \gamma_1 \|z_1\| \|\hat{v}_1 - u_1\| - b_{2,1}(\hat{u}_1 - \hat{v}_1; z_2). \tag{2.12}$$

To obtain the estimate for $\|z_1\|$, we need to estimate $X \equiv -b_{2,1}(\hat{u}_1 - \hat{v}_1; z_2)$. Since $\hat{u}_1 - \hat{v}_1 = z_1 + u_1 - \hat{v}_1$, then we have $X = -b_{2,1}(z_1; z_2) - b_{2,1}(u_1 - \hat{v}_1; z_2)$. Since $z_2 = \hat{u}_2 - \hat{v}_2 + \hat{v}_2 - u_2$, then $X = -b_{2,1}(z_1; \hat{u}_2 - \hat{v}_2) - b_{2,1}(z_1; \hat{v}_2 - \hat{u}_2) - b_{2,1}(u_1 - \hat{v}_1; z_2)$ and

$$X \leq Y + \kappa_1 \|z_1\| \|\hat{v}_2 - \hat{u}_2\| + \kappa_1 \|u_1 - \hat{v}_1\| \|z_2\|, \tag{2.13}$$

$$Y \equiv -b_{2,1}(z_1; \hat{u}_2 - \hat{v}_2). \tag{2.14}$$

We transform Y making use of (2.10). Thus,

$$Y = -\alpha b_{2,2}(z_2; z_2) - \alpha b_{2,2}(z_2; u_2 - \hat{v}_2). \tag{2.15}$$

Now we make use of the conditions imposed on $b_{2,2}$ and introduce a semi-inner product $(u_2, v_2) \equiv b_{2,2}(u_2; v_2)$. We then see that

$$\|z_2\|_{L_{2,2}}^2 + |(z_2; u_2 - \hat{v}_2)_{L_{2,2}}| \leq \|z_2\|_{L_{2,2}}^2 + \frac{1}{4}\|u_2 - \hat{v}_2\|_{L_{2,2}}^2,$$

and (2.15) leads to

$$Y \leq \frac{\alpha\gamma_3}{4}\|u_2 - \hat{v}_2\|^2. \tag{2.16}$$

Thus, (2.12)–(2.16) imply that

$$I_0 \leq \gamma_1 \|z_1\| \|\hat{v}_1 - u_1\| + \kappa_1 (\|z_1\| \|\hat{v}_2 - u_2\| + \|\hat{v}_1 - u_1\| \|z_2\|)$$

$$+ \frac{\alpha\gamma_3 \|u_2 - \hat{v}_2\|^2}{4}. \tag{2.17}$$

Choosing $\hat{v}_1$ and $\hat{v}_2$ as the best approximations to v_1 and v_2, we conclude from (2.17) that

$$\|z_1\|^2 \leq \frac{\gamma_1\rho_1 + \kappa_1\rho_2}{\gamma_0}\|z_1\| + \frac{1}{\gamma_0}\left(\kappa_1\rho_1\|z_2\| + \frac{\alpha\gamma_3}{4}\rho_2^2\right). \tag{2.18}$$

To obtain the estimate for $\|z_1\|$, we increase the right-hand side of (2.18) by virtue of (2.7) and (1.12), obtaining the inequality

$$\|z_1\|^2 - 2K_1\|z_1\| - K_2 \le 0, \qquad (2.19)$$

which leads directly to (2.6). □ [11]

2.3. General theorem on convergence.

Theorem 3. Suppose that the operator $L \in \mathcal{L}(G)$ in problem (1.13) is invertible and, for projective operators $\hat{L}$ from (2.3), there exists a constant K_0 independent of h such that (2.5) holds. Denote by P an orthoprojector of G onto $\hat{G}_h$. Then error estimates (1.17), (1.1.18) are valid.

Proof. We apply Theorem 1.1.3 by introducing a basis for the space $\hat{G}_h$ and defining the corresponding grid operator Λ (see (0.4.3)). In this case, (2.5) and the boundedness of L lead to (0.4.7) and (0.4.8), respectively. □ [12]

§ 3. Theorems on divergence of vector field

3.1. Differential case (normal solvability of divergence operator).

Theorem 1. Let Ω be a bounded domain in $\mathbf{R}^d$ such that

$$\Omega = \cup_{k=1}^m \Omega^{(k)}, \ \ where \ |\Omega^{(k)} \cap \Omega^{(k+1)}| > 0, \ k \in [1, m-1]$$

and each $\Omega^{(k)}$, $k \in [1, m]$, has the Lipschitz piecewise smooth boundary. Suppose that the Hilbert spaces G_1 and G_2 are defined by (1.30). Then (1.31) holds.

Proof. If, in the definition of G_1 and G_2, we replace Ω by $\Omega^{(k)}$, then (1.31) (with $\sigma = \sigma_k > 0$, $k \in [1, m]$,) is well-known fact. Suppose that (1.31) is not valid for our Ω. Then in G_2 there exists $\{p_n\}$ such that $|p_n|_{0,\Omega} = 1, n = 1, \ldots,$ and $s_n \to 0$, where $s_n \equiv s(p_n, \Omega)$ denotes the left-hand side of (1.31) for $p = p_n, n = 1, \ldots$. Consider the restrictions $p_n^{(k)} \in L_2(\Omega^{(k)})$ to $\Omega^{(k)}$ represented by $p_n^{(k)} = g_n^{(k)} + c_n^{(k)}$, where

$$c_n^{(k)} \equiv (p_n^{(k)}, 1)_{0,\Omega^{(k)}}, \quad (g_n^{(k)}, 1)_{0,\Omega^{(k)}} = 0, \ k \in [1, m], n = 1, \ldots. \qquad (3.1)$$

Analogously to $s(p_n, \Omega)$, we define $s_n^{(k)} \equiv s(p_n, \Omega^{(k)}) = s(g_n^{(k)}, \Omega^{(k)})$, $k \in [1, m], n = 1, \ldots,$ and observe that $0 \le s_n^{(k)} \le s_n$ (the functions from $\overset{o}{W_2^1}(\Omega^{(k)})$ are restrictions of certain functions in $\overset{o}{W_2^1}(\Omega)$, $k \in [1, m]$).

---

[11] These estimates obtained are independent of α (see (1.12)) and imply that $\|z\| \le K_3(\rho_1 + \rho_2)$, where $K_3 \equiv K_3(\gamma_0, \gamma_1, \kappa_1, \hat{\sigma}_0, \gamma_3, \alpha_0)$.

[12] Another proof is possible, based on the proof of Theorem 1.1.3 for relations of type (2.1), (2.5) written for $\hat{u} - Pu$. This theorem does not need conditions (1.14), (1.15), and (2.4), but its assumptions are difficult to verify.

Thus, $s(g_n^{(k)}, \Omega^{(k)}) \to 0$ and $g_n^{(k)}|_{0,\Omega^{(k)}} \to 0$ as $n \to \infty$ for each $k \in [1, m]$. The set of constants $c_n^{(k)}$ in (3.1) is bounded, so we may redefine $\{p_n\}$ and write $c_n^{(k)} \to c_k$ and $p_n^{(k)}|_{0,\Omega^{(k)}} \to c_k$ as $n \to \infty$ for each $k \in [1, m]$. This convergence to constants and the condition $|\Omega^{(k)} \cap \Omega^{(k+1)}| > 0$, $k \in [1, m-1]$ imply that all c_k coincide and $\{p_n\}$ converges in $L_2(\Omega)$ to a constant c. Since $(p_n, 1)_{0,\Omega} = 0$, then $c = 0$ and we arrive at the contradiction that $|p_n| = 1$, $n = 1, \ldots$. $\square$

Theorem 2. *Let the conditions of Theorem 1 be satisfied. Suppose we approximate $\bar{\Omega}^{(k)}$ by a family of regions $\hat{\Omega}_h^{(k)}$ such that $\lim_{h \to 0} |\Omega^{(k)} \triangle \hat{\Omega}_h^{(k)}| = 0$ and that there exists a constant $\sigma_k > 0$, independent of h, for which (1.31) holds with $\Omega \equiv \hat{\Omega}_h^{(k)}$ and $\sigma \equiv \sigma_k$, where $k \in [1, m]$. Suppose also that $\hat{\Omega} = \cup_{k=1}^m \hat{\Omega}_h^{(k)}$ and that*

$$|p|_{0,\hat{\Omega}_h} = 1, \quad (p, 1)_{0,\hat{\Omega}_h} = 0. \tag{3.2}$$

Then the inequality $s(p, \hat{\Omega}_h) \geq \sigma$ holds for all $\hat{\Omega}$ with a constant $\sigma > 0$ independent of h and the Hilbert spaces G_1 and G_2 defined by (1.30) with $\hat{\Omega}$ instead of Ω.

Proof. Identifying functions $p \in L_2(\hat{\Omega}_h)$ and their finite extensions enables us to apply the reasoning from the proof of Theorem 1. We thus deal with $\{h_n\}$ and $\{p_n\}$ such that $|p_n|_{0,\hat{\Omega}_h} = 1$, $(p_n, 1)_{0,\hat{\Omega}_h} = 0$, and $\lim_{n \to \infty} s(p_n, \hat{\Omega}_h) = 0$ (here and below, we write h instead of h_n). Now we make use of the sums $p_n^{(k)} = g_n^{(k)} + c_n^{(k)}$, where $c_n^{(k)} \equiv (p_n^{(k)}, 1)_{0,\hat{\Omega}_h^{(k)}}$, $(g_n^{(k)}, 1)_{0,\hat{\Omega}_h^{(k)}} = 0$, $k \in [1, m]$, $n = 1, \ldots$, and conclude that $|p_n^{(k)} - c_k|_{0,\hat{\Omega}_h^{(k)}} \to 0$ and $|p_n^{(k)} - c_k|_{0,\hat{\Omega}^{(k)}} \to 0$ as $n \to \infty$. The proof then follows from that of Theorem 1. $\square$ [13]

Now we consider the case of (1.35).

Lemma 1. *Let the domain Ω be as in Theorem 1. Suppose that the Hilbert spaces G_1 and G_2 are defined by (1.35). Suppose also that $\Gamma_1 \equiv \Gamma \backslash \Gamma_0$ contains a part Γ^* belonging to a smooth curve if $d = 2$ or surface if $d = 3$ and that $|\Gamma_{(d-1)}^*| > 0$. Then there exist a constant K^* and $\vec{v}^* \in G_1$ such that*

$$(\vec{v}^*, \vec{n})_{0,\Gamma_1} = 1 \text{ and } \|\vec{v}^*\| \leq K^* |\text{div } \vec{v}^*|_{0,\Omega}. \tag{3.3}$$

Proof. We make use of the possibility of constructing the desired vector field $\vec{v}^*$ in a suitable Descartes coordinate system. For $d = 2$, we choose

---

[13] The region Ω in Theorems 1 and 2 may have non-Lipschitz boundary. Moreover, the existence of the desired σ_k was proved in [191, 310]; they are determined (for the appropriate blocks $\Omega_h^{(k)}$) by the constants in the Lipschitz conditions for the function corresponding to the curvilinear side or face of such a block.

a smooth arc $\Gamma_f \subset \Gamma_1$ that, in a Descartes coordinate system $[y_1, y_2]$, is defined by the equation $y_2 = f(y_1) > 0$, with $|y_1| \leq a$ and

$$f'(0) = 0, \quad |f'(y_1)| \leq \kappa_f a \tag{3.4}$$

(the y_2-axis contains the normal to Γ at the point of its intersection with Γ_f, namely, the point $[0, f(0)]$; $O \equiv [0,0] \in \Omega$). We can choose $a_0 \leq 1$ so small that, for all $a \in (0, a_0]$, we have

$$\int_{-a/2}^{a/2} \cos \alpha(t_1) dt_1 \geq \kappa_f' a, \quad \kappa_f' > 0, \tag{3.5}$$

where $\alpha(t_1)$ is the angle between the y_2-axis and the normal to Γ at the point of intersection of the straight line $y_1 = t_1$ and Γ_f, κ_f' is defined by κ_f, and both these constants are independent of a.

Next, consider a quasirectangle $\Pi \subset \bar{\Omega}$ whose three sides are the straight line segments $A_1 A_2$, $A_2 A_2$, and $A_1 A_4$, with $A_1 \equiv [-a, 0]$, $A_2 \equiv [a, f(a)]$, $A_3 \equiv [-a, f(-a_1)]$, and the fourth side is our arc Γ_f. (Its part with the points y such that $|y_1| \leq a/2$ is denoted by $\Pi_{1/2}$.) We will construct the desired vector field $\vec{u} \equiv [u_1(y)], u_2(y)$ (corresponding to $\vec{v}^*$) such that $u_1 = 0$ and $u_2 \in W_2^1(\Omega; \Gamma_0)$ vanishes outside Π. With this in mind, we specify the construction of the function $u_2(y) \equiv u(y)$ defined on Π.

Define $u(y)$ on Γ_f by $u(y)|_{\Gamma_f} = g(y_1)$, where

$$g(y) = 1 \ (|y_1| \leq a/2), \quad g(y_1) = [y_1(|y_1| - a)]^2(a/2)^{-4} \leq 1 \ (|y_1| > a/2). \tag{3.6}$$

(Note that $g(-a) = g(a) = 0$ and $|g'(y_1)| \leq \kappa_g/a$.) We then define $u(y)$ on Π as $u(y_1, y_2) \equiv y_2 g(y_1)/f(y_1)$ (which is linear with respect to y_2 and vanishes on all sides of Π but Γ_f). Without loss of generality, we assume that $f(0) \equiv b = a(1 + \kappa_f)$, which implies that $a \leq f(y_1) \leq a(1 + 2\kappa_f) \equiv f^*$. It then is easy to see that

$$|D_2 u|_{0, \Pi_{1/2}}^2 \geq |\Pi_{1/2}| \frac{1}{a^2(1 + 2\kappa_f)^2} \quad \text{and} \quad |D_2 u|_{0, \Pi}^2 \leq |\Pi| \frac{1}{a^2}. \tag{3.7}$$

Next, $|D_1 u| = |g'(y_1) y_2/f(y_1) + y_2 g(y_1) f'(y_1)/(f(y_1))^2|$ and

$$|D_1 u|_{0, \Pi}^2 \leq |\Pi| \frac{(\kappa_g + a\kappa_f)^2}{a^2}. \tag{3.8}$$

From (3.7) and (3.8), we define the constant K^* by the condition $|u|_{1, \Pi}^2 \leq (K^*)^2 |D_2 u|_{0, \Pi}^2$, or

$$|\Pi| \frac{1}{a^2} + |\Pi| \frac{(\kappa_g + a\kappa_f)^2}{a^2} \leq (K^*)^2 \left(|\Pi_{1/2}| \frac{1}{a^2(1 + 2\kappa_f)^2} \right). \tag{3.9}$$

Since $|\Pi|/|\Pi_{1/2}| \leq 4(1 + 2\kappa_f)$, then K^* is fully defined by the constants a_0, κ_f, and κ_f. But (3.9) implies that $|\vec{u}|^2 \leq (K^*)^2 |\text{div } \vec{u}|_{0,\Omega}^2$ and (3.5) implies that $(\vec{u}, \vec{n})_{0,\Gamma_1} \geq \kappa_f'a$. To satisfy (3.3) it suffices to define a new vector field $\vec{u}^* \equiv [(\vec{u}, \vec{n})_{0,\Gamma_1}]^{-1}\vec{u}$, which corresponds to the desired $\vec{v}^*$.

For $d = 3$, the proof is analogous. Γ_f is now a piece of a smooth surface and is defined in a Descartes coordinate system $[y_1, y_2, y_3]$ by the equation $y_3 = f(y_1, y_2) > 0$, where $|y_s| \leq a, s = 1, 2$, and grad $f |_0 = 0$, $|D_s f'(y_1, y_2)| \leq \kappa_f a, s = 1, 2$, $a \leq f(y_1, y_2) \leq a(1 + 4\kappa_f)$. Then all considerations are related to a quasiparallelepiped P with only one possibly curvilinear face Γ_f. The new function $g(y_1, y_2)$ may be chosen as $g(y_1, y_2) \equiv g_1(y_1)g_2(y_2)$, where $g_1 = g_2$ is the same function defined in (3.6). The vector field $\vec{u} \equiv [0, 0, u_3]$; $u_3 \equiv u$ is actually constructed on P by the formula $u(y_1, y_2, y_3) \equiv y_3 g(y_1, y_2)[f(y_1, y_2)]^{-1}$. All estimates become more cumbersome but the reasoning is much the same. $\square$

Theorem 3. *Let the conditions of Lemma 1 be satisfied. Then* (1.31) *holds.*

Proof. Let $p^* \equiv \text{div } \vec{v}^*$ (see Lemma 1). Then, for each $p \in G_2$, we write $p = p - cp^* + cp^*$, where $c \equiv (p, 1)_{0,\Omega}$, $(p - cp^*, 1)_{0,\Omega} = 0$, $|c| \leq |\Omega|^{1/2}|p|_{0,\Omega}$. For $p - cp^*$, in accordance with Theorem 1 and Lemma 1.2, we choose $\vec{u} \in (\overset{0}{W_2^1}(\Omega))^d$ such that div $\vec{u} = p - cp^*$ and

$$|\vec{u}|_{G_1} \leq 1/\sigma'|p - cp^*|_{0,\Omega} \leq 1/\sigma'[1 + |\Omega|^{1/2}|\text{div } \vec{v}^*|_{0,\Omega}]|p|_{0,\Omega},$$

where σ' denotes σ from Theorem 1 and Lemma 1.2. Lemma 1 allows us to make use of the estimate

$$|\text{div } \vec{v}^*|_{0,\Omega} \leq \kappa_0/a \equiv \kappa_1. \tag{3.10}$$

This implies that $|\vec{u}|_{G_1} \leq \kappa_2|p|_{0,\Omega}$ and, for $\vec{v} = \vec{u} + c\vec{v}^* \in G_1$, that div $\vec{v} = p$ and $|\vec{v}|_{G_1} \leq (\kappa_2 + K^*\kappa_1|\Omega|^{1/2})|p|_{G_2} \equiv 1/\sigma|p|_{G_2}$. $\square$

Lemma 2. *Let the conditions of Lemma 1 and Theorem 2 be satisfied. Instead of* (1.35), *consider a family of Hilbert spaces $\tilde{G}_1$ and $\tilde{G}_2$, where*

$$\tilde{G}_1 \equiv (W_2^1(\hat{\Omega}_h; \hat{\Gamma}_0))^d, \quad \tilde{G}_2 \equiv L_2(\hat{\Omega}_h). \tag{3.11}$$

Suppose that the spaces $\tilde{G}_1$ are such that the standard blocks Π and P considered in Lemma 1 correspond to standard blocks $\hat{\Pi}$ and $\hat{P}$ that differ from Π and P in equations for the curvilinear side or face: this part of $\hat{\Gamma}_1$ is denoted by $\hat{\Gamma}_f$ and corresponds to the equations $y_2 = f_h(y_1) > 0$ or $y_3 = f_h(y_1, y_2) > 0$, which approximate the old ones as $h \to 0$. Let these continuous functions be piecewise smooth so $|f - f_h| = O(h^2)$, where $a/2 \leq f_h \leq \kappa_3 a$ and almost everywhere $|f_h'(y_1)| \leq \kappa_f a$ or $|D_s f_h'(y_1, y_2)| \leq \kappa_f a, s = 1, 2$. Then there exists a constant K^* and small enough $h_0 > 0$ for *which (3.3) holds with* $\vec{v}^* \in \tilde{G}_1$ *and* $\vec{n}, \Gamma_1, \Omega$ *replaced by their corresponding approximations, provided* $h \leq h_0$.

Proof. We apply the proof of Lemma 1, with the only difference that our function $u(y_1)$ (for $d = 2$) is continuous along with $D_2 u$, but $D_1 u$ is defined everywhere except at, say k_h, vertical lines. Smallness of h is actually needed for obtaining the corresponding analog of (3.5). Analogous modifications in the proof for $d = 3$ are quite simple, although in this case we need to specify where $D_1 u$ and $D_2 u$ are defined (for our purposes, it suffices to assume that $D_s f_h(y_1, y_2), s = 1, 2$ exist inside some triangles in the square $[-a, a]^2$). □

Theorem 4. Let the conditions of Lemma 2 be satisfied. Then there exists a constant σ independent of h such that

$$\sup_{\vec{u} \in \tilde{G}_1} \frac{(\operatorname{div} \vec{u}, p)_{0,\hat{\Omega}_h}}{\|\vec{u}\|_{\tilde{G}_1}} \geq \sigma |p|_{0,\hat{\Omega}_h}, \ \sigma > 0, \ \forall p \in \tilde{G}_2. \tag{3.12}$$

Proof. It suffices to modify slightly the proof of Theorem 3, making use of Lemma 2 instead of Lemma 1. □

3.2. Projective-grid (mixed finite element) methods for $d = 2$. We start by considering quasiuniform triangulations $T_h(\hat{\Omega}_h)$ and the spline spaces $\hat{G}_1 \subset \tilde{G}_1 \equiv W_2^1(\hat{\Omega}_h; \hat{\Gamma}_0)$ (see Lemma 2 and § 5.1). Since the Steklov averagings $Y_\rho u(x) \equiv w$ (see (2.3.2) with $\rho \asymp h$) are essential for our analysis, we assume that Γ and $\hat{\Gamma}$ are Lipschitz boundaries of Ω and $\bar{\Omega}_h$, respectively, and introduce $\hat{B}_\rho \equiv \{P : 0 < \operatorname{dist}\{P; \Omega\} \leq \rho\}$ and $\hat{B}_\rho^0 \equiv \{P : P \in \hat{B}_\rho; \operatorname{dist}\{P; \hat{\Gamma}_0\} \leq \operatorname{dist}\{P; \hat{\Gamma}_1\}\}$. We assume that we may apply extension theorems for $u \in \tilde{G}_1$ so that $Eu \in W_2^1(\hat{\Omega}_h \cup \hat{B}_\rho)$, vanishes on $\hat{B}_\rho^0$, and there exists a constant K_E independent of h such that

$$|Eu|_{1,\hat{\Omega}_h \cup \hat{B}_\rho} \leq K_E |u|_{1,\hat{\Omega}_h}, \quad \forall u \in \tilde{G}_1, \ \forall h \leq h_0. \tag{3.13}$$

(For example, when $\Gamma_0 = \Gamma$, we can use the finite extension, and when $\hat{\Gamma}_1 = \Gamma_1$, we can define the desired extension as in § 3.6 (see the proof of Theorem 3.6.8)).

Lemma 3. Suppose h_0 is so small that (3.13) holds. For $v \in \tilde{G}_1$, define a piecewise linear function $\hat{w}^* \in \hat{G}_1 \equiv V$ that coincides with $w \equiv Y_\rho Ev$ at every node $P_i \in \Omega_h$, where $\rho \asymp h$. Then there exist constants K_0 and K_1 such that

$$\|\hat{w}^*\|_V \leq K_0 \|v\|_V \text{ and } \|v - \hat{w}^*\|_{0,\hat{\Omega}_h} \leq K_1 h \|v\|_V, \quad \forall v \in \tilde{G}_1. \tag{3.14}$$

Proof. Proof of the inequality involving K_0 is much the same as that for Theorem 3.6 (see (3.6.21)) and Lemma 5.2.5. It is based on (2.3.4) with $\gamma = 0$ and the inequalities

$$|w(P_i)| \leq \frac{|S_\rho(P_i)|}{(2\rho)^2}|v|_{0,S_\rho(P_i)} \leq K_4|v|_{1,C_{2^{1/2}\rho}(P_i)}, \quad P_i \in \hat{\Gamma}_0 \qquad (3.15)$$

($S_\rho(P_i)$ are the square and circle with centers at P_i and areas $4\rho^2$ and $2\pi\rho^2$, respectively, and v vanishes on a triangle inside the circle; see (5.2.18)). We denote $\mathrm{int}_1 w$ by $\hat{w}$ (which coincides with w at the nodes) and consider the triangles $T \in T_h(\hat{\Omega}_h)$ that have a vertex in $\hat{\Gamma}_0$. If ω is a set of such triangles, then (3.15) implies that

$$\sum_{T\in\omega} |\hat{w} - \hat{w}^*|_{1,T}^2 \leq K_5 \sum_{P_i\in\hat{G}_)} w(P_i)^2 \leq K_6|v|_{1,.\hat{\Omega}_h}^2, \qquad (3.16)$$

which is equivalent to $|\hat{w}-\hat{w}^*|_{1,;0_h} \leq K_6^{1/2}|v|_{1,\hat{\Omega}_h}$. We can obtain the desired estimate by making use of the inequality

$$|\hat{w}^*|_{1,\hat{\Omega}_h} \leq |\hat{w} - \hat{w}^*|_{1,\hat{\Omega}_h} + |\hat{w}|_{1,\hat{\Omega}_h},$$

as in the proof of Theorem 3.6.5. Along the same lines we obtain

$$\sum_{T\in\omega} |\hat{w} - \hat{w}^*|_{0,T}^2 \leq K_7h^2 \sum_{P_i\in\hat{G}_)} w(P_i)^2 \leq K_8h^2|v|_{1,\hat{\Omega}_h}^2$$

and $|\hat{w} - \hat{w}^*|_{0,\hat{\Omega}_h} \leq K_8^{1/2}h|v|_{1,\hat{\Omega}_h}$. Thus,

$$|v - \hat{w}^*|_{0,\hat{\Omega}_h} \leq |\hat{w} - \hat{w}^*|_{1,\hat{\Omega}_h} + |v - \hat{w}|_{1,\hat{\Omega}_h},$$

which allows us to apply (2.3.4) with $\gamma = 0$ and the standard inequalities for the interpolant of the function in $W_2^2(\Omega)$. $\square$

Here we use the above quasiuniform triangulations $T_h(\hat{\Omega}_h)$ of the same type as in § 5.1 in our PGMs for construction of the subspace $\hat{G}_2'$, which approximates the space $L_2(\hat{\Omega}_h)$ and consists of the piecewise constant functions on our triangulations. The basis for $\hat{G}_2'$ is just a set of characteristic functions $\chi_T(x)$ of the triangles $T \in T_h(\hat{\Omega}_h)$, and after numbering them, we have the basis $\chi_1(x), \ldots, \chi_{N_2}(x)$. We will identify $\hat{G}_2'$ and $\hat{G}_2$ for the choice (1.35) and (1.34), and for the choice $G_2 = G_2' \setminus 1$ (see (1.30)), we deal with the subspace $\hat{G}_2$ of $\hat{G}_2'$ consisting of the functions $\hat{u}_2$ such that

$$(\hat{u}_2, 1)_{0,\hat{\Omega}_h} = 0 \qquad (3.17)$$

(whose basis is needed only for theoretical study; dim $\hat{G}_2 = N_2 - 1$). In approximating $\tilde{G}_1$ and $W_2^1(\hat{\Omega}_h; \hat{G}_0)$, we make use of the spline spaces $\hat{G}_{(1)} \subset W_2^1(\hat{\Omega}_h; \hat{\Gamma}_0)$ of continuous piecewise linear functions on triangulations $T_h^{(2)}(\hat{\Omega}_h)$ (recall that these triangulations are obtained by drawing midlines in each $T \in T_h(\hat{\Omega}_h)$, and the number of these smaller triangles is $4N_2$). This space was extensively used in Chapter 5 (dim $\hat{G}_{(1)} \equiv N_1$ and $\hat{\psi}_1(x), \ldots, \hat{\psi}_{N_1}(x)$ is its nodal basis). Hence, we approximate $\tilde{G}_1 = (W_2^1(\hat{\Omega}_h; \hat{G}_0))^2$ by $\hat{G}_1 \equiv (\hat{G}_{(1)})^2$. Its basis is $\vec{\psi}_1(x), \ldots, \vec{\psi}_{2N_1}(x)$, where

$$\vec{\psi}_i(x) \equiv [\hat{\psi}_i(x), 0]^T \text{ and } \vec{\psi}_{N_1+i}(x) \equiv [0, \hat{\psi}_i(x)]^T, \quad i \in [1, N_1].$$

Theorem 5. For (1.30) we use the approximating family of subspaces $\hat{G} \equiv \hat{G}_1 \times \hat{G}_2$ defined above. Suppose that the conditions of Theorem 2 are satisfied. Then there exists a constant σ_0^* independent of h such that

$$\sup_{\vec{u} \in \hat{G}_1} \frac{(\operatorname{div} \vec{u}, \hat{p})_{0,\hat{\Omega}_h}}{\|\vec{u}\|_{\hat{G}_1}} \geq \sigma_0^* |\hat{p}|_{0,\hat{\Omega}_h}, \quad \sigma_0^* > 0, \ \forall \hat{p} \in \hat{G}_2. \tag{3.18}$$

Proof. For each $\hat{p} \in \hat{G}_2 \subset \tilde{G}_2$, we apply Theorem 2 and find $\vec{v} \in \tilde{G}_1$ such that

$$|(\operatorname{div} \vec{v}, \hat{p})_{0,\hat{\Omega}_h} (\|\vec{v}\|_{\tilde{G}_1})^{-1} \geq \sigma |\hat{p}|_{0,\hat{\Omega}_h}. \tag{3.19}$$

We now construct a projection operator P of $\tilde{G}_1$ onto $\hat{G}_1$ such that

$$(\operatorname{div} \vec{v}_1 - \operatorname{div} (P\vec{v}_1), \hat{v}_2)_{0,\hat{\Omega}_h} = 0, \quad \forall \hat{v}_2 \in \hat{G}_2, \tag{3.20}$$

$$\|P\vec{v}_1\| \leq K_9 \|\vec{v}_1\|, \quad \forall v_1 \in \tilde{G}_1. \tag{3.21}$$

Now (3.20) holds if

$$\int_{a(T)} P\vec{v}_1 ds = \int_{a(T)} \vec{v}_1 ds, \tag{3.22}$$

where $a(T)$ refers to each side of triangle $T \in T_h(\hat{\Omega}_h)$. Of course, it suffices to consider only $a(T)$ not belonging to $\hat{\Gamma}_0$ and we need to prescribe only values of $P\vec{v}$ at the nodes not belonging to $\hat{\Gamma}$. With this in mind for $\vec{v}_1$, we define $\vec{v}_1^* \equiv [\hat{w}_1^*, \hat{w}_2^*] \in \hat{G}_1$ whose components are constructed in accordance with Lemma 3 for the components of $\vec{v}_1$. We require that $P\vec{v}_1 = \vec{v}_1^*$ at all vertices of the triangles $T \in T_h(\hat{\Omega}_h)$ and, at the midpoints $a_{1/2}(T)$ of the above indicated sides $a(T)$, we obtain the remaining values from the vector conditions (3.22). We rewrite them in terms of $\hat{z}_1 \equiv P\vec{v}_1 - \vec{v}_1^* \equiv [\hat{z}_{1,1}, \hat{z}_{1,2}]$

and $z_1 \equiv \vec{v}_1 - \vec{v}_1^*$, making use of the representation $P\vec{v}_1 = \vec{v}_1 - z_1 + \hat{z}_1 = \vec{v}_1^* + \hat{z}_1$. Note that $\hat{z}_{1,r}$ vanishes at vertices of T and that the vector condition (3.22) is equivalent to the linear equations

$$\hat{z}_{1,r}(a_{1/2}(T))|a(T)| = \int_{a(T)} z_{1,r}\, ds, \quad r = 1, 2. \tag{3.23}$$

In accordance with Lemma 3, we have

$$\|z_1\| \le (1 + K_0)\|\vec{v}_1\|. \tag{3.24}$$

We see also that $\|P\vec{v}_1\| \le \|\vec{v}_1^*\| + \|\hat{z}_1\|$. For estimating the last term, we apply (3.23) and see that, for $r = 1, 2$,

$$|\hat{z}_{1,r}(a_{1/2}(T))|^2 \le \frac{K_{10}}{h}|z_{1,r}|_{0,a(T)}^2 \le K_{11}\left(|z_{1,r}|_{1,T}^2 + \frac{1}{h^2}|z_{1,r}|_{0,T}^2\right), \tag{3.25}$$

(see (0.5.17)). Hence,

$$|\hat{z}_{1,r}|_{1,\hat{\Omega}_h}^2 \le K_{12}[|z_{1,r}|_{1,\Omega_h}^2 + 1/h^2|z_{1,r}|_{0,\Omega_h}^2], \quad r = 1, 2,$$

and it is possible to apply (3.14). This leads to (3.21) with

$$K_9 = K_0 + K_{12}^{1/2}[(1 + K_0)^2 + K_1^2]^{1/2}.$$

We now make use of (3.19) and take $\vec{u} \equiv P\vec{v}$. This implies that (3.18) holds with $\sigma_0^* = \sigma/K_9$. □ [14]

Theorem 6. *For (1.35), (1.34), we use the approximating family of the subspaces $\hat{G} \equiv \hat{G}_1 \times \hat{G}_2$ with $\hat{G}_2 \equiv \hat{G}_2'$. Suppose that the conditions of Lemma 2 are satisfied. Then there exists a constant σ_0^* independent of h such that (3.18) holds.*

Proof. It suffices to modify slightly the proof of Theorem 5, making use of Theorem 4 instead of Theorem 2 for (3.19). Next, in constructing $P\vec{v}_1$, we need to prescribe only values of $P\vec{v}$ at the nodes not belonging to $\hat{\Gamma}_0$, and $a(T)$ in (3.22) must not belong to $\hat{\Gamma}_0$. □

---

[14] No convexity of Ω was required in our proof (see [191]), in contrast with proofs based on estimates of the solutions in the norm $(W_2^2(\Omega))$ (see [127, 242, 293, 356]). We emphasize also that the proof was given for a family of approximating regions $\hat{\Omega}_h \subseteq \bar{\Omega}$ (later related results can be found in [261, 411, 465]). Besides, in [127] the space $\hat{G}_{(1)}$ was defined as one of piecewise quadratic functions with respect to $T_h(\hat{\Omega}_h)$. Our proof can be applied to this space as well, but on the whole our simpler choice of piecewise linear functions enables us to apply the theory and algorithms developed in Chapter 3; we will pay special attention to composite grids with local refinement below.

3.3. Three-dimensional problems. To simplify the presentation, we assume that Ω is a bounded three-dimensional domain with Lipschitz boundary such that $\bar{\Omega}$ is a union of several simplexes and Γ_0 consists of a finite number of their faces. Suppose that triangulations $T_h^{(3)}$ are obtained from quasiuniform triangulations $T_h(\bar{\Omega})$ by global refinement with ratio 3. Then our spline spaces $\vec{G}_1$ are subspaces of $G = (W_2^1(\Omega; \Gamma_0))^3$ that consist of piecewise linear (with respect to $T_h^{(3)}$) vector fields. The subspace $\hat{G}_2'$ consists of the piecewise constant functions with respect to $T_h(\bar{\Omega})$ (its basis is a set of characteristic functions $\chi_T(x)$ of the simplexes $T \in T_h(\hat{\Omega}_h)$). The spline subspaces $\hat{G}_h \subset W_2^1(\Omega; \Gamma_0) \equiv G$ are associated with our triangulations $T_h(\bar{\Omega})$. Suppose finally that $\hat{G}_2$ is either $\hat{G}_2'$ (if (1.35) and (1.34) hold) or its subspace defined by (3.17) (if (1.30) holds).

Theorem 7. For (1.35), (1.34) *or* (1.30), *we use the approximating family of the subspaces* $\hat{G} \equiv \hat{G}_1 \times \hat{G}_2$, *with* $\hat{G}_1$ *and* $\hat{G}_1$ *defined above. Then there exists a constant* σ_0^* *independent of* h *such that* (3.18) *holds.*

Proof. We indicate the necessary alterations in the proofs of Theorems 5 and 6. Note that (3.19) holds because Theorems 2 and 4 apply. We have to specify only the operator P with properties (3.20) and (3.21). The values of $P\vec{v}_1$ are defined by the values of $\vec{v}_1^* \equiv [\hat{w}_{1,1}^*, \hat{w}_{1,2}^*, \hat{w}_{1,2}^*]$ at all nodes but those that are inner points of faces $a(T)$ of simplexes $T \in T_h(\bar{\Omega})$ ($a(T)$ does not belong to Γ_0). The values at these points (we denote them by $a_{1/2}(T)$) are defined by the vector condition $(P\vec{v}_1, \vec{1})_{0,a(T)} = (\vec{v}_1, \vec{1})_{0,a(T)}$, where $\vec{1} \equiv [1,1,1]$ (see (3.22)). This leads to three linear equations, analogous to (3.23), and the proof of (3.21) is based on the inequality

$$|\hat{z}_{1,r}(a_{1/2}(T))|^2 \leq \frac{K_{13}}{h^2}|z_{1,r}|_{0,a(T)}^2 \leq K_{14}[|z_{1,r}|_{1,T}^2 + \frac{1}{h^2}|z_{1,r}|_{0,T}^2], \quad (3.26)$$

where $r = 1, 2, 3$. Finally, we may make use of (3.14) because Lemma 3 can be generalized to the case $d = 3$ in a straightforward manner. Necessary facts were given in § 3.6 and 5.2; now we may apply the estimate

$$|w(P_i)| \leq |S_\rho(P_i)|(2\rho)^{-3}|v|_{0,S_\rho(P_i)} \leq K_{15}/h^{1/2}|v|_{1,S'(P_i)},$$

where $P_i \in \Gamma_0$, $S_\rho(P_i)$ is the corresponding cube, and $S'(P_i)$ denotes a union of several simplexes in the corresponding vicinity of P_i. $\square$

3.4. Composite triangulations. As is indicated in § 2.1, composite grids with local refinement can yield good approximations to the solution of an elliptic problem even when this solution may have strong variations in different subdomains. This is especially important for many hydrodynamics problems, but now the property of PGMs defined by (3.18) deserves special attention (see [205]). To simplify the presentation, we assume that Ω is a bounded three-dimensional domain with Lipschitz boundary such that $\bar{\Omega}$ is a union of several triangles and that Γ_0 consists of a finite number of their sides.

We start by considering the triangulation $T_h(\bar{\Omega})$ from Theorems 5 and 6 extended to a composite triangulation $T_{c,h}(\bar{\Omega})$ by choosing subsets $\bar{\Omega}_{p_1} \subset \bar{\Omega}_{p_1-1} \subset \cdots \bar{\Omega}_0 \equiv \bar{\Omega}$. We carry out p_1 consecutive refinements with ratios $t_1, \ldots, t_{p_1}$ in $\bar{\Omega}_1, \ldots, \bar{\Omega}_{p_1}$, obtaining the composite triangulations $T_c^{(1)}(\bar{\Omega}), \ldots, T_c^{(p_1)}(\bar{\Omega}) \equiv T_{c,h}(\bar{\Omega})$ (each is the result of local refinement applied to a previous triangulation). Moreover, we assume that: each refinement region is a union of previously obtained triangles; $t_k \in [2, t^*], k \in [1, p_1]$; each triangle in $T_c^{(k)}(\bar{\Omega})$, which is new with respect to $T_c^{(k-1)}(\bar{\Omega})$, has no common points with the closure of $\Omega_{k-1} \setminus \Omega_k, k \geq 2$; and a similar condition holds for $k = 1$ if $\Omega_0 \equiv \Omega$. This implies that, in analyzing the structure of our composite triangulation, we can work with each level of refinement separately. For example, in Figure 1 we have $T \equiv \triangle P_0 P_1 P_2 \in T_c^{(k-1)}(\bar{\Omega})$, the triangles $P_1 P_{1,2} P_{1,3}, P_{1,2} P_{2,3} P_2, P_{1,2} P_{1,3} P_{2,3}$ (they have common points with T) are in $T_c^{(k)}(\bar{\Omega})$ (with refinement ratio 2), and they are new with respect to $T_c^{(k-1)}(\bar{\Omega}), k \geq 1$. Our space $\hat{G}_2'$, which is associated with $T_{c,h}(\bar{\Omega})$,

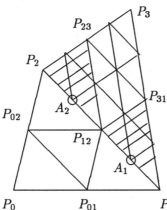

P_3

P_{23}

P_2

P_{31}

P_{02} A_2

P_{12}

A_1

P_0 P_{01} P_1 Figure 1: Augmented triangle.

will be specified below because it needs data about $\hat{G}_1$. For construction of $\hat{G}_1$, we refine each triangle $T \in T_{c,h}(\bar{\Omega})$ with ratio 2 to obtain a composite triangulation $T_{c,h}^{(2)}(\bar{\Omega})$. We define $\hat{G}_{(1)} \subset W_2^1(\Omega; \Gamma_0)$ as the spline space of piecewise linear (with respect to $T_{c,h}^{(2)}(\bar{\Omega})$) functions and $\hat{G}_1 \equiv (\hat{G}_{(1)})^2$. $\hat{G}_{(1)}$ is the space used in § 5.1, and the notion of seminodes is important for its elements (in Figure 1, we have 2 seminodes A_1 and A_2 marked by circles). Consider a triangle $T \in T_{c,h}(\bar{\Omega})$ such that it contains seminodes and none of its sides is part of a side of another triangle in $T_{c,h}(\bar{\Omega})$ (see $T = \triangle P_0 P_1 P_2$ in Figure 1). For such a T, we define its *augmented triangle* T' as the union of triangles in $T_{c,h}(\bar{\Omega})$ that have common seminodes with T (in Figure 1, $T' \equiv \triangle P_0 P_1 P_2 \cup \triangle P_1 P_{1,2} P_{1,3} \cup \triangle P_{1,2} P_{2,3} P_2$ and the added triangles are striped). If T contains no seminodes, then its augmented triangle T' is just T itself. Hence, $\bar{\Omega}$ is partitioned into a union of augmented triangles, and we define $\hat{G}'_2$ as a subspace of piecewise constant functions with respect to this partition. We then define $\hat{G}_2$ as either $\hat{G}'_2$ (if (1.35) and (1.34) hold) or its subspace defined by (3.17) (if (1.30) holds).

Theorem 8. *With the above family of subspaces $\hat{G} \equiv \hat{G}_1 \times \hat{G}_2$ and $p_1 \leq p^*$, there exists a constant σ_0^* independent of h such that (3.18) holds.*

Proof. We indicate the necessary alterations in the proofs of Theorems 5 and 6. Note that (3.19) holds because Theorems 2 and 4 apply. We have to specify only the operator P with properties (3.20) and (3.21). The values of $P\vec{v}_1$ are defined by the values of $\vec{v}_1^* \equiv [\hat{w}_{1,1}^*, \hat{w}_{1,2}^*, \hat{w}_{1,2}^*]$ at all nodes but those that are inner points of sides $a(T)$ of augmented triangles T' ($a(T)$ does not belong to Γ_0; T' in Figure 1 has 6 sides). The values at these points (denoted by $a_{1/2}(T)$) are defined by the vector condition (3.22) and lead to the estimate (3.25) with T replaced by T'. The main problem is the generalization of Lemma 3. Under our assumptions, we actually deal with a quasiuniform composite triangulation, and it is possible to use Y_ρ with $\rho \asymp h$ as in Lemma 3. A straightforward argument thus establishes (3.14). This implies that (3.18) holds. $\square$ [15]

3.5. Use of domain symmetry and other generalizations.

Theorem 9. *Let the conditions of Theorem 2 be satisfied and $\hat{\Gamma}_0 = \hat{\Gamma}$. Suppose that $\hat{G}_1$ is defined as in Theorem 2 and that, in the definition of $\hat{G}_2$ in Theorem 5, the condition*

$$(\hat{u}_2, g)_{0, \hat{\Omega}_h} = 0 \tag{3.27}$$

is used instead of (3.17), where $g \in L_2(\hat{\Omega}_h)$, $(g, 1)_{0, \hat{\Omega}_h} \geq c > 0$, and c is independent of h. Then there exists a constant σ_0^* independent of h such that (3.18) holds.

Proof. To prove (3.19), we have to generalize Theorem 2. If $\hat{v}_2$ satisfies (3.27), then $\hat{u}_2 \equiv \hat{v}_2 - c_2$ with $c_2 = (\hat{v}_2, 1)_{0, \hat{\Omega}_h} |\hat{\Omega}_h|^{-1}$ satisfies (3.17). For $\hat{u}_2$, we may apply Theorem 5 and find $\vec{u}_1 \in \hat{G}_1$ with div $\vec{u}_1 = \hat{u}_2$ and $\|u\| \leq 1/\sigma$

---

[15] When p_1 may increase, it is reasonable to use not one Y_ρ but averagings Y_{ρ_k}, $k \in [1, p_1]$, for different types of nodes. Then rather cumbersome estimates lead to the constant K_0 in (3.14) independent of p_1; the remaining inequality in (3.14) must be replaced by a more involved one (see [205]). The estimates from [448] can probably be applied as well. Theorem 8 can be generalized for $d = 3$ when each simplex in $T_{c,h}(\bar{\Omega})$ is refined with ratio 3, and the augmented simplexes are used in the definition of G'_2.

with constant $\sigma > 0$ independent of h. Observe that $(\mathrm{div} u_1, \hat{v}_2)_{0,\hat{\Omega}_h} = (\mathrm{div} u_1, \hat{u}_2)_{0,\hat{\Omega}_h}$ and that (3.27) yields $c_2 = -(\hat{v}_2, 1)_{0,\hat{\Omega}_h}(g,1)_{0,\hat{\Omega}_h}^{-1}$ and $|c_2| \le |\hat{v}_2|_{0,\hat{\Omega}_h} |\Omega_h|^{1/2}/c$. This implies that $|\hat{v}_2|_{0,\hat{\Omega}_h} \le \kappa(g)(1 + |g|_{0,\hat{\Omega}_h}/c|\hat{u}_2|_{0,\hat{\Omega}_h})$, where $\kappa(g) \equiv (1 + |g|_{0,\hat{\Omega}_h}|\Omega_h|^{1/2})/c$. Thus, we conclude that (3.18) holds with $\sigma_0^* = \sigma(\kappa(g))^{-1}$. $\square$

Condition (3.27), with g the characteristic function of a subdomain in Ω, is important for applications (see, e.g., [486]). Moreover, practically all of the above results can be generalized to certain domains with non-Lipschitz boundaries because our basic Theorems 1–4 were proved for such domains. The main difference is connected with the use of Steklov averaging, which now must be applied for separate subdomains in a manner analogous to that in § 3.6.

We indicate now very briefly that a region's symmetry may be taken into account in proving all of the above theorems. We restrict ourselves to Theorem 7 and the mirror symmetry with respect to the plane $x_1 = 0$, under the assumption that Ω, Γ_0 and our triangulations are symmetric with respect to the plane $x_1 = 0$ and that $p \equiv \hat{v}_2 \in G_2$ is either an even or odd function with respect to x_1 (the corresponding subspaces of $\hat{G}_2$ are denoted by $\hat{G}_{2,e}$ and $\hat{G}_{2,o}$, respectively). For $\vec{v} \in \hat{G}_1$, we define the symmetry operator S by (5.3.45) and the corresponding symmetric subspaces

$$\hat{G}_{(1,e)} \equiv \{\vec{v} : \vec{v} \in \hat{G}_1, S\vec{v} = \vec{v}\}, \quad \hat{G}_{(1,o)} \equiv \{\vec{v} : \vec{v} \in \hat{G}_1, S\vec{v} = -\vec{v}\},$$

(both symmetries imply that the components of $\vec{v}$ are either odd or even functions with respect to x_1, and that div $\vec{v}$ is either odd or even in the same sense). Next, we define

$$\hat{G}_e \equiv \hat{G}_{(1,e)} \times \hat{G}_{(2,e)} \text{ and } \hat{G}_o \equiv \hat{G}_{(1,o)} \times \hat{G}_{(2,o)}. \tag{3.28}$$

Theorem 10. *Let the conditions of Theorem 8 be satisfied. Suppose that Ω, Γ_0, and our triangulations are symmetric with respect to the plane $x_1 = 0$. Suppose that $\hat{G}$ coincides with $\hat{G}_e$ or $\hat{G}_o$ defined by (3.28). Then there exists a constant σ_0^* independent of h such that (3.18) holds.*

Proof. In accordance with (1.8), it suffices to prove that, for a prescribed $p \in \hat{G}_{(2,e)}$ or $p \in \hat{G}_{(2,o)}$, the problem $(\hat{u}_1, \hat{v}_1)_{1,\hat{\Omega}_h} = (\mathrm{div}\ \hat{v}_1, p)_{0,\hat{\Omega}_h}$, $\forall \hat{v}_1 \in \hat{G}_1$, has a solution $\hat{u}_1$ such that $\hat{u}_1 \in \hat{G}_{(1,e)}$ or $\hat{u}_1 \in \hat{G}_{(1,o)}$, respectively. The original problem is equivalent to finding $\hat{u}_1 = \arg\min \Phi(\hat{v}_1)$, with $\Phi(\hat{v}_1) \equiv |\hat{v}_1|^2_{1,\hat{\Omega}_h} - 2(\mathrm{div}\ \hat{v}_1, p)_{0,\hat{\Omega}_h}$. Now it is easily checked that $\Phi(S\hat{v}_1) = \Phi(\hat{v}_1)$, $\forall \hat{v}_1$, where the symmetry operator S is chosen as in (5.3.45), provided $p \in \hat{G}_{(2,e)}$. Hence, $\hat{u}_1 \in \hat{G}_{(1,e)}$. Analogously, if $p \in \hat{G}_{(2,o)}$, we see that

$\Phi(-S\hat{v}_1) = \Phi(\hat{v}_1)$, $\forall \hat{v}_1$, and $\hat{u}_1 \in \hat{G}_{(1,o)}$. $\square$ [16]

§ 4. Asymptotically optimal algorithms for Stokes type problems

4.1. Construction of grid problems and error estimates. For $d = 2$ or $d = 3$, we deal with spaces (1.30) or (1.35) (if (1.34) holds). We construct triangulations (possibly composite) $T_h(\hat{\Omega}_h)$ of the same type as in § 5.1 and 5.2 (see, e.g., Figure 0.2.9, where topologically equivalent grids typical for many hydrodynamics problems are represented) and define spline subspaces $\hat{G}$ from Theorems 2.5–2.7. We assume that the corresponding Theorem 5, 6, or 7 is valid and consider the original problem defined by (1.23), (1.24), with $l_2(v_2) \equiv 0$ and

$$b_{2,1}(u_1; v_2) \equiv (\mathrm{div}\ u_1, v_2)_{0,\Omega}, \quad b_{2,2}(u_2, v_2) \equiv (u_2, v_2)_{0,\Omega} \qquad (4.1)$$

(if (1.30) holds, $b_{1,1}(u_1, v_1) = (\vec{u}_1, \vec{v}_1)_{1,\Omega})$, and $\alpha = 0$, then we have the Dirichlet problem for the Stokes system).

In the case of (1.35), (1.34), we approximate the original problem by the projective problem of finding $\hat{u} \in \hat{G}$ such that [17]

$$\left.\begin{array}{l} b_{1,1}(\hat{u}_1; \hat{v}_1) + b_{1,2}(\hat{u}_2; \hat{v}_1) = l_1(\hat{v}_1), \quad \forall \hat{v}_1 \in \hat{G}_1 \\ b_{2,1}(\hat{u}_1; \hat{v}_2) - \alpha b_{2,2}(\hat{u}_2; \hat{v}_2) = 0, \quad \forall \hat{v}_2 \in \hat{G}_2'. \end{array}\right\} \qquad (4.2)$$

In the case of (1.30), we use (4.2) and the additional condition

$$(\hat{u}_2, 1)_{0,\hat{\Omega}} = 0 \qquad (4.3)$$

((4.2) and (4.3) are equivalent to (4.2) with $\forall \hat{v}_2 \in \hat{G}_2'$ replaced by $\forall \hat{v}_2 \in \hat{G}_2$, because $(\mathrm{div}\ \hat{u}_1, 1)_{0,\hat{\Omega}} = 0$). In accordance with Theorems 3.5–3.7, these projective problems are correct. To study convergence of our PGMs, we assume that the solution of the original problem is

---

[16] In the same fashion, we can deal with symmetries with respect to several coordinate planes. Finally, it should be mentioned that other choices of PGMs are possible. For example, it is not hard to study PGMs with certain curvilinear triangles in the corresponding triangulations $T_h(\hat{\Omega})$; more cumbersome are the study of methods associated with piecewise polynomial approximations for velocity and pressure, for which (3.18) remains valid (see [195, 261, 356]). Of special interest are PGMs dealing with continuous piecewise linear (with respect to $T_h(\hat{\Omega}_h)$) functions in $\hat{G}_2$. Then, for $d = 2$, elements of $\hat{G}_{(1)}$ can be defined as piecewise quadratic (with respect to $T_h^{(2)}(\hat{\Omega}_h)$), or they may contain functions that are defined using $T_h^{(4)}(\hat{\Omega}_h)$ and, on each $T \in T_h(\hat{\Omega}_h)$, are polynomials of degree not greater than 4.

[17] We actually deal here with the bilinear forms $b_{r,l}(\hat{u}_l; \hat{v}_l) \equiv b_{r,l}(\hat{\Omega}_h; \hat{u}_l; \hat{v}_l)$, where $r = 1, 2$, $l = 1, 2$, as in § 5.1.

$$[\vec{u}, p] \equiv [u_{1,1}, \ldots, u_{1,d}, p] \in (W_2^{1+\gamma}(\Omega))^d \times W_2^\nu(\Omega), \quad 0 < \gamma \leq 1, \quad (4.4)$$

and that $\hat{G}$ is a subspace of G (that is, that $\hat{\Omega}_h \subset \bar{\Omega}$ and $\hat{\Gamma}_1 = \Gamma_1$; see § 5.1, 5.2). In this case, Theorem 2.2 applies. To estimate ρ_1 and ρ_2 in this Theorem, we may make use of all of our results dealing with approximation of Sobolev spaces (see § 2.3, 5.1, 5.2) and write

$$\|\hat{u}_1 - u_1\|_{G_1} + \|\hat{u}_2 - u_2\|_{G_2} \leq K^* h^\gamma, \quad 0 < \gamma \leq 1. \quad (4.5)$$

In the same manner, we can prove (4.5) under the conditions

$$\|u_{1,s}\|_{\gamma,\Omega_r} \leq K_r^*, \ \|p\|_{\gamma,\Omega_r} \leq K_r^*, \ r \in [1,p]', s \in [1,d], \quad 0 < \gamma \leq 1, \quad (4.6)$$

of type (5.1.3). This can be done even for the analogous conditions of type (1.3) involving certain known singular vectors describing the asymptotic behavior of $\vec{u}_1$ in the neighborhoods of some singular points, provided we include these functions in the basis for $\hat{G}_1$ (see Lemma 2.1). When $\hat{\Gamma}_1$ approximates Γ_1, we need a more involved analysis that unites the proofs of Theorems 2.2 and 5.1.1.

4.2. Modified classical iterative methods. In accordance with standard extensions of $\hat{u}_1 \in \hat{G}_1$ and $\hat{u}_2 \in \hat{G}_2'$ with respect to the bases for these subspaces, we obtain the grid systems

$$Lu \equiv \begin{bmatrix} L_{1,1} & L_{1,2} \\ L_{2,1} & -\alpha L_{2,2} \end{bmatrix} \begin{bmatrix} u_1 \\ u_2 \end{bmatrix} = \begin{bmatrix} f_1 \\ f_2 \end{bmatrix}, \quad (4.7)$$

where: $u_1 \equiv \mathbf{u_1} \in H_1$; $u_2 \equiv \mathbf{u_2} \in H_2'$; $\dim H_1 = \dim \hat{G}_1 = dN_1$; $\dim H_2' = \dim \hat{G}_2' = N_2$; H_1 and H_2' have inner products as in § 0.3 and 5.3; $L_{2,1}$ is such that $(L_{2,1}u_1, v_2) = (\text{div } \hat{u}_1, \hat{v}_2)_{0,\hat{\Omega}_h}$, $\forall \hat{u}_1 \in \hat{G}_1$, $\forall \hat{u}_2 \in \hat{G}_2'$; and $L_{1,2} = L_{2,1}^*$. We also note that

$$L_{2,1} = [(\text{div } \bar{\psi}_r, \chi_k)_{0,\hat{\Omega}_h}], \quad r = 1, \ldots, dN_1, \ k = 1, \ldots, N_2,$$

where $\bar{\psi}_r$ corresponds to a basis function in $\hat{G}_1$ (see § 5.3); the operator $L_{1,1}$ is determined by the bilinear form $b_{1,1}(\hat{u}_1; \hat{u}_2)$ (in the simplest case of the Dirichlet problem for Stokes system, it is just the block diagonal operator with diagonal blocks equal to Λ_Ω (see § 5.1 and 5.3)); but, under general conditions (1.24), $L_{1,1}$ may be nonsymmetric.

For (1.30), (4.7) is supplemented by

$$(u_2, \bar{1}_2)_{J_2} = 0 \quad (4.8)$$

$(\bar{1}_2 \equiv [1, \ldots, 1]^T \in H_2')$, which is equivalent to (4.3) (here, $J_2 = L_{2,2}$ is the diagonal Gram matrix for the basis for $\hat{G}_2'$; its elements are determined by $|T|$, where $T \in T_h(\Omega_h)$). In this case, $\bar{1}_1 \equiv [1, \ldots, 1] \in H_1$ is the basis for Ker $L_{1,2}$.

Lemma 1. *Let the conditions* (1.24) *be satisfied. Suppose also that the bilinear forms $b_{1,1}$ and $b_{2,2}$ are symmetric. Then*

$$L = L^*, \ \gamma_0 J_1 \le L_{1,1} \le \gamma_1 J_1, \ \gamma_2 J_2 \le L_{2,2} \le \gamma_3 J_2, \qquad (4.9)$$

$$\|J_2^{-1/2} L_{2,1} J_1^{-1/2}\| \le \kappa_1, \ \|J_2^{-1/2} L_{2,1} L_{1,1}^{-1/2}\| \le \kappa_1 \gamma_0^{-1/2} \equiv \kappa_{1,2}, \qquad (4.10)$$

and

$$0 \le L_{2,1} L_{1,1}^{-1} L_{1,2} \le \kappa_{1,2}^2 J_2. \qquad (4.11)$$

Proof. In accordance with standard relations (see § 0.3), we have

$$(L_{r,l} u_l, v_r) = b_{r,l}(\hat{u}_1; \hat{v}_2), \ \forall \hat{u}_1 \in \hat{G}_1, \forall \hat{u}_2 \in \hat{G}_2', r \in [1, d], l \in [1, d].$$

Hence, (4.9) is evident. Also, we see that

$$|(L_{2,1} u_1, v_2)| \le \kappa_1 \|u_1\|_{J_1} \|v_2\|_{J_2} \le \kappa_{1,2} \|u_1\|_{L_{1,1}} \|v_2\|_{J_2},$$

which leads directly to (4.10) (see (0.4.4)). Finally, (4.11) is equivalent to

$$0 \le (J_2^{-1/2} L_{2,1} L_{1,1}^{-1/2})(L_{1,1}^{-1/2} L_{1,2} J_2^{-1/2}) \le \kappa_{1,2}^2 I_2,$$

which follows from (4.10). □ [18]

For L from (4.7), we define its Schur matrices $S_2(L) \equiv -S_2$ and $S_1(L) \equiv S_1$, that is,

$$S_2 = R_2 + \alpha L_{2,2}, \quad S_1 = L_{1,1} + 1/\alpha R_1, \qquad (4.12)$$

where $R_2 \equiv L_{2,1} L_{1,1}^{-1} L_{1,2}$, $R_1 \equiv L_{1,2} L_{2,2}^{-1} L_{2,1}$, and S_1 is defined for $\alpha > 0$. These matrices will be used in Subsection 4.4 for construction of a class of iterative methods, where at least one of the operators $L_{1,1}$ or $L_{2,2}$ is simple enough. For now, we consider a class of effective iterative methods that require no such assumption (they were considered in [198, 199, 200, 205]) and are based on symmetrization (0.3.17) with $B_1 = B_2 = B$ and inequalities (0.3.19).

With this in mind, we introduce the block diagonal operator $J \in \mathcal{L}^+(H)$, whose diagonal blocks $J_1 \in \mathcal{L}^+(H_1)$ and $J_2 \in \mathcal{L}^+(H_2')$ were used in (4.9)–(4.11) and $\|u\|_J^2 = |\hat{u}_1|_{1,\hat{\Omega}_h}^2 + |\hat{u}_2|_{0,\hat{\Omega}_h}^2$. Then, for (1.35) and (1.34) (when $\hat{G}_2' = \hat{G}_2$ and $H_2' = H_2$) the proof of (0.3.19), with J instead of $B \asymp J$ is

---

[18] The estimates from below will be sharpened in Lemma 2.

a simple consequence of the correctness of (4.2), as well as a simplification of our obtaining analogous estimates for the case (1.30) below.

For (1.30), we deal with (4.4) (recall that it is equivalent to (4.3)) and define a Euclidean space H_3 as the subspace of the Euclidean space $H_2'(J_2)$ whose elements satisfy (4.8).

Our wish to obtain estimates uniform in $\alpha \in [0, \alpha_0]$, and the special role of H_3 make it natural to introduce the new operators

$$L_{2,2}^0 \equiv L_{2,2}P, \tag{4.13}$$

where $Pu_2 \equiv u_2 - (u_2, \bar{1}_2)_{L_{2,2}}(\|\bar{1}_2\|_{L_{2,2}})^{-2}\bar{1}_2$, and

$$L^0 u \equiv \begin{bmatrix} L_{1,1} & L_{1,2} \\ L_{2,1} & -\alpha L_{2,2}^0 \end{bmatrix} \begin{bmatrix} u_1 \\ u_2 \end{bmatrix} = \begin{bmatrix} f_1 \\ f_2 \end{bmatrix}, \tag{4.14}$$

which replaces (4.7). [19] For (4.14), we introduce $S_2(L^0) \equiv -S_2^0$ with $S_2 = R_2 + \alpha L_{2,2}^0$. The indicated symmetry of P implies that

$$(L_{2,2}^0 u_2, v_2) = (Pu_2, v_2)_{L_{2,2}} = (u_2, L_{2,2}^0 v_2),$$

that is, $L_{2,2}^0 = (L_{2,2}^0)^* \geq 0$ in the sense of $\mathcal{L}(H_2))$ and $\mathrm{Ker}\, L_{2,2}^0 = \mathrm{Ker}\, L_{1,2}$. Hence,

$$\mathrm{Ker}\, L^0 = \{[0, v_2] : v_2 \in \mathrm{Ker}\, L_{1,2}\}.$$

In the sequel, we agree not to distinguish between an operator and its restriction to its invariant subspace.

Lemma 2. Let the conditions of Lemma 1 and Theorem 3.5 be satisfied. Then H_3 is an invariant subspace of $J_2^{-1}R_2$ and $J_2^{-1}S_2^0$; these operators, as elements of $\mathcal{L}(H_2'(J_2))$, are spectrally equivalent to the operator I_2:

$$\delta_0^{(2)} I_2 \leq J_2^{-1} R_2 \leq \delta_1^{(2)} I_2, \quad \delta_{0,2} I_2 \leq J_2^{-1} S_2^0 \leq \delta_{1,2} I_2, \tag{4.15}$$

where

$$\delta_0^{(2)} \equiv (\sigma_0^*)^2/\gamma_1, \quad \delta_1^{(2)} \equiv \kappa_{1,2}^2, \quad \delta_{0,2} \equiv \delta_0^{(2)} + \alpha\gamma_2, \quad \delta_{1,2} \equiv \delta_1^{(2)} + \alpha\gamma_3.$$

Proof. The operators R_2 and $L_{2,2}^0$ are symmetric in the sense of $\mathcal{L}(H_2')$. Thus, both operators $J_2^{-1}R_2$ and $J_2^{-1}S_2^0$ are symmetric in the sense of $\mathcal{L}(H_2'(J_2))$ and have the same kernel $\mathrm{Ker}\, L_{1,2}$. Hence, $H_3 = \mathrm{Im}\, J_2^{-1}R_2 =$

---

[19] $Pu_2 \equiv v_2$ corresponds to $\hat{v}_2 \in \hat{G}_2'$ such that $b_{2,2}(\hat{v}_2; 1) = 0$; our basic idea in constructing effective iterative methods is analogous to the orthogonalization procedure in Subsection 1.3.6. We emphasize that P is an orthoprojector, in the sense of the semiinner product in $H_2'(L_{2,2})$, and is symmetric as an element of $\mathcal{L}(H_2'(L_{2,2}))$ (usually, $L_{2,2} = J_2$, but the case of $L_{2,2} \neq 0$ is very important for problems associated with (3.27) and Theorem 3.9).

Im $J_2^{-1} S_2^0$ is invariant with respect to these operators. Further, it is easy to see that the desired inequalities are equivalent to

$$\delta_0^{(2)} J_2 \le R_2 \le \delta_1^{(2)} J_2 \text{ and } \delta_{0,2} J_2 \le S_2^0 \le \delta_{1,2} J_2.$$

They are verified analogously to (1.18) but need some additional new steps. For example, we have $(L_{1,1}^{-1} L_{1,2} v_2, L_{1,2} v_2) \ge 1/\gamma_1 \|L_{1,2} v_2\|_{J_1^{-1}}^2$ (see (4.9) and Lemma 0.4.3) and

$$\|L_{1,2} v_2\|_{J_1^{-1}} = \sup_{v_1 \ne 0} \frac{|(v_1, L_{1,2} v_2)|}{\|v_1\|_{J_1}} = \sup_{\hat{v}_1 \ne 0} \frac{|(\text{div } \hat{v}_1, \hat{v}_2)_{0,\hat{\Omega}_h}|}{|\hat{v}_1|_{1,\hat{\Omega}_h}}$$

(see Lemma 0.4.5). This and (3.18) show that

$$(L_{1,1}^{-1} L_{1,2} v_2, L_{1,2} v_2) \ge \frac{(\sigma_0^*)^2}{\gamma_1} \|v_2\|_{J_2}^2,$$

$\forall v_2 \in H_3$, and lead to the inequalities in (3.15) involving $\delta_0^{(2)}$ and $\delta_{0,2}$. The estimates from above are simpler and hold on H_2'. □ [20]

   Lemma 3. Let the conditions of Lemma 1 and Theorem 3.5 be satisfied. Suppose that we consider problem (4.14), (4.8) with prescribed $f_1 \in H_1$ and $f_2 \in H_3$. Then it has a unique solution $u \equiv [u_1, u_2]^T$ and

$$\|u\|_J \le \kappa_2 \|L^0 u\|_{J^{-1}}, \quad \forall u. \tag{4.16}$$

Proof. For each solution of (4.14) we have

$$J_2^{-1} S_2^0 u_2 = -J_2^{-1}(f_2 - L_{2,1} L_{1,1}^{-1} f_1) \equiv g_2.$$

Hence, it has a unique solution with $u_2 \in H_3$ if and only if $g_2 \in H_3$ (see Lemma 2). Since $(L_{2,1} L_{1,1}^{-1} f_1, \bar{1}_2) = 0$, then $g_2 \in H_3 \Leftrightarrow f_2 \in H_3$. Therefore, L^0 is an one-to-one mapping of the space $V' \equiv H_1 \times H_3$ onto itself. Further, for each solution of (4.14) and (4.8), (4.15) leads to the estimate $\|u_2\|_{J_2} \le \gamma_1 (\sigma_0^*)^{-2} \|g_2\|_{J_2}$. We now see that

$$\|g_2\|_{J_2} \le \|f_2\|_{J_2^{-1}} + \|J_2^{-1} L_{2,1} L_{1,1}^{-1/2}\| \|f_1\|_{L_{1,1}^{-1}}.$$

---

[20] We can take $\delta_{0,2} \equiv \delta_0^{(2)}$, $\delta_{1,2} \equiv \delta_1^{(2)} + \alpha_0 \gamma_3$ uniformly in $\alpha \in [0, \alpha_0]$ and, on the whole space $H_2'(J_2)$, we have $J_2^{-1} S_2 \ge \alpha I_2$, that is, for $\alpha > 0$, the operators S_2 and L (see (4.12) and (4.7)) are invertible.

   Since $S_2(L^0) = R_2 + \alpha L_{2,2}^0$, then it is easy to verify that solutions of (4.7), (4.8) and (4.15), (4.8) (with $f_2 = 0$) exist and differ only in their second components by the term $c\bar{1}_2$. Here, we use K and κ for positive constants independent of h and $\alpha \in [0, \alpha_0]$.

Hence, Lemma 2 leads to the estimates $\|u_2\|_{J_2} \leq \kappa_3\|f_2\|_{J_2^{-1}} + \kappa_4\|f_1\|_{J_1^{-1}}$, $|u_1\|_{J_1} \leq 1/\gamma_0[\|f_1\|_{J_1^{-1}} + \kappa_1\|u_2\|_{J_2}$ (see (1.24)), and (4.16). □ [21]

We now consider model operators

$$B_1 \asymp J_1 \text{ and } B_2 \asymp J_2. \tag{4.17}$$

For our PGMs, we may even take $B_2 = J_2$ (which is a diagonal matrix) and B_1 as a block diagonal operator with diagonal blocks spectrally equivalent to the operator Λ_Ω (see Chapters 5 and 3). [22]

Theorem 1. Let the conditions of Lemma 1 and Theorem 3.5 be satisfied. Suppose also that $\gamma_4 \leq b_{r,l}(\hat{\Omega}_h; 1; 1) \leq \gamma_5$, $\gamma_4 > 0$. Then, for the above defined operators L^0 and B, there exist constants $\delta_1 \geq \delta_0 > 0$, independent of h and $\alpha \in [0, \alpha_0]$, such that

$$\delta_0\|v\|_B^2 \leq \|Lv\|_{B^{-1}}^2 = \|B^{-1}Lv\|_B^2 \leq \delta_1\|v\|_B^2, \quad \forall v \in V, \tag{4.18}$$

where $V \equiv H_1 \times V_2$ and V_2 consists of vectors $v_2 \in H_2'$ orthogonal in the Euclidean space $H_2'(B_2)$ to Ker $L_{1,2}$.

Proof. The inequality involving δ_1 is proved as in § 5.1 on the basis of the estimates $|(Lu, v)| = |b(\hat{u}; \hat{v})| \leq \kappa_5\|\hat{u}\|\|\hat{v}\| \leq \kappa_6\|u\|_B\|v\|_B$. But now we have an additional term

$$X \equiv \alpha(L_{2,2}(Pu_2 - u_2), v_2) = \alpha(u_2, \bar{1}_2)_{L_{2,2}}(\|\bar{1}_2\|_{L_{2,2}})^{-2}(\bar{1}_2, v_2)_{L_{2,2}}$$

in (L^0u, v). We rewrite it as

$$X = \alpha b_{2,2}(\hat{\Omega}_h; \hat{u}_2; 1)b_{2,2}(\hat{\Omega}_h; \hat{v}_2; 1)[b_{2,2}(\hat{\Omega}_h; 1; 1)]^{-1}.$$

Hence,

$$|X| \leq \alpha_0 b_{2,2}(\hat{\Omega}_h; \hat{u}_2; \hat{u}_2)^{1/2} b_{2,2}(\hat{\Omega}_h; \hat{v}_2; \hat{v}_2)^{1/2}\gamma_5/\gamma_4$$

$$\leq \alpha_0\gamma_5|\hat{u}_2|_{0,\hat{\Omega}_h}|\hat{v}_2|_{0,\hat{\Omega}_h} = \alpha_0\gamma_5\|u_2\|_{J_2}\|v_2\|_{J_2} \leq \kappa_7\|u_2\|_{B_2}\|v_2\|_{B_2}.$$

Thus, it suffices to prove the remaining inequality in (4.18). To this end, for each $v \in V$, we define (see the proof of Theorem 3.9) $u \equiv [v_1, u_2]$, where $u_2 \equiv v_2 - c_2\bar{1}_2 \in H_3$. Since u_2 and v_2 are orthogonal to $\bar{1}_2$ in the sense of

---

[21] Here we have used the symmetry and positivity of $L_{1,1}$; more general results will be obtained in § 6.

[22] Other choices $B_1 \asymp L_{1,1}$ (see § 5.4) and $B_2 \asymp S_2^0$ are also possible.

$H_2'(J_2)$ and $H_2'(B_2)$, respectively, then $\|u_2\|_{J_2} \leq \|v_2\|_{J_2}$ (see (0.1.7)) and $\|v_2\|_{B_2} \leq \|v_2\|_{B_2}$. This implies that

$$\kappa_8\|u_2\|_{B_2} \leq \|v_2\|_{B_2} \leq \|u_2\|_{B_2}. \tag{4.19}$$

Now we observe that $L^0 u = L^0$ and that

$$\|L^0 u\|_{B^{-1}} = \|L^0 u\|_{B^{-1}}. \tag{4.20}$$

In accordance with (4.16) and the fact that $J \asymp B$, we see that $\|u\|_B \leq \kappa_9\|L^0 u\|_{B^{-1}}$, which together with (4.20) and (4.19) leads to (4.18). □

Theorem 1 implies that

$$\|L^0\|_{H(B) \mapsto H(B^{-1})} \leq \delta_1^{1/2}, \quad \|(L^0)^{-1}\|_{H(B^{-1}) \mapsto H(B)} \leq \delta_0^{1/2},$$

and sp $(B^{-1}(L^0)^* B^{-1} L^0) \subset \{0 \cup [\delta_0, \delta_1]\}$ (see Lemma 1.3.2). Therefore, for solving (4.7), it is reasonable to use iterations

$$B u^{n+1} = B u^n - \tau_n (L^0)^* B^{-1}(L^0 u^n - f), \tag{4.21}$$

convergence of which is determined by the constants δ_0 and δ_1 (see (0.3.19), Lemma 1.1.2, and Theorems 1.1.14 and 1.1.19). In case of the modified Richardson method, the adaptation procedure from § 1.3 for these constants is available; the modified conjugate gradient method can also be used.

Theorem 2. *Let the conditions* (1.24), (1.35), (1.34) *be satisfied. Suppose also that the bilinear forms $b_{1,1}$ and $b_{2,2}$ are symmetric and Theorem 3.6 or 3.7 applies. Let the operator L in* (4.7) *be denoted by L^0. Then, for the model operators B defined by* (4.17), *there exist constants $\delta_1 \geq \delta_0 > 0$, independent of h and $\alpha \in [0, \alpha_0]$, such that* (4.18) *holds with $V \equiv H_1 \times H_2'$.*

Proof. Now we deal with a simpler case where there is no need of (4.8), and we may take $H_3 \equiv H_2' \equiv H_2$. Hence, instead of (4.13), we deal with $P_2 = I_2, L_{2,2}^0 = L_{2,2}$, and $S_2^0 = S_2$, and (4.15) holds in the sense of $\mathcal{L}(H_2(J_2))$. Hence, $L^0 = L$ is a one-to-one mapping of the space $V = H_1 \times H_2$ onto itself and (4.16) holds. Further, the proof of (4.18) is much simpler because there is no X and $u = v$ in (4.20). □

4.3. Strengthened variant of the Kolmogorov-Bakhvalov hypothesis for Stokes type problems.

Theorem 3. *Consider a boundary value problem of type* (1.23), (1.24) *in the Hilbert space G defined by* (1.35), (1.34) *(or* (1.30)*) and such that its solution satisfies* (4.6). *Suppose that PGM* (4.2) *(or* (4.2) *and* (4.3)*) deals with the family of subspaces $\hat{G} \subset G$ and that the conditions of the respective Theorem 3.5–3.7 are satisfied. Then, for the resulting systems* (4.7) *(or* (4.7) *and* (4.8)*), iterative method of type* (4.21) *yields an ε-approximation to the solution of the original problem with computational* *work $W(\varepsilon) = O(\varepsilon^{-d/\gamma})|\ln \varepsilon|$, application of the multigrid acceleration to this basic iterative algorithm yields an ε-approximation with computational work $W(\varepsilon) = O(\varepsilon^{-d/\gamma})$.*

Proof. Theorem 2.2 and (4.6) yield error estimate (4.5), so $\varepsilon \asymp h^\gamma$. Theorem 2 guarantees effectiveness of modified iterative methods (4.21) with an asymptotically optimal preconditioner B. For (4.7), (4.8), these iterations define one of the possible solutions of (4.7) (differing in the term $c_1 \bar{1}_2$ in u_2). To define the desired c_2, it suffices to apply (4.8). So the proof of the first assertion is practically the same as for elliptic problems in Chapter 5.

The second assertion is analogous to Theorem 5.1.4. We use again the same sequence of triangulations $T_{h(l)}^{(l)}(\Omega^l)$ (in the role of our $T_h(\hat{\Omega}_h)$ for approximation of u_1) and their refinements $T_{h(l,d)}^{(l)}(\Omega^l)$ (in the role of our $T_h^{(d)}(\hat{\Omega}_h)$ for approximation of u_2). Choice of the number $t \asymp 1$ of iterations (4.21) on each level and estimation of computational work in this algorithm are much the same as in the proof of Theorem 1.1.4 (where $\hat{\Omega}_h = \bar{\Omega}$) or Theorem 5.1.4 where approximation of the domain must be taken into account. In the latter case for (1.30), there is one noteworthy difference. The coarse grid continuation now deals with the components $\hat{v}_1^l$ and $\hat{v}_2^l$ (see (5.1.20); I_l^{l+1} is now slightly different from the one in (5.1.20)). There is nothing new for $\hat{u}_1^l$, but for $\hat{u}_2^l$, we must project the extended function $\hat{w}_2^{(l+1)}$ into the subspace $V_2^{(l+1)}$ corresponding to the subspace V_2 in Theorem 1 on the level $l+1$, $l \in [0, p-1]$. This is equivalent to finding

$$a \equiv (B_2^{(l+1)} w_2^{(l+1)}, \bar{1}_2^{(l+1)}) \text{ and } P_2 w_2^{(l+1)} \equiv w_2^{(l+1)} - a\|\bar{1}_2^{(l+1)}\|_{B_2^{(l+1)}}^{-2} \bar{1}_2^{(l+1)},$$

where the index $(l+1)$ signifies that we deal with the level $l+1$, $l \in [0, p-1]$. The model operator $B_2^{(l+1)}$ is a diagonal matrix even for the case of composite triangulations. The number a can be easily found and is $O(h)$ because $v_2^{(l)} \in V_2^{(l)}$ and $|\Omega^l \triangle \Omega^{l+1}| \leq \kappa_{10} h_l^2$. We can thus prove a simpler modification of (5.1.25) in which the norms with index 1 are replaced by the norms with index 0. $\square$ [23]

---

[23] Theorem 3 established the strengthened variant of the Kolmogorov-Bakhvalov hypothesis for a wide class of boundary problems associated with Stokes type systems (see Subsection 1.6). Moreover, the given estimates of computational work are uniform in the parameter α when it is present in our systems. We emphasize also that these results were obtained for all $\gamma \in (0,1]$ and very general assumptions on the geometry of the original domain Ω (even certain domains with non-Lipschitz boundaries are allowed). Some relevant results dealing with multigrid methods can be found in [57, 270, 368]. Difference methods on composite grids were considered in [223]. A posteriori estimates were studied in [56, 492].

4.4. Systems with Schur matrices. If Ω is a rectangle or a paral-lelepiped and regular triangulations are used, certain simple problems (e.g., the Dirichlet problem for Stokes system) yield systems (4.7) with the operator $L_{1,1}$ such that fast direct methods from § 3.1 apply. Then a class of effective iterative methods like

$$B_2 u_2^{n+1} = B_2 u_2^n - \tau_n(S_2^0 u_2^n + g_2), \qquad (4.22)$$

(with $g_2 \equiv f_2 - L_{2,1}L_{1,1}^{-1}f_1$) can be constructed. [24] Their study with $B_2 \asymp J_2$ or $B_2 \asymp S_2^0$ is based on the theory given in § 1.3 and Lemma 2 and its simplification with $H_3 = H_2'$ (see the proof of Theorem 2).

We now consider a very interesting case of systems with $S_1(L) \equiv S_1$.

Lemma 4. *The operator* $L_{1,1}^{-1}R_1 \in \mathcal{L}(H_1(L_{1,1}))$ *is symmetric and*

$$\mathrm{sp}\,(L_{1,1}^{-1}R_1) \subset \{0 \cup [\delta_0^{(2)}, \delta_1^{(2)}]\}. \qquad (4.23)$$

Proof. The first part of the assertion is evident due to the symmetry of $R_1 \in \mathcal{L}(H_1)$. We see also that $R_1 \geq 0$ and Ker $L_{2,1} = $ Ker $L_{1,1}^{-1}R_1$. For non-vanishing $\lambda(L_{1,1}^{-1}R_1)$, we have $\lambda(L_{1,1}^{-1}R_1) = \lambda(L_{1,1}^{-1/2}L_{1,2}L_{2,2}^{-1}L_{2,1}L_{1,1}^{-1/2}) = \lambda(L_{2,2}^{-1}L_{2,1}L_{1,1}^{-1}L_{1,2})$ (see Lemma 0.4.6), which together with (4.15) yields (4.23). $\square$

Lemma 5. *The orthogonal complement of Ker* $L_{2,1}$ *in the sense of the space* $H_1(L_{1,1})$ *is* Im $L_{1,1}^{-1}R_1$ *and*

$$\delta_0^{(2)}\|u_1\|_{L_{1,1}}^2 \leq (R_1 u_1, u_1) \leq \delta_1^{(2)}\|u_1\|_{L_{1,1}}^2, \quad \forall u_1 \in \mathrm{Im}\,(L_{1,1}^{-1}R_1). \quad (4.24)$$

Proof. The first part of the assertion is a consequence of the symmetry of $L_{1,1}^{-1}R_1$ in the sense of $\mathcal{L}(H_1(L_{1,1}))$ and the fact that Ker $L_{1,1}^{-1}R_1 = $ Ker $L_{2,1}$. The second part of the assertion is a consequence of Lemma 4. $\square$

From Lemma 5, it follows that convergence of iterations (seed86)

$$u_1^{n+1} - u_1^n = -\tau_n \alpha L_{1,1}^{-1}(S_1 u_1^n - f_1) \qquad (4.25)$$

is independent of h provided they are considered on Im $L_{1,1}^{-1}R_1$. [25]

Now we tackle the problem of constructing nearly asymptotically opti-mal model operators $D_1 \asymp S_1$ with estimates of spectral equivalence inde-pendent in $\alpha \in (0, \alpha_0]$ (see [191, 197]). We assume that an asymptotically optimal model operator $B_1 \asymp L_{1,1}$ is given and

---

[24] See [191, 195, 327] and references therein; relevant iterations in Hilbert spaces were analyzed probably for the first time in [126], where the fundamental role of (1.31) was emphasized.

[25] In [42], it was observed that analogous methods show good practical convergence in several initial iterations when the rounding errors are nonsignificant.

$$\delta_{0,1}B_1 \le L_{1,1} \le \delta_{1,1}B_1 \le L_{1,1}, \ \delta_{0,1} > 0. \tag{4.26}$$

For the symmetric bilinear form $b_{2,2}$, it is required that $\gamma_2 > 0$ (see (1.24)) and that the respective systems with $L_{2,2} \equiv B_2 \ge \gamma_2 J_2$ are easy to solve. We introduce operators

$$\Lambda_2 \equiv \alpha B_2 + L_{2,1}B_1^{-1}L_{1,2}, \quad S_2' \equiv B_2^{-1}\Lambda_2 \tag{4.27}$$

and the linear subspace

$$W_2 \equiv \{v_2 : (v_2, \bar{1}_2)_{B_2} = 0\} = \{v_2 : (v_2, \bar{1}_2)_{\Lambda_2} = 0\}.$$

We make use of two inner products in W_2 determined by the inner products in $H_2'(B_2)$ and $H_2'(\Lambda_2)$ and denote these new Euclidean spaces by W_3 and W_4, respectively.

Lemma 6. Let the conditions of Lemma 2 and Theorem 3.5 be satisfied. Then W_2 is an invariant subspace of the operator S_2' and

$$\delta_{0,4}\|v_2\|_{L_{2,2}}^2 \le (S_2'v_2, v_2)_{L_{2,2}} \le \delta_{1,4}\|v_2\|_{L_{2,2}}^2, \quad \forall v_2 \in W_3, \tag{4.28}$$

where $\delta_{0,4} \equiv \alpha + \delta_0^{(2)}\delta_{0,1}/\gamma_3$ and $\delta_{1,4} \equiv \alpha + \delta_1^{(2)}\delta_{1,1}/\gamma_3$.

Proof. The proof is based on the proof of Lemma 2 because $S_2' = \alpha I_2 + B_2^{-1}R_2'$ with $R_2' \equiv L_{2,1}B_1^{-1}L_{1,2}$ (see (4.12); $B_2^{-1}R_2' \in \mathcal{L}(W_3)$ is symmetric and its kernel is just $\mathrm{Ker}\, L_{1,2}$ so $W_2 = \mathrm{Im}\, B_2^{-1}R_2'$). Further, since $\delta_{0,1}R_2 \le R_2' \le \delta_{1,1}R_2$ (see (4.26) and Lemma 0.4.3), then (4.28) is equivalent to $\delta_{0,4}/\delta_{0,1}(L_{2,2}v_2, v_2) \le (R_2v_2, v_2) \le \delta_{1,4}/\delta_{1,1}(L_{2,2}v_2, v_2), \forall v_2 \in W_2$. We prove only the estimate from below. For $v_2 \in W_2$, we consider $u_2 \equiv v_2 - c_2\bar{1}_2 \in H_3$. Since u_2 and v_2 are orthogonal to $\bar{1}_2$ in the sense of $H_2'(J_2)$ and $H_2'(B_2)$, respectively, then we have the same inequalities connecting the norms of u_2 and v_2 in $H_2'(J_2)$ and $H_2'(B_2)$ as in the proof of Theorem 1. For u_2, we apply (4.15), so that, for all $v_2 \in H_4$, we have

$$(R_2u_2, u_2) \ge \delta_0^{(2)}(J_2u_2, u_2) \ge \delta_0^{(2)}/\gamma_3(B_2u_2, u_2) \ge \delta_0^{(2)}/\gamma_3(B_2v_2, v_2).$$

The inequality involving $\delta_{1,4}$ is simpler and holds on H_2'. $\quad\square$

Lemma 7. Suppose, for the system

$$S_2'v_2 = B_2^{-1}g_2 \equiv g_2' \in W_3, \tag{4.29}$$

with $v_2 \in W_3, g_2 \equiv L_{2,1}B_1^{-1}\varphi_1$, that m iterations

$$L_{2,2}(v_2^{k+1} - v_2^k) = -\tau_k^{(2)}(\Lambda_2 v_2^k - g_2), \quad k = 0, \ldots, m-1, \qquad (4.30)$$

are applied with $v_2 = 0$ and with the reduction error operator Z_m such that
$\|Z\|_{W_3} \le q < 1$. *Then*

$$\alpha \mathcal{A}_{2,2} \equiv S_2'[(I_2 - Z_m)^{-1} - I_2] + \alpha I_2 \equiv \alpha \mathcal{A}_{2,2} \in \mathcal{L}^+(W_3) \qquad (4.31)$$

is such that $u_2 \in W_3$ defined by the system

$$Bu \equiv \begin{bmatrix} B_1 & L_{1,2} \\ B_2^{-1} L_{2,1} & -\alpha \mathcal{A}_{2,2} \end{bmatrix} \begin{bmatrix} u_1 \\ u_2 \end{bmatrix} = \begin{bmatrix} \varphi_1 \\ 0 \end{bmatrix} \qquad (4.32)$$

is just v_2^m.

Proof. Z_m is a polynomial with respect to S_2' and is symmetric in the sense of $\mathcal{L}(W_3)$ and $\mathcal{L}(W_4)$ (see § 0.3). The same holds for $\mathcal{A}_{2,2}$. Positivity of $\mathcal{A}_{2,2}$ follows from (4.31) and (4.28). Hence, (4.31) is valid. Further, we have $v_2^m - v_2 = Z_m v_2$, and $S_2'(I_2 - Z_m)^{-1} v_2^m = L_{2,2} g_2$, where matrix $S_2'(I_2 - Z_m)^{-1} \in \mathcal{L}^+(W_3)$ in accordance with the above properties of S_2' and Z_m. From (4.31) and (4.27), it follows that

$$S_2'(I_2 - Z_m)^{-1} = \alpha \mathcal{A}_{2,2} + L_{2,2}^{-1} L_{2,1} B_1^{-1} L_{1,2},$$

that is, $S_2'(I_2 - Z_m)^{-1} = -S_2(B)$ (see 4.31). Thus $u_2 = v_2^m$. $\square$ [26]

Lemma 8. *Let, for iterations (4.30), the reduction error operator Z_m, symmetric in the sense of $\mathcal{L}(W_3)$ and $\mathcal{L}(W_4)$, be such*

$$0 \le Z_m \le q_2 I_2, \qquad (4.33)$$

where $q_2 \equiv \alpha \kappa^{(0)} \le \alpha_0 \kappa^{(0)} < 1$. Then in $\mathcal{L}(W_3)$ we have

$$\alpha I_2 \le \alpha \mathcal{A}_{\epsilon,\epsilon} \le \alpha \kappa^{(1)} I_2, \qquad (4.34)$$

where $\kappa^{(1)} \equiv 1 + \kappa^{(0)}(1 - \alpha \kappa^{(0)})^{-1} \delta_{1,4}$.

Proof. From (4.31) and (4.33) it follows that

$$\alpha I_2 \le \alpha \mathcal{A}_{2,2} \alpha I_2 + \frac{q_2}{1 - q_2} S_2'$$

in the sense of $\mathcal{L}(W_3)$. This and (4.28) lead to (4.34). $\square$

---

[26] The operator B in (4.32) is analogous to the cooperative operators considered in § 1.5; its elements $B_2^{-1} L_{2,1}$ and $-\alpha \mathcal{A}_{2,2}$ in the second row can be replaced by $L_{2,1}$ and $-\alpha \mathcal{A}_{2,2}$.

Theorem 3. Suppose that the conditions of Lemma 8 are satisfied. Let

$$D_1 \equiv B_1 + 1/\alpha A_1, \tag{4.35}$$

where B_1 satisfies (4.26) and $A_1 \equiv L_{1,2}A_{2,2}^{-1}B_2^{-1}L_{2,1}$. Let $\delta_0 \equiv \min\{\delta_{0,1}; 1\}$ and $\delta_1 \equiv \max\{\delta_{1,1}; \kappa^{(1)}\}$. Then $D_1 \in \mathcal{L}^+(H_1)$ and

$$\delta_0 D_1 \leq S_1 \leq \delta_1 D_1. \tag{4.36}$$

Proof. Since $(B_2^{-1}L_{2,1}u_1, \bar{1})_{W_3} = (\operatorname{div} \hat{u}_1; 1)_{0,\hat{\Omega}_h} = 0$, then

$$u_{1,2} \equiv B_2^{-1}L_{2,1}u_1 \in W_3, \quad \forall u_1 \in H_1.$$

Further,

$$(A_1u_1, v_1) = (A_{2,2}^{-1}u_{1,2}, v_{1,2})_{W_3} = (u_1, A_1v_1)$$

and

$$(A_1u_1, u_1) \geq \frac{1}{\kappa^{(1)}}\|u_{1,2}\|_{H_3}^2 = \frac{1}{\kappa^{(1)}}\|L_{2,1}u_1\|_{L_{2,2}^{-1}}^2$$

(see (4.34)). Hence,

$$(D_1u_1, u_1) \geq \frac{1}{\delta_{1,1}}(L_{1,1}u_1, u_1) + \frac{1}{\alpha\kappa^{(1)}}(R_1u_1, u_1) \geq \frac{1}{\delta_1}(S_1u_1, u_1).$$

In the same way, we obtain $D_1 \leq 1/\delta_0 S_1$. □

Theorem 4. For finding the solution of system

$$D_1v_1 = \varphi_1, \tag{4.37}$$

with D_1 defined by (4.35), it suffices: to find $g_2 \equiv L_{2,1}B_1^{-1}\varphi_1$, to carry out m iterations (4.30) with $v_2^0 = 0$; and to find $v_1 \equiv B_1^{-1}(\varphi_1 - L_{1,2}v_2^m)$.

Proof. Consider (4.32) with the above φ_1 and g_2. Observe that $g_2' \in W_3$. The solution of (4.32), in accordance with Lemma 7, is just the final iterate v_2^m in (4.30). On the other hand, elimination of u_2 in (4.32) leads to the system $D_1u_1 = \varphi_1$ and, hence, $u_1 = v_1$. □

Theorems 3 and 4 imply that the modified Richardson method

$$D_1(u_1^{n+1} - u_1^n) = -\tau_n(S_1u_1^n - f_1) \tag{4.38}$$

or modified conjugate gradient method can be used effectively for solving systems with operator S_1. To satisfy (4.33), we may apply Theorem 3.1.14 with the set of iteration parameters determined by $\{t_i^+\}$ and constants in (4.28). Then $m = O(|\ln \alpha|)$ and the computational work for solving a system with D_1 in (4.35) is $W = O(h^{-d}|\ln \alpha|)$ because we solve $m + 2$ systems with operator B_1. [27]

For (1.35), (1.34), we have $W_2 = H_2' \equiv H_2$, so our study becomes much simpler. (4.28) holds in the sense of $\mathcal{L}(H_2(B_2))$, and in Lemma 8 and Theorem 3 we deal with $W_3 \equiv H_2(B_2)$ and

$$D_1 \equiv B_1 + 1/\alpha L_{1,2} A_{2,2}^{-1} L_{2,1}. \tag{4.39}$$

4.5. Generalizations for other problems and PGMs. Our results can be generalized for certain other boundary value problems dealing with subspaces of $\vec{V}$ (see (5.4.1) that were mentioned at the end of § 5.4. There is also a wide variety of grid methods for which the above asymptotically optimal iterations (first of all, iterations (4.21)) are useful. For example, we mention that PGMs associated with continuous piecewise linear approximations for the pressure lead to diagonal model operators B_2 (see, e.g., (2.4.9) and (2.4.10)). When we work with nonconforming finite element methods (see § 2.6 and [99, 98, 127]) the situation is even simpler because respective analogs of (3.18)—in the case of (1.30) and a polygonal domain—have been known since the seventies; (see [127]). Their generalizations, dealing with $\hat{\Omega}_h$, non-Lipschitz boundary, and (1.34), (1.35), follow from Theorems 3.1–3.4. [28] Theorem 3.10 enables us to construct asymptotically optimal algorithms that exploit symmetry of the domain.

Finally, we mention problems where Theorem 3.9 is of fundamental importance (such problems are typical when both compressible and incompressible mediums are present). For example, suppose we deal with (1.30) and the new bilinear form $b_{2,2}(u_2; v_2) \equiv (g, u_2 v_2)_{0,\Omega}$, where $g(x)$ is a characteristic function of a subdomain Ω_1 and the interface between Ω_1 and $\Omega_2 \equiv \Omega \setminus \Omega_1$ is a set of faces of simplexes in $T_h(\hat{\Omega}_h)$. For its solution, it is natural to assume that (4.6) holds. Then our PGMs are associated with the old subspaces $\hat{G}_1$ and $\hat{G}_2'$ and new subspace $\hat{G}_2$ determined by the condition $(1, \hat{u}_2)_{0,\Omega_1 \cap \hat{\Omega}_h} = 0$. This implies that (4.8) must be replaced by (4.41), but again the finding of its solution is reduced to finding of a solution of (4.14).

---

[27] Iterations of type (4.38) with $D_1 \asymp L_{1,1}$ were considered in [298]. Certain model operators based on factorization of S_1 (for analogous PGMs) were used in [244] for solution of important practical problems; each iteration for $d = 2$ required computational work $W = O(h^{-3})$, in contrast with our theoretical estimate $W = O(h^{-2}|\ln \alpha|)$. We emphasize that the constants in (4.36) are uniform in h and α. Our condition $Z_m \geq 0$ in (4.33) is not necessary but, as the analysis shows, turns out to be more profitable in comparison with the standard modified Richardson method when α and σ_0^* are small. We note also that (4.38) are actually two-stage iterations with inner iterations (4.30) and that the operators like D_1 will be used in constructing nearly asymptotically optimal algorithms for complicated eigenvalue problems in § 9.7.

[28] Iterations of type (4.22) were applied in [193]. Recent results are connected with multigrid methods (see [93, 94]), and also lead to convergence uniform in h and α.

§ 5. Iterative methods with model saddle operators

5.1. Basic classes of methods. In what follows, we deal with the Euclidean space $H \equiv H_1 \times H_2$ and the operator equations

$$\tilde{L}(u) \equiv Lu + \mathcal{P}(u) = f. \tag{5.1}$$

Here, $H_2 \equiv H_2'$ and $H_2 \equiv W_2$ if we deal with the cases (1.34)–(1.35) and (1.30), respectively, L is a strongly saddle operator in H that satisfies (4.9)–(4.11),

$$\delta_{0,2} J_2 \le L_{2,1} L_{1,1}^{-1} L_{1,2} \le \delta_{1,2} J_2, \ \delta_{0,2} > 0, \tag{5.2}$$

and $\mathcal{P}$ is a typically nonlinear perturbation operator satisfying

$$\mathcal{P}(u) \equiv [\mathcal{P}_1(u_1), \mathcal{P}_2(u_2)]^T. \tag{5.3}$$

A saddle operator A is called *model operator* if, for solving

$$Au = g, \tag{5.4}$$

there exist asymptotically optimal algorithms, and iterations

$$A(u^{n+1} - u^n) = f - \tilde{L}(u^n) \tag{5.5}$$

(with u^0 near u) converge with factor $q \in (0,1)$ independent of h. (For linear problems, u^0 is arbitrary.) [29]

5.2. Convergence for linear problems. Convergence of (5.5) for linear problems is characterized by the operator $Z \equiv I - A^{-1}\tilde{L}$. We assume that $A = A^*$ is of form

$$A = \begin{bmatrix} A_1 & L_{1,2} \\ L_{2,1} & -A_2 \end{bmatrix}, \tag{5.6}$$

$$\alpha_0^{(s)} A_s \le L_{s,s} \le \alpha_1^{(s)} A_s, \ s = 1, 2, \quad \alpha_0^{(3)} A_2 \le J_2 \le \alpha_1^{(3)} A_2, \tag{5.7}$$

and $\alpha_0^{(1)} > 0, \alpha_0^{(1)} > 0, \alpha_0^{(2)} \ge 0$, and $\alpha_0^{(3)} > 0$. Note that

$$\|Z_s\|_{A_s} \le \max\{|1 - \alpha_0^{(s)}|; |1 - \alpha_1^{(s)}|\} \equiv q_s, \tag{5.8}$$

---

[29] Sometimes, A_n instead of A in (5.4) might be useful; this class of methods is wide (see [198, 199]), containing, e.g., methods (4.22) and (4.38) and those in [309, 486].

where $Z_s \equiv I_s - A_s^{-1}\tilde{L}_s$ and $\tilde{L}_s \equiv L_{s,s} + \mathcal{P}_s, s = 1, 2$. We will show that Z is a contraction operator in a special Euclidean space $H(D)$, where the block diagonal D is such that

$$Du \equiv [p_1 A_1 u_1; p_2 A_2 u_2]^T, p_1 > 0, p_2 > 0. \tag{5.9}$$

The simplest case $p_1 = p_2 = 1$ yields D_0 and is used in Lemma 1.

Lemma 1. Let L be a strongly saddle operator in H and satisfy (4.9), (4.10). Suppose $\tilde{L} = L$ (see (5.1)). Then

$$\|Z\|_{D_0} \leq \max\{q_1; q_2\}. \tag{5.10}$$

Proof. Let $z \equiv [z_1, z_2]^T, v \equiv [v_1, v_2]^T$, and $z = Zv$. Then

$$A_1 z_1 + L_{1,2} z_2 = (A_1 - L_{1,1})v_1, \text{ and } L_{2,1} z_1 + A_2 z_2 = (L_{2,2} - A_2)v_2, \tag{5.11}$$

which yields $\|z_1\|_{A_1}^2 + \|z_2\|_{A_2}^2 = (A_1 - L_{1,1})v_1, z_1 + ((A_2 - L_{2,2})v_2, z_2)$. Since

$$\|((A_s - L_{s,s})v_s, z_s)\| = \|((I_s - A_s^{-1/2} L_{s,s} A_s^{-1/2})A_s^{1/2} v_s, A_s^{1/2} z_s)\|$$

$$\leq q_s \|v_s\|_{A_s} \|z_s\|_{A_s}, \tag{5.12}$$

then

$$\sum_{s=1}^{2} \|z_s\|_{A_s}^2 \leq \sum_{s=1}^{2} q_s \|v_s\|_{A_s} \|z_s\|_{A_s}, \tag{5.13}$$

which leads directly to (5.10). □

Theorem 1. Suppose the conditions of Lemma 1 are satisfied and

$$q_1 < q_2 \leq 1. \tag{5.14}$$

Let $\lambda_1(t) \equiv q_1(1 + 2t)[2(1 - t)]^{-1}\{1 + [1 + 4t(1 - t)(1 + 2t)^{-2}]^{1/2}\}, \lambda_2(t) \equiv q_2(1 + t\sigma^2 \alpha_0^{(1)} \alpha_0^{(3)})^{-1}$ and $t_0 \in (0, (1 - q_1)(1 + q_1)^{-2})$ be the root of the equation $\lambda_1(t) = \lambda_2(t)$. Let $p_1 = (1 - t_0)\{1 + (1 + 2t_0)(1 + 8t_0)^{-1/2}\}^{-1}, p_2 = 2^{-1}(1 + \sigma^2 \alpha_0^{(1)} \alpha_0^{(3)} t_0)$, and D be defined by (5.9). Then

$$\|Z\|_D \leq \lambda_2(t_0) < 1. \tag{5.15}$$

Proof. From (5.11) and (1.8), we see that

$$\|z_2\|_{J_2} \leq 1/\sigma_0^* \left(\|A_1 z_1\|_{L_{1,1}^{-1}} + \|(A_1 - L_{1,1})v_1\|_{L_{1,1}^{-1}} \right).$$

Hence, $\|z_2\|_{J_2} \leq 1/\sigma_0^*(\alpha_0^{(1)})^{-1/2} \left(\|z_1\|_{a_1} + \|(A_1 - L_{1,1})v_1\|_{A_1^{-1}} \right)$. This implies that $\sigma^2 \alpha_0^{(1)} \alpha_0^{(3)} \|z_2\|_{A_2}^2 \leq \|z_1\|_{A_1}^2 + 2q_1 \|z_1\|_{A_1} \|v_1\|_{A_1} + q_1^2 \|v_1\|_{A_1}^2$, which, upon multiplication by $t \in (0, 1)$ and addition to (5.13), yields

$$(1-t)\|z_1\|^2_{A_1} + (1+\sigma^2\alpha_0^{(1)}\alpha_0^{(3)}t)\|z_2\|^2_{A_2}$$

$$\leq q_1(1+2t)\|z_1\|_{A_1}\|v_1\|_{A_1} + tq_1^2\|v_1\|^2_{A_1} + q_2\|z_2\|_{A_2}\|v_2\|_{A_2}. \qquad (5.16)$$

Since $\|z_s\|_{A_s}\|v_s\|_{A_s} \leq \epsilon_s/2\|z_s\|^2_{A_s} + 2/\epsilon_s\|v_s\|^2_{A_s}$, $(\forall \epsilon_s > 0,\ s = 1,2)$, then

$$[1 - t - \epsilon_1 2^{-1}q_1(1+2t)]\|z_1\|^2_{A_1} + [1 + t\sigma^2\alpha_0^{(1)}\alpha_0^{(3)} - 2^{-1}\epsilon_2 q_2]\|z_2\|^2_{A_2}$$

$$\leq [\frac{q_1(1+2t)}{2\epsilon_1} + tq_1^2]\|v_1\|^2_{A_1} + \frac{q_2}{2\epsilon_2}\|v_2\|^2_{A_2}. \qquad (5.17)$$

We choose ϵ_1 and ϵ_2 in (5.17) so that the coefficients of $\|z_1\|^2_{A_1}$ and $\|z_2\|^2_{A_2}$ on the left-hand side of (5.17) are positive and minimize maximum of the two ratios of the coefficients of $\|v_s\|^2_{A_s}$ and $\|z_s\|^2_{A_s}$ in the obtained inequalities. For $a \geq 0$, $\min_{x\in(a,b)} [(1+ax)x^{-1}(b-x)^{-1}]^{1/2}$ is just $1/b[1+(1+ab)^{1/2}] \equiv x_0^{-1}$, which is achieved when $x = x_0$, then this choice leads to

$$\epsilon_1 = 1/q_1 2(1-t)\{1 + 2t + [(1+2t)^2 + 4t(1-t)]^{1/2}\}^{-1}$$

and $\epsilon_2 = 1/q_2(1 + \sigma^2\alpha_0^{(1)}\alpha_0^{(3)}t)$. Moreover, the relevant ratios are $(\lambda_s(t))^2$, respectively, $s = 1,2$. Now $\lambda_1(t)$ and $\lambda_2(t)$ are monotonically increasing and decreasing functions, respectively, and $\lambda_1(0) = q_1$ and $\lambda_2(0) = q_2$. Hence, there exists a unique root $t_0 \in (0, (1-q_1)(1+q_1)^{-2})$ of the equation $\lambda_1(t) = \lambda_2(t)$. We then use our chosen $t, \epsilon_1, \epsilon_2$ to determine the coefficients of $\|z_1\|^2_{A_1}$ and $\|z_2\|^2_{A_2}$ on the the left-hand side of (5.17), denoting them by p_1 and p_2. Then (5.15) is a simple consequence of (5.17). □ [30]

5.3. Convergence theorems for nonlinear problems.

Theorem 2. Let L be a strongly saddle operator in H that satisfies (4.9), (4.10), and (5.7). Suppose $\tilde{L}, \mathcal{P}$, and A (see (5.6)) satisfy

$$\|Z_s(v_s) - Z_s(v_s')\|_{A_s} \leq q_s\|v_s - v_s'\|_{A_s}, \qquad \forall v_s, \forall v_s',\ s = 1,2. \qquad (5.18)$$

Let $q_1 < q_2 \leq 1$ and D and $\lambda_2(t_0)$ be defined as in Theorem 1. Then

$$\|Z(v) - Z(v')\|_D \leq \lambda_2(t_0)\|v - v'\|_D, \qquad \forall v, \forall v'. \qquad (5.19)$$

Proof. For $z = Z(v) - Z(v')$, we have $A_1 z_1 + L_{1,2} z_2 = Z(v_1) - Z(v_1')$ and $L_{2,1} z_1 - A_2 z_2 = -\|Z_2(v_2) - Z_2(v_2')$. Observe also that $|(Z_s(v_s) - $

---

[30] When $q_1 \to 0$, we obtain $t = 0$, $p_1 = 0$, and $p_2 = 1$, that is, transition to the method (4.22) is reasonable. It is also important to note that (5.14) is not necessary and convergence can be proved for $q_2 \leq 1 + \xi$ with $\xi > 0$ small enough.

$Z_s(v_s'), z_s)| \leq q_s \|v_s - v_s'\|_{A_s} \|z_s\|_{A_s}$, $s = 1, 2$. We may thus generalize (5.13) by replacing v_s in (5.13) by $v_s - v_s'$. Analogously, we modify (5.16) because $\|Z_1(v_1) - Z_1(v_1')\|_{L_{1,1}^{-1}} \leq (\alpha_0^{(1)})^{-1/2} q_1 \|v_1 - v_1'\|_{A_1}$. The remaining reasoning is much the same as that used in proving (5.15). $\square$

Theorem 2 establishes that Z is a contraction mapping, and it allows several generalizations connected with the theory in § 1.2. We confine ourselves to operators $\tilde{L}$ that are typical for the grid Navier-Stokes systems (see § 6).

Theorem 3. Let L be a strongly saddle operator in H that satisfies (4.9), (4.10). Suppose that $P_2 = 0$ and that the continuous operator P_1 satisfies the antisymmetry property (see (1.2.5)) and

$$\|P_1(v_1)\|_{L_{1,1}^{-1}} \leq \kappa_2(\|v_1\|_{L_{1,1}}), \quad \forall v_1, \tag{5.20}$$

where $\kappa_2(t)$ is a continuous nondecreasing function on $[0, \infty)$. Then problem (5.1) with $f_2 = 0$ has a solution, and each solution u is such that

$$\|u_1\|_{L_{1,1}} \leq \|f_1\|_{L_{1,1}^{-1}} \equiv r_1^0, \quad \|u_2\|_{J_2} \leq \sigma^{-1}\{2r_1^0 + \kappa_2(r_1^0)\}. \tag{5.21}$$

Proof. In accordance with Theorem 1.2.3, it suffices to prove (5.21) for the solutions $u \equiv u(\nu)$ of problems (5.1) with $\tilde{u} \equiv L + \nu P$, where $\nu \in [0, 1]$. For all v, we have $(L_{1,1}u_1, u_1) + \alpha(L_{2,2}u_2, u_2) = (f_1, u_1)$ and $\|u_2\|_{J_2} \leq \sigma^{-1}\|L_{1,1}u_1 + \nu P_1(u_1) - f_1\|_{L_{1,1}^{-1}}$, which lead to (5.21). $\square$

Theorem 4. Let the conditions of Theorem 3 and (5.6) be satisfied and yield the estimates $\|u_s\|_{A_s} \leq r_s, s = 1, 2$. Suppose that (5.18), with the functions $q_s(r_s, r_s')$ instead of q_s, holds for all v_s with $\|v_s\|_{A_s} \leq r_s$ and all v_s', where the functions $q_s(r_s, r_s')$ are nondecreasing in $r_s' \equiv \|v_s - v_s'\|_{A_s}, s = 1, 2$. Suppose finally that u^0 in (5.5) satisfies $\|u_1 - u_1^0\|_{A_1}^2 + \|u_2 - u_2^0\|_{A_2}^2 \leq r_0^2$. Suppose also that $q_s = q_s(r_s, r_s'), s = 1, 2$, and define D and $\lambda_2(t_0)$ as in Theorem 1. Then iteration (5.5) converges and

$$\|u^{n+1} - u\|_D \leq \lambda_2(t_0)\|u^n - u\|_D. \tag{5.22}$$

Proof. The proof is much the same as for Theorem 2 with $z = u - u^{n+1}$, $v = u$, $v' = u^n$. If $u^0 = 0$, $r_0^2 = r_1^2 + r_2^2$. $\square$ [31]

5.4. Convergence of iterative methods with cooperative model operators. We now pay special attention to the choice of model saddle operators such that (5.4) is easily solved by block elimination, that is, one of the systems

---

[31] Bounds (5.15) and (5.22) can be replaced by slightly different estimates based on the use of $\kappa_{1,2}$ in (4.10) (see [199]).

$$-(A_2 - L_{2,1}A_1^{-1}L_{1,2})v_2 = g_2 - L_{2,1}A_1^{-1}g_1, \quad A_1v_1 + L_{1,2}v_2 = g_1, \quad (5.23)$$

$$(A_1 + L_{1,2}A_2^{-1}L_{2,1})v_1 = g_1 + L_{1,2}A_2^{-1}g_2, \quad L_{2,1}v_1 - A_2v_2 = g_2, \quad (5.24)$$

is solved with computational work $W(A) = O(N)$, where $N = \dim H_1 + \dim H_2$. [32] This is the case for (5.23) if

$$A_1 = 1/\tau_1 B_1, \quad A_2 + L_{2,1}A_1^{-1}L_{1,2} = 1/\tau_2 B_2, \quad (5.25)$$

where τ_1 and τ_2 refer to positive iterative parameters and $B_1 \in \mathcal{L}^+(H_1)$ and $B_2 \in \mathcal{L}^+(H_1)$ are optimal preconditioners from § 4. Along the same line, for system (5.24), an appropriate choice is

$$A_2 = \frac{1}{\tau_2}B_2, \quad A_1 + L_{1,2}A_2^{-1}L_{2,1} = \frac{1}{\tau_1}B_1, \quad (5.26).$$

Lemma 2. *Let L be a strongly saddle operator in H that satisfies (4.9), (4.10). Suppose that operators $B_s \in \mathcal{L}^+(H_s)$, $s = 1, 2$, satisfy*

$$\delta_0^{(s)}B_s \leq L_{s,s} \leq \delta_1^{(s)}B_s, \quad s = 1, 2, \quad \delta_0^{(0)} > 0, \quad \delta_0^{(2)} \geq 0, \quad (5.27)$$

$$\delta_0^{(3)}B_2 \leq J_2 \leq \delta_1^{(3)}B_2, \quad \delta_0^{(3)} > 0. \quad (5.28)$$

Let (5.25) hold and $1 - \tau_1\tau_2\delta_1 > 0$, with $\delta_1 \equiv \delta_1^{(1)}\delta_1^{(2)}\kappa_{1,2}^2$. Then $A_s \in \mathcal{L}^+(H_s)$, $s = 1, 2$, and (5.7) holds with $\alpha_j^{(1)} = \tau_1\delta_j^{(1)}$, $j = 0, 1$, $\alpha_0^{(2)} = 0$, $\alpha_0^{(3)} = \tau_2\alpha_0^{(3)}$, $\alpha_1^{(3)} = \tau_2\delta_1^{(2)}(1 - \tau_1\tau_2\delta_2)^{-1}$, and $\alpha_1^{(2)} = \alpha_1^{(3)}/\delta_0^{(2)}$.

Proof. From (5.25) and (5.27), we have $\frac{1}{\tau_1\delta_1^{(1)}}L_{1,1} \leq A_1 \leq \frac{1}{\tau_1\delta_0^{(1)}}L_{1,1}$ and $A_2 \leq \frac{1}{\tau_2\delta_0^{(3)}}J_2$. Further, (5.27) and (4.11) lead to $L_{2,1}B_1^{-1}L_{1,2} \leq \delta_1^{(1)}L_{2,1}L_{1,1}^{-1}L_{1,2} \leq \delta_1^{(1)}\kappa_{1,2}^2J_2$. This implies that

$$A_2 \geq \tau_2^{-1}B_2 - \tau_1\delta_1^{(1)}\kappa_{1,2}J_2 \geq \tau_2^{(-1)}(1 - \tau_1\tau_2\delta_1)B_2,$$

which leads to the desired $\alpha_1^{(3)}$ and, consequently, to $\alpha_1^{(2)}$. □

Theorem 5. *Let $\tilde{L} = L$ in (5.1) be a strongly saddle operator such that (4.9), (5.26) are satisfied. Suppose that iterations (5.5) are applied with*

---

[32] It is easy to see that we deal with the block triangular factorizations of A mentioned in § 1.5.

A, A_1, and A_2 from (5.6) and (5.25). *Suppose also that the conditions of Lemma 2 are satisfied,* $0 < \tau_1 < 2(\delta_1^{(1)})^{-1}$, *and* $0 < \tau_2 < 2(\delta_1^{(2)} + 2\tau_1\delta_1)^{-1} \equiv \tau_2'$. *Then method (5.5) converges and (5.22) holds.*

Proof. Lemma 2 and the above assumptions made on the iteration parameters imply that $\alpha_1^{(2)} \le 2$ and $q_1 = \max\{|1 - \tau_1\delta_0^{(0)}|; |1 - \tau_1\delta_1^{(1)}|\} < 1$. We then apply Theorem 1 with $q_2 = 1$. $\square$ [33]

Theorem 6. Let (5.7), (5.27) and the conditions of Theorem 3 be satisfied with estimates $\|u_1\|_{B_1} \le a_1$, $\|u_2\|_{J_2} \le a_2$ *(see (5.21)). Suppose that* $\tilde{L}_{1,1}$ *in (5.1) is such that, for the given u and all z with $\|z\| \le t$, we have*

$$(\tilde{L}_{1,1}(u_1 + z_1) - \tilde{L}_{1,1}(u_1), z_1) \ge \delta_0\|z_1\|_{B_1}^2, \quad \delta_0 > 0, \qquad (5.29)$$

$$\|\tilde{L}_{1,1}(u_1 + z_1) - \tilde{L}_{1,1}(u_1)\|_{B_1^{-1}}^2 \le \delta_1(t)\|z_1\|_{B_1}^2. \qquad (5.30)$$

Suppose also that the conditions of Lemma 2 are satisfied and that

$$a_{1,0} \equiv [a_1 + a_2^2(2\delta_1\delta_0^{(3)})^{-1}]^{1/2}, \quad \tau \equiv \delta_0[\delta_1(a_{1,0})]^{-1},$$

$$q_1 \equiv \{1 - \delta_0^2[\delta_1(a_{1,0})]^{-1}\}^{1/2}, \quad r_0 \equiv \alpha_{1,0}\tau_1^{-1/2}, \quad \tau_2 \equiv (\delta_1^{(2)} + 2\tau_1\delta_1)^{-1}.$$

Suppose, finally, that D and $\lambda_2(t_0)$ are defined as in Theorem 1, with $q_1 < 1$ and $q_2 = 1$, and that $u^0 \in S_{D_0}(u; r_0) \equiv S$. Then method (5.5) converges and (5.22) holds.

Proof. Uniqueness of u in (5.1) follows from (5.29). The choice of τ_2 implies $\|Z_2\|_{A_2} \le 1$ (see (5.18)). The choice of τ_1, in accordance with Theorem 1.3.1, implies that, for all z_1 with $\|z_1\|_{B_1} \le a_{1,0}$, we have

$$\|(I - \tau_1 B_1^{-1}\tilde{L}_{1,1})(u_1 + z_1) - (I - \tau_1 B_1^{-1}\tilde{L}_{1,1})(u_1)\|_{B_1} \le q_1\|z_1\|_{B_1}, \quad (5.31)$$

$$\|Z_1(u_1 + z_1) - Z_1(u_1)\|_{A_1} \le q_1\|z_1\|_{A_1}. \qquad (5.32)$$

---

[33] With regard to further specification of the iteration parameters it seems that the simple choice $\tau_1 = 2(\delta_0^{(1)}) + \delta_1^{(1)})^{-1}$ is reasonable since it is based on minimization of $\|I_1 - \tau_1 B_1^{-1}L_{1,1}\|_{B_1}$ and the respective modified method of the simple iteration for system $\tilde{L}_{1,1}(u_1) = g_1$, where $\tilde{L}_{1,1} = L_{1,1}$. Then it is natural to use $\tau_2 = 1/2\tau_2'$ or even to determine $\tau_2 \in ((\tau_1\delta_1 + \delta_1^{(2)})^{-1}, (\tau_1\delta_0 + \delta_0^{(2)})^{-1})$ by the condition $1 - \tau_2\delta_1^{(2)}(1 - \tau_1\tau_2\delta_0)^{-1} = \tau_2\delta_1^{(2)}(1 - \tau_1\tau_2\delta_1)^{-1}$, equivalent to a quadratic equation. (These parameters can be further improved by numerical minimization of $\lambda_2(t_0)$ as a function of admissible τ_1 and τ_2.) We give here an analogous choice for nonlinear problems when $\tilde{L}_{1,1} \equiv L_{1,1} + \mathcal{P}_1$.

Therefore, (5.32) holds for all $u + z \in S$ and $\|Z(u+z) - Z(u)\|_{D_0} \leq \|z\|_{D_0}$. Thus, for all iterates in (5.5), $z_1^n = u_1^n - u_1 \subset S_{B_1}(u_1; a_{1,0})$ and (5.31) holds with $z_1 = z_1^n$. We observe now that our choice of $a_{1,0}$ is determined by the fact that, for $u_0 = 0$, we have

$$\|z^0\|_{D_0}^2 \leq \frac{a_1^2}{\tau_1} + \frac{a_2^2}{\tau_2 \delta_0^{(3)}}, \quad \|z_1^0\|_{B_1} \leq \{a_1 + \frac{\tau_1 a_2^2}{\tau_2 \delta_0^{(3)}}\}^{1/2} \leq a_{1,0}, \text{ and } \|z^0\|_{D_0} \leq r_0.$$

Hence, the reasoning in the proof of Theorem 2 leads also to (5.22) for our nonlinear problem. $\square$

5.5. Possible generalizations. Clearly, iterations considered above can be applied to various types of grid methods, including, e.g., nonconforming finite element and difference methods, provided fundamental estimate (3.18) holds. The study of methods (5.5) with model saddle operators A defined by (5.6) and (5.26) is much the same. We have paid special attention to iterations (5.5), with A defined by (5.6) and (5.25), because they contain concrete iterative methods (see [309, 486]) that were applied for solving some practical problems in hydrodynamics and elasticity (for relatively simple regions, difference discretizations, special B_1 and B_2). However, they were devised heuristically on the basis of an analogy with certain implicit difference schemes. Moreover, their convergence analysis was carried out in a normed linear space with the norm dependent on iterations parameters which prevented parameter optimization.

It is very important that our results can be extended to certain more general nonlinear problems (5.1)–(5.3) that contain small additional perturbations of type

$$\tilde{P}(u) \equiv [P_1(u_1; u_2), P_2(u_1; u_2)]^T, \quad \|\tilde{P}(u+z) - \tilde{P}(u)\|_{J^{-1}} \leq \gamma^* \|z\|_J, \quad (5.33)$$

where $\|u\|_J \leq a^*$ and γ^* and a^* are small enough and z is arbitrary (see sited87). Moreover, the respective contractions operators can be useful in proving existence of solutions of the given nonlinear grid problems. Our analysis can be generalized to the case when H_1 and H_2 are Hilbert spaces. It then provides a means even of proving existence theorems for the respective differential problems.

§ 6. Asymptotically optimal algorithms for nonlinear Navier-Stokes systems

Here we extend asymptotically optimal algorithms obtained in § 4 to the important class of elliptic problems associated with two and three-dimensional Navier-Stokes systems in general regions. We impose assumptions on the parameters that are typical for mathematical analysis of the correctness of such problems (see [242, 244, 261, 323, 475]). Certain results,

e.g., related to construction of effective iterative methods, are obtained under more general conditions.

   6.1. Projective-grid (mixed finite element) methods. Instead
of linear problem (1.13) associated with (1.30) and (4.1) and $\alpha = 0$, we
consider its nonlinear perturbation. To be specific, on $(G_1)^3$ (recall that
G_1 is defined by (1.30)) we introduce the trilinear forms

$$\tilde{t}(u_1; w_1; v_1) \equiv \sum_{s=1}^{d} (D_s(u_{1,s}w_1), v_1)_{0,\Omega} \tag{6.1}$$

and

$$2t(u_1; w_1; v_1) \equiv \tilde{t}(u_1; w_1; v_1) - \tilde{t}(u_1; v_1 : w_1) =$$

$$\sum_{s,r=1}^{d} (u_{1,s}, v_{1,r} D_s w_{1,r} - w_{1,r} D_s v_{1,r})_{0,\Omega}, \tag{6.2}$$

where $u_1 \equiv \vec{u}_1 \in G_1, v_1 \equiv \vec{v}_1 \in G_1$, and $w_1 \equiv \vec{W}_1 \in G_1$. It is well known
(see [323, 475] and Lemma 5.5.7) that these forms are continuous and that
there exist constants $K^{(1)}$ and $K^{(2)}$ such that

$$|t(u_1; w_1; v_1)| \leq K^{(1)} \|u_1\| \|v_1\| \|w_1\|, \tag{6.3}$$

$$t(u_1; w_1; v_1) = -t(u_1; v_1; w_1), \quad t(w_1; u_1; u_1) = 0, \tag{6.4}$$

$$|t(v_1; v_1; w_1) - t(u_1; u_1; w_1)| \leq K^{(2)} \max\{\|u_1\|; \|v_1\|\} \|v_1 - u_1\| \|w_1\|, \tag{6.5}$$

where u_1, v_1, w_1 are arbitrary elements of G_1. Moreover, if div $u_1 = 0$, then
$t(u_1; u_1; v_1) = \sum_{s,r=1}^{d}(u_{1,s} D_s u_{1,r})_{0,\Omega}$, which corresponds to the common
way of writing the nonlinear terms in the Navier-Stokes system

$$-\nu\Delta u_1 + \sum_{s=1}^{d} u_{1,s} D_s u_1 - \text{grad } u_2 = \tilde{f}_1, \quad \text{div } u_1 = 0.$$

   As in § 4, we formulate this and other more general problems in the
modern setting: we seek $u \equiv [u_1, u_2]^T \in G = G_1 \times G_2'$ such that

$$\left.\begin{array}{c} \tilde{b}_1(u_1; v_1) + b_{2,1}(v_1; u_2) = l_1(v_1), \quad \forall v_1 \in G_1, \\ b_{2,1}(u_1; v_2) = 0, \quad \forall v_2 \in \hat{G}_2', \\ (u_2, 1)_{0,\Omega} = 0, \end{array}\right\} \tag{6.6}$$

where $\tilde{b}_1(u_1; v_1) \equiv b_{1,1}(u_1; v_1) + t(u_1; u_1; v_1)$. We also use the equivalent operator formulation

$$Lu \equiv \left[\begin{array}{c} \tilde{L}_{1,1}(u_1) + L_{1,2}u_2 \\ L_{2,1}u_1 \end{array} \right] = \left[\begin{array}{c} f_1 \\ 0 \end{array} \right],$$

under the condition t $(u_2, 1)_{0,\Omega} = 0$; here, $\tilde{L}_{1,1}(u_1) \equiv L_{1,1}u_1 + \mathcal{P}(u_1)$, where the linear operators $L_{1,1}$, $L_{1,2}$, and $L_{2,1}$ are the same as in § 4 and the non-linear operator $\mathcal{P}_\infty$ is defined by the relation $(\mathcal{P}u_1, v_1) = t(u_1; u_1; v_1)$, $\forall u_1 \in G_1, \forall v_1 \in G_1$ (see (6.2)).

Our PGM for (6.6) is a simple generalization of the PGM described in § 4 (which is associated with the subspaces $\hat{G} \equiv \hat{G}_1 \times \hat{G}'_2$): we seek $\hat{u} \equiv [\hat{u}_1, \hat{u}_2]^T \in \hat{G}'_2$ such that

$$\left. \begin{array}{r} \tilde{b}_1(\hat{u}_1; \hat{v}_1) + b_{2,1}(\hat{v}_1; \hat{u}_2) = l_1(\hat{v}_1), \ \forall \hat{v}_1 \in \hat{G}_1, \\ b_{2,1}(\hat{u}_1; \hat{v}_2) = 0, \ \forall \hat{v}_2 \in \hat{G}'_2, \\ (\hat{u}_2, 1)_{0,\Omega} = 0. \end{array} \right\}$$

We also use the equivalent operator formulation in the Euclidean space $H = H_1 \times H'_2$

$$L_h \mathbf{u} \equiv L\mathbf{u} \equiv \left[\begin{array}{c} \tilde{L}_{1,1}(\mathbf{u}_1) + L_{1,2}\mathbf{u}_2 \\ L_{2,1}\mathbf{u}_1 \end{array} \right] = \left[\begin{array}{c} \mathbf{f}_1 \\ 0 \end{array} \right], \tag{6.7}$$

under the condition

$$(\mathbf{u}_2, \bar{1}_2)_{J_2} = 0. \tag{6.8}$$

Here we use the same symbols both for original differential operators and their grid approximations. The linear operators $L_{1,1}$, $L_{1,2}$, and $L_{2,1}$ are the same as in § 4 (see (4.7), (4.8)), and the nonlinear operator $\mathcal{P}_\infty \in \mathcal{L}(H_1)$ is defined by the relation $(\mathcal{P}\mathbf{u}_1, \mathbf{v}_1) = t(\hat{u}_1; \hat{u}_1; \hat{v}_1)$, for all $\mathbf{u}_1$ and $\mathbf{v}_1$ in H_1 (see (6.2)). For this grid operator $\mathcal{P}_\infty$ and arbitrary $\mathbf{u}_1, \mathbf{v}_1$, and $\mathbf{u}_1$, properties (6.3)–(6.5) and the common relations from § 0.4 imply that

$$\left. \begin{array}{r} |(\mathcal{P}\mathbf{u}_1; \mathbf{z}_1)| \leq K^{(1)} \|\mathbf{u}_1\|_{J_1}^2, \\ (\mathcal{P}\mathbf{u}_1, \mathbf{u}_1) = 0, \\ |(\mathcal{P}\mathbf{v}_1, \mathbf{z}_1) - ((\mathcal{P}\mathbf{u}_1, \mathbf{z}_1)| \leq K^{(2)} \|\mathbf{u}_1\|_{J_1} \|\mathbf{z}_1\|_{J_1} \|\mathbf{v}_1 - \mathbf{u}_1\|_{J_1}. \end{array} \right\} \tag{6.9}$$

Properties (6.3)–(6.5) and Theorem 3.1 imply that, for each solution of the differential problem, we have the a priori estimates

$$\|\bar{u}_1\| \leq \|l_1\|/\gamma_0 \equiv a_1, \quad \|u_2\|_{J_2} = \|\hat{u}_2\| \leq a_2(\sigma_0^*)^{-2}, \tag{6.10}$$

where $a_2 \equiv a_2(a_1, K^{(1)}, \gamma_1)$ (this follows by a slightly modified proof of Lemma 1.5). This implies (see [244, 323] and Theorem 3.1) that at least one solution of our problem exists, which is unique when

$$\gamma_0 - K^{(2)} a_1 \equiv \delta_0^{(1)} > 0. \qquad (6.11)$$

This follows from the strong monotonicity inequality

$$\tilde{b}_1(u_1 + z_1; \hat{z}_1) - \tilde{b}_1(u_1; \hat{v}_z) \geq \delta_0^{(1)} \|z_1\|^2, \quad \forall z_1 \in \hat{G}_1, \qquad (6.12)$$

which was discussed in Subsection 5.5. Along the same lines, (6.9), (4.9), and Theorem 3.5 imply that, for each solution of (6.7), (6.8), we have the priori estimate

$$\|\hat{u}_1\| = \|u_1\|_{J_1} \leq a_1, \|\hat{u}_2\| \leq (\sigma_0^*)^{-2} a_2 \equiv a_3. \qquad (6.13)$$

Moreover, (6.12) holds with u_1 replaced by $\hat{u}_1$.

Theorem 1. Let conditions (1.14), (1.15), (6.11), *and of Theorem 3.5 be satisfied. Then error estimate* (2.14) *holds.*

Proof. We indicate only the necessary alterations in the proof of Theorem 2.2. Instead of (2.9), we now have $\tilde{b}_1(\hat{u}_1; \hat{v}_1) - \tilde{b}_1(u_1; \hat{v}_1) = -b_{2,1}(\hat{v}_1; z_2)$, $\forall \hat{v}_1 \in \hat{G}_1$. Hence (see (2.13) and (6.12)),

$$\delta_0^{(1)} \|z_1\|^2 \leq (\gamma_1 + K^{(1)} a_1) \|z_1\| \|\hat{v}_1 - u_1\| + \kappa_1 \{\|z_1\| \|\hat{v}_2 - u_2\| + \|\hat{v}_1 - u_1\| \|z_2\|\}$$

and $\|z_2\| \leq \|\hat{v}_2 - u_2\| + \frac{1}{\sigma_0^*} \{\kappa_1 \|\hat{v}_2 - u_2\| + (\gamma_1 + K^{(1)} a_1) \|z_1\|\}$ (see 2.11). For obtaining inequalities of type (2.6) and (2.7), we apply the same reasoning but with $\delta_0^{(1)}$ and $\gamma_1 + K^{(1)} a_1$ instead of γ_0 and γ_1, respectively. □ [34]

6.2. Properties of linearized operators. It is easy to verify that the Fréchet derivative of the differential operator $\mathcal{P}_1$ at a point w_1 is defined by the bilinear form $t_{w_1}(u_1; v_1) = t(w_1; u_1; v_1) + t(u_1; w_1; v_1)$. Analogously, the derivative of the grid operators $\mathcal{P}_\infty$ and $\tilde{L}_{1,1}$ in (6.7) at a point $\mathbf{w}_1 \equiv w_1 \in H_1$ are such that

$$(\mathcal{P}'_{1,w_1} u_1, v_1) = t_{\hat{w}_1}(\hat{u}_1; \hat{v}_1), \quad \tilde{L}'_{1,1,w_1} = L_{1,1} + \mathcal{P}'_{1,w_1}. \qquad (6.14)$$

In what follows, $\tilde{L}'_{1,1,w_1} \equiv M_1$,

$$L'_w \equiv \begin{bmatrix} M_1 & L_{1,2} \\ L_{2,1} & 0 \end{bmatrix}, \qquad (6.15)$$

---

[34] Theorem 1 can be generalized on the basis of Theorem 2.2 to problems with condition (1.33). Relevant results dealing with error estimates can be found in [242, 249, 293, 476].

and all constants are independent of h. We emphasize that

$$(M_1 z_1, z_1) \geq (\gamma_0 - K^{(2)}\|\hat{w}_1\|)\|\hat{z}_1\|^2 \geq \delta_0^{(1)}\|z_1\|_{J_1}^2, \qquad (6.16)$$

provided

$$\|\hat{w}_1\| \leq a_1 \qquad (6.17)$$

(see (4.9) and (6.11)). Hence,

$$\mathrm{Ker}\, L'_w = \mathrm{Ker}\, (L'_w)^* = \mathrm{Ker}\, L^0, \qquad (6.18)$$

where $L^{(0)}$ coincides with L in (4.7)

Lemma 1. *Suppose that (6.17) holds. Then there exists $\kappa^{(2)}$ such that*

$$\|\mathcal{P}'_{1,w_1}\|_{H_1(J_1) \mapsto H_1(J_1^{-1})} \leq \kappa^{(2)}. \qquad (6.19)$$

Proof. It suffices to combine the proof of (0.4.8) and the fact that

$$|(\mathcal{P}'_{1,w_1} u_1, v_1)| \leq 2K^{(1)}\|\hat{w}_1\|\|\hat{u}_1\|\|\hat{v}_1\| \leq \kappa^{(2)}\|u_1\|_{J_1}\|v_1\|_{J_1}.$$

Lemma 2. *Suppose that (6.11), (6.17) hold and that $\mathcal{R}_2 \equiv L_{2,1} M_1^{-1} L_{2,1}$. Let $\delta \equiv (\sigma_0^*)^2 \delta_0^{(1)}(\gamma_1 + \kappa^{(2)})^{-2}$ and W_2 be the subspace of H'_2 defined by (4.8). Then*

$$(\mathcal{R}_2 u_2, u_2) \geq \delta\|u_2\|_{J_2}^2, \quad \forall u_2 \in W_2. \qquad (6.20)$$

Proof. (6.17) yields (6.16). Thus, Lemma 1.4 with $A = J^{-1/2} M_1 J_1^{-1/2}$ and $\alpha_0 = \delta_0^{(1)}$ applies. Since

$$\|A\| \leq \|J_1^{-1/2} L_{1,1} J_1^{-1/2}\| + \|J_1^{-1/2} \mathcal{P}'_{1,w_1} J_1^{-1/2}\| \leq \gamma_1 + \kappa^{(2)},$$

then $(A^{-1}v_1, v_1) \geq \delta_{0,0}\|v_1\|^2$ and $(M_1^{-1}v_1, v_1) \geq \delta_{0,0}\|v_1\|_{J_1^{-1}}^2$, where $\delta_{0,0} \equiv \delta_0^{(1)}(\gamma_1 + \kappa^{(2)})^{-2}$. Hence, $(\mathcal{R}_2 u_2, u_2) \geq \delta_{0,0}\|L_{1,2} u_2\|_{J_1^{-1}}^2$. Note that $\mathcal{R}_2$ here is the analog of R_2 in Lemma 4.2. Thus, the right-hand side of the above inequality, for $u_2 \in W_2$, can be estimated from below as in (4.15) with $\gamma_0 = \gamma_1 = 1$. $\square$

Lemma 3. *Suppose that the conditions of Lemma 2 are satisfied. Then, for each solution of the system $L'_w u = f$ with $f \equiv [f_1, f_2]^T, f_2 \in W_2$ and $u_2 \in W_2$, we have the a priori estimate*

$$\|u\|_J \leq \kappa'_2 \|L'_w u\|_{J^{-1}}. \qquad (6.21)$$

Proof. The proof is a generalization of the proof of Lemma 4.3 (see (4.16)). We start by considering

$$\mathcal{R}_2 u_2 = f_2 - L_{2,1} M_1^{-1} f_1 \in W_2, \tag{6.22}$$

which follows from our system. Note that $\|J_1^{1/2} M_1^{-1} J_1^{1/2}\| \leq 1/\delta_0^{(1)}$ and $\|L_{2,1} M_1^{-1} f_1\|_{J_2^{-1}} \leq \kappa_1 (\delta_0^{(1)})^{-1} \|f_1\|_{J_1^{-1}}$ (see (6.16) and (4.10)). Hence, $\|u_2\|_{J_2} \leq 1/\delta \{\|f_2\|_{J_2^{-1}} + \kappa_1/\delta_0^{(1)} \|f\|_{J_1^{-1}}\}$. Since $\|J_1^{-1/2} L_{1,2} J_2^{-1/2}\| = \|J_1^{-1/2} L_{2,1} J_2^{-1/2}\| \leq \kappa_1$ (see (4.10)), then we have $\delta_0^{(1)} \|u_1\|_{J_1}^2 \leq (f_1, u_1) + |(u_1, L_{1,2} u_2)| \leq \{\|f_1\|_{J_1^{-1}} + \kappa_1 \|u_2\|_{J_2}\} \|u_1\|_{J_1}$, which yields (6.21). $\square$

The model operator $B \asymp J$ is the same as in Theorem 4.1, but now

$$\gamma_{0,1} B_1 \leq J_1 \leq \gamma_{1,1} B_1, \quad \gamma_{0,1} > 0. \tag{6.23}$$

Theorem 2. *Suppose that the conditions of Lemma 2 are satisfied and that the operator L_w' is denoted by L^0. Then the assertion of Theorem 4.1 is valid.*

Proof. It suffices in the proof of Theorem 4.1 to make use of (6.21) instead of (4.16). $\square$

We emphasize that, for $A \equiv B^{-1}(L^0)^* B^{-1} L^0$, we have Ker $A =$ Ker L^0 and Im $A = V$ (see Theorem 4.1). Hence,

$$\text{sp } A \subset \{0 \cup [\delta_0, \delta_1]\} \tag{6.24}$$

(see (4.18) and Lemma 1.3.2), which is of fundamental importance for construction of effective iterative methods. It is noteworthy that (6.17) and (6.11) were necessary for (6.24) and that (6.11) is rather restrictive. We can obtain better results by considering another linearization for $\tilde{L}_{1,1}$. This is obtained by using $M_1' \equiv M_{1,w_1}'$ instead of M_1, where

$$(M_1' u_1, v_1) = (L_{1,1} u_1, v_1) + t(\hat{w}_1; \hat{u}_1; \hat{v}_1), \quad \forall u_1 \in H_1, \forall v_1 \in H_1. \tag{6.25}$$

It is important to note that $M_{1,w_1}' w_1 = \tilde{L}_{1,1}(w_1)$ and that

$$(M_1' - L_{1,1})^* = -(M_1' - L_{1,1}), \quad (M_1 z_1, z_1) \geq \gamma_0 \|z_1\|_{J_1}^2, \forall z_1 \in H_1.$$

Moreover, for $\mathcal{R}_2' \equiv L_{2,1}(M_1')^{-1} L_{1,2}$, we obtain

$$(\mathcal{R}_2' u_2, u_2) \geq (\sigma_0^*)^2 \gamma_0 (\gamma_1 + K^{(1)} a_{1,1})^{-2} \|u_2\|_{J_2}^2, \quad \forall u_2 \in W_2, \tag{6.26}$$

provided $\|\hat{w}_1\| \leq a_{1,1}$.

Theorem 3. Let $\|\hat{w}_1\| \leq a_{1,1}$ and

$$L^0 \equiv L_w^0 \equiv \begin{bmatrix} M'_{1,w_1} & L_{1,2} \\ L_{2,1} & 0 \end{bmatrix}. \qquad (6.27)$$

Then the assertion of Theorem 4.1 is valid and (6.24) holds with positive constants $\delta_0 \equiv \delta_0(a_{1,1})$ and $\delta_1 \equiv \delta_1(a_{1,1})$.

Proof. It suffices to make use of (6.26) instead of (4.16) in the proof of Theorem 4.1 . □

Finally, we remark that

$$\|L_w^0 - L_v^0\|_{H(J) \mapsto H(1/J)} \leq \kappa^{(2)} \|w_1 - v_1\|_{J_1}, \qquad (6.28)$$

which follows from (6.19) for both linearizations; but, for (6.27), it is possible to take $\kappa^{(2)} = K^{(1)}$.

6.3. Effective iterative methods and minimization of computational work. Now we consider certain iterative methods based on a linearization of L in (6.7) (see § 3.8) and an approximate solving, via inner iterations of type (4.21), of the respective linearized systems. We make use of the spaces H_1, H_2', V and model operators B (see (6.23)) from Theorem 4.1. Our nonlinear system (6.7) is supplemented now by the condition $u_2 \in V_2$ instead of (6.8) as in Theorem 4.1. Then $\|u_1\|_{J_1} \leq a_1$ (see (6.13)). We obtain our new iterate u^{n+1} recursively assuming that $u^n \in V$ with $\|u_1^n\|_{J_1} \leq a_{1,1}$ and considering the system

$$L^0 \tilde{v} = f - L(u^n) \equiv g, \quad \tilde{v} \in V \qquad (6.29)$$

(with $L^0 \equiv L_{u^n}^0$ and defined by (6.27)). Inner iterations are defined by

$$B(v^{m+1} - v^m) = -\tau_m (L^0)^* B^{-1}(L^0 v^m - g), \quad m = 0, 1, \ldots, k_n - 1, \quad (6.30)$$

with $v^0 = 0 \in V$. Here, $g \perp \mathrm{Ker}\,(L^0)^* = \mathrm{Ker}\,L^0$ and (6.29) has a unique solution. Moreover, $V = \mathrm{Im}\,A$ and the constants in (6.24) are determined in accordance with Theorem 3 as functions of $a_{1,1}$. Hence, V is an invariant space of the error reduction operator

$$Z_n \equiv (I - \tau_0 A) \cdots (I - \tau_{k_n - 1} A) \qquad (6.31)$$

and, for its restriction on V, estimate (3.8.7) with $q_n < 1$ can be obtained. For $u^{n+1} - u^n \equiv v^{k_n}$, we obtain the known relation

$$L^0 (I - Z_n)^{-1}(u^{n+1} - u^n) = f - L(u^n), \quad u^{n+1} \in V, \qquad (6.32)$$

(see (3.8.6)) for our two-stage iterative method, which is equivalent to

$$L^0 z^{n+1} = g_0 - g_1, \quad z^{n+1} \in V, \tag{6.33}$$

where $g_0 \equiv L(u) - L(u^n) - L^0 u$, $g_1 \equiv L^0[(I - Z_n)^{-1} - I](u^{n+1} - u^n)$ (see (3.8.11)), and $z^{n+1} \equiv u^{n+1} - u$.

Lemma 4. *Suppose that (6.10) holds, that w in (6.27) is such that $w = u^n \in V$, that $\|u^n - u\|_B \leq r_0$, and that (6.24) holds with constants $\delta_0 \equiv \delta_0(a_{1,1}) > 0$ and $\delta_1 \equiv \delta_1(a_{1,1}) > 0$, where $a_{1,1} \equiv a_1 + r_0 \gamma_{1,1}^{1/2}$. Let Z_n be defined by (6.31) with these δ_0 and δ_1. Then, for g_0 and g_1 from (6.33), we have*

$$\|g_0\|_{B^{-1}} \leq K^{(3)} \|z_1^n\|_{B_1}, \quad \|g_1\|_{B^{-1}} \leq K^{(4)} \|u^{n+1} - u^n\|_B, \tag{6.34}$$

where $K^{(3)} \equiv a_1 K^{(1)} \gamma_{1,1}^{1/2} \gamma_{0,1}^{-1/2}$ *and* $K^{(4)} \equiv \delta_1^{1/2} q_n (1 - q_n)^{-1}$.

Proof. Observe that

$$(g_0, v) = (\mathcal{P}_1(u_1) - \mathcal{P}_1(u_1^n) + (L_{1,1} - M'_{1,u_1^n}) z_1^n, v_1), \quad \forall v_1 \in H_1.$$

Hence, $(g_0, v) = t(\hat{u}_1; \hat{u}_1; \hat{v}_1) - t(\hat{u}_1^n; \hat{u}_1^n; \hat{v}_1) + t(\hat{u}_1^n; \hat{z}_1^n; \hat{v}_1)$ (see (6.14)) and $(g_0, v) = -t(\hat{z}_1^n; \hat{u}_1; \hat{v}_1)$. Thus, (6.3) yields

$$|(g_0, v)| \leq K^{(1)} \|\hat{z}_1^n\| \|\hat{u}_1\| \|\hat{v}_1\| = K^{(1)} \|z_1^n\|_{J_1} a_1 \|v_1\|_{J_1}.$$

This and Lemma 0.4.5 lead to $\|g_0\|_{J_1^-} \leq K^{(1)} a_1 \|z_1^n\|_{J_1}$, from which the desired estimate for g_0 is a straightforward consequence (see (6.23)). The estimate for g_1 was already proved in Theorem 3.8.3. $\square$

Lemma 5. *Let the conditions of Lemma 4 be satisfied. Suppose also that a_1, q_n, and r_0 are small enough so that*

$$K^{(4)} \delta_0^{-1/2} \leq \bar{q} < 1, \quad [\delta_0^{-1/2} K^{(3)} + \bar{q}](1 - \bar{q})^{-1} \leq \rho < 1. \tag{6.35}$$

Then one iteration (6.32) leads to estimates

$$\|z^{n+1}\|_B \leq \rho \|z^n\|_B \leq r_0 \quad \|u_1^{n+1}\|_{J_1} \leq a_{1,1}. \tag{6.36}$$

Proof. From (6.33), (6.34), and (6.25) we see that

$$\|z^{n+1}\|_B \leq \delta_0^{-1/2} [K^{(3)} \|z^n\|_B + K^{(4)}(\|z^{n+1}\|_B + \|z^n\|_B)].$$

This implies that the first inequality in (6.36) holds, which yields the second because $\|u_1^{n+1}\|_{J_1} \leq \|u_1\|_{J_1} + \|z^{n+1}\|_{J_1}$. $\square$

Theorem 4. *Let*

$$u^0 \in V, \quad \|u^0 - u\|_B \leq r_0. \tag{6.37}$$

Suppose that (6.35) *holds and that all* $u^{n+1}, n = 0, \ldots,$ *are defined recursively by* (6.32). *Then two-stage iterative method* (6.32) *converges and*

$$\|z^n\|_B \leq \rho^n r_0, \quad \rho < 1. \tag{6.38}$$

Proof. Condition (6.37) and Lemma 5 imply that all iterates belong to V and satisfy (6.36), which leads directly to (6.38). □ [35]

Theorem 5. Let, for boundary value problem (6.6), *conditions* (1.24), (6.11) *be satisfied. Suppose that its solution satisfies* (4.6) *and that the PGM, leading to systems* (6.7), (6.8), *makes use of the same family of subspaces* $\hat{G} \subset G$ *as for the respective linear problem in* § 4. *Suppose also that B is an asymptotically optimal preconditioner. Then iterative method* (6.32) *yields the desired ε-approximation to the solution of the original problem with computational work* $W(\varepsilon) = O(\varepsilon^{-d/\gamma} |\ln \varepsilon|)$; *application of the multigrid acceleration to this basic iterative algorithm yields the desired ε-approximation with computational work* $W(\varepsilon) = O(\varepsilon^{-d/\gamma})$.

Proof. Theorem 1 and (4.6) yield error estimate (4.5), so $\varepsilon \asymp h^\gamma$. Theorem 4 guarantees effectiveness of iterative method (6.32) (with an asymptotically optimal preconditioner B). Thus, the remaining proof of the first assertion is the same as in § 4. More cumbersome is the proof of the second assertion, but those difficulties are analogous to those for elliptic problems with bounded power nonlinearity (see § 5.5). □

6.4. Other effective algorithms and possible generalizations. We outline briefly other possible asymptotically optimal algorithms. It is easy to see that the case of $\alpha L_{2,2}$ instead of 0 in (6.7) requires a simple combination of the above analysis of our nonlinear problem with $\alpha = 0$ and respective constructions and proofs for the linear problem in Theorem 4.1.

For iterative algorithms based on the Newton-Kantorovich and two-stage methods, the study (in the case of (6.11)) can be carried out on the basis of Theorem 3.8.4. However, it is important that we can now prove (6.24), with $L^0 \equiv L'_w$, only for small enough $a_{1,1}$. This results from the lack of antisymmetry and the fact that (6.26) does not hold. But, in Lemma 4, we can obtain a better estimate for $\|g_0\|_{B^{-1}}$ because now $\|g_0\|_{B^{-1}} \leq K^{(5)} \|z_1^n\|_{B_1}^2$, (see (1.2.19) with $\alpha = 1$). Such algorithms should perhaps be applied in combination with several initial iterations (6.29). Also important is to note that it is possible to obtain respective results without assumption (6.11), provided we carry out our iterations (6.32) with $u^0 \in V \cap S_B(u; r_0)$, where u is a solution of (6.7), (6.8) and, for $L^0 \equiv L'_v$ with

---

[35] Our analysis applies provided problem (6.7), (6.8) has a solution and (6.24) holds. Therefore, methods like (6.29) might be useful for situations when (6.11) does not hold but for (6.11) locally we have a proof of (6.24) (see Theorem 3). Note also that the respective adaptation of the constants in (6.24) (in inner iterations (6.30)) might be used (see § 1.3).

$v \in S_B(u; r_0)$, (6.24) holds (r_0 must be small enough). Such assumptions are natural if (6.24) holds for $L^0 \equiv L'_u$.

We can also apply iterations (1.3.12) as an analog of (4.21) (Theorem 1.3.11 can be generalized to our case because $V = \text{Im } B^{-1}(L'_{u^n})^*$ and the conditions of this theorem are satisfied for all $z' \in V$ (see (6.21) and (6.27)), provided $\|\hat{u}_1^n\| \leq a_1$ (see (6.17)). But again we have to assume that r_0 is small enough. Iterations from § 5 can be applied to our nonlinear problem as well.

Some generalizations are quite transparent for certain nonlinear problems of thermohydrodynamics arising in meteorology, when the incompressibility condition is preserved. For example, the boundary value problem considered in [194] exactly corresponds to problems for which our general Theorem 3 applies (with $u_1 \equiv [\vec{u}, t]^T$, where t is related to the temperature and, in the original differential problem, $t \in W_2^1(\Omega; \Gamma_{0,t})$).

It is very important that our results are extendable to more general nonlinear problems (5.1)–(5.3) indicated at the end of § 4 (see (5.33)). It therefore stands to reason that algorithms analogous to (5.5) and (6.32) can be applied for more involved hydrodynamics problems related to viscous compressible fluid (see [230, 244]). However, the very important side of all of the above problems, determined by the smallness of $\nu = 1/Re$, remains outside our analysis. Analogous methods might be useful in solving some magnetohydrodynamics problems (see [262]). We mention also that generalizations of (4.38) are possible (if $\alpha > 0$) because (6.12) clearly leads to

$$(\tilde{L}_{1,1}(\mathbf{u}_1 + \mathbf{z}_1), \mathbf{z}_1) - (\tilde{L}_{1,1}(\mathbf{u}_1), \mathbf{z}_1)) \geq \delta_0^{(1)} \|\mathbf{z}_1\|_{J_1}^2, \quad \forall \mathbf{z}_1 \in H_1,$$

and an analogous property for $\tilde{S}_1 \equiv S_1 + \mathcal{P}_1$. Then the conditions of Theorem 1.3.1 are satisfied with $\delta_1 = O(\alpha)$. Better convergence is obtained for $B_1 = S_1$, but each such iteration becomes rather expensive (the computational work is $O(h^{-3})$ for $d = 2$). The same comment applies to the methods based on minimization of $\Phi(v_1) \equiv \|\tilde{L}_1 v_1 - f_1\|_{S_1^{-1}}^2$, which were used for the solution of important applied problems (see [242, 244, 411]). Recent results related to multigrid methods can be found in [368].

Chapter 8

Asymptotically optimal algorithms for fourth-order elliptic problems

Fourth-order elliptic boundary value problems can be reduced to operator equations in Hilbert spaces G that are certain subspaces of the Sobolev space $W_2^2(\Omega) \equiv G^{(2)}$. Construction of asymptotically optimal grid approximations and, most particularly, asymptotically optimal algorithms are very difficult now because the associated spline subspaces are not of Lagrangian type (for relevant results, see [18, 117, 399, 470]). These difficulties evoked a series of attempts to reduce such problems to second-order differential equations (see [45, 81, 92, 117, 244, 443]), but with no essential progress in the construction of asymptotically optimal algorithms. Appearance of asymptotically optimal algorithms for Stokes type problems made it natural to focus on an approach that considers rot $w \equiv [D_2 w, -D_1 w] \equiv \vec{u}$ (for $d = 2$) as a new unknown vector-function, which automatically satisfies the condition div $\vec{u} = 0$ (see [192, 308]).[1]

---

[1] We will show in § 2 and 3 that this and the results of Chapter 7 enable one to obtain asymptotically optimal algorithms for finding ε-approximations in $W_2^1(\Omega)$ and $L_2(\Omega)$, respectively, to the first and second derivatives of the solution of the original problem $w \in W_2^{1+\gamma}(\Omega)$, where $0 < \gamma \leq 1$ and Ω is a simply connected domain (precisely these approximations are needed in many applied problems). Moreover, these results are generalized to multiconnected domains (see [196, 197]) where it is necessary to carry out all investigations on the kernels of certain bounded functionals. This means, in terms

§ 1. Boundary value problems in a rectangle

1.1. Projective-grid method for the first boundary problem in a rectangle. In this section, we consider a rectangle $\bar{Q} \equiv [0, l_1] \times [0, l_2]$ and variable $x \equiv [x_1, x_2] \in Q$. As an illustration, we consider the variational problem of finding $w = \arg \min_{w \in W} \Phi(w)$, where the space of admissible functions is $W \equiv \overset{o}{W_2^2}(Q)$ and the energy functional is defined by $\Phi(w) \equiv I_2(w) - 2l(w)$, where

$$I_2(w) \equiv \sum_{s=1}^{2} \sum_{r=1}^{2} (a_{s,r}, (D_s D_r w)^2)_{0,Q} + 2(a_0, D_1^2 w D_2^2 w)_{0,Q}, \qquad (1.1)$$

$$\left. \begin{array}{l} a_{1,2} = a_{2,1}, \quad a_{s,r}(x) \geq \kappa_0 > 0, \ s = 1, 2, \ r = 1, 2, \\ a_{1,1}(x) a_{2,2}(x) - a_0^2(x) \geq \kappa_1 > 0, \ \forall x \in Q, \ l \in W^*. \end{array} \right\} \qquad (1.2)$$

We use a rectangular grid with shifted nodes $P_i \equiv [(i_1+1/2)h_1, (i_2+1/2)h_2]$, where $h_s \equiv l_s(N_s + 2)^{-1}$, $s = 1, 2$. We make also use of the set

$$Q_h \equiv \{P_i : i_s \in [1, N_s], \ s = 1, 2\} \qquad (1.3)$$

of nodes corresponding to the basis functions

$$\varphi_i^{(2)}(x) \equiv \prod_{s=1}^{2} \varphi^{(2)}\left(\frac{x_s - (i_s + 1/2)h_s}{h_s}\right), \qquad (1.4)$$

where $\varphi^{(2)}(t) \equiv Y_1 Y_1 \chi(t)$, $\chi(t)$ is the characteristic function of the line segment $[-1/2, 1/2]$, and $Y_1 f(t) \equiv \int_{-1/2}^{1/2} f(t+\xi) d\xi$ (see (2.3.2)). It is easy to verify (see [18]) that $\varphi^{(2)}(t) \in W_2^2(\mathbf{R}^1)$, that it vanishes when $t \geq 3/2$, and that it takes the form $\varphi^{(2)}(t) = 3/4 - t^2$ if $|t| \leq 1/2$, $\varphi^{(2)}(t) = 1/2(3/2 - |t|)^2$ if $1/2 \leq |t| \leq 3/2$. Functions $\varphi_i^{(2)}(x)$ with $P_i \in Q_h$ constitute a basis for the subspace $\hat{G} \in W$. The resulting PGM was studied in [18].

1.2. Two-stage iterative methods. We write the corresponding grid system in the standard form $L_h u = f$ (see (0.1.28) and Lemma 0.4.1) with $L_h \in \mathcal{L}^+(H)$. We also consider an operator M_h that is just a particular case of L_h determined by the choices $a_0 = a_{1,2} = a_{2,1} = 0$ and $a_{1,1} = a_{2,2} = 1$.

---

of grad w instead of rot w, we give the precise description of the space consisting of grad w, $w \in W_2^2(\Omega)$, for multiconnected Ω (classical results for simply connected Ω and $w \in W_2^1(\Omega)$ can be found in [460]). Certain generalizations are obtained for systems involving several functions $w_s \in W_2^{1+\gamma}(\Omega)$. In § 1, we consider a basically simpler approach for a rectangle (see [170]).

Lemma 1. L_h *and* M_h *are spectrally equivalent operators. Moreover,*

$$(M_h u)_i = h_1 h_2 \left(Y_2^{(2)} \Delta_1^2 u + Y_1^{(2)} \Delta_2^2 u \right)_i, \quad P_i \in Q_h, \qquad (1.5)$$

where

$$u_i = 0 \text{ if } P_i \notin Q_h, \qquad (1.6)$$

and

$$Y_r^{(2)} u \mid_i \equiv \frac{1}{120} \{ u(P_i + 2h_r e_r) + u(P_i - 2h_r e_r)$$

$$+26[u(P_i + h_r e_r) + u(P_i - h_r e_r)] + 66 u(P_i) \}, \quad P_i \in Q_h, \; r = 1, 2. \quad (1.7)$$

Proof. Conditions (1.2) and Lemma 0.4.2 yield the desired inequalities $\delta_0^{(1)} M_h \leq L_h \leq \delta_1^{(1)} M_h$ with $\delta_0^{(1)} > 0$. Relation (1.7) is verified in a straightforward manner by observing that the coefficients on the right-hand side of (1.7) are determined by the respective Gram matrix for the functions $\varphi^{(2)}(x_r - (i_r + 1/2)h_r)/h_r)$ of one variable $x_r, r = 1, 2$ (see [170]). □

Now, as in § 6.4, we consider the operator $\Lambda \equiv \Lambda_{(h)} = \Lambda_{(1)} + \Lambda_{(2)}$, where $\Lambda_{(1)}$ and $\Lambda_{(2)}$ are the corresponding one-dimensional grid operators (see § 3.1 and § 6.4) defined by

$$(\Lambda_{(s)} u)_i \equiv h_1 h_2 \left(\Delta_s^2 u \right)_i, \quad P_i \in Q_h, \; s = 1, 2, \qquad (1.8)$$

under the boundary conditions (1.6).

Lemma 2. *For* L_h *and* Λ_h, *we have* $\delta_0 \Lambda_h \leq L_h \leq \delta_1 \Lambda_h$, *where* $\delta_0 \equiv 2/15 \delta_0^{(1)}$ *and* $\delta_1 \equiv \delta_1^{(1)}$.

Proof. Formulas (1.7) and (1.6) define one-dimensional operators $Y_r \in \mathcal{L}^+(H), r = 1, 2$. It is easy to see that $Y_r \leq I, r = 1, 2, Y_2 \Lambda_{(1)} = \Lambda_{(1)} Y_2$, and $Y_1 \Lambda_{(2)} = \Lambda_{(2)} Y_1$. It suffices to prove that $Y_r \geq \frac{2}{15} I$, which reduces to

$$S(u) \equiv \sum_k \{ 66 u_k^2 + 26 u_k [u_{k-1} + u_{k+1}] + u_k [u_{k-2} + u_{k+2}] \} \geq 16 \sum_k u_k^2 \quad (1.9)$$

on the respective one-dimensional grid. For $z_k \equiv (-1)^{k+1} u_k$, we have $S(u) = \{ \sum_k 60 z_k^2 - 22 z_k [z_{k-1} + z_{k+1}] \} + S_4(z)$, where

$$S_4(z) \equiv \sum_k \{ 6 z_k^2 - 4 z_k [z_{k-1} + z_{k+1}] + z_k [z_{k-2} + z_{k+2}] \}$$

$$= \sum_k [2 z_k - z_{k-1} - z_{k+1}]^2 \geq 0.$$

This yields (1.9). □

A nearly optimal model operator $B \asymp L_h$ can be taken in the form $B \equiv \Lambda_h (I - Z_k)^{-1}$ (see § 3.4 and § 6.4), where Z_k is the error reduction operator for k iterations of the ADI method applied to the equation $\Lambda u = g$. It is easy to see that sp $\Lambda_{(s)} \subset [\kappa_{0,s} h^2, \kappa_{1,s} h^{-2}], s = 1, 2$. This means that solving a system with this model operator $B \asymp \Lambda_h$ requires computational work $W = O(h^{-2} |\ln h|)$, and that we can find an ε-approximation (in the sense of $H(B)$ or $H(M_h)$) to the solution of the given grid system by, e.g., modified Richardson method with computational work $W(\varepsilon, h) = O(|\ln h \ln \varepsilon|/h^2)$. Finally, we note that the above results can be generalized to the case of essentially more general problems of the type considered in § 6.4 and 6.5.

§ 2. Reduction to problems of Stokes type

2.1. Spaces of rotors of functions in $W_2^2(\Omega)$. If an original elliptic boundary value problem is formulated in a Hilbert space W that requires existence of second-order derivatives of the functions $w_r, r \in [1, k]$, then a very natural decrease in the order of the necessary derivatives may be achieved by using $2k$ unknown functions

$$[u_{r,1}, u_{r,2}] \equiv \vec{u}_r \equiv [\frac{\partial w_r}{\partial x_2}, -\frac{\partial w_r}{\partial x_1}] \equiv [D_2 w_r, -D_1 w_r] \equiv \text{rot } w_r \qquad (2.1)$$

such that div $\vec{u}_r = 0, r \in [1, k]$. [2]

Theorem 1. *Suppose a $(p + 1)$-connected domain Ω is obtained from a simply connected domain Ω_0 by cutting out p simply-connected domains Ω_i. Assume that the boundary Γ_i of Ω_i, for $i \in [0, p]$, is a piecewise smooth curve that does not intersect other $\Gamma_j, j \neq i$, and that it may only contain corner points with angles in $(0, 2\pi)$. Let $W \equiv G^{(2)}$. Then the space rot W coincides with the subspace of $(G^{(1)})^2$ defined by restrictions*

---

[2] This condition explains why we prefer to use rot w_r instead of grad w_r. It is well known that, for an arbitrary solenoidal $\vec{u} \in (G^{(1)})^2 \equiv (W_2^1(\Omega))^2$, there exists a function $w \in W_2^2(\Omega)$ such that rot $w = \vec{u}$ (this is proved in [460] in terms of gradients), provided Ω is simply connected with Lipschitz piecewise smooth boundary. This implies that, in this case, the set consisting of rot $w, w \in W = W_2^2(\Omega)$, can be easily described by means of the Hilbert space rot W. But, if Ω is a multiconnected domain and W is a subspace of $W_2^2(\Omega)$, then, for describing the Hilbert space rot W, we need additional conditions of the kind $\phi_i(\vec{u}) = 0, i \in [1, k_1]$; here, the values $\phi_i(\vec{u})$ correspond to increments of w on some arcs or contours, depending on the type of boundary conditions and connectedness of the domain (see [196, 197]). The presence of these restrictions is a new serious mathematical problem, which will be in the center of our attention. We will show that arising complications can be overcome, so we will obtain effective numerical methods of the same type that were studied in Chapter 7. In this section, the vector $\vec{n}$ refers to the outward normal on the boundary Γ of the domain.

$$\text{div } \vec{u} = 0, \tag{2.2}$$

$$\varphi_i(u) \equiv \oint_{\Gamma_i} \vec{u}\vec{n}d\Gamma_i = \oint_{\Gamma_i} (-u_2 dx_1 + u_1 dx_2) = 0, \quad i = 0, 1, \dots, p. \tag{2.3}$$

Proof. For $w \in G^{(2)}$ extended onto the whole plane $\mathbf{R}^2$ (see § 0.1), we define its Sobolev averaging $Y_h w \equiv Y'_h w$ (see § 2.3), with a small enough h. Then, obviously, rot $Y_h w$ satisfies (2.2) and (2.3). In the limit with respect to h, the same holds for $\vec{u} \equiv \text{rot } w$. Conversely, consider an arbitrary $\vec{u} \in (G^{(1)})^2$ such that (2.2) and (2.3) hold. We may assume that u_r is an element of $W_2^1(\mathbf{R}^2)$ and $Y_h u_r$ is its averaging, $r = 1, 2$, $Y_h \vec{u} \equiv \vec{v} \equiv [v_1, v_2]^T$. It is known that the components of $\vec{v}$ are smooth functions and

$$\|v_s\|_{W_2^1(\mathbf{R}^2)} \le K\|u_s\|_{1,\Omega}, \ \lim_{h \to 0} \|u_r - v_r\|_{1,\Omega} = 0.$$

This implies that $\lim_{h\to 0} |\varphi_i(\vec{u}) - \varphi_i(\vec{v})| = 0$, $i \in [0, p]$. For small enough h, we define the domain $\Omega_h \subset \Omega$ containing the points whose distance from Γ is more than h. Its boundary is $\Gamma_h = \cup_{i=0}^p \Gamma_{i,h}$, where $\Gamma_{i,h}$ is a contour that approximates Γ_i and does not intersect the other $\Gamma_{j,h}$. We have div $\vec{v} = 0$ on Ω_h due to the commutability of averaging and differentiation operators. We construct contours $\hat{\Gamma}_{i,h}$ coinciding with $\Gamma_{i,h}$ everywhere but at h-neighborhoods of the corner points with the inner angles greater than π, where the arcs of circles of radius h are replaced by tangents of types $A_1 A_0$ and $A_0 A_2$ (see Figure 1).

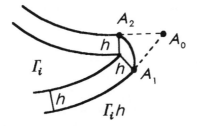

Figure 1. Shifted boundary in a vicinity of corner point with the inner angle greater than π.

Contours $\hat{\Gamma}_{i,h}$, $i \in [0, p]$, constitute the boundary of $\hat{\Omega}_h$, which differs from Ω_h only in a set of curvilinear triangles of the type $A_1 A_0 A_2$. Hence, $\hat{\Omega}_h \subset \Omega$ and we have div $\vec{v} = 0$ in $\hat{\Omega}_h$. Next, we define the subsets $\Gamma_{i,h}^{(0)} \equiv \Gamma_{i,h} \cap \hat{\Gamma}_{i,h}$, $\Gamma_{i,h}^* \equiv \hat{\Gamma}_{i,h} \setminus \Gamma_{i,h}^0$ of $\hat{\Gamma}_{i,h}$ along with the subsets Γ_i^* and

$\Gamma_i^{(0)} \equiv \Gamma_i \setminus \Gamma_i^*$. Here, Γ_i^* refers to a part of Γ_i consisting of points whose distance from $\Gamma_{i,h}$ is greater than h and is a union of a finite number of arcs of contours Γ_i of length $O(h)$ coming out of the corner points of Γ_i, with inner angles less than π. Suppose that the functionals $\hat{\varphi}_i, \hat{\varphi}_i^{(0)}, \hat{\varphi}_i^*, \varphi_i^{(0)}$, and φ_i^* are obtained from φ_i (see (2.3)) by replacing Γ_i by $\hat{\Gamma}_{i,h}, \hat{\Gamma}_{i,h}^{(0)}, \hat{\Gamma}_{i,h}^*, \Gamma_i^{(0)}$, and Γ_i^*, respectively. Then

$$|\hat{\varphi}_i(\vec{v}) - \varphi_i(\vec{u})| \leq X_1 + X_2 + X_3 + X_4, \quad i \in [0, p],$$

where $X_1 \equiv |\varphi_i^{(0)}(\vec{v} - \vec{u})|$, $X_2 \equiv |\hat{\varphi}_i^{(0)}(\vec{v}) - \varphi_i^{(0)}(\vec{v})|$, $X_3 \equiv |\hat{\varphi}_i^*(\vec{v})|$, and $X_4 \equiv |\varphi_i^*(\vec{u})|$. All indicated terms tend to 0 as $h \to 0$: X_1 according to the properties of averaging; X_2 by the standard estimates effective in theorems on traces of functions in $G^{(1)}$; and those remaining ones because the lengths of the respective curves tend to zero and the uniform estimates (see [67, 283]) of type

$$\|u_s\|_{L_1(\Gamma_i^*)} \leq |\Gamma_i^*|_{(1)}^{1/2} \|u_s\|_{L_2(\Gamma_i^*)} \leq K h^{1/2} \|u_s\|_{1,\Omega}, \quad s = 1, 2,$$

hold. Hence, $\lim_{h \to 0} \hat{\varphi}_i(\vec{v}) = 0$. Consider now some $x_i \equiv [x_{i,1}, x_{i,2}] \in \Omega_i$ and $r_i \equiv |x - x_i|$, $i \in [1, p]$. Let

$$\vec{w}_i \equiv -\frac{1}{2\pi} [\frac{x_1 - x_{i,1}}{r_i^2}, \frac{x_2 - x_{i,2}}{r_i^2}]^T.$$

It is known that div $\vec{w}_i = 0$ in Ω and $\hat{\varphi}_i(\vec{w}_j) = \delta(i, j), i \in [1, p], j \in [1, p]$, where $\delta(i, j)$ is the Kronecker symbol. For $\vec{u}_h \equiv [u_{1,h}, u_{2,h}]^T \equiv \vec{v} - \sum_{i=1}^p \hat{\varphi}_i(\vec{v}) \vec{w}_i$, we have div $\vec{u}_h = 0$ and $\oint_R (-u_{2,h} dx_1 + u_{1,h} dx_2) = 0$ for any closed piecewise smooth contour $R \subset \hat{\Omega}_h$. Now, fixing a point $x_0 \equiv [x_{0,1}, x_{0,2}]^T \in \hat{\Omega}_h$ and integrating along a broken line $R(x)$ that connects x_0 with $x \in \hat{\Omega}_h$ and consists of several straight line segments contained in the closure of $\hat{\Omega}_h$, we can construct a single-valued smooth function in $\hat{\Omega}_h$:

$$w_h(x) = \int_{R(x)} (-u_{2,h} dx_1 + u_{1,h} dx_2)$$

such that $D_1 w_h = -u_{2,h}$, $D_2 w_h = u_{1,h}$. It can be readily shown that the number of segments of $R(x)$ is less than some $k(\Omega)$. Therefore, making use again of uniform estimates of integrals along $R(x)$, we see that $\|w_h\|_{0,\hat{\Omega}_h} \leq K' \|\vec{u}_h\|_{1,\hat{\Omega}}$. Since the embedding operator of $W_2^1(\hat{\Omega}_h)$ into $L_2(\hat{\Omega}_h)$ is compact, for each $\{w_h\}$ and arbitrary domain Ω' contained strictly inside Ω, we can find a subsequence of functions converging to $w \in W_2^1(\Omega')$, where $\|w\|_{1,\Omega'} \leq K$, rot $w = \vec{u}$, and K is independent of Ω'.

This implies that $w \in G^{(1)}$. Since its derivatives also belong to $G^{(1)}$, we conclude that $w \in G^{(2)}$. Therefore, the space rot W coincides with the subspace of functions in $(G^{(1)})^2$ satisfying (2.2) and (2.3). $\square$ [3]

Our next theorems derive relations between certain subspaces W of $G^{(2)}$ and the corresponding rot W, assuming that the conditions of Theorem 1 are satisfied.

Theorem 2. Suppose that $W = \overset{o}{W}_2^2 (\Omega)$. Then the space rot W consists of vector functions $\vec{u} \in (\overset{o}{W}_1^2 (\Omega))^2$ satisfying (2.2) and conditions

$$\varphi_i^{(0)}(\vec{u}) \equiv \int_{R_i} (-u_2 dx_1 + u_1 dx_2) = 0, \ i \in [1, p], \qquad (2.4)$$

where $R_i, i \in [1, p]$ are piecewise smooth curves belonging to $\bar{\Omega}$ that connect points of Γ_0 with points of $\Gamma_i, i \in [1, p]$.

Proof. The proof of this theorem, as well as others given below, repeats in many ways the proof of Theorem 1. It suffices to observe that: for $\vec{u} \in (\overset{o}{W}_2^1 (\Omega))^2$, all conditions (2.3) are satisfied; the first derivatives of the function w, obtained as in Theorem 1, vanish on all Γ_i; hence, w is constant on each Γ_i; by conditions (2.4), all of these constants are just c, and $w - c$ is the desired function. $\square$

Theorem 3. Suppose $W \equiv W(\Omega; \Gamma^0)$ consists of $w \in G^{(2)}$ that, with their first derivatives, vanish on the set $\Gamma^0 \subset \Gamma$ consisting of $p_0 + 1$ nonintersecting arcs of contours $\Gamma_i, i \in [0, p]$. Then the space rot W consists of vector functions $\vec{u} \in (G^{(1)})^2$ that vanish on Γ^0 and satisfy (2.2), (2.3), and conditions

$$\varphi_i^{(1)}(\vec{u}) \equiv \int_{R_i} (-u_2 dx_1 + u_1 dx_2) = 0, \quad i \in [1, p_0]. \qquad (2.5)$$

Here, the curves $R_i, i \in [1, p_0]$, connecting pairs of points of disconnected parts of Γ^0, are chosen so that $\Gamma^0 \cup (\cup_{i=1}^{p_0} R_i)$ is a connected set.

This theorem is an easy generalization of Theorem 2: curves R_i can be chosen so that they have no points of intersection and, if Γ_i belongs to Γ^0, then the corresponding condition in (2.3) is redundant.

Theorem 4. Suppose that $\Gamma^0 \subset \Gamma$ is defined as in Theorem 3 and that $\Gamma^{1,0} \subset (\Gamma \setminus \Gamma^0)$ consists of p_1 nonintersecting closed arcs. Suppose also that

---

[3] Note that one of the conditions in (2.3) can be omitted because it follows from (2.2) and the remaining conditions in (2.3). Also, (2.3) may be regarded as orthogonality conditions on $\vec{u}$ with respect to some $\vec{g}_0, \ldots, \vec{g}_p$ in $(G^{(1)})^2$. This is due to the Riesz representation $\varphi_i(\vec{u}) = (\vec{g}_i, \vec{u})_{(G^{(1)})^2}$ and the fact that, for a simply connected domain, conditions (2.3) become redundant. The following is a corollary: a vector field $\vec{u} = [u_1, u_2]^T$ with components u_r continuous in $\bar{\Omega}$ and continuously differentiable in Ω, is a potential field if and only if $D_2 u_1 = D_1 u_2$ and rotations of $\vec{u}$ along all $\Gamma_i (i > 1)$ vanish.

$W \equiv W(\Omega; \Gamma^0; \Gamma^1) \subset W(\Omega; \Gamma^0)$ *contains functions w that vanish on $\Gamma^{1,0}$.*
Then the space rot W *is a subspace of* $(G^{(1)})^2$ *specified by the restrictions of Theorem 3 and by conditions*

$$\vec{u}\vec{n} \mid_{\Gamma^{1,0}} = 0, \tag{2.6}$$

$$\varphi_i^{(1)}(\vec{u}) = 0, \quad i = p_0 + 1, \ldots, p_0 + p_1. \tag{2.7}$$

Here, the functionals $\varphi_i^{(1)}$ in (2.7) are of the same form as in (2.5), and the additional curves $R_i, i > p_0$, together with R_i for $i \le p_0$, make the set of all curves in Γ^0 and $\Gamma^{1,0}$ connected.

Proof. It suffices to apply Theorem 3 and observe that conditions (2.6) correspond to the condition $\frac{\partial w}{\partial s} = 0$ on $\Gamma^{1,0}$. $\square$

The choice of curves R_i in these theorems is ambiguous: allowing some to belong to the boundary contours Γ_i, we may assume that $\Omega' = \Omega \setminus \cup_{i=1}^{p_0+p_1} R_i$ is connected.

Theorem 5. *Let $G_1 \subset (G^{(1)})^2$ be a Hilbert space with inner product $(\vec{u}, \vec{v}) \equiv (\vec{u}, \vec{v})_{1,\Omega}$, whose elements $\vec{u}$ vanish on Γ^0 and satisfy (2.6) on $\Gamma^{1,0}$ and nonlocal conditions (2.3), (2.5), (2.7). Let $G_2 = L_2(\Omega) \setminus 1$ be the subspace of functions in $L_2(\Omega)$ that are orthogonal to 1. Then there exists $\sigma_0 > 0$ such that (7.1.31) holds.*

Proof. Our space G_1 contains the space G_1^0 of vectors $\vec{u}$ that vanish on all Γ_i and R_i, and coincide with $(\overset{o}{W_2^1}(\Omega'))^2$, where Ω' may be assumed to be connected. The domain Ω' is not, in general, a domain with the Lipschitz boundary but can be presented as a finite sum of domains as in Theorem 3.7.1. Thus, this theorem applies and (7.1.31) holds. $\square^4$ $\square$

2.2. Examples of resulting boundary value problems (in a variational setting). In what follows, $(u, v)_0 \equiv (u, v)_{0,\Omega}$.

Theorem 6. *Suppose problem (0.1.4) is considered in the Hilbert space $V \equiv W \equiv \overset{o}{W_2^2}(\Omega)$, where Ω is the domain from Theorem 2, the energy functional is defined by $\Phi(w) \equiv I_2(w) - 2l(w)$,*

$$I_2(w) \equiv \sum_{s=1}^{2}\sum_{r=1}^{2}(a_{s,r}, (D_s D_r w)^2)_0 + 2(a_0, D_1^2 w D_2^2 w)_0, \tag{2.8}$$

and conditions (1.2) are satisfied (in (1.2), $Q \equiv \Omega$). Suppose that

$$l(w) = (f_{1,1}, D_2 w)_0 - (f_{1,2}, D_1 w)_0, \tag{2.9}$$

---

[4] The above results can be easily generalized to the case when $\Gamma^0 = \emptyset$.

where $f_{1,r} \in L_2(\Omega), r = 1, 2$. Then the rotor of its solution is the first component of the solution of problem (7.1.23), where G_1 and G_2 are defined as in Theorem 5 for $W = \overset{\circ}{W_2^2}(\Omega)$, $\alpha = 0$, $b_{2,1}(u_1; v_2) \equiv (div\ u_1, u_2)_0$,

$$b_{1,1}(u_1; v_1) \equiv (a_{1,1}, D_1 u_{1,2} D_1 v_{1,2})_0 + (a_{2,2}, D_2 u_{1,1} D_2 v_{1,1})_0$$

$$+(a_{1,2}, D_1 u_{1,1} D_1 v_{1,1} + D_2 u_{1,2} D_2 v_{1,2})_0 - (a_0, D_1 u_{1,2} D_2 v_{1,1} + D_1 v_{1,2} D_2 u_{1,1})_0,$$

and

$$l_1(v_1) = (f_{1,1}, v_{1,1})_0 + (f_{1,2}, v_{1,2})_0. \tag{2.10}$$

Proof. We have

$$\Phi(w) = \Phi^1(u_1) \equiv I_2^1(u_1) - 2l_1(u_1), \tag{2.11}$$

where

$$I_2^1(u_1) \equiv \sum_{s,r=1}^{2} (\tilde{a}_{s,r}, (D_s u_{1,r})^2)_0 - 2(a_0, D_1 u_{1,2} D_2 u_{1,1})_0, \tag{2.12}$$

$\tilde{a}_{1,1} = \tilde{a}_{2,2} = a_{1,2}$, $\tilde{a}_{1,2} = a_{2,2}$, and $\tilde{a}_{2,1} = a_{1,1}$. It is easy to see that $I_2^1(u_1) = b(u_1; u_1)$, where the bilinear form $b(u_1; v_1)$ (defined on $(\text{rot } W)^2$) satisfies the conditions of Theorem 0.1.3 in accordance with (1.2). Hence, minimization of $\Phi(w)$ on rot W is a correctly posed problem. Elements of rot W are elements of G_1 such that $b_{2,1}(u_1; v_2) = 0$, $\forall v_2 \in G_2$. Theorem 5 implies that (7.1.15) holds. Hence, problem (7.1.23) is also well posed (see Theorem 7.1.1), and Lemma 7.1.1 leads to the desired assertion. □

Theorem 7. Consider variational problem (0.1.4), *which differs from the one considered in Theorem 6 only in the choice of the Hilbert space* $V \equiv W \equiv W(\Omega; \Gamma^0)$ *(see Theorem 3). Then the rotor of its solution is the first component of the solution of problem* (7.1.23), *which differs from the one considered in Theorem 6 only in the choice of the Hilbert space* G_1, *now defined as in Theorem 5 for the space* $W \equiv W(\Omega; \Gamma^0)$.

Proof. The proof repeats the proof of Theorem 6 (Theorem 3 replaces Theorem 2 for describing the space rot W). □ [5]

Theorem 8. Suppose variational problem (0.1.4) *considered in Theorem 7 is such that* Ω *is a simply connected domain,* Γ^0 *consists of a single arc, and* $|\Gamma \setminus \Gamma^0|_{(1)} > 0$. *Then the rotor of its solution is the first component of*

---

[5]By virtue of (2.3), G_2 is the same as in Theorem 6, even when $|\Gamma \setminus \Gamma^0|_{(1)} > 0$. Moreover, for some cases, it is possible to use even the simpler choice $G_2 = L_2(\Omega)$.

the solution of problem (7.1.23), which differs from the one considered in Theorem 6 only in the choice of the Hilbert spaces G_1 and G_2, now defined as $G_1 \equiv (W_2^1(\Omega; \Gamma^0))^2$ and $G_2 = L_2(\Omega)$.

Proof. The space rot W is just the subspace of solenoidal vector fields in the indicated G_1 (there is no need for conditions (2.5), and the single condition (2.3) follows from (2.2)). We observe now that div $\{G_1\} = L_2(\Omega)$ and apply Theorem 7.3.3 (instead of Theorem 5) to ensure that (7.1.31) holds. $\square$

2.3. Examples of resulting nonsymmetric and nonlinear problems. We restrict ourselves to the nonlinear problems of shell theory (von Karman's type systems) considered in Subsection 6.5.3. But now we deal with general curvilinear domains Ω and, instead of u_1 and u_2 in (5.10)–(5.14), we write w_1 and w_2, respectively. We look for $w_1 \in W \equiv \overset{o}{W_2^2}(\Omega)$ and $w_2 \in W$ such that, for all $w_1' \in W$ and $w_2' \in W$, we have

$$\frac{(1-\mu^2)(w_1, w_1')_{2,\Omega}}{12} + \frac{1}{4}\{\kappa_1(D_2 w_2, D_2 w_1')_0 + \kappa_2(D_1 w_2, D_2 w_1')_0\}$$

$$-([w_1; w_2], w_1')_0 = l_1(w_1'), \qquad (2.13)$$

$$2(w_2, w_2')_{2,\Omega} - \frac{1}{4}\{\kappa_1(D_2 w_1, D_2 w_2')_0 + \kappa_2(D_1 w_1, D_1 w_2')_0\}$$

$$+([w_1; w_1], w_2')_0 = l_2(w_2'). \qquad (2.14)$$

Here,

$$-([w_1; w_2], w_1')_0 = (D_2^2 w_1, D_1 w_2 D_1 w_1')_0 + (D_1^2 w_1, D_2 w_2 D_2 w_1')_0$$

$$-(D_1 D_2 w_1, D_2 w_2 D_1 w_1' + D_1 w_2 D_2 w_1')_0, \qquad (2.15)$$

which follows from (6.5.12) and the fact that $[w_1; w_2]$ is just $D_1(D_2^2 w_1 D_1 w_2 - D_1 D_2 w_1 D_2 w_2) + D_2(D_1^2 w_1 D_2 w_2 - D_1 D_2 w_1 D_1 w_2)$ in the case of smooth functions. Problem (2.13)–(2.14) is equivalent to a nonlinear operator equation in the Hilbert space $\vec{W} \equiv W \times W$.

We now reformulate this problem in terms of $\vec{u} \equiv$ rot w_1, $\vec{v} \equiv$ rot w_2, $\vec{u}' \equiv$ rot w_1', and $\vec{v}' \equiv$ rot w_2', which are elements of the Hilbert space rot W (e.g., $\vec{u} = [D_2 w_1, -D_1 w_1]$ and rot W is defined by Theorem 2.2). We suppose that

$$l_1(w_1') = (\vec{g}_1, \vec{u}')_{1,\Omega} = (\vec{F}_1, \vec{u}')_{\text{rot } W}, \qquad (2.16)$$

$$l_2(w_2') = (\vec{g}_2, \vec{v}')_{1,\Omega} = (\vec{F}_2, \vec{v}')_{\text{rot } W}, \qquad (2.17)$$

and replace $-([w_1; w_2], w_1')_0$ (see (2.15)) by the trilinear form $t(\vec{u}; \vec{v}; \vec{u}') \equiv T$, where

$$T \equiv (D_2u_1, v_2u_2')_0 - (D_1u_2, v_1u_1')_0 - 1/2(D_1u_1 - D_2u_2, v_1v_2' - v_2v_1')_0. \quad (2.18)$$

Theorem 9. *Let Ω be a simply connected domain and $[w_1, w_2]$ be the solution of problem (2.13)–(2.14). Then rot $w_1 \equiv \vec{u}$ and rot $w_2 \equiv \vec{u}$, together with some $p_1 \in P \equiv L_2(\Omega) \setminus 1$ and $p_2 \in P$, determine a solution of the problem of finding $\vec{u} \in V \equiv (\overset{o}{W_2^1}(\Omega))^2$, $\vec{v} \in V$, and $p_1 \in P$, and $p_2 \in P$ such that, for all $\vec{u}' \in V, \vec{v}' \in V, p_1' \in P$, and $p_2' \in P$, we have*

$$\frac{1-\mu^2}{12}(\vec{u}, \vec{u}')_{1,\Omega} + \frac{1}{4}\{\kappa_1(v_1, u_1')_0 + \kappa_2(v_2, u_2')_0\} + t(\vec{u}; \vec{v}; \vec{u}')$$

$$+(div\ \vec{u}', p_1)_0 = (\vec{g}_1, \vec{u}')_0, \quad (2.19)$$

$$2(\vec{v}, \vec{v}')_{1,\Omega} - \frac{1}{4}\{\kappa_1(u_1, v_1')_0 + \kappa_2(u_2, v_2')_0\} - t(\vec{u}; \vec{u}; \vec{v}')$$

$$+(div\ \vec{v}', p_2)_0 - (\vec{g}_2, \vec{v}')_0 = 0, \quad (2.20)$$

and

$$(div\ \vec{u}, p_1')_0 = 0,\ (div\ \vec{v}, p_2')_0 = 0. \quad (2.21)$$

Proof. We rewrite (2.13) and (2.16) in the form

$$(A_{1,1}\vec{u}, \vec{u}')_V + (A_{1,2}\vec{v}, \vec{u}')_V + (N_1(\vec{u}, \vec{v}), \vec{u}')_V - (\vec{F}_1, \vec{u}')_V = 0, \quad (2.22)$$

where the linear operators $A_{1,1}, A_{1,2}$ and nonlinear operator N_1 (it maps V^2 into V) are defined in a natural way (e.g., $(N_1(\vec{u}, \vec{v}), \vec{u}')_V = t(\vec{u}; \vec{v}; \vec{u}')$ for all $\vec{u}, \vec{v}, \vec{u}'$ in V). Then (2.22) implies that

$$(A_{1,1}\vec{u} + A_{1,2}\vec{v} + N_1(\vec{u}, \vec{v}) - \vec{F}_1, \vec{u}')_V = 0,\ \forall \vec{u}' \in \text{rot } W. \quad (2.23)$$

Here, $\vec{u} \in \text{rot } W, \vec{v} \in \text{rot } W$, and rot W is the subspace of V determined by the condition div $\vec{u}' = 0$. This condition we rewrite as $L_{2,1}\vec{u}' = 0$, where $L_{2,1} \in \mathcal{L}(V; P)$ and $(L_{2,1}\vec{u}', p)_P = (\text{div } \vec{u}', p)_0, \forall \vec{u}' \in V, p \in P$. From (2.23), it then follows that, for some $p_1 \in P$, we have

$$A_{1,1}\vec{u} + A_{1,2}\vec{v} + N_1(\vec{u}, \vec{v}) - \vec{F}_1 - (L_{2,1})^*p_1 = 0.$$

Hence, we obtain (2.19) and the first condition in (2.21) in a straightforward manner. From (2.14) and (2.17), we obtain (2.20) and the second condition in (2.21). □ [6]

Lemma 1. *The trilinear form $t(\vec{u}; \vec{v}; \vec{u}')$ (see (2.18)) is bounded on $(V)^3$,*

$$t(\vec{u}; \vec{v}; \vec{u}') = t(\vec{u}; \vec{u}'; \vec{v}), \quad \forall \vec{u} \in V, \forall \vec{v} \in V, \forall \vec{u} \in V, \tag{2.24}$$

and

$$t(\vec{u}; \vec{v}; \vec{u}) - t(\vec{u}; \vec{u}; \vec{v}) = 0, \quad \forall \vec{u} \in V, \forall \vec{v} \in V. \tag{2.25}$$

Proof. Boundedness of t follows directly from boundedness of the separate terms in (2.18), e.g., $|(D_2 u_1, v_2 u_2')_0| \leq |D_2 u_1|_0 |v_2|_{L_4(\Omega)} |u_2'|_{L_4(\Omega)}$ (see (5.5.10)). It then suffices to make use of the fact that $\overset{\circ}{W}_2^1(\Omega)$ is embedded in $L_4(\Omega)$. Relation (2.24) follows directly from (2.18) and corresponds to the analogous property for $([w_1; w_2], w_1')_0$; (2.25) is a simple consequence of (2.24). □

Next, we formulate problem (2.19)–(2.21) as an operator equation in the Hilbert space $G \equiv G_1 \times G_2$, where $G_1 \equiv V \times V$ and $G_2 \equiv P \times P$: we seek $u \equiv [\bar{u}_1, \bar{u}_2]^T \in G = G_1 \times G_2$ such that

$$\left.\begin{array}{ll} \tilde{b}_1(\bar{u}_1; \bar{u}_1') + b_{2,1}(\bar{u}_1'; \bar{u}_2) = l_1(\bar{u}_1'), & \forall \bar{u}_1' \in G_1, \\ b_{2,1}(\bar{u}_1; \bar{u}_2') = 0, & \forall \bar{u}_2' \in G_2, \end{array}\right\} \tag{2.26}$$

where $\bar{u}_1 \equiv [\vec{u}; \vec{v}]^T$, $\tilde{b}_1(\bar{u}_1; \bar{u}_1') \equiv b_{1,1}(\bar{u}_1; \bar{u}_1') + t(\vec{u}; \vec{v}; \vec{u}') - t(\vec{u}; \vec{u}; \vec{v}')$, $\bar{u}_1' \equiv [\vec{u}'; \vec{v}']^T$, $b_{2,1}(\bar{u}_1; \bar{u}_2') \equiv (\text{div } \vec{u}, p_1')_0 + (\text{div } \vec{v}, p_2')_0$, $\bar{u}_2 \equiv [p_1; p_2]^T$, $\bar{u}_2' \equiv [p_1'; p_2']^T$. [7]

---

[6] In the classical setting, instead of (2.19)–(2.21), we have the system

$$-\frac{1-\mu^2}{12}\Delta^2 u_1 + \frac{\kappa_1}{4}v_1 - v_1 D_1 u_2 + \frac{1}{2}v_2(D_1 u_1 - D_2 u_2) - D_1 p_1 = g_{1,1},$$

$$-\frac{1-\mu^2}{12}\Delta^2 u_2 + \frac{\kappa_2}{4}v_2 - v_2 D_2 u_1 - \frac{1}{2}v_1(D_1 u_1 - D_2 u_2) - D_2 p_1 = g_{1,2},$$

$$-2\Delta^2 v_1 - \frac{\kappa_1}{4}u_1 + u_1 D_1 u_2 - \frac{1}{2}u_2(D_1 u_1 - D_2 u_2) - D_1 p_2 = g_{2,1},$$

$$-2\Delta^2 v_2 - \frac{\kappa_2}{4}u_2 - u_2 D_2 u_1 + \frac{1}{2}u_1(D_1 u_1 - D_2 u_2) - D_2 p_2 = g_{2,2},$$

$$D_1 u_1 + D_2 u_2 = 0, \quad D_1 v_1 + D_2 v_2 = 0,$$

with homogeneous Dirichlet boundary conditions for u_1, u_2, v_1, and v_2.

[7] It is easy to see that $b_{1,1}(\bar{u}_1; \bar{u}_1) \geq \gamma_0 \|\bar{u}_1\|^2$, for all $\bar{u}_1$, where $\gamma_0 \equiv \min\{(1 - \mu^2)/12; 2\}$. Lemma 1 implies that the nonlinear operator (which is associated with $t(\vec{u}; \vec{v}; \vec{u}') - t(\vec{u}; \vec{u}; \vec{v}'))$ satisfies antisymmetry property (1.2.5) or (7.6.9).

Lemma 2. Let the conditions of Theorem 9 be satisfied. Then

$$\sup_{\bar{v}_1 \neq 0} \frac{b_{2,1}(\bar{v}_1; \bar{v}_2)}{\|\bar{v}_1\|} \geq \sigma \|\bar{v}_2\|, \quad \sigma > 0, \forall \bar{v}_2 \in G_2. \tag{2.27}$$

Proof. Let $\bar{u}_2 \equiv [p_1; p_2]^T \in P^2$. Then, for $p \equiv p_1$ (or $p \equiv p_2$), (7.1.31) holds and we can find $\vec{u} \in V$ and $\vec{v} \in V$ such that div $\vec{u} = p_1$, div $\vec{v} = p_2$, $\|\vec{u}\| \leq 1/\sigma\|p_1\|$, and $\|\vec{v}\| \leq 1/\sigma\|p_2\|$. If we take $\bar{v}_1 \equiv [\vec{u}, \vec{v}]$, then it is easy to show that $b_{2,1}(\bar{v}_1; \bar{v}_2)/\|\bar{v}_1\| \geq \sigma\|\bar{v}_2\|$, that is, that (2.27) holds with the same $\sigma > 0$ as in (7.1.31). $\square$

Lemmas 1 and 2 indicate that our nonlinear problem (2.26) is of the same type as that considered in § 7.6. Hence, it always has a solution and all solutions can be easily localized (see (7.6.10)). Conditions of type (7.6.11) enable us to prove problem correctness, and the solution of (2.26) determines rot w_1 and rot w_2, where $[w_1, w_2]$ is the solution of original problem (2.13)–(2.14).

We conclude with a few remarks about possible generalizations in applying this approach. For multiconnected domains, conditions of type (2.4) are necessary. The case of periodic conditions is very simple, but the case for Theorem 4 is more complicated. The most difficult situation arises when the original equation contains not only the derivatives of w, but also the function w itself. We must then express w via rot w. For example, we may use

$$w(A) \equiv w(x_1, x_2) = \int_{A_0}^{A} (-u_{1,2}dx_1 + u_{1,1}dx_2) \equiv S_1 u_1,$$

where A_0 belongs to Γ^0 and the integral is taken along a broken line connecting A_0 and A. Hence, the bilinear form $b_{1,1}^{(1)}(u_1; v_1)$, which corresponds to $(w, w')_0$, takes the form

$$b_{1,1}^{(1)}(u_1; v_1) = \int_{\Omega} S_1 u_1 S_1 v_1 dx.$$

Consequently, the algorithm of computing residuals in the iterative methods for corresponding grid systems (we will consider them in § 3) becomes more complicated; but this apparently is immaterial for the asymptotic behavior of computational work, because we can use (2.28) recursively for the nodes belonging to the broken line $A_0 A$. In these computations, the choice of $A_0 A$ affects the obtained value of the integral, unlike the case when div $u_1 = 0$ and the integrals are the same along any contour connecting A_0 and A. These discrepancies can be used to reveal the actual accuracy of the method. It is even possible to make use of the representation $w(A) = \Delta^{-1}(-D_1 u_2 + D_2 u_1)$ (or its grid analog) when $w \in W_2^1(\Omega; \Gamma^0)$, but error estimates for resulting approximations deserve special study.

§ 3. Asymptotically optimal algorithms for problems with linear constraints

3.1. Projective-grid methods.[8] Not to distract attention from our essentially new mathematical problem, we confine ourselves to domains from Theorem 1 under the additional assumption that each $\Gamma_1, i \in [0, p+1]$, is a closed broken line. We can then apply triangulations $T_h(\bar{\Omega})$ (possibly composite) of the same type as in § 5.1 and 7.4 and make use of new spline spaces $\hat{G}_1$ and $\hat{G}_2$, which are just the old spaces $\hat{G}_1$ and $\hat{G}_2'$ from § 7.4 for the respective problem with the single restriction (2.2). The new spline space $\hat{G}_1$ is defined as a subspace of functions $\hat{u}_1 \equiv \vec{u} \in \hat{G}_1'$ such that (2.3), (2.5), and (2.7) hold. We rewrite these constraints in the form

$$\varphi_i(\hat{u}_1) = 0, \quad i \in [1, k], \tag{3.1}$$

and assume that the lines, along which integration in (2.3), (2.5), and (2.7) is carried out, consist of some sides of the triangles in $T_h(\bar{\Omega})$ such that the conditions of Theorem 2.5 or 2.8 are satisfied. [9]

With respect to the space $\hat{G}_2$ we assume that $\hat{G}_2$ is defined as in Theorem 3.5 when $\Gamma = \Gamma^0$ and is just $\hat{G}_2'$ when $|\Gamma^{(1)}|_{(1)} > 0$, where $\Gamma^1 \equiv \Gamma \setminus (\Gamma^0 \cup \Gamma^{1,0})$ (see Theorem 3.8).

Theorem 1. For the spline spaces $\hat{G}_1$ and $\hat{G}_2$, there exists a constant σ_0^*, independent of h such that (7.3.18) holds.

Proof. If $\hat{G}_2 \subset L_2(\Omega) \setminus 1$, then it suffices to choose the same Ω' as in Theorem 2.5 and apply the proof of Theorem 7.3.5 (or 7.3.8 for composite triangulations). If we deal with the problem considered in Theorem 2.8 then Theorem 7.3.6 can be applied. $\square$

Now it is clear that the convergence of our PGMs can be analyzed in accordance with Theorems 7.2.2 and 5.2.4 because Theorem 5.2.4 reduces approximation problems of Sobolev spaces with linear constraints to standard ones considered in § 2.3. We can thus obtain error estimates (7.4.5)

---

[8]The boundary value problems that deal with Hilbert spaces $G = G_1 \times G_2$, where $G_1 = \text{rot } W$ is defined in accordance with (2.2), (2.3), (2.5), and (2.7), will be referred as *elliptic problems with linear constraints* (see [196, 197]). The simplest way to treat these restrictions is to apply the penalty method (see § 5.5). It might be useful for practical solution of such problems when a modest precision is needed, but such an approach does not lead to asymptotically optimal algorithms. On the other hand, since these problems contain those from § 7.4 and 7.6 (where only (2.2) was present), it is natural to look for possible generalizations of algorithms investigated in Chapter 7.

[9]It is important to note that conditions (3.1) are essentially different from (2.2) because $\dim \hat{G}_1' = \dim \hat{G}_1 + k$ and k is a small integer determined by Γ^0.

under the assumption that (7.4.4) or even (7.4.6) holds (the same applies to conditions of type (5.1.3) for $\vec{u}$ involving certain known singular vectors if we include them in the basis for $\hat{G}_1$ (see Lemma 7.2.1)). [10] Note also that error estimates for the problem considered in Theorem 2.9 are easily obtained (even for a curvilinear domain) under the assumption that

$$[\bar{u}_1, \bar{u}_2] \in (W_2^{1+\gamma}(\Omega))^4 \times (W_2^\nu(\Omega))^2, \quad 0 < \gamma \le 1. \tag{3.2}$$

3.2. Construction of asymptotically optimal preconditioners. Let vector-functions $\vec{\psi}_1', \ldots, \vec{\psi}_{N+k}'$ constitute the standard nodal basis for $\hat{G}_1'$, as in § 7.4 ($N + k = 2N_1$). Assume without loss of generality that

$$\varphi_s(\vec{\psi}_{N+s}') \ne 0, \quad \varphi_s(\vec{\psi}_{N+r}') = 0 \text{ for } r \ne s, \ s \in [1, k]. \tag{3.3}$$

Then, in the role of a basis for $\hat{G}_1$, we can take the functions

$$\vec{\psi}_i \equiv \vec{\psi}_i' - \sum_{s=1}^k \alpha_{i,s} \vec{\psi}_{N+s}', \quad i \in [1, N], \tag{3.4}$$

where $\alpha_{i,s} \equiv \varphi_s(\vec{\psi}_i')(\varphi_s(\vec{\psi}_{N+s}'))^{-1}$, $i = 1, \ldots, N$. The operators $J_1' \in \mathcal{L}^+(H_1')$ and $J_1 \in \mathcal{L}^+(H_1)$ ($H_1' \equiv \mathbf{R}^{(N+k)\times(N+k)}$ and $H_1 \equiv \mathbf{R}^{N\times N}$) for these bases correspond to the matrices

$$J_1' = [(\vec{\psi}_j', \vec{\psi}_i')_{1,\Omega}] \in H_1', \quad J_1 = [(\vec{\psi}_j, \vec{\psi}_i)_{1,\Omega}] \in H_1. \tag{3.5}$$

The main problem now is construction of an asymptotically optimal preconditioner $B_1 \asymp J_1$. Fortunately, this can be reduced to the analogous problem of finding $B_1' \asymp J_1'$ in the conventional Euclidean space H_1' (sometimes we can even choose B_1' as a block diagonal matrix with diagonal blocks $B_{(1)}' \asymp \Lambda_{1,Q}$; effective algorithms for solving systems with $B_{(1)}'$ were discussed in Chapters 3, 5, and 7). To describe this important reduction, along with the routine basis $e_1, \ldots, e_{N+k}$ for H_1', we use the new basis $e_1^*, \ldots, e_{N+k}^*$, where

$$e_i^* \equiv e_i' - \sum_{s=1}^k \alpha_{i,s} e_{N+s}', \quad i \in [1, N], \tag{3.6}$$

and $e_{N+i}^* \equiv e_{N+i}'$, $i \in [1, k]$. It is easy to see that $e_1^*, \ldots, e_N^*$ is a basis for the subspace $H_{1,0}'$ (which is isomorphic to H_1). We then apply model operators $B \in \mathcal{L}^+(H_1)$ defined as

---

[10] Note that condition $w \in W_2^{2+\gamma}(\Omega)$ directly implies that (7.4.4) holds for $u_1 = \vec{u}$; the condition on $u_2 = p$ is very natural but its justification deserves special consideration if $0 < \gamma < 1$.

$$B_1 = [(e_j^*, e_i^*)_{B_1'}], \quad i \in [1, N], \ j \in [1, N]. \tag{3.7}$$

Lemma 1. *Suppose that* $\kappa_0 B_1' \le J_1' \le \kappa_1 B_1'$. *Then* $\kappa_0 B_1 \le J_1 \le \kappa_1 B_1$.
Proof. It suffices to apply Lemma 0.4.2. □

Now we describe algorithms for solving systems with matrix B_1, assuming that $k \ge 2$ (the case $k = 1$ is slightly different but simpler) and that we have already found vectors

$$\mathbf{u}_s' \equiv (B_1')^{-1} \mathbf{f}_s', \ s \in [1, k], \tag{3.8}$$

where

$$\mathbf{f}_s' \equiv [\alpha_{1,s}, \dots, \alpha_{N,s}, \underbrace{0, \dots, 0}_{s-1}, 1, \underbrace{0, \dots, 0}_{k-s-1}], \ s \in [1, k].$$

We also suppose that the matrix C_k^{-1} is known, where $C_k \in \mathbf{R}^{k \times k}$ has elements $c_{r,s} \equiv \varphi_r(\hat{u}_s')$, $r \in [1, k], s \in [1, k]$.[11]

Theorem 2. *The solution of the system*

$$B_1 \mathbf{u}_1 = \mathbf{f}_1, \tag{3.9}$$

with $\mathbf{f}_1 \equiv [f_{1,1}, \dots, f_{1,N}]^T$, *can be obtained as follows:*

1. *Find*
$$\mathbf{u}_0' \equiv (B_1')^{-1} \mathbf{f}_0', \ \text{where} \ \mathbf{f}_0' \equiv [\mathbf{f}_1^T, \underbrace{0, \dots, 0}_{k}]^T.$$

2. *Evaluate* $\mathbf{d} \equiv -[\varphi_1(\hat{u}_0'), \dots, \varphi_k(\hat{u}_0')]^T$ *and* $\mathbf{c}^0 \equiv C_k^{-1} \mathbf{d} \equiv [c_1^0, \dots, c_k^0]^T$.

3. *Evaluate*
$$\mathbf{u}_1' \equiv \mathbf{u}_0' + \sum_{s=1}^{k} c_s^0 \mathbf{u}_s'. \tag{3.10}$$

Proof. We consider (3.9) as a particular case of system (0.2.8). The vector $\mathbf{u}_1 \in H_1$ in (3.9) is formed by the coefficients of the expansion of the grid function $\mathbf{u}_1' \in H_{1,0}'$ with respect to the basis $e_1^*, \dots, e_N^*$. Therefore, $(\mathbf{u}_1', e_i^*)_{B_1'} = f_{1,i}$, $i = 1, \dots, N$ (see (0.2.10)) and according to (3.4) and (3.6), we have

$$(\mathbf{u}_1', e_i)_{B_1'} = f_{1,i} + \sum_{s=1}^{k} \alpha_{i,s}(\mathbf{u}_1', e_{N+s})_{B_1'}, \quad i \in [1, N].$$

---

[11] Linear independence of vectors $\mathbf{u}_s'$, $s \in [1, k]$, and of the respective functions $\hat{u}_s' \in \hat{G}_1'$, is easy to verify, implying that C_k^{-1} does exist.

Denote $(\mathbf{u}'_1, e_{N+s})_{B'_1}$ by c^0_s. Then $\mathbf{u}'_1$ corresponds to the solution of

$$B'_1 \mathbf{u}'_1 = \mathbf{f}', \tag{3.11}$$

where

$$\mathbf{f}' \equiv [f_{1,1} + \sum_{s=1}^{k} \alpha_{1,s} c^0_s, \ldots, f_{1,N} + \sum_{s=1}^{k} \alpha_{N,s} c^0_s, c^0_1, \ldots, c^0_k]^T. \tag{3.12}$$

If, in this vector, we take variable parameters c_s instead of c^0_s, then the solution of (3.11) yields a vector $\mathbf{u}'$ corresponding to the grid function $\mathbf{u}' = \mathbf{u}'_0 + \sum_{s=1}^{k} c_s \mathbf{u}'_s \in H'_1$. The conditions for $\mathbf{u}'$ to be in the space $H'_{1,0}$ (they imply that $\hat{u}'_1 \in \hat{G}_1$ (see (3.2)), together with (3.3), (3.4), and (3.6), lead to the system $C_k \mathbf{c} = \mathbf{d}$. This proves that representation (3.10) for the solution of (3.9) is valid. $\square$ [12]

We now indicate alterations in our PGMs and model operators for dealing with Theorem 4 and condition (2.6) on $\Gamma^{1,0}$ (an analogous condition was discussed in § 5.4). The main difference then is that each node $A \in \Gamma^{1,0}$ (writing the normal to Γ at A as $\vec{n} \equiv [n_1, n_2]$) determines a single basis function

$$\vec{\psi}_A(x) \equiv [-n_2 \hat{\psi}_A(x), n_1 \hat{\psi}_A(x)]^T \in \hat{G}_1. \tag{3.13}$$

To avoid other problems, connected, e.g., with corner points of $\Gamma^{1,0}$, we may assume that $\Gamma^{1,0}$ is a set of inner points of a line segment and take G_1 as in Theorem 2.5. Then Theorem 1 applies, including the case when $|\Gamma^{(1)}|_{(1)} > 0$. Theorems 1, 7.2.2, and 5.2.4 enable us to obtain error estimates (7.4.5). For construction of the model operator B'_1, we can apply the results of § 5.4. W can also make use of the form

$$B'_1 u_1 \equiv \begin{bmatrix} B_{1,1} & 0 & B_{1,3} \\ 0 & B_{2,2} & B_{2,3} \\ B^*_{1,3} & B^*_{2,3} & B_{3,3} \end{bmatrix} \begin{bmatrix} u_{1,1} \\ u_{1,2} \\ u_{1,3} \end{bmatrix}, \tag{3.14}$$

---

[12] When systems (3.9) are repeatedly solved for different right-hand sides, it is advisable to compute functions $\mathbf{u}'_s$, $1 \le s \le k$, and matrix C_k^{-1} beforehand. Then, for small k, solving (3.9) requires approximately the same computational work as solving (3.11) and some additional memory for storing $u'_{h,j}$. Biorthogonality conditions (3.3), which are usually easy to specify, can certainly be used under different numbering of the functions in the basis. For instance, for the spline spaces in Theorem 1, we may assume that intersections of broken lines, along which integrals in conditions (3.1) are taken, consist only of separate points. Then, for basis functions $\vec{\psi}_{N+s}$, one may take functions corresponding to nodes located sufficiently far from these points. When the above broken lines contain a common segment, conditions (3.1) can be replaced by the corresponding equivalent conditions in which new lines are of the desired type.

where the vector $u_{1,3}$ is determined by the coefficients of the functions in (3.13) and $u_{1,1}$ and $u_{1,2}$ correspond to the nodal values (at the remaining nodes) of the first and second components of the vector-function $\hat{u}_1 \in \hat{G}_1$, respectively. (3.14) makes it natural to apply methods from § 3.5, with $\Gamma^{1,0}$ being a part of the cutting lines (see [195], where the case of curvilinear $\Gamma^{1,0}$ was also considered).

3.3. Effective iterative methods. Theorems 1 and 2 enable application of modified classical iterations (7.4.21) because estimates of type (7.4.18) cause no additional difficulties. But now we must pay attention to the problem of evaluating the matrix-vector products $L_h\mathbf{v}$ and $L_h^*\mathbf{w}$, where

$$(L_h\mathbf{u}, \mathbf{v})_H = b(\hat{u}; \hat{v}), \quad \forall \hat{u} \in \hat{G}, \forall \hat{v} \in \hat{G}, \tag{3.15}$$

$H \equiv H_1 \times H_2$, and b is the bilinear form associated with our posing of the problem as an operator equation in the Hilbert space $G = G_1 \times G_2$ (see Theorems 2.5 and 2.8).

Lemma 2. *Suppose that vectors* $\mathbf{v} \in H$ *and* $\mathbf{w} \in H$ *are known. Then evaluation of the matrix-vector products* $L_h\mathbf{v}$ *and* $L_h^*\mathbf{w}$ *requires computational work* $W \le \kappa(N + N_2)$.

Proof. It is clear that L_h has the form (7.4.7), with $\alpha = 0$, and its blocks $L_{1,1}$ and $L_{1,2}$ differ from the analogous ones $L'_{1,1}$ and $L'_{1,2}$ associated with the old spaces $\hat{G}'_1$ and H'_1. More precisely, we have

$$L_{1,1} \equiv [b_{1,1}(\vec{\psi}_j^*; \vec{\psi}_i^*)] \in \mathbf{R}^{N \times N}, \; L'_{1,1} \equiv [b_{1,1}(\vec{\psi}_j; \vec{\psi}_i)] \in \mathbf{R}^{(N+k) \times (N+k)}, \tag{3.16}$$

$$L_{2,1} \equiv [(\mathrm{div}(\vec{\psi}_j^*; \chi_i)_0] \in \mathbf{R}^{N_2 \times N}, \; L'_{2,1} \equiv [(\mathrm{div}(\vec{\psi}_j; \chi_i)_0] \in \mathbf{R}^{N_2 \times (N+k)}, \tag{3.17}$$

where $\chi_1(x), \ldots, \chi_{N_2}(x)$ is our old basis for $\hat{G}'_2$ (which is just the set of characteristic functions of the triangles in our triangulation). It is easy to see that the number of $\vec{\psi}_j^* \ne \vec{\psi}_j$ is $O(k/h)$. Hence, nonzero elements of $L_{1,1}$ (see (3.16)), different from certain elements of $L'_{1,1}$, may appear only in $O((k/h)^2)$ positions. Analogously, new nonzero elements of $L_{2,1}$ (see (3.17)) may appear only in $O(k/h)$) positions. Since $L_{1,2} = L_{2,1}^T$, Lemma 2 must hold. $\square$ [13]

---

[13] These results imply that iterations (7.4.21) with optimal B_1 require computational work of the same type as for problems with no linear constraints (3.1). For nonlinear problems, the same is true for iterations (7.6.32). Application of the multigrid acceleration for these basic iterations requires no new proofs if we deal with nested spline subspaces (see Theorem 1.4.4). For general domains, its study is much the same as in § 5.1. Throughout this section, we have been mostly interested in the new aspects connected with the study of PGMs and iterative algorithms in the presence of linear constraints. More traditional questions, dealing with such issues as the use of quadrature

§ 4. Asymptotically optimal algorithms for stiffened plates and shells

In this Section, we show that the approach described in § 2 and 3 can also be developed for an important class of problems from the theory of plates and shells with stiffeners. Such problems are often found in the engineering applications, e.g., to plates and shells with stiffened edges.

4.1. Variational formulations in strengthened Sobolev spaces. We start by considering generalizations of the problems considered in Theorem 2.8 (plates without stiffeners) dealing with variational problem (0.1.4) in the Hilbert space $V \equiv W \equiv W_2^2(\Omega; \Gamma^0)$, where the energy functional is defined as in Theorem 2.8, $|\Gamma^0|_{(1)} > 0$, $|\Gamma^1|_{(1)} > 0$, and $\Gamma^1 \equiv \Gamma \setminus \Gamma^0$. Next, we consider a subset S of $\bar{\Omega}$ consisting of straight line segments (stiffeners or stringers) $S_1, \ldots, S_m$. For simplicity, we assume that the end points of each stiffener belong to Γ. Thus (considered as cuttings lines), they define a partition of $\bar{\Omega}$ into a set of blocks (panels) $P_1, \ldots, P_{m'}$. We also assume that, if an inner point of S_r belongs to Γ, then S_r belongs to Γ^1 (note that $m' = 1$ if $S \subset \Gamma$). [14] We replace $I_2(w)$ by

$$\bar{I}_2(w) \equiv I_2(w) + \sum_{r=1}^{m} \int_{S_r} [c_{r,1}(D_s^2 w)^2 + c_{r,2}(D_s D_n w)^2] ds, \qquad (4.1)$$

where $c_{r,1}$ and $c_{r,2}$ are positive constants ($r \in [1, m]$), s and $n \equiv \bar{n}$ refer to the respective arclength parameter and normal with respect to $S_r, r \in [1, m]$, and the Hilbert space W consists of functions in $W_2^2(\Omega, \Gamma^0)$ with special traces of $D_s w$ and $D_n w$ on each S_r. These traces must belong to $W_2^1(S_r), r \in [1, m]$, so we may define the inner product $(w, w')_W$ by[15]

$$(w, w')_{2,\Omega} + \sum_{r=1}^{m} [(c_{r,1}, D_s^2 w D_s^2 w')_{0,S_r} + (c_{r,2}, D_s D_n w D_s D_n w')_{0,S_r}]. \quad (4.2)$$

---

formulas, curvilinear triangles, and piecewise polynomial and singular basis functions can be considered by analogy to investigations carried out for standard elliptic boundary value problems. We have not mentioned variants of algorithms related to periodic conditions with respect to one or two variables because this case is very simple. Of course, we can make use of any symmetry of the domain as indicated in Theorem 7.3.10, provided we choose conditions (3.1) in an appropriate manner.

[14] $\Gamma' \equiv \Gamma \cup S$ corresponds to the union of the panel boundaries.

[15] It can be shown that W, with the inner product (4.2), is the completion of the preHilbert space of smooth functions that, with their first derivatives, vanish on Γ^0.

If the end points of a stiffener S_r belong to Γ^0, then these traces must belong to $\overset{o}{W}{}^1_2(S_r)$. The case with only one end point of S_r on Γ^0 is fairly similar. (We already considered a very simple example of such problems in § 6.5.) Also, we may replace $l(w)$ in (2.8) by

$$\bar{l}(w) \equiv l(w) + \sum_{r=1}^{m} [(f'_{r,1}, D^2_s w)_{0,S_r} + (f'_{r,2}, D_s D_n w)_{0,S_r}], \qquad (4.3)$$

where $f'_{r,1} \in L_2(S_r), f'_{r,2} \in L_2(S_r)$, $r \in [1, m]$. This implies that we deal with the original variational problem [16]

$$w = \arg \min_{w' \in W} [\bar{I}_2(w') - 2\bar{l}(w')]. \qquad (4.4)$$

4.2. Reduction to Stokes type systems. Let $\vec{s} \equiv [\cos\alpha_r, \sin\alpha_r] \equiv \vec{s}_r$ determine the direction of $S_r, r \in [1, m]$. Then $\vec{n} \equiv \vec{n}_r \equiv [-\sin\alpha_r, \cos\alpha_r]$ and, in accordance with (2.1), on S_r, we have

$$D_s w = -\cos\alpha_r u_2 + \sin\alpha_r u_1 \equiv I_{r,s}(\vec{u})$$

and

$$D_n w = \sin\alpha_r u_2 + \cos\alpha_r u_1 \equiv I_{r,n}(\vec{u}), \ r \in [1, m].$$

With the Hilbert space W in (4.4), we associate a Hilbert space rot W. This we describe by introducing a Hilbert space $G_1 \subset (W^1_2(\Omega; \Gamma^0))^2$, whose elements are vector fields $\vec{u}$ belonging to $(W^1_2(\Omega; \Gamma^0))^2$ and such that the traces of $I_{r,s}(\vec{v})$ and $I_{r,s}(\vec{u})$ on S_r (they exist in the sense of traces of functions in $W^1_2(\Omega)$) satisfy

$$I_{r,s}(\vec{u}) \in W^1_2(S_r), \ I_{r,n}(\vec{u}) \in W^1_2(S_r), \ r \in [1, m]. \qquad (4.5)$$

The inner product in G_1 is defined by

$$(\vec{u}, \vec{v})_{G_1} \equiv (\vec{u}, \vec{v})_{1,\Omega}$$

$$+ \sum_{r=1}^{m} [(1, I_{r,s}(\vec{u}) I_{r,s}(\vec{v}))_{1,S_r} + (1, I_{r,n}(\vec{u}) I_{r,n}(\vec{v}))_{1,S_r}] \qquad (4.6)$$

(if the end points of a stiffener S_r belong to Γ^0, then the above traces must belong to $(\overset{o}{W}{}^1_2(S_r))^2$; the case with only one end point of S_r on Γ^0 is fairly similar). Then rot $W \subset G_1$ is a subspace of solenoidal vector fields.

---

[16] First use of analogous problems in preHilbert spaces dates back to the paper of S. P. Timoshenko in 1915; see also [122].

We replace (4.4) by the problem of finding $u \in G \equiv G_1 \times G_2$ (with $G_2 \equiv L_2(\Omega)$) such that

$$\left. \begin{array}{r} \bar{b}_{1,1}(u_1; u_1') + b_{1,2}(u_2; u_1') = \bar{l}_1(u_1'), \ \forall u_1' \in G_1 \\ b_{2,1}(u_1; u_2') = 0, \ \forall u_2' \in G_2, \end{array} \right\} \tag{4.7}$$

where

$$\bar{b}_{1,1}(u_1; u_1') \equiv b_{1,1}(u_1; u_1') +$$

$$\sum_{r=1}^{m} [c_{r,1}(1, I_{r,s}(\vec{u})I_{r,s}(\vec{u}'))_{1,S_r} + c_{r,2}(1, I_{r,n}(\vec{u})I_{r,n}(\vec{u}'))_{1,S_r}] \tag{4.8}$$

(see (2.9)) and

$$\bar{l}_1(u_1') \equiv l_1(u_1') + \sum_{r=1}^{m}[(f_{r,1}, D_s I_{r,s}(\vec{u}'))_{1,S_r} + (f_{r,2}, D_s I_{r,n}(\vec{u}'))_{1,S_r}]$$

(see (2.10)) . The following generalization of Lemma 7.3.1 for each of our panels is fundamental.

Lemma 1. *Let P be a domain with piecewise smooth boundary ∂P. Suppose that ∂P contains a straight line segment $\Gamma^*(P) \equiv S^*$ and let $\Gamma^0(P) \equiv \partial P \setminus S^*$. Suppose also that the Hilbert space $G_1(P)$ is defined as in (4.6) with only one stiffener S^*. Then there exist a constant K^* and $\vec{v}^* \in G_1(P)$ such that*

$$(\vec{v}^*, \vec{n})_{0,S^*} = 1 \text{ and } [|\vec{v}^*|_{1,P}^2 + |D_s\vec{v}^*|_{0,S^*}^2]^{1/2} \le K^* |\text{div } \vec{v}^*|_{0,P}. \tag{4.9}$$

Proof. It is easy to see that we may apply the proof of Lemma 7.3.1 with $\Omega = P, \Gamma_f = S^*$, and a rectangle

$$\Pi \equiv [-a, a] \times [0, b] \subset P,$$

that is, $\kappa_f = 0$ and $b = a$. We make use of the same function

$$u_2(y) \equiv u(y_1, y_2) \equiv y_2 g(y_1)/f(y_1)$$

as before (g_1 is defined by (7.3.6)). Then $|D_1 u|_{0,S^*}^2 = |D_1 g|_{0,S^*}^2$. Thus, $|D_1 u|_{0,S^*}^2 \le \kappa_g^2/a$, which implies that K^* is defined now by (7.3.9) with the additional term κ_g^2/a on its left-hand side. $\square$

Theorem 1. *Let the Hilbert space G_1 be defined as above (see (4.6)) and let $G_2 = L_2(\Omega)$. Suppose also that $S \subset \Gamma^1$. Then there exists $\sigma_0 > 0$ such that (7.1.31) holds.*

Proof. Since all stiffeners belong to Γ, we have only one panel $P_1 = \bar{\Omega}$. Let $p^* \equiv \text{div } \vec{v}^*$ (see Lemma 1). Then, for every $p = p - cp^* + cp^* \in G_2$, we apply the proof of Theorem 7.3.3 and find $\vec{v} = \vec{u} + c\vec{v}^* \in G_1$ such that $\overset{o}{\vec{u}} \in (W_1^2(\Omega))^2$, div $\vec{v} = p$, and $|\vec{v}|_{G_1} \le 1/\sigma |p|_{G_2}$. □

Theorem 2. *Let the Hilbert space G_1 be defined by (4.6) and $G_2 = L_2(\Omega)$. Suppose also that the partition of $\bar{\Omega}$ into a set of panels $P_1, \ldots, P_{m'}$ is such that each pair P_i and P_{i+1}, $i \in [1, m'-1]$, has a common side $S_{i,i+1}^* \in S$ and P_m has a side on Γ^1 (which might belong to S). Then there exists $\sigma_0 > 0$ such that (7.1.31) holds.*

Proof. We denote by $p_1, \ldots, p_{m'-1}$ the restrictions of p to our panels (their finite extensions are denoted by $p_1^0, \ldots, p_{m'-1}^0$, so $p = p_1^0 + \cdots + p_{m'-1}^0$). For $p_1 \equiv p_1' + c_1 p_1^* ((p_1', 1)_{0, P_1} = 0)$, we construct $\vec{v}_1 \equiv \vec{v}_{1,1} = \vec{u}_1 + c_1 \vec{v}_1^*$ as in Theorem 1 with $\vec{u}_1$ vanishing on ∂P_1 and $\vec{v}_1^*$ vanishing on the complement of the rectangle $\Pi_1 \subset P_1$ (see the proof of Lemma 1) in P_1 (one side of Π_1 belongs to $S_{1,2}$). More precisely, we assume that a in the proofs of Lemmas 1 and 7.3.1 is so small that $\Pi_{1,2} \subset P_2$, where $\Pi_{1,2}$ denotes the mirror reflexion of Π_1 with respect to $S_{1,2}$. This reflexion for p_1^* and $\vec{v}_1^*$ yields their symmetric images $I_{1,2} p_1^*$ and $I_{1,2} \vec{v}_1^*$ (in the coordinate system $[y_1, y_2]$, the extended second component of $\vec{v}_1^*$ is an even function with respect to $y_2 - b$, the other component vanishes, and div $\vec{v}^* = p_1^*(y_1, y_2)$ is odd with respect to $y_2 - b$).

Next, we represent p_2 as $p_2 = c_1 I_{1,2} p_1^* + p_2'$ (we use finite extensions for $I_{1,2} p_1^*$ if necessary). For $p_2' = (p_2' - c_2 p_2^*) + c_2 p_2^* \in L_2(P_2)$ ($c_2 \equiv c_2(p_2')$), Theorem 1 again applies: we construct $\vec{v}_1 \equiv \vec{v}_{1,2} = \vec{u}_1 + c_2 \vec{v}_1^* \equiv \vec{u}_{1,2} + c_2 \vec{v}_{1,2}^*$ with $\vec{u}_1$ vanishing on ∂P_2 and $\vec{v}_1^*$ vanishing on the complement of the rectangle $\Pi_2 \subset P_2$ (see the proof of Lemma 1) in P_2 (one side of Π_2 belongs to $S_{2,3}$). We repeat this procedure for all remaining panels and finally define

$$\vec{v}_1 \equiv \vec{u}_{1,1} + c_1 E_{1,2}(\vec{v}_{1,1}^*) + \vec{u}_{1,2} + c_2 E_{2,3}(\vec{v}_{1,2}^*) + \cdots + \vec{u}_{1,m'} + c_{m'} \vec{v}_{1,m'}^*. \quad (4.10)$$

Here, $E_{1,2}(\vec{v}_{1,1}^*) \in (\overset{o}{W_2^1}(\Omega))^2$ coincides with $\vec{v}_{1,1}^*$ and $I_{1,2} \vec{v}_1^*$ on Π_1 and $\Pi_{1,2}$, respectively, and vanishes at other points; $E_{1,2}(\vec{v}_{1,1}^*), \ldots, E_{m'-1,m'}(\vec{v}_{1,m'-1}^*)$ are defined analogously. It is easy to see that div $\vec{v}_1 = p$. Moreover, we have $|c_1| \le K_{1,1} |p_1|_{0, P_1}$ (see the proof of Theorem 7.3.1) and

$$|\vec{u}_{1,1} + c_1 E_{1,2}(\vec{v}_{1,1}^*)|_{1,\Omega} + |c_1 E_{1,2}(\vec{v}_{1,1}^*)|_{1, S_{1,2}} \le K_{1,2} |p|_0.$$

Next, we see that $|p_2'|_{0, P_2} \le K_{2,0}(|p_1|_{0, P_1} + |p_2|_{0, P_2}$, $|c_2| \le K_{2,1}(|p_1|_{0, P_1} + |p_2|_{0, P_2}$, and

$$|\vec{u}_{2,1} + c_2 E_{2,3}(\vec{v}_{2,1}^*)|_{1,\Omega} + |c_2 E_{2,3}(\vec{v}_{2,1}^*)|_{1, S_{2,3}} \le K_{2,2} |p|_0.$$

Along the same lines, we obtain the desired estimates for the terms on the right-hand side of (4.10). This yields $\|\vec{v}_1\|_{G_1} \leq K|p|_0$, $\forall p$. $\square$ [17]

Theorem 3. Consider variational problem (4.4) *replaced by* (4.7). *Suppose that S is such that the respective spaces G_1 and G_2 lead to* (7.1.31). *Then the rotor of the solution of* (4.4) *is the first component of the solution of* (4.7).

Proof. In accordance with the Theorems in § 7.1, (4.7) is a correct problem and has unique solution. It is easy to show that rot w is a solution of a variational problem similar to that considered in Theorem 2.8 (they differ only in the choices of G_1 and $b_{1,1}$). Lemma 7.1.1 now applies. $\square$

We consider now the more difficult problem that differs from (4.7) in choices of G_1 and G_2. Elements of $G_1 \equiv G_1^0 \subset (\overset{o}{W_2^1}(\Omega))^2$ are vector fields $\vec{u}$, belonging to $(\overset{o}{W_2^1}(\Omega))^2$, such that the traces of $I_{r,s}(\vec{v})$ and $I_{r,s}(\vec{u})$ on S_r satisfy (4.5); the inner product in G_1 is defined by (4.6); and $G_2 \equiv L_2(\Omega) \setminus 1 \equiv G_2^0$. This problem is associated with (4.4) under the choice $W = (\overset{o}{W_2^2}(\Omega))^2$ (the inner product is defined by (4.2); see also Theorem 2.7 for a simpler case). It is easy to see that we can obtain a modification of Theorem 3 if we prove that (7.1.31) holds for the indicated pair of spaces.

Theorem 4. Let $G_1 = G_1^0$ and $G_2 = G_2^0$. Suppose also that the partition of $\bar{\Omega}$ into a set of panels $P_1, \ldots, P_{m'}$ is such that each pair P_i and P_{i+1}, $i \in [1, m'-1]$, has a common side $S_{i,i+1}^* \in S$. Then there exists $\sigma_0 > 0$ such that (7.1.31) *holds.*

Proof. The proof is a modification of the proof of Theorem 2. The essential difference is connected with the final step where we deal with $P_{m'}$, because now its boundary contains no arc on $\Gamma^1(\Gamma^1 = \emptyset)$ and we cannot make use of $c_{m'}\vec{v}_{1,m'}^*$ in (4.10). Fortunately there is no need in such a

---

[17] We observe that the proof of Theorem 2 can be easily generalized in two directions. First, if by drawing certain cutting lines belonging to S we can partition $\bar{\Omega}$ into several subregions, such as in Theorem 2, then it is possible to construct the desired $\vec{v}$ for each region independently (it vanishes on the cutting lines). This yields several independent chains of panels with different ends (different panels for which Theorem 1 can be applied directly). Second, it is possible to take the end panel in certain of these chains as an inner panel in a chain provided different sides of this panel connect it with the panels in different chains (the analogy with rivers and trees is obvious). For example, a triangular panel may belong to two chains. Thus, speaking in terms of graph theory, our set of panels is partitioned into subsets such that each is defined by a separate tree (a directed graph) whose vertices correspond to the panels in the subset. It is also very important to note that, in proving (7.1.31), we may deal with refinement of the original partition. Thus, without loss of generality, we may assume that all our panels are triangles—their curvilinear sides may be only some parts of Γ^0. For concrete triangulations, there is no problem to determining necessary chains of panels. For example, all triangulations considered in Chapter 4 exhibit the desired property. Thus, (7.1.31) can be proved for fairly general partitions of $\bar{\Omega}$ into a set of panels.

term, because now we obtain $p'_{m'} \equiv p_{m'} - c_{m'} I_{m'-1,m'} p^*_{m'-1} \in L_2(P_{m'} \setminus 1)$. Indeed, using finite extensions, we see that $p = p_1 + \cdots + p_{m'-1} \in L_2(\Omega \setminus 1)$ and that the functions $c_1(p^*_1 + I_{1,2}p^*_1), c_2(p^*_2 + I_{2,3}p^*_2), \ldots, c_{m'-1}(p^*_{m'-1} + I_{m'-1,m'}p^*_{m'-1})$ are orthogonal to 1 in $L_2(\Omega)$. We also have

$$p = [p_1 - c_1 p^*_1] + c_1(p^*_1 + I_{1,2}p^*_1) + [p'_2 - c_2 p^*_2] + c_2(p^*_2 + I_{2,3}p^*_2) + \cdots + p'_{m'-1}.$$

All terms on the right-hand side here are orthogonal to 1 in $L_2(\Omega)$, with the exception of the last one (for the terms in brackets, see the definition of the constants $c_1, \ldots, c_{m'-1}$). Hence,

$$(p'_{m'}, 1)_0 = 0 = (p'_{m'}, 1)_{0, P_{m'}},$$

and the desired $\vec{v}_1$ is defined by (4.10) with $\vec{v}^*_{1,m'} \equiv 0$. $\square$ [18]

4.3. Projective-grid (mixed finite element) methods. There is nothing essentially new in construction of subspaces $\hat{G}_1$ and $\hat{G}_2$ in comparison with the case when $S = \emptyset$ (see § 2) because all elements of the old $\hat{G}_1$ belong to the new one. (Of course, it is natural to construct original triangulations by refinement of the original panels (see also § 4.2), so we will assume that triangulations $T_h(\bar{\Omega})$ yield triangulations $T_h(\bar{P}_i)$ for all panels $P_1, \ldots, P_{m'}$.) The most serious difficulty is connected with generalizations of the theorems in § 7.3. Fortunately, we can obtain (7.3.18), without making use of Lemma 7.3.1, by analogy with the above proofs.

Lemma 2. *Let the conditions of Lemma 1 be satisfied and let P be a domain with a piecewise smooth boundary ∂P. Suppose that we deal with quasiuniform triangulations $T_h(\bar{P})$, where $h \leq h_0$ and h_0 is small enough. Then there exist a constant K^*, independent of h, and a $\vec{v}^* \in \hat{G}_1(P)$ such that (4.9) holds.*

Proof. We construct $\vec{v}^* \in G_1(P)$ as in the proof of Lemma 1 (it vanishes outside of the rectangle II) and define $\vec{v}_h \in \hat{G}_1(P)$ as an interpolant for our piecewise smooth $\vec{v}^*$. It is easy to see that this $\vec{v}_h \in \hat{G}_1(P)$ may be taken in the role of the desired $\vec{v}^*$ in (4.9). $\square$

---

[18] The above proofs of (7.1.31) are new even for the classical case of this inequality considered in § 7.3. It is easy to see how to modify them for the case when the panels correspond to vertices of a single tree of general form (the tree in the proof of Theorem 1 contained no branches). Moreover, it can be proved that, for each partition of $\bar{\Omega}$ (e.g., for its composite triangulation), it is possible to indicate the desired tree with a prescribed root panel. It is also important that we can prove (7.1.31) for an approximating sequence $\hat{\Omega}_h$ for Ω with the constant independent of h (see § 7.3), and we may assume that S belongs to cutting lines in some partition of $\bar{\Omega}$ (one or both end points of a stiffener belong to Ω). The case of a multiconnected region needs more cumbersome descriptions and proofs (see § 2).

Theorem 5. Let the conditions of Theorem 2 be satisfied. Suppose also that h_0 is so small that Lemma 2 applies for each panel. Then there exists $\sigma_0^* > 0$ independent of h such that (7.1.18) holds.

Proof. It suffices to modify slightly the proof of Theorem 2. $\square$ [19]

Having obtained (7.3.18), we can apply the theorems of § 7.2. For (4.7), it is natural to make assumptions of the form

$$\|u_1\|_{1+\gamma,P_i} \leq K_{1,i}, \quad \|u_1\|_{1+\gamma,S_r} \leq K_{1,r}, \quad \|u_2\|_{\gamma,P_i} \leq K_{2,i}, \quad (4.11)$$

where $i \in [1, m'], r \in [1, m]$ (see § 7.4 and (0.5.4)). Then it is easy to prove that asymptotic approximation properties of the strengthened Sobolev spaces (see (4.6)) are the same, and we can obtain the error estimates

$$\|\hat{u}_1 - u_1\|_{G_1} + \|\hat{u}_2 - u_2\|_0 \leq Kh^\gamma \quad (4.12)$$

(estimates for N-widths are the same as in § 0.5.)

4.4. Multigrid construction of asymptotically optimal preconditioners. Our PGM yields grid systems of type (7.4.14) such that $(L_{1,1}u_1, v_1)_{H_1} = \bar{b}_{1,1}(\hat{u}_1, \hat{v}_1)$, for all $u_1 \in H_1$ and $v_1 \in H_1$. The resulting system differs from the case when $S = \emptyset$ only for the nodes on S. We have already commented about possible preconditioners for $L_{1,1}$ (see (0.4.21)), and we may confine ourselves to the case when $\bar{b}_{1,1}(u_1, v_1) = (\vec{u}, \vec{v})_{G_1}$ and $(\vec{u}, \vec{v})_{G_1}$ is just

$$(\vec{u}, \vec{v})_{1,\Omega} + \sum_{r=1}^{m} [c_{r,1}(1, I_{r,s}(\vec{u})I_{r,s}(\vec{v}))_{1,S_r} + c_{r,2}(1, I_{r,n}(\vec{u})I_{r,n}(\vec{v}))_{1,S_r}] \quad (4.13)$$

(see (4.6)). Here, $c_{r,1}$ and $c_{r,2}$ are nonnegative numbers, $r \in [1, m]$. Moreover, we may consider Ω as a model region Q and apply constructions very similar to those in § 5.4 under the same splitting of the spline space $\hat{G}_1 \equiv \vec{G}_1$ (see (5.4.6) and (5.4.7) for the space $\vec{G}^{(l+1)}$) and under the same form (5.4.8) of the new Gram matrices.

Lemma 3. *The angle α between the subspaces $\vec{G}_2^{(l+1)} = \vec{G}^{(l)}$ and $\vec{G}_1^{(l+1)}$ is not smaller than the angle between the respective subspaces when $S = \emptyset$.*

Proof. In accordance with Lemma 2.5.1 (see (2.5.5)), it suffices to introduce the semiinner product

$$(\vec{u}, \vec{v})_S \equiv \sum_{r=1}^{m} [c_{r,1}(1, I_{r,s}(\vec{u})I_{r,s}(\vec{v}))_{1,S_r} + c_{r,2}(1, I_{r,n}(\vec{u})I_{r,n}(\vec{v}))_{1,S_r}] \quad (4.14)$$

---

[19] Now $E_{1,2}(\vec{v}_{1,1}^*), \ldots, E_{m'-1,m'}(\vec{v}_{1,m'-1}^*)$ are symmetric vector fields (with respect to $S_{1,2}, \ldots, S_{m'-1,m'}$) only in the case of local symmetry of $T_h(\bar{\Omega})$. This is essential for the case of Theorem 3—otherwise, some additional explanations are needed.

(see (4.6)), to observe that $(\vec{u}^{(l+1)}, \vec{u}^{(l)})_S = 0$ if $\vec{u}^{(l+1)} \in \vec{G}_1^{(l+1)}$, and to use triangulations with refinement ratio 2. $\square$ [20]

Now in accordance with the theory of optimal model operators given in § 3.7, we need to approximate the block $\bar{L}_{1,1}^{(l+1)} = L_{1,1}^{(l+1)}$ (see (5.4.8)).

Lemma 4. *Suppose that the basis functions for $G_1^{(l+1)}$ are indexed so that the two basis functions associated with each node on S have consecutive numbers. Then there exists a block diagonal matrix $A_{1,1}^{(l+1)} \in \mathcal{L}(\vec{H}_1^{(l+1)})$, with blocks in $\mathbf{R}^{2\times 2}$ or $\mathbf{R}^{1\times 1}$ and constants $\sigma_{0,1} > 0$ and $\sigma_{1,1} > 0$, independent of l and coefficients $c_{r,1}$ and $c_{r,2}$ ($r \in [1, m]$), such that*

$$\sigma_{0,1} A_{1,1}^{(l+1)} \leq L_{1,1}^{(l+1)} \leq \sigma_{1,1} A_{1,1}^{(l+1)}, \quad l+1 \in [1, p].$$

Proof. Our inequalities are the same as (5.4.29). We may take $A_{1,1}^{(l+1)} = A_{\emptyset,1,1}^{(l+1)} + A_{S,1,1}^{(l+1)}$, where, for all $\vec{u}^{(l+1)} \in G_1^{(l+1)}$, we have

$$(A_{\emptyset,1,1}^{(l+1)} \mathbf{u}^{(l+1)}, \mathbf{u}^{(l+1)})_{\vec{H}_1^{(l+1)}} = |\vec{u}^{(l+1)}|_{1,\Omega}^2$$

and

$$(A_{S,1,1}^{(l+1)} \mathbf{u}^{(l+1)}, \mathbf{u}^{(l+1)})_{\vec{H}_1^{(l+1)}} = \|\vec{u}^{(l+1)}\|_S^2.$$

Note that $A_{\emptyset,1,1}^{(l+1)}$ is a positive diagonal matrix (essentially the same as in § 5.4 and 3.7; its elements are uniformly bounded) and $A_{S,1,1}^{(l+1)}$ is a nonnegative block diagonal matrix (its elements are $O(1/h^{(l+1)})$).$\square$[21]

Theorem 6. *Let the operator $\Lambda_{1,1}$ be the Gram matrix for the basis functions in $\hat{G}_1$ (see (4.13)). Then there exists an asymptotically optimal model operator $B_1 \asymp \Lambda$ such that the constants of spectral equivalence and the estimates of the required computational work in solving systems with B_1 are independent of $c_{r,1}$ and $c_{r,2}$ ($r \in [1, m]$).*

Proof. It suffices to apply construction of the model cooperative operators $\bar{B}^{(l+1)}$ and $B^{(l+1)}$ from § 3.7 (see (3.7.30)), in combination with (5.4.8) and Lemmas 3 and 4. $\square$ [22]

---

[20] If we deal only with isosceles rectangular triangles in $T_h(\bar{\Omega})$, then $\alpha \geq \pi/4$.

[21] If $c_{r,1} = c_{r,2}$, ($r \in [1, m]$), then $A_{S,1,1}^{(l+1)}$ is diagonal.

[22] The use of multigrid acceleration of the obtained optimal iterative methods requires nothing new compared with the study in § 1.4 and 5.1. This implies that, for problems (4.7), we can determine asymptotically optimal algorithms under assumptions (4.11). Similar algorithms can be obtained for certain shell problems.

Chapter 9

Effective algorithms for spectral problems

In this chapter, we are interested in finding a few eigenvalues and corresponding eigenfunctions of eigenvalue problems involving symmetric elliptic operators. Special attention is paid to the algorithms that require computational work of the same type as for the corresponding boundary value problems provided the smoothness properties of eigenfunctions are of the same nature as those of the boundary problems solutions (see [184, 185, 188, 197, 208, 212, 213]). In § 1 we develop necessary properties of spectral problems with symmetric compact operators, including the minimax Courant-Fisher principle, basic properties of gaps between subspaces, fundamentals of the Rayleigh-Ritz method. A priori and a posteriori error estimates for this method are given in § 2 and § 3, with emphasis on the generalized Temple inequalities. Efficient preconditioned iterative methods have lately been gaining acceptance for solving such partial symmetric grid sparse eigenvalue problems; their study is the main subject in § 4-6 (see also [253, 348, 305, 307]). It should be noted that, in contrast to preconditioning for systems (see § 0.3), the basic theorems about such methods require significant mathematical effort, especially when the attempt is to find several eigenvalues and their eigenspaces. The latter uses approximate orthogonalization with respect to those eigenvectors that have already been computed; simultaneous calculation of a group of the eigenvectors is also very important (see § 6). The final § 7 is devoted to examples of elliptic spectral problems such that the above algorithms can be designed for them; triangulations and spline spaces are the same as before; special attention is paid to problems with linear constraints, including those obtained from

467

problems involving fourth-order operators by a reduction procedure similar to that in § 8.2.

§ 1. The Rayleigh-Ritz method for spectral problems

1.1. Operator formulations for spectral problems in mathematical physics. In a Hilbert space H, we consider the spectral (eigenvalue) problem

$$Lu = \lambda M u, \tag{1.1}$$

where $L \in \mathcal{L}^+(H)$ and $M = M^* > 0$, and the more general

$$Mu = sLu, \tag{1.2}$$

where $L \in \mathcal{L}^+(H)$ and $M = M^*$. [1] If we introduce bilinear forms $b_L(u; v)$ and $b_M(u; v)$ associated with our operators L and M, respectively, then, e.g., (1.1) is actually a problem of finding λ and $u \neq 0$ such that

$$b_L(u; v) = \lambda b_M(u; v), \quad \forall v \in H. \tag{1.3}$$

It is clear that (1.2) is equivalent to the spectral problem

$$Au = su, \tag{1.4}$$

where

$$A \equiv L^{-1}M. \tag{1.5}$$

In what follows, we consider only elliptic spectral problems that can be reduced to problems (1.4) in a Hilbert space G for which A is symmetric and compact. If $L \in \mathcal{L}^+(H)$, then in the role of G we consider the energy space $G \equiv H(L)$ with inner product

$$(u; v)_G \equiv (u; v)_L \equiv b_L(u; v) = (Lu, v). \tag{1.6}$$

Thus, $A \in \mathcal{L}(G)$ is symmetric and it is compact when M is compact.

---

[1] $u \neq 0$ is an eigenfunction of (1.1) or (1.2) if there exists a number λ or s such that (1.1) or (1.2) hold. Analogously, a number λ or s is called an eigenvalue of (1.1) or (1.2) if there exists $u \neq 0$ such that (1.1) or (1.2) holds. For a given eigenvalue s of (1.2), $U(s) \equiv \mathrm{Ker}\,(M - sL)$ is the *eigenspace* associated with the eigenvalue s and $\{s, U(s)\}$ is referred to as the *spectral pair* of (1.2). It is clear that these notions apply to (1.1) as well. Each spectral pair $\{\lambda, U(\lambda)\}$ for (1.1) is a spectral pair $\{\lambda^{-1}, U(\lambda^{-1})\}$ for (1.2). There are two reasons why we prefer to deal with (1.1) and (1.2) separately. First, problems of type (1.1) frequently appear in applications. Second, some important formulas can be simplified if we use an appropriate choice of (1.1) or (1.2).

We briefly recall the most important facts related to spectral problem (1.4) with symmetric and compact A (see [341, 449]) [2]:

1. Eigenvalues s for problem (1.4) are real and $s \in [\alpha_0; \alpha_1]$, where

$$\alpha_0 \equiv \min_{\|v\|_G=1} (Av, v)_G, \quad \alpha_1 \equiv \max_{\|v\|_G=1} (Av, v)_G. \qquad (1.7)$$

2. The norm of A satisfies

$$\|A\| = \max\{|\alpha_0|; |\alpha_1|\}. \qquad (1.8)$$

3. The set sp $A \setminus 0$ has no limit points.

4. dim $U(s) < \infty$ if $s \neq 0$.

5. The eigenspaces $U(s)$ and $U(s')$ are orthogonal if $s \neq s'$.

6. The Hilbert space G is an orthogonal sum of Ker A and Im A.

7. Im A and Ker A have orthonormal bases composed of eigenfunctions of problem (1.4).

8. The Hilbert space G has an orthonormal basis composed of eigenfunctions of problem (1.4), that is, for arbitrary $v \in G$, we have

$$v = v_0 + \sum_{i=1}^{\infty} c_i u_i, \quad v_0 \in \text{ Ker } A, \qquad (1.9)$$

where $Au_i = s_i u_i$ if $i \geq 1$ and $|s_{i+1}| \leq |s_i| > 0$, $i \geq 1$. Moreover, $(u_i, u_j)_G = \delta(i; j)$, if $i \geq 1, j \geq 1$ and $c_i = (v, u_i)_G$ if $i \geq 1$.

The statement associated with (1.9) is known as the *Hilbert-Schmidt Theorem*. Sometimes, we use representations

$$v = v_0 + \sum_{i=1}^{\infty} c_i u_i + \sum_{i=-1}^{-\infty} c_i u_i, v_0 \in \text{ Ker } M, \qquad (1.10)$$

with different numerations of positive and negative eigenvalues, where

$$Mu_i = s_i Lu_i, \qquad (1.11)$$

---

[2] Such operators are the most natural generalizations of symmetric operators in an Euclidean space.

$(u_i, u_j)_L = \delta(i; j)$, $s_1 \geq s_2 \geq \cdots > 0$, $s_{-1} \leq s_{-2} \leq \cdots < 0$, and $c_i = (v, u_i)_L$. It is also possible to use splittings

$$I = \sum_{j=-\infty}^{\infty} P_j, \quad v = \sum_{j=-1}^{-\infty} v_j + v_0 + \sum_{j=1}^{\infty} v_j, \tag{1.12}$$

where

$$t_1 > t_2 > \ldots > 0 = t_0, \quad t_{-1} < t_{-2} < \ldots < 0 \tag{1.13}$$

correspond to different eigenvalues of (1.2), $v_j = P_j v$, and P_j is the ortho-projector of G onto the subspace $U(t_j)$ for any integer j. In the case of (1.1) (with $\lambda > 0$), instead of (1.10) we have

$$v = \sum_{i=1}^{\infty} c_i u_i, \tag{1.14}$$

where $Lu_i = \lambda M u_i, (i = 1, \ldots), 0 < \lambda_1 \leq \lambda_2 \leq \ldots, (u_i, u_j)_M = \delta(i; j)$, $c_i = (v, u_i)_M, \forall i$. Here we made use of the fact that, if u_j is an eigenfunction and $(v, u_j)_L = 0$, then $(Mv, u_j) = 0$. Different eigenvalues of (1.1) will be denoted by

$$\nu_1 < \nu_2 < \ldots; \nu_1 > 0. \tag{1.15}$$

Lemma 1. *Let $U \equiv \text{lin} \{u_{i_1}, \ldots, u_{i_k}\}$ be the linear span of several eigen-functions of problem (1.2). Suppose that $P_{U,L} \equiv P$ is an orthoprojector of the Hilbert space $H(L)$ onto U. Then, for arbitrary u and $z \equiv u - Pu$, we have $(Mz, Pu) = 0$ and $(MPu, Pu) + (Mz, z) = (Mu, u)$.* [3]

Proof. Without loss of generality, we assume that the indicated functions are orthogonal in $H(L)$. We have $Pu = \sum_{r=1}^{k} c_{i_r} u_{i_r}$, where $(L(u - Pu), u_{i_r}) = 0$, $r = 1, \ldots, k$. Hence, $(M(u - Pu), u_{i_r}) = 0((u - Pu), Mu_{i_r}) = s_{i_r}(L(u - Pu), u_{i_r}) = 0$, $r \in [1, k]$, and $(Mz, Pu) = 0$. The second equality follows from the first. □

1.2. Variational properties of the eigenvalues and the minimax Courant-Fisher principle. For $v \neq 0$, define the *Rayleigh quotients*

$$s(v) = \frac{(Mv, v)}{(Lv, v)}, \quad \lambda(v) = \frac{(Lv, v)}{(Bv, v)} \tag{1.16}$$

($\lambda(v)$ is defined only for (1.1)). Expansions (1.10) imply that

$$s_1 = \max_{v \neq 0} s(v), \quad s_i = \max_{v \in W_{i-1}^{\perp} \backslash 0} s(v), \ i \geq 2, \tag{1.17}$$

---

[3] If $M > 0$ then P is also an orthoprojector of the preHilbert space $H(M)$ onto U (the original Hilbert space H is embedded into $H(M)$).

where $W_{i-1} \equiv \text{lin} \{u_1, \ldots, u_{i-1}\}$, $i \geq 2$, and the orthogonal complement is taken in the sense of the Hilbert space $H(L)$ (for $M > 0$, the case of the complement in $H(M)$ is also permitted (see Lemma 1)). From (1.12) it follows that $U(t_1 \setminus 0) = \arg\max_{v \neq 0} s(v)$, $U(t_j \setminus 0) = \arg\max_{v \in S_{j-1}^\perp \setminus 0} s(v)$, $j = 2, \ldots$, where $S_{j-1} = U(t_1) \oplus U(t_2) \oplus \cdots \oplus U(t_{j-1})$, $j \geq 2$. Along the same lines, we can show that

$$\lambda_1 = \min_{v \neq 0} \lambda(v), \quad \lambda_i = \min_{v \in W_{i-1}^\perp \setminus 0} \lambda(v), \ i \geq 2. \tag{1.18}$$

Theorem 1. *Let problem (1.2) with $L \in \mathcal{L}^+(H)$ and $M = M^*$ be such that the operator $A \equiv L^{-1}M$ is compact. Then*

$$s_i = \min_{V_{i-1}} \max_{v \in V_{i-1}^\perp \setminus 0} s(v), \quad s_i = \max_{V_i} \min_{v \in V_i \setminus 0} s(v), \ i = 1, 2, \ldots, \tag{1.19}$$

where V_{i-1} denotes an arbitrary $(i-1)$-dimensional subspace of H and its orthogonal complement is taken in the sense of $H(L)$ or $H(M)$ (if $M > 0$).

Proof. It is easy to see that there exists $v \in V_{i-1} \cap W_i$, $v \neq 0$. Hence, $s(v) \geq s_i$ and $\sup s(v) \geq s_i$. But, for $V_{i-1} = W_{i-1}$, we see that s_i is attainable and the first equality in (1.19) holds. Analogously, there exists $v \in V_i \cap W_{i-1}^\perp$, $v \neq 0$. Then $s(v) \leq s_i$ and $\min s(v) \leq s_i$. For $V_i = W_i$, $s_i = \min_{v \in V_i \setminus 0} s(v)$. $\square$

(1.19) implies that, for $M > 0$, we have

$$\lambda_i = \min_{V_i} \max_{v \in V_i \setminus 0} \lambda(v), \ i \geq 1. \tag{1.20}$$

Theorem 1 (see also (1.20)) is known as *the minimax Courant-Fisher principle.* It can be used for the proof of our next theorem (see [317, 449]).

Weyl's Theorem. *Let G be a Hilbert space and suppose that operators A and A', considered as elements of $\mathcal{L}(G)$, are symmetric and compact. Let s_i and s_i' denote their ith positive eigenvalues numbered in decreasing order. Then*

$$|s_i - s_i'| \leq \|A - A'\|. \tag{1.21}$$

1.3. The Rayleigh-Ritz method. Consider the problem of finding λ_1 in (1.18). The Rayleigh-Ritz method replaces it by the problem of minimizing $\lambda(v)$ on the finite-dimensional subspace $\hat{H}$ from a chosen sequence $\{\hat{H}_h\}$ of the same type as in the preceding chapters. Since

$$\frac{\partial \lambda}{\partial v}|_u \equiv \frac{d}{dt}\lambda(u + tv)|_{t=0} = \frac{2}{\|u\|_M^2}(Lu - \lambda(u)Mu, v), \qquad (1.22)$$

then the desired $\hat{\lambda}_1$ is the smallest number among $\hat{\lambda}$ such that there exists $\hat{u} \in \hat{H} \setminus 0$ satisfying

$$b_L(\hat{u}; \hat{v}) = \hat{\lambda} b_M(\hat{u}; \hat{v}), \quad \forall \hat{v} \in \hat{H}. \qquad (1.23)$$

We emphasize that (1.23) can be also obtained from (1.3) directly, which is especially important for approximation of other eigenvalues. If

$$\hat{L} \equiv \hat{P}L\hat{P}, \quad \hat{M} \equiv \hat{P}M\hat{P}, \qquad (1.24)$$

where $\hat{P}$ is the orthoprojector of H onto $\hat{H}$, then (1.23) leads to spectral problem

$$\hat{L}\hat{u} = \hat{\lambda}\hat{M}\hat{u}. \qquad (1.25)$$

If $\hat{\psi}_1, \ldots, \hat{\psi}_N$ is a basis for $\hat{H}$, then (1.25) implies that

$$b_L(\hat{u}; \hat{\psi}_i) = \hat{\lambda} b_M(\hat{u}; \hat{\psi}_i), \quad i = 1, \ldots, N. \qquad (1.26)$$

Hence, (1.25) is equivalent to the algebraic eigenvalue problem

$$\bar{L}\bar{u} = \bar{\lambda}\bar{M}\bar{u}, \qquad (1.27)$$

where $\bar{L} \equiv [b_L(\psi_j; \psi_i)]$, $\bar{u} \equiv [u_1, \ldots, u_N]^T \in \mathbf{R}^N \equiv \bar{H}$, $\bar{\lambda} \equiv \hat{\lambda}$, $\bar{M} \equiv [b_M(\psi_j; \psi_i)]$, Among the spectral pairs $\bar{\lambda}; U(\bar{\lambda})$ for (1.27), we will usually seek those that correspond to p left eigenvalues $\hat{\lambda}_1 \leq \hat{\lambda}_2 \leq \ldots \leq \hat{\lambda}_p$ of (1.25) (they serve as possible approximations to several spectral pairs of (1.1)). In terms of (1.2), we seek $\hat{s}_1 \geq \hat{s}_2 \geq \ldots \geq \hat{s}_p$ for problems

$$\hat{M}\hat{u} = \hat{s}\hat{L}\hat{u}, \qquad (1.28)$$

$$\bar{M}\bar{u} = \bar{s}\bar{L}\bar{u}. \qquad (1.29)$$

   Lemma 2. *Let s_i and $\hat{s}_i$ denote the ith positive eigenvalues of problems (1.2) and (1.28), respectively, numbered in decreasing order. Then*

$$\hat{s}_i \leq s_i. \qquad (1.30)$$

   Proof. Due to Theorem 1, we have $\hat{s}_i = \max_{\bar{V}_i} \min_{v \in \bar{V}_i} s(\bar{v})$, where $s(\bar{v}) \equiv (M\bar{v}, \bar{v})(\bar{L}\bar{v}, \bar{v})^{-1}$, $\bar{V}_i \subset \bar{H}$, and $\dim \bar{V}_i = i$. Expansions (0.2.2) and Lemma 0.4.1 imply that $s(\bar{v}) = s(\hat{v})$ and, therefore,

$$\hat{s}_i = \max_{\hat{V}_i \subset \hat{H}} \ \min_{\hat{v} \in \hat{V}_i \backslash 0} \ s(\hat{v}). \tag{1.31}$$

Comparison of (1.19) and (1.31) yields (1.30). □

The characterization (1.20) implies that

$$\lambda_i \leq \hat{\lambda}_i, \quad i = 1, 2, \dots. \tag{1.32}$$

Moreover, (1.28) can be obtained as a projective approximation of (1.4). Indeed, let $P \equiv P_N$ be the orthoprojector of $G = H(L)$ onto $\hat{H}$ and $\hat{A} \equiv \hat{A}_N \equiv PAP$. If we approximate (1.4) (recall that A is symmetric and compact) by

$$\hat{A}u = \hat{s}u, \tag{1.33}$$

then this with $\hat{s} \neq 0$, together with the fact that $\hat{H}$ is an invariant subspace of $\hat{A}$, implies that u can be replaced by $\hat{u}$ and $(A\hat{u} - \hat{s}\hat{u}, \hat{v})_L = 0, \forall \hat{v} \in \hat{H}$. Hence, (1.33) yields (1.28). Next, observe that $\|(A - \hat{A})v\|_L^2 = \|P(A - \hat{A})v\|_L^2 + \|(I - P)(A - \hat{A})v\|_L^2$, where $\|P(A - \hat{A})v\|_L \leq \|P\|_L \|A(I - P)\|_L = \|(I - P)A\|_L$ and $\|(I - P)(A - \hat{A})\|_L = \|(I - P)Av\|_L$. Hence, (1.30) and (1.21) lead (see [317]) to the well-known estimate

$$0 \leq s_i - \hat{s}_i \leq \|A - \hat{A}\|_L \leq 2^{1/2}\|(I - P)A\|_L. \tag{1.34}$$

Let $S \equiv \{u : \|u\|_L \leq 1\}$. Then $A\{S\}$ is a compact set, which with the condition

$$\lim_{N \to \infty} (I - P_N)v = 0, \quad \forall v \tag{1.35}$$

implies that $\lim_{N \to \infty} \hat{s}_i = s_i, \ i = 1, \dots$.

1.4. Gaps between subspaces. Below, H denotes a Hilbert or Euclidean space. Let U_1 and U_2 be subspaces of H, P_k be the orthoprojector of H onto U_k, and $P_k^\perp \equiv I - P_k$, $k = 1, 2$. The *gap between subspaces* U_1 and U_2 is defined as (see [295, 317])

$$\Theta(U_1; U_2) \equiv \|P_1 - P_2\| = \|P_1^\perp - P_2^\perp\|. \tag{1.36}$$

An important point is that the gap may serve as a distance between subspaces in the metric space of all subspaces of H. We also introduce $S \equiv \{u | \|u\| = 1\}$, $S_k \equiv U_k \cap S$, and

$$\sigma_k \equiv \min_{u \in S_k} \|P_l u\|, \quad d_k \equiv \max_{u \in S_k} \|u - P_l u\| = (1 - \sigma_k^2)^{1/2}, \tag{1.37}$$

$l \neq k$, $k = 1, 2$. It is easy to see that

$$\|P_k^\perp P_l\| \leq d_l, \quad l \neq k. \tag{1.38}$$

Lemma 3. *Let*

$$dim\ U_1 = \ dim\ U_2 < \infty. \tag{1.39}$$

Then $d_1 = d_2$.

Proof. If $d_1 = 1$ and $\sigma_1 = 0$, then there exists $u_1 \in U_1$ such that $\|u_1\| = 1$ and $P_2 u_1 = 0$. This implies that $u_1 \perp U_2$. Then, in accordance with (1.39), we can find $u_2 \in U_2$ such that $u_2 \perp U_1$. Hence, $\sigma_2 = 0$ and $d_2 = 1$. Let now $\sigma_1 \sigma_2 > 0$, $u_1 \in S_1, \|P_2 u_1\| = \sigma_1 = \cos\alpha$, that is, $u_1 \in \arg\min\limits_{u_1 \in U_1} \|P_2 u_1\|^2 \|u_1\|^{-2}$ and $\min\limits_{u_1 \in U_1} \|P_2 u_1\|^2 \|u_1\|^{-2} = \sigma_1^2$. Then $(P_2 u_1 - \sigma_1^2 u_1) \perp U_1$ (see (1.22) and (1.23)) and for $u_2 = 1/\sigma_1 P_2 u_1$ we see that $P_1 u_2 = \sigma_1 u_1$. Hence, $\|P_1 u_2\| = \sigma_1$ and $\sigma_2 \leq \sigma_1$. Analogously, it can be shown that $\sigma_1 \leq \sigma_2$. Hence, $\sigma_1 = \sigma_2$ and $d_1 = d_2$. $\square$

Theorem 3. Suppose (1.39) *is satisfied. Then*

$$\Theta(U_1; U_2) = d_k = (1 - \sigma_k^2)^{1/2} \leq 1, \quad k = 1, 2. \tag{1.40}$$

Proof. It is easy to verify that $(P_2 - P_1)u = P_2 P_1^\perp u - P_2^\perp P_1 u$. Next we make use of the facts that $P_1^\perp = (P_1^\perp)^2$ and $P_1 = (P_1)^2$. Then $(P_2 - P_1)u = P_2(P_1^\perp)^2 u - P_2^\perp(P_1)^2 u$. In accordance with (0.1.7) this implies that $\|(P_2 - P_1)u\|^2 = \|P_2(P_1^\perp)^2 u\|^2 + \|P_2^\perp P_1^2 u\|^2$. Hence,

$$\|(P_2 - P_1)u\|^2 \leq \|P_2 P_1^\perp\|^2 \|P_1^\perp u\|^2 + \|P_2^\perp P_1\|^2 \|P_1 u\|^2. \tag{1.41}$$

Observe that $\|P_2^\perp P_1\| \leq d_1$ (see (1.38)) and $\|P_2 P_1^\perp\| = \|P_1^\perp P_1\| \leq d_2$. This and (1.41), for any u with $\|u\| = 1$, leads to $\|(P_2 - P_1)u\|^2 \leq d_1^2 = d_2^2$ (see Lemma 1) and to (1.40). $\square$

1.5. Symmetries of eigenfunctions. In the sequel we suppose that a Hilbert or Euclidean space $H \equiv G \equiv W(\Omega)$ is associated with a domain Ω such that it is symmetrical in a sense (see § 2.1) and a symmetry operator $S \in \mathcal{L}(H)$ is defined such that

$$S^2 = I, \quad b_M(Su; Sv) = b_M(u; v), \quad b_L(Su; Sv) = b_L(u; v), \tag{1.42}$$

$\forall u \in H, \forall v \in H$. We also denote by H_s and H_a the subspaces of H defined by the conditions $Su = u$ or $Su = -u$, respectively.

Lemma 4. *Let s_1 be the maximal eigenvalue of problem* (1.2) *(see* (1.10), (1.11)*). Suppose that* (1.42) *holds. Then either*

$$s_1 = \max_{u \in H_s \backslash 0} s(u) \text{ or } s_1 = \max_{u \in H_a \backslash 0} s(u). \tag{1.43}$$

Proof. Let v be any eigenfunction in $U(s_1)$. Consider the element Sv ($Sv \in U(s_1)$ due to (1.42)). Then $v + Sv \in U(s_1)$. If $v + Sv \neq 0$ then $u \equiv v + Sv \in H_s$ and if $v + Sv = 0$ then $v \in H_a$. $\square$

Theorem 4. Let the conditions of Lemma 4 be satisfied. Then there exists an orthonormal basis of $H(L)$ such that (1.10), (1.11) hold and each element u_i belongs to either H_s or H_a.

Proof. For u_1 applies Lemma 4. After that we consider an orthogonal complement in the sense of $H(L)$ to u_1 as a new Hilbert or Euclidean space and again apply Lemma 4. Repeating this procedure we can construct the desired u_i for all $s_i > 0$ and, along the same lines, for $s_i < 0$. $\square$ [4]

§ 2. Error estimates for the Rayleigh-Ritz and projective methods

2.1. Auxiliary key inequalities. [5]

Lemma 1. Let $L = L^* > 0$, and $M = M^* > 0$. Suppose that $Lu_i = \lambda_i M u_i$. Then, for any v with $\|v\|_M = 1$, we have

$$\|v - u_i\|_L^2 = \lambda(v) - \lambda_i + \lambda_i \|v - u_i\|_M^2. \qquad (2.1)$$

Proof. It is easy to see that $\|v - u_i\|_L^2 = (Lv, v) + (Lu_i, u_i) - 2(Lu_i, v)$. This yields $\|v - u_i\|_L^2 = (Lv, v) + \lambda_i(Mu_i, u_i) - 2\lambda_i(Mu_i, v)$ and

$$\|v - u_i\|_L^2 = \lambda(v) + \lambda_i \|u_i\|_M^2 - 2\lambda_i(Mu_i, v).$$

The obtained formula leads to (2.1). $\square$

Lemma 2. Let $L = L^* > 0$, and $M = M^*$. Suppose that $M u_i = s_i L u_i$. Then, for any v with $\|v\|_L = 1$, we have

$$(M(v - u_i), (v - u_i)) = s(v) - s_i + s_i \|v - u_i\|_L^2. \qquad (2.2)$$

Proof. It suffices to verify that $(M(v - u_i, v - u_i) = s(v) + s_i(Lu_i, u_i) - 2s_i(Lu_i, v)$, from which (2.2) is a straightforward consequence. $\square$

Lemma 3. Let $u_{i_1}, \ldots, u_{i_k}$ be eigenfunctions of (1.1) or (1.2), $U \equiv \lim \{u_{i_1}, \ldots, u_{i_k}\}$, and $P \equiv P(U)$ be the orthoprojector of $H(L)$ onto U. Suppose that $\hat{U} \equiv \lim \{\hat{u}_{i_1}, \ldots, \hat{u}_{i_k}\}$ is an approximation to the subspace U in the sense that

$$\Theta \equiv \Theta_L\{U; \hat{U}\} < 1. \qquad (2.3)$$

---

[4] We note that the similar basis for Ker M can be constructed and that each subspace $U(s_j) = U_s(s_j) \oplus U_a(s_j)$, where $U_s(s_j) \subset H_s, U_a(s_j) \subset H_a$.

[5] H again denotes a Hilbert or Euclidean space. We present proofs for the Hilbert space case only because they are easily modified for the simpler Euclidean spaces.

Then

$$\lambda(P\hat{u}) \geq \lambda(\hat{u})(1 - \Theta^2), \quad \forall \hat{u} \in \hat{U} \tag{2.4}$$

(for (1.1)), or

$$s(P\hat{u}) - s_{-1} \leq \frac{s(\hat{u}) - s_{-1}}{1 - \Theta^2}, \quad \forall \hat{u} \in \hat{U} \tag{2.5}$$

(for (1.2)), where $s_1 \leq \inf s(u)$.

Proof. In accordance with (2.3) and Lemma 1.3, the restriction of P to $\hat{U}$ yields a one-to-one correspondence between $\hat{U}$ and U. For (1.1), P is an orthoprojector in the sense of $H(M)$ (see Lemma 1.1) and $\|P\hat{u}\|_M^2 \leq \|\hat{u}\|_M^2$. Hence, it is easy to see that

$$\lambda(P\hat{u}) = \frac{\|P\hat{u}\|_L^2}{\|\hat{u}\|_L^2} \frac{\|\hat{u}\|_L^2}{\|P\hat{u}\|_M^2} \geq \lambda(\hat{u}) \min_{\hat{u} \in \hat{U}} \frac{\|P\hat{u}\|_L^2}{\|\hat{u}\|_L^2}.$$

This and (1.40) lead to (2.4). Next, for (1.2), we observe that $A \equiv M - s_{-1}L \geq 0$ and apply Lemma 1 to our problem, rewritten in the form $Au = (s - s_{-1})Lu$. Then $\|P\hat{u}\|_A^2 + \|(I - P)\hat{u}\|_A^2 = \|\hat{u}\|_A^2$ and we see that $s(P\hat{u}) - s_{-1} = \|P\hat{u}\|_A^2 \|P\hat{u}\|_L^{-2} \leq \|\hat{u}\|_A^2 \|P\hat{u}\|_L^{-2} \equiv X(\hat{u})$. We then have

$$X(\hat{u}) = \frac{\|\hat{u}\|_A^2}{\|\hat{u}\|_L^2} \frac{\|\hat{u}\|_L^2}{\|P\hat{u}\|_L^2} \leq [s(\hat{u}) - s_{-1}](\min\{\frac{\|P\hat{u}\|_L^2}{\|\hat{u}\|_L^2}\})^{-1} \leq \frac{s(\hat{u}) - s_{-1}}{1 - \Theta^2},$$

which yields (2.5). $\square$

Note that $s_{-1} = 0$ for $M \geq 0$ and $s_{-1} = s(u_{-1})$ otherwise (see (1.11)).

2.2. Estimates for approximation of the eigenvalues. We make use of $\hat{P}$—the orthoprojector of $H(L)$ onto an approximating subspace $\hat{H}_h$—and introduce $S_i \equiv \text{lin}\{u_1, \ldots, u_i\}$, where $i \leq p < \dim \hat{H}_h$ and p is a fixed integer such that $\dim \hat{P}\{S_p\} = p$. This implies that $\dim S_i = \dim \hat{S}_i'$, where $\hat{S}_i' \equiv \hat{P}\{S_i\}$, $i \leq p$. We analyze the error in approximation of the subspace S_i by $\hat{S}_i'$ using the quantity

$$\rho(S_i) \equiv \sup_{u \in S_i, \|u\|_L = 1} \|u - \hat{P}u\|_L = \Theta_L\{S_i; \hat{S}_i'\} < 1. \tag{2.6}$$

Theorem 1. *For problem (1.1) with expansions (1.14), let* $\hat{S}_p'$ *be introduced as above and suppose* $\dim \hat{S}_p' = p$. *Suppose also that* $\hat{\lambda}_1 \leq \hat{\lambda}_2 \leq \ldots \leq \hat{\lambda}_p$ *and* $\hat{u}_1, \ldots, \hat{u}_p$ *denote the first p eigenvalues and corresponding eigenfunctions of projective problem (1.25), with the eigenvectors forming an orthonormal basis for* $\hat{S}_p \equiv \hat{S}$ *in the sense of $H(M)$. Then*

$$0 \leq \hat{\lambda}_i - \lambda_i \leq \hat{\lambda}_i(\rho(S_i))^2, \quad i \in [1, p]. \tag{2.7}$$

Proof. For problem (1.25) we apply the inner Rayleigh-Ritz method with the approximating subspace $\hat{S}'_i, i \leq p$. Let $\hat{\lambda}'_1 \leq \hat{\lambda}'_2 \leq \ldots \leq \hat{\lambda}'_i$ and $\hat{u}'_1, \ldots, \hat{u}'_i$ be respective approximations to $\hat{\lambda}_1, \ldots, \hat{\lambda}_i$ and $\hat{u}_1, \ldots, \hat{u}_p$. These approximate eigenfunctions form a basis for $\hat{S}'_i$ and $\lambda(\hat{u}'_i) \equiv \hat{\lambda}'_i \geq \hat{\lambda}_i$. Moreover,

$$0 \leq \hat{\lambda}_i - \lambda_i \leq \hat{\lambda}'_i - \lambda_i. \tag{2.8}$$

In accordance with Lemma 1.3, we observe that $\rho_i < 1$ (otherwise dim $\hat{S}'_i \neq$ dim $\hat{S}_i$) and apply (2.4) with $u = \hat{u}'_i$ and $Pu \in S_i$. This yields $\lambda_i \geq \hat{\lambda}'_i(1 - \rho_i^2) \geq \hat{\lambda}_i(1 - \rho_i^2)$, which, together with (2.8), leads to (2.7). □

Theorem 2. For problem (1.2) with expansions (1.10), let $\hat{S}'_p$ be introduced as above and suppose dim $\hat{S}'_p = p$. Suppose also that $\hat{s}_1 \geq \hat{s}_2 \geq \ldots \geq \hat{s}_p$ and $\hat{u}_1, \hat{u}_2, \ldots, \hat{u}_p$ denote the first p respective eigenvalues and eigenfunctions of projective problem (1.29) which form an orthonormal basis for $\hat{S}_p \equiv \hat{S}$ in the sense of $H(L)$. Then $0 \leq s_i - \hat{s}_i \leq (s_i - s_{-1})(\rho(S_i))^2$, $i \in [1, p]$.

Proof. The proof is analogous to the proof of (2.7) and is based on the inequalities $s_i - \hat{s}_i \leq s_i - \hat{s}'_i$ and $s_i - s_{-1} \leq (\hat{s}'_i - s_{-1})(1 - \rho_i^2)^{-1}$. □

2.3. Estimates for gaps between subspaces. We make use of expansions (1.14) and (1.15).

Lemma 4. Suppose v is such that $\|v\|_M = 1$ and $\nu_1 \leq \lambda(v) \leq \nu_2$. Then

$$\bar{c}_1^2 \geq \frac{\nu_2 - \lambda(v)}{\nu_2 - \nu_1}, \tag{2.9}$$

where $\bar{c}_1^2 \equiv c_1^2 + \cdots + c_{k_1}^2$, the coefficients c_i are defined by (1.14) and dim $U(\nu_1) = k_1$.

Proof. We have $\lambda(v) = (Lv, v) = \sum_{i=1}^{\infty} \lambda_i c_i^2$. Hence, $(\lambda(v) - \lambda_1)\bar{c}_1^2 = \sum_{i>k_1}(\lambda_i - \lambda(v))c_i^2 \geq (\nu_2 - \lambda(v))(1 - \bar{c}_1^2)$ and (2.9) holds. □

In accordance with the fact that dim $U(\nu_j) = k_j$, we partition the set of $\hat{\lambda}_i$ into groups

$$\hat{\lambda}_1 \leq \hat{\lambda}_2 \leq \cdots \leq \hat{\lambda}_{k_1} \equiv \hat{\nu}_1, \quad \hat{\lambda}_{k_1+1} \leq \cdots \leq \hat{\lambda}_{k_1+k_2} \equiv \hat{\nu}_2, \cdots, \tag{2.10}$$

regarding $\hat{\nu}_l \equiv \hat{\lambda}_p$ with $p = k_1 + \cdots + k_l$ as an approximation to ν_l. Then $\hat{U}_l \equiv \text{lin}\{\hat{u}_{q+1}, \ldots, \hat{u}_p\}$ with $q \equiv k_1 + \cdots + k_{l-1}$ will serve as an approximation to $U(\nu_l)$. Analogously, we define $\hat{U}_j$ with $j \leq l$ and $\hat{S}_p \equiv \hat{U}_1 \oplus \cdots \oplus \hat{U}_l$ as an approximation to $S_p \equiv U(\nu_1) \oplus \cdots \oplus U(\nu_l)$.

Theorem 3. Suppose that the conditions of Theorem 1 are satisfied with $p = k_1 = dim \, U(\nu_1)$ *and* $\nu_1 \leq \hat{\nu}_1 \equiv \hat{\lambda}_{k_1} \leq \nu_2$. *Then* $\Theta_M(\hat{U}_1; U(\nu_1)) \leq [\hat{\nu}_1 - \nu_1]^{1/2}[\nu_2 - \nu_1]^{-1/2}$.

Proof. Consider any $\hat{v} \equiv v \in \hat{S}_p$ such that $\|v\|_M = 1$. Then (2.9) applies and

$$\text{dist}_M\{v; U(\nu_1)\} = (1 - \bar{c}_1^2)^{1/2} \leq \frac{(\lambda(v) - \nu_1)^{1/2}}{(\nu_2 - \nu_1)^{1/2}}. \qquad (2.11)$$

This, together with (1.40), yields (2.11). □

Theorem 4. Suppose that the conditions of Theorem 1 are satisfied with $p = k_1 + \cdots + k_j$ *and* $\hat{\nu}_{j-1} < \hat{\nu}_j \leq \nu_{j+1}$. *Then*

$$\Theta_M(\hat{U}_j; U(\nu_j)) \leq (\hat{\nu}_j - \nu_j)^{1//2}[\frac{1}{(\nu_{j+1} - \nu_j)^{1/2}} + \frac{1}{(\hat{\nu}_j - \hat{\nu}_{j-1})^{1/2}}]. \qquad (2.12)$$

Proof. Let $U'_{j,h} \equiv \hat{S}_p \cap [\text{lin} \{u_1, \ldots, u_q\}]^{\perp}$, where $p = q + k_j$ and the orthogonal complement is taken in $H(M)$. Then $\dim U'_{j,h} = k_j$ and

$$\Theta_M(\hat{U}_j; U(\nu_j)) \leq \Theta_M(\hat{U}'_{j,h}; U(\nu_j)) + \Theta_M(\hat{U}_j; U'_{j,h}). \qquad (2.13)$$

To estimate the first term on the right-hand side of (2.13), consider $v \in U'_{j,h}$ with $\|v\|_M = 1$. In accordance with (1.14), $v = c_{q+1}u_{q+1} + \cdots$. Observing that $\lambda(v) \in [\lambda_p, \hat{\lambda}_p] \subset [\nu_j, \nu_{j+1}]$ and applying estimates of type (2.9) and (2.11) yields

$$\Theta_M(\hat{U}'_{j,h}; U(\nu_j)) \leq \frac{(\hat{\nu}_j - \nu_j)^{1/2}}{(\nu_{j+1} - \nu_j)^{1/2}}. \qquad (2.14)$$

Next, we estimate $\Theta_M\{\hat{U}_j; U'_{j,h}\}$, where $\hat{U}_j \subset \hat{S}_p, U'_{j,h} \subset \hat{S}_p$, and $\hat{u}_1, \ldots, \hat{u}_p$ is an orthonormal basis for $\hat{S}_p$ in the sense of $H(M)$. Let $\phi(v) = X \equiv [x_1, \ldots, x_p]^T$ (with $v = x_1\hat{u}_1 + \ldots + x_p\hat{u}_p$) define an isometry between $\hat{S}_p$ and $\mathbf{H} \equiv \mathbf{R}^p$; $\phi(\hat{u}_i) \equiv X_i, i \in [1, p]$. Let $L_p \equiv [(\hat{u}_j, \hat{u}_i)_L]$ be the Gram matrix for $\hat{u}_1, \ldots, \hat{u}_p$. Then it is easy to see that $L_p X_i = \hat{\lambda}_i X_i, \lambda(v) = \lambda(X) \equiv (L_p X, X)\|X\|^{-2}$, and $\phi\{\hat{U}_j\} = \text{lin}\{X_{q+1}, \ldots, X_p\}$. If $v \in U'_{j,h}$ and $\|v\|_M = 1$, then $\lambda(v) \geq \nu_{j-1}$ and $[\lambda(v) - \hat{\nu}_{j-1}]c^2 \leq [\hat{\nu}_j - \lambda(v)](1 - c^2)$, where $c^2 \equiv x_1^2 + \cdots + x_q^2 = \text{dist}_M^2\{v; \hat{U}_j\}$. Hence, we conclude that $\Theta_M^2(\hat{U}_j; U'_{j,h}) \leq [\hat{\nu}_j - \nu_j][\hat{\nu}_j - \hat{\nu}_{j-1}]^{-1}$ which leads to (2.12). □

From (2.12) and (2.1) we can obtain an estimate for $\Theta_L^2(\hat{U}_j; U'_{j,h})$. But a better way is indicated below for problems (1.2), (1.28) when

$$\hat{s}_1 \geq \ldots \geq \hat{s}_{k_1} \equiv \hat{t}_1 \geq \hat{s}_{k_1+1} > \ldots \geq \hat{s}_{k_1+k_2} \equiv \hat{t}_2 \ldots \qquad (2.15)$$

and the subspaces $\hat{U}_j$ are defined by $\hat{t}_j$ as above.

Lemma 5. Let $\dim U(t_1) = k_1$, $\|v\|_L = 1$, and $t_2 \leq s(v) \leq t_1$. Then $\alpha_1^2 \equiv c_1^2 + \cdots + c_{k_1}^2 \geq [s(v) - t_2][t_1 - t_2]^{-1}$, where the coefficients c_i are defined by (1.10).

Proof. Let $\omega_1 \equiv [1, k_1]$. Then it is easy to see that $(t_1 - s(v))\alpha_1^2 = \sum_{i \notin \omega_1}(s_i - s(v))c_i^2 - c_0^2$, where $c_0^2 \equiv \|v_0\|_L^2$ and $s_0 = 0$, and we can apply much the same proof as for (2.9). □

Theorem 5. Suppose that the conditions of Theorem 2 are satisfied with $p = k_1 = \dim U(t_1)$ and that $t_2 \leq \hat{s}_p \equiv \hat{t}_1 \leq t_1$. Then $\Theta_L(\hat{U}_1; U(t_1)) \leq [(t_1 - \hat{t}_1)(t_1 - t_2)^{-1}]^{1/2}$.

Proof. It suffices to combine Lemma 5 and the proof of (2.11). □

Theorem 6. Suppose that the conditions of Theorem 2 are satisfied with $p = k_1 + \cdots + k_j = q + k_j$, where $k_j \equiv \dim U(t_j)$, and that $\hat{s}_p \equiv \hat{t}_j \geq t_{j+1}$ and $\hat{t}_{j-1} > \hat{t}_j$. Then $\Theta_L(\hat{U}_j; U(t_j)) \leq (t_j - \hat{t}_j)^{1/2}[(t_j - t_{j+1})^{-1/2} + (\hat{t}_{j-1} - \hat{t}_j)^{-1/2}]$.

Proof. It suffices to apply much the same proof as for Theorem 4, replacing $H(M)$ everywhere by $H(L)$. For example, instead of (2.14), we now have the estimate

$$\Theta_L\{U'_{j,h}; U(t_j)\} \leq \frac{(t_j - \hat{t}_j)^{1/2}}{(t_j - t_{j+1})^{1/2}} \tag{2.16}$$

which is obtained via (1.10) and (1.11) with $c_i = 0$, $1 \leq i \leq q$. □ [6]

2.4. Estimates of accuracy for projective-grid methods. The above estimates indicate that ε^2-approximations for eigenvalues are consistent with ε-approximations for eigenspaces. We thus speak about ε-approximations for the spectral pair $\{t_j, U(t_j)\}$ if we find $\{\hat{t}_j, \hat{U}(t_j)\}$ such that

$$|t_j - \hat{t}_{j,h}| \leq \epsilon^2, \quad \Theta_L\{\hat{U}_j; U(t_j)\} \leq \epsilon. \tag{2.17}$$

If the original region Ω is such that we can apply PGMs based on quasiuniform triangulations (possibly composite) $T_h(\bar{\Omega})$, then the estimates given in § 2 in terms of $\rho(S_i) \equiv \rho_h(S_i)$ (see (2.6) and (0.2.12)) can be rewritten in terms of $\rho_h(S_p)$, where $p = k_1 + \cdots + k_l$. We then have the estimates $|t_j - \hat{t}_j| \leq K_j^0(\rho_h(S_p))^2$, $\Theta_L\{\hat{U}_j; U(t_j)\} \leq K_j^1\rho_h(S_p)$, $j \leq l$. Thus, to obtain (2.17), we must take $\rho_h(S_p) \asymp \epsilon$ as for the respective problems $Lu = f$ in H. It should be noted that the indicated constants become large when $t_j - t_{j+1}$ and $\hat{t}_j - \hat{t}_{j+1}$ are small. It is then better to use approximations not of the eigenspaces, but their direct sums, provided they correspond to close eigenvalues (see Lemma 5.7). The case when $\bar{\Omega}$ is approximated by

---

[6]The above estimates are of the same type as in [207, 208]. Somewhat different estimates can be found in [30, 305, 449].

$\{\hat{\Omega}_h\}$ is very similar to our study in § 5.1, 5.2.

§ 3. A posteriori error estimates

3.1. Properties of residuals. In the sequel, we consider (1.1) and (1.2) in the Hilbert or Euclidean space H. Again we restrict our proofs to the more difficult Hilbert case. For $v \neq 0$, define the Rayleigh quotients by (1.16) and the residuals by

$$r(v) \equiv Lv - \lambda(v)Mv, \tag{3.1}$$

$$r_0(v) \equiv Mv - s(v)Lv. \tag{3.2}$$

Lemma 1. *For the residuals defined above, we have*

$$\|r(v)\|_{M^{-1}} = \min_\alpha \|Lv - \alpha Mv\|_{M^{-1}} \tag{3.3}$$

and

$$\|r_0(v)\|_{L^{-1}} = \min_\alpha \|Mv - \alpha Lv\|_{L^{-1}}. \tag{3.4}$$

Proof. We observe that $(r(v), Mv)_{M^{-1}} = (Lv, v) - \lambda(v)(Mv, v) = 0$ and that $Mv - \alpha Lv = r(v) - (\alpha - \lambda(v))Mv$. For proving (3.4), we use the orthogonality of $r_0(v)$ and Lv in $H(L^{-1})$. □

Lemma 2. *Suppose that expansions* (1.10) *hold,* $\|v\|_L = 1$, *and*

$$t_{j+1} \leq s(v) \leq t_j, \quad j > 0. \tag{3.5}$$

Then

$$\|r_0(v)\|^2_{L^{-1}} \geq (t_j - s(v))(s(v) - t_{j+1}). \tag{3.6}$$

Proof. The operator $A = L^{-1}M$ is symmetric as an element of $\mathcal{L}(H(L))$. If s does not satisfy (3.5), then $(s - t_{j+1})(s - t_j) > 0$. Therefore,

$$((A - t_j I)(A - t_{j+1}I)v, v)_L \geq 0, \quad \forall v, \tag{3.7}$$

which leads to

$$\|Mv\|^2_{L^{-1}} \geq (t_j + t_{j+1})(s(v) + t_j t_{j+1}) \geq 0. \tag{3.8}$$

Since $\|r_0(v)\|^2_{L^{-1}} = \|Mv\|^2_{L^{-1}} - s^2(v)$, then from (3.8) it follows that $\|r_0(v)\|^2_{L^{-1}} \geq (t_j + t_{j+1})s(v) - t_{j+1}t_j - s^2(v)$. Hence, (3.6) holds. □

Lemma 3. Suppose that expansions (1.14) hold, $\|v\|_M = 1$, and

$$\nu_j \le \lambda(v) \le \nu_{j+1}. \tag{3.9}$$

Then

$$\|r(v)\|_{L^{-1}}^2 \ge \frac{\lambda(v)(\lambda(v) - \nu_j)(\nu_{j+1} - \lambda(v))}{\nu_j \nu_{j+1}}. \tag{3.10}$$

Proof. Taking $w \equiv v/\|v\|_L = v/\lambda^{1/2}(v)$, then $s(w) = \frac{1}{\lambda(v)} \in [t_{j+1}, t_j]$ and $\|r(v)\|_{L^{-1}}^2 = \lambda^3(v)\|r_0(w)\|_{L^{-1}}^2$. Hence (3.10) holds (see (3.6)). [7] □

3.2. A posteriori error estimates.

Theorem 1. Suppose that expansions (1.10) hold, $\|v\|_L = 1$, and

$$t_j \ge s(v) > t_{j+1/2} \ge t_{j+1}. \ j > 0. \tag{3.11}$$

Let $B = B^* > 0$ be a model operator such that $L^{-1} \le 1/\delta_0 B^{-1}$. Then

$$s(v) \le t_j < s(v) + \xi_0(v), \tag{3.12}$$

$$|t_j - s(v) - \frac{\xi_0(v)}{2}| \le \frac{\xi_0(v)}{2}, \tag{3.13}$$

where $\xi_0(v) \equiv \|r_0(v)\|_{B^{-1}}^2 (\delta_0(s(v) - t_{j+1/2}))^{-1}$.
Proof. From (3.11) and (3.6), it follows that

$$\|r_0(v)\|_{L^{-1}}^2 \ge (t_j - s(v))(s(v) - t_{j+1/2}).$$

Hence, $t_j < s(v) + \|r_0(v)\|_{L^{-1}}^2 [s(v) - t_{j+1/2}]^{-1}$, which leads to (3.12) if we replace $\|r_0(v)\|_{L^{-1}}^2$ by $\delta_0^{-1}\|r_0(v)\|_{B^{-1}}^2$. Finally, (3.13) is equivalent to (3.12). □

Theorem 2. Suppose that expansions (1.15) hold, $\|v\|_M = 1$, and

$$\nu_j \le \lambda(v) < \nu_{j+1/2} \le \nu_{j+1}. \tag{3.14}$$

Let $B = B^* > 0$ be a model operator such that $L^{-1} \le 1/\delta_0 B^{-1}$. Then

$$\lambda(v)(1 + \xi(v))^{-1} < \nu_j \le \lambda(v), \tag{3.15}$$

$$|\nu_j - \frac{2\lambda(v) + \xi(v)}{2(1 + \xi(v))}| \le \frac{\lambda(v)\xi(v)}{2(1 + \xi(v))}, \tag{3.16}$$

---

[7] We refer to (3.6) and (3.10) as *generalized Temple's inequalities*; our proof is a generalization of one given by Kato (see references in [295]) for the simplest case of (3.6) with $L = I$; the proof of (3.10) based on the use of expansions (1.14) was given in [184, 213].

where $\xi(v) \equiv \|r(v)\|_{B^{-1}}^2 [\delta_0 \lambda(v)(1 - \lambda(v)/\nu_{j+1/2})]^{-1}$.
    Proof. It suffices to start from (3.10) and verify that

$$\frac{\lambda(v)}{\nu_j} - 1 \leq \frac{\|r(v)\|_{L^{-1}}^2}{\lambda(v)(1 - \lambda(v)\nu_{j+1/2}^{-1})}.$$

We then obtain (3.15) and (3.16). □
    We use the model operator B to make the estimates more practical. For
finding $\|r\|_{B^{-1}}^2$, it suffices to solve the model operator equation $Bz = r$ and
evaluate (z, r).
    Theorem 3. *Let the conditions of Theorem 1 be satisfied. Then*

$$\min\{|s(v) - t_j|; |s(v) - t_{j+1}|\} \leq \|r_0(v)\|_{B^{-1}}^2 \delta_0^{-1}. \qquad (3.17)$$

    Proof. Following a well-known proof (see [402]), it suffices to consider
only the case when $t_{j+1} < s(v) < t_j$ and the operator $A - s(v)I$ (see the
proof of Lemma 2) is invertible. Since $L^{-1}r_0(v) = (A - s(v)I)v$, then
$\|v\|_L \leq \|(A - s(v)I)^{-1}\|_L \|r_0(v)I\|_{L^{-1}}$, which leads to (3.17). □

§ 4. Modified gradient methods with model operators

    4.1. Basic computational algorithms. In the Euclidean space $H \equiv$
$\mathbf{R}^N$, we pose the algebraic eigenvalue problem (actually a discretization of
(1.1))

$$Lu = \lambda M u, \qquad (4.1)$$

with $L = L^* > 0$ and $M = M^* > 0$. We order its eigenvalues λ_i according
to

$$\lambda_1 \leq \lambda_2 \leq \ldots \leq \lambda_N \qquad (4.2)$$

and write them also in the form

$$\nu_1 < \nu_2 < \ldots < \nu_l, \qquad (4.3)$$

where $\lambda_1 = \nu_1, \ldots, \lambda_N = \nu_l$. For $\lambda_i = \nu_j$, we denote by $U(\lambda_i) \equiv U(\nu_j)$ the
corresponding eigenspace, where $\dim U(\nu_j) \equiv k_j$, $j \in [1, l]$, $k_1 + \cdots + k_l = N$,
and $\lambda_1 = \cdots = \lambda_{k_1} = \nu_1 < \lambda_{k_1+1} = \cdots = \lambda_{k_1+k_2} = \nu_2 < \cdots < \lambda_N = \nu_l$.
    There are many well-known algorithms for solving (4.1), especially for its
partial variant when several, say, the first p eigenvalues are needed (see [402,
505]). We are interested in the *modified gradient methods* for minimizing
$\lambda(u)$. The modification is of the same nature as in § 0.3—instead of the
standard gradient in H we make use of the gradient in the Euclidean space

$H(B)$, where $B \in \mathcal{L}^+(H)$ is a model operator. It is easy to see that $\frac{\partial \lambda}{\partial v}\,|_u = (2B^{-1}r(u), v)_B/\|v\|_M^2$ (see (1.22) and (3.1)), so

$$\mathrm{grad}_B \lambda(u)\,|_v = 2B^{-1}r(u)/\|u\|_M^2. \tag{4.4}$$

For a given u^0 with $\|u^0\|_M = 1$, the above methods are defined by the recursive relations

$$\tilde{u}^{n+1} = u^n - \tau_n v^n, \quad u^{n+1} = \frac{\tilde{u}^{n+1}}{\|\tilde{u}^{n+1}\|_M}, \tag{4.5}$$

where $\|u^n\|_M = 1$, $v^n \equiv B^{-1}r(u^n)$, $r(u^n) \equiv r \equiv Lu^n - \mu^n Mu^n$, and $\mu^n \equiv (Lu^n, u^n)/(Mu^n, u^n)$.[8]

In addition to (4.5) with a fixed iteration parameter $\tau > 0$, we also consider methods based on the choice

$$\tau_n = \arg\min_\tau \mu(u^n - \tau w^n), \tag{4.6}$$

which is typical for the *method of the steepest descent*. [9]

4.2. Analysis of the modified gradient method with fixed step size. Here we study convergence of (4.5) with $\tau_n = \tau > 0$ and of its generalization dealing with orthoprojection with respect to known eigenvectors. Let $\omega_0 \equiv \{1, \ldots, m\}$ be a subset of the indices i for which the eigenvalues λ_i and corresponding eigenvectors u_i have been computed. We assume that $u_1, \ldots, u_m$ form an orthonormal basis in the sense of $H(M)$ for the subspace $Q \equiv Q_m \equiv \mathrm{lin}\,\{u_1, \ldots, u_m\}$, and we denote by $P \equiv P_\omega$ the orthoprojector of $H(M)$ onto Q. We also use $P^\perp \equiv I - P$, that is, the

---

[8] We prefer here and below to write μ^n instead of $\lambda(u^n)$ (see (1.16)) to emphasize the fact that these numbers are obtained explicitly in our algorithms. We emphasize that (4.5) may be considered as a standard gradient method for the preconditioned problem $B^{-1}Lu = \lambda B^{-1}Mu$ in the Euclidean space $H(B)$.

[9] The idea of the method on the differential level can be found in [292, 441]; similar iterative methods for difference problems were applied in [247, 431]; the importance of methods (4.6) for the theory of asymptotically optimal algorithms was stressed in [175]; the first results relevant to the study of such methods for finding λ_1 under natural assumptions on the operators and initial iterate were obtained in [212, 213]; in [184, 185, 188, 189, 190, 192, 195, 197], similar algorithms were considered and studied for more difficult problems dealing with several eigenvalues and based on orthogonalization. An important class of algorithms based on iterations of subspaces was obtained in [207, 208]. Currently, iterative algorithms with model operators have attracted a lot of attention (see, e.g., [253, 305, 307, 348, 359, 381]). For grid problems, an appropriate choice for B is usually associated with the spectral equivalence of the operators L and B (as stressed in [175]), and the model operator B can be selected out of the known samples (see Chapters 3, 5, 8).

orthoprojector of $H(M)$ onto $Q_m^\perp$. By virtue of Lemma 1.1, these operators are also orthoprojectors in the sense of $H(L)$. [10] We replace (4.5) by

$$\tilde{u}^{n+1} \equiv u^n - \tau P^\perp v^n \text{ and } u^{n+1} \equiv \tilde{u}^{n+1}/\|\tilde{u}^{n+1}\|_M, \tag{4.7}$$

where $u^n \in Q_m^\perp, \|u^n\|_M = 1, v^n \equiv B^{-1}r^n$, and the model operator $B \in \mathcal{L}^+(H)$ satisfies

$$\delta_0 B \le L \le \delta_1 B, \quad \delta_0 > 0. \tag{4.8}$$

    Lemma 1. *Suppose that conditions (4.8) are satisfied,* $\|u^n\|_M = 1$, *and* $\tilde{u}^{n+1} \equiv u^n - \tau w^n$, *where* $w^n \equiv P^\perp v^n$ *and* $v^n \equiv B^{-1}r^n$. *Then* $\|\tilde{u}^{n+1}\|_L^2 \ge \delta\|u^n\|_L^2$, *where* $\delta \equiv \delta_0/\delta_1 \le 1$.
    Proof. If $m = 0$ and $P^\perp = I$, then $(Bu^n, v^n) = 0$ and $\|\tilde{u}^{n+1}\|_B^2 = \|u^n\|_B^2 + \tau^2\|r^n\|_{B^{-1}}^2$. If $m \ge 1$, then we need a more complicated proof. First, observe that $Q^\perp$ is an invariant subspace of the operator $C \equiv P^\perp B^{-1}LP^\perp$. We consider $Q^\perp$ as a Euclidean space G with the inner product defined by $H(L)$. Let C_0 be the restriction of C to G. Then it is not difficult to verify that $C_0 \in \mathcal{L}^+(G)$ and that $(C_0u, u) = \|Lu\|_{B^{-1}}^2$. Next, observe that $\delta_0 I \le C_0 \le \delta_1 I$, $L^{-1}r^n \equiv r_L^n \in G$, and $\tilde{u}^{n+1} = u^n - \tau C_0 r_L^n$. Since $(C_0^{-1}(C_0 r_L^n), u^n)_G = (r_L^n, u^n)_L = 0$, then $\|\tilde{u}^{n+1}\|_{G(C_0^{-1})}^2 = \|u^n\|_{G(C_0^{-1})}^2 + \tau^2\|r^n\|_{L^{-1}}^2$. This yields $1/\delta_0\|\tilde{u}^{n+1}\|_G^2 \ge 1/\delta_1\|u^n\|_G^2$. □
    To study convergence of the method, we simplify rather complicated formulations and the proof of Theorem 1, by introducing some conditions and notations beforehand. If

$$0 < \tau < 2/\delta_1, \tag{4.9}$$

then we write $\gamma \equiv \tau(2 - \tau\delta_1) > 0$. We also write $\lambda_{m+1} = \nu_j$, $U'(\nu_j) \equiv U(\nu_j) \cap Q^\perp$. For $u^n \in Q^\perp$, we have $\lambda(u^n) \equiv \mu^n \ge \nu_j$.
    Lemma 2. *Suppose that* $\nu_j \le \mu \le \mu^0 < \nu_{j+1}$ *and consider the function*

$$\rho(\mu) \equiv \frac{1 - \delta_0\gamma\nu_{j+1}^{-1}(\nu_{j+1} - \mu)}{1 + X(\mu)}, \text{ where } X(\mu) \equiv \frac{\delta_0\gamma(\mu - \nu_j)(\nu_{j+1} - \mu)}{\nu_j\nu_{j+1}}.$$

Then $\rho(\mu) \le q(\mu^0)$, *where* $q(\mu) \equiv \max\{\rho(\mu); \rho(\nu_j)\}$.
    Proof. Study of the function $\rho(\mu)$ on $[\nu_j, \nu_{j+1}]$ is reduced to the study of $f(t) \equiv [1 - \beta(1 - t)][1 + \beta(1 - t)(s/t - 1)]^{-1}$, where $t \equiv \mu/\nu_{j+1} \in [s, 1]$, $\beta \equiv \delta_0\gamma < 1$, $s \equiv \nu_j/\nu_{j+1} < 1$. It is not difficult to verify that the functions $f'(t)$ and $t^2 - (1 - \beta)(1 - t^2)$ have the same signs and that $f'(t) = 0$ only

---

[10] If m is small and M is sufficiently simple—this is the case very often—then finding $P^\perp u$ for a given u presents no problems.

for $t = \omega \equiv (1 - \beta)^{1/2}[1 + (1 - \beta)^{1/2}]^{-1}$. This implies that the maximal value of $f(t)$ can be only when $t = s$ or $t = 1$, which proves the lemma. $\square$

Note that $q(\mu^0) = \rho(\nu_j) \equiv \rho_0$ if $\omega > \nu_j/\nu_{j+1}$ and $\mu^0 \leq \omega\nu_{j+1}$.

Theorem 1. *For* (4.7), *suppose that conditions* (4.8) *and* (4.9) *are satisfied. Suppose also that* $u^0 \in Q_m^\perp$ *is such that* $\|u^0\|_M = 1$ *and* $\mu^0 < \nu_{j+1}$. *Then* $\lim_{n \to \infty} \mu^n = \nu_j$, $\lim_{n \to \infty} \|r^n\|_{B^{-1}} = 0$,

$$0 \leq \mu^n - \nu_j \leq \sigma(n)(\mu^0 - \nu_j), \tag{4.10}$$

and

$$\|r^n\|_{B^{-1}}^2 \leq \mu^n(\mu^n - \mu^{n+1})[\gamma\mu^{n+1}]^{-1}, \tag{4.11}$$

where $\sigma(n) \equiv \rho(\mu^0) \cdots \rho(\mu^{n-1}) \leq (q^*)^n$ *and* $q^* \equiv q(\mu^0) < 1$. *Moreover, for* $\rho_M \equiv \mathrm{dist}_M\{u^n; U'(\nu_j)\}$, *we have*

$$\rho_M \leq \left(\frac{\mu^n - \nu_j}{\nu_{j+1} - \nu_j}\right)^{1/2}, \quad \mathrm{dist}_L\{u^n; U'(\nu_j)\} \leq \nu_{j+1}^{1/2}\rho_M. \tag{4.12}$$

Proof. Let u^n satisfy the same conditions as u^0. We will show that u^{n+1} (see (4.7)) also satisfies these conditions. Indeed,

$$\mu^n - \mu^{n+1} = Z_1/Z_2, \tag{4.13}$$

where $Z_1 \equiv 2\tau(w^n, r^n) - \tau^2\|w^n\|_L^2 + \tau^2\mu^n\|w^n\|_M^2$, and $Z_2 \equiv \|\tilde{u}^{n+1}\|_M^2 = 1 - 2\tau(u^n, w^n)_M + \tau^2\|w^n\|_M^2$. First, observe that

$$(w^n, r^n) = (P^\perp v^n, r^n) = (M^{-1}r^n, P^\perp v^n)_M$$

and that, for $u^n \in Q_m^\perp$, we have $P^\perp M^{-1}r^n = M^{-1}r^n$. Hence, $(w^n, r^n) = (P^\perp M^{-1}r^n, (I - P)v^n)_M$, and we see that $(w^n, r^n) = \|r^n\|_{B^{-1}}^2$. Next, we estimate from below the term $-\|w^n\|_L^2$, making use of the evident inequality $\|P^\perp\|_L \leq 1$ (see Lemma 1.1). This yields $-\|w^n\|_L^2 \geq -\|B^{-1}r^n\|_L^2 \geq -\delta_1\|r^n\|_{B^{-1}}^2$, which leads to

$$Z_1 \geq \gamma\|r^n\|_{B^{-1}}^2 + \tau^2\mu^n\|w^n\|_M^2.$$

Now, for Z_2, we apply the estimate $Z_2 \leq 1 + \alpha + \tau^2(1 + 1/\alpha)\|w^n\|_M^2$, where $\alpha > 0$ is a free parameter. But

$$\mu^n - \mu^{n+1} \geq \max_{\alpha>0}\min\left\{\frac{\gamma\|r^n\|_{B^{-1}}^2}{1 + \alpha}; \frac{\mu^n}{1 + \alpha^{-1}}\right\}.$$

Thus, for $\alpha = \gamma\|r^n\|_{B^{-1}}^2/\mu^n$, we obtain the following key inequality:

$$\mu^n - \mu^{n+1} \geq I(u^n) \equiv \frac{\gamma \|r^n\|_{B^{-1}}^2}{1 + \gamma(\mu^n)^{-1}\|r^n\|_{B^{-1}}^2} \geq 0. \qquad (4.14)$$

(4.14) implies that u^{n+1} is such that $\nu_j \leq \mu^{n+1} \leq \mu^n$ and that the sequence $\{\mu^n\}$ generated by (4.7) is nonincreasing and has a limit. We show now that this limit is the desired λ_{m+1}, that is, ν_j. Observe that the function $f(t) \equiv t/(1 + at)$, with a positive constant a, is increasing for $t \geq 0$. Hence, we may replace $\|r^n\|_{B^{-1}}^2$ by $\delta_0 \|r^n\|_{L^{-1}}^2$ in the expression for $I(u^n)$. This yields the estimate $\mu^n - \mu^{n+1} \geq \delta_0 \gamma \|r^n\|_{L^{-1}}^2 (1 + (\mu^n)^{-1}\delta_0 \gamma \|r^n\|_{L^{-1}}^2)^{-1}$. Now we again make use of a lower bound for $\|r^n\|_{L^{-1}}^2$ (see (3.10)) and see that $\mu^n - \mu^{n+1} \geq \mu^n X(\mu^n)(1 + X(\mu^n))^{-1}$ (the function $X(\mu)$ was defined just before Lemma 2). This is equivalent to

$$\mu^{n+1} - \nu_j \leq \rho(\mu^n)(\mu^n - \nu_j). \qquad (4.15)$$

Since $\rho(\mu^n) < 1$ on $[\nu_j, \mu^0]$, then (4.15), together with the conditions of the theorem and Lemma 2, leads to (4.10). Finally, (4.11) follows from (4.14). Hence, inequalities (4.12) follow from (2.11) and (2.1). □

The rate of convergence of (4.7) depends on $\delta_0 \gamma$, ν_j, ν_{j+1}, and $\nu_{j+1} - \mu^0$, but is independent of sp L. The best value of τ is $\tau = 1/\delta_1$, since we then obtain $\gamma = 1/\delta_1$ and the least values of $\rho(\mu^0)$ and q^*. If L and B are spectrally equivalent operators, then the rate of convergence is independent of the grid. For $u^n \in Q_m^\perp$, with $\mu^n \leq \nu_{j+1}$, we have

$$\|w^n\|_M \leq \|w^n\|_L / \nu_j^{1/2} \leq \|v^n\|_L / \nu_j^{1/2} \leq \delta_1^{1/2} \|r^n\|_{B^{-1}} / \nu_j^{1/2}.$$

This, together with (4.11) and the fact that $\gamma = 1/\delta_1$, implies that

$$\|w^n\|_M \leq \frac{\delta_1(\mu^0)^{1/2}}{\nu_j}(\mu^n - \nu_j)^{1/2} \equiv Y_0(\mu^n - \nu_j)^{1/2}. \qquad (4.16)$$

Theorem 2. *Suppose that the conditions of Theorem 1 are satisfied,* $\|w^n\|_M \leq (2\tau)^{-1}$, *and* $4\tau Y_0(\mu^0 - \nu_j)^{1/2}(q^*)^{n/2} \leq \rho^n$ *for some* $\rho \in (0,1)$ *and for all* $n \geq n_0$, *where* n_0 *is large enough. Then there exists* $u^* \in U'(\nu_j)$ *such that* $\|u^*\|_M = 1$ *and*

$$\|u^n - u^*\|_M \leq \frac{\rho^n}{1 - \rho}, \quad \forall n \geq n_0. \qquad (4.17)$$

Proof. For $n > n_0$, we have $a_n \equiv \|u^n - \tau w^n\|_M \geq 1/2$ and $\|u^n - u^{n+1}\|_M = a_n^{-1}\| - \tau w^n + (1 - a_n)u^n\|_M \leq 4\tau \|w^n\|_M \leq \rho^n$ (see (4.10) and (4.16)). Hence, $\{u^n\}$ is fundamental and (4.17) holds. □

It is easy to estimate $\|u^n - u^{n+1}\|_L$ on the basis of (2.1) and (4.17).

Theorem 3. *Let the conditions of Theorem 1 be satisfied,* $\tau = 1/\delta_1$, *and* $\mu^0 \in I \equiv (\nu_j + (\nu_j - \nu_{j+1})/2, \nu_{j+1})$. *Then there exists* $\kappa(\delta, \nu_j, \nu_{j+1})$ *such that* $\mu^k \leq \nu_j + (\nu_{j+1} - \nu_j)/2$, *where* $k \equiv \kappa(\delta, \nu_j, \nu_{j+1})| \ln(\nu_{j+1} - \mu^0)|$.

Proof. From (4.15) and the inequality $\rho(\mu^n) \leq 1 - \delta/\nu_{j+1}(\nu_{j+1} - \mu^n)$ (for $\mu^n \in I$), we have

$$\nu_{j+1} - \mu^{n+1} \geq Z(\nu_{j+1} - \mu^n),$$

where $Z \equiv 1 + \delta(\nu_{j+1})^{-1}(\mu^n - \nu_j) \geq 1 + \delta(2\nu_{j+1})^{-1}(\nu_{j+1} - \nu_j)$. This leads to the desired estimate. $\square$[11]

4.3. A posteriori adaptation of iterative parameters. Below we denote by δ_1^* the smallest δ_1 in (4.8).

Theorem 4. *Suppose that the conditions of Theorem 1 are satisfied,* $\tau \equiv \tau_n = 1/\delta_1^{(n)} > 0$. *If the computeded* μ^{n+1} *satisfies* $\mu^{n+1} > \mu^n$, *then*

$$\delta_1^* > 2\delta_1^{(n)} + \mu^n \|w^n\|_M^2 \|r^n\|_{B^{-1}}^{-2}. \tag{4.18}$$

If

$$\mu^{n+1} \leq \mu^n, \quad \delta_1^* < 2\delta_1^{(n)}, \tag{4.19}$$

then

$$\delta_1^* \geq \delta_1^{(n)} + [1 - \kappa_n \|r^n\|_{B^{-1}}^{-2}]\delta_1^{(n)}, \tag{4.20}$$

where $\kappa_n \equiv \delta_1^{(n)}(\mu^n - \mu^{n+1})\mu^n/\mu^{n+1}$.

Proof. If $\mu^{n+1} > \mu^n$, then for Z_1 (see the Proof of Theorem 1) we must have $Z_1 < 0$. Combining this with the estimate from below, we see that $\tau(2 - \tau\delta_1^*)\|r^n\|_{B^{-1}}^{-2} + \tau^2 \mu^n \|w^n\|_M^{-2} < 0$. This yields (4.18). If (4.19) holds, then so does (4.11), where $\gamma \equiv \tau(2 - \tau\delta_1^*) > 0$. This implies that $\|r^n\|_{B^{-1}} \leq \kappa_n(2 - \tau\delta_1^*)^{-1}$. Hence, (4.20) holds. $\square$

On the basis of (4.18) or (4.20), we can construct better approximations δ_1^{n+1} to δ_1^* and to the optimal τ.

4.4. Modified method of the steepest descent and its generalizations. We return now to method (4.7) with the iteration parameter defined by (4.6). It is easy to find the explicit form of the function $\varphi(\tau) \equiv \lambda(u^n - \tau w^n)$ (see (1.16)) and to see that the roots of the equation $\varphi'(\tau) = 0$ are just the roots of

$$\tau^2(a_2b_1 - a_1b_2) + \tau(a_1 - \mu^n a_2) - (r^n, w^n) = 0,$$

---

[11]If $\mu^0 > \nu_{j+1}$, then μ^n converges to some ν_k for $k \geq j$. Moreover, normalization of $\tilde{u}^{n+1}$ in (4.7) is not necessary at each iteration. Generalizations of Theorem 1 to the case of the set ω_0 and the subspace Q are possible. We shall consider even the more difficult case in § 5, where this set corresponds to indices of the already computed eigenvectors.

where $a_1 \equiv \|w^n\|_L^2, a_2 \equiv \|w^n\|_M^2, b_1 \equiv (u^n, Lw^n)$, and $b_2 \equiv (u^n, Mw^n)$. So finding τ that corresponds to (4.6) is not a problem if we neglect the possibility of obtaining very large τ. In order to avoid appearance of such large values, we can use minimization of $\lambda(u^{n+1})$ as in the Raylegh-Ritz method. For the more general case, this leads to

$$\tilde{u}^{n+1} = \arg \min_{u \in S_p^{(n)}, \|u\|_M = 1} \mu(u), \tag{4.21}$$

where $\dim S_p^{(n)} \geq 2$ and $\lim \{u^n, w^n\} \equiv S_2^{(n)} \subset S_p^{(n)} \subset \bar{Q}^\perp$. For example, it seems reasonable to use $S_p^{(n)} \equiv \lim \{u^n, P^\perp B^{-1} r^n, \dots, P^\perp B^{-p+1} r^n, \}$.

   Theorem 5. Suppose the initial iterate $u^0 \in Q_m^\perp$, $\|u^0\|_M = 1$, $\mu^0 < \nu_{j+1}$, and we carry out iterations (4.21), with $n = 0, \dots$. Then for this iterative method estimate (4.10) holds, with $\delta_1 = \delta_1^* = 1/\tau$.

   Proof. It suffices to observe that $u^n - \tau w^n \in S_p^{(n)}$ (see Theorem 1). $\square$

   4.5. The general case of the symmetric operator M. Here, with $H \equiv \mathbf{R}^n$, we consider algebraic spectral problem (1.2) with eigenvalues $t_1 = s_1 = \cdots = s_{k_1} > t_2 = s_{k_1+1} = \cdots = s_{k_1+k_2} > \cdots = s_N$. We assume also that $s_N \geq t^*$ and that we know $s_1, \dots, s_m$ and the corresponding eigenvectors $u_1, \dots, u_m$, which form an orthonormal basis in the sense of $H(L)$ for the subspace $Q \equiv Q_m \equiv \lim\{u_1, \dots, u_m\}$. We denote by P the orthoprojector of $H(L)$ onto Q and define $P^\perp \equiv I - P$.

   For finding $s_{m+1} \equiv t_{j+1}$, as in [305] we apply iterations

$$\tilde{u}^{n+1} = u^n + \frac{\tau}{s^n - t^*} w^n, \qquad u^{n+1} = \frac{\tilde{u}^{n+1}}{\|\tilde{u}^{n+1}\|_L}, \tag{4.22}$$

where $\|u^n\|_L = 1$, $w^n \equiv P^\perp v^n$, $v^n \equiv B^{-1} r_0^n$,

$$r_0^n \equiv r_0(u^n) \equiv Mu^n - s^n Lu, \qquad s^n \equiv s(u^n) \equiv \frac{(Mu^n, u^n)}{(Lu^n, u^n)},$$

and the iteration parameter τ either satisfies (4.9) or may be for fastest increase, that is, $\tau = \arg\max s(u^n + \tau w^n)$.

   Theorem 6. Suppose that $u^0 \in Q_m^\perp$, $\|u^0\|_L = 1$, $s^0 > t_{j+1}$, and we carry out iterations (4.22), with τ from (4.9). Then $s^{n+1} \geq s^n$,

$$0 \leq t_j - s^n \leq g(n)(t_j - s^0), \tag{4.23}$$

and

$$\|r_0(u^n/\|u^n\|_L)\|_{B^{-1}}^2 \leq \frac{s^{n+1} - s^n}{\gamma(s^n - t^*)}, \tag{4.24}$$

where $g(n) \equiv f(s^0) \cdots f(s^{n-1}) \leq (\rho^*)^n$, $f(s) \equiv 1 - \delta_0 \gamma \frac{s - t_{j+1}}{s - t^*}$, and $\rho^* \equiv f(t_j) < 1$.

Proof. For $M_0 \equiv M - \omega L > 0$, consider spectral problem

$$Lu = \lambda M_0 u. \tag{4.25}$$

Denote the eigenvalues of (4.25) by $\lambda_1 \leq \cdots \leq \lambda_N$, $\lambda_1 > 0$. It is clear that the eigenspaces $U(t_j)$ of the original problem and $U_0(\nu_j)$ of (4.25) coincide if $\nu_j = (t_j - \omega)^{-1}$ $(\lambda_i = (s_i - \omega)^{-1})$. This implies also that the subspace Q_m spanned by the known eigenvectors is one and the same for both problems. It is important also that P defined for the original problem is the orthoprojector of the Euclidean space $H(M_0)$ onto Q (see Lemma 1.1). Thereby, if for $v^n \neq 0$ we use $\mu_0^n \equiv (Lv^n, v^n)/(M_0 v^n, v^n) = 1/(s(v^n) - \omega)$, then the basic recursive relation of method (4.7) for (4.25) may be rewritten as $v^{n+1} = v^n - \tau P^{\perp} B^{-1}(Lv^n - \mu_0^n M_0 v^n)$ (we emphasize that v^n is not normalized as in (4.7) since it is of no importance here). Hence,

$$v^{n+1} = v^n + \frac{\tau}{s(v^n) - \omega} P^{\perp} B^{-1} r_0(v^n), \tag{4.26}$$

and we have (4.22) with t^* replaced by ω, $u^n = v^n$, and $\tilde{u}^{n+1} = v^{n+1}$. For (4.26) on the basis of (4.15), we have

$$\mu_0^{n+1} - \frac{1}{t_j - \omega} \leq \rho(\mu_0^n)\left(\mu_0^n - \frac{1}{t_j - \omega}\right). \tag{4.27}$$

Note that $\tilde{u}^{n+1} = v^{n+1}$ depends on $\omega < t^*$. Since $\mu_0^{n+1} = [s(u_\omega^{n+1}) - \omega]^{-1}$, then it is not very difficult to show that (4.27) leads to

$$t_j - s(u_\omega^{n+1}) \leq [t_j - s(v^n)]\left(1 - \delta_0\gamma\frac{s(v^n) - t_{j+1}}{s(v^n) - \omega}\right). \tag{4.28}$$

Returning now to (4.22), with $u^n \equiv v^n$, we replace t^* by ω. Then, for each $\omega < t^*$, (4.28) holds. Since both sides of (4.28) depend continuously on the parameter ω and have limits, we may let ω tend to t^*. This implies that

$$t_j - s^{n+1} \leq (t_j - s^n)\left(1 - \delta_0\gamma\frac{s^n - t_{j+1}}{s^n - t^*}\right), \tag{4.29}$$

and therefore (4.23) holds. To prove (4.24), we just apply (4.11) for problem (4.25), where $r^n \equiv r(v^n/\|v^n\|_{M_0}) = [Lv^n - \mu_0^n M_0 v^n]\|v^n\|_{M_0}^{-1}$ and v^n is the same as in (4.26). □

Finally, we emphasize that Theorems 2–5 can be generalized for the case under consideration in a straightforward manner.

§ 5. Modified gradient methods under presence of perturbations

This section is devoted to the analysis of modified gradient methods (see [187, 190]) for the most difficult case where we must account for the effects of approximate orthogonalization of the current eigenvector approximations against those for other eigenvectors that have already been accepted.. We also allow these to include some perturbations due, for example, to approximate solutions of the associated systems or rounding errors. This leads to some new mathematical problems connected with the study not only of the above considered behavior of μ^n (see Theorem 4.1), but also of the possible lack of monotonicity. We show, for properly chosen iteration parameters, μ^n may only increase within an $O(\epsilon^2)$-vicinity of an eigenvalue. The methods we consider are able to find the corresponding eigenvector approximations, but convergence rates depend on the separation of the given eigenvalue λ_i from its neighbors. A possible improvement of the methods dealing with a cluster of eigenvalues will be investigated in § 6.

5.1. Basic computational algorithms. We denote by $u_1, \ldots, u_N$ an orthonormal basis for $H(M)$ such that

$$Lu_i = \lambda_i M u_i, \quad i \in [1, N], \tag{5.1}$$

where $\lambda_1 \leq \lambda_2 \leq \ldots \leq \lambda_N$. Instead of $\lambda(v)$ in (1.16) we prefer to write $\mu(v)$. Let ω_0 be a subset of the indices i for which there have already been found approximations $\bar{u}_i$ to u_i such that

$$\|\bar{u}_i - u_i\|_L \leq \epsilon_{L,i}, \quad \|\bar{u}_i - u_i\|_M \leq \epsilon_{M,i}, \quad \max_{i \in \omega_0} \max\{\epsilon_{L,i}, \epsilon_{M,i}\} = \epsilon; \tag{5.2}$$

$Q \equiv Q_{\omega_0}$ and $\bar{Q} \equiv \bar{Q}_{\omega_0}$, respectively, are the linear spans of u_i and $\bar{u}_i, i \in \omega_0$; dim $Q = m$; the respective $Q^\perp$ and $\bar{Q}^\perp$ are orthogonal complements in $H(M)$ to Q and $\bar{Q}$; the respective P and $\bar{P}$ are orthoprojectors of $H(M)$ onto Q and $\bar{Q}$; $P^\perp \equiv I - P$; and $\bar{P}^\perp \equiv I - \bar{P}$. The proximity of the subspaces Q and $\bar{Q}$ will be characterized by (5.2); the use of the gaps (see § 1) between these subspaces in $H(M)$ and $H(L)$ are denoted by

$$\Theta_M \equiv \Theta_M(Q; \bar{Q}), \quad \Theta_L \equiv \Theta_L(Q; \bar{Q}); \tag{5.3}$$

note that they are just the gaps between the orthogonal complements of the given subspaces.

The simplest generalization of (4.7) for the case under consideration is given by the recursive relations

$$\tilde{u}^{n+1} \equiv \bar{u}^n - \tau \bar{P}^\perp B^{-1} r(\bar{u}^n) + \xi^n, \quad \bar{u}^{n+1} = \tilde{u}^{n+1}[\|\tilde{u}^{n+1}\|_M]^{-1}, \quad (5.4)$$

where $\bar{u}_0 \in \bar{Q}^\perp, \|\bar{u}_0\|_M = 1, B$ is a model operator (see (4.8)), and

$$\bar{P}^\perp \xi^n = \xi^n, \qquad (5.5)$$

$$\sup_n \max\{\|\xi^n\|_L; \|\xi^n\|_M\} \le \epsilon_\xi \le \kappa_1 \epsilon^2, \qquad (5.6)$$

$$\bar{P}^\perp u \equiv u - \sum_{i \in \omega_0} (u, \bar{u}_i)_M u_i. \qquad (5.7)$$

The vector

$$\bar{w}^n \equiv \bar{P}^\perp B^{-1} r(\bar{u}^n) + \bar{\xi}^n, \qquad (5.8)$$

with $\bar{\xi}^n \equiv -\tau^{-1}\xi^n$, may be considered as the final outcome of algorithms for evaluating Lu^n, for solution of systems with the model operator B, and so on. [12] Hence,

$$\tilde{u}^{n+1} = \bar{u}^n - \tau \bar{w}^n. \qquad (5.9)$$

In addition to iterative methods (5.9) with $\tau \approx 1/\delta_1$ (for small enough ϵ), we consider the methods based on the choice

$$\tau = \arg\min_{\tau \ge 0} \mu(\bar{u}^n - \tau \bar{w}^n), \qquad (5.10)$$

or even the more general

$$\tilde{u}^{n+1} = \arg\min_{u \in S_p^{(n)}; \|u\|_M = 1} \mu(u), \qquad (5.11)$$

where

$$\lim\{u^n; w^n\} \subset S_p^{(n)} \subset \bar{Q}^\perp. \qquad (5.12)$$

5.2. Auxiliary inequalities.
Lemma 1. *Suppose (5.2) holds and* dim $Q =$ dim $\bar{Q} = m$. *Then*

$$\Theta_M \le [\sum_{i \in \omega_0} \epsilon_{M,i}^2]^{1/2} \le m^{1/2}\epsilon, \qquad (5.13)$$

---

[12] If M is simple enough, as is often the case, then finding $\bar{P}^\perp u$ for a given u is not difficult. It is possible to use algorithms with orthogonalization in $H(L)$ or even in $H(B)$ (see [305]) when algorithms for obtaining Bv are available.

$$\|(P - \bar{P})u\|_L \le \epsilon[m^{1/2} + \sum_{i \in \omega_0} \lambda_i^{1/2}]\|u\|_M \le \epsilon\kappa_L\|u\|_L, \quad \forall u, \qquad (5.14)$$

where $\kappa_L \ge \lambda_1^{-1/2}[m^{1/2} + \sum_{i \in \omega_0} \lambda_i^{1/2}]$ and

$$\|u - P^\perp u\|_L \le [\sum_{i \in \omega_0} \epsilon_{M,i}\lambda_i^{1/2}]\|u\|_M, \quad \forall u \in \bar{Q}^\perp. \qquad (5.15)$$

Proof. Let $u = \sum_{i \in \omega_0} c_i\bar{u}_i \in \bar{Q}$ and $\|u\|_M = 1$. Then $u - Pu = \sum_{i \in \omega_0} c_i(\bar{u}_i - P\bar{u}_i)$. Hence, $\|u - Pu\|_M \le [\sum_{i \in \omega_0} \epsilon_{M,i}^2]^{1/2}$. This and (1.36) lead to (5.13). Next, we see that

$$X \equiv \|(P - \bar{P})u\|_L = \|\sum_{i \in \omega_0}[(u, u_i)_M u_i - (u, \bar{u}_i)_M \bar{u}_i]\|_L$$

and $X \le \sum_{i \in \omega_0} |(u, \bar{u}_i)_M|\|u_i - \bar{u}_i\|_L + \|u\|_M \sum_{i \in \omega_0} \|u_i - \bar{u}_i\|_M\|u_i\|_L$. Hence,

$$\|(P - \bar{P})u\|_L \le [[\sum_{i \in \omega_0} \epsilon_{L,i}^2]^{1/2} + \sum_{i \in \omega_0} \lambda_i^{1/2}\epsilon_{M,i}]\|u\|_M, \quad \forall u \in \bar{Q}^\perp,$$

which yields (5.14). Since $u - P^\perp u = \sum_{i \in \omega_0}(u, \bar{u}_i - u_i)_M u_i$, then (5.15) holds. $\square$

Let H_p be a p-dimensional subspace of H and

$$\lambda_p^\perp \equiv \min_{H_p \subset Q^\perp} \max_{u \in H_p} \mu(u), \quad \bar{\lambda}_p^\perp \equiv \min_{H_p \subset \bar{Q}^\perp} \max_{u \in H_p} \mu(u), \qquad (5.16)$$

where $p = 1, \ldots, N - m$. It is easily seen that $\lambda_p^\perp$ coincides with the pth eigenvalue λ_i (in increasing order) that has remained after eliminating all λ_i with $i \in \omega_0$ from sp $M^- L$; if $\lambda_i \ge \lambda_p^\perp$ for all $i \in \omega_0$, then $\lambda_p^\perp = \lambda_p$; and it is this case that will be most important in what follows. Of course, $\lambda_1^\perp \le \lambda_2^\perp \le \ldots \le \lambda_{N-m}^\perp$ and similar inequalities hold for $\bar{\lambda}_i^\perp, i = 1, \ldots, N$. Let $\nu_k^\perp$ be $\lambda_i^\perp$ closest to $\lambda_k^\perp$ and strictly greater than $\lambda_k^\perp$.

Lemma 2. Let $\lambda_p^\perp$ and $\bar{\lambda}_p^\perp$ be defined by (5.16) and $q \equiv N - m - p + 1$. Then

$$\lambda_p^\perp = \max_{H_q \subset Q^\perp} \min_{u \in H_q} \mu(u), \quad \bar{\lambda}_p^\perp = \max_{H_q \subset \bar{Q}^\perp} \min_{u \in H_p} \mu(u), \qquad (5.17)$$

where $p = 1, \ldots, N - m$.

Proof. The proof follows from Theorem 1.1. $\square$

Lemma 3. Suppose that (5.2) holds and that $1 - m\epsilon^2 \ge \kappa_0 > 0$ and $(\Theta_L)^2 \le \kappa_0'\epsilon^2$. Then there exist numbers $K_{0,p} > 0$ and $K_{1,p} > 0$ such that

$$-K_{0,p}\epsilon^2 \le \lambda_p^\perp - \bar{\lambda}_p^\perp \le K_{1,p}\epsilon^2. \qquad (5.18)$$

Moreover, if $\lambda_i \leq \lambda_p^\perp$ for all $i \in \omega_0$, then $\bar{\lambda}_p^\perp \leq \lambda_p^\perp$; if $\lambda_i \geq \lambda_p^\perp$ for all $i \in \omega_0$, then $\bar{\lambda}_p^\perp \geq \lambda_p^\perp$; in particular, if $\omega_0 = \{1, \ldots, m\}, p = m + 1$, and $\bar{\lambda}_{m+1} = \bar{\lambda}_p^\perp$, then

$$0 \leq \lambda_{m+1} - \bar{\lambda}_{m+1} \leq \sum_{i=1}^m \epsilon_{M,i}^2 (\lambda_{m+1} - \lambda_i). \tag{5.19}$$

Proof. Bounds (5.13) imply that $P^\perp$ is a one-to-one mapping of $\bar{Q}^\perp$ onto $Q^\perp$, and each d-dimensional subspace of $\bar{Q}^\perp$ corresponds to the respective d-dimensional subspace of $Q^\perp$. Let $u \in \bar{Q}^\perp$ and $\|u\|_M = 1$. Then $u = \sum_{i=1}^N c_i u_i$, where $|c_i| \leq \epsilon_{M,i}$ if $i \in \omega_0$. Hence,

$$\|P^\perp u\|_M^2 \geq 1 - \sum_{i \in \omega_0} \epsilon_{M,i}^2, \quad \mu(P^\perp u) \leq (1 - m\epsilon^2)^{-1} \mu(u).$$

This and (5.16) imply that $\lambda_p^\perp \leq \bar{\lambda}_p^\perp (1 - m\epsilon^2)^{-1}$ and $K_{1,p} \geq \bar{\lambda}_p^\perp m(1 - m\epsilon^2)^{-1}$. To find $K_{0,p}$ in (5.18), we apply (2.4) to the above function u. This and (5.16) imply that $\lambda_p^\perp \geq \bar{\lambda}_p^\perp (1 - \kappa_0' \epsilon^2)$ and $K_{0,p} \geq \kappa_0' / \bar{\lambda}_p^\perp$. The remaining inequalities follow from simpler proofs. $\square$

In what follows, for $u \neq 0$, we write $\mu(u) \equiv \mu$ and $r(u) \equiv r \equiv Lu - \mu M u$ and make use of expansion $u = \sum_{i=1}^N c_i u_i$ (see (5.1)). The set of indices not belonging to ω_0 is partitioned into sets $\omega_1(\mu) \equiv \omega_1$ and $\omega_2(\mu) \equiv \omega_2$, where ω_1 contains only $\lambda_i \leq \mu$ and ω_2 contains only $\lambda_i > \mu$. We let

$$\lambda \equiv \lambda(\mu) \equiv \max_{i \in \omega_1} \lambda_i, \quad \nu \equiv \nu(\mu) \equiv \min_{i \in \omega_2} \lambda_i.$$

Lemma 4. *Suppose that conditions (5.2) are satisfied and that, for $u \in \bar{Q}^\perp$ with $\|u\|_M = 1$, the sets ω_1 and ω_2 are defined as above. Then*

$$\sum_{i \in \omega_1} (\nu - \lambda_i) c_i^2 \geq \nu - \mu + \sum_{i \in \omega_0} (\lambda_i - \nu) c_i^2 \tag{5.20}$$

and

$$\|r(u)\|_{L^{-1}}^2 \geq \frac{\mu(\mu - \lambda)(\nu - \mu)}{\lambda \nu} - \frac{\mu^2}{\lambda \nu} \sum_{i \in \omega_{0,0}} \epsilon_{M,i}^2 \frac{(\nu - \lambda_i)(\lambda_i - \lambda)}{\lambda_i}, \tag{5.21}$$

where $\omega_{0,0} \equiv \omega_{0,0}(\mu)$ is a subset of the indices $i \in \omega_0$ such that $\lambda < \lambda_i < \nu$.
Proof. From (5.1), it follows (see the proof of Lemma 2.4) that

$$\sum_{i \in \omega_1} (\nu - \lambda_i) c_i^2 = \sum_{i \in \omega_2} (\lambda_i - \nu) c_i^2 + \sum_{i \in \omega_0} (\lambda_i - \nu) c_i^2 \tag{5.22}$$

and

$$\sum_{i \in \omega_1} (\nu - \lambda_i) c_i^2 \geq (\nu - \mu) \sum_{i \in \omega_2} c_i^2 + \sum_{i \in \omega_0} (\lambda_i - \nu) c_i^2. \tag{5.23}$$

Adding the term $(\nu - \mu) \sum_{i \in \omega_0 \cup \omega_1} c_i^2$ to both parts of (5.23) leads to (5.20).
To prove (5.21) we observe that

$$\|r\|_{L^{-1}}^2 = -\nu(Lu - \mu Mu, L^{-1}Mu) = \mu \sum_{i=1}^{N} \frac{\nu - \lambda_i}{\lambda_i} c_i^2.$$

Next, according to the definition of the sets ω_1 and ω_2, we have

$$\|r\|_{L^{-1}}^2 \geq \sum_{i \in \omega_0 \cup \omega_1} \frac{\mu - \lambda_i}{\lambda_i} c_i^2 - \frac{\mu}{\nu} \sum_{i \in \omega_2} (\lambda_i - \mu) c_i^2.$$

Here, the term involving $\sum_{i \in \omega_2}$ can be rewritten by way of (5.22). Hence,
in accordance with the definition of ν and λ, we see that

$$\|r\|_{L^{-1}}^2 \geq \frac{\mu}{\nu} \sum_{i \in \omega_1} \frac{(\mu - \lambda_i)(\nu - \lambda_i)}{\lambda_i} c_i^2 + \frac{\mu}{\nu} \sum_{i \in \omega_0} \frac{(\mu - \lambda_i)(\nu - \lambda_i)}{\lambda_i} c_i^2$$

and

$$\|r\|_{L^{-1}}^2 \geq \frac{\mu(\mu - \lambda)}{\lambda\nu} \sum_{i \in \omega_1} (\nu - \lambda_i) c_i^2 + \frac{\mu}{\nu} \sum_{i \in \omega_0} \frac{(\mu - \lambda_i)(\nu - \lambda_i)}{\lambda_i} c_i^2.$$

Now, making use of (5.20), we have

$$\|r\|_{L^{-1}}^2 \geq \frac{\mu(\mu - \lambda)(\nu - \mu)}{\lambda\nu} + \frac{\mu}{\nu} \sum_{i \in \omega_0} c_i^2 (\nu - \lambda_i) \left(\frac{\mu - \lambda_i}{\lambda_i} - \frac{\mu - \lambda}{\lambda} \right),$$

which leads directly to (5.21). □

Lemma 5. Suppose $L = L^* > 0$ and $M = M^* > 0$. Then, for all $u \neq 0$
and $v \neq 0$,

$$|\mu(u) - \mu(v)| \leq \frac{[\mu^{1/2}(u) + \mu^{1/2}(v)][\|u - v\|_L + \mu^{1/2}(v)\|u - v\|_M]}{\|u\|_M}. \tag{5.24}$$

Proof. We have $\mu(u) - \mu(v) = [\|u\|_L^2 \|v\|_M^2 - \|v\|_L^2 \|u\|_M^2] \|u\|_M^{-2} \|v\|_M^{-2}$.
Denote $\mu(u) - \mu(v)$ by X. Then

$$|X| \leq [|\|u\|_L\|v\|_M - \|v\|_L\|u\|_M|[(\|u\|_M\|v\|_M)^{-1}[\mu^{1/2}(u) + \mu^{1/2}(v)].$$

Note that $|\|u\|_L\|v\|_M - \|v\|_L\|u\|_M| \leq \|u - v\|_L\|v\|_M + \|v\|_L\|u - v\|_M$. Hence, (5.24) holds. $\square$

Lemma 6. *Suppose that* (4.8) *holds and that* $L \geq \lambda_1 M$, *where* $\lambda_1 > 0$. *Let* $A_L \equiv L^{1/2}B^{-1/2}$, $A_M \equiv M^{1/2}B^{-1/2}$, *and* $A \equiv M^{1/2}B^{-1/2}$. *Then*

$$\|A_L\| = \|A_L^*\| \leq \delta_1^{1/2}, \ \|A_M\| = \|A_M^*\| \leq \lambda_1^{-1/2}, \ \|A\| = \|A^*\| \leq \delta_1^{1/2}\lambda_1^{-1/2}.$$

Proof. It suffices to apply Lemmas 0.4.4 and 0.4.6. $\square$

Lemma 7. *Suppose that the conditions of Lemma 4 are satisfied and* $U(\mu; \omega_3)$ *is the direct sum of some eigenspaces* $U(\lambda_i)$ *with* $i \in \omega_3$, *where* ω_3 *is a subset of* $\omega_1 \cup \omega_2$. *Let* $\omega_4 \equiv (\omega_1 \cup \omega_2) \setminus \omega_3$. *Then*

$$dist_L^2\{u; U(\mu; \omega_3)\} \leq \frac{K_{\mu,\omega_3}}{\delta_0}\|r(u)\|_{B^{-1}}^2 + \sum_{i \in \omega_0} \lambda_i \epsilon_{M,i}^2, \tag{5.25}$$

and $dist_M^2\{u; U(\mu; \omega_3)\} \leq K'_{\mu,\omega_3}/\delta_0\|r(u)\|_{B^{-1}}^2 + \sum_{i \in \omega_0} \epsilon_{M,i}^2$, *where*

$$K_{\mu,\omega_3} \geq [\min_{i \in \omega_4}[1 - \mu(u)/\lambda_i]^2]^{-1}, \ K'_{\mu,\omega_3} \equiv K_{\mu,\omega_3}[\min_{i \in \omega_4}\{\lambda_i\}]^{-1}. \tag{5.26}$$

If $\omega_0 = \emptyset$ *and* $\omega_3 = \omega_1$ *then*

$$\text{dist}_M^2\{u; U(\mu; \omega_1)\} \leq \frac{\mu - \lambda}{\nu\mu}, \ \text{dist}_L^2\{u; U(\mu; \omega_1)\} \leq \nu\frac{\mu - \lambda}{\nu\mu}. \tag{5.27}$$

Proof. Using the same expansion as in Lemma 4 we see that

$$\|r(u)\|_{L^{-1}}^2 = \sum_{i=1}^N \frac{(\lambda_i - \mu)^2}{\lambda_i}c_i^2 \geq \sum_{i \in \omega_4} \frac{(\lambda_i - \mu)^2}{\lambda_i^2}\lambda_i c_i^2.$$

Hence, $\|r(u)\|_{L^{-1}}^2 \geq a\sum_{i \in \omega_4} \lambda_i c_i^2$ and

$$\|r(u)\|_{L^{-1}}^2 \geq a \sum_{i \in \omega_4 \cup \omega_0} \lambda_i c_i^2 - a \sum_{i \in \omega_0} \lambda_i c_i^2,$$

where $a \equiv \min_{i \in \omega_4}[1 - \mu/\lambda_i]^2$. This yields

$$\|r(u)\|_{L^{-1}}^2 \geq a\text{dist}_L^2\{u; U(\mu; \omega_3)\} - a \sum_{i \in \omega_0} \lambda_i \epsilon_{M,i}^2$$

and (5.25). The estimate for $\text{dist}^2_M\{u; U(\mu; \omega_3)\}$ is obtained in the same manner. Finally, if $\omega_0 = \emptyset, \omega_3 = \omega_1$, and $k_1 = \dim U(\mu; \omega_1)$, then it is easy to see that $\bar{c}_1^2 \geq (\nu - \mu)(\nu - \lambda)^{-1}$, where $\bar{c}_1^2 \equiv c_1^2 + \cdots + c_{k_1}^2$ and the coefficients c_i are defined by (1.14) for the given u (see the proof of (2.9)). This yields the desired estimate for $\text{dist}^2_M\{u; U(\mu; \omega_1)\}$ (see also Theorem 2.3) and, together with (2.1), leads to (5.27). $\square$

A reasonable choice for ω_3 in this lemma is defined by indices i such that λ_i are close to $\mu(u)$. In the simplest case, we have $U(\mu; \omega_3) = U(\lambda^*)$, where λ^* corresponds to $\min\limits_{i \in \omega_1 \cup \omega_2} |\mu(u) - \lambda_i|$. In this case we write simply K_μ instead of K_{μ, ω_3}. [13]

5.3. Convergence of the method. Here we study convergence of method (5.4). For a given $\bar{u}^n \in \bar{Q}^\perp$ with $\|\bar{u}^n\|_M = 1$, we use the notation $\mu(\bar{u}^n) \equiv \bar{\mu}^n$, define $\tilde{u}^{n+1}$ by (5.4), and write $\hat{u}^{n+1} \equiv \tilde{u}^{n+1} - \xi^n$.

Theorem 1. Let conditions (4.8), (4.9), (5.2), and (5.6) be satisfied. Suppose ϵ is small enough, $\|\bar{u}^n\|_M = 1$, and

$$\bar{\mu}^n \leq \bar{\beta}^0. \tag{5.28}$$

Then there exist numbers $\kappa_2 > 0$ and $\kappa_3 > 0$ such that

$$\|\hat{u}^{n+1}\|_M \geq \kappa_2, \quad \|\tilde{u}^{n+1}\|_M \geq \kappa_3. \tag{5.29}$$

Proof. We define $u^n \equiv P^\perp \bar{u}^n \in Q^\perp$ and $\zeta^n \equiv u^n - \bar{u}^n = (P^\perp - \bar{P}^\perp)u^n$, and estimate $\|\zeta^n\|_M$ and $\|\zeta^n\|_L$ from above by (5.13) and (5.15). Hence, $\|u^n\|_M^2 \geq 1 - m\epsilon^2 \geq \kappa_0 > 0$ and $\mu(u^n) \leq \bar{\mu}^n(1 - m\epsilon^2)^{-1} \leq \bar{\mu}^n/\kappa_0$ (see the proof of Lemma 3). For $u^{n+1} \equiv u^n - \tau P^\perp B^{-1} r(u^n)$, we have the estimates $\|u^{n+1}\|_L^2 \geq \delta\|u^n\|_L^2$ (see Lemma 4.1) and $\mu(u^{n+1}) \leq \mu(u^n)$, provided $\tau \in (0, 2/\delta_1)$ (see (4.14)). This implies that

$$\|u^{n+1}\|_M^2 \geq \delta\mu(u^n)\kappa_0/\mu(u^{n+1}) \geq \sigma_0 > 0.$$

Defining $\hat{u}^{n+1} \equiv \tilde{u}^{n+1} - \xi^n$, for $\hat{\zeta}^{n+1} \equiv \hat{u}^{n+1} - u^{n+1}$ we then have

$$\|\hat{\zeta}^{n+1}\|_M \leq \|\zeta^n\|_M + \tau(X_1 + X_2 + X_3 + X_4),$$

where $X_1 \equiv \|(\bar{P}^\perp - P^\perp)B^{-1}r(u^n)\|_M$, $X_2 \equiv \|\bar{P}^\perp B^{-1}L\zeta^n\|_M$, $X_3 \equiv \mu(u^n)\|\bar{P}^\perp B^{-1}M\zeta^n\|_M$, and $X_4 \equiv |\mu(u^n) - \mu(\bar{u}^n)|\|\bar{P}^\perp B^{-1}M\bar{u}^n\|_M$. It can be verified (via Lemmas 1,5,6) that there exist constants $\kappa_i^*, i \in [1, 4]$, such that $X_i \leq \kappa_i^*\epsilon, i \in [1, 4]$. Thus, $\|\hat{\zeta}^{n+1}\|_M \leq K_1^*\epsilon$ and

$$\|\hat{u}^{n+1}\|_M \geq -\|\hat{\zeta}^{n+1}\|_M \geq \sigma_0^{1/2} - K_1^*\epsilon \equiv \kappa_2 > 0.$$

---

[13] If φ is the angle between u and $U(\mu; \omega_3)$, then $\sin\varphi = \text{dist}^2_L\{u; U(\mu; \omega_3)\}\mu^{-1/2}$ and $\text{dist}^2_L\{u/\|u\|_L; U(\mu; \omega_3) \cap S_L(1)\} = 2\sin\varphi/2$.

Finally, $\|\tilde{u}^{n+1}\|_M \geq \kappa_2 - \kappa_1 \epsilon^2 \equiv \kappa_3 > 0$ (see (5.6)). $\square$

In what follows, we write

$$\bar{\mu}^{n+1} \equiv \mu(\tilde{u}^{n+1}) = \mu(\bar{u}^{n+1}), \ \hat{\mu}^{n+1} \equiv \mu(\hat{u}^{n+1}), \ \Delta\mu \equiv \bar{\mu}^n - \hat{\mu}^{n+1}), \quad (5.30)$$

$$\bar{r}^n \equiv \bar{r} \equiv L\bar{u}^n - \bar{\mu}^n M\bar{u}^n; \ \bar{w}^n \equiv \bar{w} \equiv \bar{P}^{\perp}B^{-1}\bar{r}. \quad (5.31)$$

Theorem 2. Let the conditions of Theorem 1 be satisfied. Then there exist numbers $\bar{\gamma} > 0$ and $K_2 \geq 0, \kappa_4 \geq 0$ such that

$$\bar{\mu}^n - \bar{\mu}^{n+1} \geq I(\bar{u}^n) - K_2 \epsilon^2 \quad (5.32)$$

and

$$|\bar{\mu}^{n+1} - \hat{\mu}^{n+1}| \leq \kappa_4 \epsilon_\xi \leq \kappa_4 \kappa_1 \epsilon^2 \equiv \epsilon_\mu, \quad (5.33)$$

where

$$I(\bar{u}^n) \equiv \bar{\gamma}\|\bar{r}\|^2_{B^{-1}}[1 + \bar{\gamma}\|\bar{r}\|^2_{B^{-1}}(\bar{\mu}^n)^{-1}]^{-1}.$$

Proof. From (5.4) and (5.30), we obtain

$$\Delta\mu = [2\tau(\bar{w}, \bar{r}) - \tau^2\|\bar{w}\|^2_L + \tau^2\bar{\mu}^n\|\bar{w}\|^2_M]\|\hat{u}^{n+1}\|^{-2}_M. \quad (5.34)$$

As in § 4, we estimate this number from below. First, $\bar{P}^{\perp} = I - P + \bar{P}^{\perp} - P^{\perp}$ leads to

$$(\bar{w}, \bar{r}) = (\bar{P}^{\perp}B^{-1}\bar{r}, \bar{r}) = \|\bar{r}\|^2_{B^{-1}} - (PB^{-1}\bar{r}, \bar{r}) + ((\bar{P}^{\perp} - P^{\perp})B^{-1}\bar{r}, \bar{r}).$$

Next, $|(PB^{-1}\bar{r}, \bar{r})| \leq \|PM^{-1}\bar{r}\|_M\|B^{-1}\bar{r}\|_M \leq \|\bar{r}\|_{M^{-1}}\delta_1^{1/2}\lambda_1^{-1/2}\|\bar{r}\|_{B^{-1}}$ and

$$|(PB^{-1}\bar{r}, \bar{r})| \leq 2K_\omega \epsilon\|\bar{r}\|_{B^{-1}} \leq a_1\|\bar{r}\|^2_{B^{-1}} + (K_\omega \epsilon)^2 a_1^{-2},$$

where $2K_\omega \geq \delta_1^{1/2}/\lambda_1^{1/2}[\sum_{i \in \omega_0}(\lambda_i - \bar{\mu}^n)^2]^{1/2}$ and $a_1 \in (0,1)$ is a free parameter. Moreover, $|((\bar{P}^{\perp} - P^{\perp})B^{-1}\bar{r}, \bar{r})| \leq \|P - \bar{P}\|_B\|\bar{r}\|^2_{B^{-1}} \leq \kappa_L \epsilon\delta^{-1/2}\|\bar{r}\|^2_{B^{-1}}$ (see (5.14)). Hence,

$$(w, \bar{r}) \geq \|\bar{r}\|^2_{B^{-1}}[1 - \epsilon\kappa_L\delta^{-1/2} - a_1] - K_\omega^2\epsilon^2/a_1.$$

Now we estimate from below the term $-\|\bar{w}\|^2_L$, making use of (5.14) and the evident inequality $\|P^{\perp}\|_L \leq 1$ (see Lemma 1.1). This yields $-\|\bar{w}\|^2_L \geq -(1 + \kappa_L \epsilon)^2\|B^{-1}\bar{r}\|^2_L \geq -\delta_1(1 + \kappa_L \epsilon)^2\|\bar{r}\|^2_{B^{-1}}$. Hence,

$$2\tau(\bar{w}, \bar{r}) - \tau^2\|\bar{w}\|^2_L \geq \gamma'\|\bar{r}\|^2_{B^{-1}} - 2\tau K_\omega \epsilon^2/a_1,$$

where $\gamma' \equiv 2\tau[1 - \kappa_L \epsilon \delta^{-1/2} - a_1] - \tau^2 \delta_1 (1 + \epsilon \kappa_L)^2 \geq \bar{\gamma} > 0$ for $\tau < 2/\delta_1$ and small enough ϵ and a_1. These inequalities, (5.34), and (5.29) lead to

$$\Delta \mu \geq \frac{\bar{\gamma} \|\bar{r}\|^2_{B^{-1}} + \tau^2 \bar{\mu}^n \|\bar{w}\|^2_M}{1 - 2\tau(\bar{u}^n, \bar{w})_M + \tau^2 \|\bar{w}\|^2_M} - K_2' \epsilon^2, \text{ where } K_2' \equiv 2\tau K_\omega^2 (a_1 \kappa_2^2)^{-1}$$

(we have used the same technique in estimating the right-hand side in (4.13)). In a straightforward manner, we thus obtain

$$\Delta \mu \geq \max_{\alpha > 0} \min \left\{ \frac{\bar{\gamma} \|\bar{r}\|^2_{B^{-1}}}{1 + \alpha}; \frac{\bar{\mu}^n}{1 + \alpha^{-1}} \right\} - K_2' \epsilon^2 \text{ and } \hat{\mu}^n - \bar{\mu}^{n+1} \geq I(\bar{u}^n) - K_2' \epsilon^2$$

(for $\alpha = \bar{\gamma} \|\bar{r}\|^2_{B^{-1}} / \bar{\mu}^n$). This implies that, for $\bar{\mu}^n$ from (5.28), we have $\hat{\mu}^{n+1} \leq \bar{\beta}^0 + K_2' \epsilon^2 \leq R^2$ with some $R > 0$. Let $\epsilon_1(\xi^n) \geq \|\xi^n\|_L + R \|\xi^n\|_M$ and apply (5.24) with $u \equiv \bar{u}^{n+1}$ and $v \equiv \hat{u}^{n+1}$. This yields

$$\Delta \mu \leq \frac{(\bar{\mu}^{n+1})^{1/2} + (\hat{\mu}^{n+1})^{1/2}}{\kappa_3} \epsilon_1(\xi^n) \text{ and } (\bar{\mu}^{n+1})^{1/2} - (\hat{\mu}^{n+1})^{1/2} \leq \frac{\epsilon_1(\xi^n)}{\kappa_3}.$$

Thus (see (5.24)),

$$|\mu(\bar{u}^{n+1}) - \hat{\mu}^{n+1}| \leq \frac{(2R + \kappa_3^{-1} \epsilon_1(\xi^n)) \epsilon_1(\xi^n)}{\kappa_3}.$$

Hence, (5.33) holds with $\kappa_4 \geq [2R + \kappa_3^{-1} \epsilon_\xi (1 + R)](1 + r)/\kappa_3$. Next, we obtain (5.32) (with $K_2 \equiv K_2' + \kappa_1 \kappa_4$) as a consequence of (5.33) and the above estimate for $\bar{\mu}^n - \hat{\mu}^{n+1}$. $\square$

This theorem implies that

$$\bar{\mu}^{n+1} \leq \bar{\mu}^n + K_2 \epsilon^2 + \epsilon_\mu \leq \bar{\mu}^n + K_2 \epsilon^2 \tag{5.35}$$

and

$$\|\bar{r}^n\|^2_{B^{-1}} \leq \frac{\bar{\mu}^n (\bar{\mu}^n - \bar{\mu}^{n+1} + K_2 \epsilon^2)}{\bar{\gamma}(\bar{\mu}^{n+1} - K_2 \epsilon^2)} \tag{5.36}$$

((5.36) with $\epsilon = 0$ yields (4.11)).

Theorem 3. *Suppose that conditions (4.8), (4.9), (5.2), and (5.6) are satisfied, that ϵ is small enough, and, for $\bar{u}^0 \equiv u$ with $\|u^0\|_M = 1$, that λ and ν are defined as in Lemma 4 so that $\lambda \leq \bar{\mu}^0 \leq \beta^0 < \nu$. Suppose also that the constants $K_2 \equiv K_2(\bar{\beta}^0)$ and κ_4 are defined as in Theorem 2 on the basis of condition (5.28) with some $\bar{\beta}^0 \geq \beta^0$ and $q \equiv \max\limits_{\mu \in [\lambda, \beta^0]} \rho(\mu)$, where*

$$\rho(\mu) \equiv \frac{1 - \nu^{-1} \delta_0 \bar{\gamma}(\nu - \mu)}{1 + (\lambda \nu(\nu - \mu))^{-1} \delta_0 \bar{\gamma}} < 1. \tag{5.37}$$

Suppose further that constants K_3, K_4, and K_5 satisfy

$$K_3 \geq \frac{\bar{\gamma}\delta_0\bar{\beta}^0}{\lambda\nu} \sum_{i \in \omega_{0,0}} \frac{(\nu - \lambda_i)(\lambda_i - \lambda)}{\lambda_i}, \quad K_4 \equiv K_3\beta^0,$$

$K_5 \equiv K_2 + K_3$, *and* $K_6 \geq K_5/(1 - q)$. *Suppose finally that* $\lambda, \beta^0, \nu, \epsilon, \kappa_1$, *and* κ_4 *are such that*

$$\lambda + K_6\epsilon^2 < \beta^0 - K_2\epsilon^2. \tag{5.38}$$

Then (5.28) holds for all $\bar{\mu}^n$ with $n = 1, 2, \ldots$ and

$$\bar{\mu}^n - \lambda \leq q \max\{0; \bar{\mu}^n - \lambda\} + K_5\epsilon^2. \tag{5.39}$$

Proof. If $\bar{\mu}^n \leq \bar{\beta}^0 < \nu$, then (5.32) holds. To estimate $I(\bar{u}^n)$ (see Theorem 2) from below, we proceed in the same manner as for (4.14). Hence, we replace $\|\bar{r}^n\|_{B^{-1}}^2$ in the expression for $I(\bar{u}^n)$ by $\delta_0\|\bar{r}^n\|_{L^{-1}}^2$ and, in turn by a smaller term from (5.21). Therefore, from (5.21) for $\bar{\mu}^n \in [\lambda, \nu)$ and $\mu^* \equiv \bar{\mu}^{n+1} - K_2\epsilon^2$, we obtain

$$\bar{\mu}^n - \mu^* \geq \bar{\mu}^n Y(\bar{\mu}^n)[1 + Y(\bar{\mu}^n)]^{-1}, \tag{5.40}$$

where $Y(\bar{\mu}^n) \equiv X - K_3\epsilon^2$ and

$$X \equiv [\delta_0\bar{\gamma}(\bar{\mu}^n - \lambda)(\nu - \bar{\mu}^n)][\lambda\nu]^{-1}.$$

Hence, $(\bar{\mu}^n - \mu^*)(1 + X) \geq \bar{\mu}^n X - K_3\epsilon^2\mu^*$ and

$$\bar{\mu}^n - \mu^* \geq \frac{\bar{\mu}^n X}{1 + X} - \frac{K_3\epsilon^2\mu^*}{1 + X} \geq \frac{\bar{\mu}^n X}{1 + X} - K_4\epsilon^2.$$

This implies (see (4.15)) that $\mu^* - \lambda \leq \rho(\bar{\mu}^n)(\bar{\mu}^n - \lambda) + K_4\epsilon^2$ and

$$\bar{\mu}^{n+1} - \lambda \leq \rho(\bar{\mu}^n)(\bar{\mu}^n - \lambda) + K_5\epsilon^2. \tag{5.41}$$

Thus, (5.39) follows from (5.41) and, for $\bar{\mu}^n \in [\lambda + K_6\epsilon^2, \beta_0]$, we have $\bar{\mu}^{n+1} \leq \bar{\mu}^n$. Also, for $\bar{\mu}^n < \lambda + K_6\epsilon^2$, we have (5.35), and an increase of $\bar{\mu}^n$ is possible only until it remains smaller than $\lambda + K_6\epsilon^2$. $\square$ [14]

---

[14]Note that (5.38) always holds for small enough ϵ, that $q = 1 - \delta_0\bar{\gamma}[1 - \lambda/\nu]^{-1}$ if $\lambda/\nu > [1 + (1 - \delta_0\bar{\gamma})^{-1/2}]^{-1}$ and $\beta^0/\nu \leq [1 + \lambda((1 - \delta_0\bar{\gamma})(\nu - \lambda))^{-1}]^{-1}$, and that $q = \rho(\beta^0)$ otherwise. Our estimates contain the free parameter $a_1 \in (0, 1 - \kappa_L\epsilon\delta^{-1/2})$ and parameter τ (see (4.9)); we may take $a_1 = 1 - \kappa_L\epsilon\delta^{-1/2}$ (if $\epsilon \leq \epsilon_0 < \kappa_L/\delta^{1/2}$) and $\tau \approx 1/\delta_1$, since we then obtain the least values of $\rho(\beta^0)$ and q for small enough ϵ.

Theorem 4. Let the conditions of Theorem 3 be satisfied, with the exception of (5.38). *Suppose that*

$$q_{\lambda,\nu} \equiv \max_{\lambda \le \mu \le \lambda + 2^{-1}(\nu - \lambda)} \rho(\mu); \quad \hat{q}_{\lambda,\nu} \equiv \frac{\delta_0 \bar{\gamma}(\nu - \lambda)}{2\nu} > 0, \quad K_7 \equiv \frac{2K_5}{1 - q_{\lambda,\nu}},$$

and $K_8 \equiv 2K_5[\hat{q}_{\lambda,\nu}]^{-1}$, *where* $\max\{K_7; K_8\}\epsilon^2 \le (\nu - \lambda)/2$. *Suppose also that* s *iterations of* (5.4) *are computed with the initial iterate* $\bar{u}^0$ *such that* $\bar{\mu}^0 \in [\lambda, \nu)$, *where* s *is so large that* $s \equiv M_{\lambda,\nu}(\epsilon) = O(|\log_{\nu-\lambda}\epsilon|)$. *Then either* $\bar{\mu}^s < \lambda$, *or else, for some* $n \le s$, *we have*

$$|\bar{\mu}^{n+1} - \bar{\mu}^n| \le K_2\epsilon^2. \tag{5.42}$$

Proof. (5.36), (5.25), and the inequality $\bar{\mu}^{n+1} \ge \bar{\mu}^n$ lead to (5.42) and

$$\text{dist}_L\{\bar{u}^n; U(\lambda)\} = O(\epsilon(\Delta\lambda)^{-1}), \tag{5.43}$$

where $\Delta\lambda \equiv \min|\lambda_i - \lambda_j|$ with the minimum taken with respect to distinct λ_i and λ_j such that either they are not greater than $\bar{\mu}^0$ or $\lambda_i = \nu(\bar{\mu}^0)$ ($\nu(\bar{\mu}^0)$ was defined above Lemma 4). Hence, if $\bar{\mu}^{n+1} \ge \bar{\mu}^n$, then the iterations should be stopped and $\bar{u}^n$ should be taken as a new approximate eigenvector; the set of ω_0 may be extended and new iterations of the same type may be used for determining another eigenvalue. Hence, it suffices to prove the theorem only when $\bar{\mu}^n - \bar{\mu}^{n+1} \ge K_2\epsilon^2$ for all n. If $\lambda + (\nu - \lambda)/2 \le \bar{\mu}^0 \le \nu$, then a finite number k_1 of iterations yields $\bar{\mu}^{k_1} \le \nu - K_8\epsilon^2$. Observe that, for $\lambda + (\nu - \lambda)/2 \le \bar{\mu}^0 \le \nu - K_8\epsilon^2$, in accordance with (5.41), we have $\nu - \bar{\mu}^{n+1} \ge (1 + \hat{q}_{\lambda,\nu}/2)(\nu - \bar{\mu}^n)$. This implies that, for large enough $k_2 = O(|\log_{\nu-\lambda}\epsilon|)$, we have $\bar{\mu}^{n+k_2} \le \lambda + (\nu - \lambda)/2$. For $\lambda + K_7\epsilon^2 \le \bar{\mu}^n \le \lambda + (\nu - \lambda)2$, in accordance with Theorem 3, (5.39) holds. This yields

$$\bar{\mu}^{n+1} - \lambda \le (1/2 + q_{\lambda,\nu}/2)(\bar{\nu}^n - \lambda).$$

Observe also that

$$\max_{\mu \in [\lambda, \bar{\mu}^0]} = \max\{\rho(\lambda); \rho(\bar{\mu}^0)\} = 1 - \kappa(\nu - \lambda),$$

with $\kappa > 0$, where $\kappa \approx \delta_0\bar{\gamma}/\nu$ if $\bar{\mu}^0 \approx \lambda$. Hence, for $\lambda + K_7\epsilon^2 \le \bar{\mu}^m \le \lambda + (\nu - \lambda)/2$ and large enough $k_3 = O(|\log_{\nu-\lambda}\epsilon|)$, we obtain $\bar{\mu}^{m+k_3} \le \lambda + K_7\epsilon^2$ and (after a finite number of further iterations) $\bar{\mu}^s \le \lambda$. $\square$ [15]

---

[15] Inequality (5.42) and, in particular, an increase in $\bar{\mu}^n$ may occur only for $\bar{\mu}^n \in [\lambda, \lambda + K_7'\epsilon^2] \cup [\nu - K_8'\epsilon^2, \nu)$, where $K_7' \equiv [K_2 + K_5][1 - q_{\lambda,\nu}]^{-1}$, $K_8' \equiv (K_2 + K_5)/\hat{q}_{\lambda,\nu}$; that is, when $\bar{\mu}^n$ is an $O(\epsilon(\Delta\lambda)^{-1})$-approximation to λ_i.

5.4. Determination of eigenvectors.

Theorem 5. Let $H_p^0 \subset \bar{Q}^\perp$ be a p-dimensional subspace such that

$$\max_{u \in H_p^0} \mu(u) \equiv \bar{\beta}^0 < \nu(\bar{\beta}^0) \tag{5.44}$$

with $\bar{\beta}^0$ independent of ϵ. Suppose that there are exactly p eigenvalues λ_i such that $i \notin \omega_0$ and $\lambda_i \leq \bar{\beta}^0$. Suppose that $\Delta\lambda \equiv \min\limits_{\lambda_i \neq \lambda_j, \lambda_i \leq \bar{\beta}^0, \lambda_j \leq \bar{\beta}^0} |\lambda_i - \lambda_j|$ and that the conditions of Theorem 4 with respect to ϵ are satisfied with $\nu - \lambda \equiv \Delta\lambda$. Then an $O(\epsilon(\Delta\lambda)^{-1})$-approximation to one of the desired eigenvalues and an $O(\epsilon(\Delta\lambda)^{-1/2})$-approximation to the corresponding eigenvector are obtained from s iterations of (5.4) with an initial iterate $\bar{u}_0 \in H_p^0$ such that $\|\|\bar{u}_0\|\|_M = 1$, provided that s is large enough and $s = O(|\log_{\Delta\lambda}\epsilon|)$.

Proof. It suffices to apply Theorem 4 and (5.25). □

The initial iterate may be chosen by way of the Rayleigh-Ritz method associated with a basis for H_p^0. For example, if vectors $v_1, \ldots, v_p$ form an orthonormal basis for H_p^0 in the sense of the Euclidean space $H(M)$, then we obtain the algebraic spectral problem $\bar{L}_p \bar{X} = \xi \bar{X}$, where $\bar{X} \in \mathbf{R}^p$, $\bar{L}_p \equiv [(Lv_j, v_i)]$. Its eigenvalues $\xi_1 \leq \ldots \leq \xi_p$ are such that $\xi_i \geq \bar{\lambda}_i^\perp, i = 1, \ldots, p$, and $\xi_p = \bar{\beta}^0$ (see (5.18)). One of the corresponding eigenvectors $\bar{X}_j \equiv [x_{1,j}, \ldots, x_{p,j}]^T$, with $\|\bar{X}_j\| = 1$, defines the possible initial iterate $\bar{u}^0 \equiv x_{1,j}v_1 + \cdots + x_{p,j}v_p$.

Theorem 6. Suppose that the conditions of Theorem 5 are satisfied. Then $O(\epsilon(\Delta\lambda)^{-1})$-approximations to all of the desired eigenvalues and $O(\epsilon(\Delta\lambda)^{-1/2})$-approximations to corresponding eigenvectors are obtained by a total of $\bar{s} = O(|\log_{\Delta\lambda}\epsilon|)$ iterations of type (5.4) dealing in consecutive order with the initial subspaces $H_p^0, H_{p-1}^0, \ldots, H_1^0$, where $H_1^0 \subset H_2^0 \subset \cdots \subset H_p^0$.

Proof. For a given $H_p^0 \subset \bar{Q}^\perp$, we apply Theorem 5. The iterations considered above lead to an $O(\epsilon(\Delta\lambda)^{-1/2})$-approximation u^* to the corresponding eigenvector. We then include it as an additional basis vector for a new subspace $\bar{Q}_{\omega_0}$. Clearly, conditions (5.2) for the new set of indices are also satisfied with a possible increase of the constants in (5.2) (for the new vector) in a finite times. Defining H_{p-1}^0 as a subset of elements of H_p^0 orthogonal in $H(M)$ to u^* and noting that

$$\max_{u \in H_{p-1}^0} \mu(u) \leq \max_{u \in H_p^0} \mu(u),$$

then condition (5.44) allows us again to use a new cycle of iterations (5.4). This procedure is repeated in consecutive order, and the theorem clearly follows. □

In accordance with Theorems 6 and 1.3, then given the dimensions k_i of the eigenspaces U_{λ_i} associated with all $\lambda_i < \bar{\beta}^0$, we can construct subspaces $\tilde{U}_i$ (they are linear spans of k_i approximate to eigenvectors) such that

$$\Theta_M(\tilde{U}_i; U(\lambda_i)) = O(\epsilon), \quad \Theta_L(\tilde{U}_i; U(\lambda_i)) = O(\epsilon). \qquad (5.45)$$

Hence, our algorithms with orthogonalization are stable in a certain sense, and the obtained results may differ only slightly when $\epsilon \to 0$. [16]

5.5. Eigenvalue clustering. The above estimates depend strongly on $\Delta\lambda$ when we deal with several closely-spaced λ_i, $i \leq p$. This is the case, e.g., when the above λ_i correspond to approximations of a multiple eigenvalue of a differential problem and $\Delta\lambda = O(\epsilon^2)$. Then it is reasonable (see [213]) to replace the original problem (5.1) by

$$\tilde{L}u = \lambda M u, \qquad (5.46)$$

where $\tilde{L}$ has the same eigenvectors u_i, but $\tilde{L}u = \tilde{\lambda}_i M u_i$, $i = 1, \ldots, N$, and, for $i \in \omega_1$ associated with the indicated close eigenvalues, $\tilde{\lambda}_i$ coincides with, say, the minimal eigenvalue in the cluster. Hence, $0 \leq \lambda_i - \tilde{\lambda}_i \leq K_\lambda \epsilon^2$, and we may treat iterations (5.4) as iterations for (5.46) with the additional perturbation

$$\bar{\xi}^n \equiv -\tau \bar{P}^\perp B^{-1} \left((L - \tilde{L})\bar{u}^n - (\bar{\mu}^n - \tilde{\mu}^n)M\bar{u}^n \right), \qquad (5.47)$$

where $\tilde{\mu}^n \equiv (\tilde{L}\bar{u}^n, \bar{u}^n)$. Then $0 \leq \bar{\mu}^n - \tilde{\mu}^n \leq K_\lambda \epsilon^2$, $(1 - K_\lambda \epsilon^2/\lambda_1)L \leq \tilde{L} \leq L$, and
$\|\bar{\xi}^n\|_L \leq \tau \|\bar{P}^\perp\|_L [\|L^{1/2}B^{-1}L^{1/2}\| \|(L - \tilde{L})\bar{u}^n\|_{L^{-1}} + K_\lambda \epsilon^2 \|L^{1/2}B^{-1}M^{1/2}\|]$.
Hence, $\|\bar{\xi}^n\|_L \leq 2\tau \delta_1 \lambda_1^{-1/2} \|\bar{P}^\perp\|_L K_\lambda \epsilon^2$ and for $\tilde{L}$ we have $\Delta\tilde{\lambda} \asymp 1$, that is, Theorem 5 applies to a problem without the original cluster.

§ 6. Modified subspace iteration method

This section is devoted to the analysis of preconditioned subspace iterations for simultaneous calculation of a group of the eigenvalues and their eigenvectors (see [207, 208]). Such methods are especially well suited for parallel computations and for problems with eigenvalue cluster. They also

---

[16] In practical applications the choice of ω_0 and $\bar{Q}$ should be made on the basis of information about μ^0 and the already computed λ_i and u_i. More precisely, we should include in ω_0 only i for which $\mu(\bar{u}_i) \leq \mu^0$ and possibly some i for which $\mu(\bar{u}_i) \geq \mu^0$ and $\mu(\bar{u}_i)$ close to μ^0 (here, $\bar{u}_i$ is the approximation to u_i). After all $\bar{u}_1, \ldots, \bar{u}_p$ are found, it is possible to improve the first $l \leq p$ of them by the Rayleigh-Ritz method for the respective p-dimensional subspace with the basis $\bar{u}_1, \ldots, \bar{u}_p$. This may yield better choices of ω_0 and the respective adaptations of our original iterations. Some numerical experiments with such methods can found in [208].

enable one to find the corresponding approximation to U_{λ_p} in the new subspace. Using orthogonalization with respect to previously computed eigenvectors, one can construct similar iterative methods of finding all $\lambda_i, i \leq p$, and corresponding subspaces U_{λ_i}. Convergence of these methods is geometric with ratio $q_i \leq q_p = q$. Thus, the presence of some closely-spaced λ_i, $i \leq p$ does not decrease the convergence rate of the iterative methods being discussed. As in § 5 special attention is paid to the errors involved in orthogonalization with respect to previously computed eigenvectors. We therefore preserve the notation of § 5 dealing with $\bar{Q}, \bar{P}$, etc.

6.1. Basic computational algorithms. As in the classic subspace iterations (see, e.g., [402]), the modified method begins with a given p-dimensional subspace H_p^n and constructs a new p-dimensional subspace H_p^{n+1}. The aim is to determine better approximations to λ_p and U_{λ_p}, where λ_p is the pth eigenvalue of problem (5.1) in increasing order. More precisely, the new subspace is defined by

$$H_p^{n+1} \equiv R_{\beta_n}\{H_p^n\}, \tag{6.1}$$

where $R_{\beta_n} \equiv I - \tau B^{-1}(L - \beta^n M)$ and B is an operator from (4.8). (Such an iteration requires solution of systems $Bv_i^{n+1} = a_i^n, i \in [1, p]$, where $a_1^n, \ldots, a_p^n$ is a basis for H_p^n.)

For each n, we consider the algebraic spectral problem

$$\bar{L}_p^n \bar{X} = \lambda \bar{M}_p^n \bar{X}, \tag{6.2}$$

where $\bar{X} \equiv [x_1, \ldots, x_p]^T$, $\bar{L}_p^n \equiv [(La_j^n, a_i^n)] \in \mathbf{R}^{p \times p}$, $\bar{M}_p^n \equiv [(Ma_j^n, a_i^n)] \in \mathbf{R}^{p \times p}$, and λ_i^n and $\bar{X}^{(i)} \equiv [x_1^{(i)}, \ldots, x_p^{(i)}]^T$, $i \in [1, p]$, are the respective eigenvalues and eigenvectors of (6.2) which form an orthonormal system in accordance with the inner product $(\bar{M}_p^n \bar{X}, \bar{Y}) = \bar{Y}^T \bar{M}_p^n \bar{X}$. Then the vectors

$$v_i \equiv \sum_{j=1}^{p} x_j^{(i)} a_j^n, \quad i \in [1, p], \tag{6.3}$$

form an orthonormal basis for H_p^n in $H_p^n(\bar{M}_p^n)$ and

$$\mu(v_i) = \lambda_i^n = \min_{H_i \subset H_p^n} \max_{u \in H_i} \mu(u) \geq \lambda_i, \quad \lambda_p^n = \beta(H_p^n) \equiv \beta^n. \tag{6.4}$$

Moreover, orthogonality of $\bar{X}^{(1)}, \ldots, \bar{X}^{(p)}$ in the sense of inner product $(\bar{L}_p^n \bar{X}, \bar{Y})$ entails orthogonality of the vectors $v_1, \ldots, v_p$ in $H_p^n(\bar{L}_p^n)$. Note that $\beta^n = \max_{u \in H_p^n} \mu(u)$, which serves as an approximation to the desired $\bar{\lambda}_p^\perp$

(see (5.17)). The new approximation β^{n+1} is defined in the same way, but for the subspace

$$H_p^{n+1} \equiv \bar{R}_{\beta^n}\{H_p^n\}, \tag{6.5}$$

where

$$\bar{R}_{\beta^n} \equiv I - \tau \bar{P}^\perp B^{-1}(L - \beta^n M). \tag{6.6}$$

Lemma 1. *Let (4.8) be satisfied and suppose that, for a given $H_p^n \in \bar{Q}^\perp$, the new subspace H_p^{n+1} is defined by (6.5) and (6.6), where $0 < \tau < 4/\delta_1$. Then*

$$H_p^{n+1} \in \bar{Q}^\perp, \quad \dim H_p^{n+1} = p. \tag{6.7}$$

Proof. The assertion $H_p^{n+1} \in \bar{Q}^\perp$ is evident, and we are left with proving only that $\dim H_p^{n+1} = p$. Let $u \in H_p^n$ be such that $\|u\|_{\bar{M}_p^n} \equiv \|u\|_M = 1$ and let $t \equiv \|Mu\|_{B^{-1}}$. Then, for $\beta^n \equiv \beta$, we have

$$I \equiv (\bar{R}_\beta u, u)_M = 1 - \tau(B^{-1}Lu, Mu) + \tau\beta t^2$$

and $I \geq 1 - \tau\|B^{-1/2}L^{1/2}\|\|L^{1/2}u\|t + \tau\beta t^2$. Observe that $\|B^{-1/2}L^{1/2}\| \leq \delta_1^{1/2}$ (see Lemma 5.6). Since $u \in H_p^n$, then $\|L^{1/2}u\| \leq \beta^{1/2}$. Hence, $X \equiv (\bar{R}_\beta u, u)_M \geq 1 - \tau\delta_1^{1/2}\beta^{1/2}t + \tau\beta t^2$, and $X \geq 1 - 1/4\tau\delta_1 \geq K_0 > 0$. This implies that $\|\bar{R}_\beta u\|_M > 0$, $\forall u \in H_p^n, u \neq 0$, which leads to (6.7). $\square$ [17]

To find the basis $a_1, \ldots, a_p$ for H_p^{n+1} it suffices to solve

$$Bz_i = -\tau(Lv_i - \beta Mv_i), \quad i \in [1, p], \tag{6.8}$$

and set $a_i = v_i - \bar{P}^\perp z_i, i \in [1, p]$. [18]

6.2. Convergence of the method. Here we study convergence of $\beta^n \equiv \beta(H_p^n)$ (see (6.4)) to $\lambda_p^\perp$, with H_p^n determined by (6.5) for $n = 0, 1, \ldots$.

Lemma 2. *Let the conditions of Lemma 5.4 and bounds (4.8) be satisfied. Let $\omega_{0,0}$ be a subset of indices $i \in \omega_0$ such that $\lambda < \lambda_i < \nu$, and suppose that*

$$r_\beta(u) \equiv r_\beta \equiv Lu - \beta Mu. \tag{6.9}$$

Then

$$\|r_\beta\|_{L^{-1}}^2 \geq \frac{\beta^2}{\mu}\left(\frac{(\mu - \lambda)(\nu - \mu)}{\lambda\nu} + 1 - \frac{\mu}{\beta}\right)$$

---

[17] If ω_0 is empty, then $R_\beta = I - \tau B^{-1}(L - \beta M) \in \mathcal{L}(H(B))$ is symmetric and $(R_\beta u, u)_B \geq \tau\beta\|u\|^2$, for all $u \in H$ and $\tau \in (0, 1/\delta_1)$; if we also have $u \in H_p^n$, then $(R_\beta u, u)_B \geq \|u\|_B^2$ for all $\tau \geq 0$.

[18] We can again construct the problem of type (6.2) for H_p^{n+1} and find an orthonormal basis for $H_p^n(\bar{M}_p^n)$ consisting of the vectors $v_1^{n+1}, \ldots, v_p^{n+1}$ (see (6.3)). Note that if $a_1^n, \ldots, a_p^n$ is an orthonormal basis for H_p^n in $H(\bar{M}_p^n)$, then $\bar{M}_p^n = I$ and problem (6.2) is simplified.

$$-\frac{\beta^2}{\lambda\nu}\sum_{i\in\omega_{0,0}}\epsilon_{M,i}^2\frac{(\nu-\lambda_i)(\lambda_i-\lambda)}{\lambda_i},\tag{6.10}$$

Proof. It suffices to modify the proof of Lemma 5.4. $\square$

Lemma 3. Suppose that (4.8) and (5.2) hold and that

$$0<\tau<2/\delta_1.\tag{6.11}$$

Let $u\equiv u^n\in H_p^n$, $\|u\|_M=1$, $\mu(u)\equiv\mu^n\equiv\mu$, $t(u^n)\equiv\bar{\gamma}\|r_\beta\|_{B^{-1}}^2+\beta-\mu$, and $u^{n+1}\equiv R_\beta u^n$, where $\beta\equiv\beta^n$ and $\mu^{n+1}\equiv\mu(u^{n+1})$. Then, for small enough $\epsilon>0$, there exist numbers $\bar{\gamma}>0$ and $K_2>0$ such that

$$\beta^n-\mu^{n+1}\geq\frac{t(u^n)}{1+\beta^{-1}t(u^n)}-K_2\epsilon^2.\tag{6.12}$$

Proof. Let $w\equiv\bar{P}^\perp B^{-1}r_\beta$. Then

$$\mu^n-\mu^{n+1}=[2\tau(w,r_\beta)-\tau^2\|w\|_L^2+\tau^2\mu\|w\|_M^2+2\tau(\beta-\mu)(u,w)_M]\|u^{n+1}\|_M^{-2},$$

where $(w,r_\beta)=\|r_\beta\|_{B^{-1}}^2-(PB^{-1}r_\beta,r_\beta)+((\bar{P}^\perp-P^\perp)B^{-1}r_\beta,r_\beta)$. By analogy with the estimate for $(\bar{w},\bar{r})$ in the proof of Theorem 5.2, we have

$$|(PB^{-1}r_\beta,r_\beta)|\leq\|PM^{-1}r_\beta\|_M\|B^{-1}r_\beta\|_M\leq 2K_\omega\epsilon\|r_\beta\|_{B^{-1}}$$

$$\leq a_1\|r_\beta\|_{B^{-1}}^2+(K_\omega\epsilon)^2a_1^{-2},$$

where $2K_\omega\geq\delta_1^{1/2}\lambda_1^{-1/2}[\sum_{i\in\omega_0}(\lambda_i-\beta)^2]^{1/2}$ and $a_1\in(0,1)$ is a free parameter. Next, as in the proof of Theorem 5.2, we have

$$|((\bar{P}^\perp-P^\perp)B^{-1}r_\beta,r_\beta)|\leq\|P-\bar{P}\|_B\|r_\beta\|_{B^{-1}}^2\leq\kappa_L\epsilon\delta^{-1/2}\|r_\beta\|_{B^{-1}}^2,$$

where $\kappa_L\geq\lambda_1^{-1/2}[m^{1/2}+\sum_{i\in\omega_0}\lambda_i^{1/2}]$ and $\delta\equiv\delta_0/\delta_1$ (see (4.8)). Hence, $(w,r_\beta)\geq\|r_\beta\|_{B^{-1}}^2[1-\epsilon\kappa_L\delta^{-1/2}-a_1]-K_\omega^2\epsilon^2/a_1$. We estimate $-\|w\|_L^2$ as $-\|\bar{w}\|_L^2$ in the proof of Theorem 2 and obtain

$$-\|w\|_L^2\geq-(1+\kappa_L\epsilon)^2\|B^{-1}r_\beta\|_L^2\geq-\delta_1(1+\kappa_L\epsilon)^2\|r_\beta\|_{B^{-1}}^2.$$

When combined with the above inequality, this implies that $2\tau(w,r_\beta)-\tau^2\|w\|_L^2\geq\gamma\|r_\beta\|_{B^{-1}}^2-K_1\epsilon^2$, where $K_1\equiv2\tau K_\omega^2/a_1$ and $\gamma\equiv2\tau[1-a_1-\kappa_L\epsilon\delta^{-1/2}]-\tau^2\delta_1(1+\kappa_L\epsilon)^2]\geq\bar{\gamma}>0$ for $\tau<2/\delta_1$ and small ϵ and a_1. Hence,

$$\mu - \mu^{n+1} \geq \frac{\bar{\gamma}\|r_\beta\|_{B^{-1}}^2 + \tau^2\mu\|w\|_M^2 + 2\tau(\beta - \mu)(u, w)_M}{1 - 2\tau(u, w)_M + \tau^2\|w\|_M^2} - K_1\epsilon^2.$$

In straightforward manner we obtain

$$\mu - \mu^{n+1} \geq \frac{\bar{\gamma}\|r_\beta\|_{B^{-1}}^2 + \tau^2\mu\|w\|_M^2 + \beta - \mu}{1 - 2\tau(u, w)_M + \tau^2\|w\|_M^2} + \mu - \beta - K_1\epsilon^2.$$

Note that $\|u^{n+1}\|_M = \|R_\beta u\|_M \geq K_0 > 0$ (see Lemma 1). Hence, for $K_2 \equiv K_1/K_0$, we have

$$\beta - \mu^{n+1} \geq \frac{\bar{\gamma}\|r_\beta\|_{B^{-1}}^2 + \tau^2\mu\|w\|_M^2 + \beta - \mu}{1 - 2\tau(u, w)_M + \tau^2\|w\|_M^2} - K_2\epsilon^2.$$

We increase the denominator on the right-hand side by replacing the term $-2\tau(u, w)_M$ with the upper bound $a\|u\|_M^2 + a^{-1}\|w\|_M^2, a > 0$. Since $\beta \geq \mu$, then $\beta - \mu^{n+1} \geq \min\{t(u^n)(1+a)^{-1}; \beta(1+a^{-1})^{-1}\} - K_2\epsilon^2$. We may assume that $r_\beta \neq 0$ and take $a = t(u^n)/\beta$. In view of this, we obtain the lower bound maximal in a for $\beta - \mu^{n+1}$, which leads directly to (6.12). $\square$

Theorem 1. *Let $\omega_0, \bar{Q}_{\omega_0}$ and $p > 1$ in (6.5) be such that $\lambda_i \geq \lambda_p \equiv \lambda, \forall i \in \omega_0, \Theta_L(Q; \bar{Q}) \leq \epsilon$, and $\Theta_M(Q; \bar{Q}) \leq \epsilon$. Suppose that, for an initial p-dimensional subspace $H_p^0 \subset \bar{Q}^\perp$,*

$$\beta(H_p^0) \equiv \beta^0 \leq \bar{\beta} < \nu_p^\perp \equiv \nu, \tag{6.13}$$

where $\nu_p^\perp$ coincides with the eigenvalue $\lambda_i, i \notin \omega_0$, closest to $\lambda_p \equiv \lambda$ and strictly greater than λ. Suppose that (4.8), (5.2), and (6.11) hold, that ϵ is small enough, and that $\bar{\gamma}\delta_0 \leq 1$ (see Lemma 3). Suppose that the constants K_ω, K_1, and K_2 in Lemma 3 are independent of $\beta \in [\lambda_1, \bar{\beta}]$, and that

$$K_3 \geq \frac{\bar{\beta}}{\lambda\nu} \sum_{i \in \omega_0, \lambda \leq \lambda_i \leq \nu} \frac{(\lambda_i - \lambda)(\nu - \lambda_i)}{\lambda_i}, \quad K_4 \geq K_3\bar{\beta} + K_2(1 + K_3\epsilon^2).$$

Then

$$\beta^{n+1} - \lambda \leq \rho(\beta^n)(\beta^n - \lambda) + K_4\epsilon^2, \tag{6.14}$$

where

$$\rho(\beta) \equiv \frac{1 - \bar{\gamma}\delta_0(1 - \beta/\nu)}{1 + \bar{\gamma}\delta_0(\beta/\lambda - 1)(1 - \beta/\nu)}.$$

Proof. It is possible to choose $u \equiv u^n \in H_p^n$ in (6.12) such that $\mu(u^{n+1}) = \beta(H_p^{n+1}) \equiv \beta^{n+1}$. Then, by monotonicity in t of the function $t(1 + at)^{-1}$ with $a > 0$ and $t > 0$, we have

$$\beta^n - \beta^{n+1} \geq \frac{\varphi}{1 + \beta^{-1}\varphi} - K_2\epsilon^2, \quad 0 < \varphi \leq \min_{u \in H_p^n} t(u). \tag{6.15}$$

Now we determine the explicit form of φ. In view of (4.8), we have $\|r_\beta\|_{B^{-1}}^2 \geq \delta_0\|r_\beta\|_{L^{-1}}^2$. The lower bound of the right-hand side of this inequality can be found in accordance with (6.10). Then, using the fact that $\lambda_k^\perp = \lambda_k$ for $k \leq p$, we obtain

$$t(u) \geq \bar{\gamma}\delta_0 \left(\frac{\beta^2(\lambda_k + \nu_k - \mu)}{\lambda_k \nu_k} + \mu - 2\beta \right) + \beta - \mu \equiv g_k(\mu)$$

(for $\lambda_k \leq \mu < \nu_k \leq \lambda_p \equiv \lambda$) and

$$t(u) \geq \frac{\bar{\gamma}\delta_0[\beta^2(\lambda_k + \nu - \mu)]}{\lambda\nu} + \mu - 2\beta] + \beta - \mu - K_3\epsilon^2 = g_p(\mu) - K_\beta\epsilon^2$$

(for $\mu \geq \lambda$), where we use the assumption on ω_0, the form of $\omega_{0,0}$ (see(6.10)), and (5.2). Since $\bar{\gamma}\delta_0[1 - \beta^2(\lambda_k\nu_k)^{-1}] < 1$ and $g_k(\lambda_{k+1}) = g_{k+1}(\lambda_{k+1})$, then $\min_{k \leq p} \min_{\mu \in [\lambda_k, \min\{\nu_k, \beta\}]} g_k(\mu) = g_p(\beta)$ and $t(u) \geq \varphi \equiv \gamma\delta_0\beta(1 - \beta/\nu)(\beta/\lambda - 1) - K_3\epsilon^2$ which, together with (6.15), leads to (6.14). $\square$

Theorem 2. *Suppose that the conditions of Theorem 1 are satisfied, that*

$$q_p \equiv \max_{\lambda \leq \beta \leq \beta^0} \rho(\beta) < 1,$$

and that $K_5 \geq K_4(1 - q_p)^{-1}$. *Then method* (1.5) *enables use to obtain* $\beta^r \in [\lambda_p, \lambda_p + \epsilon^2(K_4 + K_5)]$ *in*

$$s \equiv [log_{q_p}(\epsilon^2 K_4(\beta^0 - \lambda)^{-1})] + 1 \tag{6.16}$$

iterations for a certain value of $r \leq s$.

Proof. It is easy to see that $\beta^n \geq \bar{\lambda}_p^\perp \geq \lambda_p$. Moreover, (6.14) implies that $\beta^n - \lambda \leq q_p^n(\beta^0 - \lambda) + K_5\epsilon^2$ and, hence, β^s will have the required accuracy. In fact, such an accuracy could have been reached previously, since there could be a violation in the monotonic decrease of β^n, which could occur only for $\beta^r \in [\lambda_p, \lambda_p + K_5\epsilon^2)$ (see (6.14)). $\square$ [19]

---

[19] For small ϵ, the best value of τ is $\tau \approx 1/\delta_1$. In the particular cases when $\epsilon = 0$ ($P = \bar{P}$) and when ω_0 is empty, the investigation of the method is significantly simplified. Then $\beta^n - \lambda_p \leq \sigma(n)(\beta^0 - \lambda_p)$, where $\sigma(n) \equiv \prod_{i=0}^{n-1} \rho(\beta^i) \leq q_p^n$, $\tau = 1/\delta_1$, $\rho(\beta) \equiv [1 - \delta(1 - \beta/\nu_p)][1 + \delta(\beta/\lambda_p - 1)(1 - \beta/\nu_p)]^{-1}$, and $q_p \equiv \max\{\rho(\beta_0); \rho(\lambda_p)\}$.

6.3. Eigenspaces and their direct sums.

Lemma 4. *Let $H_p^r \subset \bar{Q}^\perp$ be a subspace such that, from Theorem 1, we have*

$$0 \leq \beta^r - \lambda_p \leq K_6 \epsilon^2 \tag{6.17}$$

and $\lambda_i \geq \beta^r$ for any $i \in \omega_0$. Let p' be the dimension of the direct sum of the eigenspaces U_{λ_i} of problem (5.1) associated with all $\lambda_i < \lambda_p$. Let $K_7 \equiv [K_6(\nu_p^\perp - \lambda_p)^{-1} + m]^{1/2} \geq 1$. Then there exists a $(p - p')$-dimensional subspace $\tilde{U}^r \subset H_p^r$ such that

$$\Theta_M^2(\tilde{U}^r; U(\lambda_p)) \leq \frac{\beta^r - \lambda_p}{\nu_p^\perp - \lambda_p} + m\epsilon^2 \leq K_7^2 \epsilon^2. \tag{6.18}$$

Proof. Consider $\tilde{U}^r \subset U' \equiv H_p^r \cap \operatorname{lin}\{u_{p'+1}, \ldots, u_N\}$. Then, for each $u = \sum_{i=1}^N c_i u_i \in \tilde{U}^r$ with $\|u\|_M = 1$, we have $\mu(u) \equiv \mu \in [\lambda_p, \beta^r]$ and

$$\sum_{i=p'+1}^N c_i^2(\lambda_i - \mu) = 0, \quad \sum_{i \in \omega_2}(\lambda_i - \mu)c_i^2 = \sum_{i=p'+1}^p c_i^2(\mu - \lambda_i)c_i^2$$

(see (5.1) and the proof of Lemma 2.4), where $\omega_2 \equiv \{p + 1, \ldots, N\} \setminus \omega_0$. Note that $\lambda_i - \mu \geq \nu_p^\perp - \beta^r$ for $i \in \omega_2$ and that $\mu - \lambda_i \leq 0$ for $i \in \omega_0$. Then $\sum_{i \in \omega_2} c_i^2 \leq \frac{\beta^r - \lambda_p}{\nu_p^\perp - \beta^r}$, which with $\sum_{i \in \omega_0} c_i^2 \leq m\epsilon^2$ yields

$$\{dist\}_M^2(u; U_{\lambda_p}) = \sum_{i \in \omega_0 \cup \omega_2} c_i^2 \leq \frac{\beta^r - \lambda_p}{\nu_p^\perp - \beta^r} + m\epsilon^2. \tag{6.19}$$

This, (1.40), and (6.17) give (6.18). □

For H_p^r and algebraic problem (6.2), consider the eigenvalues $\lambda_1^r \leq \lambda_2^r \leq \ldots \leq \lambda_p^r$. It is clear (see (2.7)) that $\lambda_i \leq \lambda_i^r, i \in [1, p]$, and that $\lambda_p \leq \lambda_{p'+1}^r \leq \cdots \leq \lambda_p^r = \beta^r$.

Theorem 3. *Suppose that (6.17) holds and that ϵ is so small that $\varphi_1 \equiv \bar{\gamma}\delta_0\beta^r[1 - \beta^r(\nu_p^\perp)^{-1}][\beta^r(\lambda_{p'})^{-1} - 1] - K_3\epsilon^2 > 0$. Then*

$$\lambda_{p'}^{r+1} \leq \beta^r[1 + \varphi_1/\beta^r] - 1 + K_2\epsilon^2. \tag{6.20}$$

Proof. Consider $S^r \equiv H_p^r \cap \operatorname{lin}\{u_1, \ldots, u_{p'}, u_{p+1}, \ldots, u_N\}$ and $S_{r+1} \equiv \bar{R}_{\beta^r}\{S^r\}$. Observe that S^r is a set of elements of H_p^r orthogonal in the sense of $H(M)$ to $u_{p'+1}, \ldots, u_p$ and, hence, $\dim S^r = \dim S^{r+1} \geq p'$. If $u \in S_r, \|u\|_M = 1$ and $\mu(\bar{R}_{\beta^r}u) \equiv \mu^{r+1} = \beta(S^{r+1})$, then $\mu^{r+1} \geq \lambda_{p'}^{r+1} \geq \lambda_{p'}$, and we may apply (6.12). This yields

$$\lambda_{p'}^{r+1} \leq \beta^r - \frac{t(u)}{1 + t(u)/\beta^r} + K_2 \epsilon^2,$$

$$\lambda_{p'}^{r+1} \leq \mu^{r+1} \leq \beta^r - \frac{\tilde{\varphi}_1}{1 + \tilde{\varphi}_1/\beta^r} + K_2 \epsilon^2 = \frac{\beta^r}{1 + \tilde{\varphi}_1/\beta^r} + K_2 \epsilon^2, \qquad (6.21)$$

where $\tilde{\varphi}_1 \geq 0$ is the lower bound for $t(u)$ ($u \in S^r, \|u\|_M = 1$). We show now that $\tilde{\varphi}_1$ may be replaced by φ_1. If we drop the values $\lambda_i^\perp$, $i = p' + 1, \ldots, p$, from the set $\lambda_1^\perp, \ldots, \lambda_{N-m}^\perp$ (see (5.17)) then orthogonality of S^r to $u_{p'+1}, \ldots, u_p$ implies that (6.10) holds, and the same argument as in the proof of Theorem 1 applies to the lower bound of $t(u)$. Hence, we conclude that $t(u) \geq \tilde{\phi}_1 = \phi_1$. Combining the lower bound with (6.21) we obtain (6.20). $\square$

From (6.20) it follows that, for small enough ϵ, we have

$$\lambda_{p'}^r < \lambda_p \leq \lambda_{p'+1}^r \leq \cdots \leq \lambda_p^r = \beta^r \leq \lambda_p + K_6 \epsilon^2. \qquad (6.22)$$

Hence, $\lambda_{p'+1}, \ldots, \lambda_p^r$, with $K_6 \epsilon^2$ accuracy can be united into a separate group that is a finite distance apart from $\lambda_{p'}^r$. From the practical point of view, the number p' can be determined from the results of the calculations.

Theorem 4. *Suppose that the conditions of Lemma 4 are satisfied and that $\lambda_{p'}^r < \lambda_p$. Suppose also that algebraic problem (6.2) corresponds to H_p^r and that $\bar{X}^{(i)}, i \in [1, p]$, are its eigenvectors orthonormalized in accordance with inner product $(\bar{Y}, \bar{X})_{\bar{M}_p^r} \equiv \bar{Y}^T \bar{M}_p^r \bar{X}$. Let them be in correspondence with the vectors $v_i^r \equiv v_i$ (see (6.3)), forming an orthonormal basis for $H_p^r(\bar{M}_p^r)$, and let U^r be the linear span of $v_{p'+1}, \ldots, v_p$. Then*

$$\Theta_M(U^r; U(\lambda_p)) \leq K_7 \epsilon + \left[\frac{\beta^r - \lambda_p}{\beta^r - \lambda_{p'}^r}\right]^{1/2} \leq K_8 \epsilon \qquad (6.23)$$

and

$$\Theta_L(U^r; U(\lambda_p)) \leq [\beta^r - \lambda_p + \lambda_p \Theta_M^2(U^r; U(\lambda_p))]^{1/2} \leq K_9 \epsilon. \qquad (6.24)$$

Proof. We estimate $\Theta_M(\tilde{U}^r; U^r)$ ($\tilde{U}^r$ was defined in Lemma 4). Both of these $(p - p')$-dimensional subspaces $\tilde{U}^r$ and U^r belong to H_p^r. To estimate $\text{dist}_M\{z; U^r\}$, where $z \in \tilde{U}^r$ and $\|z\|_M = 1$, we perform an isometric transition in $H(M)$ from H_p^n to the p-dimensional Euclidean space $\bar{H}_p^r(\bar{M})$ consisting of $\bar{X} \equiv [x_1, \ldots, x_p]^T$ with inner product $(\bar{Y}, \bar{X})_{\bar{M}}$, where $\bar{M} \equiv \bar{M}_p^r$ (see (6.2) and (6.3)). In this case, for $z = \sum_{i=1}^p z_i v_i$ with $\|z\|_M = 1$, we have

$$\bar{Z} = \sum_{i=1}^{p} z_i \bar{X}^{(i)}, \ \text{dist}_M^2\{z; U^r\} = \sum_{i=1}^{p'} z_i^2.$$

Hence, applying the proof of Lemma 2.4, we obtain $\sum_{i=1}^{p} z_i^2(\lambda_i^r - \mu) = 0$, where $\mu \equiv \mu(z) \geq \lambda_p$, $\lambda_i^r \geq \lambda_i$ and $\sum_{i=1}^{p'} z_i^2(\mu - \lambda_i^r) = \sum_{i=p'+1}^{p} z_i^2(\lambda_i^r - \mu)$. Thus,

$$(\mu - \lambda_{p'}^r) \sum_{i+1}^{p'} z_i^2 \leq (\beta^r - \mu)(1 - \sum_{i=1}^{p'} z_i^2) \ \text{and} \ \text{dist}_M^2\{z; U^r\} \leq \frac{\beta^r - \mu}{\beta^r - \lambda_{p'}^r}.$$

This implies that

$$\Theta_M^2(\tilde{U}^r; U^r) \leq \frac{\beta^r - \lambda_p}{\beta^r - \lambda_{p'}^r}. \tag{6.25}$$

Because the gap between the subspaces is a metric, then (6.23) follows from (6.18) and (6.25). Now it is easy to estimate $\Theta_L(U^r; U_{\lambda_p})$, because, for any eigenvector $u \in U_{\lambda_p}$ with $\|v\|_M = 1$, (2.1) holds with $i = p$. This leads to (6.24). □

6.4. Eigenspaces for smaller eigenvalues. The algorithm considered above for finding approximations to the spectral pair $\{\lambda_p; U(\lambda_p)\}$ works under the assumption that the initial p-dimensional subspace $H_p^0 \subset \bar{Q}^\perp$ satisfies (6.13) (see Theorem 1). Recall (see § 2) that the spectral pair $\{\lambda_p; U(\lambda_p)\}$ was also denoted by $\{\nu_j; U(\nu_j)\}$, where

$$p = \dim U(\nu_1) + \cdots + \dim U(\nu_j), \tag{6.26}$$

$\dim U(\nu_i) \equiv k_i, i \in [1, j]$ (see (1.15)), and $k_j = p - p'$. We emphasize, that for small enough ϵ, the multiplicity k_j can be practically determined from the results of the calculations when the distance between approximations to λ_p and λ_p' remain sufficiently large relative to the distances between the approximate eigenvalues with indices $p' + 1, \ldots, p$ (see (6.22)). However, from the theoretical point of view, no such calculations can guarantee that $\dim U(\lambda_p) = p - p'$ (algorithms with two-side approximations to eigenvalues are of special interest but beyond the scope of this book). Finding $\{\nu_j; U(\nu_j)\}$ may be regarded as the first stage of the general algorithm for finding all $\{\nu_i; U(\nu_i)\}$ with smaller $i \leq p$.

Having found U_{ν_j} with $O(\epsilon)$-accuracy (see (6.23) and (6.24)), we are in a position to change ω_0 and $Q_{\omega_0}^\perp$ and turn to the determination of $\lambda_{p'}$ and $U_{\lambda_p'}$, that is, to the next spectral pair $\{\nu_{j-1}; U(\nu_{j-1})\}$. By virtue of (6.20), we may take the linear span of the vectors $v_1, \ldots, v_{p'}$ belonging to H_p^r as an initial iterate $H_{p'}^0$. If $\lambda_{p'}^r$ is sufficiently far away from $\lambda_{p'+1}^r \approx \lambda_p$, then orthogonalization may be unnecessary, but if $\lambda_{p'}^r$ differs from $\lambda_{p'+1}^r$ only slightly, then orthogonalization with respect to U^r can essential. Inserting the indices $p' + 1, \ldots, p$ into ω_0, with possible elimination of some higher indices, enables one to obtain the convergence rate estimates that are no worse than for the first stage. Thus, this provides the possibility to determine ν_i and U_{ν_i} with the accuracy required for all $i \leq j$, given, e.g., an initial iterate H_p^0 with

$$\beta(H_p^0) \leq \bar{\beta} < \lambda_{p+1} = \nu_{j+1} \tag{6.27}$$

(then we can even start with $\omega = \emptyset$ in the first stage of the general algorithm). Moreover, in situations when there are difficulties in finding any multiplicity k_i, it is natural to look not for the approximate eigenspaces, but for an orthonormal (in the sense of $H(M)$) basis $\bar{w}_1, \ldots, \bar{w}_p$ for $U_{\nu_1} \oplus \cdots U_{\nu_j}$, that is an ϵ-approximation (see (5.2)) to an orthonormal basis $w_1, \ldots, w_p$ (instead of $u_1, \ldots, u_p$ in (1.14)) and such that $0 \leq \mu(\bar{w}_i) - \lambda_i \leq K_\lambda \varepsilon^2$, $i = 1, \ldots, p$. We can then specify an effective algorithm consisting of p stages. Each stage produces only one vector: $\bar{w}_p$ in the first stage (iterations (6.5) with $\omega_0 = \emptyset$), $\bar{w}_{p-1}$ in the second (iterations (6.5) but with $(p-1)$-dimensional subspaces, $\omega_1 = \lin \{\bar{w}_p\}$), and the initial iterate

$$H_{p-1}^0 = \lin \{v_1, \ldots, v_{p-1}\}$$

found at the final iterate of the first stage, and so on. Thus, the presence of a cluster is no hindrance provided we can construct good model operators B. Numerical experiments can be found in [208]. Finally, we remark that for many grid problems in practice, it is possible to construct H_p^0 in (6.27) as a linear span of p functions obtained from the first p eigenfunctions of a certain model operator associated with a model domain Q by transformations of type (2.1.4) (sometimes we can even use piecewise linear mappings as in § 4.2–4.4). Numerical experiments on coarse grids can also produce the desired subspace H_p^0. Generalizations to problems (1.2) are fairly transparent.

§ 7. Estimates of computational work for spectral elliptic problems

In this section, we show that for obtaining ε-approximations to several spectral pairs (see (2.17)) for symmetric problems with elliptic operators, algorithms can be developed with computational work estimates identical to those for asymptotically optimal algorithms for boundary value problems with the same operators. This assumes that the desired eigenfunctions satisfy the same smoothness conditions as do the solutions of the corresponding boundary value problems. We confine ourselves to the case when quasiuniform triangulations (possibly composite) $T_h(\hat{\Omega}_h)$ are constructed and the respective PGMs are actually just special cases of the Rayleigh-Ritz method (see § 1 and 2). Resulting algebraic problems (4.1) or (1.2) contain symmetric and positive grid operators $L \equiv L_h$ exactly the same as in Chapters 5–8, and the problem of choosing optimal model operator $B \equiv B_h \asymp L_h$ requires nothing new. The desired algorithms are then just a combination of the iterative methods considered in § 4–6, with coarse grid continuation which is defined by analogy with § 1.4. [20]

7.1. Spectral problems with second-order operators. In what follows, we consider: problem (1.3) in the Hilbert space $G \equiv W_2^1(\Omega; \Gamma_0)$, where Ω is the domain considered in § 5.1 or 5.2; the bilinear form $b_L(u; v) \equiv b_L(\Omega; u; v) \equiv b(\Omega; u; v)$ ($b(\Omega; u; v)$ was defined in § 5.1 and 5.2 and corresponds to the operator $L \in \mathcal{L}^+(G)$ (see (5.2.1)–(5.2.4)));

$$b_M(u; v) \equiv b_M(\Omega; u; v) \equiv (d, uv)_{0,\Omega}, \qquad (7.1)$$

$$(d, u^2)_{0,\Omega} \geq \kappa |u|_{0,\Omega}^2, \quad \kappa > 0; \qquad (7.2)$$

$M = M^* > 0$ (but not positive definite); and $A \equiv L^{-1}M \in \mathcal{L}(G(L))$ is symmetric and compact (see Lemma 5.2.1). In this section we return to our use of H only for Euclidean spaces. For a prescribed j, we look for ε-approximations to the spectral pairs $\{\nu_i, U(\nu_i)\}$, with $i \leq j$ and p defined by (6.26). [21]

We confine ourselves to the case when quasiuniform triangulations (possibly composite) $T_h(\hat{\Omega}_h)$ are constructed (of the same type as those in § 5.1, 5.2) and the respective PGMs are actually just special cases of the Rayleigh-Ritz method (see § 1 and 2).

With each triangulation $T_h(\hat{\Omega}_h)$ we associate the same subspace $\hat{G}_h \equiv \hat{G}$ used in § 5.1 or 5.2 for the respective boundary value problem associated with the prescribed bilinear form $b_L(u; v)$. We approximate our problem by the standard projective problems

---

[20] Our exposition is based on the results obtained in [175, 178, 184, 187, 195, 197, 205, 213, 208]; relevant results dealing largely with multigrid methods can be found in [14, 90, 265, 348].

[21] This means that we deal with partial differential eigenproblems such as those that arise, e.g., in structural analysis and reactor physics, and especially in optimization problems (see, e.g., [340]). Sometimes, when eigenvalues are badly separated, we prefer to deal with the problem of obtaining an orthonormal (in $G(M)$) basis $w_1, \ldots, w_p$ for $U_p \equiv U(\nu_1) \oplus \cdots \oplus U(\nu_j)$ as in Subsection 6.6. We assume that all $u_1, \ldots, u_p$ in (1.14) and, therefore, all elements in U_p meet our standard smoothness conditions, say, (5.1.3).

$$b_L(\hat{\Omega}; \hat{u}; \hat{v}) = \lambda b_M(\hat{\Omega}; \hat{u}; \hat{v}), \quad \forall v \in \hat{G}_h.$$

If $|\hat{\Omega}_h \setminus \Omega| > 0$, then those problems do not conform to the Rayleigh-Ritz method, so we prefer to use approximations defined by

$$b_L(\Omega; \hat{u}; \hat{v}) = \lambda b_M(\Omega; \hat{u}; \hat{v}), \quad \forall v \in \hat{G}_h. \tag{7.3}$$

They may be regarded as small perturbations of the above scheme. [22]

For (7.3), the theory given in § 2 applies. We can thus attain the desired ε-approximations if we take $h^\gamma \asymp \varepsilon$ (the arising algebraic problems (4.1), with $L \equiv L_h$, $M \equiv M_h$, and $\lambda \equiv \lambda_h$, are formulated in the Euclidean space $H \equiv \mathbf{R}^N$ with $N \asymp h^{-d} \asymp \varepsilon^{-d/\gamma}$).

To apply methods from § 5 and 6, we assume that

$$\mu(\hat{u}) \equiv \lambda(\hat{u}) \leq \beta^0 < \nu_{j+1}, \quad \forall u \in \hat{S}_p^0 \setminus 0, \tag{7.4}$$

where $\hat{S}_p^0 \subset \hat{G}$ is given and dim $\hat{S}_p^0 = p$. We emphasize that

$$\mu(\hat{u}) = \frac{b_L(\Omega; \hat{u}; \hat{u})}{b_M(\Omega; \hat{u}; \hat{u})} = \frac{(L_h u, u)}{(M_h u, u)}, \quad \forall \hat{u} \in \hat{G} \setminus 0, \tag{7.5}$$

where the vector $u \in H$ corresponds to $\hat{u} \in \hat{G} \setminus 0$ (see (0.2.2)). Then (6.27) follows from (7.5) if H_p^0 corresponds to $\hat{S}_p^0$ (we note that $p << N$).

Theorem 1. *Let the original partial spectral problem (1.3) be approximated by (7.3). Suppose that (6.26) holds and that all $u_1, \ldots, u_p$ in (1.14) satisfy smoothness condition (5.1.3) with $n_2 = 0$ and $\gamma \in (0, 1]$. Suppose also that (7.4) holds for $\hat{S}_p^0 \subset \hat{G}$. Then the algorithm indicated in Subsection 6.4 (with the initial iterate H_p^0 determined by $\hat{S}_p^0$ and $\omega = \emptyset$ in the first stage) yields an orthonormal (in $G(M)$) basis $\hat{w}_1, \ldots, \hat{w}_p$ for U_p such that*

$$0 \leq \mu(\hat{w}_i) - \lambda_i \leq K_\lambda \varepsilon^2, \quad i \in [1, p], \tag{7.6}$$

and the computational work is $W(\varepsilon) = O(\varepsilon^{-d/\gamma} |\ln \varepsilon|)$.

Proof. Since problems (7.3) yield $O(\varepsilon)$-approximations to the desired spectral pairs, it suffices to find ε-approximations to the eigenvalues of the arising algebraic problems (4.1) in the Euclidean space $H \equiv \mathbf{R}^N$ with $N \asymp h^{-d} \asymp \varepsilon^{-d/\gamma}$ on the basis of the algorithm at hand provided that we have an optimal model operator $B \equiv B_h \asymp L_h$. But such operators were already used in § 5.1 and 5.2. Hence, our iterations yield an orthonormal (in $H(M_h)$) system $w_1, \ldots, w_p$ such that $\hat{w}_1, \ldots, \hat{w}_p$ constitute an orthonormal (in $G(M)$) basis $\hat{w}_1, \ldots, \hat{w}_p$ for U_p and $0 \leq \mu(\hat{w}_i) - \hat{\lambda}_i \leq \bar{K}_\lambda \varepsilon^2, \quad i \in [1, p]$.

---

[22] We used a similar approach in § 5.1; both formulations are identical under the simplifying but reasonable assumption that $\hat{\Omega}_h \subset \bar{\Omega}$.

This and (2.17) imply (7.6). Each stage of our algorithm requires $O(|\ln \varepsilon|)$ iterations because the recursive orthogonalization procedure requires also $O(|\varepsilon|)$-approximations to the subspace $\bar{Q}$ on each of the stages (ϵ in § 6 is our ε under consideration). This yields the desired estimates for $W(\varepsilon)$. $\square$

Theorem 2. Let the conditions of Theorem 1 be satisfied and assume $k_i \equiv \dim U(\nu_i), i \in [1, j]$, *are known. Then the algorithm indicated in Subsection 6.4 (with the subspace H_p^0 determined by $\hat{S}_p^0$ and $\omega = \emptyset$ in the first stage) yields an orthonormal (in $G(M)$) basis $\hat{w}_1, \ldots, \hat{w}_p$ for U_p such that*

$$\|\hat{w}_i - u_i\|_L \le K_L \varepsilon, \quad |\mu(\hat{w}_i) - \lambda_i| \le K_\lambda \varepsilon^2, \quad i \in [1, p], \qquad (7.7)$$

where $u_1, \ldots, u_p$ is an orthonormal (in $G(M)$) basis for U_p indicated by (1.14).

Proof. The proof is much the same as for Theorem 2 but now the known structure of the eigenspaces assures us that $w_i \in H$ and $\hat{w}_i \in \hat{G}$ are $O(|\varepsilon|)$-approximations to the respective eigenvectors and eigenfunctions due to the known structure of eigenspaces. $\square$ [23]

Now we eliminate the multiplier $\ln \varepsilon$ in these estimates and obtain optimal estimates $W(\varepsilon) = O(\varepsilon^{-2/\gamma})$ (see (5.1.19)) for $d = 2$ and for the general case of curvilinear boundaries under the simplifying assumption that $\hat{\Omega}_h \subset \bar{\Omega}$ (see (5.1.5) and (5.1.20)) (for $d \ge 3$, we can obtain estimate $W(\varepsilon) = O(\varepsilon^{-d/\gamma})$ when $\bar{\Omega}$ consists of a finite number of simplexes (see § 5.2)). [24]

The acceleration procedure deals with $l^* = O(|\ln h|) = O(|\ln \varepsilon|)$ grid problems

$$L^l u^l = \lambda^{(l)} M^l u^l, \quad u^l \in H^l \setminus 0, \quad l \in [0, l^*],$$

associated with triangulations $T^{(l)}(\hat{\Omega}_{h_l})$ and subspaces $\hat{G}_{h_l} \equiv \hat{V}^{(l)}, l \in [0, l^*]$, described in § 5.1 (see (1.4.17); note that p instead of $l*$ was used in § 1.4, 5.1, and 5.2, and that $h_0 \asymp 1$, which is small enough). For simplicity, we confine ourselves to the case when $\lambda_1 < \lambda_2 < \cdots < \lambda_p$ and

---

[23] Theorem 2 yields the desired $O(|\varepsilon|)$-approximations to the spectral pairs in the sense of (2.17). Also we might apply iterations from Theorem 5.6 with the difference that now the order of obtained vectors might be different from the order $w_p, w_{p-1}, \ldots, w_1$ in Theorem 1 because convergence to smaller eigenvalues might occur. The more general smoothness conditions present no additional trouble but the case $|\hat{\Omega}_h \setminus \Omega| = O(h^2)$ requires more cumbersome descriptions and proofs.

[24] This is done by applying *multigrid acceleration of the basic iterative algorithm* similar to that considered in § 1.4 and 5.1 (seed-o77,d-o80; closely related approaches were developed in [90, 359]). Recall that the desired result is attained by moving progressively from coarse to finer grids to provide a good starting guess, for the fine grid iteration, and that the routine correspondence (0.2.2) between functions and vectors is used.

$\lambda_i^{(0)} < \lambda_{i+1}, i \in [0,p]$ (then $\lambda_i \leq \lambda_i^{(l)} < \lambda_{i+1}, \ i \in [0,p], \ l \in [0,l^*]$). The respective eigenvectors, forming an orthonormal system in $H^{(l)}(M^{(l)})$, are denoted by $u_1^{(l)}, \ldots, u_p^{(l)}$. We assume that

$$\lambda_i^{(l)} - \lambda_i \leq K_0 h_l^{2\gamma} \leq \Delta\lambda/3, \ \|\hat{u}_i^{(l)} - u_i\|_L \leq K_0' h_l^\gamma, \ i \in [1,p], \ l \in [0,l_*], \quad (7.8)$$

$(\lambda_{i+1} - \lambda_i \geq \Delta\lambda, i \in [1, p-1])$ and that $S_p^0 \subset \hat{V}^{(0)}$ (see (7.4)). Thus, on level $l = 0$, we may apply Theorem 2 with $t \asymp 1$ iterations and obtain vectors $\hat{v}_i^{(0)}, i \in [1,p]$, that form an orthonormal system in $G(M)$ and satisfy

$$\lambda_i \leq \mu(\hat{v}_i^{(l)}) \leq \lambda_i + 2K_\lambda h_l^{2\gamma} \leq \lambda_i + 1/2\Delta\lambda, \ \|\hat{v}_i^{(l)} - u_i\|_L \leq \bar{K}_L h_l^\gamma, \quad (7.9)$$

where $i \in [1,p], l = 0$. This indicates that the vectors $w_i^{(l+1)} \equiv I_l^{l+1} v_i^{(l)} \in H^{(l+1)}, i \in [1,p]$ (see (5.1.20)), may serve as good initial iterates for the methods developed in § 4 and 5. When $\hat{V}^{(l)} \subset \hat{V}^{(l+1)}$ the study simplifies; but, for the general case, we need some lemmas similar to those in § 5.1.

Lemma 1. Let our triangulations be as in § 5.1. Then there exists K_0, independent of l, such that $|\hat{v}^{(l)} - \hat{w}^{(l+1)}|_{1,\Omega} \leq K_0 h_l^{1/2}, \forall \hat{v} \in \hat{V}^{(l)}$ with $\|\hat{v}\|_{1,\Omega} = 1$.
Proof. It suffices to make use of (5.5.22). □
Lemma 2. Let $L = L^* > 0, M = M^* > 0$ and

$$\|u\|_M = 1, \ \|u - v\|_M \leq \epsilon_M < 1, \ \|u - v\|_L \leq \epsilon_L.$$

Then $\mu^{1/2}(v) \leq [\mu^{1/2}(u) + \epsilon_L]/(1 - \epsilon_M)$ and, in the particular case when $Lu_i = \lambda_i M u_i, \|u\|_M = 1, \|z\|_M < 1$, we have

$$\mu(u_i + z) - \lambda_i \leq [\|z\|_L^2 - \lambda_i\|z\|_M^2](1 - \|z\|_M)^{-2}.$$

Proof. The first inequality follows from (5.24). The second is an easy consequence of the fact that $\mu(u_i + z) - \lambda_i = [\|z\|_L^2 - \lambda_i\|z\|_M^2][1 + 2(u_i, z)_M + \|z\|_M^2]^{-1}$. □
Lemma 3. Suppose that $\hat{\Omega}_h \subset \bar{\Omega}$ and that (7.9) holds for a given level $l \leq l^* - 1$, where $u_1, \ldots, u_p$ satisfy the smoothness conditions $u_i \in W_2^{1+\gamma}(\Omega)$ with $\gamma \in (0,1]$. Suppose also that h_0 is small enough. Then the functions $\hat{w}_1^{(l+1)}, \ldots, \hat{w}_p^{(l+1)}$ form a basis for $\hat{W}_p^{(l+1)}$ and

$$\lambda_i \leq \mu(\hat{w}_i^{(l+1)}) \leq \lambda_i + \bar{K}_\lambda h_l^{2\gamma}, \|\hat{v}_i^{(l)} - u_i\|_L \leq \bar{K}_L h_l^\gamma, \quad i \in [1,p],$$

where the constants are independent of l.

Proof. The functions $\hat{w}_1^{(l+1)}, \ldots, \hat{w}_p^{(l+1)}$ are linearly independent in accordance with Lemma 1 and the smallness of h_0. Next, for each i, we may apply (5.1.26), which implies that

$$\|u_i - \hat{w}_i^{(l+1)}\|_L^2 \le \sigma_0 \|u_i - \hat{v}_i^{(l)}\|_L^2 + \sigma_1 h_l^{2\gamma}, \quad i \in [1, p],$$

and leads to the desired inequalities (see Lemma 2). □

Theorem 3. *Let the conditions of Theorem 2 and Lemma 3 be satisfied and suppose that (7.8) holds. Suppose also that, on each level $l \in [1, p]$, we perform t iterations considered in § 4–6 with the initial iterate taken from level $l - 1$. Then it is possible to choose the number $t \asymp 1$ of iterations on each level in such a way that the final iteration on level $l*$ yields an $O(\varepsilon)$-approximation to the desired spectral pair with the computational work $W(\varepsilon) = O(\varepsilon^{-2/\gamma})$.*

Proof. Consider, e.g., iterations (4.5) for $i = 1$. Then

$$X_{l+1} \equiv \mu(v_1^{(l+1)}) - \lambda_1 = \lambda_1^{(l)} - \lambda_p + \mu(v_1^{(l+1)}) - \lambda_1^{(l)}$$

and $X_{l+1} \le K_0 h_l^{2\gamma} + q^t |\mu(w_1^{(l+1)}) - \lambda_1^{(l)}|$, where $q \equiv q^* < 1$ (see Theorem 4.1). Hence,

$$X_{l+1} \le K_0 h_l^{2\gamma} + q^t X_l.$$

If $\hat{V}^{(l)} \subset \hat{V}^{(l+1)}$, then the proof is simplest and the computational work is estimated as in § 1.4. For the general case, assume that (7.9) holds on level l and that $\bar{K}_\lambda h_l^{2\gamma} \le 2/3\Delta\lambda$ (it suffices to use small enough h_0). Then Lemma 3 implies that we may start the iterations (on level $l + 1$) with the initial iterate w_1^{l+1} and obtain, after t iterations, $\hat{v}_1^{(l+1)}$ such that $\mu(\hat{v}_1^{(l+1)}) - \lambda_1 \le q^t \bar{K}_\lambda h_l^{2\gamma}$. It is easy to prove (7.9) for level $l + 1$ if t is large enough and $t \asymp 1$. This proves our theorem for λ_1. For the remaining λ_i ($i \in [2, p - 1]$), we apply iterations from § 6 dealing with i-dimensional subspaces and $\omega_0 = \emptyset$. Then we can avoid orthogonalization and obtain analogous recursive estimates. □ [25]

---

[25] Even for a nonempty set ω_0, theorems given in § 6 lead to the desired estimates under the assumption that we use $O(\varepsilon_l)$-approximations to the orthoprojectors on the level l, where $\varepsilon_l \asymp h_l^\gamma$. Algorithms with the above properties can be also constructed for problems involving strongly elliptic operators L such as those considered in § 5.3. (M might correspond to certain first-order differential operators.) Some of the above algorithms have been used with success for two- and three-dimensional practical problems (see, e.g., relevant results in [75, 90, 305, 359]. Some classes of problems are actually determined by the geometry of the domain and several parameters (e.g., this is the case for eigenvalue problems involving the Laplace operator or elasticity operator problems in § 5.4). Then the estimates for N-widths given in § 1.5 are badly suited for such classes and, hence, the optimization problem becomes more difficult. Nonetheless, algorithms

7.2. Spectral elliptic problems with fourth-order operators.
We consider only those algorithms based on the approach discussed in
§ 8.2, which apply to fairly general regions and boundary conditions. As the
simplest example, we consider the minimization problem. Such problems,
as well as the problems of finding the maximum of $\tilde{I}_1(w)(\tilde{I}_2(w))^{-1}$,
are important, for example, for analysis of plates stability and their oscillations.
The non-sign-definite matrices $[b_{i,j}]$ correspond to problems of type (1.2).

$$\lambda_1 \equiv \min_{w \in W \backslash 0} \Phi(w), \tag{7.10}$$

where $W \equiv W(\Omega; \Gamma^0)$ (see Theorem 8.2.8),

$$\Phi(w) \equiv I_2(w)/I_1(w), \tag{7.11}$$

where

$$I_1(w) \equiv \sum_{r=1}^{2} \sum_{r=1}^{2} (b_{r,l}, D_r w D_l w)_{0,\Omega},$$

$I_2(w) \equiv b_1(w; w)$, $b_1(u; v)$ is defined by (8.2.9), and $[b_{i,j}]^T = [b_{i,j}] \geq \nu I$.
Let $G = G_1 \times G_2$ be the Hilbert space in Theorem 8.2.8. Consider now the
eigenvalue problem

$$b_1^{(1)}(u_1; v_1) + b(v_1; u_2) = \lambda b_M(u_1; v_1), \quad \forall v_1, \tag{7.12}$$

$$b(u_1; v_2) = 0, \quad \forall v_2, \tag{7.13}$$

where $b_1^{(1)}(u_1; v_1) \equiv b_{1,1}(u_1; v_1)$ is the same as in Theorem 8.2.8, $b(u_1; v_2) =$
$(\text{div } u_1, v_2)_{0,\Omega}$, and

$$b_M(u_1; v_1) \equiv (b_{1,1}, u_{1,2} v_{1,2})_{0,\Omega}$$

$$+(b_{2,2}, u_{1,1} v_{1,1})_{0,\Omega} - (b_{1,2}, u_{1,2} v_{1,1} + u_{1,1} v_{1,2})_{0,\Omega}. \tag{7.14}$$

Theorem 4. λ_1 *in* (7.10) *coincides with the minimal* λ *in problem*
(7.12)–(7.14).

Proof. Problem (7.12)–(7.14) can be rewritten in operator form as

$$Lu \equiv \begin{bmatrix} L_{1,1} & L_{1,2} \\ L_{2,1} & 0 \end{bmatrix} \begin{bmatrix} u_1 \\ u_2 \end{bmatrix} = \lambda M u \equiv \lambda \begin{bmatrix} M_1 u_1 \\ 0 \end{bmatrix} \tag{7.15}$$

---

producing $O(\varepsilon^2)$-approximations to the mentioned eigenvalues were constructed in [395]
that require $W(\varepsilon) = O(\varepsilon^{-(1+\kappa)})$, $\forall \kappa > 0$; the Dirichlet conditions in two-dimensional
domains with smooth boundary were considered; and the PGM made use of a nonuniform
grid with the number of the nodes $N(\varepsilon) = O(\varepsilon^{-(1+\kappa')})$, $\kappa' < \kappa$, and of iterations similar
to (4.5). These results were further specified for simpler domains in [307].

in the Hilbert space $G = G_1 \times G_2$, where L is a strongly saddle operator in G, $M_1 = M_1^* > 0$ ($M_1 \in \mathcal{L}(G_1)$), $M = M^* \geq 0$, and $L^{-1}M$ is a compact operator (see Lemma 5.2.1). Along the same lines, we can identify all λ_i and the stationary points of the functional $\Phi(w)$ (the eigenvalues of the respective problem with the fourth-order elliptic operator).

To approximate (7.12)–(7.14), we make use of the same subspaces $\hat{G} = \hat{G}_1 \times \hat{G}_2$ as in § 8.3 (we consider them later for more general cases (see (7.25)).[26] The resulting algebraic problems in the respective Euclidean spaces $H \equiv H_h = H_{1,h} \times H_{2,h}$ have the same form (7.15) but with $L \equiv L_h, M \equiv M_h$, and $\lambda \equiv \lambda_h$. To obtain effective algorithms for them, we suggest the penalty method, yielding the problem

$$\begin{bmatrix} L_{1,1} & L_{1,2} \\ L_{2,1} & -\alpha J_2 \end{bmatrix} \begin{bmatrix} u_1 \\ u_2 \end{bmatrix} = \lambda \begin{bmatrix} M_1 u_1 \\ 0 \end{bmatrix}, \quad \alpha > 0 \qquad (7.16)$$

(see J_2 in Lemma 7.4.1), which is reduced to the standard problem

$$S_1 u_1 \equiv (L_{1,1} + 1/\alpha L_{1,2} J_2^{-1} L_{2,1}) u_1 = \lambda M_1 u_1 \qquad (7.17)$$

in the Euclidean space H_1 with $S_1 \in \mathcal{L}(H_1)$ (see Theorem 7.4.3 and (7.4.12) with $L_{2,2} = J_2$). The effect of introducing $\alpha > 0$ will be investigated later (see Theorem 7 for more general problems) and yields the estimate $\alpha = O(\varepsilon)^{2\gamma}$ (see assumptions (7.4.5) with respect to eigenfunctions of (7.15)). We emphasize that model operators $D_1 \in \mathcal{L}^+(H_1)$ for S_1 in (7.17) were constructed in Theorems 7.4.3 and 7.4.4 such that (7.4.36) holds with constants δ_0 and δ_1 in (4.36) independent of h and $\alpha \in (0, \alpha_0]$ and computational work $W(\alpha, h) = O(|\ln \alpha|/h^2)$ for solving a system with D_1.

Next, consider the special case of problems considered in Theorem 2.3.1 (see also (0.5.8)). Let T be the triangle with vertices $(0,0), (1,0)$, and $(0,1)$, and let W consist of functions $w \in W_2^2(T)$ that vanish at these vertices. We seek

$$\lambda_1 \equiv \min_{w \in W \backslash 0} |w|_{2,T}^2 |w|_{1,T}^{-2}. \qquad (7.18)$$

The Hilbert space G_1 is now a subspace of functions in $(W_2^1(T))^2$ such that

$$\phi_1(u) \equiv \int_0^1 u_{1,1}(0, x_2) dx_2 = 0, \quad \phi_2(u) \equiv \int_0^1 u_{1,2}(x_1, 0) dx_1 = 0, \quad (7.19)$$

and the space rot W is defined by

---

[26] The convergence of such projective approximations were studied in [27, 311, 369].

$$\text{rot } W \equiv \{u_1 : u_1 \in G_1 \text{ and div } u_1 = 0\}. \tag{7.20}$$

Theorem 5. Problem (7.18) is equivalent to finding

$$\lambda_1 \equiv \min_{u_1 \in \text{rot } W \backslash 0} |u_1|^2_{1,T} |u|^{-2}_{0,T} \tag{7.21}$$

and is reduced to a particular case of (7.12)–(7.14).

Proof. It suffices to rewrite (7.18) in terms of rot w. It is also simple to obtain (7.1.31) for the indicated G_1 and $G_2 \equiv L_2(T)$. □

It is not difficult to show that, for the indicated problem we may apply PGMs, based on quasiuniform triangulations and local refinements around the vertices of T. Moreover, an optimal model operator B can be constructed as in § 5.4 and 8.4. This implies that it is even possible to indicate asymptotically optimal algorithms for problems of this type and solve them with high accuracy if need be. But, when a moderate accuracy is required, a simpler approach based on the penalty method might be more useful. For example, in [453], it was suggested that we replace (7.21) by

$$\lambda_1(\alpha) \equiv \min_{u_1 \in G_1 \backslash 0} \frac{|u_{1,1}|^2_{1,T} + |u_{1,2}|^2_{1,T} + 1/\alpha |\text{div } u_1|^2_{0,T}}{|u_{1,1}|^2_{0,T} + |u_{1,2}|^2_{0,T}}. \tag{7.22}$$

Moreover, the symmetry of the problems with respect to the bisector $x_1 = x_2$ was taken into account and good practical approximations to λ_1 were obtained even on grids with $h = 1/8$ (they were specified in § 0.5 and were in agreement with the known approximations (see [12]) and control computations with $h = 1/21$ and $\alpha = 10^{-5}$).[27]

7.3. Spectral problems with linear constraints. Problems (7.15) are examples of more general *spectral problems with linear constraints* of type

$$Lu \equiv \begin{bmatrix} L_{1,1} & L_{1,2} \\ L_{2,1} & -\alpha L_{2,2} \end{bmatrix} \begin{bmatrix} u_1 \\ u_2 \end{bmatrix} = \lambda \begin{bmatrix} M_1 u_1 \\ 0 \end{bmatrix} \equiv \lambda M u, \ \alpha > 0, \tag{7.23}$$

where $M_1 = M_1^* > 0$ and L is a strongly saddle operator (see § 7.1) with $L_{2,2} \in \mathcal{L}^+(G_2)$. By virtue of Theorem 7.1, the symmetric operator L has a bounded inverse L^{-1}. In the sequel, we assume that $A = L^{-1}M$ is compact

---

[27] The results obtained in § 8.4 indicate that our approach can be generalized to certain spectral problems in the strengthened Sobolev spaces. For strongly elliptic operators L, description of the corresponding PGMs and iterative processes necessarily becomes more cumbersome. Some examples, which model certain problems of the shell stability theory, were given in [197].

in G; but for many concrete problems, e.g., for (7.15), this property of A is a simple consequence of Lemma 5.2.1. We will deal even with more general problems

$$Au = su \tag{7.24}$$

when $M_1 = M_1^*$ and no positivity is assumed. Note that A might be nonsymmetric. Thus, for an eigenvalue s of (7.24) with the corresponding eigenspace $U(s)$, we introduce a generalized eigenspace $K(s) \equiv \{u$: there exists k such that $(A - sI)^k u = 0\}$ (if $s \neq 0$, then dim $K(s) < \infty$).

Lemma 4. If $s \neq 0$, then $K(s) = U(s)$. [28]

Proof. It is not difficult to verify that $L^{-1} = [R_{i,j}], i = 1, 2, j = 1, 2$, where $R_{1,1} = (L_{1,1} + \alpha^{-1} L_{1,2} L_{2,2}^{-1} L_{2,1})^{-1} > 0$ (if $\alpha > 0$); if $\alpha = 0$ then

$$R_{1,1} = L_{1,1}^{-1/2} (I_1 - \Lambda_1^* (\Lambda_1 \Lambda_1^*)^{-1} \Lambda_1) L_{1,1}^{-1/2},$$

where $\Lambda_1 = L_{2,1} L_{1,1}^{-1/2}$ and $\Lambda_1^* = L_{1,1}^{-1/2} L_{1,2}$. Hence, $R_{1,1} = R_{1,1}^* \geq 0$ in the sense of $\mathcal{L}(G_1)$ (it is easy to obtain a generalization of Lemma 0.4.6). Moreover, $R_{1,1} M_1$ is compact. Further, the nonnegativeness of $R_{1,1}$ and symmetry of M_1 are sufficient for any generalized eigenspace $K(\nu)$ corresponding to an eigenvalue $\nu \neq 0$ of $R_{1,1} M_1$ to be identical to the eigenspace. Hence, if $(A - \nu I)^2 u = 0$, then $(R_{1,1} M_1 - \nu I_1)^2 u_1 = 0$ implies that $(R_{1,1} M_1 - \nu I_1) u_1 = 0$. This and the fact that $-\nu R_{2,1} M_1 u_1 + \nu^2 u_2 = 0$ lead to $(A - \nu I) u = 0$. $\square$

We are interested in several first spectral pairs for (7.24) when the positive s in (7.24) are numbered in decreasing order (see (1.10) and (1.13); t_i is of multiplicity k_i; see (6.27)). The projective approximation of (7.24) (see (7.2.1), (7.2.2)) is defined by

$$\left. \begin{array}{l} b_{1,1}^{(1)}(\hat{u}_1; \hat{v}_1) = \hat{s}[b_{1,1}(\hat{u}_1; \hat{v}_1) + b_{2,1}(\hat{v}_1; \hat{u}_2), \quad \forall \hat{v}_1 \in \hat{G}_1 \equiv \hat{G}_{1,h}, \\ 0 = b_{2,1}(\hat{u}_1; \hat{v}_2) - \alpha b_{2,2}(\hat{u}_2; \hat{v}_2), \quad \forall \hat{v}_2 \in \hat{G}_2 \equiv \hat{G}_{2,h} \end{array} \right\} \tag{7.25}$$

and its positive eigenvalues are

$$\hat{t}_1 \equiv \hat{s}_1 \geq \hat{s}_2 \geq \cdots \geq \hat{s}_{k_1} \geq \hat{s}_{k_1+1} \equiv \hat{t}_2 \geq \cdots.$$

Its operator form (see (7.2.3)) is

$$\hat{M}_1 \hat{u}_1 = \hat{s}[\hat{L}_{1,1} \hat{u}_1 + \hat{L}_{1,2} \hat{u}_2], \quad 0 = \hat{L}_{2,1} \hat{u}_1 - \alpha \hat{L}_{2,2} \hat{u}_2, \tag{7.26}$$

---

[28] Consequently, we need not distinguish hereafter between algebraic or geometric multiplicity of s in algebraic problems (7.24).

where $\hat{L}_{r,l} \in \mathcal{L}(\hat{G}_l; \hat{G}_r)$, $(r = 1, 2,\ l = 1, 2)$, $\hat{M}_1 \in \mathcal{L}(\hat{G}_1; \hat{G}_1)$. We approximate t_i and the eigenspace $U(t_i)$ by $\hat{t}_i$ and $\hat{U}_i$, respectively, where

$$t_1 \equiv (\hat{s}_1 + \cdots + \hat{s}_{k_1})/k_1, \quad t_i \equiv (\hat{s}_{k_1 + \cdots + k_{i-1} + 1} + \cdots + \hat{s}_{k_1 + \cdots + k_i})/k_i$$

if $i \geq 2$ and

$$\hat{U}_i \equiv \lim \{\hat{u}_{k_1 + \cdots + k_{i-1} + 1}, \ldots, \hat{u}_{k_1 + \cdots + k_i}\}.$$

Theorem 6. *Let problems* (7.23) *with* $\alpha \in [0, \alpha_0]$ *be approximated by* (7.25). *Suppose that condition* (7.2.4) *is satisfied and let*

$$\rho_i \equiv \sup_{u \in U(t_i), \|u\| = 1} dist_G\{u; \hat{G}_h\}.$$

Then there exist constants $K(i)$ *and* $K'(i)$, *independent of* h *and* α, *such that*

$$|t_{i,h} - t_i| \leq K(i)\rho_i^2, \quad \Theta \equiv \Theta_{H_1(L_{1,1})}(\hat{U}_i; U(t_i)) \leq K'(i)\rho_i. \qquad (7.27)$$

Proof. If $\alpha = 0$, then estimates (7.27) were obtained in [369] (see also [27, 311]). For $\alpha > 0$, we make use of estimates (7.1.16) and (7.2.5) uniform in α. Then it is easy to show that

$$\|(A - \hat{P}_h A)G\| \leq \kappa(i)\rho_i$$

for all $g \in U(t_i)$ such that $\|g\| = 1$. This enables us to apply the analysis in [369] and obtain (7.27). $\square$

7.4. Regularization with respect to α and effective iterative methods. We concentrate now on algorithms for algebraic problems (7.23), with $L_{2,2} \in \mathcal{L}^+(H_2)$, in Euclidean spaces $H \equiv H_1 \times H_2$. If $\alpha > 0$, this reduces to

$$L_1 u_1 \equiv (L_{1,1} + 1/\alpha L_{1,2} L_{2,2}^{-1} L_{2,1}) u_1 = \lambda^{(\alpha)} M_1 u_1, \qquad (7.28)$$

where $R_1 \equiv L_{1,2} L_{2,2}^{-1} L_{2,1}$. Then

$$\lambda_1^{(\alpha)} = \min_{u_1} \mu^{(\alpha)}(u_1),$$

where $\mu^{(\alpha)}(u_1) \equiv (L_1 u_1, u_1)\|u_1\|_{M_1}^{-2}$. If $\alpha = 0$, then (7.23) can be reduced to the problem

$$\tilde{L}_1 \tilde{u}_1 = \lambda^{(0)} \tilde{M}_1 \tilde{u}_1 \qquad (7.29)$$

on the subspace $\tilde{H}_1 \equiv \text{Ker } L_{2,1}$, where $\tilde{L}_1 \equiv PL_{1,1}P$, $\tilde{M}_1 \equiv PM_1P$, and P is an orthoprojector of H_1 onto $\tilde{H}_1$. Then

$$\lambda_1^{(0)} = \min_{u_1 \in \tilde{H}_1} \mu^{(0)}(u_1).$$

We enumerate the eigenvalues in both problems (7.28) and (7.29) in increasing order, and we retain this numbering for the corresponding eigenvectors that form an orthonormal basis in $H_1(M_1)$. Denote by $S_m^{(\alpha)}$ the span of the first m eigenvectors; if $m \equiv k_1$ is the multiplicity of $\lambda_1^{(0)} \equiv \nu_1$, then $S_m^{(0)}$ is just the corresponding eigenspace. In the sequel, K refers to nonnegative constants independent of $\alpha \in [0, \alpha_0]$.

Theorem 7. *Let the conditions of Theorem 7.2.1 for $L = L^*$ in (7.23) be satisfied. Suppose that $L_{2,2} = J_2$ and $M_1 = M_1^* > 0$. Then*

$$0 \leq \nu_1 - \lambda_k^{(\alpha)} \leq K\alpha, k \in [1, m],$$

and

$$\Theta_{H_1(L_{1,1})} \equiv \Theta_{H_1(L_{1,1})}(S_1^{(\alpha)}; S_1(\nu_1) \leq K\alpha^{1/2}, \tag{7.30}$$

where m is the multiplicity of $\nu_1 \equiv \lambda_1^{(0)}$ and $\Theta_{H_1(L_{1,1})}$ is the gap between the indicated subspaces in $H_1(L_{1,1})$.

Proof. If $u_1 \in S(\lambda_1^{(0)})$, then $L_{2,1}u_1 = 0$ and $L_1u_1 = L_{1,1}u_1$. Hence, $\nu_1 = \mu^{(\alpha)}(u_1)$ and the Fisher-Courant principle for (7.28) yields

$$\lambda_k^{(\alpha)} \leq \nu_1, \quad k \in [1, m].$$

Further, let $k \leq m$ and $w_1 = u_k^{(\alpha)}$ be an eigenfunction of (7.28) corresponding to $\lambda_k^{(\alpha)}$ and $\|w_1\|_{M_1} = 1$. Then $\|w_1\|_{L_{1,1}}^2 \leq \nu_1$ and

$$\|w_1\|_{R_1} = \|J_2^{-1}L_{2,1}w_1\|_{J_2} \leq \frac{1}{\hat{\sigma}_0} \sup_{v_1} \frac{(J_2^{-1}L_{2,1}w_1, L_{2,1}v_1)}{\|v_1\|_{J_1}}$$

$$= \frac{\alpha}{\hat{\sigma}_0} \sup_{v_1} \frac{(L_{1,1}w_1 - \lambda_1^{(\alpha)}M_1w_1, v_1)}{\|\hat{v}_1\|_{J_1}} \leq K_2\alpha. \tag{7.31}$$

Let P_1 be the orthoprojector of $H_1(L_{1,1})$ onto $\text{Ker } L_{2,1}$. Then

$$w_1 - P_1w_1 \in \text{Im } (L_{1,1}^{-1}(R_1))$$

and

$$\|w_1\|_{R_1} = \|w_1 - P_1w_1\|_{R_1} = O(\alpha), \tag{7.32}$$

which with (7.4.24) implies that

$$\|w_1 - P_1 w_1\|_{L_{1,1}} = O(\alpha), \quad \|w_1 - P_1 w_1\|_{M_1} = O(\alpha),$$

and

$$\|w_1 - P_1 w_1\|_{L_1} = O(\alpha^{1/2}). \tag{7.33}$$

Take $v_1 = \|P_1 w_1\|_{M_1}^{-1} P_1 w_1$. Then (2.1), (7.32), and (7.33) lead to

$$\|v_1 - w_1\|_{L_1}^2 = O(\alpha). \tag{7.34}$$

Hence, $\mu(v_1) - \lambda_k^{(\alpha)} = O(\alpha)$. It suffices now to show that

$$\|w_1 - P_1 w_1\|_{L_{1,1}} = O(\alpha^{1/2}), \quad \forall w_1 \in S_m^{(\alpha)}, \; \|w_1\|_{L_{1,1}} = 1. \tag{7.35}$$

It therefore suffices to obtain (7.34) for all eigenvectors w_1 of (7.28) associated with $\lambda_k^{(\alpha)}, k \in [1, m]$, and such that $\|w_1\|_{M_1} = 1$. Inequalities (7.33) and (7.35) hold for these w_1 and imply that

$$\|w_1 - v_1\|_{L_{1,1}} = O(\alpha^{1/2})$$

(see (9.2.1)). Hence, if $u_1^{(0)}$ is the function in $S_{H_1}(v_1)$ nearest to v_1 in the sense of $H_1(L_{1,1})$, then (9.2.1) implies that, for problem (7.28), we have

$$\|u_k^{(\alpha)} - u_1^{(0)}\|_{L_{1,1}} = O(\alpha^{1/2}), k \in [1, m],$$

which yields (7.30).□ [29]

We propose to apply to (7.28) the modified gradient method discussed in § 4–6 with a model operator $B_1 \equiv D_1 \in S^+(H_1)$, where $\delta_0 D_1 \leq L_1 \leq \delta_1 D_1$ and $\delta_0 > 0$ and $\delta_1 > 0$ are independent of h and $\alpha \in (0, \alpha_0)$. Then their convergence can be made uniform in h and α. Precisely such D_1 were defined in Theorems 7.4.3 and 7.4.4 (see (7.4.35)). The proposed iterative procedure is actually a two-stage iterative method with inner iterations of type (7.4.30). Since, in the cases of interest to us, $L_{2,2}$ is a diagonal matrix, then those iterations are simple enough. Hence, the computational work of one of our basic iterations is estimated as in § 7.4 for a prescribed α. When assumptions of type (7.4.5) with respect to eigenfunctions of the differential problem (7.23) with $\alpha > 0$ are satisfied uniformly in $\alpha \in (0, \alpha_0]$, then ε^2-accuracy for λ_1 is attained for $h^\gamma \asymp \varepsilon$ and $N \asymp \varepsilon^{-2/\gamma}$ (see (7.27)).

---

[29] Theorem 7 can be generalized to the case of other v_j; the proof requires surmounting certain technical difficulties connected with the presence of additional orthogonalization conditions (see [197], where also the case of non-sign-definite M_1 was considered).

Thus, for the resulting algebraic problems, $\alpha > 0$ remains the same and obtaining ε^2-approximations requires $O(h^{-2}|\ln \varepsilon \ln \alpha|$ (or $O(\varepsilon^{-2/\gamma}|\ln \varepsilon \ln \alpha|))$ arithmetic operations. These estimates can be replaced by $O(\varepsilon^{-2/\gamma}|\ln \alpha|)$ if the multigrid acceleration procedure of the basic iterations is applied.

For differential problem (7.23) with $\alpha = 0$, Theorem 7 implies that we should take the regularization with $\alpha \asymp h^\gamma$, and the required computational work is also easily estimated.

7.5. Final comments. The above given results clearly indicate how the Kolmogorov-Bakhvalov hypothesis (see § 0.5) is specified for spectral problems, and that its strengthened variant can be proved for certain types of such problems. Summarizing all relevant results, we see that this hypothesis was proved in this book for a fairly large class of elliptic problems. Nonetheless, we stress again that there are certain important classes of problems, such as those associated with variational inequalities, for which no asymptotically optimal algorithms have yet been found. In this respect, we should conclude that this remarkable hypothesis can be proved only for prescribed types of problems, and new types will serve as a formidable mathematical challenge. (This hypothesis may not even be correct for some problem types.)

P. Chebyshev, one of the founders of optimization in approximation theory, remarked more than a hundred years ago that the history of mathematics consists of two periods: the first is characterized by the fact that the mathematical problems were set by Gods (duplication of the cube, trisection of angle, and so on), and in the second they were formulated by semigods (Fermat, Pascal, and others). He wrote that "we are entering a third one, when mathematical problems are dictated by vital necessity." He actually formulated three features that characterize a really great mathematical problem: it must be natural and beautiful, it must not be an easily solved one, and its applications must be fairly significant. All of these features are present in the theory of optimization of numerical methods. As a consequence of them, an international character of this theory should be stressed: many important steps have been made by many different people at different times. There is no doubt that many young readers of this book will make further contribution to this extremely important and compelling branch of mathematics.

Bibliography

[1] Abramov, A. A., On the separation of the principal part of some algebraic problems, *USSR Comp. Math. and Math. Phys.*, **2**, 147–151, 1962.

[2] Abramov, A. A., and Neuhaus, M., Bemerkungen uber eigenwert probleme von matrizen hoherer ordnung, in *Comptes Rendus du Congres International des Mathematiques de l'Ingenier, Mons et Bruxelles, 9–14 juni, 1958*, 176–179, 1958.

[3] Adams, R. A., *Sobolev Spaces*, Academic Press, New York, 1975.

[4] Agapov, V. K., and Kuznetsov, Yu. A., On some versions of the domain decomposition method, *Sov. J. Numer. Anal. Math. Modelling*, **3**, 245–265, 1988.

[5] Ahlberg, J. H., Nilson, E. N., and Walsh, S. L., *The Theory of Splines and Their Applications*, Academic Press, New York and London, 1967.

[6] Aitken, A. C., Studies in Practical Mathematics. On the Iterative Solution of a System of Linear Equations, *Proc. Roy. Soc., Edinburgh, A*, **63**, 652–660, 1950.

[7] Albert, A., *Regression and the Moor–Penrose Pseudoinverse*, Academic Press, New York and London, 1972.

[8] Allgower, E. L., and George, K., *Computational Solution of Nonlinear Systems of Equations*, American Mathematical Society, Providence, Rhode Island, 1990.

[9] Andreev, V. B., On the convergence of difference schemes approximating the second and third boundary value problems for elliptic equations, *USSR Comp. Math. and Math. Phys.*, **8**, 44–62, 1968.

[10] Andreev, V. B., Stability of difference schemes for elliptic equations with respect to the boundary Dirichlet conditions, *USSR Comp. Math. and Math. Phys.*, **12**, 35–52, 1972.

[11] Ansorge, R., and Lei, J., The convergence of discretization methods if applied to weakly formulated problems. Theory and Examples, *ZAMM*, **71**, 207–221, 1991.

[12] Arbenz, P., Computable finite element error bounds for Poisson's equation, *IMA J. Numer. Anal.*, **2**, 475–479, 1982.

[13] Astrakhantsev, G. P., An iterative method for solving grid elliptic problems, *USSR Comp. Math. and Math. Phys.*, **11**, 171–182, 1971.

[14] Astrakhantsev, G. P., The iterative improvement of eigenvalues, *USSR Comp. Math. and Math. Phys.*, **16**, 171–182, 1976.

[15] Astrakhantsev, G. P., Method of fictitious domains for a second-order elliptic equation with natural boundary conditions, *USSR Comp. Math. and Math. Phys.*, **18**, 114–121, 1978.

[16] Astrakhantsev, G. P., On numerical solution of mixed boundary value problem on the basis of difference analogs of potentials of simple and double layers, in *Variational-difference Methods in Mathematical Physics*, Bakhvalov, N. S., and Kuznetsov, Yu. A., Eds., Otdel. Vychisl. Matem. Akad. Nauk SSSR, Moscow, 1984, 26–34 (in Russian).

[17] Astrakhantsev, G. P. and Rukhovets, L. A., Fictitious component method for solving grid equations used to approximate higher-order elliptic equations with natural boundary conditions, *Sov. J. Numer. Anal. Math. Modelling*, **1**, 37–46, 1986.

[18] Aubin, J. P., *Approximation of Elliptic Boundary Value Problems*, Wiley, New York, 1972.

[19] Aubin, J. P., and Ekeland I., *Applied Nonlinear Analysis*, Wiley–Interscience Publication, New York, 1984.

[20] Axelson, O., *Iterative Solution Methods*, Cambridge University Press, Cambridge, 1994.

[21] Axelson, O., and Gustafson, I., Preconditioning and two-level multigrid methods of arbitrary degree of approximation, *Math. Comp.*, **40**, 219–242, 1983.

[22] Axelson, O., and Vassilevski, P. S., Algebraic multilevel preconditioning methods, I, *Numer. Math.*, **56**, 157–177, 1989.

[23] Axelson, O., and Vassilevski, P. S., Algebraic multilevel preconditioning methods. II, *SIAM J Numer. Anal.*, **27**, 1569–1590, 1990.

[24] Axelson, O., and Vassilevski, P. S., A survey of multilevel preconditioned iterative methods, *BIT*, **29**, 769–793, 1989.

[25] Babushka, I., The finite element method with penalty, *Math. Comput.*, **27**, 221–228, 1973.

[26] Babushka, I., The finite element method with Lagrangian multipliers, *Numer. Math.*, **20**, 179–193, 1987.

[27] Babushka, I., and Aziz, A. K., Survey lectures on the mathematical foundations of the finite element method, in *The Mathematical Foundations of the Finite Element Method with Applications to Partial Differential Equations*, Aziz, A. K., Ed., Academic Press, New York and London, 3–360, 1972.

[28] Babushka, I., Craig, A., Mandel, J., and Pitkaranta, J., Efficient preconditioning for the p-version finite element method in two dimensions, *SIAM J. Numer. Anal.*, **28**, 624–661, 1991.

[29] Babushka, I., and Miller, A., A feedback finite element method with a posteriori error estimation: Part I. The finite element method and some basic properties of the a posteriori error estimator, *Computer Methods in Appl. Mechanics and Engineering*, **61**, 1–40, 1987.

[30] Babushka, I., and Osborn, J. M., Finite element-Galerkin approximation of the eigenvalues and eigenvectors of selfadjoint operator, *Math. Comput.*, **52**, 275–297, 1989.

[31] Babushka, I., and Rosenzweig, M. B., A finite element scheme for domains with corners, *Numer. Math.*, **20**, 1–21, 1972.

[32] Babushka, I., and Sobolev, S. L., Optimization of numerical methods, *Appl. Mat.*, **10**, 96–130, 1965 (in Russian).

[33] Babushka, I., and Suri, M., The optimal convergence rate of p-version of the finite element method, *SIAM J. Numer. Anal.*, **24**, 750–776, 1987.

[34] Babushka, I., Zienkiewichz, O. C., Gago, J., and de A. Oliveira, E. R., Eds., *Accuracy Estimates and Adaptive Refinements in Finite Element Computations*, John Wiley, New York, 1986.

[35] Baiocchi, C., and Capelo, A., *Variational and Quasivariational Inequalities Applications to Free Boundary Problems*, John Wiley and Sons, Chichester, 1984.

[36] Baker, G. A., Jureidini, W. N., and Karakashian, O. A., Piecewise solenoidal vector fields and the Stokes problem, *SIAM J. Numer. Anal.*, **27**, 1466–1485, 1990.

[37] Bakhvalov, N. S., About convergence of a relaxation method under natural constraints on the elliptic operator, *USSR Comp. Math. and Math. Phys.*, **6**, 101–135, 1966.

[38] Bakhvalov, N. S., On optimization of methods for solving boundary problems in the presence of a boundary layer, *Z. Vychisl. Mat. i Mat. Fiz.*, **9**, 841–859, 1969 (in Russian).

[39] Bakhvalov, N. S., About optimization of numerical methods, in *Actes Congress Intern. Math., Nice–September 1970*, **3**, 289–295, 1970.

[40] Bakhvalov, N. S., On the optimality of linear methods for operator approximation in convex classes of functions, *USSR Comp. Math. and Math. Phys.*, **11**, 244–249, 1971.

[41] Bakhvalov, N. S., *Numerical Methods*, Mir Publishers, Moscow, 1977.

[42] Bakhvalov, N. S., Knyazev, A. V., and Kobel'kov, G. M., Iterative methods for solving equations with highly varying coefficients, in *Fourth International Symposium on Domain Decomposition Methods for Partial Differential Equations*, Glowinski, R., Kuznetsov, Yu., Meurant, G., Periaux, J., and Widlund, O., Eds, SIAM, Philadelphia, 1991, 197–205.

[43] Bakhvalov, N. S., and Orehov, M. Yu., On fast methods of solving Poisson's equation, *Z. Vychisl. Mat. i Mat. Fiz.*, **22**, 1386–1392, 1982 (in Russian).

[44] Bakhvalov, N. S., and Panasenko, G., *Homogenization: Averaging Processes in Periodic Media. Mathematical Problems in the Mechanics of Composite Materials*, Kluwer Academic Publishers, Dordrecht, Boston, London, 1989.

[45] Balasundaram, S. and Bhattacharyya, P. K., A mixed finite element method for fourth order partial differential equations, *ZAMM*, **66**, 489–499, 1986.

[46] Bank, R., An automatic scaling procedure for D'yakonov–Gunn iteration scheme, *Linear Algebra and its Applications*, **28**, 17–33, 1979.

[47] Bank, R., Analysis of a multilevel inverse iteration procedure for eigenvalue problems, *SIAM J. Numer. Anal.*, **19**, 886–898, 1982.

[48] Bank, R., and Benbourenane, M., The hierarchical basis multigrid method for convection-diffusion equations, *Numer. Math.*, **61**, 7–37, 1992.

[49] Bank, R., and Dupont, T., An optimal order process for solving finite element equations, *Math. Comput.*, **36**, 35–51, 1981.

[50] Bank, R., Dupont, T., and Yserentant, H., The hierarchical basis multigrid method, *Numer. Math.*, **52**, 427–458, 1988.

[51] Bank, R., and Rose, D. J., Marching algorithms for elliptic boundary value problems, *SIAM J. Numer. Anal.*, **14**, 792–829, 1977.

[52] Bank, R., and Rose, D. J., Analysis of a multilevel iterative method for nonlinear finite element equations, *Math. Comp.*, **39**, 453–465, 1982.

[53] Bank, R., and Rose, D. J., Some error estimates for the box method, *SIAM J. Numer. Anal.*, **24**, 777–787, 1987.

[54] Bank, R., and Scott, L. R., On the conditioning of finite element equations with highly refined meshes, *SIAM J. Numer. Anal.*, **26**, 1383–1394, 1989.

[55] Bank, R., and Weiser, A., Some a posteriori error estimators for elliptic partial differential equations, *Math. Comp.*, **44**, 283–301, 1985.

[56] Bank R., and Welfert, B. D., A posteriory error estimates for the Stokes problem, *SIAM J. Numer. Anal.*, **28**, 591–623, 1991.

[57] Bank R., Welfert, B. D., and Yserentant, H., A class of iterative methods for solving saddle point problems, *Numer. Math.*, **56**, 645–666, 1990.

[58] Barrett, J. W., and Elliotte, C. M., Finite element approximation of the Dirichlet problem using the boundary penalty method, *Numer. Math.*, **49**, 343–366, 1986.

[59] Baumgardner, J. R., and Frederickson, P. O., Icosahedral discretization of the two-sphere, *SIAM J. Numer. Anal.*, **22**, 1107–1115, 1985.

[60] Baylis, A., and Goldstein, C. I., An iterative method for the Helmholtz equation, *J. of Comput. Phys.*, **49**, 443–457, 1983.

[61] Bensousan, A., Lions, J. L. and Papanicolaou, G., *Asymptotic Analysis for Periodic Structures*, North Holland, Amsterdam, 1978.

[62] Belii, M. V., and Bulgakov, V. E., On the multigrid technique for solving three-dimensional boundary value engineering problems, *Internat. J. Numer. Methods Engrg.*, **33**, 753–764, 1992.

[63] Belinskii, P. P., Godunov, S. K., Ivanov, Yu. B., and Yanenko, I. K., An application of a class of quasiconformal mappings for constructing grids in regions with curved boundaries, *Z. Vychisl. Mat. i Mat. Fiz.*, **15**, 1499–1511, 1975 (in Russian).

[64] Berezin, I. S., and Zhidkov, N. P., *Computing Methods*, Addison-Wesley, Reading, Mass., 1965.

[65] Berger, M. S., Bifurcation theory for nonlinear elliptic differential equations and systems, in *Bifurcation Theory and Nonlinear Eigenvalue Problems*, Keller, J. B., and Antman, S., Eds, W. A. Benjamin, Inc., New York, 71–113, 1969.

[66] Bernardi, C., Optimal finite element interpolation on curved domains, *SIAM J. Numer. Anal.*, **26**, 1212–1240, 1989.

[67] Besov, O. V., Il'in, V. P., and Nikol'skii, S. M., *Integral Representation of Functions and Embedding Theorems*, v. 1, Winston and Sons, Washington, 1978; v. 2, A Halsted Press Book, John Wiley, New York, 1979.

[68] Besov, O. V., Kudryavtsev, L. D., Lizorkin, P. I., and Nikol'skii, S. M., Investigations in the theory of spaces of differentiable functions of several variables, in *Proceedings of the Steklov Institute of Mathematics*, 73–139, 1990.

[69] Bhardwaj, K. K., Kadalbajov, M. K., and Sanakav, R., Symmetric marching technique for the Poisson equation. I. Dirichlet boundary conditions, *Appl. Math. and Computation*, **15**, 137–149, 1984.

[70] Birkhoff, G. and Varga, R. S., Implicit alternating direction methods. *Trans. Amer. Math. Soc.*, **92**, 13–24, 1959.

[71] Bjorstad, P. E., and Widlund, O. B., Iterative methods for the solution of elliptic problems on regions partitioned into substructures, *SIAM J. Numer. Anal.*, **23**, 1097–1120, 1986.

[72] Blum, H., Numerical treatment of corner and crack singularities, *Cours and Lect. CISN, Intern. Cent. Mech. Sci.*, n. 301, 171–212, 1988.

[73] Blum, H., and Rannacher, R., Extrapolation techniques for reducing the pollution effect of reentrant corners in the finite element method, *Numer. Math.*, **52**, 539–564, 1988.

[74] Bourlard, M., Dauge, M., Lubuma, M. S., and Nicaise, S., Coefficients of the singularities for elliptic boundary value problems on domains with conical points, III, Finite element methods on polygonal domains, *SIAM J. Numer. Anal.*, **29**, 136–155, 1992.

[75] Bourquin, F., Component mode synthesis and eigenvalues of second order operators: discretization and algorithm, *Mathematical Modelling and Numer. Anal.*, **26**, 385–423, 1992.

[76] Braess, D., On the combination of the multigrid methods and conjugate gradients, *Lecture Notes in Mathematics*, **1228**, Springer-Verlag, 52–64, 1986.

[77] Braess, D., and Blomer, C., A multigrid method for parameter dependent problem in solid mechanics, *Numer. Math.*, **57**, 747–761, 1990.

[78] Bramble, J. H., *Multigrid methods, Pitman Research Notes in Mathematics Series*, **294**, 1993.

[79] Bramble, J. H., Ewing, R. E., Parashkevov, R. E., and Pasciak, J. E., Domain decomposition methods for problems with partial refinement, *SIAM J. Sci. Statist. Comput.*, **13**, 397–410, 1992.

[80] Bramble, J. H., Ewing, R. E., Pasciak, J. E., and Schatz, A. H., A preconditioning technique for the efficient solution of problems with local grid refinements, *Computer Methods in Applied Mechanics and Engineering*, **67**, 149–159, 1988.

[81] Bramble, J. H., and Falk, R. S., A mixed Lagrange multiplier finite element method for the polyharmonic equation, *Math. Modelling and Numer. Anal.*, **19**, 519–557, 1985.

[82] Bramble, J. H., and Pasciak J. E., Preconditioned iterative methods for nonselfadjoint or indefinite elliptic boundary value problems, in *Unific. Finite Element Meth.*, North Holland, Amsterdam, 167–184, 1984.

[83] Bramble, J. H., and Pasciak, J. E., New convergence estimates for multigrid algorithms, *Math. Comp.*, **49**, 311–329, 1987.

[84] Bramble, J. H., Pasciak, J. E., and Schatz, A. N., An iterative method for elliptic problems on regions partitioned into substructures, *Math. Comp.*, **46**, 361–369, 1986.

[85] Bramble, J. H., Pasciak, J. E., and Schatz, A. H., The construction of preconditioners for elliptic problems by substructuring. I, *Math. Comput.*, **47**, 103–134, 1986.

[86] Bramble, J. H., Pasciak, J. E., and Schatz, A. H., The construction of preconditioners for elliptic problems by substructuring. IV, *Math. Comput.*, **53**, 1–24, 1989.

[87] Bramble, J. H., Pasciak, J. E., and Xu, J., Parallel multilevel preconditioners, *Math. Comput.*, **55**, 1–22, 1990.

[88] Bramble, J. H., and Zlamal, M., Triangular elements in the finite element method, *Math. Comp.*, **24**, 809–821, 1970.

[89] Brandt, A., Algebraic multigrid theory: the symmetric case, *Appl. Math. Comput.*, **19**, 23–56, 1986.

[90] Brandt, A. S., McCormick, S., and Ruge, J., Multigrid methods for differential eigenproblems, in *SIAM J. Sci. Statist. Comput.*, **4**, 244–260, 1983.

[91] Brenner, S. C., An optimal-order multigrid method for $P1$ nonconforming finite elements, *Math. Comp.*, **52**, 1–15, 1989.

[92] Brenner, S. C., An optimal-order nonconforming multigrid method for the biharmonic equation, *SIAM J. Numer. Anal.*, **26**, 1124–1138, 1989.

[93] Brenner, S. C., A nonconforming mixed multigrid method for the stationary Stokes equations, *Math. Comp.*, **55**, 411–437, 1990.

[94] Brenner, S. C., Multigrid methods for parameter dependent problems, Report, the 6th Maunt. Copper Conf., 1993.

[95] Brenner, S. C., and Scott, L. R., *The Mathematical Theory of Finite Element Methods (Texts in Applied Mathematics, **15**)*, Springer, 1994.

[96] Brezis, H., and Sibony, M., Me'thodes d' Approximation et d' Iteration pour les operateurs monotones, *Arch. Ration. Mech. and Analysis*, **28**, 59–82, 1968.

[97] Brezzi, F., On the existence, uniqueness and approximation of saddle-point problems arising from Lagrangian multipliers, *Rev. Fran. Automat. Inf. Recherche Operationelle. Ser. Rouge. Anal. Numer.*, **2**, 129–151, 1974.

[98] Brezzi, F., and Douglas, J., Stabilized mixed methods for the Stokes problem, *Numer. Math.*, **53**, 225–235, 1988.

[99] Brezzi, F., and Fortin, M., *Mixed and Hybrid Finite Element Methods*, Springer-Verlag, 1991.

[100] Brezzi, F., Rappaz, J., and Raviart, P. A., Finite dimensional approximation of nonlinear problems, *Numer. Math.*, **38**, 1–30, 1981.

[101] Browder, F. E. and Petryshyn, W. V., Construction of fixed points of nonlinear mappings in Hilbert spaces, *J. Math. Anal. Appl.*, **20**, 197–228, 1967.

[102] Buleev, N. I., A numerical method for solving two-dimensional and three-dimensional diffusion equations, *Mat. Sb*, **51(2)**, 227–238, 1960 (in Russian).

[103] Burenkov, V. I., On the additivity of the classes $W_p^{(r)}(\Omega)$, *Proc. Steklov Inst. Math.*, **89**, 32–62, 1967.

[104] Burenkov, V. I., and Goldman, M. L., On interconnection between norms of operators for periodic and non-periodic function spaces, *Proc. Steklov Inst. Math.*, **161**, 53–112, 1983.

[105] Cai, X. C., Gropp, W. D., and Keyes, D. E., Convergence estimate for a domain decomposition method, *Numer. Math.*, **61**, 153–169, 1992.

[106] Cai, Z., and McCormick, S., On the accuracy of the finite volume element method for diffusion equations on composite grids, *SIAM J. Numer. Anal.*, **27**, 636–655, 1990.

[107] Canuto, C., Parallelism in spectral methods, *Calcolo*, **25**, 53–74, 1988.

[108] Canuto, C., Hussaini, M. Y., Quarteroni, A., and Zang, T. A., *The Spectral Methods in Fluid Dynamics*, New York, 1987.

[109] Cea, J., *Optimisation. Theorie et Algorithmes*, Dunod, Paris, 1971.

[110] Cesari, L., Sulla risolutzione del sistemi di equazioni lineari per approssimazioni successivi, *Atti Accad. Naz. Lincei, Rend. Cl. Sci. Fiz. Mat. Natur.*, VI, **25**, 422–428, 1937.

[111] Chan, R. H., Iterative methods for overflow queuing models, I, *Numer. Math.*, **51**, 143–180, 1987.

[112] Chan, R., and Jin, X., Circulant and skew-circulant preconditioners for skew-Hermitian type Toeplitz systems, *BIT*, **31**, 632–646, 1991.

[113] Chan, R. H., and Yeung, M. C., Circulant preconditioners constructed from kernels, *SIAM J. Numer. Anal.*, **29**, 1093–1103, 1992.

[114] Chang, S. C., Solution of elliptic PDEs by fast Poisson solvers using a local relaxation factor, *J. Comput. Phys.*, **67**, 91–123, 1986.

[115] Chinn, W. G., and Steenrod, N. E., *First Concepts of Topology*, Random House, New York, 1966.

[116] Chow, S. S., Finite element error estimates for nonlinear elliptic equations of monotone type, *Numer. Math.*, **54**, 373–393, 1989.

[117] Ciarlet, P., *The Finite Element Method for Elliptic Problems*, North-Holland Publishing Company, Amsterdam, 1975.

[118] Ciarlet, P. G, and Rabier, P., *Les Equations de von Karman*, Springer-Verlag, Berlin, 1980.

[119] Concus, P., and Golub, G. H., Use of fast direct methods for the efficient numerical solution of nonseparable elliptic equations, *SIAM J. Numer. Anal.*, **10**, 1103–1120, 1973.

[120] Cooley J. W., and Tukey, J. W., An algorithm for the machine calculation of complex Fourier series, *Math. Comput.*, **19**, 297–301, 1965.

[121] Cosserat, Eugene et Francois, Sur les equations de la theorie de l'elastite, *C. R. Acad. Sci. (Paris)*, **126**, 1089–1091, 1898.

[122] Courant, R., Variational methods for the solution of problems of equilibrium and vibrations, *Bull. of Amer. Math. Soc.*, **49**, 1–23, 1943.

[123] Courant, R., Friedriechs, K. O., and Lewy, H., Uber die partiellen differenzengleichungen der physic, *Math. Ann.*, **100**, 32–74, 1928–1929.

[124] Courant, R., and Hilbert, D., *Methods of Mathematical Physics*, Wiley, New York, 1953.

[125] Coxeter, H. S. M., Discrete groups generated by reflections, *Ann. Math.*, **35**, 588–621, 1934.

[126] Crouzeuz, M., Etude d'une methode de linearisation; resolution des equations de Stokes stationaries; application aux equations des Navier–Stokes stationares, *Cahiere de l' IRIA*, n. 12, 139–244, 1974.

[127] Crouzeux, M. and Raviart, P. A., Conforming and nonconforming finite element methods for solving the stationary Stokes equations, *RAIRO*, n. 3, 33–75, 1973.

[128] Dahmen, W., and Elsner, L., Algebraic multigrid methods and the Schur complement, in *Robust Multi-Grid Methods: Proc. GAMM-Semin., Kiel, Yan. 22–24,1988*, Braunschweig, Wiesbaden, 1989, 58–68.

[129] Demianovich, Yu. K., On convergence of grid methods for elliptic problems, *Dokl. Akad. Nauk SSSR*, **170**, 27–30, 1966 (in Russian).

[130] De Roeck, Y. H., and Le Tallec, P., Analysis and test of a local domain-decomposition preconditioner, in *Fourth International Symposium on Domain Decomposition Methods for Partial Differential Equations*, Glowinski, R., Kuznetsov, Yu., Meurant, G., Periaux, J., and Widlund, O. B., Eds, SIAM, Philadelphia, 112–128, 1991.

[131] Deuflhard, P., and Engquist, B., Eds., *Large Scale Scientific Computing*, Birkhauser, Boston, 1987.

[132] Douglas, C. C., A tupleware approach to domain decomposition methods, *Applied Numerical Mathematics*, **8**, 353–373, 1991.

[133] Douglas, C. C., and Douglas, J., A unified convergence theory for abstract multi-grid or multilevel algorithms, serial and parallel, *SIAM J Numer. Anal.*, **30**, 136–158, 1993.

[134] Douglas, C. C., and Smith, B. F., Using symmetries and antisymmetries to analyze a parallel multigrid algorithms: the elliptic boundary value problem case, *SIAM J. Numer. Anal.*, **26**, 1439–1461, 1989.

[135] Douglas, J., Alternating direction iteration for mildly nonlinear elliptic difference equations, *Numer. Math.*, **3**, 92–98, 1963.

[136] Douglas, J., Alternating direction methods for three space variables, *Numer. Math.*, **4**, 41–63, 1962.

[137] Douglas, J., and Dupont, T., Alternating direction Galerkin methods on rectangles, in *Numerical Solution of Partial Differential Equations, II*, Bramble, J. H., Ed., Academic Press, New York, 1971.

[138] Douglas, J., and Rachford, H. H., On the numerical solution of heat conduction problems in two and three variables, *Trans. Amer. Math. Soc.*, **82**, 421–439, 1956.

[139] Dryja, M., On the stability in W_2^2 of schemes with splitting operator, *USSR Comp. Math. and Math. Phys.*, **7**, 71–80, 1967.

[140] Dryja, M., Prior estimates in W_2^2 in a convex domain for systems of difference elliptic equations, *USSR Comp. Math. and Math. Phys.*, **12**, 291–300, 1972.

[141] Dryja, M., A capacitance matrix method for Dirichlet problem on polygonal region, *Numer. Math.*, **39**, 51–64, 1982.

[142] Dryja, M., and Widlund, O., Towards a unified theory of domain decomposition algorithms for elliptic problems, in *The Third International Symposium on Domain Decomposition Methods for Partial Differential Equations*, Chan, T. F., Glowinski, R., Periaux, J., and Widlund, O. B., Eds, SIAM, Philadelphia, 3–21, 1989.

[143] Dryja, M., and Widlund, O. B., Multilevel additive methods for elliptic finite element problems, *Notes Numer. Fluid Mech.*, **31**, 58–69, 1991.

[144] Dryja, M., Smith, B. F., and Widlund, O. B., Schwarz analysis of iterative sub-structuring algorithms for elliptic problems in three dimensions, *SIAM J. Numer. Anal.*, **31**, 1662–1694, 1994.

[145] Dubinskii, Yu. A., Quasilinear elliptic and parabolic equations of an arbitrary order, *Uspekhi Mat. Nauk*, **23**, 45–90, 1968 (in Russian).

[146] Duff, I. S., Erisman, A. M., and Reid, J. K., *Direct Methods for Sparse Matrices*, Clarendon Press, New York, 1989.

[147] Dupont, T., A factorization procedure for the solution of elliptic difference equations, *SIAM J. Numer. Anal.*, **5**, 753–782, 1968.

[148] Dupont, T., and Scott, R., Polynomial approximation of functions in Sobolev spaces, *Math. Comput.*, **34**, 441–463, 1980

[149] Duran, R., Maschietti, M. A., and Rodriguez, R., On the asymptotic exactness of error estimates for linear triangular finite elements, *Numer. Math.*, **59**, 107–128, 1991.

[150] D'yakonov, E. G., On an iterative method for solving finite difference equations, *Soviet Math. Dokl.*, **2**, 647–650,1961.

[151] D'yakonov, E. G., A method for solving Poisson's equation, *Soviet Math. Dokl.*, **3**, 320–323, 1962.

[152] D'yakonov, E. G., On the application of disintegrating difference operators, *USSR Comp. Math. and Math. Phys.*, **3**, 511–515, 1963.

[153] D'yakonov, E. G., Methods with majorizing operator for solving difference analogs of strongly elliptic systems, *Uspekhi Mat. Nauk*, **19**, 385–386, 1964 (in Russian).

[154] D'yakonov, E. G., The use of spectrally equivalent operators in solving difference analogs of strongly elliptic systems, *Soviet Math. Dokl.*, **6**, 1105–1109, 1965.

[155] D'yakonov, E. G., The construction of iterative methods based on the use of spectrally equivalent operators, *USSR Comp. Math. and Math. Phys.*, **6**, 14–46, 1966.

[156] D'yakonov, E. G., On convergence of an iterative process, *Uspekhi Mat. Nauk*, **21**, 181–183, 1966 (in Russian).

[157] D'yakonov, E. G., On a class of differential partial equations arising in closing difference methods based on the use of split operators, in *Vychisl. Metody i Programmir.*, Izdat. Moskov. Gos. Univ., Moscow, 144–152, 1967 (in Russian).

[158] D'yakonov, E. G., Economical difference methods based on the use of split difference operator, in *Vychisl. Metody i Programmir.*, Izdat. Moskov. Gos. Univ., Moscow, 76–120, 1967 (in Russian).

[159] D'yakonov, E. G., On certain iterative methods for solving elliptic difference equations, *Lect. Notes Math.*, **109**, 7–22, 1969.

[160] D'yakonov, E. G., The solution of some nonlinear systems of difference equations, *Soviet Math. Dokl.*, **10**, 1216–1220, 1969.

[161] D'yakonov, E. G., On a problem of net shells theory and its difference analog, *USSR Comp. Math. and Math. Phys.*, **9**, 124–141, 1969.

[162] D'yakonov, E. G., *Iterative Methods for Solving Difference Analogs of Boundary Value Problems for Elliptic Equations. Materials of International School on Numerical Methods, June, 1966, Kiev*, Inst. Cybernetics Akad Nauk Ukr., 1970 (in Russian).

[163] D'yakonov, E. G., On the solution of some elliptic difference equations, *JIMA*, **7**, 1–20, 1971.

[164] D'yakonov, E. G., *Difference Methods for Solving Boundary Value Problems, 2 (Stationary Problems)*, Moscow State University, Moscow, 1971 (in Russian).

[165] D'yakonov, E. G., On approximate methods for the solution of operator equations, *Soviet Math. Dokl.*, **12**, 826–830, 1971.

[166] D'yakonov, E. G., On some operator inequalities and their applications, *Soviet Math. Dokl.*, **12**, 921–925, 1971.

[167] D'yakonov, E. G., Questions of numerical solving nonlinear elasticity and plasticity problems, in *Proceedings of the Second All-Union Conference on Numerical Methods for Problems of Elasticity and Plasticity, 1970, Trakai*, Yanenko, N. N., Ed., Vychisl. Tsentr Sibirsk. Otdel. Akad. Nauk SSSR, Novosibirsk, 95–104, 1971 (in Russian).

[168] D'yakonov, E. G., *Difference Methods for Solving Boundary Value Problems, 2 (Nonstationary Problems)*, Moscow State University, Moscow, 1972 (in Russian).

[169] D'yakonov, E. G., Difference analogs of embedding theorems and their applications to investigation of difference schemes and iterative processes, in *Application of Functional Analysis Methods to Boundary Value Problems of Mathematical Physics*, Izd. Sibirsk. Otdel. Akad. Nauk SSSR, Novosibirsk, 55–62, 1972 (in Russian).

[170] D'yakonov, E. G., On some methods for solving difference and projective-difference schemes, in *Comput. Methods of Linear Algebra*, Marchuk, G. I., Ed., Vychisl. Tsentr Sibirsk. Otdel. Akad. Nauk SSSR, Novosibirsk, 25–58, 1972 (in Russian).

[171] D'yakonov, E. G., On stability of difference schemes for some nonstationary problems, in *Topics in Numerical Analysis, II*, Miller, J. J. H., Ed., Academic Press, London, 63–87, 1973.

[172] D'yakonov, E. G., Projective–difference methods for quasilinear problems, in *Variationally-difference Methods in Mathematical Physics*, Marchuk, G. I., Ed., Vychisl. Tsentr Sibirsk. Otdel. Akad. Nauk SSSR, Novosibirsk, 34–46, 1973 (in Russian).

[173] D'yakonov, E. G., About convergence of difference methods for nonlinear problems, *Acta Universitatis Carolinae, Math. et Phys.*, 11–15, 1974 (in Russian).

[174] D'yakonov, E. G., About an iterative method for solving discretized elliptic systems, *Compte rendue de l' Academie Bulgare des Sciences*, **28**, 295–297, 1975 (in Russian).

[175] D'yakonov, E. G., Projective-difference and difference methods for solving nonlinear stationary problems of elasticity and plasticity, *Chisl. Metody Mekh. Sploshnoi Sredy*, **2**, 14–78, 1976 (in Russian).

[176] D'yakonov, E. G., Classes of spectrally equivalent operators and their applications, in *Variationally-difference Methods in Mathematical Physics*, Marchuk, G. I., Ed., Sibirsk. Otdel. Akad. Nauk SSSR, Novosibirsk, 1976, 49–61 (in Russian).

[177] D'yakonov, E. G., On the triangulations in the finite element and efficient iterative methods, in *Topics in Numerical Analysis, III*, Miller, J. J. H., Ed., Academic Press, London, 103–124, 1977.

[178] D'yakonov, E. G., On selecting a triangulation in the projective-difference method, in connection with minimizing computational work, *Soviet Math. Dokl.*, **18**, 1029–1033, 1977.

[179] D'yakonov, E. G., Some modifications of projective-difference schemes, *Vestnik Moskov. Univ., Ser. Vychisl. Mat. i Kibernet.*, **1**, 1–14, 1977.

[180] D'yakonov, E. G., Some topological and geometrical problems associated with triangulation of the region in projective-difference methods, *Math. Notes of the Academy of Sciences of the USSR*, **21**, 238–245, 1977.

[181] D'yakonov, E. G., Use of consequences of grids for solving strongly elliptic systems, in *Computational Methods of Linear Algebra*, Marchuk, G. I., Ed., Sibirsk. Otdel. Akad. Nauk SSSR, Novosibirsk, 146–162, 1977 (in Russian).

[182] D'yakonov, E. G., Solving of projective-difference systems with nonnegative operators, in *Computational Methods of Linear Algebra*, Marchuk, G. I., Ed., Sibirsk. Otdel. Akad. Nauk SSSR, Novosibirsk, 51–60, 1977 (in Russian).

[183] D'yakonov, E. G., Asymptotic minimization of computational work in applying projective-difference methods, in *Variationally-difference Methods in Mathematical Physics*, Marchuk, G. I., Ed., Sibirsk. Otdel Akad. Nauk SSSR, Novosibirsk, 149–164, 1978 (in Russian).

[184] D'yakonov, E. G., Modified iterative methods for eigenvalue problems, in *Comput. Methods of Linear Algebra*, Marchuk, G. I., Ed., Sibirsk. Otdel. Akad. Nauk SSSR, Novosibirsk, 149–164, 1978 (in Russian).

[185] D'yakonov, E. G., Rate of convergence of iterative methods for obtaining the smallest eigenvalues, in *Numerical Methods in Mathematical Physics*, Marchuk, G. I., Ed., Sibirsk. Otdel. Akad. Nauk SSSR, Novosibirsk, 155–157, 1979 (in Russian).

[186] D'yakonov, E. G., About some direct and iterative methods based on bordering of the matrix, *Numerical Methods in Mathematical Physics*, Marchuk, G. I., Ed., Sibirsk. Otdel. Akad. Nauk SSSR, Novosibirsk, 45–68, 1979 (in Russian).

[187] D'yakonov, E. G., Asymptotic minimization of computational work in solving strongly elliptic boundary value problems, in *Theory of Numerical Integration and Computational Mathematics*, Sobolev, S. L., Ed., Sibirsk. Otdel. Akad. Nauk SSSR, Novosibirsk, 31–37, 1980 (in Russian).

[188] D'yakonov, E. G., Estimates of computational work in solving grid eigenvalue problems, in *Computational Methods of Linear Algebra*, Marchuk, G. I., Ed., Sibirsk. Otdel. Akad. Nauk SSSR, Novosibirsk, 78–92, 1980 (in Russian).

[189] D'yakonov, E. G., On solving some singular eigenvalue problems, in *Computations with Sparse Matrices*, Kuznetsov, Yu. A, Ed., Sibirsk. Otdel. Akad. Nauk SSSR, Novosibirsk, 45–53, 1981 (in Russian).

[190] D'yakonov, E. G., On certain iterative methods for solving eigenvalue problems, *Math. Notes of the Academy of Sciences of the USSR*, **34**, 945–953, 1983 .

[191] D'yakonov, E. G., Estimates of computational work for boundary value problems with the Stokes operators, *Soviet Math. (Iz. VUZ)*, **27**, 57–71, 1983.

[192] D'yakonov, E. G., On solving grid eigenvalue problems, in *Computational Methods of Linear Algebra*, Voevodin, V. V., Ed., Otdel. Vychisl. Matem. Akad. Nauk SSSR, Moscow, 72–86, 1983 (in Russian).

[193] D'yakonov, E. G., On a nonconformal projective-grid method for Stokes system, *Vestnik Moskov. Univ., Ser. Vychisl. Mat. i Kibernet.*, **2**, 17–24, 1984.

[194] D'yakonov, E. G., Asymptotic minimization of the computational work for steady-state problems of meteorology, *Soviet Math. Dokl.*, **30**, 258–262, 1984.

[195] D'yakonov, E. G., Estimates of computational work for modified projective-grid methods, in *Variational-difference Methods in Mathematical Physics*, Bakhvalov, N. S., and Kuznetsov, Yu. A., Eds., Otdel. Vychisl. Matem. Akad. Nauk SSSR, Moscow, 58–71, 1984 (in Russian).

[196] D'yakonov, E. G., Methods of solving fourth order elliptic problems that are asymptotically optimal with respect to labor consumption, *Soviet Math. Dokl.*, **32**, 128–132, 1985.

[197] D'yakonov, E. G., Effective methods for solving eigenvalue problems with fourth-order elliptic operators, *Soviet J. Numer. Anal. Math. Modelling*, **1**, 59–82, 1986.

[198] D'yakonov, E. G., On iterative methods with saddle operators, *Soviet Math. Dokl.*, **35**, 166–170, 1987.

[199] D'yakonov, E. G., On classes of iterative methods with saddle operators, in *Computational Processes and Systems, 5*, Marchuk, G. I., Ed., Nauka, Moscow, 101–115, 1987 (in Russian).

[200] D'yakonov, E. G., Asymptotically optimal algorithms for correct elliptic problems, in *Computational Processes and Systems, 6*, Marchuk, G. I., Ed., Nauka, Moscow, 121–133, 1988 (in Russian).

[201] D'yakonov, E. G., *Minimization of Computational Work. Asymptotically Optimal Algorithms for Elliptic Problems*, Nauka, Moscow, 1989 (in Russian).

[202] D'yakonov, E. G., On some iterative methods for nonlinear grid systems, in *Computational Processes and Systems, 8*, Marchuk, G. I., Ed., Nauka, Moscow, 95–111, 1991 (in Russian).

[203] D'yakonov, E. G., On some modern approaches to constructing spectrally equivalent grid operators, in *Fourth International Symposium on Domain Decomposition Methods for Partial Differential Equations,* Glowinski, R., Kuznetsov, Yu. A., Meurant, G., Periaux, J., and Widlund, O. B., Eds, SIAM, Philadelphia, 35–40, 1991.

[204] D'yakonov, E. G., On increasing the efficiency of grid methods for solution of elasticity problems, in *Computer Mechanics of Solids,* Moscow, n. 2, 133–157, 1991 (in Russian).

[205] D'yakonov, E. G., Composite grids and asymptotically optimal algorithms for problems of Stokes and Navier-Stokes type, *Russian Acad. Sci. Dokl. Math.,* **47**, 221–227, 1993.

[206] D'yakonov, E. G., Iterative methods based on linearization for nonlinear elliptic grid systems, in *Numerical Methods and Applications,* Marchuk, G. I., Ed., CRC Press, Boca Raton, 1–43, 1994.

[207] D'yakonov, E. G., and Knyazev, A. V., On a group iterative method for finding lower eigenvalues, *Vestnik Moskov. Univ., Ser. Vychisl. Mat. i Kibernet.,* n. 2, 29–34, 1982 (in Russian).

[208] D'yakonov, E. G., and Knyazev, A. V., On an iterative method for finding lower eigenvalues, *Russ. J. Numer. Anal. Math. Modelling,* **7**, 473–486, 1992.

[209] D'yakonov, E. G., and Lebedev, V. I., Split operator method in solving the third boundary value problem for parabolic equations, in *Vychisl. Metody i Program-mir.,* Izdat. Moskov. Gos. Univ., Moscow, 121–143, 1967 (in Russian).

[210] D'yakonov, E. G., and Nikolaev, I. K., An application of a grid method for modelling of shells of rotation, *Z. Vychisl. Mat. i Mat. Fiz.,* **13**, 938–951, 1973 (in Russian).

[211] D'yakonov, E. G., and Nikolaev, I. K., On numerical solution of nonlinear problems of net shells problems, in *Proceedings of the Third All-Union Conference on Numerical Methods for Problems of Elasticity and Plasticity, 1973, Kishinev,* Yanenko, N. N., Ed., Vychisl. Tsentr Sibirsk. Otdel. Akad. Nauk SSSR, Novosibirsk, 85–101, 1974 (in Russian).

[212] D'yakonov, E. G., and Orehov, M. Yu., On minimization of computational work in solving eigenvalue problems, *Soviet Math. Dokl.,* **18**, 1074–1077, 1977.

[213] D'yakonov, E. G., and Orehov, M. Yu., On minimization of computational work in finding smallest eigenvalues of differential operators, *Math. Notes of the Academy of Sciences of the USSR,* **27**, 382–391, 1980.

[214] D'yakonov, E. G., and Stolyarov, N. N., On solving nonlinear stationary problems of theory of plates and shells, *Chisl. Metody Mekh. Sploshnoi Sredy,* **10**, n. 5, 39–62, 1979 (in Russian).

[215] Eaves, B. C., A course in triangulations for solving equations with deformations, *Lect. Notes in Economics and Math. Systems,* **234**, Springer-Verlag, Berlin, 1984.

[216] Ehrlich, L., Numerical analogs to the Schwarz alternating procedure and SOR, *SIAM J. Sci. Statist. Comput.,* **7**, 989–993, 1986.

[217] Eisenstat, S. C., Efficient implementation of a class of preconditioned conjugate gradient methods, *SIAM J. Sci. Statist. Comput.*, **2**, 1–4, 1981.

[218] Eisenstat, S. C., Elman, H. C., and Schultz, M. H., Variational iterative methods for nonsymmetric systems of linear equations, *SIAM J. Numer. Anal.*, **20**, 345–357, 1983.

[219] Eisenstat, S. C., Schultz, M. H., and Scherman, A. H., The application of sparse matrix methods to the numerical solution of nonlinear elliptic partial differential equations, *Lect. Notes Math.*, **430**, 131–153, 1974.

[220] Ekeland, I., and Temam, R., *Convex Analysis and Variational Problems*, North-Holland Publishing Company, Amsterdam, 1976.

[221] Elman, W. C., and Schultz, M. H., Preconditioning by fast direct methods for nonself-adjoint nonseparable elliptic equations, *SIAM J. Numer. Anal.*, **23**, 44–57, 1986.

[222] Erikson, K., High-order local rate of convergence by mesh refinement in the finite element methods, *Math. Comput.*, **171**, 109–142, 1991.

[223] Ewing, R. E., Lazarov. R. D., and Vassilevski, P. S., Local refinement techniques for elliptic problems on cell-centured grids, I, Error analysis, *SIAM J. Numer. Anal.*, **56**, 437–461, 1991.

[224] Ewing, R. E., Lazarov. R. D., and Vassilevski, P. S., Mixed finite element solutions of second order elliptic problems with regular grid refinement, in *Fourth International Symposium on Domain Decomposition Methods for Partial Differential Equations*, Glowinski, R., Kuznetsov, Yu. A., Meurant, G. A., Periaux, J., and Widlund, O., Eds., SIAM, Philadelphia, 206–212, 1991.

[225] Faber, V., Manteuffel, T. A., and Parter, S. A., On the theory of equivalent operators and application to the numerical solution of uniformly elliptic partial differential equations, *Advances in Applied Mathematics*, **11**, 109–163, 1990.

[226] Faddeev, D. K., and Faddeeva, V. N., *Computational Methods of Linear Algebra*, Freeman, San Francisco, 1963.

[227] Fairweather, G., Finite element Galerkin methods for differential equations, *Lecture Notes in Pure and Appl. Math.*, **34**, 1978.

[228] Fedorenko, R. P., The speed of convergence of one iterative process, *USSR Comp. Math. and Math. Phys.*, **4**, 227–235, 1964.

[229] Filippov, A. F., and Rjaben'kii, V. S., *Uber die Stabilitat von Differenzengleichungen*, VEB, Deutscher Verlag Wiss., Berlin, 1960.

[230] Fix, G. J., Finite element approximations to compressible flow problems, in *Elliptic Problem Solvers. II; Proc. Conf. Monterley, Calif., 10-12 Jan., Orlando e.a.,*, 369–393, 1984.

[231] Fix, G. J., Gulati, S., and Wakoff, G. I., On the use of singular functions with finite element approximations, *J. Comput. Phys*, **13**, 209–228, 1973.

[232] Forsythe, G. E., and Wasov, W. R., *Finite Difference Methods for Partial Differential Equations*, John Wiley, New York, 1960.

[233] Fortin, M., An analysis of the convergence of mixed finite element methods, *R.A.I.R.O. Numer. Anal.*, **11**, 341–354, 1977.

[234] Friedrichs, K. O., and Keller, H. B., A finite difference scheme for generalized Neumann problems, in *Numerical Solution of Partial Differential Equations*, New York, 1–19, 1967.

[235] Fuchik. S., and Kufner, S., *Nonlinear Differential Equations*, Elsevier Scientific Publishing Company, Amsterdam, 1980.

[236] Funaro, D., Quarteroni, A., and Zanolli, P., An iterative procedure with interface relaxation for domain decomposition methods, *SIAM J. Numer. Anal.* , **25**, 1213–1236, 1988.

[237] Gagliardo, E., Caracterizzazione delle tracce sulla frontiera relative ad alcune classi di funzioni in *n* variabili, *Rend. Semin. Mat. Univ. Padova*, **27**, 284–305, 1957.

[238] Gaejewski, H., Groger, K., and Zacharias, K., *Nicht lineare Operatorgleichungen*, Akademie-Verlag, Berlin, 1974.

[239] Gantmacher, F. R., *The Theory of Matrices*, Chelsea Publishing Company, 1989.

[240] George, A., and Liu, J. W. H., *Computer Solution of Large Sparse Positive Definite Systems*, Prentice-Hall, Inc., Englewood Cliffs, New Jersey, 1981.

[241] Geymonat, G., and Oswald, P., Some remarks on approximation by finite element methods, in *Banach Center Publications, 22*, PWN Polish Scientific Publishers, Warsaw, 137–164, 1989.

[242] Girault, V., and Raviart, P. A., *Finite Element Methods for Navier–Stokes equations. Theory and Algorithms*, Springer, Berlin, 1986.

[243] Glazman, I., and Liubitch, Y., *Analyse Lineaire dans les Espaces de Dimensions Finies*, Mir, Moscow, 1972.

[244] Glowinski, R., *Numerical Methods for Nonlinear Variational Problems*, Springer-Verlag, New York, 1983.

[245] Glowinski, R., Keller, H. B., and Reinhart, L., Continuation-conjugate gradient methods for the least squares solution of nonlinear boundary value problems, *SIAM J. Sci. Statist. Comput.*, **6**, 793–832, 1985.

[246] Glowinski, R., Lions, J. L., and Tremolieres, R., *Analyse Numerique des Inequations Variationnelles*, Dunod, Paris, 1976.

[247] Godunov, S. K., Ogneva, V. V., and Prokopov, G. P., On convergence of the modified method of the steepest descent for computing of eigenvalues, in *Differential Partial Equations*, Bizadze, A. V., Ed., Nauka, Moscow, 77–80, 1970 (in Russian).

[248] Godunov, S. K., and Ryaben'kii, V. S., *Difference Schemes*, North Holland, Amsterdam, 1987.

[249] Goering, H., and Tobiska, L., Finite element methods for singularly perturbed elliptic boundary value problems and its application to the stationary Navier-Stokes equations, *Z. Anal. Anwendungen*, **8**, 13–23, 1989.

[250] Goldstein, C. I., Analysis and application of multigrid preconditioners for singularly perturbed boundary value problems, *SIAM J. Numer. Anal.*, **26**, 1090–1123, 1989.

[251] Goldstein, C. I., Multigrid analysis of finite element methods with numerical integration, *SIAM J. Numer. Anal.*, **56**, 409–436, 1991.

[252] Golub, G. H., and Van Loan, C. F., *Matrix Computations*, Johns Hopkins, Baltimore, 1989.

[253] Golub, G. H., Greenbaum, A., and Luskin, M. F., Eds, *Recent Advances in Iterative Methods*, (The IMA Volumes in Mathematics and its Applications), **60**, 1994.

[254] Goodsell, G., and Whiteman, J. R., Superconvergence of recovered gradients of piecewise quadratic finite element approximations. Part I: L_2-error estimates, *Numerical Methods for Partial Differential Equations*, **1**, 61–83, 1991.

[255] Griffiths, D. F., Ed., *The Mathematical Basis of Finite Element Methods with Applications to Partial Differential Equations*, Clarendon Press, London, 1984.

[256] Grisvard, P., *Elliptic Problems in Nonsmooth Domains*, Pitman, Boston, 1985.

[257] Gropp, W. D., and Keyes, D. E., Domain decomposition on parallel computers, in *Domain Decomposition Methods for Partial Differential Equations. II*, Chan, T., Glowinski, R., Meurant, G. A., Periaux, J., and Widlund, O., Eds., SIAM, Philadelphia, 260–268, 1989.

[258] Gunn, J. E., The numerical solution of $\nabla a \nabla u = f$ by a semi-explicit alternating direction iterative techniques, *Numer. Math.*, **6**, 181–184, 1965.

[259] Gunn, J. E., On the two-stage iterative method of Douglas for mildly nonlinear elliptic difference equations, *Numer. Math.*, **6**, 243–249, 1964.

[260] Gunn, J. E., The solution of elliptic difference equations by semi-explicit iterative techniques, *SIAM J. Numer. Anal.*, **2**, 24–45, 1965.

[261] Gunzburger, M. D., *Finite Element Methods for Viscous Incompressible Flow: A Guide to Theory, Practice and Algorithms*, Academic Press, Boston, 1989.

[262] Gunzburger, M. D., Meir, A. J., and Peterson, J. S., On the existence, uniqueness, and finite element solutions of the equations of stationary incompressible magnetohydrodynamics, *SIAM J. Numer. Anal.*, **56**, 523–563, 1991.

[263] Haase, G., Langer, U., and Meyer, A., Domain decomposition preconditioners with inexact subdomain solvers, *Journal of Numerical Linear Algebra with Applications*, **1**, 27–41, 1991.

[264] Habetler, G. J., and Wachspress, E. L., Symmetric successive overrelaxation in solving diffusion difference equations, *Math. Comput.*, **15**, 356–362, 1961.

[265] Hackbusch, W., On the computation of approximate eigenvalues and eigenfunctions by means of a multi-grid method, *SIAM J. Numer. Anal.*, **16**, 201–215, 1979.

[266] Hackbusch, W., On the regularity of difference schemes, *Ark. Mat.*, **19**, 71–95, 1981.

[267] Hackbusch, W., *Multi-Grid Methods and Applications*, Springer-Verlag, Berlin, 1985.

[268] Hackbusch, W., On first and second order box schemes, *Computing*, **41**, 277–296, 1989.

[269] Hackbusch, W., *Iterative Solution of Large Sparse Systems of Equations*, Springer–Verlag, Berlin, 1994.

[270] Hackbusch, W., and Trottenberg, V., Eds., *Multi-Grid Methods, Proceedings, Koln–Porz, Nov. 1981, Lecture Notes Math.*, **960**, 1982.

[271] Hageman, L. A., and Young, D. M., *Applied Iterative Methods*, Academic Press, New York, 1981.

[272] Halmos, P. K., *A Hilbert Space Problem Book*, D. Van Nostrand Company, Inc., Toronto, 1967.

[273] Han, W., The p-version penalty finite element method, *IMA Journal of Numerical Analysis*, **12**, 47–56, 1992.

[274] Hardy, G. H., Littlewood, J. E., and Polya, G. *Inequalities*, Cambridge University Press, Cambridge, 1934.

[275] Heinrich, B., Finite difference methods on irregular networks, *ISNM*, **82**, 1987.

[276] Heinrichs, W., Improved condition numbers for spectral methods, *Math. Comput*, **53**, 103–119, 1989.

[277] Hestenes, M. R., The conjugate-gradient method for solving linear systems, *Proc. Symp. Appl. Math. 6, Providence P.I*, Amer. Math. Soc., 83–102, 1956.

[278] Hestenes, M. R., *Conjugate Direction Methods in Optimization*, Springer, New York, 1980.

[279] Hlavacek, I., and Necas, J., On inequalities of Korn's type, *Archiv Ration. Mech. Anal.*, **36**, 312–334, 1970.

[280] Hlavacek, I., Necas, J., and Lovisek, J., *Solving Variational Inequalities in Mechanics*, Mir, Moscow, 1986, (in Russian).

[281] Hockney, R. W., The potential calculations and some applications, *Methods in Computational Physics, 9*, Adler, B., Fernbach, S., and Rotenberg, S., Eds., 131–211, 1969.

[282] Hughes, T. J. R., *The Finite Element Method: Linear Static and Dynamic Finite Element Analysis*, Prentice-Hall, Englewood Cliffs, New Jersey, 1987.

[283] Il'in, V. P., Properties of some classes of differentiable functions defined on n-dimensional region, *Amer. Math. Soc. Transl., Ser.2*, **81**, 91–256, 1969.

[284] Ishihara, K., Monotone explicit iterations of the finite element approximations for the nonlinear boundary value problem, *Numer. Math.*, **43**, 419–438, 1984.

[285] Jerom, J., On the N-width in Sobolev spaces and applications to elliptic boundary value problems, *J. Math. Anal. Appl.*, **29**, 201–215, 1970.

[286] Jerri, A. J., *Integral and Discrete Transforms with Applications and Error Analysis*, Marcel Dekker, Inc., New York, 1992.

[287] Joe, R., Delaunay triangular meshes in convex polygons, *SIAM J. Sci. Statist. Comput.*, **7**, 514–539, 1986.

[288] Johnson, S. L., Saad, Y., and Schultz, M. H., Alternating direction methods on multiprocesors, *SIAM J. Sci. Statist. Comput.*, **8**, 686–700, 1987.

[289] Jung, M., Langer, U., and Semmer, U., Two-level hierarchically preconditioned conjugate gradient methods for solving linear elasticity finite element equations, *BIT*, **29**, 748–768, 1989.

[290] Kaasschieter, E. F., A general finite element preconditioning for the conjugate gradient methods, *BIT*, **29**, 824–849, 1989.

[291] Kacvara, M., and Mandel, J., A multigrid method for three-dimensional elasticity and algebraic convergence estimates, *Appl. Math. Comput.*, **23**, 121–135, 1987.

[292] Kantorovich, L. V., and Akilov, G. P., *Functional Analysis in Normed Spaces*, Pergamon, London, 1964.

[293] Karakashian, O. A., On a Galerkin–Lagrange method for the stationary Navier–Stokes equations, *SIAM J. Numer. Anal.*, **19**, 909–923, 1982.

[294] Karchevskii, M. M., Nonlinear problems of theory of plates and shells and their grid approximations, *Izv. Vuzov, Matem.*, n.10, 17–30, 1985 (in Russian).

[295] Kato, T., *Perturbation Theory for Linear Operators*, Springer Verlag, Berlin, 1966.

[296] Kato, T., Notes on some inequalities for linear operators, *Math. Ann.*, **125**, 208–213, 1952.

[297] Kaufman, L., Matrix methods for queeneing problems, *SIAM J. Sci. Statist. Comput.*, **4**, 525–552, 1982.

[298] Kaushilaite, D., On solving a nonlinear problem of hydrodynamics, *Dokl. Akad. Nauk. SSSR*, 198, 555–558, 1971 (in Russian).

[299] Kavanagh, J. P., and Neumann, M., Consistency and convergence of the parallel multisplitting method for singular M-matrices, *SIAM J. Matrix Anal. Appl.*, **10**, 210–218, 1989.

[300] Keller, J. B., and Antman, S., Eds., *Bifurcation Theory and Nonlinear Eigenvalue Problems*, W. A. Benjamin, Inc., New York, 1969.

[301] Kelly, D. W., Gago, J. P. de S. R., Zienkiewicz, O. C., and Babushka, I., A posteriori error analysis and adaptive processes in the finite element method, *Internat. J. Numer. Meths. Engrg.*, **19**, 1593–1656, 1983.

[302] Khoromskij, B. N., and Wendland, W. L., *Spectrally Equivalent Preconditioners in Substructuring Techniques*, *East-West J. Numer. Anal.*, **1**, 1–27, 1992.

[303] Kightley, J. R., and Thompson, C. P., On the performance of some rapid elliptic solvers on a vector processor, *SIAM J. Sci. Statist. Comput.*, **8**, 701–715, 1987.

[304] Knupp, P. M., and Steinberg, S,. *The Fundamentals of Grid Generation*, CRC Press, Boca Raton, 1993.

[305] Knyazev, A. V., *Computing Eigenvalues and Eigenvectors in Grid Problems: Algorithms and Estimates of Errors*, Otdel Vychisl. Matem. Akad. Nauk SSSR, Moscow, 1986 (in Russian).

[306] Knyazev, A. V., Convergence rate estimates for iterative methods for a mesh symmetric problem, *Sov. J. Numer. Anal. Math. Modelling*, **2**, 371–396, 1987.

[307] Knyazev, A. V., and Skorokhodov, A. L., Preconditioned gradient-type iterative methods in a subspace for partial generalized symmetric eigenvalue problems, *SIAM J. Numer. Anal.*, **31**, 1226–1239, 1994.

[308] Kobel'kov, G. M., On reducing of a boundary value problem for a biharmonic equation to one with an operator of Stokes type, *Izv. Vuzov., Matem.*, n.10, 39–46, 1985 (in Russian).

[309] Kobel'kov, G. M., Efficient methods for solving elasticity theory problems, *Sov. J. Numer. Anal. Math. Modelling*, **6**, 361–375, 1991.

[310] Kobel'kov, G. M., and Valedinsky,V. D., On the inequality $\|p\|_{L_2} \leq c\|\nabla p\|_{-1}$ and its finite dimensional analog, *Sov. J. Numer. Anal. Math. Modelling*, **1**, 189–200, 1986.

[311] Kolata, W. G., Eigenvalue approximation by the finite element method: the method of Lagrange multipliers, *Math. Comput.*, **33**, 63–76, 1979.

[312] Kondrat'iev, V. A., and Oleinik, O. A., Boundary value problems for partial differential equations in nonsmooth domains, *Russian Math. Surveys*, **38**, 1–86, 1983.

[313] Kopachevsky, N. D., Krein, S. G., and Ngo, Z. C., *Operator Methods in Hydrodynamics; Evolution and Spectral Problems*, Nauka, Moscow, 1989 (in Russian).

[314] Korneev, V. G., *Schemes of Finite Elements of High Order of Accuracy*, Izdat. Leningrad. Univ., Leningrad, 1977.

[315] Koshelev, A. I., *Smoothness of Solutions of Elliptic Equations and Systems*, Nauka, Moscow, 1986 (in Russian).

[316] Krasnosel'kii, M. A., *Topological Methods in the Nonlinear Integral Equations*, Pergamon Press, New York, 1964.

[317] Krasnosel'kii, M. A., Vainikko, G. M., Zabreiko, P. P., and Stezenko, V.Ya., *Approximate Solution of Operator Equations*, Wolters-Noordorf, Groningen, 1969.

[318] Krein, S. G., *Linear Differential Equations in Banach Spaces*, Nauka, Moscow, 1967 (in Russian).

[319] Krizek, M., and Neittaanmaki, P., On superconvergence techniques, *Acta Appl. Math.*, **9**, 175–198, 1987.

[320] Krizek, M., Neittaanmaki, P., and Stenberg, R., Eds, *Finite Element Methods*, Marcel Dekker, Inc., New York, 1994.

[321] Kuznetsov, Yu. A., Block relaxation methods in subspace; optimization and application, in *Variationally-difference Methods in Mathematical Physics*, Marchuk, G. I., Ed., Sibirsk. Otdel Akad. Nauk SSSR, Novosibirsk, 178–212, 1978 (in Russian).

[322] Kuznetsov, Yu. A., Algebraic multigrid domain decomposition methods, *Soviet J. Numer. Meth. Math. Modelling*, **4**, 351–380, 1989.

[323] Ladyzenskaya, O. A., *Mathematics of the Dynamics of Viscous Noncompressible Liquid*, Gordon and Breach, London, 1963.

[324] Ladyzenskaya, O. A., *Boundary Value Problems of Mathematical Physics*, Nauka, Moscow, 1973, (in Russian).

[325] Ladyzenskaya, O. A., and Solonnikov, V. A., On some problems of vector analysis and generalized settings of boundary problems for Navier–Stokes equations, *Zap. Nauchn. Seminar. Leningrad. Otdel. Matem. Inst. Steklov Akad. Nauk SSSR*, **59**, 81–118, 1976 (in Russian).

[326] Ladyzenskaya, O. A., and Ural'tzeva, N. N., *Linear and Quasilinear Elliptic Equations*, Academic Press, New York, 1968.

[327] Langer, V., and Queek, W., Preconditioned Uzawa-type iterative methods for solving mixed finite element equations: Theory - Applications - Software, *Wissenschaftliche Schriftenreihe der Technischen Universitat Karl-Marx Stadt*, 1987, n. 3.

[328] Lankaster, P., *Theory of Matrices*, Academic Press, New York, 1969.

[329] Lapin, A. V., and Lyashko, A. D., Investigation of the grid method for nonlinear problems of any order, *Izv. Vuzov, Matem.*, n.10 (101), 37–43, 1970 (in Russian).

[330] Lau, W. T. F., A new method of numerical calculation for stagnation point flows, *ZAMP*, **15**, 581–588, 1964.

[331] Lebedev, V. I., On Zolotarev's problem in ADI methods, *Z. Vychisl. Mat. i Mat. Fiz.*, **17**, 349–366, 1977, (in Russian).

[332] Lebedev, V. I., and Finogenov, S. A., On the order of choosing of iterative parameters in Chebyshev cyclic iterative methods, *Z. Vychisl. Mat. i Mat. Fiz.*, **11**, 425–438, 1971 (in Russian).

[333] Lenoir, M., Optimal isoparametric finite elements and error estimates for domains involving curved boundaries, *SIAM J. Numer. Anal*, **23**, 562–580, 1986.

[334] Levin, N., Superconvergent recovery of the gradient from piecewise linear finite element approximations, *IMA J. Numer. Anal.*, **5**, 407–427, 1985.

[335] Li, R. H., Generalized difference methods for a nonlinear Dirichlet problem, *SIAM J. Numer. Anal.*, **24**, 77–88, 1987.

[336] Lin, Q., and Xie, R., Error expansions for finite element approximations and their applications, *Lect. Notes Math.*, **1297**, 98–112, 1987.

[337] Lions, J. L, *Quelque Methodes de Resolution des Problem aux Limites non Lineare*, Dunod, Paris, 1969.

[338] Lions, J. L, and Magenes, E., *Problemes aux Limites non Homogeenes et Applications*, Dunod, Paris, 1968.

[339] Lions, P. L., On the Schwartz alternating method, in *Domain Decomposition Methods for Partial Differential Equations. I*, Glowinski, R., Golub, G. H., Meurant, G. A., and Periaux, J., Eds., SIAM, Philadelphia, 1–41, 1988.

[340] Litvinov, V. G., *Otimization in Elliptic Boundary Value Problems with Applications to Mechanics*, Nauka, Moscow, 1989 (in Russian).

[341] Liusternik, L. A., and Sobolev, V. I., *Elemente der Functionalanalysis*, Academie-Verlag, Berlin, 1955.

[342] Lyashko, A. D., and Karchevskii, M. M., Difference methods of solving nonlinear filtration problems, *Soviet Math. (Iz. VUZ)*, **27**, 34–56, 1983.

[343] Lysche, T., and Schumaker, L. L., *Mathematical Methods in Computer Aided Geometric Design*, 1989, ISBN:0-12-460515-X.

[344] Maitre, J. P., and Musy, F., The contraction number of a class of two-level methods; an exact evaluation for some finite element subspaces and model problems, *Lect. Notes Math.*, **960**, 535–544, 1982.

[345] Mandel, J., Two-level domain decomposition preconditioning for the p-version finite element method in three dimensions, *International Journal for Numerical Methods in Engineering*, **29**, 1095–1108, 1990.

[346] Mandel, J., and Brezina, M., Balancing domain decomposition: Theory and computations in two and three dimensions, *UCD/CCM, Report 2*, Center for Computational Mathematics, University of Colorado at Denver, 1993.

[347] Mandel, J., and McCormick, S., Iterative solution of elliptic equations with refinement: the two-level case, in *Domain Decomposition Methods for Partial Differential Equations. II*, Chan, T., Glowinski, R., Meurant, G. A., Periaux, J., and Widlund, O., Eds., SIAM, Philadelphia, 81–92, 1989.

[348] Mandel, J., and McCormick, S., A multilevel variational method for $Au = \lambda Bu$ on composite grids, *J. Comput. Phys.*, **80**, 442–452, 1989.

[349] Mandel, J., McCormick, S., and Bank, R., Variational multigrid theory, in *Multigrid Methods*, McCormick, S., Ed., SIAM, Philadelphia, 131–177, 1987.

[350] Mandel, J., McCormick, S., and Ruge, J., An algebraic theory for multigrid methods for variational problems, *SIAM J. Numer. Anal.*, **25**, 91–110, 1988.

[351] Marchuk, G. I., *Methods of Numerical Mathematics*, Springer, New York, 1975.

[352] Marchuk, G. I., Splitting and Alternating Direction Methods, in *Handbook of Numerical Analysis*, Ciarlet, P. G., and Lions, J. L., Eds, v. Finite Difference Methods (Part 1), North-Holland, Amsterdam, 198–462, 1990.

[353] Marchuk, G. I., and Kuznetsov Yu. A., *Iterative Methods and Quadratic Functionals*, Nauka, Novosibirsk, 1975 (in Russian).

[354] Marchuk, G. I., Kuznetsov Yu. A., and Matsokin, A. M., Fictitious domain and domain decomposition methods, *Soviet J. Numer. Anal. Math. Modelling*, **1**, 3–35, 1986.

[355] Marchuk, G. I., and Lebedev, V. I., *Numerical Methods in Theory of Neutrons Transport*, Harwood Academic Publishers, Chur, Switzerland, 1986.

[356] Mansfield, L., Finite element subspaces with optimal rates of convergence for the stationary Stokes problem, *R. A. I. R. O. Numer. Anal.*, **16**, 49–66, 1982.

[357] Manteuffel, T. A., and Parter, S. V., Preconditioning and boundary conditions, *SIAM J. Numer. Anal.*, **27**, 656–694, 1990.

[358] Manteuffel, T., An incomplete factorization technique for positive definite linear systems, *Math. Comput.*, **38**, 114–123, 1980.

[359] McCormick, S., A mesh refinement method for $Ax = \lambda Bx$, *Math. Comput.*, **36**, 485–498, 1981.

[360] McCormick, S., Multigrid methods for variational problems: general theory for the V-cycle, *SIAM J. Numer. Anal.*, **22**, 634–643, 1985.

[361] McCormick, S., Ed., *Multigrid Methods*, SIAM, Philadelphia, 1987.

[362] McCormick, S., Ed., *Multigrid Methods: Theory, Applications, and Supercomputing*, *Lecture Notes in Pure and Applied Mathematics*, **110**, 1988.

[363] McCormick, S., *Multilevel Adaptive Methods for Partial Differential Equations*, SIAM, Philadelphia, 1989.

[364] McCormick, S., and Ruge, J. W., Multigrid methods for variational problems: further results, *SIAM J. Numer. Anal.*, **21**, 255–263, 1984.

[365] McCormick, S., and Thomas, J., The fast adaptive composite grid (FAC) method for elliptic boundary value problems, *Math. Comput.*, **46**, 439–456, 1986.

[366] McNeal, An asymmetric finite difference network method, *Quart. Appl. Math.*, **11**, 295–310, 1953.

[367] Melson, N. D., Manteuffel, T. A., and McCormick, S., Eds, *Sixth Copper Mountain Conference on Multigrid Methods*, NASA Conference Publication 3224, Part 1, 1993.

[368] Melson, N. D., Manteuffel, T., and McCormick, S., Eds, *Sixth Copper Mountain Conference on Multigrid Methods*, NASA Conference Publication 3224, Part 2, 1993.

[369] Mercier, B., Osborn, J., Rappas, J., and Raviart, P. A., Eigenvalue approximation by mixed and hybrid methods, *Math. Comput.*, **36**, 427–454, 1981.

[370] Michavila, F., Gavete, L., and Diez, F.,Two different approaches for treatment of boundary singularities, *Numerical Methods for Partial Differential Equations*, **4**, 255–282, 1988.

[371] Mikhaylov, V. P., *Partial Differential Equations*, Mir, Moscow, 1978.

[372] Mikhlin, S. G., *Variational Methods in Mathematical Physics*, Maxmillan Co., New York, 1964.

[373] Mikhlin, S. G., *Numerical Performance of Variational Methods*, Wolters-Noordorf, Groningen, 1971.

[374] Mikhlin, S. G., *Error Analysis in Numerical Processes*, Wiley, Chichester, 1991.

[375] Mikhlin, S. G., The spectrum of the pencil of elasticity theory operators, *Uspekhi Mat. Nauk*, **28**, 43–82, (1973) (in Russian).

[376] Miller, K., Numerical analogs to the Schwarz alternating procedure, *Numer. Math.*, **7**, 91–103, 1965.

[377] Miller, J. J. H., and Wang, S., A triangular mixed finite element method for the stationary semiconductor device equations, *Math. Modelling and Numer. Anal.*, **25**, 441–463, 1991.

[378] Mitchell, A., *Computational Methods in Partial Differential Equations*, Wiley, London, 1969.

[379] Mitchell, A. R, and Wait, R., *The Finite Element Method in Partial Differential Equations*, Wiley, Chichester, 1984.

[380] Mittelman, H. D., and Roose, D., Eds, *Continuation Techniques and Bifurcation Problems*, Birkhauser Verlag, Basel e.a., 1990.

[381] Morgan, R. B., Computing interior eigenvalues of large matrices, *Linear Algebra and its Applications*, n. 154–156, 289–309, 1991.

[382] Necas, J., *Les Methodes Directes en Theorie des Equations Elliptiques*, Masson, Paris, 1967.

[383] Necas, J., and Hlavacek, I., *Mathematical Theory of Elastic and Elastic-plastic Bodies: an Introduction*, Elsever, Amsterdam, 1981.

[384] Nepomnyaschikh, S. V., Application of domain decomposition to elliptic problems with discontinuous coefficients, in *Fourth International Symposium on Domain Decomposition Methods for Partial Differential Equations*, Glowinski, R., Kuznetsov, Yu. A., Meurant, G. A., Periaux, J., and Widlund, O., Eds., SIAM, Philadelphia, 242–251, 1991.

[385] Nichols, N. K., On the convergence of two-stage iterative processes for solving linear equations, *SIAM J. Numer. Anal.*, **10**, 460–469, 1973.

[386] Niethammer, W., Iterative solution of nonsymmetric systems of linear equations, *International Series of Numerical Mathematics*, **86**, 381–390, 1988.

[387] Nikol'skii, S. M., *Approximation of Functions of Several Variables and Embedding Theorems*, Springer-Verlag, Berlin, 1975.

[388] Nikol'skii, S. M., *Quadrature Formulas*, Nauka, Moscow, 1974 (in Russian).

[389] Nikol'skii, S. M., *Course of Mathematical Analysis, I*, Nauka, Moscow, 1983 (in Russian).

[390] Oden, J. T., *Finite Elements of Nonlinear Continua*, McGraw-Hill, New York, 1972.

[391] Oden, J. T., *Applied Functional Analysis*, Prentice-Hall, Englewood Cliffs, New Jersey, 1979.

[392] O'Donnell, S. T., and Rokhlin, V., A fast algorithm for the numerical evaluation of conformal mappings, *SIAM J. Sci. Statist. Comput.*, 10, 475–487, 1989.

[393] Oganesjan, L. A., and Ruhovec, L. A., Investigation of the convergence rate of variationally-difference schemes for elliptic second order equations in a two dimensional domain with a smooth boundary, *USSR Comp. Math. and Math. Phys.*, 9, 153–188, 1969.

[394] Oganesjan, L. A., and Ruhovec, L. A., *Variationally-difference Methods for Solving Elliptic Equations*, Izdat. Armyansk. Akad. Nauk, Erevan, 1979 (in Russian).

[395] Orehov, M. Yu., On estimates of the computational work in finding the first eigenvalues of some differential operators, *Soviet Math. Dokl.*, 22, 29–33, 1980.

[396] Ortega, J. M., Efficient implementation of certain iterative methods, *SIAM J. Sci. and Statist. Comput.*, 9, 882–891, 1988.

[397] Ortega, J. M., and Rheinboldt, W. C., *Iterative Solution of Nonlinear Equations in Several Variables*, Academic Press, New York, 1970.

[398] Ortega, J. M., and Voigt, R. G., Solution of partial differential equations on vector and parallel computers, *SIAM Review*, 27, 49–213, 1985.

[399] Oswald, P., A hierarchical finite element method for the biharmonic equation, *SIAM J. Numer. Anal.*, 29, 1610–1625, 1992.

[400] Otroschenko, I. V, and Fedorenko, R. P., A relaxation method of solving difference biharmonic equations, *USSR Comp. Math. and Math. Phys.*, 23, 57–63, 1983.

[401] Pan, V. Ya., New fast algoritms for matrix operations, *SIAM J. Comput.*, 9,321–342, 1980.

[402] Parlett, B. N., *The Symmetric Eigenvalue Problem*, Prentice-Hall, Inc., Englewood Cliffs, New Jersey, 1980.

[403] Parter, S. V., Mildly nonlinear elliptic partial differential equations and their numerical solution, *Numer. Math.*, 7, 113–128, 1965.

[404] Parter, S. V., On an estimate for the three-grid MGR multigrid method, *SIAM J. Numer. Anal.*, 24, 1032–1045, 1987.

[405] Peaceman, D. W., and Rachford, H. H., The numerical solution of parabolic and elliptic differential equations, *J. Soc. Ind. Appl. Math.*, 3, 28–41, 1955.

[406] Peronnet, A., The club MODULEF. A library of subroutines for finite element analysis, *Lecture Notes Math.*, **704**, 127–153, 1979.

[407] Petryshyn, W. V., Direct and iterative methods for the solution of linear operator equations, *Trans. Amer. Math. Soc.*, **105**, 136-175, 1962.

[408] Petryshyn, W. V., On the extension and solution of nonlinear operator equations, *Illinois. J. Math.*, **10**, 255–274, 1966.

[409] Pinkus, A., *n*-Width and Approximation Theory, Springer-Verlag, New York, 1985.

[410] Pironneau, O., *Optimal Shape Design for Elliptic Systems*, Springer-Verlag, New York, 1983.

[411] Pironneau, O., *Finite Element Methods for Fluids*, John Wiley, New York, 1990.

[412] Polak, E., *Computational Methods in Optimization Mathematics in Science and Engineering*, Academic Press, New York, 1971.

[413] Polya, G., *Mathematical Discovery. On Understanding, Learning and Teaching Problem Solving*, John Wiley, New York, 1965.

[414] Quarteroni, A., *Numerical Approximations of Partial Differential Equations*, Springer Series in Computational Mathematics, **23**, 1994.

[415] Rao, C. R., and Mitra, S. K., *Generalized Inverse of Matrices and Its Application*, John Wiley, New York, 1971.

[416] Rao, K. R., and Yip, P., *Discrete Cosine Transform. Algorithms, Advantages, Applications*, Academic Press, Boston, 1990.

[417] Raviart, P. A., Mixed finite element methods, in *The Mathematical Basis of Finite Element Methods*, Oxford, Clarendon Press, 123–156, 1984.

[418] Rectorys, K., *Variational Methods in Mathematics, Science, and Engineering*, Publishers of Technical Literature, Prague, 1980.

[419] Rheinhardt, H. J., *Analysis of Approximation Methods for Differential and Integral Equations*, Springer-Verlag, New York, 1985.

[420] Ribbence, C. J., A fast adaptive grid scheme for elliptic partial differential equations, *ACM Transactions on Mathematical Software*, **15**, 179–197, 1989.

[421] Rice, J. R., and Boisver, R. F., *Solving Elliptic Problems Using ELL PACK*, Springer-Verlag, New York, 1985.

[422] Richardson, L. F., The approximate arithmetical solution by finite differences of physical problems involving differential equations, with an application to the stresses in a masonry dam, *Trans. Roy, Soc. London*, A210, 307–357, 1911.

[423] Riesz, F., and Nagy, B. S., *Functional Analysis*, Frederic Unger Publishing Co., New York, 1955.

[424] Rivkind, V. Ya., On estimating the rate of convergence of homogeneous difference schemes for elliptic and parabolic equations with discontinuous coefficients, in *Probl. Matem. Anal.*, Izdat. Leningrad. Univ., 1966, 110–119 (in Russian).

[425] Rivkind, V. Ya., A finite element method for solving obstacle problems, in *Proceedings of the Third All-Union Conference on Numerical Methods for Problems of Elasticity and Plasticity*, Yanenko, N. N., Ed., Vychisl. Tsentr Sibirsk. Otdel. Akad. Nauk SSSR, Novosibirsk, 74–82, 1974 (in Russian).

[426] Rivkind, V. Ya., and Ural'tzeva, N. N., Projective- difference schemes for solving quasilinear equations with the bounded non linearity, *Vestnik Leningrad. Univ.*, **19**, 65–70, 1972, (in Russian).

[427] Rogers, C. A., *Packing and Covering*, Cambridge University Press, Cambridge, 1964.

[428] Romanova, S. E., On solving of a mixed boundary value problem for difference Laplace equation on by method of approximate solving difference Laplace equation on step regions by the Schwartz alternating method, *Soviet Math. Dokl.*, **21**, 689–693, 1980.

[429] Romanova, S. E., An economical method of approximate solving difference Laplace equation on rectangular regions, *USSR Comp. Math. and Math. Phys.*, **23**, 92–101, 1983.

[430] Rourke, C., and Sanderson, B. J., *Introduction to Piecewise-Linear Topology*, Springer-Verlag, Berlin, 1972.

[431] Rue, A., Iterative eigenvalue algorithms based on convergent splittings, *J. Comp. Physics.*, **19**, 110–120, 1970.

[432] Saad, Y., Krylov subspace methods on supercomputers, *SIAM J. Sci. Statist. Comput.*, **10**, 1200–1232, 1989.

[433] Saad, Y., and Shultz, M. H., GMRES: A generalized minimal residual algorithm for solving nonsymmetric linear systems, *SIAM J. Sci. Statist. Comput.*, **7**, 856–869, 1986.

[434] Sachs, A., Iterationsverfahren fur monotone, nicht notwending Lipschitz-beschrankte operatoren in Hilbert-raum, *Numer. Math.*, **20**, 356–364, 1973.

[435] Samarskii, A. A., *Theory of Difference Schemes*, Nauka, Moscow, 1977, (in Russian).

[436] Samarskii, A. A., *Introduction into Numerical Methods*, Nauka, Moscow, 1987, (in Russian).

[437] Samarskii, A. A., and Andreev, V. B., *Difference Methods for Elliptic Equations*, Nauka, Moscow, 1976, (in Russian).

[438] Samarskii, A. A., and Gulin, A. V., *Stability of Difference Schemes*, Nauka, Moscow, 1973, (in Russian).

[439] Samarskii, A. A., Lazarov, R. D., and Makarov, V. L., *Difference Schemes for Differential Equations with Generalized Solutions*, Izdat. Vyssh. Shkoly, Moscow, 1987, (in Russian).

[440] Samarskii, A. A., and Nikolaev, E. S., *Numerical Methods for Grid Equations, Vol. II: Iterative methods,* Birkhauser, Basel, 1989.

[441] Samokish, B. A., Method of the steepest descent in eigenvalue problems with semibounded operators, *Izv. Vuzov, Matem.,* **5,** 105–114, 1958 (in Russian).

[442] Sapagovas, M. P., The method of finite differences for the solution of quasilinear elliptic equations with discontinuous coefficients, *USSR Comp. Math. and Math. Phys.,* **5,** 72–85, 1965.

[443] Scapolla, T., A mixed finite element method for the biharmonic problem, *R.A.I.R.O. Numer. Anal.,* **14,** 55–79, 1980.

[444] Schellbach, K., Probleme der Variationsrechnung, *J. Reine Angew. Math.,* **41,** 293–363, 1851.

[445] Schroeder, W. J., and Shephard, M. S., Geometry-based fully automatic mesh generation and the Delaunay triangulation, *Internat. J. for Numer. Meth. in Engineering,* **26,** 2503–2515, 1988.

[446] Schultz, M. H., Elliptic spline functions and the Rayleigh-Ritz-Galerkin method, *Math. Comput.,* **24,** 65–80, 1970.

[447] Schultz, M. H., Ed., *Elliptic Problem Solvers,* Academic Press, New York, 1981.

[448] Scott, L. R., and Zhang, S,, Finite element interpolation of nonsmooth functions satisfying boundary conditions, *SIAM J. Numer. Anal.,* **54,** 483–493, 1990.

[449] Shatelin, F., *Spectral Approximation of Linear Operators,* Academic Press, New York, 1983.

[450] Sheshenin, S. V., and Kuz', I. S., On applied iteration methods, *Computer Mechanics of Solids,* Moscow, 2, 63–74, 1991 (in Russian).

[451] Shishkin, G. I., Grid approximation of singularly perturbed boundary value problems with a regular boundary layer, in *Sov. J. Numer. Anal. Math. Modelling,* **4,** 397–417, 1989.

[452] Shishov, V. S., Block partitioning for finding eigenvalues of large order matrices, *USSR Comp. Math. and Math. Phys.,* **1,** 186–190, 1962.

[453] Siganevich, G. L., On the best error estimate of the linear interpolation on a triangle for functions in $W_2^2(T)$, *Soviet Math. Dokl.,* **37,** 745–748, 1988.

[454] Siganevich, G. L., A version of the fictitious-domain method, *USSR Comp. Math. and Math. Phys.,* **28,** 134–139, 1988.

[455] Siganevich, G. L., *Increasing Efficiency of Grid Methods for Multidimensional Elliptic Problems on the Basis of Usage of Fictitious Grid Regions,* Thesis, Moscow Aviation Institute, 1992.

[456] Smirnov, V. I., *A Course of Higher Mathematics, 5,* Addison-Wesley, Reading, Massachusetts, 1964.

[457] Smith, B. F., A domain decomposition algorithm for elliptic problems in three dimensions, *Numer. Math.,* **60,** 219–234, 1991.

[458] Sobolev, S. L., On estimates of some sums for functions defined on a grid, *Izvest. Akad. Nauk SSSR, Matem.*, **4**, 5–16, 1940 (in Russian).

[459] Sobolev, S. L., *Applications of Functional Analysis in Mathematical Physics*, Math. Monographs, 7, Amer. Math. Soc., Providence, R. I., 1963.

[460] Sobolev, S. L., *Intoduction into Theory of Numerical Integration*, Nauka, Moscow, 1974 (in Russian).

[461] Sobolevskii, P. E., On equations with operators forming acute angles, *Dokl. Akad. Nauk SSSR*, **116**, 754–757, 1957 (in Russian).

[462] Sobolevskii, P. E., and Vasil'ev, V.V., On an ε-approximation of the Navier-Stokes equations, *Chisl. Metody Mekh. Sploshnoi Sredy*, **9**, n. 5, 115–139, 1978 (in Russian).

[463] Sommerville, D. M. Y., *An Introduction to the Geometry of N dimensions*, New York, 1958.

[464] Starke, G., Alternating direction preconditioning for nonsymmetric systems of linear equations, *SIAM J. Sci. Comput.*, **15**, 369–384, 1994.

[465] Stenberg, R., Error analysis of some finite element methods for the Stokes problem, *Math. Comp.*, **54**, 495–508, 1990.

[466] Stevenson, R., Discrete Sobolev spaces and regularity of elliptic difference schemes, *Math. Modelling and Numer. Anal.*, **25**, 607–640, 1991.

[467] Stolyarov, N. N., Large deformations of layered elastic-plastic plates and shells, *Prikl. Mekh.*, **20**, 72–77, 1984 (in Russian).

[468] Strang, G., *Linear Algebra and Its Applications*, Academic Press, New York, 1976.

[469] Strang, G., *Introduction to Applied Mathematics*, Wellesley-Cambridge Press, Wellesley, Massachusetts, 1986.

[470] Strang, G., and Fix, G., *An Analysis of the Finite Element Method*, Prentice-Hall, Englewood Cliffs, New Jersey, 1973.

[471] Strelkov, N. A., Simplicial extensions of grid functions and their application to the solution of problems of mathematical physics, *USSR Comp. Math. and Math. Phys.*, **11**, 190–207, 1971.

[472] Strelkov, N. A., On the relationship between difference and projection-difference methods, *Banach Center Publications*, PWN-Polish Scientific Publishers, Warsaw, **24**, 355–377, 1990.

[473] Strelkov, N. A., Projection-grid n-widths and grid packings, *Math. USSR Sbornik*, **74**, 251–269, 1993.

[474] Swarztrauber, P. N., Approximate cyclic reduction for solving Poisson's equation, *SIAM J. Sci. Statist. Comput.*, **8**, 199–209, 1987.

[475] Temam, R., *Navier–Stokes Equations, Theory and Numerical Analysis*, North-Holland Publishing Co., Amsterdam, 1979.

[476] Thomasset, F., *Implementation of Finite Element Methods for Navier–Stokes Equations*, Springer-Verlag, New York, 1981.

[477] Tikhomirov, V. M., *Some Problems of Approximation Theory*, Izdat. Moskov. Univ., Moscow, 1976 (in Russian).

[478] Tompson, J. F., Warsi, U. A., and Mastin, C. W., Boundary-fitted coordinate systems for numerical solution of partial differential equations–a review, *J. Comp. Phys.*, **47**, 1–108, 1982.

[479] Traub, J. F., *Analytic Computational Complexity*, Academic Press, New York, 1976.

[480] Traub, J. F., and Wozniakowski, H., *A General Theory of Optimal Algorithms*, Academic Press, New York, 1980.

[481] Trenogin, V. A., *Functional Analysis*, Nauka, Moscow, 1980, (in Russian).

[482] Tyrtyshnikov, E., On symmetrizable preconditioners with low rank updates, *J. Numer. Linear Algebra with Appl.*, **1**, 227–235, 1992.

[483] Tyrtyshnikov, E., Optimal and super-optimal circulant preconditioners, *SIAM J. Matrix Anal. Appl.*, **13**, 459–473, 1992.

[484] Vainberg, M. M., *Variational Method and Method of Monotone Operators in Theory of Nonlinear Equations*, Nauka , Moscow, 1972 (in Russian).

[485] Vainikko, H. J., *Funktionalanalysis der Diskretisierungsmethoden*, Teubner, Leipzig, 1976.

[486] Valedinsky, V. D., Numerical solution of an elasticity problem concerning contact of compressible and incompressible materials, *Vestnik Moskov. Univ., Ser. Vychisl. Mat. i Kibernet.*, **4**, 3-12, 1980.

[487] Van der Vorst, H. A., High performance preconditioning, *SIAM J. Sci. Statist. Comput.*, **10**, 1174–1183, 1989.

[488] Varga, R. S., *Matrix Iterative Analysis*, Prentice-Hall, Englewood Cliffs, New Jersey, 1962.

[489] Varga, R. S., *Functional Analysis and Approximation Theory in Numerical Analysis*, SIAM, Philadelpia, 1971.

[490] Vassilevski, P. S., Preconditioning nonsymmetric and indefinite finite element matrices, *J. of Numer. Linear Algebra with Applications*, **1**, 59–76, 1992.

[491] Verfurth, R., The contraction number of a multigrid method with mesh ratio 2 for solving Poisson's equation, *Linear Algebra Appl.*, **60**, 332–348, 1984.

[492] Verfurth, R., A posteriori error estimators for the Stokes equations, *Numer. Math.*, **55**, 309–325, 1989.

[493] Vishik, M. I., Quasilinear strongly elliptic systems of differential equations in divergence form, *Transactions of the Moscow Mathematical Society for the Year 1963*, 140–208, 1963.

[494] Volkov, E. A., The method of composite grids for bounded and unbounded poly-gons and error bounds in terms of known quantities, *Proc. Steklov Inst. Math.*, **96**, 117–148, 1970.

[495] Volkov, E. A., An asymptotically fast approximate method of finding the solution of the difference Laplace equation on certain line segments, *Proc. Steklov Inst. Math.*, **173** , n. 4, 71–92, 1987.

[496] Wachspress, E. L., Extended application of alternating direction implicit iteration model problem theory, *SIAM J. Numer. Anal.*, **11**, 994–1016, 1963.

[497] Wachspress, E. L., *Iterative Solution of Elliptic Systems*, Prentice-Hall, New-York, 1966.

[498] Wait, R., Audish, S. E., and Willis, C. J., Finite element analysis on a highly parallel multiprocessor architecture, *International Series on Numerical Mathematics*, **86**, 507–518, 1988.

[499] Wesseling, P., Multigrid methods in computational fluid dynamics, *ZAAM*, **70**, 337–347, 1990.

[500] Weyl, H., *Symmetry*, Princeton University Press, Princeton, New Jersey, 1952.

[501] Whiteman, J. R., Singularities in three-dimensional elliptic problems and their treatment with finite element methods, *Lect. Notes Math.*, **1066**, 264–278, 1984.

[502] Widlund, O. B., On the use of fast methods for separable finite difference equations for the solution of general elliptic problems, in *Sparse Matrices and Applications*, Eds. Rose, D. S., and Willoughby, K. A., Plenum Press, New York, 121–131, 1972.

[503] Widlund, O., Capacitance matrix methods for Helmholtz equation on general bounded region, *Lecture Notes Math.*, **631**, 209–219, 1978.

[504] Widlund, O., Optimal iterative refinement methods, in *Domain Decomposition Methods for Partial Differential Equations. II*, Chan, T., Glowinski, R., Meurant, G. A., Periaux, J., and Widlund, O., Eds., SIAM, Philadelphia, 114–125, 1989.

[505] Wilkinson, J. H., *The Algebraic Eigenvalue Problem*, Clarendon Press, Oxford, 1965,

[506] Yakovlev, G. N., Traces of functions in the space W_p^l on piecewise smooth surfaces, *Math. USSR Sb.*, **3**, 481–498, 1967.

[507] Yakovlev, M. N., On some methods for solving nonlinear equations, *Proc. Steklov Inst. Math.*, **84**, 5–41, 1965.

[508] Yanenko, N. N., *Fractional Steps Method for Solving Multi-Dimensional Problems of Mathematical Physics*, Nauka, Novosibirsk, 1967 (in Russian).

[509] Young, D. M., and Gregory, R. T., *A Survey of Numerical Mathematics, II*, Addison-Wesley Publishing Company, Reading, 1973.

[510] Young, D. M., Melvin, R. G., Johnson, F. T., Bussoletti, J. E., Wighton, L. B., and Samant, S. S., Application of sparse matrix solvers as effective preconditioners, *SIAM J. Sci. Statist. Comput.*, **10**, 1186–1199, 1989.

[511] Yosida, K., *Functional Analysis*, Academic Press, New York, 1965.

[512] Yserentant, H., On the multilevel splitting of finite element spaces, *Numer. Math.*, **49**, 379–412, 1986.

[513] Yserentant, H., Preconditioning indefinite discretization matrices, *Numer. Math.*, **54**, 719–734, 1988.

[514] Zenisek, A., The finite element method for nonlinear elliptic equations with discontinuous coefficients, *Numer. Math.*, **58**, 51–77, 1990.

[515] Zhang, S., Optimal order nonnested multigrid iterations for solving finite element equations, *SIAM J. Numer. Anal.*, **55**, 23–36, 1990.

[516] Zienkiewicz, O. C., and Craig, A. W., A posteriori error estimation and adaptive mesh refinement in the finite element methods, in *The Mathematical Basis of Finite Element Methods*, Griffiths, D. F., Ed., Clarendon Press, Oxford, 71–89, 1984.

[517] Zienkiewicz, O. C., Cago, J., Babushka, I., Hierarchical finite element approaches, error estimates and adaptive refinement, in *The Mathematics of Finite Elements and Applications*, Whiteman, J. R., Ed., Academic Press, London, 1982.

[518] Zienkiewicz, O. C., and Morgan, K., *Finite Elements and Approximation*, John Wiley, New York, 1983.

[519] Zlotnik, A. A., Convergence rate estimates of finite element methods for second-order hyperbolic equations, in *Numerical Methods and Applications*, Marchuk, G. I., Ed., CRC Press, Boca Raton, 155–220, 1994.

Index